North American Thrust-Faulted Terranes

Selected papers reprinted from the AAPG Bulletin and other geological journals

Compiled by
William J. Perry
Dietrich H. Roeder
David R. Lageson

Published by
The American Association of Petroleum Geologists
Tulsa, Oklahoma U.S.A.

Published July 1984
Second Printing, July 1985

North American thrust-faulted terranes.

Includes bibliographies.
1. Thrust faults (Geology)--North America--Addresses, essays, lectures. 2. Geomorphology--North America--Addresses, essays, lectures. 3. Petroleum--Geology--North America--Addresses, essays, lectures. I. Perry, William J. II. Roeder, Dietrich H. III. Lageson, David R. IV. AAPG bulletin.
QE606.5.N7N67 1984 551.8'7'097 84-45319
ISBN 0-89181-548-1

Contents:

INTRODUCTION

Developments in knowledge and concepts over the last quarter century have revolutionized our understanding of thrust-faulted terranes in North America. For this reason, it is timely to reprint widely cited reports from the *AAPG Bulletin* that led up to this revolution, as well as significant reports from the *AAPG Bulletin* and other journals that document our rapid expansion in knowledge concerning thrust-faulted regions in North America and the mechanics of thrusting. Our volume is divided into two sections: (1) reports of general interest concerning thrust mechanics, particularly thrust-fault geometries, fabric elements, relation of thrusts to folds, cross-sectional modelling, and relation of overthrust faulting to abnormally high fluid pressures; and (2) reports of more regional importance from the Appalachian thrust belt, Ouachita thrust belt, the various segments of the Cordilleran thrust belt, and the first comprehensive report on low-angle thrust faulting involving crystalline basement in the Rocky Mountain foreland.

We call attention to the special reprint volume published in 1982 by the Canadian Society of Petroleum Geologists, consisting of three landmark reports on the Canadian Rockies' portion of the Cordilleran thrust belt (Shaw, 1963; Bally et al, 1966; Dahlstrom, 1970), which should be considered a companion to our volume. We express our appreciation to the Geological Society of America, the Geological Society of London, the Canadian Society of Petroleum Geologists, and the National Research Council of Canada for permission to reprint selected reports from their publications. We appreciate the encouragement of the late David Elliott, who first proposed this volume and that of E.A. Beaumont and Kaspar Arbenz who made it possible. This volume is dedicated to the memory of David Elliott and Leonard D. Harris, who, in different ways, made significant and lasting contributions to advances in the knowledge of North American thrust-faulted terranes. Helpful comments by Steven S. Oriel, John Rodgers, and Russell L. Wheeler, particularly concerning the historical sketch which follows, are appreciated.

A BRIEF HISTORY OF THE ANALYSIS OF NORTH AMERICAN THRUST-FAULTED TERRANES

Twenty-five years ago, Hubbert and Rubey (1959) published their landmark report on the role of fluid pressure in overthrust faulting. In a companion report, Rubey and Hubbert (1959) introduced abnormally high pore-fluid pressures as an explanation for a gravity sliding (also called gravitational sliding or gravity gliding) origin of overthrust belts; sliding was introduced in the Alpine literature early in this century. Coupled with the theory of reduced effective normal stress by elevated fluid pressures, the gravity sliding hypothesis became a ruling paradigm among many structural geologists.

Other geologists, particularly in the petroleum industry, who realized that gravity sliding could not explain the three-dimensional geometries of thrust belts obtained from analysis of deep drilling and seismic-reflection surveying as well as surface geology, became a nonvocal minority, without a theory to explain their observations. Finally, Price and Mountjoy (1970), modifying a concept originally proposed by Bucher (1956), introduced the theory of gravity spreading. The principal difference between sliding and spreading is that spreading accounts for the general dip of the basal detachment toward and beneath the orogen, observed by seismic-reflection surveying. The two theories are contrasted by Price (1971) who pointed out other inadequacies in the gravity-sliding paradigm. A problem with both theories was that neither considered that plate tectonics, specifically plate collisions, might play a key role in the origin of thrust-faulted terranes. Roeder (1973) and Bally (1975, 1981) argued that gravity gliding is not an important mechanism in the origin of thrust belts; they convincingly made the case for compression and subduction as governing factors in the origin of thrust belts. In the mean time, Chapple (1978) laid the mathematical framework for the compressive origin of thin-skin fold and thrust belts. Price (1981, reprinted this volume) extended the geodynamic theory to explain the Canadian foreland fold and thrust belt, leaving only scattered adherents of the older paradigms.

The entire subject of the Cordilleran hinterland is currently being revolutionized by "collage" or accretionary tectonics (Coney et al, 1980), spearheaded by biostratigraphers (Jones et al, 1980; and others), Canadian geologists (Monger, 1977; and others) and workers in paleomagnetics (Irving, 1979; and others). Although this work does not bear directly on the analysis of foreland thrusting, the timing of accretion at convergent plate margins versus thrusting within overthrust belts is a topic of intensive current research.

Thrust faulting, involving Precambrian crystalline basement in the Rocky Mountain foreland, was first comprehensively documented by Berg (1962, reprinted this volume). Following analyses that focused on vertical tectonics (Prucha et al, 1965) and wrench fault tectonics (Sales, 1968) in this region, Lowell (1974) placed Rocky Mountain foreland thrust faulting in a geodynamic context. More recently, Smithson et al (1979) presented the results of COCORP deep-reflection seismic surveying and gravity modelling, proving to many that the Wind River uplift is underlain and generated by a major low-angle thrust involving a significant part of the subjacent continental crust. Problems with the original interpretation were resolved by Zawislak and Smithson (1981). The recent synthesis of deep drillhole and proprietary seismic reflection data by Gries (1983) shows that this style of deformation is widespread throughout the Rocky Mountain foreland and provides exciting targets for oil and gas exploration.

In concentrating on the recent rapid expansion in our knowledge of thrust-faulted terranes, we should not overlook the past. Reverse-faulted and folded structural terranes have been recognized in North America for nearly 150 years. Rodgers (1949, reprinted this volume) outlined the pioneering work by Professors W.B. and H.D. Rogers, beginning in the 1830s in what is now known as the Appalachian thrust belt, and provided an excellent historical summary of later structural studies in this belt. As discussed by Owen (1975, p. 62-63), E.B. Andrews (1861) documented the first known association of oil with anticlines, within "fissures" on the flanks of the Burning Springs anticline in West Virginia. This fold, at the western limit of detachment within the thrust-faulted central Appalachians, is developed above blind (nonsurfacing) thrusts that branch from a decollement in the Upper Silurian salt beds (Rodgers, 1963, reprinted this volume).

Following the pioneering work of W.B. and H.D. Rogers in Virginia and Pennsylvania and J.M. Safford in eastern Tennessee (Rodgers, 1949), a remarkable expansion in knowledge of thrust faulting took place in the last years of the nineteenth century. This expansion began, for many North American geologists, in northwest Scotland in the 1880s: Archibald Geikie, Director of the Geological Survey of Great Britain, sent a remarkable team of geologists into the region of northwestern Sutherland to settle a controversy concerning the stratigraphic succession in that region. This controversy stemmed from the work of Sir Roderick Murchison, a former director and mentor, who had not recognized irregularities in the succession (Challinor, 1972). In spite of his assumption that Murchison's views would prove correct, Geikie (1884, p. 29) provided the classic "injunction to the officers now intrusted with the detailed survey of the region to divest themselves of any prepossessions in favor of published views and to map the actual facts in entire disregard of theory." In the process, the first major thrust system documented in the English speaking world, the Moine system of northwestern Scotland, was deciphered.

Geikie's introduction to the preliminary report by Peach and Horne (1884, his "officers") contained the following highly quotable remarks, which echo in the work still continuing in the world's thrust-faulted regions (Geikie, 1884, p. 30):

> *"No one cursorily visiting the ground could form any notion of its extraordinary complication, which could only be satisfied by patient detailed mapping* [now aided by the tools of geophysics and remote sensing] *such as had never yet been bestowed upon it. ... By a system of reversed faults, a group of strata is made to cover a great breadth of ground and actually overlie higher members of the same series. The most extraordinary dislocations, however, are those to which for distinction we have given the name of* **Thrust-plane**. *They are strictly reversed faults, but with so low a hade that the rocks on their up-throw side have been, as it were, pushed horizontally forward. The distance to which this horizontal displacement has reached is almost incredible. ... When a geologist finds vertical beds of gneiss overlying gently inclined sheets of fossiliferous quartzite, shale, and limestone, he may be excused if he begins to wonder whether he himself is not really standing on this head."*

Although thrust faults were recognized earlier (Hubbert and Rubey, 1959, p. 119-120), both in the Alps and in the Moine system, the significance and even the existence of such faults was severely questioned and the earlier work was not widely accepted.

The pioneering work of Törnebohm (1883), showing large-scale horizontal thrust displacements in the Scandinavian Peninsula, was apparently unknown in North America. McConnell (1887), in his classic traverse of the Canadian Rocky Mountains along the route of the Canadian Pacific railway, and Hayes (1891), working in the southern Appalachians, were the first North American geologists to describe major low-angle thrust faults. Although both were influenced by earlier workers in the Appalachians, particularly J.M. Safford, McConnell and Hayes were strongly influenced by Geikie's (1884) eloquent introduction and the preliminary report by Peach and Horne (1884). McConnell shortened Geikie's term "thrust-plane" to *thrust* and *thrust fault*, and was the first in North America to recognize long-distance, nearly horizontal transport along a thrust, namely, that along the mountain front in the Bow Valley area, later named the McConnell thrust in his honor. Hayes (1891, p. 142) was the first to recognize such long distance transport in the Appalachians, caused by "*overthrust faults* which are developed in northwestern Georgia, comparable in magnitude with those of the Scottish highlands and the Rocky Mountains as described by Geikie (1884) and McConnell (1887)."

As discussed by Rodgers (1949, reprinted this volume), Hayes was a member of a team of geologists from the U.S. Geological Survey who were mapping the Appalachians under the direction of Bailey Willis. Their work culminated in Willis' (1893) report on the "Mechanics of Appalachian Structure." This report along with the later reports by Peach, Horne, and their co-workers on the Moine thrust system of Scotland provided the basis for discussion on thrust faulting in the first American textbooks on structural geology (Leith, 1913; Willis, 1923). Some of Willis' conclusions were inappropriate, particularly that major overthrusts were preceded by major asymmetric anticlines caused in part by unequal syndepositional subsidence of bounding synclines and that thrusts broke out of the steeper limbs — the "break-thrust" concept (Willis, 1893). This concept was later revived by Cooper (1964), who questioned thin-skin Appalachian foreland tectonics. The cornerstone of Cooper's argument, the Bane dome of southwestern Virginia, is now known to be a folded thrust complex, with no evidence of basement involvement (Perry et al, 1979, reprinted this volume).

In a remarkably intensive two-months reconnaissance "during the summer of 1901...of a strip 180 miles long" in the Front ranges of northwestern Montana, Willis (1902), assisted by Stuart Weller and G.I. Finlay, defined the great mass of Precambrian sedimentary rocks of Glacier National Park, which were later assigned to the Belt Supergroup. More important to us, they defined and mapped the Lewis overthrust that displaced the Precambrian Belt rocks "7 miles or more" eastward over Upper Cretaceous shales (Willis, 1902, p. 307), an estimate now realized to be much too conservative (Price, 1965; Mudge and Earhart, 1980). Willis (1902, Figures 5 and 6) clearly portrayed the structural unit composed of roof thrust, floor thrust, and enclosed imbricate stack that Dahlstrom (1970, p. 352-353) would later term "duplex fault zone." Elliott and Johnson (1980) would later show the importance of this type of structure in the Moine thrust system, completing the circle beginning with the work of Peach and Horne (1884). Although much of Willis' (1893, 1902, 1923) work is now dated, he firmly ingrained into the North American literature the concept of low-angle thrusting as a major structural style.

We will not trace the entire history of North American thrust-fault concepts in this brief report; Hubbert and Rubey's report (1959, p. 119-122) and many of the papers reprinted in this volume offer further details.

Although Willis (1893) briefly mentioned without citation the work of several Alpine geologists including Heim, to whom he attributed the "stretch-thrust" concept, as well as Margerie and Suess, the discoveries of European geologists concerning thrust faulting in the Alps had little impact on the analysis of thrust systems in North America. The comprehensive and perceptive treatment of Alpine tectonics by Collet (1927), published in Great Britain, received only brief mention by Nevin (1931, p. 277) and has seldom been cited elsewhere in North American literature.

In the first edition of his influential textbook on structural geology, Nevin (1931) recognized the importance of "bedding faults" (decollements) and low-angle thrust faults, corrected earlier misconceptions (particularly the use of the strain ellipsoid to characterize fault planes), illustrated folded thrusts, briefly described the structure of the Alps and Scottish Highlands, and discussed the remarkable experimental and analytical work by Chamberlin and Miller (1918) on low-angle faults. Workers prior to Willis (1893) and Chamberlin and Miller (1918) had recognized the importance of the alternation of mechanically strong and weak layers in the development of thrust-faulted terranes (specifically Cadell, 1890). The modern concept of ramps (steps or treads) and flats was first illustrated by Chamberlin and Miller (1918, Figure 10), who horizontally compressed a multilayer composed of alternating strong and weak layers of different proportions of plaster and clay in the laboratory, arriving at results similar to those for which Rich (1934, reprinted this volume) is widely recognized, that is, that the inclination of the plane of thrust faulting to bedding is controlled by the stiffness of the layers, steeper through brittle layers (ramps) and often flat in weak layers (flats or decollements). Hubbert (1951) would later outline the mechanical basis for thrust faulting and would correct inadequacies in the analysis by Chamberlin and Miller (1918) with respect to the relation of faults to orientation of principal stresses. Rich (1934) provided a major specific field example, the Pine Mountain thrust fault (Cumberland overthrust block), and Rich's analysis laid much of the groundwork for modern analysis of thrust-faulted regions. Bailey's (1935) influential essays on Alpine tectonics would not appear for another year.

Beginning with the studies of coal fields and oil accumulations in southwestern Wyoming (Schultz, 1907; Veatch, 1907), reverse faults were recognized in the area, but cross sections were conceptually similar to those by Safford and Rogers in the Appalachians; no low-angle thrusting was shown or described in the 1907 studies.

Blackwelder (1910) discovered and described the Willard overthrust near Ogden, Utah, and was the first to recognize low-angle thrusting in what later became known as the Overthrust belt. Richards and Mansfield (1912) recognized low-angle thrusting in southeastern Idaho, but these workers incorrectly linked major overthrusts formed at different times (Rubey, 1958; Armstrong and Cressman, 1963). Later, Schultz (1914) correctly portrayed the Darby fault of southwestern Wyoming as a warped low-angle thrust. Work by the U.S. Geological Survey in this region is summarized by Armstrong and Cressman (1963) and Armstrong and Oriel (1965, reprinted this volume), who, with W.W. Rubey and J.I. Tracy, pioneered our present understanding of this region of great economic importance with respect to oil and gas in the present overthrust play. R.L. Armstrong (1968, reprinted this volume) was the first to correctly interpret the Cordilleran thrust belt of Utah, south of the Uinta Mountains, and Nevada, although low-angle thrusts had earlier been identified in southeastern Nevada by Longwell (1922) and in central Nevada by Merriam and Anderson (1942).

Although earlier workers, beginning with Veatch (1907), had discussed the oil and gas potential and resources of the region, Crittendon (1974, reprinted this volume) predicted the exploration targets of northern Utah that resulted in the Pineview discovery, the first major field discovered in the "Western Overthrust belt." This topic had received much unpublished attention — geologic and geophysical studies by the petroleum industry — prior to Crittendon's work. Royse et al (1975), who built from an impressive data base of previous geologic interpretations and proprietary geophysical data, used concepts derived from Rich (1934) and Dahlstrom (1970) to define the presently understood elaborate geologic framework of this region.

Following the discovery of the Appalachian and Cordilleran thrust belts, King (1937), working in the Marathon Mountains of west Texas, and Hendricks et al (1947), working in Oklahoma, defined segments of the major Ouachita thrust belt, in which the timing of the last major episode of thrust deformation is better documented than in the Appalachians. Nicholas and Rozendal (1975, reprinted this volume) documented the complex structural style of a portion of this major thrust belt, of which only a few segments are exposed.

The classic study of folded thrusts in southwestern Alberta by Douglas (1950), who adapted concepts from Rich's Appalachian analysis, gave the Canadians a clear lead in thrust-belt studies. Economic incentives played a major role. The Turner Valley oil and gas field, southwest of Calgary, Alberta, was discovered in 1924. By the mid-1930s, T.A. Link recognized that the Turner Valley structure was controlled by thrust faulting (Link, 1949, reprinted this volume). This led to the world's first major overthrust hydrocarbon play, beginning with exploration in the 1930s that led to the discovery of the Jumping Pound gas and condensate field in 1944. This play involved the first systematic combined use of geologic, seismic, and gravity-meter surveys to predict subsurface thrust geometries. The early efforts to predict these geometries, and the economic implications of these predictions, are carefully documented by Link (1949).

At the same time Douglas' (1950) work was being published, the importance of the analysis of folded thrust faults in deciphering the sequence of structural events was becoming known to the petroleum industry (Scott, 1951, reprinted this volume). Fox (1959, reprinted this volume) summarized the progress in development of tectonic models and in petroleum exploration in the southern Foothills structural province of Alberta in the decade following Link's work. In the process of exploration, a series of rules and constraints were developed or first applied to thrust-belt analysis by petroleum geologists in Calgary, perhaps the most important of which is the constraint of material balance (Dahlstrom, 1969, reprinted this volume). Simply stated, "A balanced section is one in which there is geometric compatibility from one level to another... rock units which occur on one side of a thrust fault must have matching counterparts on the other side" (Price, 1981, p. 432).

Norris (1958, 1964) developed the important concept of contraction and extension faults, a concept that uses layering rather than the present horizon as the frame of reference for fault analysis, a critical step in the analysis of folded faults. In his landmark study of mesoscopic deformation, Price (1967, p. 39-40, reprinted this volume) stated, "The pronounced mechanical anisotropy inherent in this layering has exerted a fundamental control on the development of a characteristic structural style that has come to be considered typical of the Canadian Rockies," the basis for Norris' concept.

Jones (1971, reprinted this volume) summarized the importance of correct analysis of folded faults in determining the sequence of deformation and in hydrocarbon exploration, building on the work of Douglas, Scott, and others. Using the rules, constraints, and concepts that they and previous workers in the Canadian Rockies had defined or applied, Shaw (1963), Bally et al (1966), Roeder (1967), Dahlstrom (1970), Price and Mountjoy (1970), and others deciphered the framework and sequence of deformation of the Canadian segment of the Cordilleran thrust belt, with the support of seismic reflection profiles.

As a result of the first terrain-corrected gravity profile in the central Appalachians, Hammer and Heck (1941) suggested, but did not seriously consider, that the crystalline basement was structurally decoupled from the deformation of the sedimentary cover. Earlier, White (1924) interpreted isostatic anomalies in the eastern United States "as though the heavy Piedmont mass was overthrust on the Valley trough zone." Following Rich's (1934) analysis of the Cumberland overthrust block, Rodgers (1949, 1950, and 1953) extended the concept of thin-skin tectonics to explain foreland deformation of both the central and southern Appalachians. Rodgers (oral communication, 1982) considered the results and interpretation of structural detachment between basement and cover by Joesting et al (1949), critical to this hypothesis.

Seismic reflection surveying and deep exploratory drilling in the central Appalachians during the period 1958 through 1964 by the petroleum industry, was accompanied by a fervor and variety of interpretations (Woodward et al, 1959; Skinner et al, 1963, and Perry, 1964) reminiscent of exploration in the Alberta foothills 20 years earlier. Rodgers (1963) and Gwinn (1964) pulled the results of this exploration together to show that folding at the surface was controlled by structural detachment (decollement) and branching thrust faults in the subsurface. Important studies of thrust faulting by Harris (1967, 1970) and Harris and Milici (1977) in the southern Appalachians, based in part on the analysis of deep exploratory wells, independently derived many concepts and constraints derived over a decade earlier by the Canadians and used in the analysis of the Canadian segment of the Cordilleran thrust belt.

The rules and constraints, collectively developed or refined first by geologists in western Canada, have now been applied to the analysis of a number of thrust systems: (1) the Appalachians (Gwinn, 1970, Perry, 1978; Roeder et al, 1978; Harris et al, 1981; and Boyer and Elliott, 1982, reprinted this volume), (2) the Idaho, Utah, Wyoming systems (Royse et al, 1975; Dixon, 1982), (3) the Alpine system (Boyer and Elliott, 1982), and (4) the Moine system (Elliott and Johnson, 1980).

William J. Perry, Jr.
Dietrich H. Roeder
David R. Lageson

REFERENCES

Andrews, E.B. 1861, Rock oil, its geologic relations and distribution: American Journal of Science, 2nd Series, v. 32, p. 85-93.

Armstrong, F.C., and E.R. Cressman, 1963, The Bannock thrust zone, southeastern Idaho: U.S. Geological Survey Professional Paper 374-J, 22 p.

———, and S.S. Oriel, 1965, Tectonic development of Idaho-Wyoming thrust belt: AAPG Bulletin, v. 49, p. 1847-1866.

Armstrong, R.L., 1968, Sevier orogenic belt in Nevada and Utah: Geological Society of America Bulletin, v. 79, p. 429-458.

Bailey, E.B., 1935, Tectonic essays, mainly Alpine: Oxford, Clarendon Press, 200 p.

Bally, A.W., 1975, A geodynamic scenario for hydrocarbon occurrences: Tokyo, 9th World Petroleum Congress, v. 2, p. 23-44.

———, 1981, Thoughts on the tectonics of folded belts, *in* K.R. McClay, and N.J. Price, eds., Thrust and nappe tectonics: Geological Society of London, Special Publication Number 9, p. 13-32.

———, P.L. Gordy, and G.A. Stewart, 1966, Structure, seismic data, and orogenic evolution of southern Canadian Rocky Mountains: Bulletin of Canadian Petroleum Geology, v. 14, p. 337-381.

Berg, R.R., 1962, Mountain flank thrusting in Rocky Mountain Foreland, Wyoming and Colorado: AAPG Bulletin, v. 46, p. 2029-2032.

Blackwelder, E., 1910, New light on the geology of the Wasatch Mountains, Utah: Geological Society of America Bulletin, v. 21, p. 517-542.

Boyer, S.E., and D. Elliott, 1982, Thrust systems: AAPG Bulletin, v. 66, p. 1196-1230.

Bucher, W.H., 1956, The role of gravity in orogenesis: Geological Society of America Bulletin, v. 67, p. 1295-1318.

Cadell, H.M., 1890, Experimental researches in mountain building: Transactions of the Royal Society of Edinburgh, v. 35, p. 337-357.

Challinor, J., 1972, Geikie, Archibald, *in* C.C. Gillespie, ed., Dictionary of scientific biography: New York, Scribners, v. 5, p. 333-338.

Chamberlin, R.T., and W.Z. Miller, 1918, Low-angle faulting: Journal of Geology, v. 26, p. 1-44.

Chapple, W.M., 1978, Mechanics of thin-skinned fold and thrust belts: Geological Society of America Bulletin, v. 89, p. 1189-1198.

Collet, L.W., 1927, The structure of the Alps: London, Edward Arnold, 289 p.

Coney, P.J., D.L. Jones, and J.W.H. Monger, 1980, Cordilleran suspect terranes: Nature, v. 288, p. 329-333.

Cooper, B.N., 1964, Relation of stratigraphy to structure in the southern Appalachians, *in* Tectonics of the southern Appalachians: Virginia Polytechnic Institute, Department of Geological Sciences Memoir 1, p. 81-114.

Crittenden, M.D., Jr., 1974, Regional extent and age of thrust near Rockport Reservoir and relation to possible exploration targets in northern Utah: AAPG Bulletin, v. 58, p. 2428-2435.

Dahlstrom, C.D.A., 1969, Balanced cross sections: Canadian Journal of Earth Sciences, v. 6, p. 743-757.

———, 1970, Structural geology in the eastern margin of the Canadian Rocky Mountains: Bulletin of Canadian Petroleum Geology, v. 18, p. 332-406.

Dixon, J.S., 1982, Regional structural synthesis, Wyoming salient of Western Overthrust belt: AAPG Bulletin, v. 66, p. 1560-1580.

Douglas, R.J.W., 1950, Callum Creek, Langford Creek, and Gap map-areas, Alberta: Geological Survey of Canada Memoir Number 255.

Elliott, D., and M.R.W. Johnson, 1980, Structural evolution in the northern part of the Moine thrust belt, NW Scotland: Trans. Royal Soc. Edinburgh, v. 71, p. 69-96.

Fox, F.G., 1959, Structures and accumulation of hydrocarbons in southern Alberta foothills, Alberta, Canada: AAPG Bulletin, v. 52, p. 641-664.

Geikie, A., 1884, The crystalline rocks of the Scottish highlands: Nature, v. 31, p. 29-31.

Gries, R., 1983, Oil and gas prospecting beneath Precam-

brian of foreland thrust plates in Rocky Mountains: AAPG Bulletin, v. 67, p. 1-28.
Gwinn, V.E., 1964, Thin-skinned tectonics in the Plateau and northwestern Valley and Ridge provinces of the central Appalachians: Geological Society of America Bulletin, v. 75, p. 863-900.
———, 1970, Kinematic patterns and estimates of lateral shortening, Valley and Ridge and Great Valley provinces, central Appalachians, south-central Pennsylvanian, *in* G.W. Fisher, et al, eds., Studies of Appalachian geology; central and southern: New York, Wiley-Interscience, p. 127-146.
Hammer, S., and E.T. Heck, 1941, A gravity profile across the central Appalachians; Buckhannon, West Virginia to Swift Run Gap, Virginia: American Geophysical Union Transactions, p. 353-362.
Harris, L.D., 1967, Geology of the L.S. Bales well, Lee County, Virginia; a Cambrian and Ordovician test, *in* Kentucky Oil and Gas Association 29th Annual Meeting Proceedings, Technical Session: Kentucky Geological Survey, Series 10, Special Publication 14, p. 50-55.
———, 1970, The Valley and Ridge and Appalachian Plateau, structure and tectonics; details of thin-skinned tectonics in parts of Valley and Ridge and Cumberland Plateau provinces of the southern Appalachians, *in* G.W. Fisher, et al, eds., Studies of Appalachian geology - central and southern: New York, Wiley-Interscience, p. 217-259.
———, A.G. Harris, W. DeWitt, Jr., and K.C. Bayer, 1981, Evaluation of southern Eastern Overthrust belt beneath Blue Ridge-Piedmont thrust: AAPG Bulletin, v. 65, p. 2497-2505.
———, and R.C. Milici, 1977, Characteristics of thin-skinned style of deformation in the southern Appalachians, and potential hydrocarbon traps: U.S. Geological Survey Professional Paper 1018, 40 p.
Hayes, C.W., 1891, The overthrust faults of the southern Appalachians: Geological Society of America Bulletin, v. 2, p. 141-152.
Hendricks, T.A., L.S. Gardiner, and M.M. Knechtel, 1947, Geology of the western part of the Ouachita Mountains of Oklahoma: U.S. Geological Survey Oil and Gas Investigations Map 66, 3 sheets.
Hubbert, M.K., 1951, Mechanical basis for certain familiar geologic structures: Geological Survey of America Bulletin, v. 62, p. 355-372.
———, and W.W. Rubey, 1959, Role of fluid pressure in overthrust faulting, I. Mechanics of fluid-filled porous solids and its applications to overthrust faulting: Geological Society of America Bulletin, v. 70, p. 115-166.
Irving, E., 1979, Paleopoles and paleolatitudes of North America and speculations about displaced terranes: Canadian Journal of Earth Sciences, v. 16, p. 669-694.
Joesting, H.R., F. Keller, Jr., and E. King, 1949, Geologic implications of aeromagnetic survey of Clearfield-Pittsburgh area, Pennsylvania: AAPG Bulletin, v. 33, p. 1747-1766.
Jones, D.L., et al, 1980, Age and structural significance of ophiolite and adjoining rocks in the Upper Chulitna District, south-central Alaska: U.S. Geological Survey Professional Paper 1121A, 21 p.
Jones, P.B., 1971, Folded faults and sequences of thrusting in Alberta foothills: AAPG Bulletin, v. 55, p. 292-306.
King, P.B., 1937, Geology of the Marathon region, Texas: U.S. Geological Survey Professional Paper 187, 148 p.
Leith, C.K., 1913, Structural geology: New York, Henry Holt and Company, 169 p.
Link, T.A., 1949, Interpretations of Foothills structures, Alberta, Canada: AAPG Bulletin, v. 33, p. 1475-1501.
Longwell, C.R., 1922, The Muddy Mountain overthrust in southeastern Nevada: Journal of Geology, v. 30, p. 63-72.
Lowell, J.D., 1974, Plate tectonics and foreland basement deformation: Geology, v. 2, p. 275-278.
McConnell, R.G., 1887, Report on the geologic structure of a portion of the Rocky Mountains: Geological Survey Canada, Annual Report for 1886, part D, p. 31D-40D.
Merriam, C.W., and C.A. Anderson, 1942, Reconnaissance survey of the Roberts Mountains, Nevada: Geological Society of America Bulletin, v. 53, p. 1675-1728.
Monger, J.W.H., 1977, Upper Paleozoic rocks of the western Canadian Cordillera and their bearing on Cordillera evolution: Canadian Journal of Earth Sciences, v. 24, p. 1832-1859.
Mudge, M.R., and R.L. Earhart, 1980, The Lewis thrust fault and related structures in the Disturbed belt, northwestern Montana: U.S. Geological Survey Professional Paper 1174, 18 p.
Nevin, C.M., 1931, Principles of structural geology: New York, John Wiley and Sons, Incorporated, 303 p.
Nicholas, R.L., and R.A. Rozendal, 1975, Subsurface positive elements within Ouachita foldbelt in Texas and their relation to Paleozoic cratonic margin: AAPG Bulletin, v. 59, p. 193-216.
Norris, D.K., 1958, Structural conditions in Canadian coal mines: Bulletin of Canadian Petroleum Geology 44, 54 p.
———, 1964, Microtectonics of the Kootenay Formation near Fernie, British Columbia: Bulletin of Canadian Petroleum Geology, v. 12, Special Issue, p. 512-535.
Owen, E.W., 1975, Trek of the oil finders; a history of exploration for petroleum: AAPG Memoir 6, 1,647 p.
Peach, B.N., and J. Horne, 1884, Report on the geology of the northwest of Sutherland: Nature, v. 31, p. 31-35.
Perry, W.J., Jr., 1964, Geology of the Ray Sponaugle well, Pendleton County, West Virginia: AAPG Bulletin, v. 48, p. 659-669.
———, 1978, Sequential deformation in the central Appalachians: American Journal Science, v. 278, p. 518-542.
———, A.G. Harris, and L.D. Harris, 1979, Conodont-based reevaluation of Bane Dome—structural reevaluation of Allegheny Frontal zone: AAPG Bulletin, v. 63, p. 647-654.
Price, R.L., 1965, Flathead map-area, British Columbia and Alberta: Geological Survey of Canada Memoir 336, 221 p.
———, 1967, The tectonic significance of mesoscopic subfabrics in the southern Rocky Mountains of Alberta and British Columbia: Canadian Journal of Earth Sciences, v. 4, p. 39-70.
———, 1971, Gravitational sliding and the foreland thrust and fold belt of the North American Cordillera, discussion: Geological Society of America Bulletin, v. 82, p. 1133-1138.
———, 1981, The Cordilleran foreland thrust and fold

belt in the southern Canadian Rocky Mountains, *in* Thrust and Nappe tectonics: Geological Society of London, p. 427-448.

———, and E.W. Mountjoy, 1970, Geologic structure of the Canadian Rocky Mountains between Bow and Athabasca Rivers - a progress report: Geological Association of Canada, Special Paper Number 6, p. 7-25.

Prucha, J.J., J.A. Graham, and R.P. Nickelson, 1965, Basement controlled deformation in Wyoming province of Rocky Mountain foreland: AAPG Bulletin, v. 49, p. 966-992.

Rich, J.L., 1934, Mechanics of low-angle overthrust faulting as illustrated by Cumberland thrust block, Virginia, Kentucky, and Tennessee: AAPG Bulletin, v. 18, p. 1584-1596.

Richards, R.W., and G.R. Mansfield, 1912, The Bannock overthrust, a major fault in southeastern Idaho and northeastern Utah: Journal of Geology, v. 20, p. 681-709.

Rodgers, J., 1949, Evolution of thought on structure of middle and southern Appalachians: AAPG Bulletin, v. 33, p. 1643-1654.

———, 1950, Mechanics of Appalachian folding as illustrated by Sequatchie anticline, Tennessee and Alabama: AAPG Bulletin, v. 34, p. 672-681.

———, 1953, The folds and faults of the Appalachian Valley and Ridge province: Kentucky Geological Survey, Series 9, Special Publication 1, p. 150-166.

———, 1963, Mechanics of Appalachian foreland folding in Pennsylvania and West Virginia: AAPG Bulletin, v. 47, p. 1527-1536.

Roeder, D.H., 1967, Rocky Mountains, Der Geologische Aufbau des Kanadischen Falsengebirges; Beitrage zur Regionalen Geologie der Erge, Band 5: Berlin, Gebruder Borntraeger, 318 p.

———, 1973, Subduction and orogeny: Journal of Geophysical Research, v. 78, p. 5005-5024.

———, O.E. Gilbert, Jr., and W.D. Witherspoon, 1978, Evolution and macroscopic structure of Valley and Ridge thrust belt, Tennessee and Virginia: Knoxville, University of Tennessee, Department of Geological Sciences, Studies in Geology 2, 25 p.

Royse, F., Jr., M.A. Warner, and D.L. Reese, 1975, Thrust belt of Wyoming, Idaho, and northern Utah; structural geometry and related problems: Rocky Mountain Geological Association, Symposium on Deep Drilling Frontiers in Central Rocky Mountains, p. 41-54.

Rubey, W.W., 1958, Geologic map of the Bedford Quadrangle, Wyoming: U.S. Geological Survey, Geological Quadrant Map GQ-109, scale 1:62,500.

———, and M.K. Hubbert, 1959, Role of fluid pressure in mechanics of overthrust faulting; II. Overthrust belt in geosynclinal area of western Wyoming in light of fluid-pressure hypothesis: Geological Society of America Bulletin, v. 70, p. 167-206.

Sales, J.K., 1968, Crustal mechanics of Cordilleran foreland deformation; a regional and scale-model approach: AAPG Bulletin, v. 52, p. 2016-2044.

Schultz, A.R., 1907, Coal fields in a portion of central Uinta County, Wyoming, *in* Contributions to economic geology, pt. 2; coal, lignite, and peat, M.R. Campbell, geologist in charge: U.S. Geological Survey Bulletin 316, p. 212-240.

———, 1914, Geology and geography of a portion of Lincoln County, Wyoming: U.S. Geological Survey Bulletin 543, 141 p.

Scott, J.C., 1951, Folded faults in Rocky Mountain foothills of Alberta, Canada: AAPG Bulletin, v. 35, p. 2316-2347.

Shaw, E.W., 1963, Canadian Rockies - orientation in space and time, *in* O.E. Childs, ed., Backbone of the Americas: AAPG Memoir 2, p. 231-242.

Skinner, W., et al, 1963, Road log, *in* Guidebook, tectonics and Cambrian-Ordovician stratigraphy in the central Appalachians of Pennsylvania: Pittsburgh Geological Society with the Appalachian Geological Society, p. 31-129.

Smithson, S.B., et al, 1979, Structure of the Laramide Wind River uplift, Wyoming, from COCORP deep reflection data and gravity data: Journal of Geophysical Research, v. 84, p. 5955-5972.

Törnebohm, A.E., 1883, Om Dalformationens geologiska ålder: Geol. Fören. Stockholm Förhandl., v. 6, p. 622-661.

Veatch, A.C., 1907, Geography and geology of a portion of southeastern Wyoming with special reference to coal and oil: U.S. Geological Survey Professional Paper 56, 178 p.

White, D., 1924, Isostatic anomalies for the United States east of 110°W: Geological Society of America Bulletin, v. 35, p. 207-278.

Willis, B., 1893, Mechanics of Appalachian structure: U.S. Geological Survey, 13th Annual Report, part 2, p. 211-281.

———, 1902, Stratigraphy and structure, Lewis and Livingstone ranges, Montana: Geological Society of America Bulletin, v. 13, p. 305-351.

———, 1923, Geologic structures: New York, McGraw Hill, 1st edition, 295 p.

Woodward, H.P., et al, 1959, A symposium on the Sandhill deep well, Wood County, West Virginia: West Virginia Geological Survey Report of Investigations 18, 182 p.

Zawislak, R.L., and S.B. Smithson, 1981, Problems and interpretation of COCORP deep seismic reflection data, Wind River range, Wyoming: Geophysics, v. 46, p. 1684-1701.

BULLETIN OF THE AMERICAN ASSOCIATION OF PETROLEUM GEOLOGISTS
VOL. 18, NO. 12 (DECEMBER, 1934), PP. 1584-1596, 9 FIGS.

MECHANICS OF LOW-ANGLE OVERTHRUST FAULTING AS ILLUSTRATED BY CUMBERLAND THRUST BLOCK, VIRGINIA, KENTUCKY, AND TENNESSEE[1]

JOHN L. RICH[2]
Cincinnati, Ohio

ABSTRACT

The structural relations of the Cumberland overthrust block are such as would occur if gliding on the thrust plane took place parallel with the bedding along certain shale beds in such a way that the thrust plane followed a lower shale bed for some distance, then sheared diagonally up across the intervening beds to a higher shale, followed that for several miles, and again sheared across the bedding to the surface.

Reasons are given for the belief that subsidiary faults and folds within the block are superficial and do not extend below the thrust plane. This possibility should be borne in mind when exploration of such structures for oil or gas is contemplated.

Study of the Cumberland block throws new light on the broader problems of the nature of folding and faulting in the sedimentary rocks bordering great mountain ranges and on the function of friction in setting limits to the distance through which overthrust blocks can be moved.

INTRODUCTION

It is now generally recognized that low-angle overthrust faults are common and that the movement on some of them has amounted to several miles, but in most cases the manner in which such thrusts are produced, the limits of movement set by friction, and the function of the nature and attitude of the rocks in guiding and controlling the faulting have not been clearly understood.

A clue to the solution of one such problem has been revealed by a study of the Cumberland fault block, a rectangular overthrust mass about 125 miles long and 25–30 miles wide, located at the common corners of the states of Virginia, Kentucky, and Tennessee. This area, which marks the outer limit of the Appalachian overthrusting in that district, shows the essential mechanics of the thrust faulting because the process has not been carried far enough to obscure the record.

CUMBERLAND FAULT BLOCK

The Cumberland overthrust block has been described in consider-

[1] Presented before the Association at the Houston meeting, March 24, 1933 Manuscript received, May 4, 1934.

[2] Department of geology, University of Cincinnati.

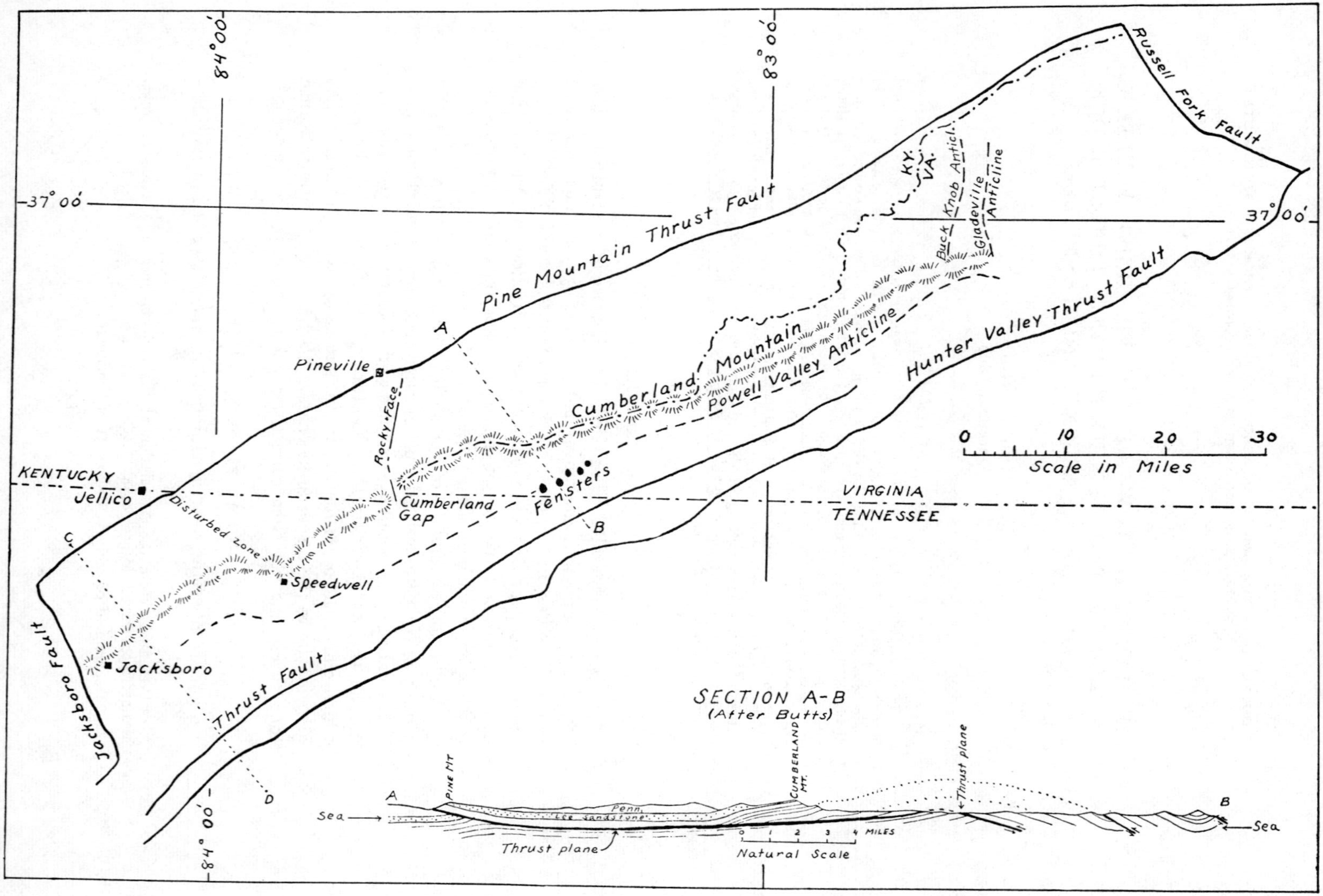

FIG. 1.—Cumberland overthrust block.

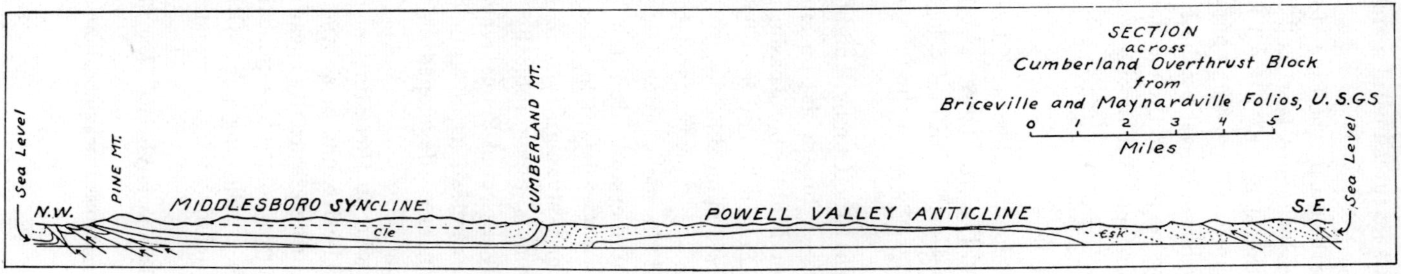

FIG. 2.—Topographic and geologic cross section of western part of Cumberland overthrust block along line *CD* (Fig. 1), showing flat-topped Powell Valley anticline and flat-bottomed Middlesboro syncline; also showing earlier, now disproved, conception of relations of Pine Mountain thrust fault. From Briceville and Maynardville folios, United States Geological Survey.

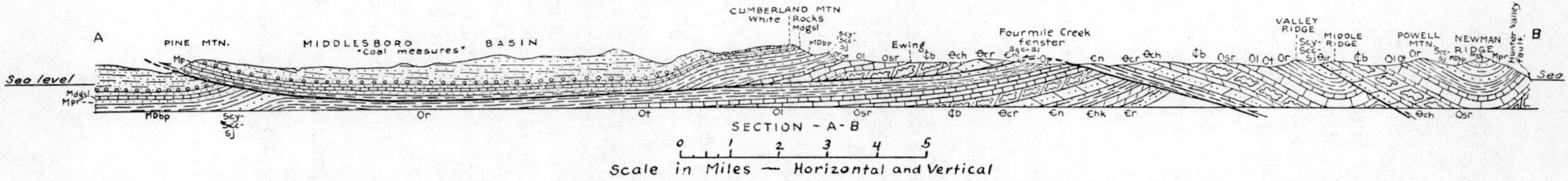

FIG. 3.—Cross section of Cumberland overthrust block showing fensters and newer conception of relations of Pine Mountain fault. Courtesy of Charles Butts and Virginia Geological Survey.

able detail by Wentworth,[3] and certain of its features in still more detail by Butts.[4]

The block (Fig. 1) is bounded on the northwest by the Pine Mountain thrust fault; on the southwest by the Jacksboro tear fault, upthrown toward the southwest; on the northeast by the Russell Fork tear fault, along which there has been some overthrusting, but not so much as on the Jacksboro fault; and on the southeast by the Hunter Valley thrust fault. The latter fault plays no part in the mechanics of the block—it merely marks it off from other fault blocks farther southeast.

STRUCTURE WITHIN BLOCK

The structure within the Cumberland block is illustrated by the two accompanying cross sections (Fig. 2 and Fig. 3). Figure 2 is taken from the Briceville folio (No. 33) of the United States Geological Survey, and Figure 3, from Butts' paper referred to in the foregoing paragraphs.

From these two sections it is apparent that the Cumberland block is divided lengthwise into two major units—a broad, flat-bottomed syncline called the Middlesboro syncline or Middlesboro basin, and a broad, flat-topped anticline, called the Powell Valley anticline—separated by the sharp Cumberland Mountain monocline.

The topographic features of the block are expressions of these structures and of the varying resistances of the rocks brought to the surface by them. Pine Mountain and Cumberland Mountain are monoclinal ridges formed by the outcrop of the resistant basal sandstone (part of the Lee formation) of the Pennsylvanian system. The Middlesboro synclinal basin between them is topographically expressed as a mountainous region of maturely dissected Pennsylvanian rocks and the Powell Valley anticline is expressed as an anticlinal lowland—part of the Great Valley of Tennessee—developed on the relatively non-resistant Ordovician limestones.

The Cumberland block is broken crosswise by three lines of disturbance (Fig. 1), trending roughly parallel with one or another of the tear faults at its ends and therefore bearing essentially shearing-angle relations to the outlines of the block as a whole. The significance of these cross-disturbances is discussed on a following page. Other structural disturbances within the block include at least two strike

[3] Chester K. Wentworth, "Russell Fork Fault of Southern Virginia," *Jour. Geol.*, Vol. 29 (1921), pp. 351–69.

[4] Charles Butts, "Fensters in the Cumberland Overthrust Block in Southwestern Virginia," *Virginia Geol. Survey Bull. 28* (1927), pp 1–12.

faults near the top of the Cumberland Mountain monocline. One is on Brush Mountain about 7 miles northeast of Cumberland Gap,[5] and the other detaches Powell Mountain from Cumberland Mountain about 3 miles southwest of Cumberland Gap. No reference to the last-mentioned fault has been found, and it was not examined in detail in the field, so it is not known whether it is a normal or a thrust fault. These strike faults may have significance as adjustment phenomena—tensional or otherwise—to the movements which accompanied the overthrusting of the Cumberland block.

INTERPRETATION OF STRUCTURE

The older interpretation of the structure of the Cumberland block is indicated by the cross section (Fig. 2) taken from the Briceville and Maynardville folios. That interpretation shows a thrust fault extending indefinitely downward from the base of Pine Mountain. It presents a difficult problem as to the mechanics of the structure for it offers no explanation of the sharp Cumberland monocline or of the flatness of the Powell Valley anticline.

The interpretation of the structure of the block was fundamentally changed by the discovery of several fensters, or "windows," on the Powell Valley anticline[6] which revealed the fact that beneath the relatively flat-lying Ordovician rocks of the Powell Valley anticline is a thrust plane, below which Silurian rocks are exposed. Later drilling in these fensters proved the presence of the normal Ordovician section below the Silurian.

Butts interpreted the presence of the Silurian rocks beneath the Ordovician at the fensters as indicating that the thrust plane revealed in the fensters must be the same as that which crops out at the base of Pine Mountain and that the thrust plane must, therefore, underlie the whole of the Cumberland block at shallow depth (Fig. 3). Butts suggested that the thrust may have followed the Chattanooga shale, on or within which gliding would have been relatively easy. He estimates that the forward movement on the thrust at this point has been about 7 miles.

Butts' interpretation is believed to be correct and is accepted as a basis for the discussion which follows.

If, in accordance with conventional ideas, the thrust plane is thought of as originally a plane surface, it would have required considerable warping since the thrusting occurred to have produced its

[5] G. H. Ashley and L. C. Glenn, "Geology and Mineral Resources of the Cumberland Gap Coal Field, Kentucky," *U. S. Geol. Survey Prof. Paper 49* (1906).

[6] Charles Butts, *op. cit.*

present attitude, with the northerly dip into the Middlesboro syncline. In addition to the pronounced apparent warping of the thrust plane shown on Butts' cross section (Fig. 3), an interesting and significant feature is the fact that in the area between the fensters and Cumberland Mountain, the strata dip northward into the thrust plane. This feature finds a ready explanation in, and is a necessary consequence of, the mechanics of thrusting outlined in the following section. Besides, the explanation to be proposed accounts for the present attitude and northward dip of the thrust plane, as shown in Figure 3, without the necessity of any warping of the thrust plane since it was formed.

EXPLANATION OF CUMBERLAND THRUST

The peculiar structural features of the Cumberland block, namely, the flat-topped Powell Valley anticline, the sharp monoclinal structure at Cumberland Mountain, the flat-bottomed Middlesboro syncline between Cumberland Mountain and Pine Mountain, and the

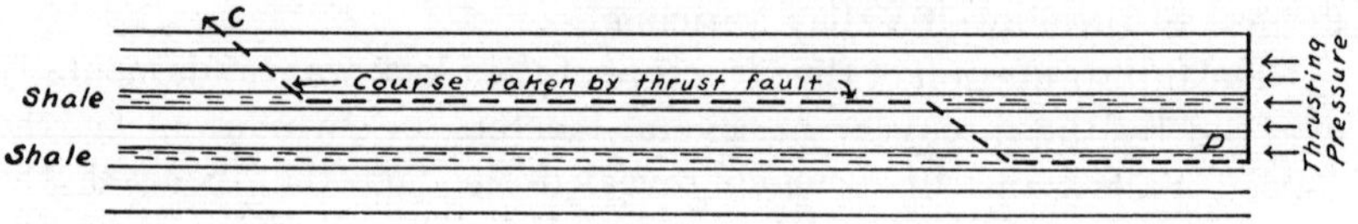

FIG. 4.—Diagram representing course of incipient thrust plane in series of sedimentary rocks. Plane follows beds of easy gliding, such as shales, and breaks diagonally across more brittle beds from one shale to another, and, finally, up to surface.

apparent warping of the thrust plane, all fit into a consistent picture when the thrusting is considered as having taken place along and across the bedding of the sedimentary rocks along some such path

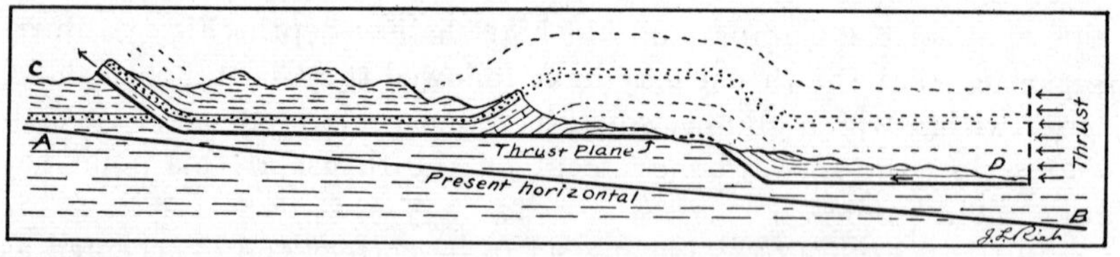

FIG. 5.—Diagram showing result of movement on thrust plane such as that shown in Figure 4. Tilting so that horizontal is represented by line *AB* and erosion down to line *CD* gives structure and topography to be compared with present cross section of Cumberland block (Fig. 3).

as that represented by the dotted line, *CD*, of Figure 4, and the whole region as having been subsequently tilted slightly toward the north.

The thrust plane may be pictured as following some zone of easy gliding such as the lower shale of Figure 4 until frictional resistance became too great; then shearing diagonally up across the bedding to another shale; following that for several miles, and finally shearing across the bedding to the surface.

Movement along such a thrust plane would inevitably produce almost an exact counterpart of the present structure of the Cumberland block, as may be seen by comparing the actual cross section (Fig. 3) with the diagram (Fig. 5)[7] and the model (Fig. 6). The model was made by piling sheets of thin paper into a form like that of Figure 4, with a sheet of paper inserted along the course of the potential thrust plane, and then pushing the paper from one side while the other was held rigid. To reproduce, almost exactly, from the diagram or model, the present structure of the Cumberland block, it is only necessary to tilt the diagram or model so that a horizontal line corresponds with the line *AB* of Figure 5 and then to erode the surface down to a line such as *CD*. The fensters then appear along the line where the thrust plane passes into the upper gliding bed after having broken across the bedding from the one below.

Examination of the diagram (Fig. 5), and the photograph of the model (Fig. 6) shows that whether a broad, flat-topped anticline or a narrow one with a rounded crest is formed where the thrust breaks

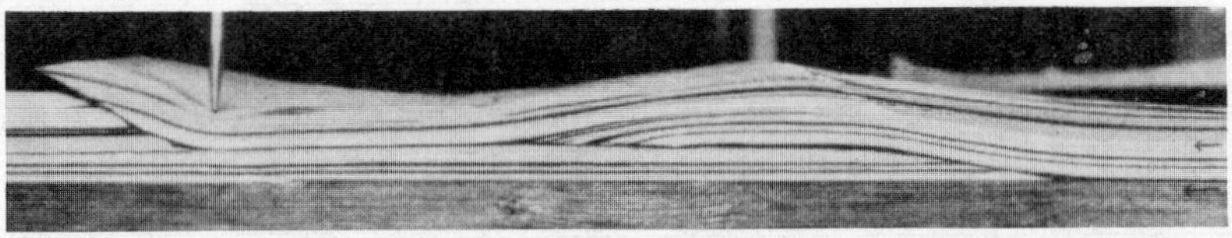

FIG. 6.—Model made by piling paper sheets as in Figure 4 and thrusting toward left. Note similarity to structure of Cumberland block.

upward from the lower gliding plane depends on the amount of thrusting after the break occurs. If the forward movement is slight, a narrow anticline is formed, whereas greater movement produces a broad, flat-topped anticline.[8]

The Powell Valley anticline seems to illustrate both conditions in different parts of its course. At the west end it is broad and flat-

[7] The diagrams, Figures 4 and 5, indicate the thrust pressure as applied entirely at one end. Without going further into the matter, attention is called to the fact that this representation is only diagrammatic, and that it seems possible, and certain features make it seem even probable, that the actual force which caused the slipping on the thrust plane was applied from above as well as from the side through the medium of an overthrust sheet now removed by erosion.

[8] This relation is easily seen by constructing a model like that shown in Figure 6 and varying the amount of thrust.

topped; toward the east it gradually narrows and assumes a rounded crest; and ends abruptly before the end of the block is reached. In accordance with the proposed interpretation, this condition indicates that at the west end, where the thrust broke across from the lower to the upper gliding bed, the strata were pushed forward for several miles; at the east end of the present Powell Valley anticline the forward movement on the lower gliding plane was sufficient to form a narrow anticline; and still farther east the thrusting occurred only along the upper of the two gliding zones, so that no anticline was formed.

Confidence in the foregoing interpretation of the nature of the thrusting is heightened by the discovery, in the southeast corner of the map in the Maynardville folio, of a structure section, reproduced as Figure 7, showing a beveled remnant of a structure such as that shown on Butts' cross section (Fig. 3), at and immediately north of

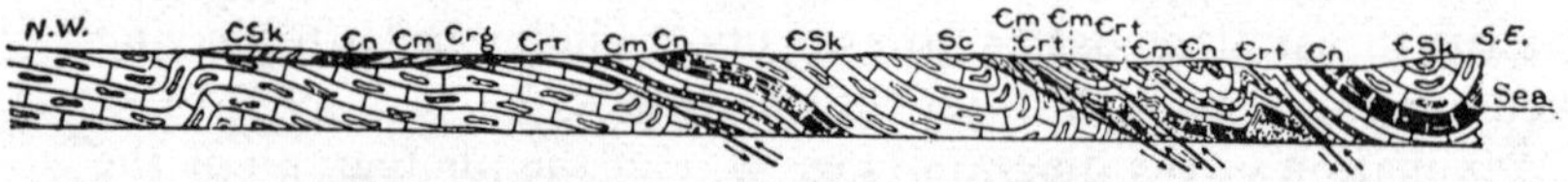

FIG. 7.—Reproduction of structure section from southeast corner of Maynardville folio. Note shape of thrust plane at left and manner in which strata dip into thrust plane along flat part of its course. Compare with cross section of Cumberland block (Fig. 3) in area northwest of fensters.

the fensters, and also on the model (Fig. 6), where the thrust breaks from the lower gliding plane up to the higher. This structure section proves that the type of thrust plane and thrust movement indicated for the Cumberland block is not unique, but occurs in connection with other faults of the Appalachian valley.

CROSS FRACTURES IN CUMBERLAND BLOCK

As has already been mentioned (Fig. 1), the Cumberland block is broken across by at least three lines of disturbance—one between Speedwell and Jellico, another between Cumberland Gap and Pineville, and a third extending from the eastern end of the Powell Valley anticline near Little Stone Gap northward toward Pine Mountain at Pound Gap.[9]

These lines of disturbance cross the block in directions which make the same angles with the Pine Mountain fault and with the block as a whole as the tear faults at its ends—essentially shearing

[9] See map accompanying the report of J. Brian Eby and others, "The Geology and Mineral Resources of Wise County and the Coal-Bearing Portion of Scott County, Virginia," *Virginia Geol. Survey Bull. 24* (1923).

angles. The most southwesterly of these lines of disturbance extends diagonally northwestward across the block from the vicinity of Speedwell toward Jellico. No adequate description of the details of its structure has been found; but, as inferred from a meager description[10] and from the geological map of Tennessee, it seems to be a line of rather sharp monoclinal dip northeast, so that the portion of the block southwest of this line stands somewhat higher structurally than that on the northeast. The trend of the disturbed zone appears to be about northwest. The zone has the appearance of an incipient tear fault separating an area of somewhat greater thrust movement in the Cumberland block west of it from one of less movement on the east.

The middle one of the three cross zones of disturbance is the Rocky Face fault zone extending from Cumberland Gap northward toward Pineville, and trending approximately parallel with the Jacksboro tear fault at the end of the block. It has been described by Ashley and Glenn[11] and certain of its implications discussed by Rich.[12] At Rocky Face it is essentially a buckled half dome on the east side of a fault upthrust toward the northwest.

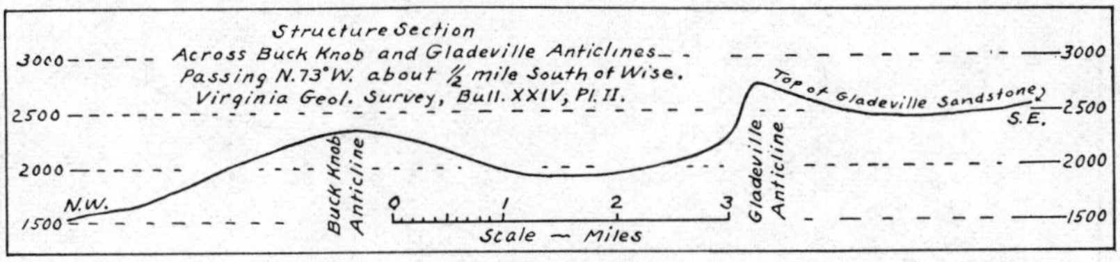

FIG. 8.—Profile across Buck Knob and Gladeville anticlines showing asymmetry of latter.

The third disturbed zone crosses the block northward from the point where the Powell Valley anticline dies out. It has been described and mapped by structure contours.[13] It consists of two parallel anticlines, the Buck Knob anticline and the Gladeville anticline, whose forms and relations to each other are shown in cross section in Figure 8. The eastern of these, the Gladeville anticline, is very unsymmetrical, being steep on the west side. It is marked by two sharp subsidiary domes along its crest.

[10] L. C. Glenn, "Northern Tennessee Coal Field," *Tennessee Geol. Survey Bull.* *33-B* (1925).

[11] G. H. Ashley and L. C. Glenn, *op. cit.*

[12] John L. Rich, "Physiography and Structure at Cumberland Gap," *Bull. Geol. Soc. America*, Vol. 44 (1933), pp. 1219–36.

[13] J. Brian Eby and others, *op. cit.*

All these cross fractures are believed to be superficial features of the Cumberland thrust block, representing a tendency toward breaking into segments as the block was pushed northwestward, and it is believed that all these cross fractures and folds will be found to end downward at the thrust plane underlying the block.

It seems that the formation of the Buck Knob and Gladeville anticlines can be accounted for by the excess northwestward shoving of that portion of the block east of the end of the Powell Valley anti-

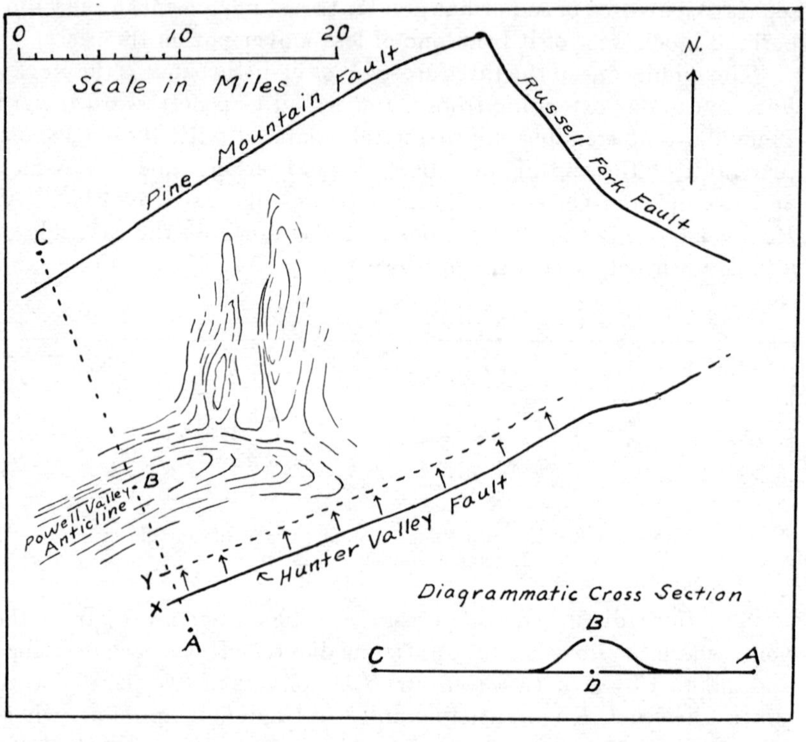

FIG. 9.—Sketch map of eastern end of Cumberland block to illustrate discussion of origin of Buck Knob and Gladeville anticlines. General form of anticlines indicated by sketch structure contours.

cline. Assuming that the Powell Valley anticline was formed during the thrusting, and as a result of the shearing up of the thrust plane from a lower bed to the horizon of the Chattanooga shale, and further assuming a uniform push from the southeast and a movement which may be diagrammatically represented by the distance *XY* (Fig. 9); then the northwestward movement of a point north of the anticline must have been less than *XY* by an amount equal to the difference between the length of a line such as *ABC* over the top of the anticline, and a straight line *ADC* (see cross section, Fig. 9), whereas east of the end of the Powell Valley anticline the northwestward movement would have been the full amount, *XY*.

The beds east of the end of the anticline, being crowded northwestward in this way farther than those on the west, must have tended to tear loose along a shearing plane or incipient tear fault, forming a block of the same general shape as the Cumberland block as a whole. In view of the trend of the incipient tear fault, the excess northward movement of the area east of the end of the Powell Valley anticline must have produced a wedging effect similar to that which caused the upthrusting along the Jacksboro tear fault at the southwestern end of the Cumberland block. In this case, the movement seems not to have gone far enough to form an actual break, but only enough to form the highly unsymmetrical Gladeville anticline and the Buck Knob anticline in front of it.

The extreme sharpness of the Gladeville anticline and its asymmetry suggest strongly that it may pass downward into a thrust fault. The anticline bears many points of similarity to the Rocky Face fault zone near Cumberland Gap, even to the presence of sharp local domes along its crest, but it differs in that it has not been so far developed.

PROBABLE SHALLOWNESS OF LOCAL STRUCTURES IN CUMBERLAND BLOCK

The exposure of the fensters in the Cumberland block (Fig. 1) seems to make inevitable Butts' conclusion that the entire Cumberland block is underlain by a low-angle thrust plane and that the Pine Mountain fault marks the line along which the thrust plane sheared upward to the surface.

If the whole block is underlain by a thrust plane, it seems probable that any local anticlines or synclines appearing within the block are only superficial and do not extend down below the thrust plane which, as nearly as can be judged at present, follows the approximate horizon of the Chattanooga shale. There seems to be no reason to expect that such features as the Rocky Face half-dome or the Buck Knob or Gladeville anticlines extend below the thrust, especially since, as in the latter case, the presence of such features in the rocks above the thrust seems to be required by the dynamics of the thrusting.

BROADER ASPECTS—POSSIBLE CORRELATION OF PINE MOUNTAIN FAULT WITH SEQUATCHIE VALLEY ANTICLINE

Analogy with the features displayed by the Cumberland overthrust block strongly suggests that the Sequatchie Valley anticline[14] is the expression of similar low-angle thrust faulting carried not so far as in the Cumberland block. The Sequatchie anticline, where eroded most deeply, reveals thrust faulting in its lower part; its distance from the western margin of the Appalachian Valley belt of strong shingle-type thrust faulting is comparable with that of the Pine Mountain fault; and the anticline west of Lookout Mountain in the southern part of the area occupies a position analogous to that of the Powell Valley anticline.

It is not difficult to picture a somewhat greater original movement along the thrust plane bringing the thrusts now exposed in the deeper parts of the Sequatchie anticline up to the surface (before erosion), shoving Walden Plateau farther westward, breaking it loose along a diagonal tear fault somewhere north of Pikeville, Tennessee, and producing an overthrust block almost identical with the Cumberland block.

BROADER BEARING ON GENERAL PROBLEM OF OVERTHRUST FAULTING

Revelation of the way in which thrust faults may follow for long distances along the bedding of strata, such as shales, on which movement is easy, completely changes the basis of calculations of the possible distance the rocks may move on overthrust faults. In the past such calculations have been based on the coefficient of friction of dry granite. The friction involved in movement along the bedding of clay shales, probably wet with water at the time of movement, is something of an entirely different order than that for granite.

Wentworth, for example, made a calculation of the force required to move the Cumberland block,[15] but he used the coefficient of friction of dry granite. His results would have been radically different had they been based on movement along the bedding of shales.

Recognition of the importance of thrusting along bedding planes throws a new light on many problems in thrust faulting. For example, the striking fact that in crossing the Appalachian valley southeast-

[14] The Sequatchie Valley anticline, as conspicuously shown on the geologic maps of Tennessee and Alabama, forms a remarkable anticlinal valley essentially in line and trend with the Pine Mountain fault, though separated from it by a gap of nearly 50 miles. The portion of the Allegheny Plateau between Sequatchie Valley and the Appalachian Valley province is known as Walden Plateau.

[15] Chester K. Wentworth, *op. cit.*

ward from the Cumberland block toward the great up- and over-thrust mass of the crystalline Appalachians one passes half a dozen or more large thrust faults which bring up repeatedly, shingle-fashion, almost the whole of the unmetamorphosed Paleozoic sedimentary series without once bringing up the crystalline basement, makes it appear likely that the thrusting in that part of the Appalachian valley is entirely confined to the sediments, which have been sheared off from the underlying crystalline basement, pushed forward, and piled up in shingle-fashion by a great plunger moving from the southeast—presumably the pre-Cambrian mass of the crystalline Appalachians—which came up diagonally along a shear plane to the base of the sediments, then rode forward horizontally, pushing and piling up the sediments before it. It is not unlikely that the pre-Cambrian basement beneath the Appalachian valley would be found to be undisturbed.

BULLETIN OF THE AMERICAN ASSOCIATION OF PETROLEUM GEOLOGISTS
VOL. 34, NO. 4 (APRIL, 1950), PP. 672-681, 4 FIGS.

MECHANICS OF APPALACHIAN FOLDING AS ILLUSTRATED BY SEQUATCHIE ANTICLINE, TENNESSEE AND ALABAMA[1]

JOHN RODGERS[2]
New Haven, Connecticut

ABSTRACT

There is controversy as to whether the folds of the Appalachian Valley and Ridge province reach to the crystalline basement and are controlled by it, or whether they are independent of the basement, all the deformation having taken place within the overlying sediments. The Sequatchie anticline of Tennessee and Alabama has been cited as evidence by proponents of both schools of thought. The theory of the second school, as developed by Rich, requires transverse faults at each end of the fold. These faults are present, though their significance has been overlooked. Further application of the theory to the Sequatchie anticline suggests the presence of two windows comparable with the oil-bearing windows of Lee County, Virginia.

INTRODUCTION

In recent decades two rather different views on the mechanics of the folding in the Appalachian Valley and Ridge province have grown up. One theory holds that the folds in the Paleozoic sedimentary rocks are reflections of adjustments in the underlying pre-Cambrian basement, the forces that produced the folds having been transmitted upward from that basement. The other theory holds that the folds are quite independent of the basement and that the folded rocks have been stripped off it by forces acting tangentially within the sedimentary prism. (For a review of the history of these ideas, see Rodgers, 1949.) The present paper is an attempt to analyze what is known about one of the more prominent Appalachian folds, in the hope of determining which theory provides a better explanation.

SEQUATCHIE ANTICLINE

The Sequatchie anticline (Fig. 1) extends southwestward from the south side of the Emory River in Morgan County, Tennessee, nearly to the forks of the Black Warrior River in Walker and Jefferson counties, Alabama, a distance of more than 200 miles. At its north end it is a low, nearly symmetrical arch in the Pennsylvanian rocks of the Cumberland Plateau, the uparched Pennsylvanian sandstone and conglomerate beds forming a prominent line of hills (the Crab

[1] Published by permission of the director of the United States Geological Survey. Manuscript received, October 29, 1949.

[2] Department of geology, Yale University. Acknowledgment is due John L. Rich, of the University of Cincinnati, to whose article on the Pine Mountain thrust sheet (Rich, 1934), the present paper is merely a sequel. The ideas in the paper have been discussed with several colleagues on the United States Geological Survey, especially Ralph L. Miller, under whose guidance the writer had the opportunity to see the Lee County windows. Philip B. King of the Survey criticized the manuscript with care and thoroughness. The manuscript has been read also by R. E. Sherrill of the University of Pittsburgh, C. R. Longwell of Yale University, B. C. Moneymaker of the Tennessee Valley Authority, H. W. Ferguson of the Tennessee Division of Geology, and R. A. Laurence, C. B. Hunt, and J. B Hadley of the United States Geological Survey, to all of whom my thanks are due for their helpful criticism. The figures were drawn by John Evans and Shirley P. Glaser.

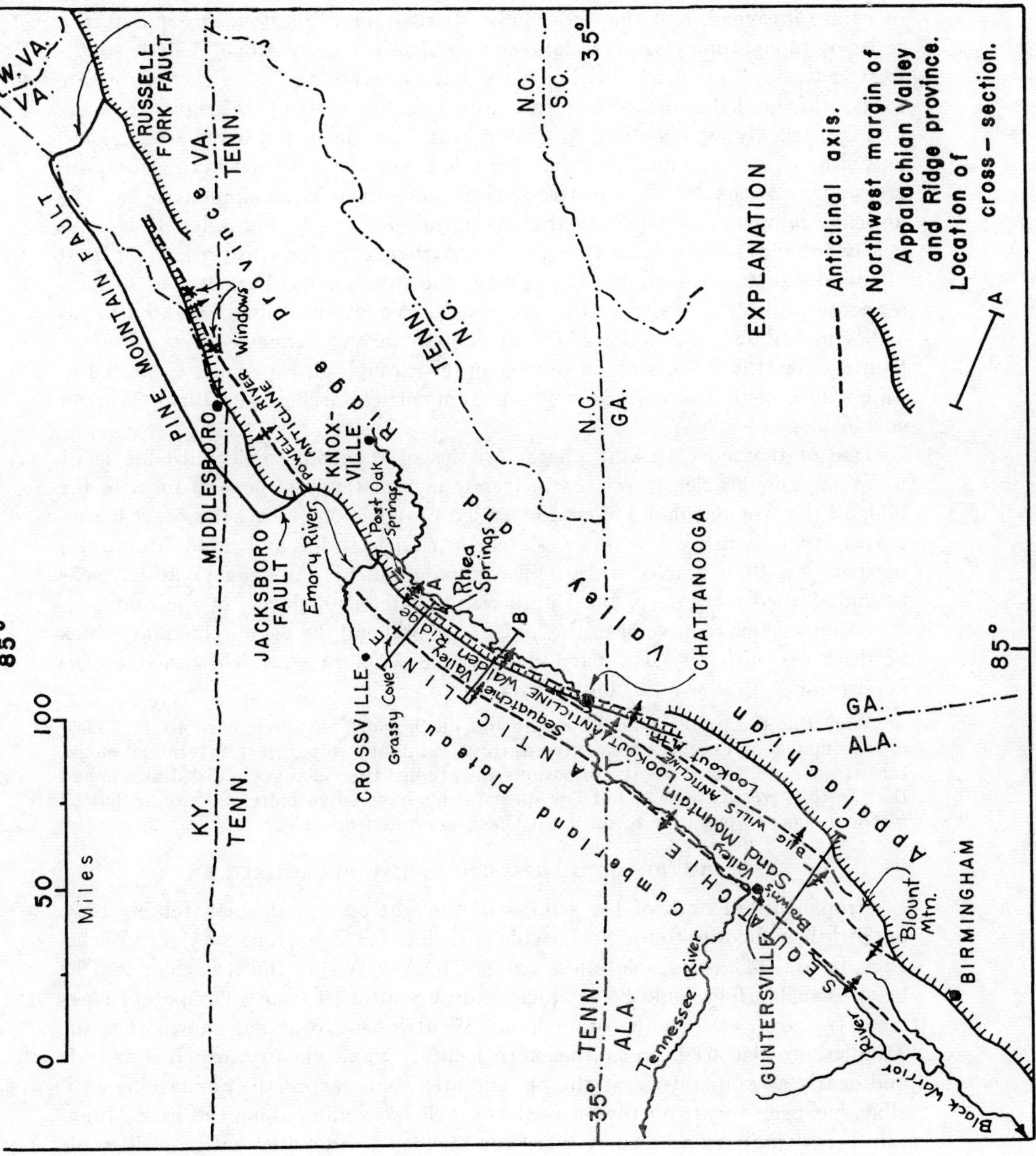

FIG. 1.—Major structural features in Cumberland Plateau province of southern Appalachians.

Orchard Mountains). Traced southwest, the fold rises so that the Pennsylvanian rocks are breached, and the underlying Mississippian limestone is exposed in a series of topographic lows, the largest of which is Grassy Cove. A little southwest of Grassy Cove, the anticline rises sharply, exposing Ordovician carbonate rocks, which are the surface rocks along its crest for most of its length. Here the fold is markedly asymmetrical, its northwest flank being much the steeper, and only a little farther southwest that flank is broken by a thrust fault, whose increasing displacement thrusts Ordovician rocks over Mississippian rocks. The anticline retains approximately this magnitude for nearly 130 miles, to a point nearly 20 miles southwest of Guntersville, Alabama; the fault is present for most of this distance. Throughout this segment the anticline has been eroded to form a narrow and fertile valley, called Sequatchie Valley in Tennessee and Browns Valley in Alabama, and followed by the Sequatchie and Tennessee rivers. Beyond Guntersville, the fold passes in reverse order through all the stages exhibited at its northeast end and ends as a gentle symmetrical arch in the Black Warrior coal field.

One of the most striking characteristics of the Sequatchie anticline is its isolation. Though clearly related in trend, asymmetry, and general form to the folds of the Appalachian Valley and Ridge province on the southeast, it is separated from them by a strip of plateau (Walden Ridge and Sand Mountain) nowhere less than 7 miles wide, which is underlain by flat Pennsylvanian rocks as undisturbed as those of the Cumberland Plateau northwest of the fold.

Safford seems to have been the first to comprehend the Sequatchie fold, which he describes with his customary clarity (1869, pp. 137–41). He concludes his description with the following remarks.

> Were this the proper place, I might add much more with reference to this most interesting fold and range. I have, myself, traversed it, and stood upon its terminal mountain at each end, traced out its formations and studied their changes, and I desire to add that, in the investigations of but few special fields have I had more pleasure, or have I met with more to point me to the great Creator, whose work it is.

APPLICATION OF THEORIES TO SEQUATCHIE ANTICLINE

Proponents of both of the schools of thought on Appalachian folding have applied their theories to the Sequatchie anticline (Rich, 1934, p. 1595; Ver Wiebe, 1936, p. 924). Thus its isolation is stressed by Ver Wiebe, and it is clear that he believes such a fold could be produced only by upthrust from a competent basement (pp. 934–35). On the other hand, Wentworth (1921) has shown that the Middlesboro coal basin in southeastern Kentucky and vicinity, which is as wide and nearly as undisturbed as the plateau area southeast of the Sequatchie anticline, has been thrust northwestward as much as 10 miles along the Pine Mountain thrust fault.[3] To be sure, Wentworth's idea was greeted with considerable

[3] This statement and similar statements which follow are phrased *as if* it were known that the faulted rocks on the southeast moved northwestward while the Cumberland Plateau on the northwest stood still. Actually, of course, all that is known is the relative movement.

skepticism when first announced (*cf.* Glenn, 1925, pp. 140–41), but the discovery by Butts (1927a) of windows in the Pine Mountain fault in Lee County, Virginia, more than 15 miles southeast of the main trace of the fault, made Wentworth's interpretation inescapable. It is a remarkable fact, noted by all observers (Safford, 1869, p. 142; Campbell, 1893, p. 21; Rich, 1934, p. 1595, footnote; Ver Wiebe, 1936, p. 924), that the Sequatchie anticline is almost exactly in line with the Pine Mountain fault, though the two are different in form and are separated by 30 miles of relatively undisturbed strata. Rich, having shown that the movement on the Pine Mountain fault was independent of the basement and was impelled by tangential forces acting within the sedimentary rocks, suggested (1934, p. 1595) that a similar movement, though carried less far, might explain the Sequatchie anticline. Clearly the facts so far adduced can be adjusted to fit either theory.

If the Sequatchie anticline and the Pine Mountain fault are to be explained by the same mechanism, then somewhere near the ends of the Sequatchie fold there should be structural features like the two cross faults that cut off the two ends of the thrust sheet above the Pine Mountain fault. There is no need to discuss those cross faults in detail here, as they have been adequately described by Wentworth (1921), but it may be noted that they differ considerably from each other, apparently because the thrust sheet in moving northwestward pivoted more or less on its northeast end. Thus the Russell Fork fault at the northeast end is chiefly a tear fault, with little stratigraphic displacement but much crushing, whereas the Jacksboro fault at the southwest end is both a tear and a thrust fault and has a large stratigraphic displacement.

STRUCTURAL FEATURES NEAR THE ENDS OF THE SEQUATCHIE ANTICLINE

Areas of exceptionally disturbed rocks near the Emory River north and northeast of the north end of the Sequatchie anticline (Fig. 2) were noted by Keith (1897, p. 3) and by Glenn (1925, pp. 334–35), but the first geologist to recognize that these disturbances are related and show a consistent pattern was Jillson (1923). Jillson stated that the observable features mark a west-northwest-trending thrust fault dipping southwest, and that at the northwest this fault crosses a west-southwest-trending (faulted) anticline. The proposed fault he called the Emory River fault and the anticline the Hatfield Mountain anticline. He recognized the similarity of the fault to the Jacksboro fault at the southwest end of the Pine Mountain fault but did not carry the comparison further.

Since Jillson's paper appeared, excellent new topographic maps of the area in question have been issued by the Tennessee Valley Authority and the United States Geological Survey, and they show that many of the structural features have strong topographic expression. Hoping to glean additional facts, the writer made a brief excursion into the area in 1947, and is convinced that a continuous line of disturbance exists approximately along the line of Jillson's proposed Emory River fault, but is inclined to think that the disturbance is mostly sharp buckling of the competent sandstone layers, broken only here and there by small thrust faults which may dip either north or south. As Jillson indicated, however, the block south of the line has risen relative to the block on the north. It is also suggested that the line of buckling does not cross the Hatfield Mountain anticline, as Jillson thought, but instead abruptly changes direction and merges into the anticline. From Hatfield Mountain, a series of anticlines continues *en échelon* west-southwest to and beyond Crossville and terminates against a marked west-northwest-trending tear fault, the Potts Creek fault (Butts, 1925, geologic map).

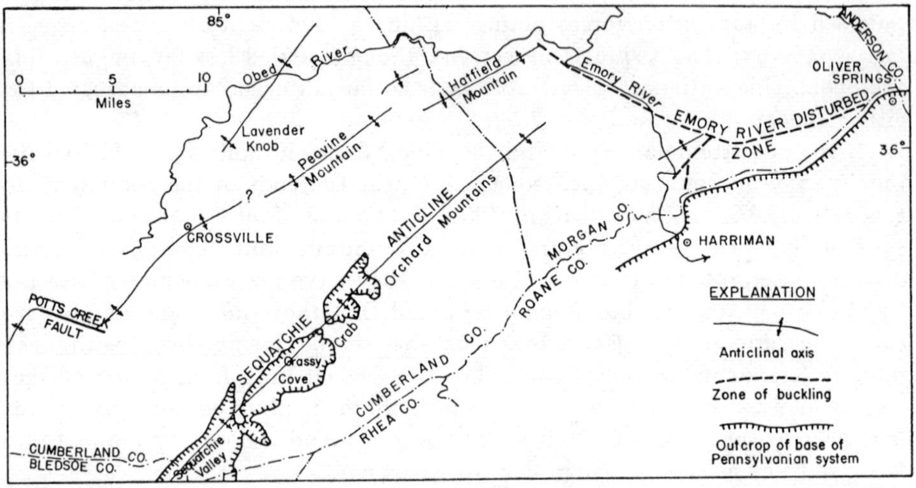

FIG. 2.—Minor structural features near north end of Sequatchie anticline, Tennessee.

Thus the Emory River disturbed zone may closely resemble the Jacksboro fault as it was at a very early stage in its development, and the abrupt change from the west-northwest line of buckling to the west-southwest line of broader folding may correspond mechanically with the abrupt change from the Jacksboro fault to the Pine Mountain fault. The Hatfield Mountain-Crossville line of folds is somewhat northwest of the plunging end of the Sequatchie anticline, but doubtless if movement had continued the latter would have carried through to the Emory River line and the other folds would have ceased to grow.

If the Emory River line at the northeast end of the Sequatchie anticline was formed in the same manner as the Jacksboro fault at the southwest end of the Pine Mountain fault, tear faults like the Russell Fork fault might be expected at the southwest end of the anticline. In the coal basin northwest of Birmingham, Alabama, Butts (1910, 1927b, economic geology maps) found north-northwest-trending and north-trending faults which he described (1910, p. 11) as "short normal faults . . . They vary in throw from 70 to 110 feet. The fault planes are generally inclined at 70° or more, but a single one is reported with an inclination of 45°." It is clear from the structure contours on Butts' maps that these faults are downthrown here on one side, there on another, and it therefore is possible that they are not normal but strike-slip or tear faults. In position and trend they accord well with what would be expected by analogy with the Russell Fork fault.

INTERPRETATION OF SEQUATCHIE ANTICLINE

The success of Rich's analogy with the Pine Mountain fault in predicting[4] structural features associated with the Sequatchie anticline creates a strong presumption that he was correct in suggesting that the anticline and the fault are mechanically similar. The remainder of this paper is therefore devoted to an interpretation of the Sequatchie anticline and structurally related features comparable with that worked out by Wentworth (1921), Butts (1927a), and Rich (1934) for the features of the Pine Mountain thrust sheet. According to Rich (1934), the Pine Mountain fault for most of its extent downdip is a bedding-plane fault that follows tectonically weaker layers or formations in the sedimentary column, and that only here and there cuts across from one such layer to another along diagonal shears through the more competent formations. The details of the cross-cutting have determined the location, size, and shape of the various structural features of the thrust sheet above the fault, such as the Middlesboro coal basin, the Powell River anticline southeast of it, and the windows along the crest of the anticline (Fig. 1).

Now throughout the long central segment of the Sequatchie anticline, Upper Cambrian and Lower and Middle Ordovician carbonate rocks are exposed along the crest of the fold, and a thrust fault on the northwest flank of the fold throws these rocks over Mississippian rocks. If the Sequatchie anticline corresponds with an early stage in the development of the Pine Mountain block, then the fold and fault represent one of the places where the fault broke across competent beds from one weak layer to another or perhaps actually to the surface of the ground. The next weak zone below the massive carbonate rocks now exposed in the core of the anticline is the Conasauga shale. Presumably, therefore, the entire mass of Walden Ridge and Sand Mountain, down to the Conasauga shale, has been shoved northwestward along a bedding-plane fault or faults in the shale. As a result of this movement, which brought Walden Ridge and Sand Mountain closer to the main Cumberland Plateau, the rocks buckled up along the line where the fault cuts upward from the Conasauga shale and formed the present Sequatchie anticline. In other words, the anticline is a mere by-product of the faulting, formed where the fault cut from one weak layer to another. At least a minimum estimate of the movement can be obtained by reconstructing the fault and fold and measuring the shortening they record. Section B in Figure 3 includes a cross section of the fold between Pikeville and Dunlap, Tennessee, and suggests a minimum displacement of $1\frac{1}{4}$ miles. If the movement had been much greater, the rocks above the fault would have lapped over the rocks of the Cumberland

[4] The features in question had been described by 1934, but were not known to Rich when he wrote his paper (personal communication).

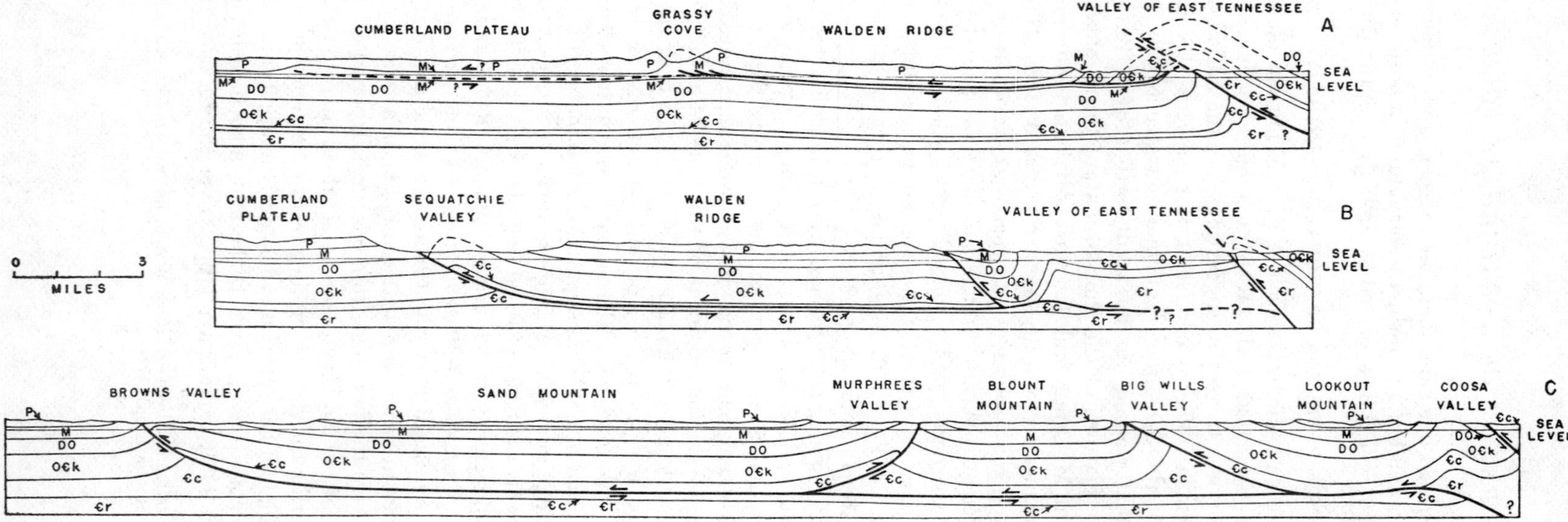

FIG. 3.—Interpretative sections across Sequatchie fold and related structural features, along lines shown in Figure 1. P—Pennsylvanian rocks. M—Mississippian rocks. DO—Devonian to Middle Ordovician rocks. O-Ꞓk—Knox group. Ꞓc—Conasauga shale. Ꞓr—Rome formation. Horizontal and vertical scales, same.

Plateau, but the marked diminution of the anticline in both directions from the central segment rules out any such large movement.

Southeast of Walden Ridge, along the same line of section (Section B, Fig. 3), is a belt of nearly vertical beds, locally cut by a thrust fault or two, beyond which older rocks are exposed in a faulted anticline; only southeast of this do the typical thrust slices of the Valley and Ridge province begin. By analogy with the Powell River anticline, this anticline may mark the point at which the fault under Walden Ridge descends from the Conasauga shale to the next lower weak layer, probably in the lower part of the Rome formation.

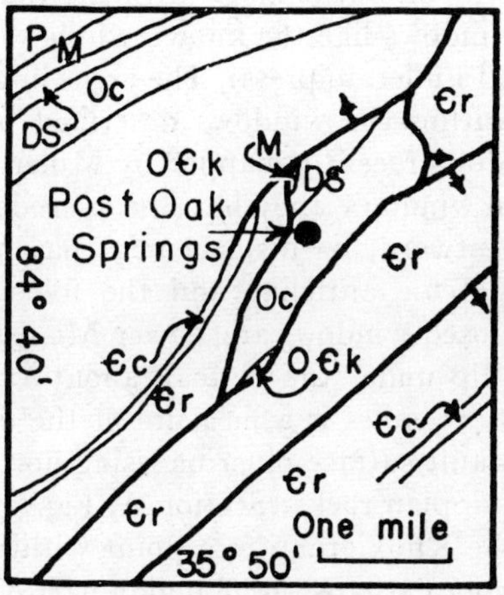

FIG. 4.—Geology around Post Oak Springs, Roane County, Tennessee. Simplified from Hayes, 1894. Heavy lines are faults; direction of their dip, as shown on Hayes' cross section, indicated by arrows. P—Pennsylvanian rocks. M—Mississippian rocks. DS—Devonian and Silurian rocks. Oc—Chickamauga limestone. O€k—Knox group. €c—Conasauga shale. €r—Rome formation.

Farther southwest, in northeast Alabama and extreme northwest Georgia, the pattern of folds becomes more complex and can no longer be compared with the relatively simple pattern behind the Pine Mountain fault. Southeast of Sand Mountain there are additional plateau-like masses of flat Pennsylvanian rocks (Lookout Mountain, Blount Mountain) separated from Sand Mountain by narrow anticlinal valleys (Big Wills Valley, Murphrees Valley) comparable with Sequatchie Valley, and only southeast of these does the typical structure of the Appalachian Valley and Ridge province appear. Section C in Figure 3 gives one possible interpretation of these features.[5]

Somewhat northeast of Section B (Fig. 3) and just south of Grassy Cove, the

[5] The writer can not follow Rich (1934, p. 1595) in drawing an analogy between the Powell River anticline and the Lookout anticline, for the latter is northwest of the plateau-like mass of Lookout Mountain. However, it will not be possible to settle the mechanics of the Lookout anticline until a detailed map is available for the peculiar doubled plunging anticline southeast of Lookout Mountain and just south of Chattanooga (Butts and Gildersleeve, 1948, p. 61).

Sequatchie anticline plunges northeastward rather abruptly, and beyond that point no rocks older than Mississippian are exposed. Northwest of the point of steep plunge, and trending more or less toward it, is the Potts Creek tear fault, which marks the southwest end of the Hatfield Mountain-Crossville line of folds (Fig. 2). Southeast of this northeastern part of the Sequatchie anticline, in the core of the first anticline southeast of Walden Ridge, are two anomalous areas of younger rocks completely surrounded by older rocks (Hayes, 1894, geologic map, areas around Post Oak Springs, Roane County, and Rhea Springs, Rhea County; see Fig. 4). Hayes interpreted these areas as bounded by thrust faults facing each other across them, but analogy with the Pine Mountain thrust sheet suggests that they are windows like the known windows in Lee County, Virginia (Butts, 1927a; Miller and Fuller, in press). The areas in question are considerably larger than the areas within the windows described by Butts, but not much larger than in other windows recently mapped by Miller, and the geology appears to be similar. If they are windows, they have been eroded through a thrust fault that disappears northwestward, its displacement having been taken up in the arching along the Sequatchie anticline and the folds farther northwest. The highest beds in the supposed windows are Lower Mississippian, and in the overlying block these beds dip under the plateau about $1\frac{1}{2}$ miles northwest. If the interpretation is correct, $1\frac{1}{2}$ miles is a measure of the displacement of the fault at this point. If so, the fault surface must be rising northwestward to some bedding plane in the Mississippian rocks (Section A, Fig. 3), whereas farther southwest it remains below the Knox group to a point within a mile or so of the axis of the Sequatchie anticline. Perhaps the plunging part of the Sequatchie fold and the Potts Creek fault on the northwest mark the position of the shift.

The writer is not familiar enough with the geology of northern Alabama to hazard any interpretation of the geology around the southwest-plunging end of the anticline, but if the analogy with the Pine Mountain fault holds, the displacement should gradually decrease and become negligible in that direction.

CONCLUDING REMARKS

If the interpretation offered in this paper seems bizarre and theoretical, field studies of the critical areas should nevertheless prove or disprove it. It is hoped therefore that those who disagree will be stimulated to study such areas and uncover additional facts bearing on the problem, for the problem is not one of merely academic interest. Oil has been discovered beneath the Pine Mountain fault in the area around the windows in Lee County, Virginia, and if the Sequatchie fold is comparable with the Pine Mountain fault, the Post Oak Springs and Rhea Springs areas are comparable with the oil-bearing areas. In more general terms, a decision on the merits of the two schools of thought about the depth of Appalachian deformation is necessary before a correct appraisal of the oil possibilities of the thrust-faulted belt of the southern Appalachians can be made.

REFERENCES CITED

BUTTS, CHARLES (1910), "Birmingham, Alabama," *U. S. Geol. Survey Geol. Atlas Folio 175.*
——— (1925), "Geology and Mineral Resources of the Crossville Quadrangle, Tennessee," *Tennessee Div. Geology Bull. 33-D.*
——— (1927a), "Fensters in the Cumberland Overthrust Block of Southwest Virginia," *Virginia Geol. Survey Bull. 28.*
——— (1927b), "Bessemer-Vandiver, Alabama," *U. S. Geol. Survey Geol. Atlas Folio 221.*
———, AND GILDERSLEEVE, BENJAMIN (1948), "Geology and Mineral Resources of the Paleozoic Area in Northwest Georgia," *Georgia Geol. Survey Bull. 54.*
CAMPBELL, M. R. (1893), "Geology of the Big Stone Gap Coal Field of Virginia and Kentucky," *U. S. Geol. Survey Bull. 111.*
GLENN, L. C. (1925), "The Northern Tennessee Coal Field," *Tennessee Div. Geology Bull. 33-B.*
HAYES, C. W. (1894), "Kingston, Tennessee," *U. S. Geol. Survey Geol. Atlas Folio 4.*
JILLSON, W. R. (1923), "Emory River Overthrust of East Tennessee," *Pan-American Geol.*, Vol. 39, pp. 373–78.
KEITH, ARTHUR (1897), "Wartburg, Tennessee," *U. S. Geol. Survey Geol. Atlas Folio 40.*
MILLER, R. L., AND FULLER, J. O. (in press), "Geology and Oil Resources of the Rose Hill District, Lee County, Virginia," *Virginia Geol. Survey Bull. 71.*
RICH, J. L. (1934), "Mechanics of Low-Angle Overthrust Faulting as Illustrated by Cumberland Thrust Block, Virginia, Kentucky, and Tennessee," *Bull. Amer. Assoc. Petrol. Geol.*, Vol. 18, pp. 1584–96.
RODGERS, JOHN (1949), "Evolution of Thought on Structure of Middle and Southern Appalachians," *Bull. Amer. Assoc. Petrol. Geol.*, Vol. 33, No. 10 (October), pp. 1643–54.
SAFFORD, J. M. (1869), *Geology of Tennessee*, Nashville.
VER WIEBE, W. A. (1936), "Geosynclinal Boundary Faults," *Bull. Amer. Assoc. Petrol. Geol.*, Vol. 20, No. 7 (July), pp. 910–38.
WENTWORTH, C. K. (1921), "Russell Fork Fault of Southwest Virginia," *Jour. Geology*, Vol. 29, pp. 351–369; *Virginia Geol. Survey Bull. 21*, pp. 53–66.

BULLETIN OF THE AMERICAN ASSOCIATION OF PETROLEUM GEOLOGISTS
VOL. 47, NO. 8 (AUGUST, 1963), PP. 1527-1536, 2 FIGS.

MECHANICS OF APPALACHIAN FORELAND FOLDING IN PENNSYLVANIA AND WEST VIRGINIA[1]

JOHN RODGERS[2]
New Haven, Connecticut

ABSTRACT

The Sandhill well on the north-south Burning Springs anticline in West Virginia found little or no deformation below the Lower Devonian. Nevertheless the shortening produced in the Carboniferous and Devonian strata by the anticline—probably not less than $\frac{1}{4}$ mile—can not simply disappear but must be taken up somewhere, both laterally, beyond the abrupt ends of the anticline, and beneath, in the underlying strata.

The pattern of fold trends south and east of the Burning Springs anticline suggests the presence of conjugate sets of strike-slip faults that have hitherto escaped recognition because, being in flat sediments, they show no stratigraphic displacement. Such faults may mark parts of the boundary of a block whose shifting produced the Burning Springs anticline; their orientation suggests that it moved roughly N.50°W. relative to the surrounding parts of the Appalachian Plateau. Complex structure reported in the Lower Devonian rocks under other foreland folds in northern West Virginia and southwestern Pennsylvania indicates that they are similar in mechanics and that the block extends northeast at least into west-central Pennsylvania. The location of the Burning Springs anticline over the western pinch-out of Upper Silurian salt suggests that the block has slipped forward on the salt layers, leaving the underlying strata relatively undisturbed.

The hypothesis is therefore presented that the Appalachian foreland folds of Pennsylvania and West Virginia result from the slipping of the upper strata of a part of the Appalachian Plateau northwestward over their substratum, as a marginal effect of the deformation of the Valley and Ridge province on the southeast. The mechanics is exactly that demonstrated by Rich for the Pine Mountain thrust sheet in a similar position not far southwest; the block in question is larger but would have moved much less far.

General Character of Folds

The folds in the rocks of the Appalachian Plateau province of Pennsylvania and West Virginia (Fig. 1) have been known since the geological surveys by the Rogers brothers in the 1830s and 1840s, and they have been repeatedly studied because of their influence on the location of the great deposits of bituminous coal and of oil and gas in the region. Most of the area has been covered by detailed geological maps, generally published on a scale of a mile to an inch and accompanied by structure-contour maps showing the attitude of the surface strata. Moreover, much subsurface information has been accumulated, though a large part of this is still unpublished.

The surface rocks in this part of the Plateau province outline a major structural basin, the Pittsburgh-Huntington basin,[3] and the folds here considered occur mostly on its southeast flank, southeast of its vaguely defined trough line (as expressed in the surface strata). They show both resemblances to and differences from the classical Appalachian folds of the Valley and Ridge province next southeast. The most notable resem-

[1] Manuscript received, October 29, 1962.

[2] Department of Geology, Yale University.

Much of the material in this paper was presented at a joint meeting of the Pittsburgh and Appalachion Geological Societies at Summit, Pennsylvania, on October 14, 1960. The writer is indebted to Addison S. Cate, H. P. Woodward, John C. Ludlum, James Affleck, and others who were there for ideas and stimulation, and he also recalls with pleasure meeting Mrs. Richard P. Sherrill and Mrs. Charles R. Fettke, whose husbands did so much of the basic work in western Pennsylvania. John C. Ludlum, Herbert P. Woodward, Robert E. Bayles, Paul H. Price, Oscar L. Haught, John M. Dennison, Ping-fan Chen, Addison S. Cate, and James Affleck kindly read the paper as prepared for publication and suggested improvements. They do not, however, necessarily accept the conclusions. Special thanks are due to Edward C. Hansen, who drew the figures.

It must be emphasized that the writer does not claim originality for the ideas in this paper. He developed the main ideas in 1959 shortly after first reading West Virginia Geological Survey Report of Investigation 14 (Bayles and others, 1956), but he found that they were already "in the air" at the Summit meeting, and evidently they occurred to many other people (e.g., Ludlum, 1958; Woodward, 1959a; Carey, 1962—a lecture given in May, 1960); moreover, Robert Bayles states (personal communication) that Finn and Fettke had similar ideas many years earlier. This paper is intended to show rather that these ideas make a consistent picture, one that deserves to be borne in mind as an alternative hypothesis by those working in the Appalachian Plateau province.

[3] Sometimes incorrectly called the Appalachian geosyncline, which however is the belt where Paleozoic strata are thick in contrast to the platform area on the west where they are thin, and which includes therefore the Valley and Ridge province if not also the Blue Ridge and much of the Piedmont. Perhaps a still better term for the basin is Allegheny synclinorium or Allegheny basin.

blance is in areal pattern; the folds of the two provinces are for the most part parallel among themselves and also from one province to the other. In northern and eastern West Virginia (less so in central West Virginia) they are nearly straight and trend about N. 30° E. in both provinces, whereas in Pennsylvanian they outline great arcs convex toward the northwest (the arc in the Valley and Ridge province seems to be a bit tighter, and its center of curvature is probably not as far southeast as that of the arc on the Plateau). Moreover, individual fold axes are notably persistent along their trend, though on the Plateau they tend to show more obvious *en échelon* offsets than in the Valley and Ridge. The wave length of the folds is also comparable in the two provinces, averaging 5–10 miles and being a little larger on the Plateau.

The most obvious differences between the two groups of folds are in amplitude and steepness. In Pennsylvania, the largest of the Plateau folds, the Chestnut Ridge anticline at Summit, has a structural relief at the surface of nearly 3,500 feet and a flank dip of 20°, but none of the others passes 2,500 feet and 10° (except very locally); on the other hand, the first fold in the Valley and Ridge province, the western anticline on the Nittany arch, has a relief of 25,000 feet, and its northwest limb is commonly overturned, locally dipping only 30° southeast. In eastern West Virginia the contrast is somewhat less sharp, for two large folds at the east edge of the Plateau, the Deer Park-Elkins Valley anticline and the Browns Mountain or Pocahontas anticline, have steep northwest flanks and intermediate amplitudes;[4] west of them, however, the folds of central West Virginia (with one exception, the aberrantly trending Burning Springs anticline) are less than 1,000 feet high and show only very low dips. Furthermore, the Valley and Ridge folds all show a striking and consistent asymmetry, the northwest flanks of the anticlines being the steeper and the axial planes hading strongly southeast, whereas the asymmetry of the Plateau folds (with the same exceptions in eastern West Virginia) is much less obvious, and many, perhaps the majority, of the anticlines are steeper on the southeast side. In the Valley and Ridge province, thrust faults are fairly common and generally dip southeast, and some have displacements measured in thousands of feet, whereas on the Plateau they have been observed only locally at the surface where they are small and may dip in either direction (again a large southeast-dipping thrust fault on the Browns Mountain anticline is an exception).

In the subsurface, the last of these differences is less strong, for faults are common beneath the Plateau at the level of the Lower Devonian Oriskany sandstone, some thousands of feet below the surface, and apparently their throws range up to 1,800 feet; likewise steep dips of bedding seem to be much less rare. Most of the faults are strike faults, probably mainly northwest-dipping reverse faults (the angle of dip is rarely known) although some pairs of faults appear to outline narrow graben near the crests of the anticlines (where the subsurface information is apt to be concentrated). On the Plateau in Clearfield and Centre Counties, Pennsylvania, several transverse faults with small and inconsistent throw cut both surface and subsurface rocks; as Fettke (1954, p. 7) pointed out, they occur precisely where the curvature of the fold trends is the tightest.

The other differences between the folds in the two provinces remain, however, most notably the abrupt change in amplitude from the relatively low folds of the Plateau to the great anticlines and synclines of the Valley and Ridge, the highest of which is precisely that closest to the Plateau—the Nittany arch in central Pennsylvania and the Wills Mountain anticline in Maryland, West Virginia, and Virginia.

Growth of Ideas

The great breadth of the Plateau folds relative to their height has been interpreted for many years as evidence that they are also deep. Thus Ashley (1908) argued that even the strongest, most "competent" layers in the sedimentary column are far too weak to support themselves in arches of this shape and size, even without considering the weight of the weaker overlying

[4] One or both of these folds are generally reckoned to the Valley and Ridge province, but the writer prefers to draw the boundary between the two provinces along the continuous zone of steep or overturned strata on the northwest flank of the Nittany arch and its southward continuation, the Wills Creek anticline, which extends across Maryland and northeastern West Virginia into Virginia where it ends against the northeast end of the St. Clair thrust fault; according to everyone, the latter forms the province boundary in southern West Virginia and adjacent Virginia. It is suggested that the boundary so drawn marks the northwest limit of decollement in Cambrian strata (Rodgers, 1953).

layers, and that therefore every layer must have been continuously supported from below, ultimately from the underlying basement. Sherrill (1934) added to this argument from his careful study of the asymmetry of the folds in Pennsylvania; he showed that, if only the regional dip at the surface is taken into account, the folds show a slight southeast-facing asymmetry, but that, if the folds are projected down to the upper surface of the basement, the asymmetry can be fully accounted for by the regional dip of that surface. He concluded therefore that all these folds are in reality basement structures, merely reflected upward through the sedimentary sequence.

Subsurface information (e.g., Fettke, 1954, Pl. 3) shows clearly that the folds do persist at least as far as the Oriskany Sandstone, where they are perhaps steeper than at the surface though not much if any higher, and where faulting is certainly commoner; below that level, published information has been inadequate.[5]

In 1955, however, the Sandhill well of the Hope Natural Gas Company in Wood County, West Virginia, was completed to the basement under the Burning Springs anticline, and the data from this well at once suggested an alternative interpretation, for that fold at least. A detailed log of the well was promply published (Bayles and others, 1956), and a symposium on the well and its stratigraphic and structural implications was held by these Societies in 1957 and published in 1959 (Woodward and others, 1959). In discussing the lessons to be learned from the well, the present writer leans heavily on this symposium, especially Woodward's contribution. Other deep wells have since been drilled on the same anticline and on other anticlines in this region; though detailed information on these wells has not been published, some well logs and other data have become available and have been used in preparing the published paper.

Burning Springs Anticline

The Burning Springs anticline is not a normal Appalachian foreland fold. It lies northwest of the main belt of low foreland folds in central West Virginia, themselves somewhat atypical because less continuous, less straight, and less parallel among themselves than the folds farther northeast. It trends north and south (bending toward the southeast at its south end) directly across the lowest part of the Pittsburgh-Huntington basin, whereas they trend roughly N. 35°–45° E. and are mainly on the basin's southeast flank, and it is far higher and steeper than any other fold within 70 miles. It is a box fold, with a flat top nearly a mile wide and with steep, locally vertical sides, and the maximum structural relief observed at the surface is about 1,650 feet.

The Sandhill well began on the flat top of the fold close to its highest point and penetrated a normal succession about 4,000 feet thick down to the Huntersville Chert, next above the Oriskany Sandstone, with little evidence of disturbance or any but low dips. It then entered a 2,000-foot zone of intensely deformed rocks, with irregular steep dips and faults, in which the Lower Devonian sequence including the Oriskany is repeated several times. Woodward (1959, p. 18) estimates the excess over the true stratigraphic thickness in this zone at 1,658 feet. Below this zone, the well reentered apparently flat Lower Devonian rocks and showed no evidence of further disturbances through more than 7,000 feet of strata to the Precambrian basement.

Thus the excess stratigraphic thickness in the well exactly matches the surface relief, and the depth at which the disturbed zone was encountered matches the surface width of the coffer-shape fold. Furthermore, if one compares the elevation of various Silurian marker beds in this well with those in the nearest wells drilled on the basin floor on either side of the anticline (cf. logs in Tucker and Price, 1943, and Martens, 1944, with log of Sandhill well in Woodward and others, 1959), one finds no evidence for an anticline in the Silurian and older rocks where they are penetrated by the Sandhill well; if such an anticline is present, it is very much lower than the surface fold. Assuming that the fold is concentric, as is probable for a fold in otherwise undeformed sedimentary rocks of varying competence, one is led to the conclusion, drawn by Woodward and others, that it is confined to the Carboniferous and Devonian strata and closes downward in a zone of disharmony in or just below the Lower Devonian (see Woodward, 1959a, Fig. 8, p. 1077).

[5] I suppose that at the Pittsburgh meeting in 1948, when I talked to the same Societies about the difference between "thick-skinned" and "thin-skinned" theories for the Valley and Ridge province (Rodgers, 1949), nobody questioned the thick-skinnedness of the Plateau folds. I remember, however, that Professor Sherrill, whom I had mentioned in my talk as belonging to the thick-skinned school, rose to remark: "I'm not as thick-skinned as I was."

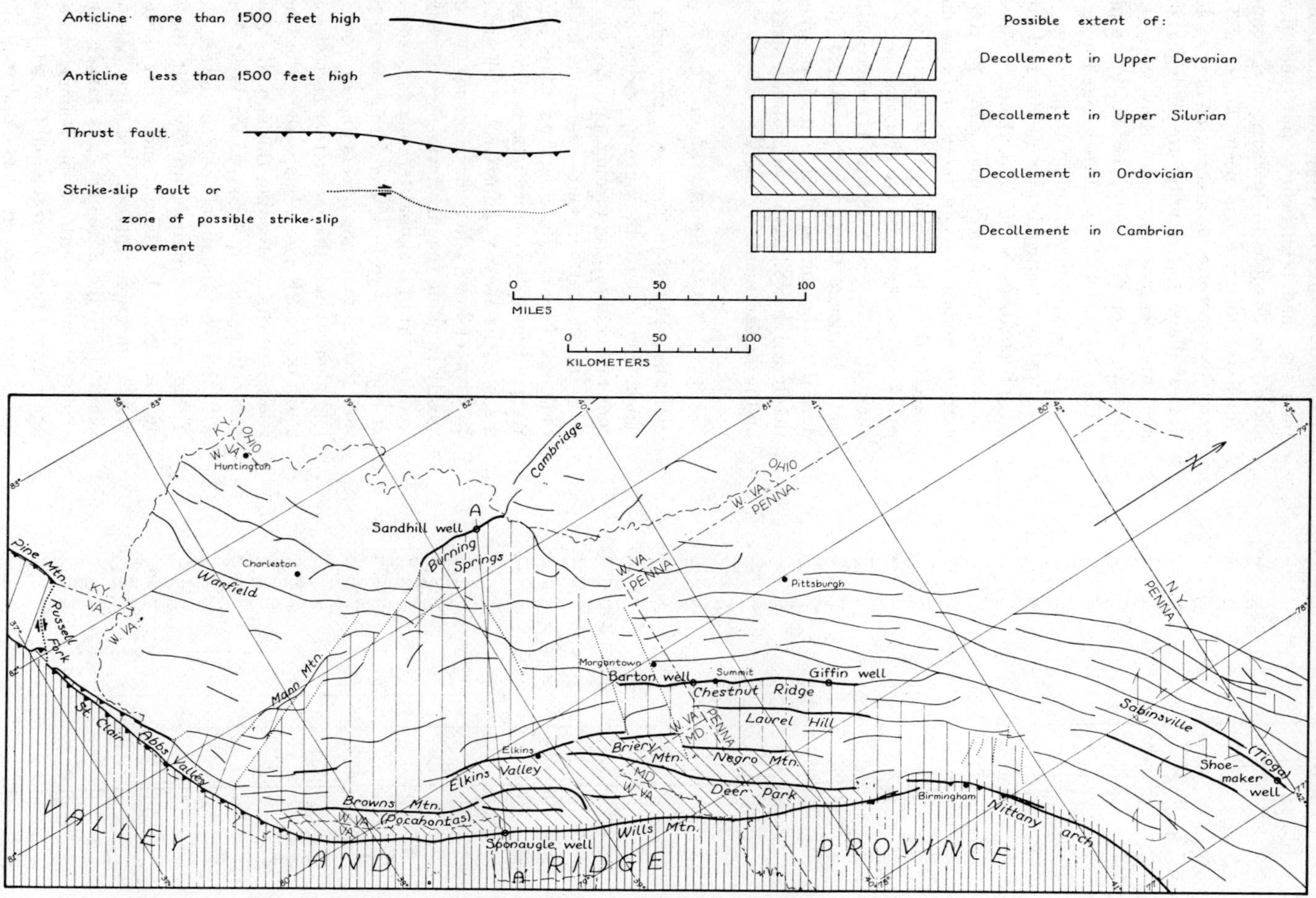

FIG. 1.—Map showing anticlines of Appalachian Plateau province in Pennsylvania, West Virginia, and adjacent states.

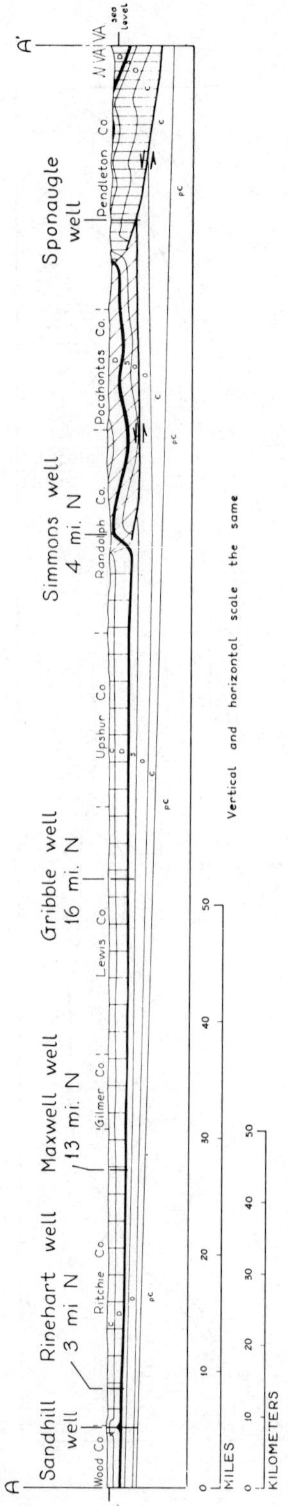

FIG. 2.—Theoretical cross section of Appalachian Plateau province in West Virginia (along line A–A′, Fig. 1), to illustrate the decollement hypothesis for folding. Vertical and diagonal lines indicate possible extent of decollements, as in Figure 1.

Letter symbols: C—Carboniferous (including Permian); D—Devonian (heavy line=Lower Devonian); S—Silurian; O—Ordovician; C—Cambrian; pC—Pre-cambrian.

The folding of the Burning Springs anticline must have caused the Carboniferous rocks on either side of the fold to approach closer together than they had been before, and theoretically it should be possible to measure the amount by which the distance has been shortened from a carefully constructed cross section of the fold. The data available to the writer are inadequate for an accurate measurement, but they suggest that, estimated conservatively, the shortening perpendicular to the fold amounts to about $\frac{1}{4}$ mile. If the actual relative movement of the rocks on either side was not perpendicular to the fold (it may have been about N. 50° W.), then the amount of movement would be considerably greater. The extreme deformation in the disturbed zone also suggests that this estimate is minimal. But if the rocks east of the fold have moved at least $\frac{1}{4}$ mile closer to those west of it, then (1) where is the rest of the border of the block that shifted? (2) how deep does the shifted block extend, or where did the underlying Silurian to Cambrian strata take up the displacement?

LIMITS OF SHIFTED BLOCK

Let us look first for the rest of the border of the block. On the south the Burning Springs anticline ends rather abruptly, turning southeastward as it does so, but farther south and east, in Roane, Calhoun, Gilmer, Braxton, and Clay Counties, West Virginia (see county reports and oil and gas maps of the West Virginia Geological Survey), is an area of low rather short basins and domes, mostly trending roughly northeast but characterized by abrupt diagonal terminations and offsets at high angles to their trends; in several places, such terminations and offsets line up from fold to fold along trends that are either east of south or south of east. Regional maps show the Chestnut Ridge anticline extending through this area and connecting, east of Charleston, West Virginia, with the Warfield anticline of southwestern West Virginia and eastern Kentucky, but the recent detailed maps by Haught show that there is no such connection. Instead, the high part of the Chestnut Ridge anticline ends rather abruptly about 20 miles south of the Pennsylvania state line; beyond that point the fold is much lower and shows several diagonal offsets, and it finally dies out in the eastern part of the area in question. Probably no individual fold extends through central West Virginia. Farther south, extending from the south edge of Clay County across Nicholas and Fayette Counties, is the low Mann Mountain anticline; its trend is irregular but generally about S. 20° E., and it is nearly in line with the Burning Springs anticline; aligned offsets with similar trends partly bridge the 20-mile gap between them. Still farther south are a few more short northeast-trending folds, with suggestions of diagonal offsets along the same trends, and finally one reaches the larger Abbs Valley anticline near the point where it loses height going eastward; it lies only a few miles northwest of and parallel with the abrupt margin of the Valley and Ridge province—the St. Clair fault, here trending N. 65° E.

The zone of structural disturbance thus outlined, trending roughly N. 20° W. across the whole state of West Virginia, seems somewhat like an oversize mirror-image of the Russell Fork fault at the northeast end of the Pine Mountain thrust block, on the Plateau about 90 miles southwest, and is even more like the Emory River fault line at the northeast end of the Cumberland Plateau thrust block in Tennessee; both these faults escaped detection for a long time though they have net slips of one or two miles, because they are strike-slip faults in essentially flat beds, so that their stratigraphic displacements are negligible. The pattern of minor folds in central West Virginia, with their abrupt diagonal ends and offsets, resembles the pattern of folds and strike-slip faults demonstrated by Stearns and Wilson (Stearns, 1954, 1955; Wilson and Stearns, 1958) on the Cumberland Plateau in Tennessee, suggesting that the diagonal trends in West Virginia likewise represent strike-slip faults, though of smaller slip. Accordingly, the Burning Springs-Mann Mountain line may well be a zone of left-handed strike-slip movement, the block east of it having moved generally northwest relative to the block on the west.

North and northeast of the Burning Springs anticline, we must look for other kinds of evidence. The duplicated Lower Devonian section found in the Sandhill well is remarkable but not unique. The Giffin well No. 4 on a dome on the crest of the Chestnut Ridge anticline in northern Westmoreland County, Pennsylvania, penetrated the Oriskany Sandstone three times in 1,800 feet (Fettke, 1951, p. 1186–1187; Shaffner, 1958, p. 75–77), and nearby wells suggest a faulted fold overturned toward the southeast; the structural relief is about 2,500 feet. Similarly the Barton well No. 8 on the Summit dome, on the same anticline not far north of the West Virginia line, penetrated nearly 2,500 feet of Upper Silurian strata, apparently without reaching the base (Fettke, 1950, p. 1055); salt was found through several hundred feet. Fettke (1954), Ludlum (1958), and Woodward (1959a) have called attention to the presence of considerable deformation in the Lower Devonian on these and other domes along the Plateau anticlines; queer graben-like structures are reported on many, on the basis of the elevations of first penetrations of the Oriskany, a steep fault being postulated where the elevations in two nearby wells are widely discrepant, but these discrepancies might equally well represent the subcrops of overturned or recumbent folds or small flap-like thrusts. One asks therefore with Ludlum (1958), Woodward (1959a), and Carey (1962, p. 120 ff.) whether these anticlines persist much deeper or whether like the Burning Springs anticline they close below in a zone of disharmony. Measurement of a non-exaggerated cross section of the three major foreland folds in southern Pennsylvania (Chestnut Ridge, Laurel Hill, Negro Mountain) suggests that taken together they represent a shortening of the order of 0.2 mile, no allowance being made for faults. Finally, the transverse faults in Clearfield and Centre Counties have some of the characters of strike-slip faults, such as steeply diagonal trends relative to the fold axes and inconsistent and generally small stratigraphic displacements.

Perhaps therefore this group of transverse faults marks the northeast end of the block whose shifting caused the Burning Springs anticline and probably the Chestnut Ridge and related anticlines as well. The northwest border of the block might lie along or just west of the Chestnut Ridge anticline from Clearfield County into northern West Virginia, as far as the point 20 miles south of the state line where that anticline loses height. A line extending N. 80° W. from this point would pass through the northern terminations of the Wolf Summit and Arches Fork anticlines, the most prominent folds between here and the Burning Springs anticline, and would be roughly parallel with several lines of offset of these and other folds; possibly right-handed strike-slip movement along this line, which would pass not far from the north end of the Burning Springs anticline, took up the shift in the interval between that anticline and Chestnut Ridge.

The block would thus consist of a blunt triangular wedge in West Virginia and a parallelogram-like strip extending northeast into central Pennsylvania. The two sides of the wedge would be conjugate zones of strike-slip movement; as they trend about N. 20° W. and N. 80° W., the movement of the wedge, relative to the rest of the Plateau, should have been about N. 50° W.

DEPTH OF SHIFTED BLOCK

If the Carboniferous and Devonian rocks on the two sides of the Burning Springs anticline have moved closer together, then somewhere the same amount of shortening should also be taken up in the underlying Silurian to Cambrian strata and in the basement beneath. Beneath the Sandhill well, these older strata appear to be flat, but if the axial plane of the fold is not vertical, the well could have passed out of the fold with depth into the horizontal rocks on one side; in other words, the fold in the lower strata might lie just at one side of the surface fold. Woodward (1959, p. 163) rejects this hypothesis; his reason is that the disturbed zone does not have the character of the steep flank of an asymmetrical anticline but of a zone of disharmonic folding. Additional arguments are the close coincidences of the surface width of the fold with the depth to the disturbed zone and of the structural relief with that zone's thickness. These are precisely what should be expected if a concentric fold of the size and shape of the Burning Springs anticline closes with depth, whereas it is difficult to draw a properly scaled cross section of the fold persisting much farther down, especially at one side, without invoking drastic thinning or changes in the amount of shortening in successive beds. Evidence from more recent nearby wells and from seismic surveys indicates, however, that on the east the zone of strong disturbance also involves the Upper Silurian and that dips of several degrees (5°–7° by dipmeter) are present even deeper.

If the main shortening in the lower beds is not taken up beneath the anticline, it must be taken up somewhere at one side, and as that much shortening, even a mile down, could hardly be accomplished without deforming the overlying rocks to the surface, one is led to look for the nearest folds of the same order of magnitude that lie approximately across strike. These folds are the Elkins Valley and Browns Mountain (Pocahontas) anticlines; moreover, surface and subsurface data (Reeves and Price, 1950) indicate that Silurian and at least some Ordovician strata are involved in both. But if shortening in the lower strata took place here, then the upper strata have slipped at least $\frac{1}{4}$ mile sidewise over the lower strata through the whole distance between those folds and the Burning Springs anticline, 70 miles at the least, and a plausible slip surface is needed. In addition the aberrant trend of the Burning Springs anticline calls for explanation.

Now the Upper Silurian strata cut by the Sandhill well contain considerable anhydrite (averaging about 10 per cent through 500 feet of section) but no rock salt; the same is true for several wells in the basin on the west and southwest (Martens, 1943; Haught, 1955, p. 7). But the Rinehart well in Ritchie County, only 4 miles east of the Sandhill well, encountered 100 feet of salt, and salt is known in nearly every well that cuts the Upper Silurian from that point eastward at least to the Chestnut Ridge anticline and northeastward entirely across Pennsylvania to the outcrop in New York (Fettke, 1955). The western limit of Upper Silurian salt has also been located with fair accuracy in Ohio (Pepper, 1947); it trends slightly west of north, following the low Cambridge arch, which is directly in line with the Burning Springs anticline. The southern limit of the salt is not well established, for south of the Rinehart well and a couple of others nearly due east of it, all of which cut 50–100 feet of salt, no wells appear to have been drilled through the pertinent strata east of the Burning Springs-Mann Mountain trend. Farther east, salt is not recorded on the Silurian outcrop at the west edge of the Valley and Ridge province, but salt crystal impressions have been found there (Ludlum, 1959).

One may well ask, however, what controlled the edge of the salt, especially its sharp west edge. Woodward (1959, p. 164 ff.), though he considers that the sharp deformation in the Burning Springs fold goes no deeper than the Lower Devonian or Upper Silurian, suggests nevertheless that the fold was localized by a low persistent basement arch, also expressed by a decrease in the thickness of the basal Silurian sandstone under the anticline. The figure of 132 feet which he gives for that thickness seems, however, entirely in line with the figures he gives (1959, Fig. 4, p. 27) for the nearest well in each of four different directions—67, 125, 178, and 247 feet—especially in view of the gradational upper boundary of the unit. If the low dips previously mentioned in the strata below the Upper Silurian were produced *before* the main Burning Springs anticline, they may have had something to do with the location of the main fold; perhaps the fold was then no higher than the present Cambridge arch at the north.

The relation of the high steep Burning Springs anticline to the edge of the salt seems strongly to support the hypothesis that the whole body of Carboniferous and Devonian strata east of the anticline slid northwestward, the slip taking place in the plastic salt layers of the Upper Silurian (Fig. 2). Where the salt layers gave out, so that the force required for the bedding slip increased greatly, the slip surface had to cut upward toward the surface, producing the aberrantly trending anticline. At the northeast the edge of the shifting block was not determined by the edge of the salt; instead, the shortening was taken up by more conventional folds coupled with strike-slip faults, the latter nearly invisible because they produced little or no stratigraphic displacement. Deformation in the Silurian to Cambrian strata beneath the main part of the shifting block was small or absent; the main part of the necessary shortening in those strata was taken up instead under the Deer Park-Elkins Valley and Browns Mountain (Pocahontas) anticlines, or under the Valley and Ridge province. That is to say, the slip surface remains in the Upper Silurian as far east as those anticlines, and only begins to cut down in the section there (perhaps to the Upper and Middle Ordovician Martinsburg shale) or at the edge of the Valley and Ridge (probably to Middle and Lower Cambrian shaly strata). The low-angle thrust faults that have recently been discovered in the subsurface beneath Cambrian rocks on the westernmost Valley and Ridge anticline both in Pennsylvania, at Birmingham, Blair County, (Moebs and Hoy, 1959), and in West Virginia, at the Sponaugle well, Pendleton County (United Fuel Gas Company information release, February, 1961), but which do not come to the surface in those areas, are probably the eastern extension of this slip surface, though the net slip may be larger there than in central West Virginia because the strata above have been folded more than those below by that anticline and others on the west.

We know little about any possible deformation in the Upper Silurian salt layers under West Virginia or southwestern Pennsylvania, except for the thickening in the Barton well No. 8 on the Summit dome, but there is more information in northern Pennsylvania and southern New York, where large lateral variations in the thickness of salt and the salt-bearing formations have long been remarked (Fettke, 1933; Finn, 1949; Cate, 1961). Some have attributed these variations to facies changes, but it is perhaps no coincidence that the greatest thicknesses are reported in the cores of anticlines. Indeed, Fettke (1933, p. 648; 1938, p. 255) suggested diapiric intrusion of salt to explain 954 feet of salt and not less than 2,887 feet of Upper Silurian strata (normal thickness probably about 1,000 feet) found in the Shoemaker well on the Tioga dome of the Sabinsville anticline in Tioga County, Pennsylvania; the structural relief of the dome, eliminating regional dip, is between 1,500 and 2,000 feet. No wells go deep enough to show whether this fold persists below the Upper Silurian, but the coincidence between structural relief and excess thickness suggests that it does not. The folds in this area are outside the limits suggested in this paper for the block whose shifting raised the Burning Springs anticline; perhaps in this part of the Plateau another block slid northward on the Silurian salt, but if so its limits are not obvious.

Summary

It is suggested, therefore, that the Appalachian foreland folds of southern Pennsylvania and West Virginia are the surface expression of the shifting of a large block of Carboniferous and Devonian strata $\frac{1}{4}$–$\frac{1}{2}$ mile northwestward on a slip surface in the Upper Silurian, greased by the salt layers at that level. In southern Pennsylvania and the northern edge of West Virginia, the shifting of the block was taken up by producing three or four broad anticlines; in central West Virginia, on the other hand, the block reached the edge of the salt basin and most of the relative shifting was concentrated in the Burning Springs anticline or distributed along two conjugate sets of strike-slip faults. A similar but perhaps smaller shifting may account for the folds of northern Pennsylvania and southern New York; if so, the boundaries of the block that shifted are still uncertain. Strata lower than Upper Silurian were entirely unaffected by this movement, except that the Silurian and some part of the Ordovician were involved in the Deer Park-Elkins Valley and Browns Mountain or Pocahontas anticlines along the east edge of the Plateau province. The movement was later than the deposition of the Dunkard group (earliest Permian); probably it was contemporaneous with the main deformation of the Valley and Ridge province on the southeast, of which it was merely a marginal effect.

years ago by Rich (1934) for the Pine Mountain block in Virginia, Kentucky, and Tennessee, and since extended, following Rich's own suggestions, to other folds and faults in the southern part of the Appalachian Plateau province (Rodgers, 1950; Stearns, 1954, 1955; Wilson and Stearns, 1958) and to the Valley and Ridge province (Miller, 1945; Rodgers, 1953; for a rebuttal see Cooper, 1961, p. 100–118). Rough comparative figures are given here for four such structures in the Plateau province; the estimates are intended to be conservative.

Block	Sequatchie—Walden Ridge	Cumberland Plateau	Pine Mountain	Burning Springs—Clearfield County
Slip surface	Cambrian Conasauga shale	Pennsylvanian shale units	Devonian Chattanooga shale (and Cambrian shale)	Upper Silurian salt
Length (miles)	250	100	125	250
Average width (miles)	20	30	25	40
Average depth (miles)	1	$\frac{1}{5}$	$1\frac{3}{5}$	1
Volume ($miles^3$)	5,000	600	5,000	10,000
Slip (miles)	$1\frac{1}{2}$ maximum say 1 mile	2 maximum, say 1 mile	2 to 12, say 6 miles	0.2 to 0.5?, say $\frac{1}{4}$ mile
Product ($miles^4$)	5,000	600	30,000	2,500

This thin-skinned hypothesis for the Plateau folds (and for those of the Valley and Ridge province) appears to receive some support from geophysical evidence presented in the papers and panel discussion at the meeting of the Societies at Summit, where this paper was first presented, but not available in published form. Sample east-west magnetic and gravity profiles across southern Pennsylvania and northern West Virginia and extending as far east as the crystalline rocks at the Blue Ridge front showed little correlation between anomalies and surface folds, strongly suggesting that no folds of any size involve the upper surface of the crystalline basement rocks. Seismic results on several structures showed major faulting at intermediate depths, the faults terminating beneath the upper folded layers. On the other hand, seismic data also indicate some deformation of the strata beneath the Upper Silurian under the Burning Springs and other anticlines. Evidently geophysical methods can provide much of the information needed to test the hypothesis.

The mechanics here suggested for the foreland folding on the Plateau is not particularly new; it is exactly comparable with that proposed some

From this tabulation it appears that the central Appalachian block is also quantitatively similar to the others; though as long as any and considerably broader, so that the volume involved is greater, the probable movement is much smaller, and the energy needed would therefore appear to be less than for either the Pine Mountain or Sequatchie blocks, even without considering that salt probably slides more easily than shale. With detailed mapping and drilling, the Pine Mountain case seems proved, and the Cumberland Plateau case seems hardly less clear. Quantatively, therefore, the proposed hypothesis for the foreland folds of the central Appalachians seems feasible; needed to test it are detailed mapping, drilling, and geophysical exploration done with the hypothesis in mind.

REFERENCES

Ashley, G. H., 1908, Studies in mechanics of Allegheny structure: Science, new ser., v. 27, p. 924–925.

Bayles, R. E., Henry, W. H., Fettke, C. R., Harris, L. D., Flowers, R. R., and Haught, O. L., 1956, Wood County deep well: W. Va. Geol. Survey Rept. Inv. 14, 62 p.

Carey, S. W., 1962, Folding: Alberta Soc. Petroleum Geologists Jour., v. 10, p. 95–144.

Cate, A. S., 1961, Stratigraphic studies of the Silurian rocks of Pennsylvania. Part I: Penna. Geol. Survey Special Bull. 10.

Cooper, B. N., 1961, Grand Appalachian field excursion: Va. Polytech. Inst. Eng. Ext. ser., Geol. Gdbk. 1, 187 p. (Geol. Soc. America Ann. Mtg. 1961, Gdbk. 1), Blacksburg, Va.

Fettke, C. R., 1933, Subsurface Devonian and Silurian sections across northern Pennsylvania and southern New York: Geol. Soc. America Bull., v. 44, p. 601–659.

——— 1938, Oriskany as a source of gas and oil in Pennsylvania and adjacent areas: Am. Assoc. Petroleum Geologists Bull., v. 22, p. 241–266.

——— 1950, (Oil and gas) Developments in Pennsylvania in 1949: Am. Assoc. Petroleum Geologists Bull., v. 34, p. 1048–1057.

——— 1951, (Oil and gas) Developments in Pennsylvania in 1950: Am. Assoc. Petroleum Geologists Bull., v. 35, p. 1177–1187.

——— 1954, Structure-contour maps of the Plateau region of north-central and western Pennsylvania: Penna. Geol. Survey Bull. G 27, 50 p., maps.

——— 1955, Preliminary report: occurrence of rock salt in Pennsylvania: Penna. Geol. Survey Prog. Rept. 145.

Finn, F. H., 1949, Geology and occurrence of natural gas in Oriskany sandstone in Pennsylvania and New York: Am. Assoc. Petroleum Geologists Bull., v. 33, p. 303–335.

Haught, O. L., 1955, Oil and gas report and map of Pleasants, Wood, and Ritchie Counties, West Virginia: W. Va. Geol. Survey Bull. 11, 21 p.

Ludlum, J. C., 1958, Relation of surface and subsurface structures in West Virginia to future drilling for natural gas: W. Va. Engineer, v. 20, no. 7.

——— 1959, Rock salt, rhythmic bedding, and salt-crystal impressions in the Upper Silurian limestones of West Virginia: Southeastern Geology, v. 1, p. 22–31.

Martens, J. H. C., 1943, Rock salt deposits of West Virginia: W. Va. Geol. Survey Bull. 7, 67 p.

——— 1944, Well-sample records: W. Va. Geol. Survey Rept., v. 17, 889 p.

Miller, R. L., 1945, Relation of overthrust faults in southwest Virginia (abs.): Geol. Soc. America Bull., v. 56, p. 1182–1183.

Moebs, N. N., and Hoy, R. B., 1959, Thrust faulting in Sinking Valley, Blair and Huntingdon Counties, Pennsylvania: Geol. Soc. America Bull., v. 70, p. 1079–1088.

Pepper, J. F., 1947, Areal extent and thickness of the salt deposits of Ohio: Ohio Jour. Sci., v. 47, p. 225–239; Ohio Geol. Survey Rept. Inv. 3 (1948), 15 p.

Reeves, Frank, and Price, P. H., 1950, Early Devonian gas in northern West Virginia and pre-Devonian oil prospects: Am. Assoc. Petroleum Geologists Bull., v. 34, p. 2095–2132.

Rich, J. L., 1943, Mechanics of low-angle overthrust faulting as illustrated by Cumberland thrust block, Virginia, Kentucky, and Tennessee: Am. Assoc. Petroleum Geologists Bull., v. 18, p. 1584–1596.

Rodgers, John, 1949, Evolution of thought on structure of middle and southern Appalachians: Am. Assoc. Petroleum Geologists Bull., v. 33, p. 1643–1654.

——— 1950, Mechanics of Appalachian folding as illustrated by Sequatchie anticline, Tennessee and Alabama: Am. Assoc. Petroleum Geologists Bull., v. 34, p. 672–681.

——— 1953, The folds and faults of the Appalachian Valley and Ridge province: Ky. Geol. Survey Special Pub. 1, p. 150–166.

Shaffner, M. N., 1958, Geology and mineral resources of the New Florence quadrangle, Pennsylvania: Penna. Geol. Survey Atlas A 57, 185 p.

Sherrill, R. E., 1934, Symmetry of northern Appalachian foreland folds: Jour. Geology, v. 42, p. 225–247.

Stearns, R. G., 1954, The Cumberland Plateau overthrust and geology of the Crab Orchard Mountains area, Tennessee: Tenn. Geol. Survey Bull. 60, 47 p.

——— 1955, Low-angle overthrusting in the central Cumberland Plateau, Tennessee: Geol. Soc. America Bull., v. 66, p. 615–628.

Tucker, R. C., and Price, P. H., 1943, Summarized records of deep wells: W. Va. Geol. Survey Rept., v. 16, 938 p.

Wilson, C. W., Jr., and Stearns, R. G., 1958, Structure of the Cumberland Plateau, Tennessee: Geol. Soc. America Bull., v. 69, p. 1283–1296.

Woodward, H. P., and others, 1959, A symposium on the Sandhill deep well, Wood County, West Virginia: W. Va. Geol. Survey Rept. Inv. 18, 182 p.

Woodward, H. P., 1959a, The Appalachian region: World Petroleum Cong., 5th, New York, 1959, Proc. Sect. 1, p. 1061–1079.

Canadian Journal of Earth Sciences
1967, v. 4, p. 39-70

The Tectonic Significance of Mesoscopic Subfabrics in the Southern Rocky Mountains of Alberta and British Columbia[1]

R. A. PRICE[2]

Abstract Deformation throughout much of the southern Rocky Mountains was characterized by brittle failure in a strongly anisotropic layered sequence of non-metamorphic rocks. On a megascopic scale, the overall structure is dominated by an interlocking system of imbricate thrust plates that have moved relatively eastward or northeastward and upward. On a mesoscopic scale, the principal elements in the fabric of these rocks are fractures that are statistically parallel or perpendicular to the bedding, or else intersect it at preferred angles of approximately 25° or 70°. During deformation many of these fractures obviously were kinematically active, as discrete surfaces of slip that became slickensided, as zones of dilation that became filled with vein minerals, or as surfaces of pressure solution that are now marked by stylolites. Each of these fractures provides a partial record of the kinematics of some stage of the deformation, even when they are considered individually rather than as components in a fracture array whose symmetry is related to that of the movement picture during deformation. Each defines a unique line of slip and axis of rotation for slip, or a unique direction of relative extension or compression. Collectively, they provide a direct and succinct record of the kinematic history of an individual fabric domain, and a sound basis for dynamic analyses of deformation.

Some preliminary results of a reconnaissance study of these mesoscopic subfabrics illustrate their tectonic significance.

A movement picture can be established for the deformation that occurs within an individual thrust plate during its development and translation. Kinematic relationships between and among the interlocking thrust plates can be studied.

Within a broad area centered along the prominent structural reentrant that crosses the Rocky Mountains near Crowsnest Pass, two different movement pictures occur in superposition. Movement about both northerly and northwesterly trending axes can be outlined on the basis of the mesoscopic subfabrics of rocks which, on a megascopic scale, have either a northwesterly or a northerly trending fabric axis. Movement patterns for the deformation associated with each of two regional structural salients converge in the vicinity of the reentrant.

The mesoscopic subfabrics associated with transverse faults in parts of the Front Ranges outline a pattern of movement which indicates that they did not originate as tear faults related to the translation of the thrust plates, but instead are probably older gravity faults, whose orientation may be controlled by the fabric of the Hudsonian basement extending beneath the mountains from the Canadian Shield.

INTRODUCTION

The southern Rocky Mountains of Alberta and British Columbia form part of a distinctive structural belt along the foreland margin of the Cordilleran orogen in western Canada. They represent a deformed westerly segment of an eastward-tapering wedge of sedimentary rocks that mantles the flank of the Canadian Shield. They consist, for the most part, of a distinctly layered mass of non-metamorphic Precambrian, Palaeozoic, and Mesozoic rocks, that is characterized by abrupt changes in lithology, and hence in mechanical properties, between thin stratigraphic sequences or individual beds. The pronounced mechanical anisotropy inherent in this layering has exerted a fundamental control on the development of a characteristic structural style that has come to be considered typical of the Canadian Rockies. However, locally, as in part of the central and western Rockies where the Cambrian and Ordovician succession is represented by a thick and relatively homogeneous shale facies, the deformation assumes a different style that is beyond the scope of this discussion and will not be considered further here.

The structure of the Rocky Mountains in southern Alberta and British Columbia is dominated by an array of interleaved thrust faults that commonly follow the bedding over large areas (see Fig. 1). These faults mark the soles of individual thrust plates, most of which have moved up toward the northeast or east relative to underlying rocks. They impart a distinct monoclinic symmetry to the deformation. The array of interleaved thrust faults represents a system of discrete, discontinuous, interlocked slip surfaces (Douglas 1958). As the displacement along any one fault decreases, that along other faults above and below it increases, so that the cumulative displacement across the system of faults as a whole changes very gradually in comparison with variations in displacement along an individual fault. The fault surfaces are generally arcuate perpendicular to the regional strike, and the bedding has commonly been rotated externally as a result of displacement on the surfaces. Concentric folds have developed in conjunction with the thrust faulting, by a combination of flexure and interstratal slip, and not uncommonly the fault surfaces themselves are involved in the folding. The imbricate thrust faults and the folds outline two broad structural salients that are convex toward the northeast and meet at a structural reentrant, near Crowsnest Pass, at about latitude 49°30′N. There, northwesterly trending and northerly trending structures occur in

---

[1]Presented at a symposium of the Royal Society of Canada, June 1965.

[2]Geological Survey of Canada

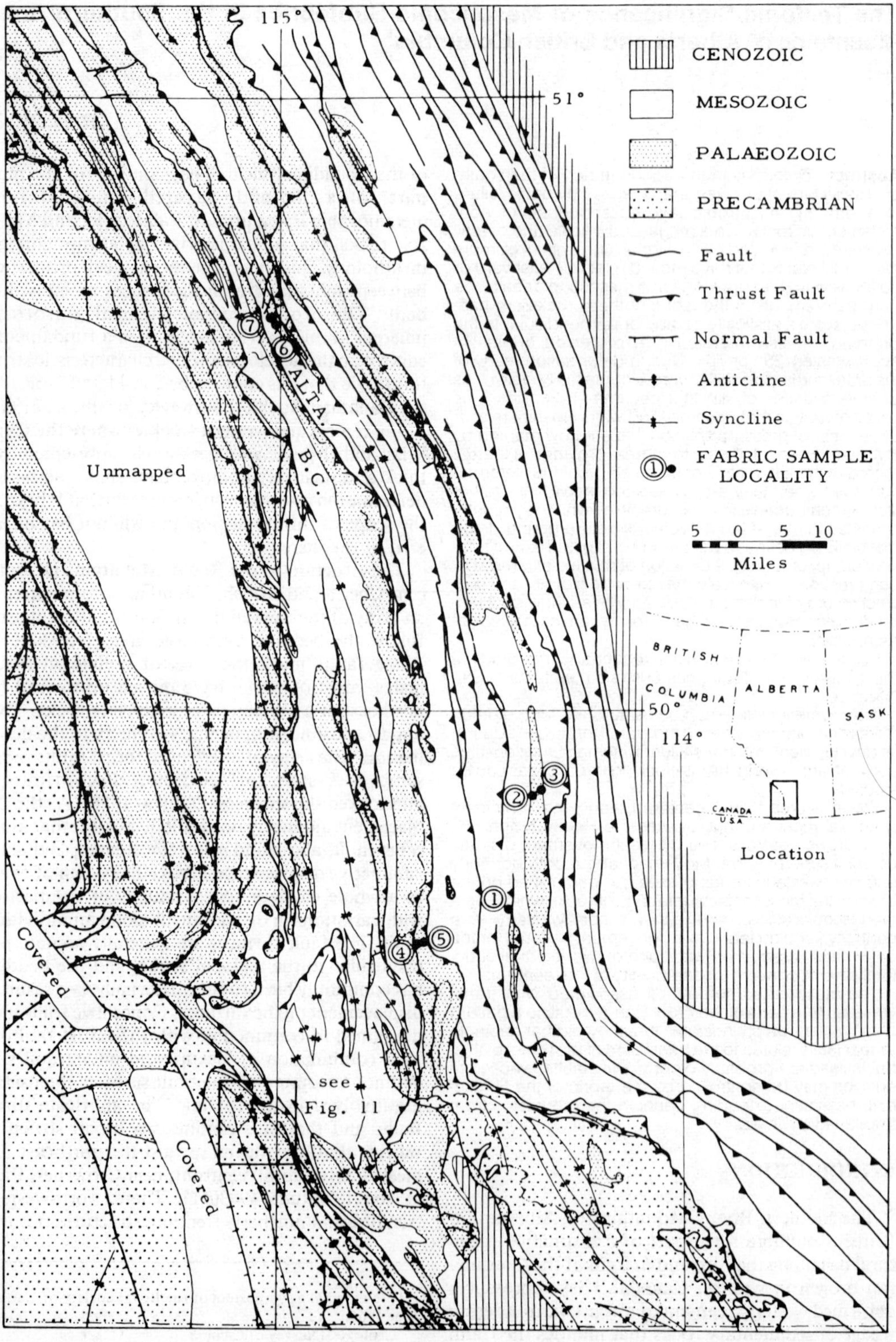
115°
51°
50°
114°
CENOZOIC
MESOZOIC
PALAEOZOIC
PRECAMBRIAN
Fault
Thrust Fault
Normal Fault
Anticline
Syncline
FABRIC SAMPLE LOCALITY
5 0 5 10
Miles
BRITISH COLUMBIA
ALBERTA
SASK
CANADA
USA
Location
Unmapped
ALTA.
B.C.
Covered
Covered
see Fig. 11
1
2
3
4
5
6
7

superposition or in close proximity to each other. The thrusting and folding across the breadth of the southern Rocky Mountains have foreshortened a portion of the western Canada sedimentary basin in a northeasterly to easterly direction by more than 50% (North and Henderson 1954; Shaw 1963), and have produced a corresponding thickening of the tabular prism of sedimentary rock occupying the upper level of the earth's crust in this region. In overall aspect the deformation can be viewed as a result of flow up and out of the interior of the Cordilleran orogen and out over the sedimentary cover of the Canadian Shield. On this scale of observation the two salients represent two frontal lobes in the allochthonous mass, and the interlocking system of thrust faults represents a pervasive array of discrete strain discontinuities.

The concepts and techniques that have assumed so fundamental a role in the structural analysis of metamorphic terranes have been applied to advantage, with some modifications, in the study of the deformation of non-metamorphic rocks (Hoeppener 1953; Bolsenkoetter 1955; Waard 1955; Bonham 1957; Lindstrom 1962; Rambow 1963). However, in the southern Rocky Mountains of Canada most structural studies have been concerned exclusively with the gross geometry of the deformation. With a few notable exceptions, little attention has been directed toward the systematic investigation of those structural features that can be observed on the scale of an individual outcrop or rock specimen ("mesoscopic scale" of Turner and Weiss 1963, pp. 15-16), or toward the analysis of aggregates of fabric data in general. Norris (1958*a*), as part of a study of structural conditions in Canadian coal mines, analyzed aggregates of faults in terms of the sense of slip on them relative to the plane of the bedding, and the angle of intersection of the fault and the bedding. He established certain relationships that are fundamental to the overall interpretation of the mechanics of deformation in the region. More recently he (Norris 1964, 1965) investigated the dynamic implications of the relationships between aggregates of several different types of small-scale structures. Roeder (1960) presented an important contribution to the structural geology of the Canadian Rockies in which aggregates of mesoscopic fabric elements are used as a basis for kinematic analyses of the deformation expressed in terms of larger scale structural features. Fitzgerald (1963) studied the geometric and kinematic relationships between mesoscopic structures and larger scale structures in a northern part of the Foothills belt of Rocky Mountains in Alberta, and Price (1964) analyzed aggregates of mesoscopic fabric elements as part of a study of the geometry and kinematics of some folds in the Rocky Mountains of southern Alberta and British Columbia.

The non-metamorphic rocks that form the imbricate thrust sheets and associated concentric folds of the southern Rocky Mountains of Alberta and British Columbia are characterized, on a mesoscopic scale, by a distinctive group of mechanically effective fabric elements that define a kinematically active mesoscopic subfabric.[3] Each of these fabric elements provides a partial record of the kinematics of some stage of the deformation, event when considered individually rather than a component of an array of elements whose symmetry is related to that of the movement picture during deformation. This paper is a discussion of the special tectonic significance of this type of subfabric in the Canadian Rockies and of some of the principles and techniques that can be applied to the study of it. It is based on preliminary results of a continuing study of tectonic fabrics in the southern Rocky Mountains of Alberta and British Columbia. It is a pleasure to acknowledge helpful discussion with R. J. W. Douglas of the Geological Survey of Canada during the course of the work, and the critical comments on an earlier version of this paper by G. B. Leech and D. K. Norris of the Geological Survey of Canada.

MESOSCOPIC FABRIC ELEMENTS

On a mesoscopic scale, deformation in the non-metamorphic rocks of the southern Canadian Rockies was dominated by brittle failure in a strongly anisotropic layered medium. The mesoscopic subfabrics of these rocks are characterized by several distinctive types of planar and curviplanar fracture surfaces. These range from simple smooth, plumose, or rough surfaces of parting, through interpenetrating stylolitic surfaces coated with insoluble residues, to fissures filled or coated with

[3]"A fabric is defined by the spatial array of all its elements. However it is frequently sufficient or convenient to consider the array of only one kind of element. This array is called by Sander a Teilgefuge and here a subfabric." (Turner and Weiss 1963, p. 34).

FIG. 1—Generalized geological map of the southern Rocky Mountains, Alberta, and British Columbia. Based on compilations by Henderson and Douglas (1954), Douglas (1958), and Dahlstrom and Henderson (1959), with additional data from Norris (1958*b*, 1958*c*, 1959), Leech (1959, 1960, 1962), and Price (1959, 1962), and from photogeological interpretations by the writer.

vein minerals, to shear surfaces marked with slickenside striae. These surfaces bound individual rock fragments which do not appear to have undergone any appreciable pervasive deformation. The fractures obviously represent discrete strain discontinuities, and the bulk deformation may have been somewhat analogous to "catalastic flow" in a granular or blocky aggregate, in which individual fragments are free to move relative to one another along their bounding surfaces during deformation.

Each fracture that is slickensided, stylolitic, or filled or coated with vein minerals is, even when considered individually, a unique kinematic indicator for some stage of the deformation of thes rocks. In contrast, the other fractures serve as kinematic indicators only when considered as components in a fracture array whose symmetry can be related to the kinematic symmetry of the deformation. A fracture filled or coated with vein minerals represents a locus of dilation and defines a unique direction of extension. A stylolitic fracture represents a locus of pressure solution and defines a unique direction of compression. A slickensided fracture represents a locus of shear and defines a unique set of three orthogonal kinematic axes comprising (1) the line of slip marked by the slickenside striae (a-axis), (2) an axis of rotation for slip (b-axis), which lies in the plane of the fracture and is perpendicular to the slickenside striae, and (3) the normal to the plane of the fracture (c-axis), which together with the direction of slip defines the plane of deformation (ac-plane). Aggregates of slickensided, stylolitic, and mineralized fractures can provide a direct and unequivocal basis for analyzing the kinematics of deformation in these rocks.

The fractures can be divided into several distinct categories on the basis of the angle at which they intersect the primary compositional and textural layering that marks the bedding in these rocks. The dynamically most important category is probably the bedding fractures, which follow the planar anisotropy inherent in the layering. These form a pervasive array of subparallel fractures, which may be slickensided, stylolitic, or coated with vein minerals. The remaining fractures fall into two main groups: faults and joints. For convenience of description these can be referred to a set of three orthogonal fabric axes *a*, *b*, and *c*, with *c* perpendicular to the bedding and *b* either parallel to an axis of cylindrical folding or else horizontal. The faults intersect the bedding at an acute angle, offset it, and except where there has been overprinting of two or more movement pictures, all tend to lie in *hOl*.[4] The joints are statistically perpendicular to the bedding and lie in *hkO*, or in the *ac* or *bc* planes. These various types of mesoscopic fabric elements are illustrated schematically in the block diagram of Fig. 2 and the stereograms of Fig. 3.

The faults can be divided into two groups on the basis of the sense of the displacement that has occurred on them with reference to the plane of the bedding (Norris 1958*a*, p. 14). Extension faults produce an elongation in the plane of the bedding, and for flat-lying beds include normal or gravity faults. Contraction faults produce a shortening in the plane of the bedding, and for flat-lying beds include reverse or thrust faults. In view of the external rotation of bedding that has occurred in conjunction with flexural-slip folding and with displacements on curved fault surfaces, it is obvious that not all extension faults are normal or gravity faults, and not all contraction faults are reverse or thrust faults. Both the extension faults and the contraction faults may be either synthetic or antithetic (Hoeppener 1955, pp. 31-34) to the interstratal slip marked by slickensided bedding, or to the sense of displacement of the thrust sheets in which they occur. Accordingly, there are four distinct potential categories of faults for any single movement picture as shown in Fig. 3, left. These categories are as follows:

(1) synthetic contraction faults, cf_s;
(2) antithetic contraction faults, cf_a;
(3) synthetic extension faults, ef_s;
(4) antithetic extension faults, ef_a.

All have a common axis of rotation (b-axis of slip) that corresponds with the axis of rotation for slip on the bedding; but the slip linears[5] vary in direction and sense from one category of fault to another, although they all fall within a plane normal to the axis of rotation. Faults formed during one phase of deformation may act as loci of dilation, pressure solution, or shear, which are marked by vein fillings, stylolites, or new slickenside striae, respectively, during a later phase of deformation. Some of the fractures which intersect the bedding at less than 70° show no obvious offset of the plane of the bedding on a mesoscopic scale, and may not be slickensided. They probably represent incipient faults rather than joints.

The joints, if considered only as fractures that are statistically perpendicular to the bedding but are otherwise undifferentiated, are unsystematic and of

---

[4]The symbols *hkl*, *Okl*, *hOl*, and *hkO* refer to orientation with respect to the fabric axes; *h*, *k*, and *l* are intercepts on the *a*, *b*, and *c* fabric axes respectively, and *O* indicates parallelism with a fabric axis.

[5]Slip linears mark the direction of slip on a shear surface. In stereographic projection they are represented by a segment of the great circle passing through both the pole to the shear surface and the slickenside striae on that shear surface. An arrow denotes the direction of displacement of the block in the hanging wall of the shear surface, if the relative displacement is known (Hoeppener 1955, pp. 37-38).

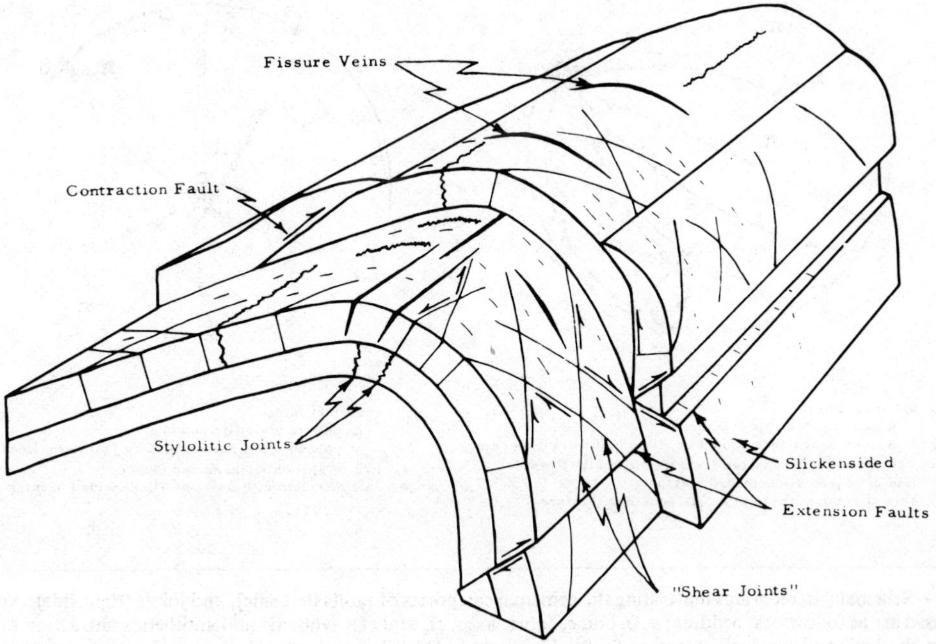

FIG. 2—Schematic block diagram of a segment of a fold illustrating the principal categories of mesoscopic fabric elements.

only limited use in kinematic analyses of the orogenic deformation. Although, locally, a relatively large planar or folded segment of a bed may be statistically homogeneous with respect to joints if the joints are referred to fabric axes that parallel both the bedding and the axis of folding (Price 1964, Fig. 2*b*), in general there is considerable heterogeneity in the spatial array of joints in even a small area of a planar segment of a single bed or of a succession of beds that comprise a planar segment of some larger structure (Price 1964, Fig. 1*d*; Norris 1964, Fig. 4, 1965, Fig. 2). This heterogeneity may be due to inadequate sampling occasioned by variations in weathering and the nature of the exposures, but complexity in jointing might be anticipated if jointing has been a continuing process involving several types of fractures and varying conditions of stress. Moreover, the full kinematic significance of an individual simple fracture forming a joint only becomes apparent when the fracture is considered as a component in an array whose overall symmetry can be related to the symmetry of the deformation. Accordingly, in the examples of mesoscopic subfabrics considered here, attention will be focused on those special categories of joints that can serve directly as kinematic indicators; joints that are slickensided, stylolitic, or coated or filled with vein minerals. Examples of these are shown schematically in Fig. 3, right.

Most of the joints in these rocks are similar to those which occur in the undeformed, flat-lying rocks that cover the Canadian Shield over much of the continental interior. It would appear that the thrusting and folding which are characteristic of the Rocky Mountains have not been a prerequisite for the development of the joints. Many presumably formed before the more conspicuous large-scale orogenic structures, and under different stress conditions. Any preexisting joint subfabric would impart a mechanical anisotropy to the rocks during orogenic deformation, and only those preexisting joints which were favorably oriented could assume a specific active kinematic role as a locus of shear, pressure solution, or dilation and precipitation of vein minerals, during the orogenic deformation. Consequently, subfabrics defined in terms of slick-

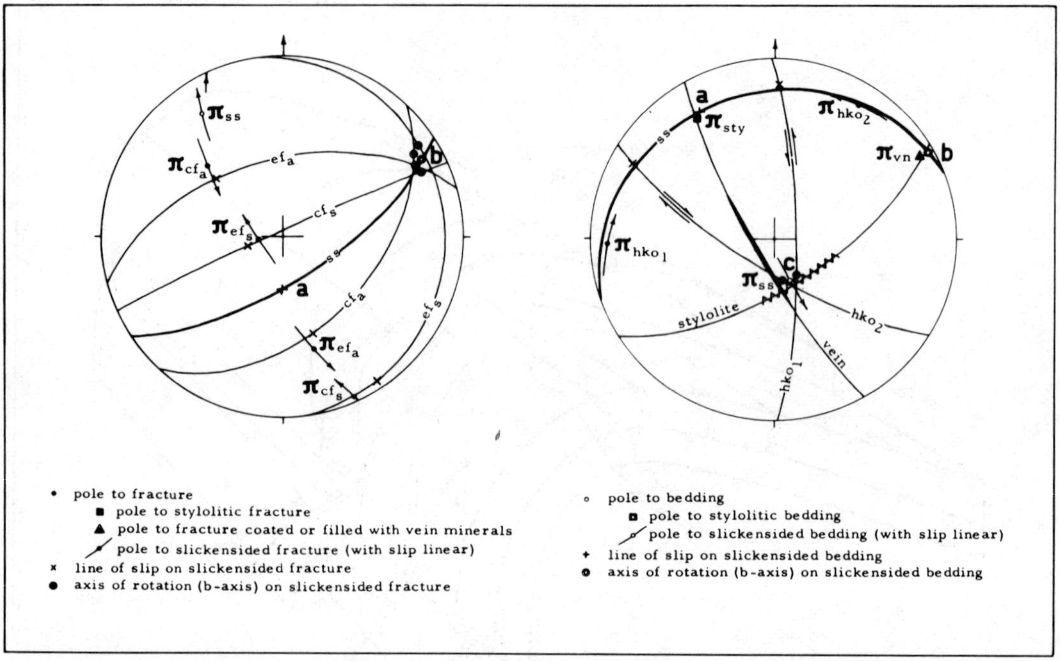

FIG. 3—Schematic stereograms illustrating the principal categories of faults (left side), and joints (right side). Abbreviations used are as follows: ss, bedding; *a*, *b*, and *c*, fabric axes; cf_s and cf_a, synthetic and antithetic contraction faults; ef_s and ef_a, synthetic and antithetic extension faults; hkO_1 and hkO_2, "shear joints". These and all subsequent stereograms are equal-area projections from the lower hemisphere. The legend is common to all subsequent sterograms but all the symbols listed do not occur in every stereogram.

FIG. 4—Locality 1, Grassy Mountain, Alberta. Left side: 600 poles to bedding; contours at 0.17%, 1%, 4%, and 10% per 1% area outline and axis of cylindrical folding (B). Right side: 134 poles to fault surfaces; contours at 0.75%, 4%, 8%, and 12% per 1% area; axis of folding for bedding, B.

ensided, stylolitic, or mineralized joints may reflect the combined influence of both a preexisting joint subfabric and the stress conditions during later orogenic deformation.

Some distinctions can be drawn among the various slickensided joints on the basis of the direction of displacement on them with reference to the plane of the bedding. The most common category of slickensided joints is characterized by slip parallel with the plane of the bedding. These joints, marked by an axis of rotation (b-axis of slip) that corresponds with the normal to the bedding (see Fig. 3, right), have been called "shear joints" (De Sitter 1956, p. 131) or "*hkO* shear surfaces" (Roeder 1960, Fig. 5). Where they occur as conjugate sets, they produce an elongation and a shortening in two mutually perpendicular directions in the plane of the bedding. A second category of slickensided joints can be distinguished by slip that is essentially perpendicular to the plane of the bedding. Joints of this type are marked by an axis of rotation that is essentially parallel with the bedding. Many of them deviate from the mean position perpendicular to the bedding and can produce an elongation or shortening in the plane of the bedding in the manner of extension faults or contraction faults. Other categories of slickensided joints are characterized by slip that is oblique to the plane of the bedding. They probably represent preexisting fractures which, although favorably oriented for slip under new conditions of stress, did not conform with the symmetry of the new stress system.

The tectonic significance of aggregates of all these various categories of mesoscopic fabric elements can be assessed in terms of several examples of mesoscopic subfabrics from the southern Rocky Mountains of Alberta and British Columbia. These examples, taken from two separate parts of this area, are intended primarily to illustrate the tectonic significance of a distinctive type of mesoscopic subfabric characteristic of deformed non-metamorphic rocks, but they also provide some insight into specific structural problems in these areas.

ANALYSES OF MESOSCOPIC SUBFABRICS

Crowsnest Pass Region

Two regional structural salients within the Rocky Mountains meet along a reentrant near Crowsnest Pass (see Fig. 1). Southeast of this reentrant the regional structural grain trends northwest. Northwest of the reentrant it trends north or northeast, and, within the vicinity of the reentrant itself, northwesterly, northerly, and northeasterly trends occur in superposition and in close proximity to each other. The kinematic significance of the change in structural trend between the two salients is a matter of considerable interest, particularly with regard to its implications concerning the direction of translation of the thrust plates (Price 1958, 1962; Norris and Black 1962; Leech 1962; Dahlstrom, Daniel, and Henderson 1962). Comparisons among examples of mesoscopic subfabrics from several different localities within the region of the reentrant provide an indication of the kinematic relationships between the two structural salients.

Grassy Mountain, Alberta

On the south slope of Grassy Mountain (locality 1, Fig. 1), a highly deformed sequence of mudstones, sandstones, and coal beds that belongs to the upper part of the Jurassic and (?)Cretaceous Kootenay Formation has been exposed extensively, as a result of strip mining operations (see Fig. 7). These rocks have been examined in some detail as part of a study of folds in this region (Price 1964). They lie within the northern structural salient, close to the Turtle Mountain fault (Norris 1955). The zone in which the structural trend changes to northwesterly lies a few miles to the south.

Within the general area shown in Fig. 7, the bedding in these rocks is statistically parallel to an axis of cylindrical folding (B) that has a plunge of less than 5° toward the south (see Fig. 4, left). The folds, although complex and disharmonic in gross aspect, are all concentric, and all terminate against thrust faults that have served as detachment surfaces for the concentric folding. Some of the thrust faults have been folded along with the bedding. Some follow the axial regions of folds and terminate along a line that serves as a focus or center of curvature for concentric folding; others rise out of a surface of interstratal slip, cut across several beds, and then pass into another surface of interstratal slip (Price 1964). All are contraction faults, and the displacements on them vary from several hundred feet to a fraction of an inch. Extension faults are common in the steeply dipping, vertical, or overturned beds (see Fig. 8). Some are truncated along surfaces of interstratal slip. Displacements on the extension faults range from a few feet to a fraction of an inch. Collectively, the fault surfaces are statistically cozonal[6] with the axis of folding (B) for the bedding, but they show greater dispersion than the bedding (see Fig. 4, right). Many of the faults are slickensided, as are many of the abundant joints in these rocks. Most of the striae on these slickensided surfaces are essentially parallel with a plane normal to the axis of folding (B) and represent slip that occurred

[6]Cozonal refers to a group of planar, surface elements that are parallel with, and may intersect parallel with, a common line that constitutes the zone axis for the group.

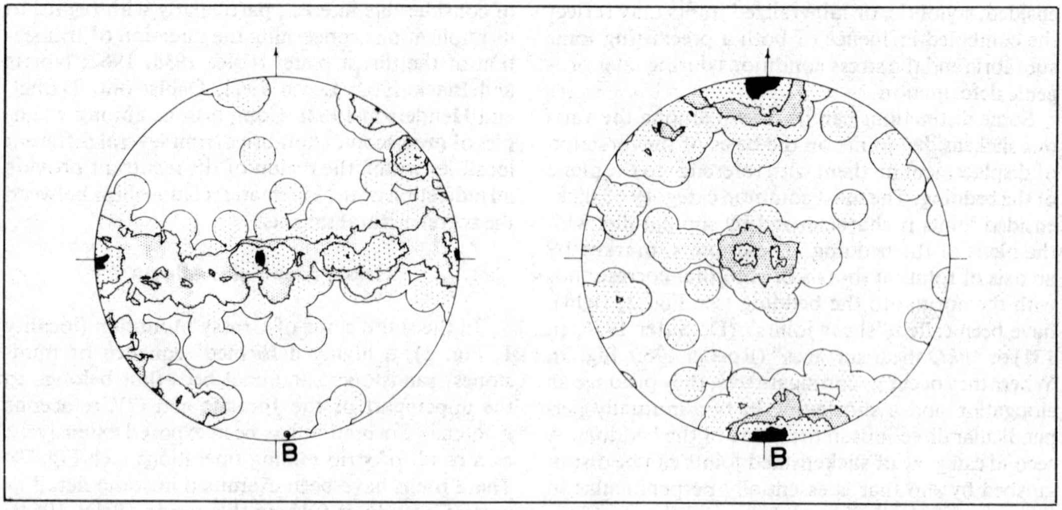

FIG. 5—Locality 1, Grassy Mountain, Alberta. Left side: 227 slickenside striae; contours at 0.44%, 2.2%, 5.3%, and 15% per 1% area; axis of folding for bedding, B. Right side: 227 axes of rotation for slip on slickensided fractures; contours at 0.44%, 2.2%, 5.3%, and 15% per 1% area; axis of folding for bedding, B.

perpendicular to the axis of folding (see Fig. 5, left). However, a significant number of the striae represent slip that occurred in a northeast-southwest direction, oblique to the axis of folding (B). Many of these oblique striae occur on *hkO* surfaces that might be described as "shear joints", but some are on faults.

The axes of rotation on all the various slickensided surfaces, when considered in aggregate, show three preferred orientations (see Fig. 5, right). By far the most conspicuous of these is an essentially horizontal, northerly trending axis that is more or less parallel with the axis of folding (B). It represents a mean kinematic b-axis for slip on an array of cozonal faults and related shear surfaces, and is parallel with the regional structural grain in this part of the more northerly structural salient. A second group of slickensided surfaces defines a mean axis of rotation that is perpendicular to this northerly trending axis and essentially vertical. These are the northeasterly trending slickensided *hkO* surfaces ("shear joints"). The third and least obvious preferred orientation is marked by an axis that plunges gently toward the northwest. It represents slip that occurred in a northeast-southwest direction, and is parallel to the structural grain in the nearest part of the more southerly structural salient. In summary, the mesoscopic subfabric of these rocks appears to outline two superimposed movement pictures. A dominant one, characterized by slip on an array of cozonal shear surfaces and external rotation of bedding and fault surfaces about an essentially horizontal, north-trending axis, appears to mask another less well defined movement picture involving slip about a northwesterly trending axis. Slip about an essentially vertical axis may have occurred in conjunction with both movement pictures.

Preferred spatial relationships between the bedding and various slickensided shear fractures on the one hand and the direction and sense of slip on them on the other, are of interest both in view of their implications for dynamic interpretations of the deformation and as a basis for the analysis of the kinematics of the deformation. The acute angle between the bedding and the surface of slip for 227 slickensided shear fractures is shown in relation to the acute angle between the bedding-fracture line of intersection and the axis of rotation (b-axis of slip) on the shear fracture in Fig. 6. On some of these fractures the sense of slip can be classified in terms of the effect it produced in the plane of the bedding, and the fractures can be designated as contraction faults or extension faults. These are shown by the larger open circles and solid circles, respectively, in Fig. 6.

The slickensided fractures fall into three main groups. In one group the axis of rotation for the slip on the fractures tends to be perpendicular to the line of intersection of the fracture and the bedding, and the fractures themselves tend to be perpendicular to the bedding. These are the slickensided joints that do not offset the bedding, that is, the "shear joints" or "*hkO* shear surfaces". In the other two groups the axis of rotation for the slip on the fracture tends to be parallel to the line of intersection of the fracture and the bedding. These are the faults that offset the bedding, and they show two preferred acute angles of intersection with the bedding. The con-

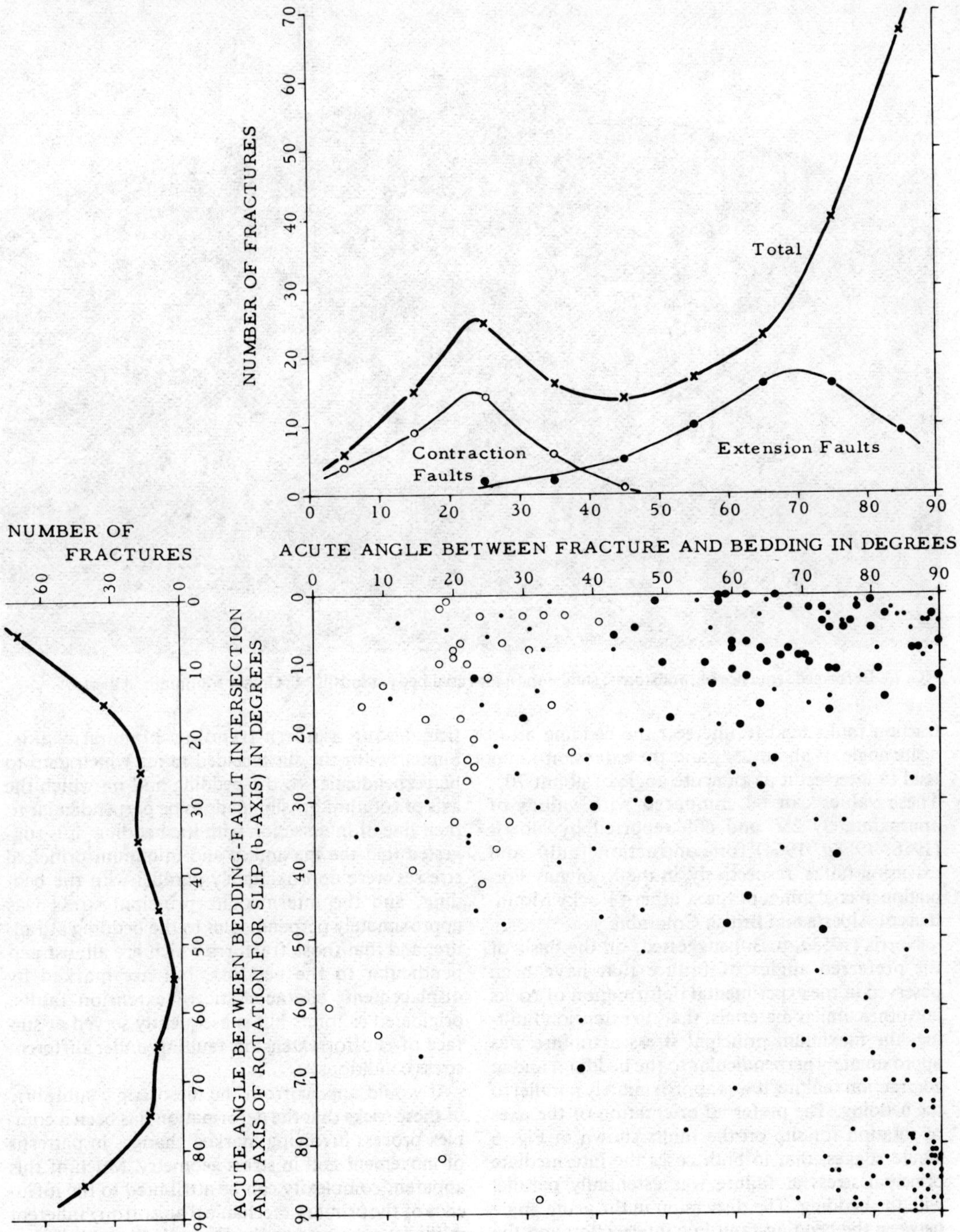

FIG. 6—The spatial relationships with reference to the plane of the bedding between axes of rotation for slip, sense of slip, and inclination for 227 slickensided fractures at locality 1, Grassy Mountain, Alberta. Large solid circles, extension faults; large open circles, contraction faults; small solid circles, sense of slip not established. The frequency distribution curves are sketched from histograms with 10° class intervals. The curve labelled "Total" refers to all slickensided fractures without differentiation, and includes those fractures upon which the curves for contraction faults and extension faults are based as well as other fractures on which the sense of slip is not established.

FIG. 7—Deformed sequence of mudstones, sandstones, and coal beds at locality 1, Grassy Mountain, Alberta.

traction faults tend to intersect the bedding at an acute angle of about 25°, and the extension faults tend to intersect it at an acute angle of about 70°. These values can be compared with values of approximately 25° and 60° reported by Norris (1956, 1958*a*, 1964) for contraction faults and extension faults, respectively, in the Kootenay Formation in coal mines of the southern Rocky Mountains of Alberta and British Columbia.

Norris (1958*a*, p. 30) suggested, on the basis of the preferred angles of failure that have been observed in the experimental deformation of rocks and other similar materials, that, in extension faulting, the maximum principal stress at failure was approximately perpendicular to the bedding, and in contraction faulting it was approximately parallel to the bedding. The preferred orientation of the axes of rotation for slip on the faults shown in Fig. 6 would suggest that in both cases the intermediate principal stress at failure was essentially parallel with the bedding. The dispersion in the acute angle between the bedding-fault line intersection and the axis of rotation for slip on the faults (see Fig. 6) may result from renewed shear movements on some of the faults under new stress conditions, such as might be expected, for example, if deformation about a northwest-trending kinematic axis were superimposed on a mesoscopic subfabric marking deformation about a north-trending kinematic axis. Similarly, for the slickensided joints which tend to be perpendicular to the bedding and on which the axis of rotation for slip tends to be perpendicular to their line of intersection with the bedding, it is suggested that the maximum and minimum principal stresses were approximately parallel with the bedding, and the intermediate principal stress was approximately perpendicular to the bedding at failure; and that those fractures which are almost perpendicular to the bedding, but are marked by displacements characteristic of extension faults, originated as joints but subsequently served as surface of slip for extension faulting under different stress conditions.

It would appear from the mesoscopic subfabric of these rocks that the deformation has been a complex process involving marked changes in patterns of movement and in stress geometry. Much of this apparent complexity can be attributed to the influence of the primary mechanical anisotropy inherent in the layering and to the flexural-slip folding that has occurred in conjunction with this anisotropy. The effect of planar anisotropy that is due to layering on the stress conditions at failure and the orientation of the plane of failure, for rocks that fail in brittle shear, has been evaluated theoretically and experimentally (Jaeger 1960; Donath 1961, 1963).

FIG. 8—Locality 1, Grassy Mountain, Alberta. Extension faults in near vertical beds.

For inclinations of the layering to the maximum principal stress over a range of 10°-60°, failure may occur, under lower stress, along a single set of shear fractures that follow the layer interfaces, instead of occurring on two conjugate sets of shear fractures that cut the layering and are inclined to the maximum principal stress at about 30°. The existence of a distinct planar anisotropy that is due to layering facilitates the interstratal slip, which, combined with buckling of individual layers, gives rise to concentric or flexural-slip folds like those characteristic of the Kootenay Formation at Grassy Mountain. In flexural-slip folding the shear stress on layer interfaces is relieved by interstratal slip, and the stress geometry within the flexed layers is modified by a strain that is due to buckling and is directly proportional to the ratio of thickness to radius of curvature for the flexed layer (Donath and Parker 1964). As the shear stress on any layer interface decreases as a result of interstratal slip, the stress trajectories will tend to become parallel with the layer interface; and the orientation of any shear fracture that develops within the layers will be governed by principal stresses that are almost parallel with, and perpendicular to, the layering. The relative magnitudes of the principal stresses that are in the plane of the layering and perpendicular to it will depend on the orientation of the layering relative to the principal stresses before the interstratal slip occurred on the layer interface. If the layering was essentially planar and inclined at a relatively low angle to the maximum principal stress, the maximum principal stress would tend to become parallel with the layering, and subsequent failure within the layers would take the form of contraction faults. If the layering had been rotated externally in the course of flexural-slip folding and was inclined at a relatively high angel to the maximum principal stress, the maximum principal stress would tend to become perpendicular to the layering, and subsequent failure within the layers would take the form of extension faults (Price 1964). The pattern of shear on the *hkO* surfaces surfaces (the "shear joints") suggests that the intermediate principal stress was perpendicular to the layering when failure occurred along them. This type of stress geometry might arise as a result of the interactions among a series of layers of different thicknesses and mechanical properties in which some of the layers had been forced to conform to the geometry of one or more dominant members during buckling.

The variations in stress conditions that have been suggested for each of the three different types of shear fractures involve changes in the relative magnitudes of the three principal stresses, but do not involve changes in the orientation of the principal stress axes with respect to the layering and axis of folding. In each case one of the principal stresses is approximately perpendicular to the layering, another lies approximately within the plane of the layering and is parallel with the axis of folding, and the third is approximately parallel with the layering and is perpendicular to the axis of folding. Accordingly, the three different types of shear fractures may be integrated dynamically in terms of this unifying factor, and the kinematics of the deformation may be considered on the basis of a single, overall movement picture involving a combination of interstratal slip, shear failure across and within the layers, and external rotation of the layering in conjunction with flexural-slip folding. Furthermore, the development of fissure veins along joints that are approximately perpendicular to the direction of elongation, and of stylolites along joints that are approximately perpendicular to the direction of shortening, within the layers as they are buckled, also can be considered as an integral part of this movement picture.

The Gap, Livingstone Range, Alberta

Livingstone Range, a north-trending, faulted anticlinorium in Mississippian beds, lies within the more northerly structural salient in the Livingstone thrust plate. At The Gap, where Oldman River

FIG. 9—Carnarvon Member of the Mount Head Formation at locality 2, the west side of The Gap. Left side: antithetic and synthetic contraction faults and joints. Two contraction faults in the thick bed at the center intersect its lower surface at about 25°. The synthetic contraction fault dips more steeply than the bedding, the antithetic fault less steeply. Right side: detail of stylolitic joint and stylolitic bedding.

crosses the range, the beds are folded about an approximately horizontal, northerly trending axis (Douglas 1950). The Carnarvon Member of the Mount Head Formation, a dark cryptocrystalline limestone in beds a few feet thick with shaly partings, is exposed along the south side of The Gap on both flanks of the range.

At the west side of The Gap (locality 2, Fig. 1), bedding in the Carnarvon Member in a long, planar segment of the limb of an anticline strikes north and dips about 50° west. Some of the bedding surfaces are stylolitic (see Fig. 9); others are polished, presumably as a result of interstratal slip, but no slickenside striae were observed on them. The beds are cut by synthetic and antithetic contraction faults of small displacement that lie in *hOl* (see Fig. 9, left), and by numerous joints, many of which are slickensided, coated or filled with white calcite, or are stylolitic (see Fig. 9, right). A sample of the mesoscopic subfabric of these rocks, comprising five measurements on bedding and 59 on other fracture surfaces, is shown in Fig. 10 (left). It can be compared with data from the same general locality reported by Roeder (1960, p. 589).

Consistent patterns for the movements during deformation can be established on the basis of these various mesoscopic fabric elements. The axes of rotation for slip on both the antithetic and synthetic *hOl* contraction faults are essentially horizontal and trend slightly west of north. They define a kinematic axis (b) that is more or less parallel with the axis of folding (*b*-axis of the fabric). The slickensided joints occur in two conjugate sets in *hkO* (shown as broken lines in Fig. 10, left) that are inclined at about 25° to a plane more or less perpendicular to the *b*-axis of the fabric. The axes of rotation for slip on the joints are parallel with the poles to the bedding (*c*-axis of the fabric). Slip on the joints was parallel with the plane of the bedding. Joints coated or filled with calcite occur in two main sets. One set is approximately perpendicular to the axis of folding and lies in the *ac* plane. The other set lies in *hkO* and intersects the bedding along a line that trends about N.30°W. The stylolitic joints also fall into two main sets. Each set is approximately perpendicular to one of the two sets of calcite-filled joints. One set of stylolitic joints lies approximately in the *bc* plane, the other set lies in *hkO* and intersects the bedding along a line that trends southwest. The contraction faults, slickensided joints, calcite-filled joints in *ac*, and stylolitic joints in *bc* outline a dominant pattern of deformation involving shortening in *a* and elongation in both *c* and *b*. The trend of the kinematic b-axis for this deformation corresponds with the regional structural trend in this part of the northern structural salient. The stylolitic and calcite-filled joints in *hkO* provide evidence of another pattern of deformation involving shortening and elongation in the plane of the bedding along kinematic axes that trend northwest and southwest, respectively. These kinematic axes are approximately parallel to and perpendicular to the structural trends in the nearest part of the southern salient.

These fabric data and the patterns of movement outlined by them are not entirely consistent with the analysis of the deformation at this locality pre-

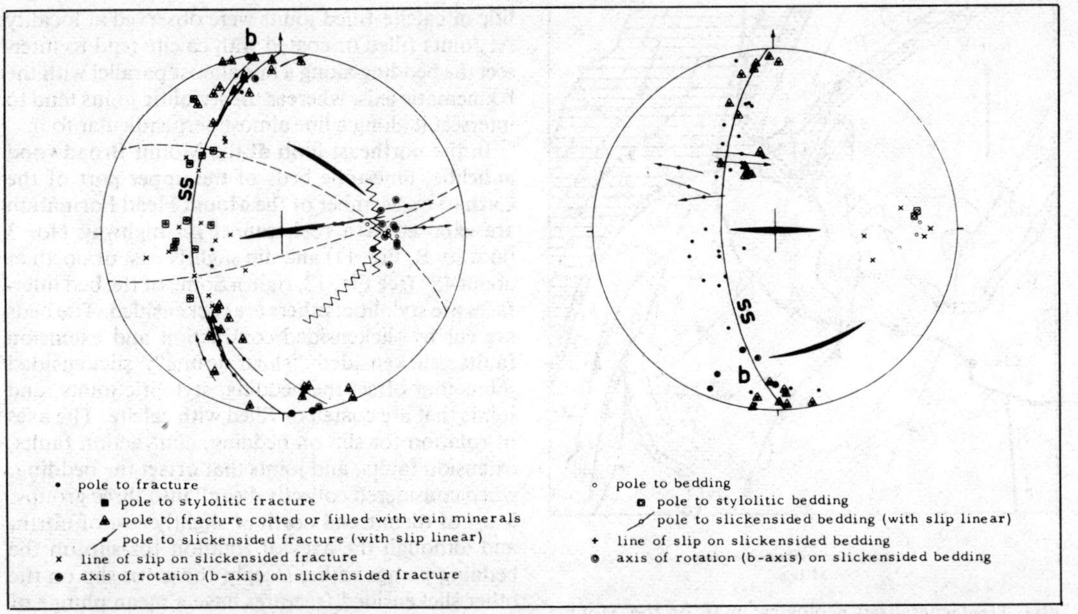

FIG. 10—Samples of mesoscopic subfabrics. The Gap, Livingstone Range, Alberta. Left side: Carnarvon Member of the Mount Head Formation at locality 2, the west side of The Gap. Right side: Carnarvon Member of the Mount Head Formation at locality 3, the east side of The Gap. Solid great circle curve labelled "ss", mean bedding; broken great circle curves, mean of individual *hkO* joint sets; short thick segments of great circle curves, mean of individual mineralized joint sets; short zigzag segments of great circle curves, mean of individual stylolitic joint sets.

sented by Roeder (1960, p. 585). Roeder concluded that a younger phase of deformation involving shortening in *a* and elongation in both *c* and *b* had been overprinted on an older phase of deformation involving elongation in *a* and shortening in *b*. He cited calcite-filled joints in *bc* and reversals in the sense of slip on the slickensided joints as evidence of the older phase of deformation. No calcite-filled *bc* joints occur in the sample of 54 joints measured in the present study, and on the few joints on which the sense of slip could be inferred no evidence of reversals in slip was observed.

At the east side of The Gap (locality 3, Fig. 1) the bedding in the Carnarvon Member strikes north and is overturned with a westerly dip of about 65°. A sample of the mesoscopic subfabric at this locality, comprising five measurements on bedding and 50 measurements on fractures, is shown in Fig. 10 (right). All of the fractures appear to be joints. A few that are inclined to the bedding at between 10° and 20° have served as surfaces of extension faulting and are slickensided. The axes of rotation for slip on these fractures plunge slightly west of south at about 15° and define a kinematic axis (b) that is more or less parallel with the *b*-axis of the fabric. Joints that are filled or coated with calcite fall into two main sets. One set lies in the *ac* plane. The other set lies in *hkO*, dips about 30° south, and intersects the bedding along a line almost parallel with the b-axis for slip on the slickensided joints. The calcite-filled *ac* joints mark an elongation in *b* similar to that noted at the west side of The Gap. The extension faulting on joints and the south-dipping, calcite-filled joints in these steeply dipping overturned beds at the east side of The Gap, mark an elongation in *a* that was not observed in the more gently dipping upright beds at the west side of The Gap. This elongation in *a* may have occurred after the bedding had been rotated externally into a steeply dipping position as a result of flexural-slip folding.

Mount Broadwood Area, British Columbia

The Mount Broadwood area (see Fig. 11) lies within the more southerly structural salient, in a region of northwest-trending thrust faults and northwest-plunging folds. Mesoscopic subfabrics were sampled at four localities within the area. Two of these localities (A and B) are in Mississippian rocks in an anticline that plunges northwest from Mount Broadwood; one (D) is in Mississippian rocks in a fault slice along the Wigwam fault; and the fourth (E) is in Purcell (Precambrian) rocks in the Wigwam fault plate. Along highway No. 3 the axis (B) of the Mount Broadwood anticline plunges

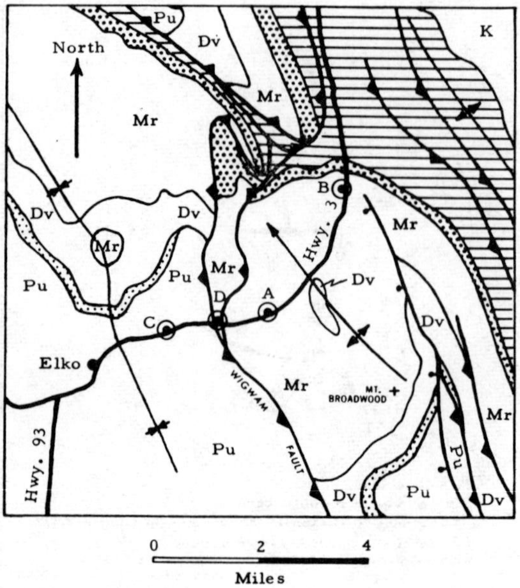

FIG. 11—Generalized geological map of the Mount Broadwood area, British Columbia (after Leech 1960; Price 1962). Heavy lines, faults (teeth indicate upper plate on thrust faults and barbs indicate downthrown block on the gravity fault); Pu, undivided Purcell (Precambrian); light stipple, undivided Cambrian; Dv, undivided Devonian; Mr, undivided Mississippian; heavy stipple, Pennsylvanian and Permian Rocky Mountain Group; horizontal ruling, Triassic Spray River Formation and Jurassic Fernie Group; K, Jurassic and (?)Cretaceous Kootenay Formation; A, B, C, and D, mesoscopic subfabric sample localities.

about N.40°W. at between 20° and 25° (see Fig. 12). Folds in the Wigwam fault plate also plunge northwest, but less steeply.

In the southwest limb of the Mount Broadwood anticline, at locality A (see Fig. 11), the bedding in the upper part of the Banff Formation dips about 20° northwest and is cut by numerous faults. Although many of these are slickensided, a lack of distinctive marker beds precludes reliable determinations of the sense of slip on them. Most intersect the bedding at a relatively low angle and are probably contraction faults (see Fig. 12, left). The axes of rotation for the slip on these slickensided faults fall into two main groups, both of which tend to be parallel with the bedding. One group plunges north or slightly east of north and defines a kinematic axis (b) that is discordant with respect to the major structures in this area, but conforms with the structural trends in the nearest part of the more northerly salient. The other group plunges northwest, parallel with the axis of the fold (B), and defines a kinematic axis (b′) that does conform with the structural trend in this part of the southern salient. Only a few stylolitic or calcite-filled joints were observed at locality A. Joints filled or coated with calcite tend to intersect the bedding along a line almost parallel with the b kinematic axis, whereas the stylolitic joints tend to intersect it along a line almost perpendicular to b.

In the northeast limb of the Mount Broadwood anticline, limestone beds of the upper part of the Carnarvon Member of the Mount Head Formation are exposed at a rock tunnel on highway No. 3 (locality B, Fig. 11) and dip slightly east of north at about 45° (see Fig. 12, right). Some of the bed interfaces are stylolitic; others are slickensided. The beds are cut by slickensided contraction and extension faults, slickensided "shear joints", slickensided joints that offset the bedding, stylolitic joints, and joints that are coated or filled with calcite. The axes of rotation for slip on bedding, contraction faults, extension faults, and joints that offset the bedding, when considered collectively, fall into three groups. Most of them trend north or slightly east of north, and although the axes of rotation for slip on the bedding plunge rather steeply, those for slip on the other slickensided fractures have a mean plunge of about 25° and define a kinematic axis (b) that is almost parallel with the b-axis for the Banff Formation at locality A. A few axes of rotation plunge northwest, more or less parallel with the axis of folding (B) and the b′-axis for the Banff Formation at locality A. The remainder plunge northeast perpendicular to the b′-axis. The calcite-filled joints occur in three sets, each of which is approximately parallel with one of the three groups of axes of rotation. The stylolitic joints tend to intersect the bedding along a line that plunges N.20°E and is inclined only slightly to the b-axis.

In the Wigwam fault plate, quartz sandstones of the Phillips Formation are exposed in a roadcut along highway No. 3 at locality C (see Fig. 11). The bedding dips west at about 20° and is cut by a number of slickensided contraction faults of small displacement (see Fig. 13, left). The axes of rotation for slip on slickensided contraction faults and bedding tend to plunge about N.30°W. at 10° or less. They define a kinematic axis (b′) that is almost parallel with the b′-axis of localities A and B. Thin quartz fissure veins and drusy coatings of quartz occur along two sets of joints at this locality. Most of the veins strike northwest parallel with the b′-axis. Drusy coatings and a few veins occur along joints that strike N.75°W. and are almost perpendicular to the b-axes of localities A and B.

A fault slice of Banff limestone that occurs along the Wigwam fault is exposed in a roadcut on highway No. 3 at locality D (see Fig. 11). Bedding in the limestone dips steeply northeast to vertical and is warped about an almost vertical axis (see Fig. 13, right). Many of the bed interfaces are marked by slickenside striae that have developed on top of a

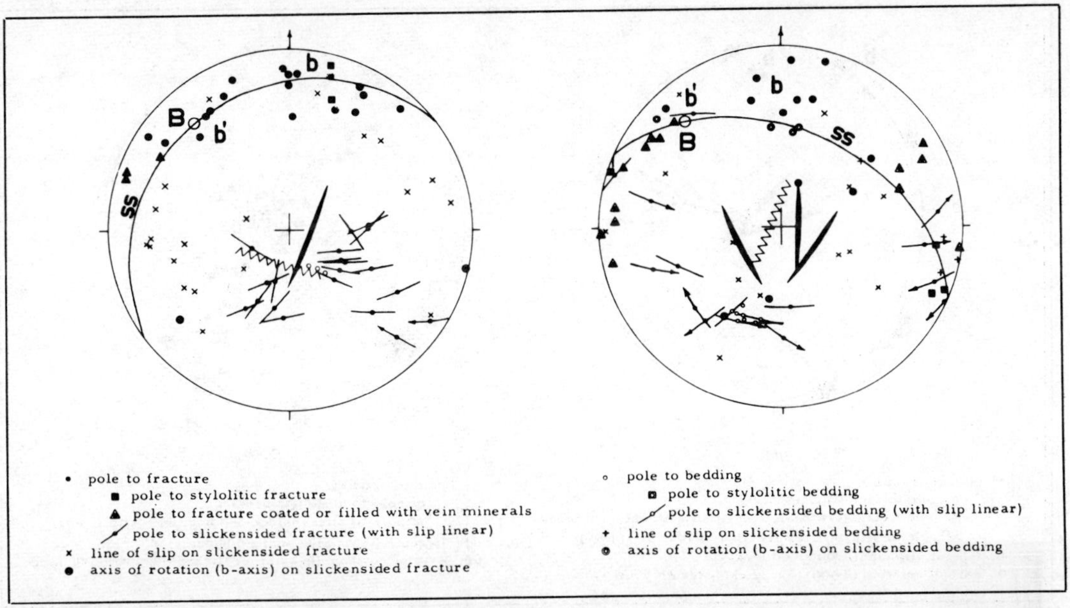

FIG. 12—Samples of mesoscopic subfabrics, Mount Broadwood area. Left side: locality A, Banff Formation, southwest limb of Mount Broadwood anticline. Right side: locality B, Carnarvon Member of the Mount Head Formation, northeast limb of Mount Broadwood anticline. B, axis of folding; solid great circle curve labelled "ss", mean bedding; short thick segments of great circle curves, mean of individual mineralized joint sets; short zigzag segments of great circle curves, mean of individual stylolitic joint sets.

coating of coarse, white calcite (see Fig. 15); some are stylolitic. Most of the axes of rotation for slip on the bedding or on contraction faults plunge very steeply and are distributed about a mean kinematic axis (b), which is more or less parallel with the axis of folding for the bedding. A couple plunge northwest or southeast and are approximately parallel with the trend of the Wigwam fault. The stylolitic joints in these rocks tend to strike about N.60°E., perpendicular to the trend of the b′-axis for localities A, B, and C, and parallel to the near vertical axis of folding for the bedding at this locality. Most of the joints that are filled or coated with calcite are almost horizontal. They tend to be perpendicular to the axis of folding for the bedding.

In spite of considerable variation in structural environment and in rock type among these four localities in the Mount Broadwood area, each sample of the mesoscopic subfabric consistently provides evidence of deformation about a northwesterly trending kinematic axis that conforms with the regional trend of the megascopic structures in this part of the more southerly salient. Furthermore, three of the samples provide significant evidence of deformation about a northerly or northeasterly trending axis that is not apparent in the megascopic structures in this area, but nevertheless conforms with the trend of megascopic structures in the structural salient that lies to the north.

Crowsnest Lake, Alberta

A faulted, west-dipping homoclinal succession of Devonian and Mississippian carbonate rocks in the lower part of the Lewis thrust plate is exposed in cliffs along the south side of Crowsnest Lake. These rocks lie within the northern structural salient, but are neat its southern margin and the area in which both northwesterly trending and northerly trending megascopic structures occur in superposition (Price 1958).

In the upper part of the Mississippian Banff Formation adjacent to highway No. 3 (locality 4, Fig. 1), the bedding strikes about N.10°W., dips between 50° and 70° west, and is offset by a conspicuous thrust fault that dips 70° west (see Fig. 14, left). Many of the bed interfaces are slickensided, and many small slickensided antithetic and synthetic contraction faults and a few slickensided extension faults cut the beds. The axes of rotation for the slip on these slickensided surfaces tend to plunge slightly west of north at bout 10°. They define a mean kinematic axis (b) that conforms with the trend of the megascopic structures in this area. The dispersion in the orientation of this group of rotation axes suggests that slip on some of the shear surfaces may be related to a second kinematic axis (b′), that plunges northwest and is parallel with the structural trends in the salient that lies to the south. The axes

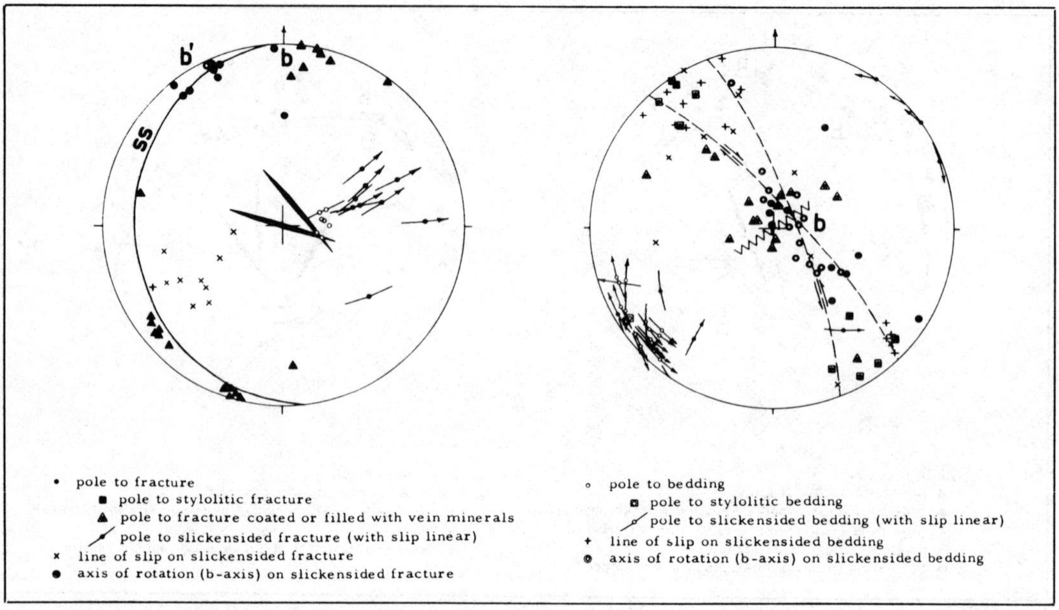

FIG. 13—Samples of mesoscopic subfabrics, Mount Broadwood area. Left side: locality C, Phillips Formation, Wigwam fault plate. Right side: locality D, Banff Formation, adjacent to Wigwam fault. Solid great circle curve labelled "ss", mean bedding; short thick segments of great circle curves, mean of individual mineralized joint sets; short zigzag segment of great circle curve, mean of stylolitic joint set; broken great circle curves, schematic examples of slickensided bedding surfaces showing sense of slip.

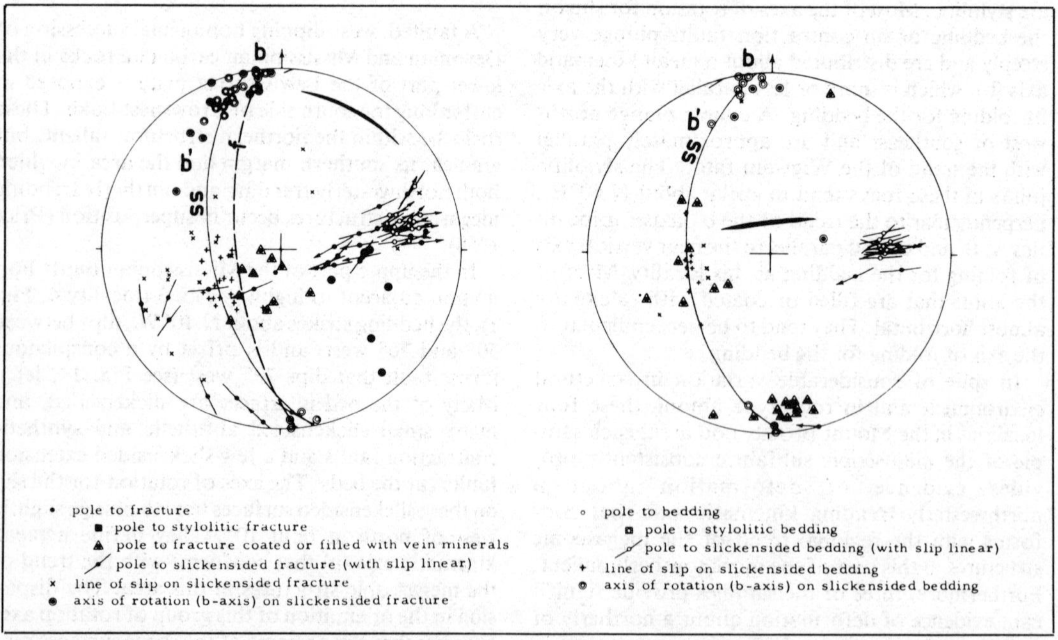

FIG. 14—Samples of mesoscopic subfabrics, Crowsnest Lake, Alberta. Left side: locality 4, Banff Formation. Right side: locality 5, Palliser Formation. Solid great circle curve labelled "ss", mean bedding; broken great circle curve labelled "f", thrust fault; short thick segments of great circle curves, mean of individual mineralized joint sets.

of rotation for slip on some of the small extension faults plunge southeast and are distributed through the interval between the ideal orientation for axes of rotation on "shear joints", parallel with the pole to the bedding, and the b- and b′-axes. Joints that are coated or filled with calcite tend to dip about 15° northeast and to intersect the bedding along a line parallel with b.

At locality 5, approximately 1,000 ft east of locality 4 (see Fig. 1), the bedded limestones in the upper part of the Devonian Palliser Formation dip about 50° west, and are cut by numerous joints that are coated or filled with calcite and by a few slickensided contraction faults (see Fig. 14, right). Many of the bedding surfaces are slickensided and a few show two sets of slickenside striae. The axes of rotation for slip on the slickensided shear surfaces tend to plunge about N.10°W. at 10° or less, but a few trend northwest and plunge more steeply. One one bedding surface, a younger set of slickenside striae perpendicular to the north-plunging kinematic axis (b) has been overprinted on an older northwest-plunging set that is perpendicular to b′. The calcite-filled joints occur in two sets that are almost perpendicular to and parallel with b.

At both localities the mesoscopic subfabric outlines a dominant pattern of deformation involving movement about a northerly trending axis that conforms with the trend of megascopic structures in this part of the northern salient; and at both there is also some evidence for movement about a northwesterly trending axis that parallels the regional structural trend in nearby parts of the more southerly salient. Thus, there appears to have been some uniformity in the patterns of movement during deformation throughout the Crowsnest Pass region. On a mesoscopic scale, deformation about both northerly and northwesterly trending axes appears to have been pervasive throughout the entire region embracing the southern part of the more northerly structural salient and the northern part of the more southerly salient. In general, where the regional structures trend north, deformation about a northerly trending axis is dominant on a mesoscopic scale; and where they trend northwest, deformation about a northwesterly trending axis is dominant on a mesoscopic scale. However, at least locally, as at locality B (see Figs. 11 and 12, right) in the northeast limb of the Mount Broadwood anticline, a dominant pattern of deformation on a mesoscopic scale, involving movement about a northerly trending axis, is discordant relative to the northwesterly trending megascopic structures in which it occurs, and appears to have been superimposed upon them. Elsewhere, mesoscopic fabric data bearing on the question of the relative ages of the two patterns of movement are, at best, suggestive. For example, at those localities where the regional structures trend north, the northwesterly trending kinematic axis tends to lie in the plane of the bedding; a relationship that might be interpreted as indicating that it represents older fabric elements which have been rotated externally along with the bedding during deformation about the northerly trending axis. Whatever their relative ages may be, the pervasive patterns of movement associated with each of the two structural salients converge in the vicinity of the reentrant near Crowsnest Pass.

Transverse Faults in Opal and Misty Ranges

Northeasterly and easterly trending transverse faults are relatively common locally in parts of the Front Ranges of the Rock Mountains in southwestern Alberta (see Fig. 1). Many of these have been interpreted as tear faults that are related to the thrusting (Douglas 1958; Dahlstrom and Henderson 1959). The stratigraphic separation across the transverse faults is generally quite small; but for a few of them it may be as much as several thousand feet. Most of the transverse faults abut the thrust faults marking the upper and lower limits of the plate in which they occur. In a few cases intervening thrust faults or folds abut a transverse fault, and the latter has clearly been active during the thrusting and folding. If all of these transverse faults originated as tear faults during the thrusting and folding, the slip that has occurred on them should consistently be parallel with the sole fault of the thrust plate in which they lie.

The mesoscopic subfabric along the transverse faults affords a basis for analyzing the movements that have occurred along them, and for establishing their kinematic relationships to the thrust faults. Accordingly, mesoscopic subfabrics have been sampled in the same stratigraphic unit on either side of several of the smaller transverse faults in the Mount Rundle thrust plate in southern Opal Range (see Fig. 16) and northern Misty Range. In this general area bedding in the Mount Rundle thrust plate dips steeply southwest, and the Mount Rundle thrust fault either follows the dip of the bedding or else dips southwest a little more steeply than the beds.

In northern Misty Range, at locality 6 (see Fig. 1) on the southwest flank of Mount Rae, the top of the Palaeozoic succession is offset with a stratigraphic separation of about 400 ft along a right-hand transverse fault that strikes about N.40°E. and dips about 80° northwest. Fractures in the upper part of the Pennsylvanian and Permian Rocky Mountain Group adjacent to the fault are commonly slickensided (see Fig. 17). Many of these fractures are more or less perpendicular to the bedding and appear to have originated as joints, but a few intersect the bedding at a low angle and appear to be contraction faults (see Fig. 17, left), and others are approxi-

FIG. 15—Slickenside striae on a bedding surface, locality D, Banff Formation, adjacent to Wigwam fault.

mately parallel with the transverse fault and presumably developed in conjunction with the movements on it. On the northwest side of the fault (see Fig. 17, right) the axes of rotation for slip on a few slickensided bedding surfaces, and on a few joints that are almost perpendicular to the fault, trend northwest, parallel with the regional structural trend in the area. Slip on these surfaces appears to have been related to the thrusting. However, the axes of rotation for slip on most of the slickensided shear surfaces tend to be parallel with the plane of the transverse fault (see Fig. 17, left). They plunge southwest or are almost vertical. The slip on these surfaces has been almost perpendicular to the plane of the bedding and the Mount Rundle thrust, whereas the slip on a tear fault related to the Mount Rundle thrust would, of necessity, be parallel to the plane of the thrust and therefore almost parallel to the bedding. The pattern of movement expressed in terms of thes slickensided shear surfaces indicates that this transverse fault is not a tear fault related to the Mount Rundle thrust, but instead, is an extension fault on which there has been slip more or less perpendicular to the plane of the bedding.

A number of small transverse faults occur in southern Opal Range, at locality 7 (see Fig. 1) on the southeast flank of Mount Elpoca. The mesoscopic fabric data pertaining to one left-hand transverse fault with a stratigraphic separation of about 35 ft at the top of the Rocky Mountain Group (see Fig. 16, right), is shown in Fig. 18. The pattern of movement on this fault, as expressed in terms of slickensided surfaces in and adjacent to the fault, is similar to that for the right-hand fault at locality 6 in Misty Range. It indicates that this fault is not a tear fault, but, instead, is an extension fault on which there has been slip more or less perpendicular to the plane of the bedding and the Mount Rundle thrust. The same relationships are implicity in the mesoscopic fabric data from a nearby right-hand transverse fault that has a stratigraphic separation of 60 ft at the level of the top of the Rocky Mountain Group (see Fig. 19).

The mesoscopic subfabrics adjacent to all three

FIG. 16—Transverse faults at locality 7, southern Opal Range, Alberta. Left side: transverse faults showing left-hand offset of the stripped surface at the top of the Rocky Mountain Group. Right side: fractures and breccia zone along a right-hand transverse fault cutting Rocky Mountain Group (see Fig. 19).

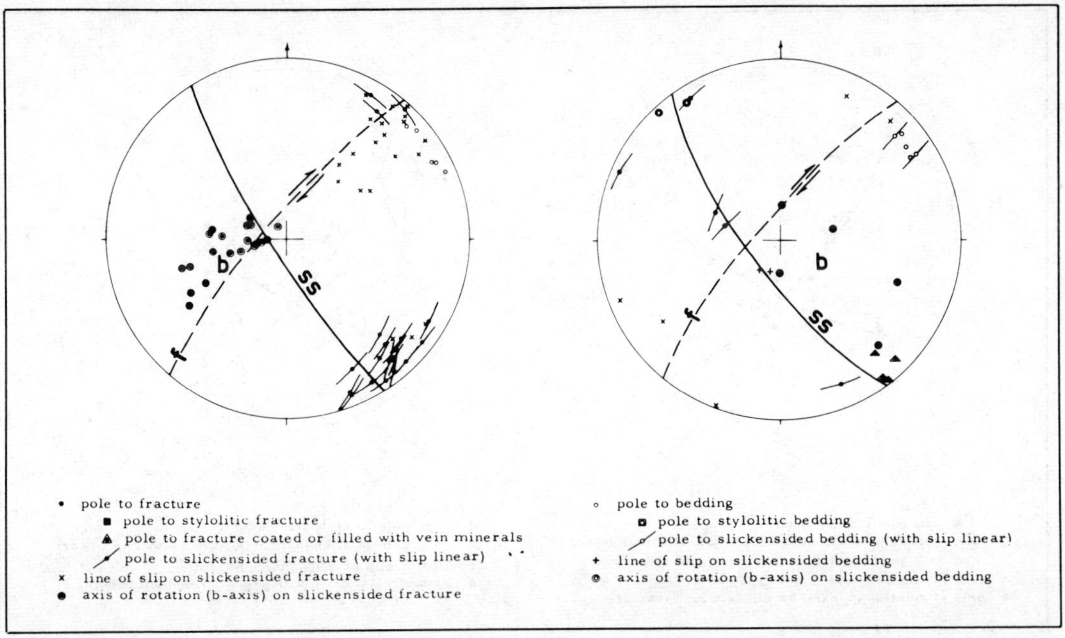

FIG. 17—Samples of the mesoscopic subfabric adjacent to the right-hand transverse fault at locality 6, northern Misty Range, Alberta. Left side: on the northern side of the fault. Right side: on the southeast side of the fault. Solid great circle curve labelled "ss", mean bedding; broken great circle curve labelled "f", transverse fault showing sense of stratigraphic separation.

northeasterly trending transverse faults outline a consistent pattern of movement that indicates that they are not tear faults related to the translation of the thrust sheets, but, instead, are extension faults on which there has been slip approximately perpendicular to the plane of the bedding. The available fabric data do not preclude the possibility that the faults are relatively young and represent a longitudinal extension within the Mount Rundle thrust plate after it had formed, but the high angle at which some of the transverse faults abut the thrust faults militates against such an interpretation. Instead, the transverse faults appear to predate the displacement on the Mount Rundle thrust and the rotation of the bedding from an initial near horizontal position into the present steeply dipping position that occurred in conjunction with this thrust displacement. Moreover, major easterly or northeasterly trending transverse structures that appear to have developed long before the period of thrusting and folding are known from parts of the western Rocky Mountains (Leech 1958), and they are essentially parallel with the northeasterly trending structural grain of the Hudsonian basement that extends westward from the Canadian Shield to the eastern margin of the Cordilleran orogen in this region. Accordingly, it is suggested that these northeasterly trending transverse faults in the Front Ranges of the Rocky Mountains originated before the thrusting and folding as gravity faults whose orientation was controlled by the fabric of the Hudsonian basement beneath them.

CONCLUSIONS

The mesoscopic subfabrics of the deformed nonmetamorphic rocks in the southern Rocky Mountains of Alberta and British Columbia are characterized by several distinctive types of fracture surfaces that were mechanically effective as strain discontinuites during deformation. Fractures that are slickensided, stylolitic, or coated or filled with vein minerals mark a locus of shear, pressure solution, or dilation, respectively. Each is a kinematic indicator for some stage of the deformation in these rocks, even when considered individually rather than as a component of a fracture array whose symmetry can be related to the symmetry of the deformation. Each defines a unique line of slip and axis of rotation for slip, a unique direction of relative compression, or a unique direction of relative extension. Collectively, they provide a direct and succinct record of the kinematic history of an individual fabric domain, and a sound basis for dynamic analyses of the deformation.

Kinematic patterns, for the deformation within

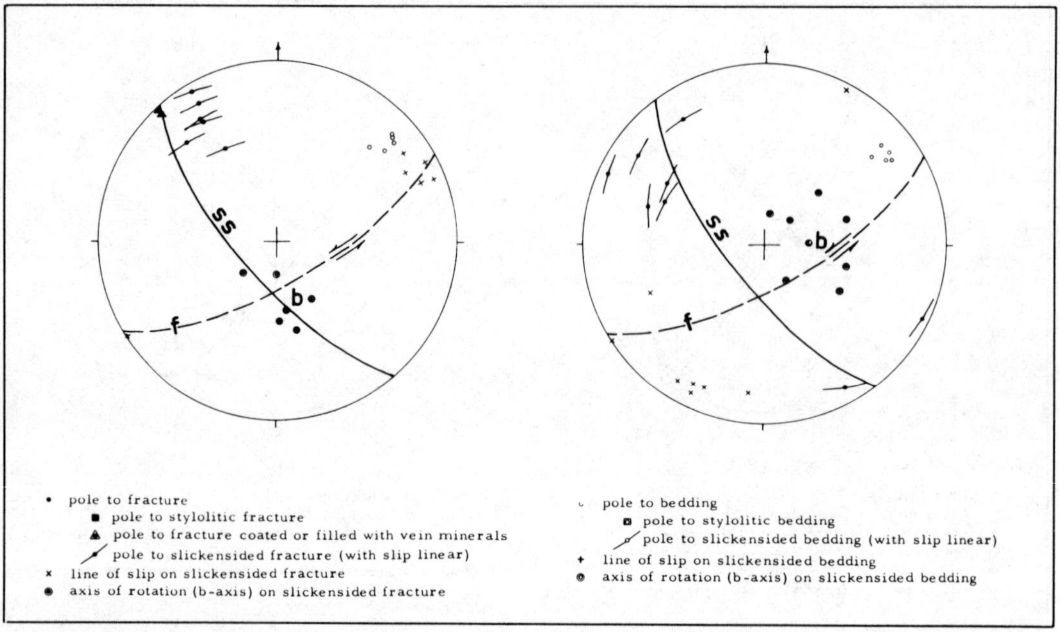

FIG. 18—Samples of the mesoscopic subfabric adjacent to a left-hand transverse fault at locality 7, southern Opal Range, Alberta. Left side: on the southeast side of the fault. Right side: on the northwest side of the fault. Solid great circle curve labelled "ss", mean bedding; broken great circle curve labelled "f", transverse fault showing sense of stratigraphic separation.

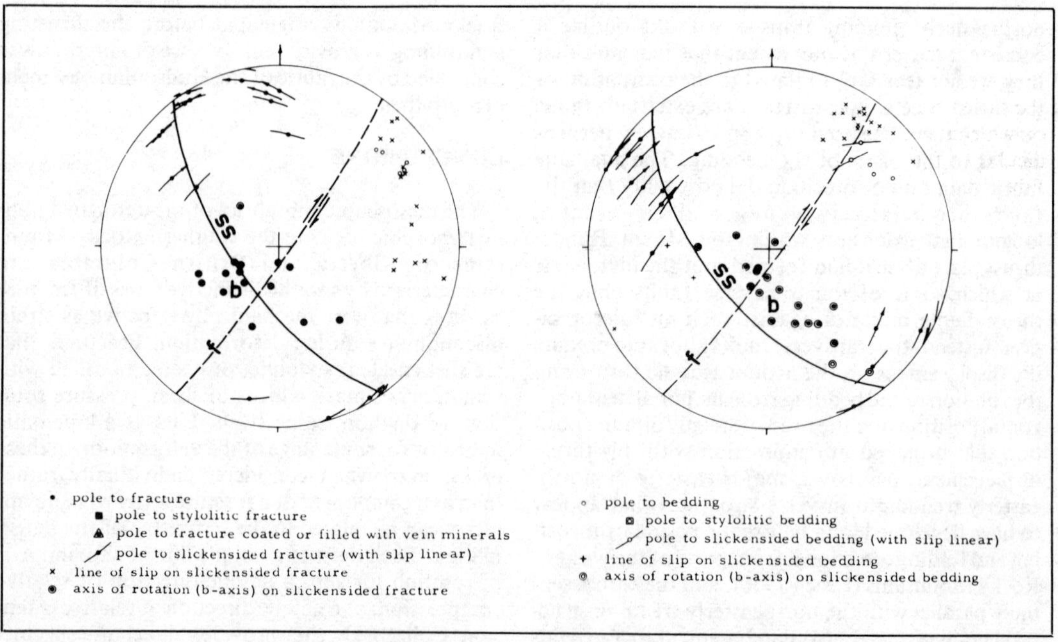

FIG. 19—Samples of the mesoscopic subfabric adjacent to a right-hand transverse fault at locality 7, southern Opal Range, Alberta. Left side: on the southeast side of the fault. Right side: on the northwest side of the fault. Solid great circle curve labelled "ss", mean bedding; broken great circle curve labelled "f", transverse fault showing sense of stratigraphic separation.

individual thrust plates during their development and translation, can be outlined in terms of samples of mesoscopic subfabrics. The kinematic relationships between and among interlocking thrust plates can be established.

Consistent spatial relationships among various categories of slickensided fractures, and the compositional and textural layering that defines the bedding, attest to the strong influence of the primary mechanical anisotropy inherent in the bedding on the character of the deformation. The orientation of the shear fractures that follow the bedding has been controlled by this anisotropy. The other slickensided fractures fall into two main groups: faults that intersect the bedding at preferred angles of 25° or 70° and are generally marked by slip perpendicular to the bedding, and "shear joints" that intersect the bedding at a preferred angle of 90° and are generally marked by slip parallel with the bedding. By analogy with data on brittle shear failure in the experimental deformation of rocks, these spatial relationships suggest that the principal stress axes were approximately parallel with and perpendicular to the bedding when slip occurred on these surfaces, but that the relative magnitude of the principal stress perpendicular to the bedding differed among each of the three categories of slickensided shear surface.

Within a broad area embracing the prominent structural reentrant that extends across the Rocky Mountains near Crowsnest Pass, deformation on a mesoscopic scale about both northerly and northwesterly trending axes was essentially pervasive. Movement about both northerly and northwesterly trending axes can be outlined on the basis of the mesoscopic subfacies of rocks which, on a megascopic scale, have either a northwesterly or northerly trending fabric axis; however, in general, the movement pattern that is parallel with the trend of the megascopic structures is dominant. Movement patterns for the deformation associated with each of two regional structural salients converge in the vicinity of the reentrant.

The mesoscopic subfabrics associated with transverse faults in parts of the Front Ranges of the Rocky Mountains outline a pattern of movement which indicates that they did not originate as tear faults related to the translation of the thrust plates, but, instead, are older gravity faults whose orientation may be controlled by the fabric of the Hudsonian basement extending under the mountains from the Canadian Shield.

REFERENCES

Bolsenkoeter, H., 1955. Feintektonische Untersuchungen an Schlechten und Kluften in Steinkohlenflozen des Ruhrgebietes. Geol. Rundschau, 44, 443.

Bonham, L. C., 1957. Structural petrology of the Pico anticline, Los Angeles County, California. J. Sediment. petrology, 27, 251.

Dahlstrom, C. D. A., and Henderson, G. G. L., 1959. Structural geology of the Moose Mountain area, Alberta. Alberta Soc. Petrol. Geologists, Guidebook, 9th Ann. Field Conf. p. 53.

Dahlstrom, C. D. A., Daniel, R. E., and Henderson, G. G. L., 1962. The Lewis thrust at Fording Mountain, British Columbia. J. Alberta soc. Petrol. Geologists, 10, 373.

De Sitter, L. U., 1959. Structural geology. McGraw-Hill Book Co., Inc., New York, p. 552.

Donath, F. A., 1961. Experimental study of failure in anisostropic rocks. Bull. Geol. Soc. Am. 72, 985.

———, 1963. Fundamental problems in dynamic structural geology. *In* The earth sciences, problems and progress in current research. Edited by T. W. Donnelly. Univ. Chicago Press, Chicago, p. 83.

Donath, F. A., and Parker, R. B., 1964. Folds and folding. Bull. Geol. Soc. Am. 75, 45.

Douglas, R. J. W., 1950. Callum Creek, Langford Creek, and Gap map-areas. Geol. Surv. Can. Mem. 255.

———, 1958. Mount Head map-area, Alberta. Geol. Surv. Can. Mem. 291.

Fitzgerald, E. L., 1963. Structural analysis along Torrens River, Alberta Foothills, Bull. Can. Petrol. Geologists, 11, 125.

Henderson, G. G. L., and Douglas, R. J. W., 1954. Southern Rocky Mountains of Canada, tectonic compilation map. Alberta Soc. Petrol. Geologists. Guidebook, 4th Ann. Field Conf.

Hoeppener, R., 1953. Faltung and Kluftung in Nordteil des Rheinischen Schiefergebirges. Geol. Rundschau, 41, 128.

———, 1955. Tektonik im Schiefergebirge. Geol. Rundschau, 44, 26.

Jaeger, J. C., 1960. Shear failure of anisotropic rocks. Geol. Mag. 98, 65.

Leech, G. B., 1958. Fernie map-area, west half, British Columbia. Geol. Surv. Can. Paper 58-10.

———, 1959. Canal Flats, British Columbia. Geol. Surv. Can. Map 24-1958.

———, 1960. Fernie, West Half, British Columbia. Geol. Surv. Can. Map 11-1960.

———, 1962. Structure of the Bull River valley near latitude 49°35′. J. Alberta Soc. Petrol. Geologists, 10, 396.

Lindstrom, M., 1962. A structural study of the southern end of the French Jura. Geol. Mag. 99, 193.

Norris, D. K., 1955. Blairmore, Alberta, map and marginal notes. Geol. Surv. Can. Paper 55-18.

———, 1956. Structural conditions in coal mines of the Canmore area, Alberta. Alberta Soc. Petrol. Geologists, Guidebook, 6th Ann. Field Conf. p. 59.

———, 1958*a*. Structural conditions in Canadian coal mines. Geol. Surv. Can. Bull. 44.

———, 1958*b*. Beehive Mountain, Alberta and British Columbia. Geol. Surv. Can. Paper 58-5.

———, 1958*c*. Livingstone River, Alberta, Geol. Surv. Can. Map 5-1958.

———, 1959. Carbondale River Alberta—British Columbia. Geol. Surv. Can. Map 5-1959.

———, 1964. Microtectonics of the Kootenay Formation near Fernie, British Columbia. Bull. Can. Petrol. Geol-

ogists, 12, 383.
———, 1965. Structural analysis of part of A-North coal mine, Michel, British Columbia. Geol. Surv. Can. Paper 64-24.
Norris, D. K., and Black, R. F., 1962. Palaeomagnetism and differential rotation in the Lewis thrust plate. J. Alberta Soc. Petrol. Geologists, 10, 13.
North, F. K., and Henderson, G. G. L., 1954. Summary of the geology of the southern Rocky Mountains of Canada. Alberta Soc. Petrol. Geologists, Guidebook, 4th Ann. Field Congr. p. 15.
Price, R. A., 1958. Structure and stratigraphy of the Flathead North map-area (east half), British Columbia and Alberta. Ph.D. Dissertation, Princeton University, Princeton, N.J.
———, 1959. Flathead, British Columbia and Alberta. Geol. Surv. Can. Map 1-1959.
———, 1962. Fernie map-area, east half, Alberta and British Columbia. Geol. Surv. Can. Paper 61-24.
———, 1964. Flexural-slip folds in the Rocky Mountains, southern Alberta and British Columbia. Seminars in tectonics—IV. Dept. Geol. Sciences, Queen's Univ., Kingston, Ont. p. 6.
Rambow, D., 1963. Scherflachen in steilstehenden Kohlenflozen. Z. Deut. Geol. Ges. 114, 85.
Roeder, D. H., 1960. Der tektonische Stil der Rocky Mountains in Alberta, Canada. Geol. Rundschau, 50, 577.
Shaw, E. W., 1963. Canadian Rockies—Orientation in time and space. Am. Assoc. Petrol. Geologists, Mem. 2, 231.
Turner, F. J., and Weiss, L. E., 1963. Structural analysis of metamorphic tectonites. McGraw-Hill Book Co., Inc., New York p. 545.
Waard, D. de., 1955. The inventory of minor structures in a simple fold. Geol. Mijnbouw, 17, 1.

Balanced cross sections

C. D. A. DAHLSTROM
Chevron Standard Limited, Calgary, Alberta

Received March 26, 1969
Accepted for publication April 30, 1969

Post-depositional concentric deformation produces no significant change in rock volume. Since bed thickness remains constant in concentric deformation, the surface area of a bed and its length in a cross-sectional plane must also remain constant. Under these conditions, a simple test of the geometric validity of a cross section is to measure bed lengths at several horizons between reference lines located on the axial planes of major synclines or other areas of no interbed slip. These bed lengths must be consistent unless a discontinuity, like a décollement, intervenes. Consistency of bed length also requires consistency of shortening, whether by folding and (or) faulting, within one cross section and between adjacent cross sections.

The number of possible cross-sectional explanations of a set of data is reduced by the fact that, in a specific geological environment, there is only a limited suite of structures which can exist. This imposes a set of local "ground rules" on interpretation. When these local restrictions are coupled with the geometric restrictions which follow from the law of conservation of volume, it is often possible to produce structural cross sections that have a better-than-normal chance of being right.

The concept of consistency of shortening can be extrapolated to a mountain belt as a whole, thereby indicating the necessity for some kind of transfer mechanism wherein waning faults or folds are compensated by waxing en echelon features. These concepts are illustrated diagrammatically and by examples from the Alberta Foothills.

Introduction

The purpose of this paper is to describe a method of checking cross sections for geometric acceptability. In the elementary schools of a bygone era children were taught to prove their answers to arithmetic problems by reversing the process: subtraction was checked by addition and division by multiplication. To apply this principle to a geological cross section, one would flatten out the deformed beds and return them to their depositional position. If this restoration could be done, one would conclude that the cross section was geometrically possible (although not necessarily true) but, if the beds could not be restored, one would conclude that the cross section was geometrically impossible. Construction of restored cross sections is tedious so they are seldom used to check structural interpretation. However, there are shortcuts which make the principle easy to use.

The method to be discussed is now being used consciously by a few geologists and unconsciously by many more to check their cross sections. The essence of the method has been discussed by Hunt (1957), mentioned by Goguel (1952), and illustrated by the published works of Carey (1962) and Bally *et al.* (1966), where it is obvious that the method is being consciously applied. However, the rules have not been set forth and discussed so that many geologists have not been exposed to the ideas. This paper is intended to remedy this oversight because the writer believes that a "balanced" cross section is a better cross section.

Source of Restrictions in Cross Section Construction

Relatively few geologists still demand for themselves the license of the artist, that is, the right to put on their cross section any interpretation which their imaginations can conceive. Aside from the data itself, most geologists recognize that other restrictions impose boundary conditions within which the imagination must be confined. These restrictions derive from two sources, (1) generalizations derived from observed facts, and (2) geometric principles.

This paper discusses one of the geometric principles and shows how specific rules of interpretation can be derived from that principle. The derivation involves simplifying assumptions which are valid only for specific structural environments. Therefore the same geometric principle will give rise to interpretive rules which vary from one structural environment to another. The writer will not attempt to develop all the possible variants. Discussion is restricted to one simple geological environment, the marginal part of an orogenic belt, and to one example of that environment, the

Canadian Journal of Earth Sciences, 6, 743 (1969)

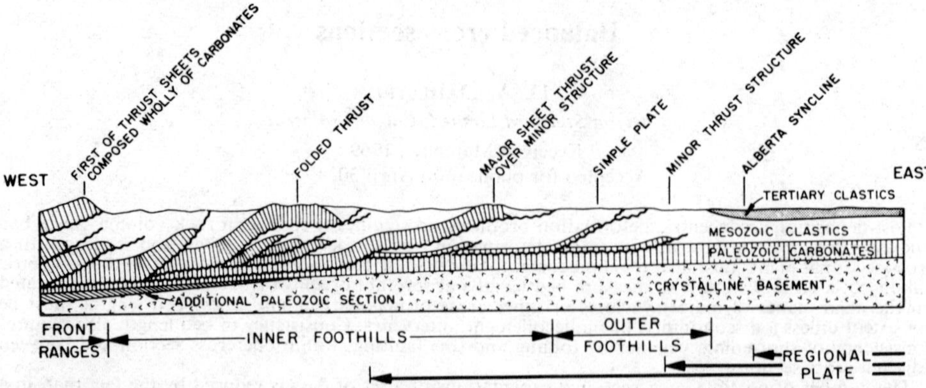

FIG. 1. Schematic cross section of Alberta Foothills (after R. E. Daniel, Chevron Standard). Note that the regional plate extends progressively farther west as older rocks are considered until, at Precambrian level, it extends completely across the section.

Alberta Foothills. To avoid misunderstanding, let it be emphasized that the interpretive rules derived in this paper are literally applicable only to the Alberta Foothills. With minor variants, they can be used in other marginal belts, but they cannot be applied to another environment, such as a salt dome province. However, the principle and method can be used to develop rules that do apply to those other environments.

Generalizations of Observed Facts

As a consequence of intensive government and industry search for oil and gas, the Alberta Foothills is well mapped (Bally *et al.* 1966, and the authors in their bibliography) and sufficient data are available for sound generalizations as to the pattern of structural behavior within this structural province. These generalizations establish the local ground rules for structural interpretation. For example, one of the local ground rules is that folds are ordinarily concentric, rarely chevron but never similar. Recognition of such local interpretive rules is an acknowledgment of the existence of "familial associations" of structures. According to this idea a specific geological environment contains a limited suite or "family" of structures. A large low-angle thrust, which would be perfectly reasonable in Foothills cross sections, would be as incongruous in a cross section through the plains of Saskatchewan as an elephant on the tundra, because thrusts are not part of the cratonic family of structures. The "foothills family of structures" comprise:

1. concentric folds,
2. decollement,
3. thrusts (usually low angle and often folded)
4. tear faults, and
5. late normal faults.

The first restriction then in constructing Foothills structure sections is that one is limited to some variant of these basic structural forms.

The area being used as an example is shown in diagrammatic cross section in Fig. 1. Of the many generalizations from observed data which can be made about the Alberta Foothills belt, the following are pertinent to the subsequent discussion.

1. The Precambrian basement extends, unbroken, beneath the Foothills structures.

2. The rocks east of the Foothills are essentially undeformed.

3. The Foothills structures were formed in the latter stages of the Laramide orogeny long after most of the rocks now preserved were deposited.

4. There is no thinning or thickening of beds by "flow" and therefore folding is concentric.

The through-going basement and the lack of deformation east of the Foothills edge limit the number of solutions which can be postulated for particular geometric problems. In an earlier and simpler day, it was the custom of geologists to draw shallow Foothills cross sections wherein geometric problems were solved by faults which dribbled off the bottom of the cross section.

Seismic and drilling data (Bally *et al.* 1966) have established the basement configuration so that now cross sections like Fig. 1 are closed systems except at the western end. Now the sophisticated geologist draws deep Foothills cross sections with a through-going basement, wherein geometric problems are solved by faults which are gathered into flat sole faults and hustled westward through the only available exit.

The Laramide orogeny moved progressively from west to east with the Foothills being deformed in the Paleocene Epoch. With the probable exception of the youngest beds in the Alberta syncline, all of the Paleozoic and Mesozoic rocks had been deposited long before the deformation. During the course of their depositional history in the miogeosyncline, these Paleozoic and Mesozoic rocks had indeed been subject to epeirogenic fluctuations and consequent erosion but they had not been orogenically deformed prior to Paleocene time. This single, post-depositional deformation of the Foothills is an important point because it simplifies structural interpretation by eliminating from consideration:

1. pronounced angular unconformities,
2. structures growing during deposition, or
3. substantial amounts of compaction during deformation.

The foregoing terse comments touch only upon those topics which are necessary background for the subsequent geometric discussion. No attempt has been made to summarize Foothills geology.

Conservation of Volume

The law of elementary physics (which Einstein amended) that matter can neither be created nor destroyed is paraphrased in geology as the "law of conservation of volume" (Goguel 1952, p. 147). A rock consists of mineral particles and voids filled with fluids. In the initial stages of clastic deposition, the proportion of voids to mineral grains is large and density is low, but this stage is brief because the compaction rate is very high during the first few hundred feet of burial. Thereafter density increases only slowly with depth. It is worth noting here for future reference that the volume change due to load compaction does reduce the thickness of a bed but it does not alter the areal extent. Volume reduction on account of deformation is thought to be negligible, particularly in the earlier stages of tectonism represented by the marginal part of the Cordillera, because no significant deformation-dependent decrease of porosity (or increase of density) is recognized. For practical purposes, one may assume that the law of volume conservation applies and that rock volume does not change during Foothills type of deformation.

Volume is three dimensional which makes the law of conservation of volume awkward to apply. However, in specific instances it can be made simpler. To start the simplification process one can use the three mutually perpendicular tectonic axes as the three dimensions: "*a*" is the direction of tectonic transport of rock during deformation, "*b*" is the direction of the fold axes, and "*c*" is the third axis which is perpendicular to the "*ab*" plane.

In the Foothills, the fold axes and the fault strikes are parallel, the folds are concentric, and the faults are dip-slip thrusts. Therefore the "action" takes place in the *ac* plane, which is a vertical transverse cross section, and virtually nothing happens in the *b* direction. By ignoring changes in the *b* direction as insignificant, the law of conservation of volume becomes a two-dimensional statement that the cross-sectional area of a bed does not change during deformation. Cross-sectional areas are manageable quantities so the law has been applied in this form to check cross sections (Hunt 1957) or to calculate depth to detachments (Bucher 1933).

Even further simplification is possible when folding is concentric because bed thickness does not change during deformation. This has been demonstrated in the Foothills, directly by detailed studies of the anatomy of folds (Price 1964), and indirectly by many stratigraphic studies, which do not detect any tectonic thinning of units whether they be on gentle upright limbs, on vertical limbs, on overturned limbs of folds, or on far-travelled thrust sheets. Because the cross-sectional area of a bed is a function of bed length and bed thickness and because the thickness remains constant, it is possible to eliminate one more dimension and to state the law of conservation of volume in terms of length alone: *In concentric regimes the cross-*

sectional length of a bed remains constant during deformation.

Consistency within Cross Sections

In simple deposition the initial areal extent (or length in cross section) of any bed is the same as that of the beds above and below it. Since bed area (or length in cross section) is not appreciably altered by either compaction or by post-depositional concentric deformation, it follows that bed lengths in a cross section must be consistent with one another (Figs. 2*a* and 2*b*), unless a discontinuity such as a sole fault or décollement intervenes (Fig. 2*c*) between the longer and shorter beds.

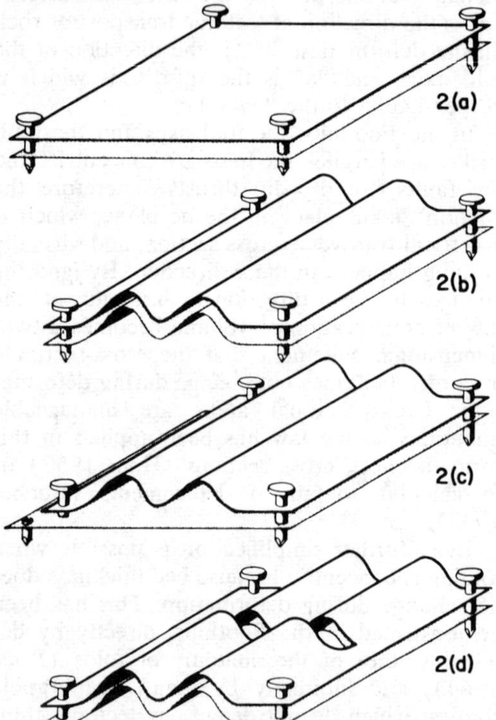

FIG. 2. Consistency of bed length.

Using the idea of consistent bed lengths to check a cross section is very simple. The first step is to establish a pair of reference lines at either end of the section in areas of no interbed slip. These reference lines should be the axial planes of major anticlines or synclines or other planes of no slip such as a plane perpendicular to the regional dip of the undisturbed "plains" section at the eastern end of a regional Foothills cross section. The second step is to measure the bed lengths of selected horizons between these reference lines. They should all be the same so that bed lengths "balance". If the cross-sectional lengths do not balance then the cross section must show a valid explanation of why they do not. In regional Foothills cross sections for instance, there is never a balance between the length of the Mesozoic and Paleozoic beds and the length of basement. The "cover" beds are always too long, which is explained in cross section (Fig. 1) by a sole fault along which the upper beds have been moved (shoved? glided?) into the cross section from the west. This is not simply a way of sloughing one's problems into an adjacent area, because the necessary implication of such a section is that the sole fault continues to the west until the bed length anomaly is resolved by shortening of the basement (compressional hypothesis), or extension of the Mesozoic and Paleozoic "cover" rocks (glide hypothesis). One should note that checking Foothills cross sections for balance by measuring bad lengths is purely a geometric test which is quite independent of the genesis of the structures.

The fundamental rule, that the cross-sectional length of a bed remains constant during concentric deformation, has led to the idea that bed lengths in a cross section ought to be consistent with one another. If this is true, then the displacement on thrust faults ought to be consistent as well. In Fig. 3 the displacement at B must be the same as at A unless it is postulated that elastic, plastic, or compactional deformation of the rock takes place between A′ and B′. None of these things happen in a concentric regime.

Despite this conclusion that thrust displacement ought to be consistent, there are many instances where the displacement can be observed to change along the fault plane. There are only two basic ways in which this paradox (Fig. 4*a*) can be resolved:

1. by interchanging fold shortening and fault displacement (Fig. 4*b*), or
2. by imbrication (Fig. 4*c*).

Thrust faulting and folding are both mechanisms for making a packet of rock shorter and

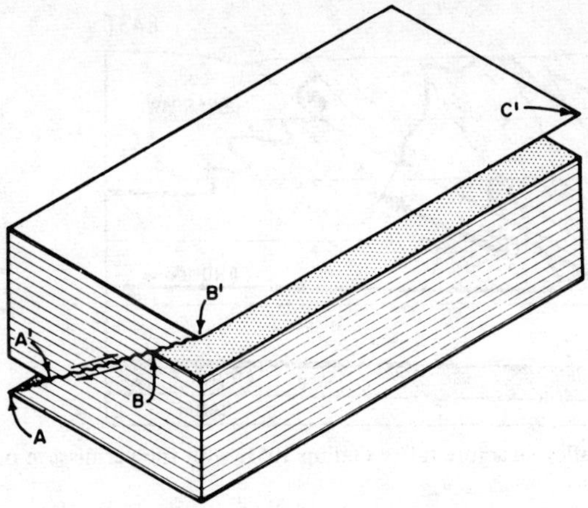

FIG. 3. Consistency of thrust displacement.

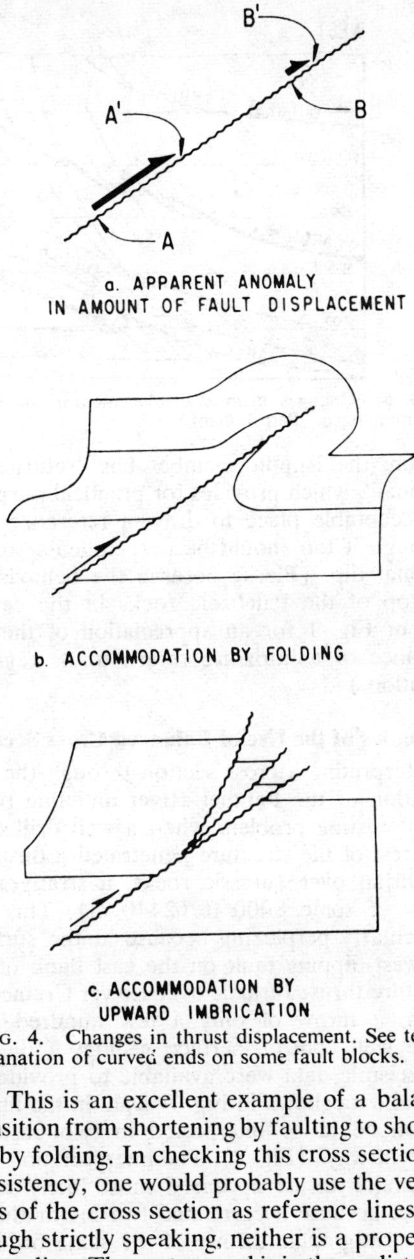

FIG. 4. Changes in thrust displacement. See text for explanation of curved ends on some fault blocks.

thicker than it was originally, so that one could expect the two mechanisms to be interchangeable. Imbrication is the distribution of displacement from one large fault to several minor ones. Figure 4*c* shows the common imbricate pattern, but imbricates can develop in the footwall rather than the hanging wall of the main thrust, and they can be in the hanging wall but dipping in the opposite direction to the principal fault ("back thrusts").

In Fig. 4 the thrust faults were arbitrarily and diagrammatically represented as simple planar features. With planar faults and constant bed length, a substantial amount of interbed slippage is required, which would alter the originally vertical ends of the blocks to the curved shapes shown. From these block diagrams it is fairly evident that:

1. faults with changing displacement are apt to be curved in cross section rather than planar (can be confirmed by observation);
2. interbed slippage is a necessary part of the thrust faulting process just as it is in concentric folding;
3. interbed slippage can contribute to the change of displacement along a fault plane and, in extreme instances, could become a species of imbrication.

The Turner Valley cross section (Fig. 5) is based on good well, seismic, and surface geological control. It shows a remarkable change from more than 2 miles (3.2 km) of fault displacement at depth to virtually none at surface and the accommodation of this change by folding. This is an excellent example of a balanced transition from shortening by faulting to shortening by folding. In checking this cross section for consistency, one would probably use the vertical ends of the cross section as reference lines even though strictly speaking, neither is a proper reference line. The eastern end is in the undisturbed plains section, but the reference line should be perpendicular to the regional dip rather than vertical. The western end of the section does not extend as far as the synclinal axial plane, which would be the proper place for the second reference line. However, at this end of the

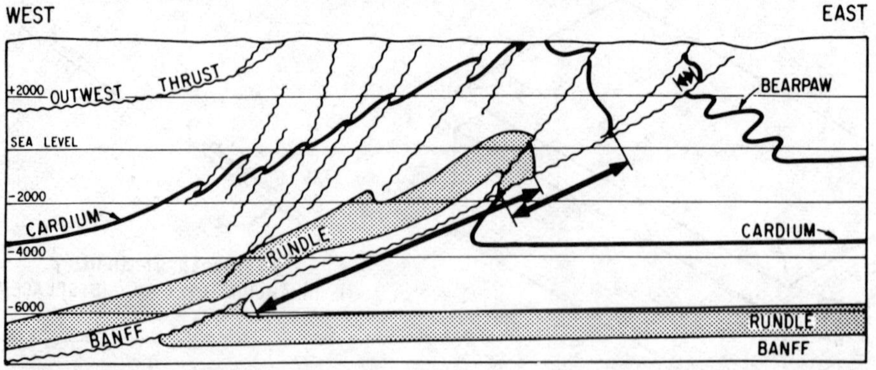

FIG. 5. Changes in thrust displacement in the Turner Valley structure (after Gallup 1951, with the permission of the Amer. Assoc. Petrol. Geol.).

section, the Rundle member has "returned to regional", which provides for practical purposes an acceptable place to draw a reference line, although it too should be perpendicular to the regional dip. (Please observe the behavior of the top of the Paleozoic rocks in the central part of Fig. 1 for an appreciation of the significance of a horizon's return to its regional elevation.)

Examples of the Use of Balanced Cross Sections

Interpreting a cross section through the culmination of the Panther River anticline posed an interesting problem when a well drilled on the crest of the structure penetrated a thrust of Cambrian over Jurassic rocks, a stratigraphic throw of some 8000 ft (2440 m). This was particularly perplexing because at the surface, the west-dipping fault on the east flank of the structure thrust Jurassic over Lower Cretaceous rocks, a throw of only a few hundred feet. Some well control, surface geology (Fig. 6), and seismic data were available to provide the critical data shown in Fig. 7. Data in the line of section could be interpreted in two basically different ways (Fig. 8). The interpretation with one major thrust appeared most likely to be correct because it correlated two major thrusts with stratigraphic throws of the same order of magnitude. The interpretation with two major thrusts seemed unreasonable because it linked a major fault in the subsurface with a minor fault at the surface (one order of magnitude difference in stratigraphic throws). Which alternative one selected had considerable economic significance, because it affected where and at what depth one could expect to find closure in the potentially hydrocarbon-bearing objective horizon at the top of the Paleozoic section.

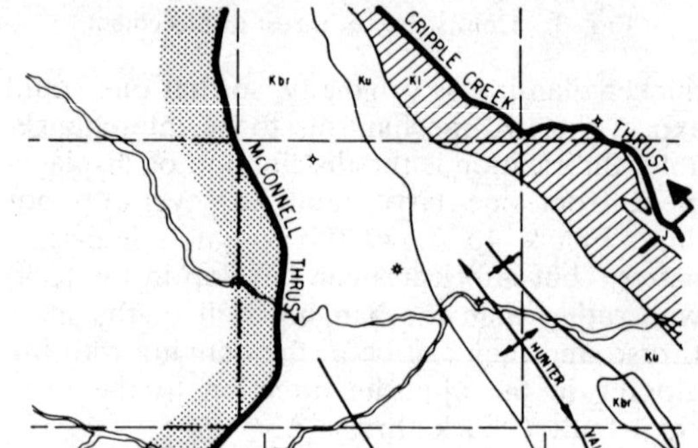

FIG. 6. General geology of the Panther River area.

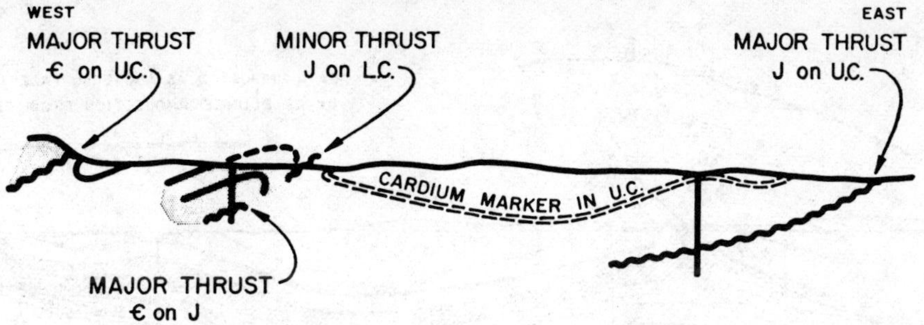

FIG. 7. Basic data available for the original interpretation of the Panther River cross section.

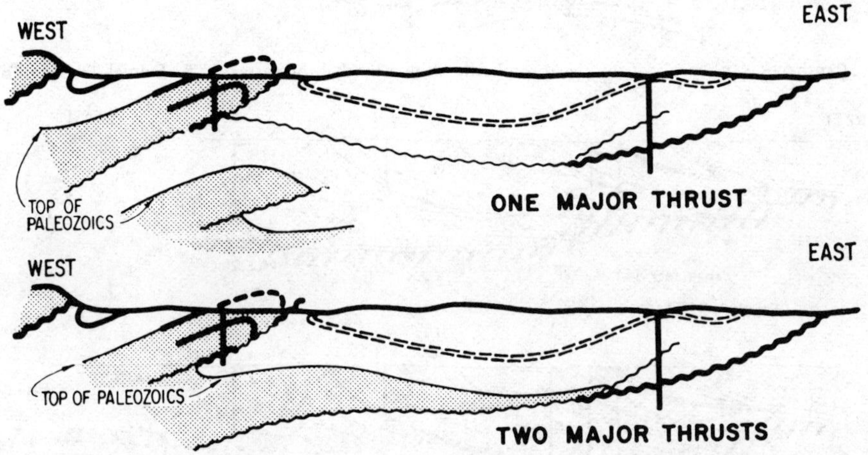

FIG. 8. Two alternate cross sections which would satisfy the data on the original line of section.

However, the geological map provided some data off the line of section, which had not been used. R. E. Daniel of Chevron Standard statistically analyzed the fold (Dahlstrom 1954), found it to be cylindrical, and calculated plunge values. The data on the geological map were projected up plunge (Stockwell 1960) into the plane of a vertical cross section passing through the culmination to provide the information shown above ground level at the west end of the section in Fig. 9. On this cross section it is apparent that the axial planes of the anticline and syncline are converging in depth, which is confirmed by the reduced length of the steep fore limbs in successively lower stratigraphic horizons. Having the fold disappear with depth would produce a geometrically unacceptable discordance in bed lengths unless the fold disappearing downward could be compensated by the fault dying out upward. This would be a repetition of the Turner Valley situation in Fig. 5. Once this fundamental point was grasped, it was apparent that the superficially unlikely looking cross section in Fig. 8 with the two major thrusts was the correct one. The cross section which R. E. Daniel constructed according to this concept was subsequently confirmed by the drilling of a well on the deep structure immediately to the east, and by the deepening of the original well.

Hypotheses of regional significance can develop from balancing cross sections. Figure 10 shows two alternative interpretations of seismic, well, and surface data. The data in the two sections are the same, but in Section A, the simply migrated seismic section, the vertical dimension is essentially two-way transit time, whereas B is the normal natural scale structural cross section. The Mesozoic and Paleozoic sections are both cut by thrusts, but even a casual inspection shows that the shortening in the Mesozoic section is far in excess of that in

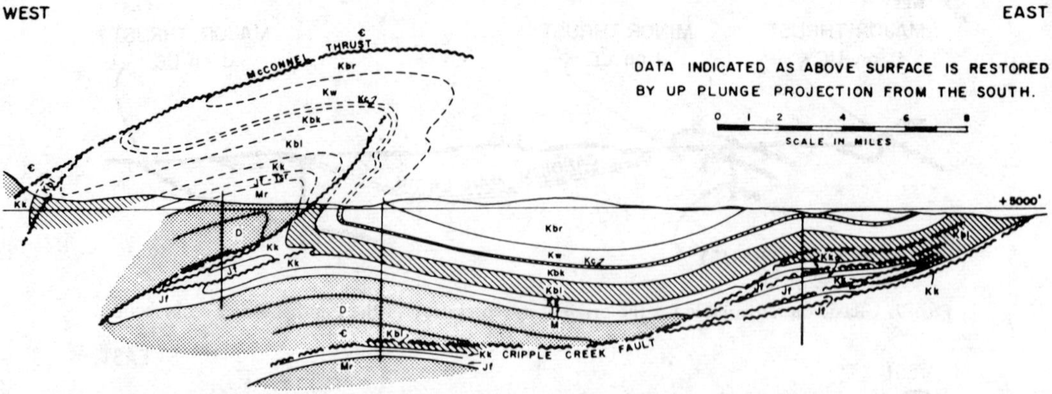

FIG. 9. Panther River cross section as subsequently shown by drilling (after R. E. Daniel, Chevron Standard).

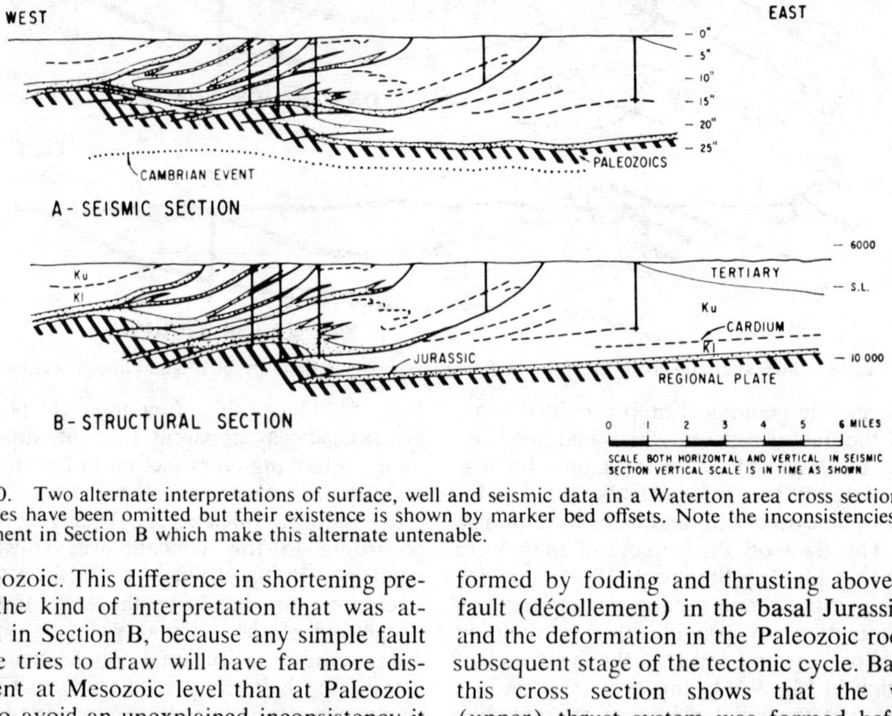

FIG. 10. Two alternate interpretations of surface, well and seismic data in a Waterton area cross section. Minor fault traces have been omitted but their existence is shown by marker bed offsets. Note the inconsistencies of fault displacement in Section B which make this alternate untenable.

the Paleozoic. This difference in shortening precludes the kind of interpretation that was attempted in Section B, because any simple fault that one tries to draw will have far more displacement at Mesozoic level than at Paleozoic level. To avoid an unexplained inconsistency it is necessary to postulate a décollement separating the Mesozoic rocks from the Paleozoic rocks. This décollement would be in the basal Jurassic beds just above the top of the Paleozoic section. Since this horizon is itself now deformed by both folding and faulting, it becomes necessary to postulate a two-stage sequence of deformation. According to this hypothesis the Mesozoic rocks were first deformed by folding and thrusting above a sole fault (décollement) in the basal Jurassic beds, and the deformation in the Paleozoic rocks is a subsequent stage of the tectonic cycle. Balancing this cross section shows that the western (upper) thrust system was formed before the eastern (lower) one, and demonstrates (Bally *et al.* 1966, Fig. 6) that, in the regional sense, deformation advanced from west to east.

Consistency between Cross Sections

The foregoing discussion has been concerned with a one-dimensional check of the internal consistency of an individual cross section by simple measurement and comparison of bed lengths. During the stage of reducing the law of conservation of volume from three dimensions to one, there was a two-dimensional stage wherein one could state that deformation changes the form but does not alter the surface area of an individual bedding plane. This stage is represented in Fig. 2*d*, where it is evident that no abrupt change can occur in fold form and (or) location unless there is a discontinuity. Similarly, in the block diagram of Fig. 3, the fault displacement between B′ and C′ cannot change abruptly unless a discontinuity intervenes. Such transverse discontinuities (tear faults) do occur in a few places in the Foothills, where they may affect either or both thrusts and folds. From such considerations of the two-dimensional version of the law of conservation of volume one can derive the rule that: *In adjacent cross sections the amount of "shortening" at a specific horizon between comparable reference lines must be nearly the same unless there is a tear fault between them.*

In this context, "shortening" is the difference between actual bed length and the horizontal distance it now occupies. This statement does not deny the possibility of a gradual change in the shortening along an individual structure nor along a mountain belt as a whole, but it does deny that these changes can occur abruptly without tear faulting.

The reader can convince himself of the reasonableness of this rule by taking a long thin strip of light paper and attempting to produce a long parallel system of folds and faults. Paper is a suitable medium for this kind of demonstration because its lack of plasticity obliges one to maintain a constant bed area. The Rocky Mountain belt is 1000 miles (1610 km) long from the Idaho batholith to the Liard River. The deformed width of say 100 miles (160 km) is, perhaps, half of the width of the rocks as originally deposited. These figures provide some approximate dimensions for the paper experiment.

Having performed the experiment actually or mentally, the reader may now have reason to suspect that there should be comparable amounts of shortening in adjacent cross sections. Figure 11 shows the north end of Turner Valley, where there is an abrupt change in structure from the simple faulted anticline of Section B–B to the sheaf of imbricates represented by Section C–C. Despite this change in structural form, the shortening at the top of the Paleozoic level in both Sections B–B and C–C in the published sections is in phenomenally good agreement at a figure of 15 000 ft (457.2 m). These two sections are consistent. Consider the third section D–D. In Section D–D the well control does not establish how far to the left (west) the regional plate extends under the Turner Valley sole fault. The natural tendency is to put the footwall "cut-off" of the top of the Paleozoic rocks at X. If this is done, the shortening in Section D–D becomes approximately 6000 ft (1830 m), some 9000 ft (2740 m) less than in Section C–C. A 60% reduction in shortening from C–C to D–D would be quite inconsistent, and prompts an interpretation where the top of the Paleozoic beds in the regional plate extends back to Y. This would also produce better internal consistency within section D–D by making the fault displacement on the Home Sand (R–S) equivalent to that on the top of Paleozoic section (Y–Z).

The two southern sections have provided a clear example of consistency of shortening in adjacent sections despite rather drastic changes in structural form, and the third section shows how the rules of consistency can be used to choose between alternate interpretations in the absence of definitive well control.

Transfer Zones

Previously it was stated that shortening on the local scale in individual structures and in the regional scale in a whole mountain belt could change, but that the change would be gradual. Using the same word "gradual" for change at both scales is not really appropriate because the rate of change is substantially greater for individual structures than it is for the mountain belt as a whole. The Lewis Thrust, for instance (Dahlstrom *et al.* 1962), has a minimum of 23 miles (37 km) of thrust displacement at the United States border and 135 airline miles (217.2 km) to the north, the fault displacement is zero. Over the same distance the overall shortening in the mountain belt as a whole may have diminished, but certainly not by 23 miles (37 km). Since the whole does not change as rapidly as its component parts, it

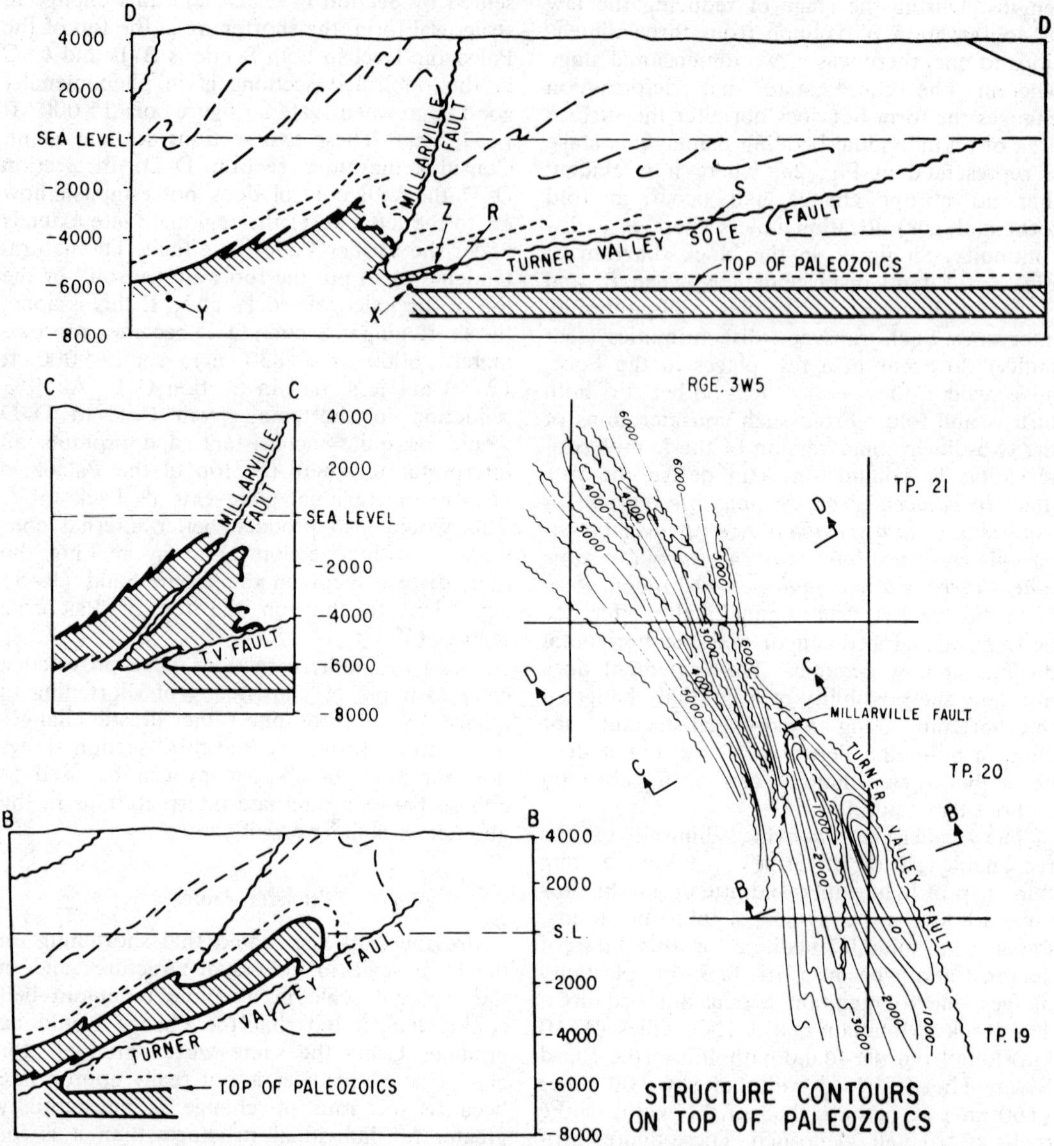

FIG. 11. North end of the Turner Valley structure (after W. B. Gallup, 1951, with the permission of the Amer. Assoc. Petrol. Geol.). Note that the cross-sectional trace of minor faults has been omitted, but their presence is indicated by offsets in the marker horizons. See text for discussion.

follows that there must be some sort of compensating mechanism at work whereby displacement is "transferred" from one structure to another. Such mechanisms have been observed in the *ac* cross-sectional plane (Fig. 5), so it is not unreasonable to expect that comparable phenomenon would function in the horizontal *ab* plane (as in Fig. 11).

The compensatory mechanism for thrusts is a kind of lap joint wherein the fault whose displacement is diminishing is replaced by an echelon fault whose displacement is increasing. Clearly such a "transfer zone" could not exist unless all of the faults involved in the transfer zone are rooted in a common sole fault. Figure 12 shows a relatively simple transfer zone consisting of three faults. In each of the five cross sections the shortening is exactly the same, although at one end virtually all of the displacement is on fault C and, at the other, on Fault A. In natural examples, the pattern is often complicated by folds and folded thrust imbricates as shown by the two mapped examples in Fig. 13.

through-going basement. Consequently, in the gross view, one would expect some transfer of displacement between the six thrust zones.

It is not proposed to discuss en echelon folding, although it is a part of the Foothills movement pattern (Fitzgerald 1968) wherein the transfer mechanism operates for folds as it does for faults. En echelon folds are tied to one another by a sole fault (décollement) and maintain consistent shortening by replacing a dying structure with an en echelon growing equivalent (Wilson 1967).

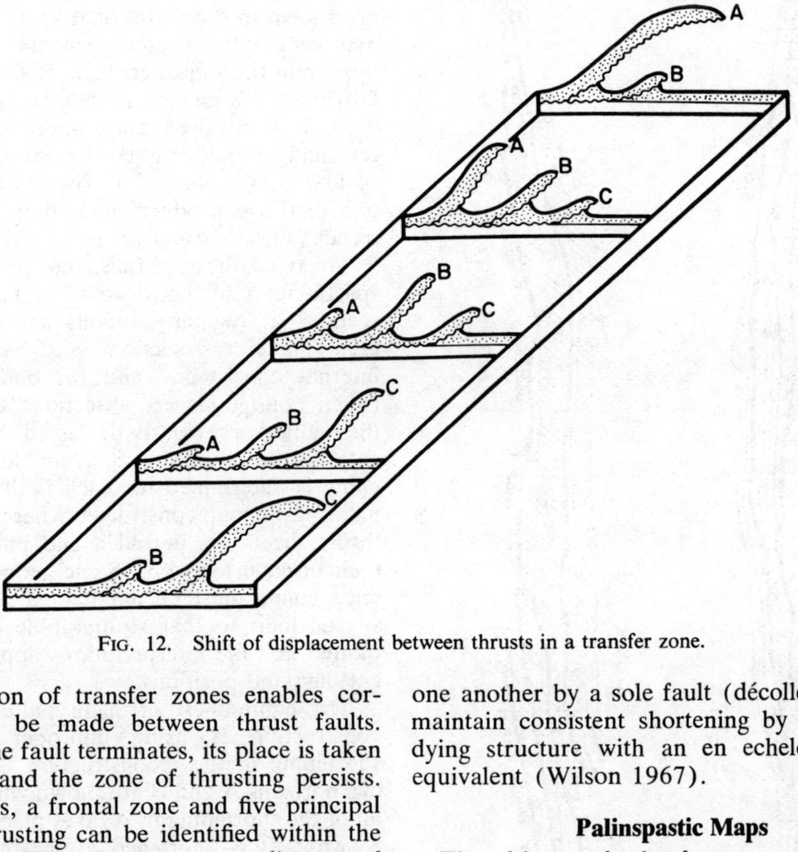

FIG. 12. Shift of displacement between thrusts in a transfer zone.

Recognition of transfer zones enables correlations to be made between thrust faults. Although one fault terminates, its place is taken by another and the zone of thrusting persists. On this basis, a frontal zone and five principal zones of thrusting can be identified within the Foothills and Front Ranges over a distance of some 400 miles (643.7 km) (Fig. 14). At the north end correlation fails because the surficial deformation was primarily folding, and faults are not prominent. The persistence and parallelism of these zones of thrusting over substantial distances lends credence to the suggestion that overall shortening is reasonably consistent along the mountain trend. In one respect, designating six zones of thrusting may be rather arbitrary because at depth all of these zones will join to a common sole fault above the

Palinspastic Maps

The ultimate check of cross sections in deformed terrane is whether or not the process can be reversed and the beds put back into their depositional position without introducing inexplicable bed length anomalies. Since the construction of restored cross sections is a tedious process, the discussion has been concerned with shortcuts which would be adequate for checking the geometric acceptability of individual cross sections and suites of cross sections. However, when stratigraphic studies are being done in deformed terrane, the shortcuts are

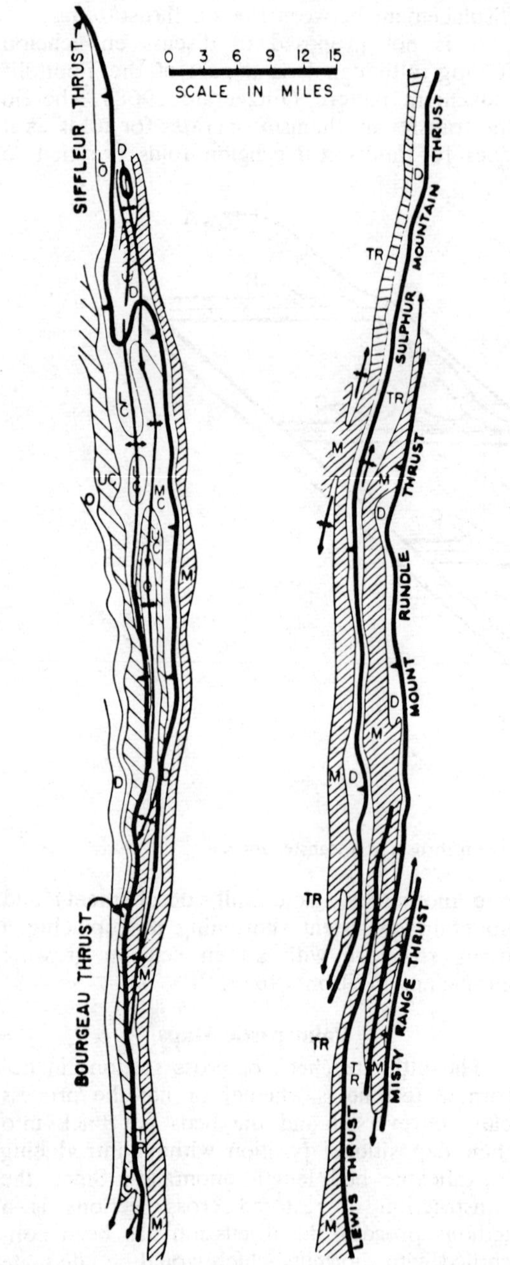

FIG. 13. Two transfer zones in the Front Ranges of Alberta. The abbreviations designate formational ages (i.e., M = Mississippian).

inadequate, and it becomes necessary to make a restored palinspastic map (Dennison and Woodward 1963), which shows beds in their depositional position.

When stratigraphic data obtained by study of deformed terrane are recorded and interpreted on an ordinary map, it is possible to make serious errors. In Fig. 15*a*, seven pieces of surface and subsurface stratigraphic data are shown in cross-sectional view. On the normal geographic presentation this data would appear in the sequence 1, 4, 5, 6, 2, 3, and 7. Obviously, what is needed for stratigraphic work is a restored map where the data are returned to their original depositional sequence of 1, 2, 3, 4, 5, 6, and 7. The normal presentation can also produce misleading stratigraphic trends (Figs. 15*b* and 15*c*).

To avoid these pitfalls, one begins a palinspastic map by constructing a suite of cross sections at regular intervals across the study area. These cross sections must be checked for internal consistency and for consistency between contiguous cross sections according to the methods previously discussed. When a consistent suite of sections is available, then shortening is determined for each fault plate and a palinspastic map constructed wherein individual thrust sheets are unfolded and pulled back to their original locations. Some geographic reference points must be maintained on the palinspastic map so that stratigraphic data can be plotted and the interpretations applied to present-day land positions.

The commonest error in palinspastic map construction is using improper values for shortening in the reconstruction. In Fig. 15*d*, two horizons A and B are shown where virtually all of the shortening in A takes place at position S, while B is shortened at R. In calculating shortening for palinspastic map construction, it would be very easy to add the shortening at R and the shortening at S together, to arrive at an answer that was twice the proper value. The best method to determine shortening is to do all the measurement on one horizon. If one is forced to change horizons, this can only be done in an area where there is no discontinuity between the reference horizons. In Fig. 15*d*, one could change from horizon A to horizon B at positions V or T, but certainly not at position U.

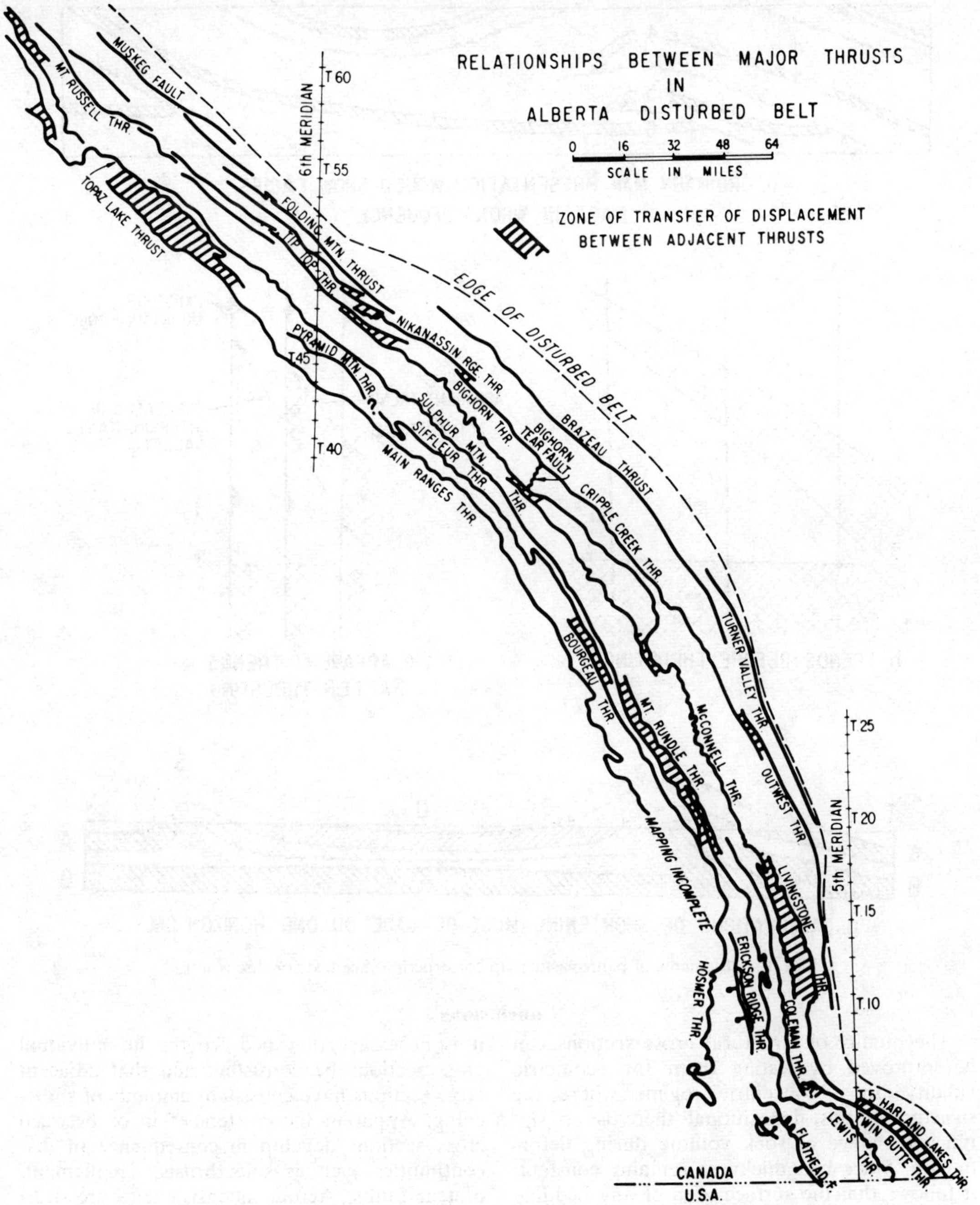

FIG. 14. Use of transfer zones in correlating zones of thrusting.

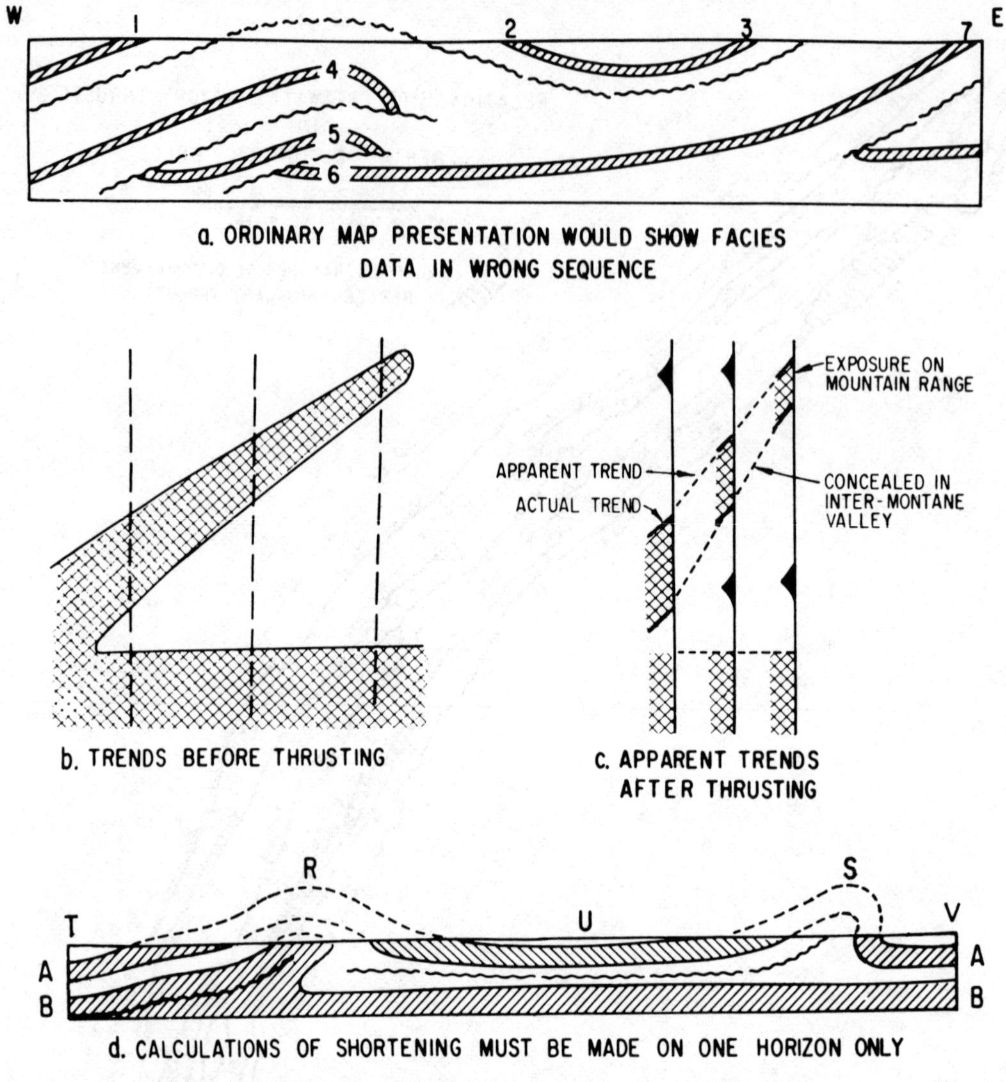

FIG. 15. Elements of palinspastic map construction. See text for discussion.

Conclusions

The quality of structural cross sections can be improved by testing them for geometric validity. In a concentric regime where the structure is post-depositional there is no significant change of rock volume during deformation. Since bed thickness remains constant, it follows that the surface area of any bedding plane remains constant during deformation. Because the area of successive beds in an undeformed depositional sequence is constant, it is necessary that bed lengths in individual cross sections be consistent and that adjacent cross sections have consistent amounts of shortening. Apparent inconsistencies in or between cross sections develop in consequence of discontinuities such as sole thrusts, décollement, or tear faults. Actual inconsistencies are ordinarily due to conceptual, observational, or drafting errors, which might pass undetected were there no tests for geometric validity.

In some lines of geological endeavor, cross sections are significant only as diagrams that are used to convey concepts. In geology applied to oil and mining exploration or to engineering projects, cross sections are used to convey predictions as to rock behavior and they must be conceptually and, more important, geometrically correct. In these areas it is important to the geologist and to his client that there be some way of checking cross-section interpretations prior to drilling. It should be emphasized that a cross section which passes the geometric tests is not necessarily correct, because completely ridiculous cross sections can be drawn which abide by the law of conservation of volume. However, if a cross section passes the geometric tests, it could be correct, and if it has been drawn with due regard for the "local ground rules" it probably is correct. On the other hand, a cross section that does not pass the geometric tests could not possibly be correct.

The rules developed in this paper pertain to a simple, concentrically deformed packet of "layer cake" geology in the Alberta Foothills. The rules themselves cannot be transported to a more complicated area, but the basic concept and method can be used to generate sets of rules which apply to other environments. Thus it is possible to devise amended rules, which apply to concentric extensional structures and to diapiric structures that grow during deposition or to "similar fold" regimes. Each of these areas requires a different set of rules, which must be derived from the basic geometric principles and observational restrictions. The interpretational rules for similar folding were discussed by Carey (1962). In other areas the geologist will have to develop his own rules, an endeavor that he should find interesting and rewarding.

Acknowledgments

The author is indebted to Chevron Standard for permission to publish this paper. It grew from his attempts to provide a rational explanation of normal interpretational practice to structural geologists, new to the ways of the Foothills geology.

BALLY, A. W., GORDY, P. L., and STEWART, G. A. 1966. Structure, seismic data, and orogenic evolution of southern Canadian rocky mountains. Bull. Can. Petrol. Geol., **14**, pp. 337–381.

BUCHER, W. H. 1933. The deformation of the Earth's crust. Princeton Univ. Press. Princeton, N.J. 518 pp.

CAREY, S. W. 1962. Folding. J. Alta. Soc. Petrol. Geol., **10**, pp. 95–144.

DAHLSTROM, C. D. A. 1954. Statistical analysis of cylindrical folds. Trans. Can. Inst. Mining Met., **57**, pp. 140–145.

DAHLSTROM, C. D. A., DANIEL, R. E., and HENDERSON, G. G. L. 1962. The Lewis thrust at Fording mountain. J. Alta. Soc. Petrol. Geol., **10**, pp. 373–395.

DENNISON, J. M. and WOODWARD, H. P. 1963. Palinspastic maps of central Appalachians. Bull. Amer. Assoc. Petrol. Geol., **47**, pp. 666–680.

FITZGERALD, E. L. 1968. Structure of British Columbia foothills, Canada. Bull. Amer. Assoc. Petrol. Geol., **52**, pp. 641–664.

GALLUP, W. B. 1951. Geology of Turner Valley oil and gas field, Alberta, Canada. Bull. Amer. Assoc. Petrol. Geol., **35**, pp. 797–821.

GOGUEL, J. 1952. Tectonics (1962 translation). Freeman and Company, San Francisco, 384 pp.

HUNT, C. W. 1957. Planimetric equation. J. Alta. Petrol. Geol., **5**, pp. 259–264.

PRICE, R. A. 1964. Flexural slip folds in the Rocky mountains, southern Alberta and British Columbia. Seminars on tectonics—IV, Dept. Geol. Sci., Queen's Univ., Kingston, Ont. (Also as Geol. Surv. Can., Reprint 78, 16 pp.)

STOCKWELL, C. H. 1960. The use of plunge in the construction of cross sections of folds. Proc. Geol. Assoc. Can., **3**, pp. 97–121.

WILSON, G. 1967. The geometry of cylindrical and conical folds. Proc. Geol. Assoc. London, **78**, pp. 179–210.

BULLETIN OF CANADIAN PETROLEUM GEOLOGY
VOL. 20, NO. 3 (SEPTEMBER, 1972), P. 583-607

THOUGHTS ON OVERTHRUST FAULTING IN A LAYERED SEQUENCE

P. E. GRETENER[1]

ABSTRACT

This paper concentrates on some aspects of thrust faulting in a layered sequence. First, the fact that thrust faulting, as the name implies, constitutes duplication of the sequence is considered. The effects of loading due to an advancing thrust sheet are studied. In particular, the effect of the load on the fluid pressure in the underlying beds is reviewed. Second, the peculiar geometric attitude of thrust faults in layered sequence; i.e., the stepping, is discussed. Further, the apparent ease with which movement takes place along faults of this type is investigated, and previously advanced theories are supplemented. Such factors as the pore pressure, the existence of bedding planes (planes of weakness) and the unequal distribution of the stresses in a layered sequence under horizontal compression are all considered important when trying to understand the mechanics of overthrust faulting.

INTRODUCTION

Overthrusts have fascinated geologists for well over a hundred years. One of the classical locations in Canada is shown in Figure 1. The sheer wall of Mount Yamnuska on the Trans-Canada highway between Calgary and Banff consists of middle Cambrian limestone thrust over upper Cretaceous clastics along the McConnell fault. Thousands of travellers pass by this place every year, all of them admiring the scenic beauty, few of them realizing the profound geological significance of the picture. Older on younger — a defiance of the principle of superposition! A unique and unequivocal demonstration of the dynamics of the earth which only the insufficient yardstick of our lifespan makes seemingly 'dead' and static. This paper represents a tribute to this — one of our truly magnificent — geological location.

Figure 2 shows an overthrust in its most simple form. On the left of the figure the thrust is a sole fault stepping through the sedimentary sequence near the middle of the figure and becoming an actual overthrust (or toe) forward of the step. On the lower portion of Figure 2 are plotted the additional uneroded load (UL) due to the thrusting process as well as the stratigraphic displacement (SD) as observed along the thrust fault. A. Escher was the first to describe such structures from the Canton of Glarus in Switzerland as early as 1841 (Staub, 1954). The term *overthrust* is a straight translation from the German 'Ueberschiebung'. The term is widely used but recently some authors have objected to it because they feel it implies an absolute sense of motion or, in terms of Figure 2, that the left-hand side has moved over

---

[1]Department of Geology, The University of Calgary, Calgary 44, Alberta.

To my colleagues Drs. Simony and Spang my thanks for many stimulating discussions on the subject. Dr. D. W. Stearns read the manuscript and has contributed a number of valuable suggestions. This work was supported by a National Research Council of Canada Grant.

Fig. 1. Mount Yamnuska, Alberta. The cliff consists of middle Cambrian Elden lime-stone thrust over upper Cretaceous Belly River sandstones and shales forming the lower slopes.

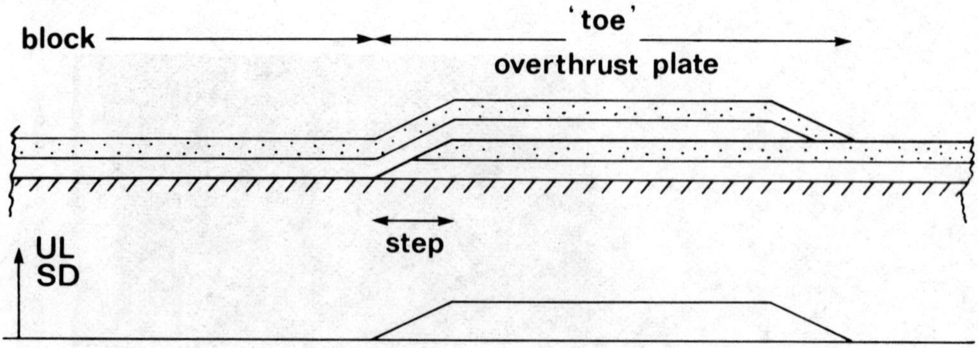

Fig. 2. Schematic diagram showing the major elements of an overthrust. The lower portion shows the additional uneroded load (UL) above, and the stratigraphic displacement (SD) along the thrust.

For convenient reference, certain notations have been adopted throughout the paper, as follows:

SS	step spacing (length of link)
SH	step height
TD	total displacement
SD	stratigraphic displacement
UL	uneroded load
p	pore pressure
S_z	total overburden pressure
σ_z	effective overburden pressure
σ_x	maximum effective horizontal stress
τ_0'	cohesion
τ_0	reduced cohesion on pre-existing fracture or bedding plane (may be zero)
n	coefficient of internal friction
μ	coefficient of sliding friction
δ	angle of fracture or bedding plane to axis of maximum compressive stress (σ_1)
λ	ratio of pore pressure to total overburden pressure (p/S_z)
E	Young's modulus
ν'	effective Poisson's ratio

1 bar = 1.02 kp/cm^2

1 bar = 14.5 psi

1 kp/cm^2 = 14.2 psi

the right-hand side, the latter remaining stationary. Of course, only a relative motion is indicated by Figure 2. To this author the question of overthrusting versus underthrusting is not of importance for the aspects discussed here, and when the term 'overthrust' is used, its connotation is simply a relative one.

Needless to say, Escher's interpretation was not readily accepted, (Heim, 1882). It was not until about the turn of the century that geologists were willing to acknowledge the existence of such fantastic structures (e.g., Staub, 1954, p. 143). The enigma of their formation remained: mechanically, these structures seemed to defy rational explanation. Only in 1959 did Hubbert and Rubey in two companion papers shed some light on the possible mechanics of these structures. Their ideas sparked much discussion, and it is the purpose of this paper to carry on the debate and point out some additional aspects which must be considered.

From Figure 2 it is apparent that the whole mass above and to the left of the fault plane has moved in regard to the sequence on the right. Only to the right of the area of the step does the 'toe' or overthrust plate exist. To the left of the step the fault plane is present, but it is no more than a bedding plane with no stratigraphic displacement. It is important to realize that only in the frontal part is the overthrust recognizable as such.

Figure 3 shows the area of the toe in more detail and clarifies the terminology used in this paper. Usually the thrust does not break to the surface in one giant step, but reaches it in a series of discrete minor steps. According to observations (Dahlstrom, 1970) the steps occur across the competent formations, whereas in the incompetent units the fault is frequently parallel or subparallel to bedding. In this paper that portion will be referred to as the link. This author accepts such behaviour of thrust faults as an observed fact (Rich, 1934; Douglas, 1950;

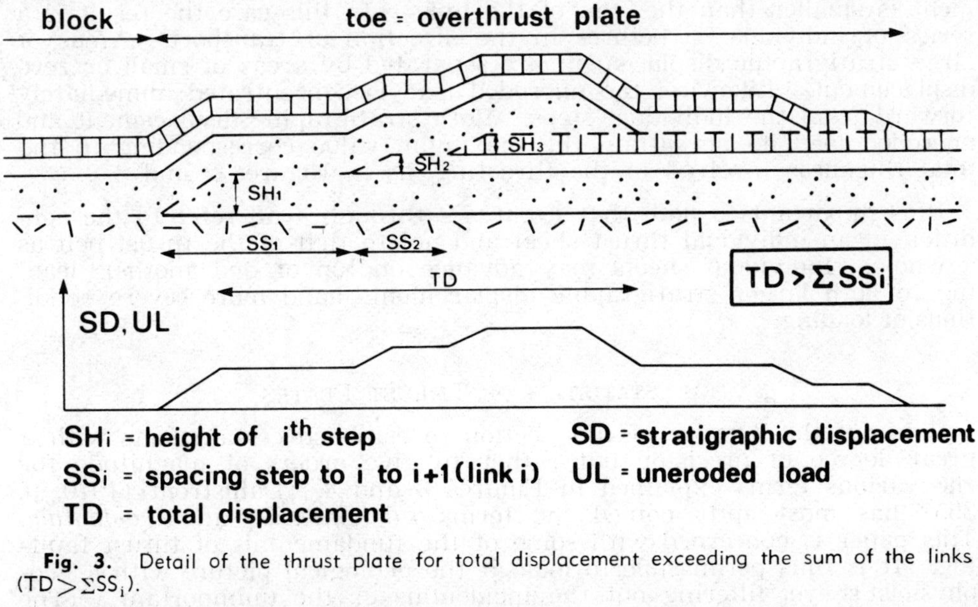

Fig. 3. Detail of the thrust plate for total displacement exceeding the sum of the links ($TD > \Sigma SS_i$).

Bally *et al.,* 1966; Dahlstrom, 1970). Thrusts become solefaults such as shown on the left in Figures 2 and 3, thus indicating that the shortening implied by thrust faulting has affected only an uppermost skin of the earth's crust in the area of observation. Thus, in one of the best explored thrust belts of the world, the southern Canadian Rockies, drilling and seismic work indicate that thrusting is confined to the sedimentary skin in the area east of the Rocky Mountain Trench and does not involve the Hudsonian basement (Bally *et al.,* 1966; Price and Mountjoy, 1969; Dahlstrom, 1970). The thrusts become observable forward from the deepest step. The thrust fault can be geometrically described by such parameters as the step height (SH), the step spacing or length of links (SS), the total displacement (TD) and the stratigraphic displacement (SD), all of which are explained in Figure 3. The uneroded load (UL) is directly proportional to the stratigraphic displacement. Uneroded load assumes that the thrust sheet remains essentially intact during the whole process of thrusting. It represents an extreme case and is not necessarily a reasonable assumption. The question of erosion during thrusting will be further discussed in the respective section.

From Figures 2 and 3 it is evident that the thrust motion has an appreciable vertical component only in the areas of the steps. On a link the motion is parallel or subparallel to bedding, so that deformation results only over the steps. This geometry was described by Rich as early as 1934 for the Appalachians. More recently it has become very important in the southern Canadian Rockies, where these structures are the loci of natural gas fields (Bally *et al.,* 1966).

The vertical extent of these structures is a function of the height of the respective buried step, whereas their lateral extent is given by the total displacement. In Figure 3 the total displacement is larger than the total length of the links (ΣSS_i). Under these conditions, maximum stratigraphic displacement is obtained and the structure formed is basically simple. In Figure 4 a case is shown where the total displacement is smaller than the sum of the links. In this case the result is a series of individual structures in the direction of transport. Areas of large stratigraphic displacement are separated by areas of small or zero displacement. Likewise, the uneroded load is concentrated immediately forward from the individual steps. Both stratigraphic displacement and uneroded load do not attain the maximum value corresponding to the total thickness involved in the thrusting, as in Figures 2 and 3.

It is important to note that Figures 2 through 4 all refer to the condition in an individual thrust sheet and not to that of the thrust belt as a whole. Individual sheets may advance on top of one another, leading to both larger stratigraphic displacements aand more severe conditions of loading.

Vital Statistics of Thrust Plates

It is not the purpose of this section to establish certain facts with a great degree of precision but rather to give orders of magnitude for the various terms explained in Figures 3 and 4. Dahlstrom (1970, p. 355) has most aptly coined the terms *fundamentals* and *incidentals.* This paper is concerned with some of the fundamentals of thrust faulting. It is thus permissible to look at the geological picture with a geophysicist's eye, filtering out the incidentals or the unimportant. (The

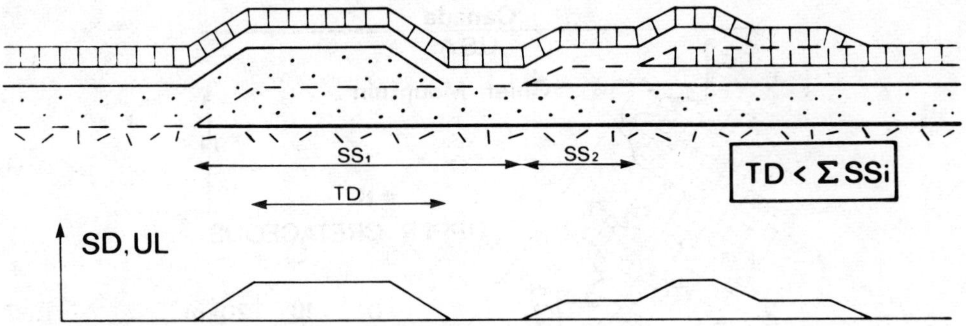

Fig. 4. Detail of the thrust plate for total displacement smaller than the sum of the links ($TD < \Sigma SS_i$).

latter term refers strictly to the objectives of this study. For those concerned with the detailed mapping or the economic exploration of such belts, the incidentals may well be all-important. For the present study they only obscure the fundamental picture or, in geophysical language they represent 'noise'.)

It may well be worth while to discuss briefly the definition of fundamentals and incidentals as used in this paper. To some extent it is of course a matter of scale, large features tending to be in the class of fundamentals whereas small ones belong to the incidentals. It may also be a matter of repetitiveness, fundamentals occurring more frequently than incidentals. More specifically, the fundamentals can be understood from a model but the incidentals cannot, being often the result of coincidence. Thus an original weakness in the rock may well locate a minor tear fault in a thrust sheet, or a facies change eliminating a thick competent unit will lead to a change in structural style.

There can be little question that in the Helvetide area of the Alps the incidentals play a much more prominent role than in the southern Canadian Rockies. The result is an apparent dissimilarity inducing some geologists to state: 'that there are no two mountain chains alike'. This statement is certainly unacceptable. Major structural provinces such as thrust belts must have certain basic features in common. It is suggested that the incidentals are largely responsible for masking those similarities.

In this paper an attempt is made to understand some fundamental features of thrust belts and, more precisely, individual thrust sheets. As a result, the details or incidentals shall not concern us or confuse the issue. It is for this reason that in the following subsection only very rough guides are established for the magnitude of certain features. Only the fact that they do exist, and to an appreciable degree, is of importance at this stage.

a) *Total Displacement*

This term is illustrated in Figure 3. According to the introduction to this section, no accurate values are sought but rather orders of magnitude. It is thus permissible to rely on rather simple techniques. Figure 5 shows the Lewis thrust in northwestern Montana just south of

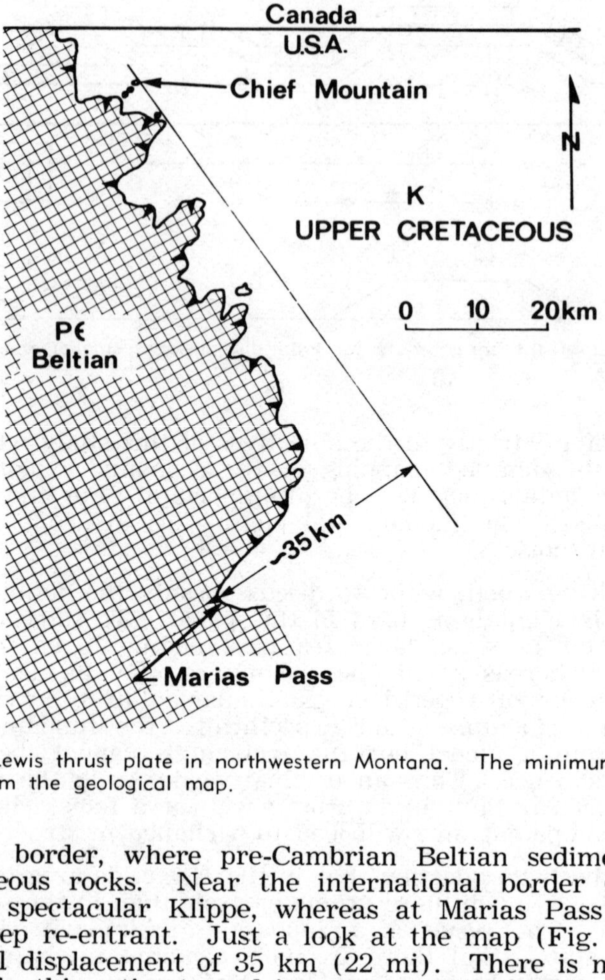

Fig. 5. The Lewis thrust plate in northwestern Montana. The minimum total displacement as read from the geological map.

the Canadian border, where pre-Cambrian Beltian sediments rest on Upper Cretaceous rocks. Near the international border Chief Mountain forms a spectacular Klippe, whereas at Marias Pass erosion has produced a deep re-entrant. Just a look at the map (Fig. 5) reveals a minimum total displacement of 35 km (22 mi). There is no interpretation involved in this estimate and it represents an absolute minimum.

Northwest of Calgary, Alberta, the McConnell thrust at the Panther River culmination has a visible displacement of 15 km (10 mi), where Cambrian rocks rest on Upper Cretaceous strata. The Glarus thrust in Switzerland shows Permian over Tertiary for a distance of 35 km (22 mi). Gansser's (1964) map of the Himalayas shows displacements of 50 km (31 mi) in several places. Maps of the Moine thrust show displacements of 10 km (6 mi).

All these displacements are simply taken from the geological maps in the manner shown for the Lewis thrust. They in no way reflect the actual total displacement and are undisputed minimum values. However, for the purpose of this study it is unimportant whether the true total displacement is one and a half, two or even three times the values stated above. It suffices to recognize that total displacements of major overthrust units in various parts of the world are measured in tens of kilometres.

b) *Stratigraphic Displacement*

At the Lewis thrust in Montana, pre-Cambrian Beltian rocks overlie Upper Cretaceous and the stratigraphic displacement is in the order of 6 km (4 mi). At Mount Yamnuska, west of Calgary (Fig. 1), the McConnell thrust brings middle Cambrian strata over Upper Cretaceous rocks with a stratigraphic displacement of more than 4 km (2.5 mi). Thickness of the Glarus thrust is estimated to have been 5 to 6 km (3 to 4 mi) at the time of emplacement (Trümpy, 1969).

Again it is not our intention to document in great detail the stratigraphic displacements on various thrusts. The above values indicate that on major thrusts the stratigraphic displacement is to be measured in kilometres.

c) *Observed Nature of Thrust Faults*

Not many detailed descriptions of thrust faults are available in the literature. However, modern mapping has led to many places where

Fig. 6. Detail of the McConnell thrust at Mount Yamnuska, west of Calgary, Alberta.

the thrusts themselves can be observed. Invariably the descriptions use such words as : 'knife sharp', 'drawn with a knife edge', etc. Figure 6 shows the McConnell thrust as exposed on Mount Yamnuska (detail, middle of Fig. 1). It is indeed possible to place one's hand on the surface separating the Middle Cambrian Eldon Formation from the Upper Cretaceous Belly River Formation. Descriptions in the literature (e.g., Gansser, 1964, photos 29 & 30, p. 132) and personal observations in both the Canadian Rockies and the Alps indicate that the picture shown may be termed 'typical' for thrust faults. The fact that this boundary is so sharp becomes more impressive when one considers that each point above the fault surface in Figure 6 has moved a minimum distance of 15 km (10 mi). It certainly points to an ease of movement and makes understandable the geologists' old postulate of a 'basal lubrication'. Until the classic companion papers of Hubbert and Rubey (1959) — that is, for over 100 years — this occurrence has remained a mystery. Today we begin to understand the mechanism of these faults and hopefully, the ideas presented further on in this paper wil contribute to that understanding.

While the sharp nature of the faults has been emphasized by many authors (Heim, 1919, p. 378, 387, 389; Holmes, 1964, p. 223, 1164; Shelton, 1966, p. 103, 323), one should not forget that the rocks above and below a thrust may show considerable deformation for distances of some hundreds of feet. Still the thrust plates as a whole have remained intact and are in no way comparable to the jumbled masses of landslides. Thus, the presumption underlying this paper — that we are dealing with virtually undistorted plates moving over the undisturbed sequence along sharp planes — is basically correct.

d) *Progession of Thrusting as Deduced from Field Evidence*

In the Canadian Rockies most authors agree that thrusting has progressed essentially from west to east; i.e., outward (Bally *et al.*, 1966; Dahlstrom, 1970). According to Rubey and Hubbert (1959, p. 190) and Armstrong and Oriel (1965), the same is true for the Idaho-Wyoming thrust belt. In the Helvetides of the Alps there is some evidence that in certain places thrusting has progressed outward, but the case cannot be as unequivocally stated as for the southern Canadian Rockies (Trumpy, *pers. comm.*)

As pointed out by Dahlstrom (1970) and Jones (1971), there are cases, particularly those of the so-called back-limb thrusts, that do not conform to this general rule. In our terminology they would fall into the class of the incidentals, their position and origin being determined by the preceding folding of a thrust plate due to an underlying step (Figs. 3, 4). These thrusts will only obscure the fundamental picture. As will be shown later, it seems mechanically very plausible that thrusting should progress basically outward. It is to be hoped that in future the field geologists will pay even more attention to this question.

e) *The Sedimentary Wedge*

The deformation in the outer thrust belts, under consideration in this paper, involves the sediments of a miogeocline. These sediments form a wedge and the case of the southern Canadian Rockies is particularly well documented. Even a casual field trip from Calgary to the Rocky Mountain Trench reveals the rapid thickening of the sedimentary section in this direction, as pointed out by Bally *et al.* (1966, Fig. 10).

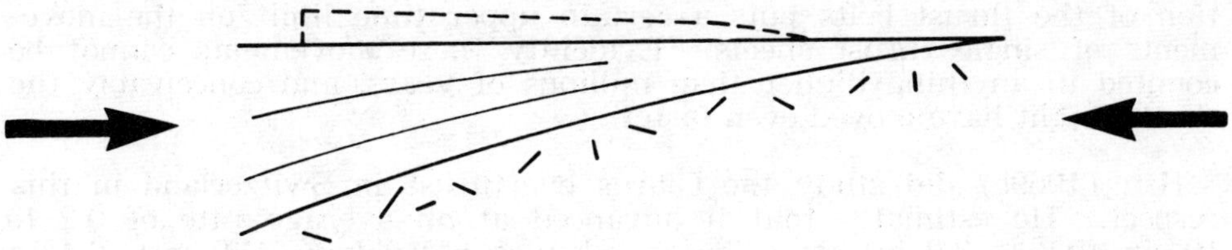

—— before shortening
---- after shortening

Fig. 7. The sedimentary wedge in compression.

The section deformed by thrusting is thus a wedge, a fact often not sufficiently appreciated. It necessarily means that more mass is shifted at the rear than is piled up at the front (Fig. 7). Furthermore, the bedding planes of such a sedimentary wedge are not completely parallel to the orientation of maximum horizontal compression during deformation.

Also, in the case of the southern Canadian Rockies, the tip of the wedge escapes deformation. In the Jura Mountains the very tip of such a wedge is strongly deformed (Laubscher, 1961), whereas the Swiss Plateau intervening between the Alps and Jura has experienced little deformation. A similar case seems to exist in the Mackenzie and Franklin mountains in the Northwest Territories of Canada. Clearly incidentals, such as the nature of the sediments involved, must play a decisive role. Thus Laubscher (1961) rightly attributes great significance to the presence of the Triassic evaporites in the subsurface of the Jura Mountains, the Swiss Plateau and the Alps proper. Incidentals may then determine to a large degree what happens to the tip of the wedge or, in other words, how far outward the deformation may reach. Whatever the case may be, certainly the fact that a wedge and not a block is deformed must be kept in sight.

The Question of the Rate of Advance

The formation of a total thrust belt is a process that must be counted in tens of millions of years. In the Austrian Alps, sizable thrust movements took place as early as pre-Coniacian and as late as post-Oligocene (Trümpy, *pers. comm.*) A total time span of at least 55 my is indicated. According to Dahlstrom (1970, p. 402) an even longer time span is indicated for the formation of the Canadian thrust belt. This is not, however, the question that is of concern in the present paper. The problem at hand is: how does an individual thrust sheet advance, what is the rate of movement? Is the process of thrusting a continuous one or does it proceed in spurts? Do individual sheets advance fast, followed by a pause until the next sheet becomes activated? It is these detailed questions of the kinematics in a thrust belt that are important in regard to the problems considered here. Not much has been written on these aspects of thrust faulting and it is certainly too early to present a final answer. The total time span covered by the formation of the thrust belts puts a certain upper time limit on the movements of single thrust sheets. Evidently these movements cannot be counted in anything higher than millions of years, and conceivably the sheets might have moved even faster.

Hsü (1969c) did study the Glarus overthrust in Switzerland in this respect. He estimates that it advanced at an average rate of 0.2 to 10 cm (0.8 to 3.9 in)/yr. For an advance of 40 km (24.9 mi), 0.4 to 20 my would be required. Thus an area 40 km wide and possibly 100 km (62 mi) long would be loaded with a plate about 3 km (1.9 mi) thick in the above time span. If we accept the minimum velocity we find that we load 3 km (1.9 mi) in 20 my, or an average of 0.15 mm/yr. For the maximum velocity we arrive at an average of 7.5 mm/yr. Certainly these values correspond to a fast rate of loading, compared with the values given for sedimentation rates by Rubey and Hubbert (1959, p. 171) and are in substantial agreement with their estimate for loading due to overthrusting in the Idaho-Wyoming thrust belt (Rubey and Hubbert, 1959, p. 192).

While it is not possible to give very precise answers to the question posed in this subsection, we may again be content; for our purposes it is sufficient to accept the advance of thrust sheets as a — geologically speaking — case of fast loading.

The Question of Contemporaneous Erosion

The thrust sheets as they present themselves today are strongly eroded remnants of initially much larger units. Since late uplift is a well-established fact in all major thrust belts (Trümpy, 1960; Bally *et al.*, 1966; Shaw, 1970), much of the erosion must postdate the thrusting and is therefore of no consequence with regard to the questions dealt with in this paper. Our problem requires the thrust sheets as they existed at the time of advance. It is almost certain that the uneroded load, as schematically given in Figures 2 to 4, does not provide the real situation. However, total erosion of the toe or thrust sheet, as assumed by Rubey and Hubbert (1959, p. 194, Fig. 9), is equally unrealistic in view of the total and stratigraphic displacements evident even today. The real question is thus to find the extent and thickness of the thrust sheets as they existed at the time of emplacement. An exact answer cannot be given at this time. Some evidence for more extensive loading than is evident today will appear in the following section. To establish the original extent of certain major thrust sheets is actually a challenging field problem that, hopefully, will be tackled in the near future.

Fortunately for this paper a precise answer is not required. It suffices to state that the evidence presently visible suggests that thrust sheets must have been extensive at the time of formation. For the Glarus thrust Trümpy (1969, p. 130) estimates the thickness of the sheet was 5 to 6 km (3.1 to 3.7 mi). Thus the shift of large loads on the earth's crust as a consequence of thrusting cannot be disputed — the only factor of real significance for our current considerations.

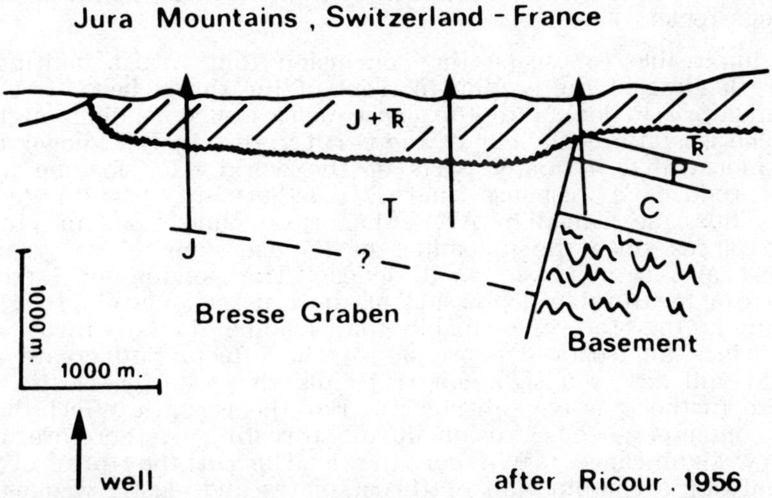

Fig. 8. Deformation of a thrust plane in the Jura Mountains. The downward deflection is interpreted as an effect of loading. The young Tertiary clastics of the Bresse graben were compacted under the load of the thrust sheet. (After Ricour, 1956).

The Effect of Loading

Loading due to thrusting must be accepted as a fact, even if it is not as severe as schematically shown in Figures 2 through 4. There are two examples that may be cited as additional evidence for loading due to thrust sheets.

The first concerns thrusting in the Jura Mountains. Recent work has shown that sizable thrusting is present. Figure 8 is after Ricour (1956) and shows one of the thrusts moving out over the Bresse Graben. According to Laubscher (1961), the graben structure was produced by pre-thrusting deformation of the basement. It is evident that the thrust plane drops about 200 m (650 ft) over the edge of the graben. This deformation of the fault plane is well documented by three wells. Laubscher (1961) attributes the deformation to the gouging action of the thrust. My preferred interpretation ascribes it to a loading effect. As the carbonates and evaporites of the thrust plate were loaded onto the Tertiary clastics of the Bresse Graben, the latter compacted under the additional load. The deformation of the thrust plane is thus a differential-compaction phenomenon and of post-thrusting age.

A second example relates to the McConnell thrust in the southern Canadian Rockies. A Cretaceous sandstone (ledge near Valley floor in Fig. 1) located under Mount Yamnuska capped by an erosional remnant of the McConnell plate was subjected to a crushing strength test. Six tests gave an average strength of 2450 kp/cm^2 with a coefficient of variation of 10 per cent. Cretaceous sandstones from more easterly positions never covered by thrust plates do not exhibit such high strength. While the idea requires further testing, the present results nonetheless suggest that Mount Yamnuska represents only a small remnant of a once much more extensive thrust sheet, and this tectonic loading may have produced the anomalously high strength in the underlying Cretaceous rocks.

It is impossible to escape the conclusion that thrust faulting leads to severe loading of the section in front of the thrust belt. It is therefore mandatory to look into the consequences such loading might produce. The isostatic behaviour of the earth's crust is well known and has been demonstrated in many parts of the world. Ice loading and unloading produces a response that is, furthermore, fast in geological terms. Thus, the Canadian Arctic has risen almost 300 m (1,000 ft) in some places since the unloading of the ice some 10,000 years ago. The crust also responds to small loads. The ponding of Lake Mead has led to a measurable depression of the crust (Longwell, 1954). The conclusion is therefore inescapable that loading due to thrust sheets must produce an isostatic depression. The effect is to create a foredeep that will act as a sediment trap, and the resulting clastic wedges will cause further passive subsidence. For the isostatic effect the question of contemporaneous erosion during thrusting is, therefore, of only secondary significance. Whether the load is in the form of thrust plates only or a combination of thrust plates and clastic wedges is immaterial. The foredeep is an essential yart of thrustbelts, an isostatic effect well explained by Price (1969) and Price and Mountjoy (1970, p. 16). The isostatic effects of thrusting may therefore be dismissed at this stage, and our attention turned to another consequence of the advancing thrust sheets more closely related to the actual progression of thrusting.

Discussion of the rate of advance has shown that thrusting represents a case of, geologically speaking, fast loading — a fact already briefly considered by Rubey and Hubbert (1959, p. 192). Under these circumstances we may expect high fluid pressures to develop under areas of maximum loading; i.e., in the front of major steps such as shown in Figure 9a. Under this condition the original fault may become inactive and a new thrust fault may develop in the area of high fluid pressure. As this new fault is activated, the old one is carried 'piggyback' and becomes folded in the process (Figure 9b). One postulates, therefore, thrusting to progress outward and deeper thrusts to be younger, for mechanical reasons. This general concept is well accepted in the Canadian Rockies (Bally *et al.*, 1966; Price, 1969; Dahlstrom, 1970). Exceptions or incidentals are the backlimb thrusts that may develop at a later stage on some thrust sheets. In the Helvetides of the Alps there is some evidence for the progression of thrusting in an outward direction, but the case cannot be as clearly stated as in the Canadian Rockies (Trümpy, pers. comm.). Theoretically one might, therefore, predict that thrusting is a somewhat selfperpetuating process (Gretener, 1969a). It is reassuring that the prediction that thrusting ought to progress in an outward direction is not at variance with present field observations.

On The Ease of Movement

The low basal friction that evidently must exist while the thrust sheets move into place has remained an enigma for many years. Only in 1959 did Hubbert and Rubey, in their companion papers discussing the role of the fluid pressure in this process, point the way to a possible solution. They effectively introduced the concept of effective stress into the geological literature. Possibly it is more accurate to say that they have put this concept in the limelight, since it had been already stated explicitly by Goguel (1962, p. 176, French edition 1952). The concept goes back to Terzaghi who developed it in 1923 (Skempton, 1960). In his classic paper (Terzaghi, 1950), he applied the idea to landslides, stating: 'that when the fluid pressure reaches a critical value the overburden may in fact be floated'. It is this principle that Hubbert and Rubey (1959) have applied to the mechanism of overthrusts.

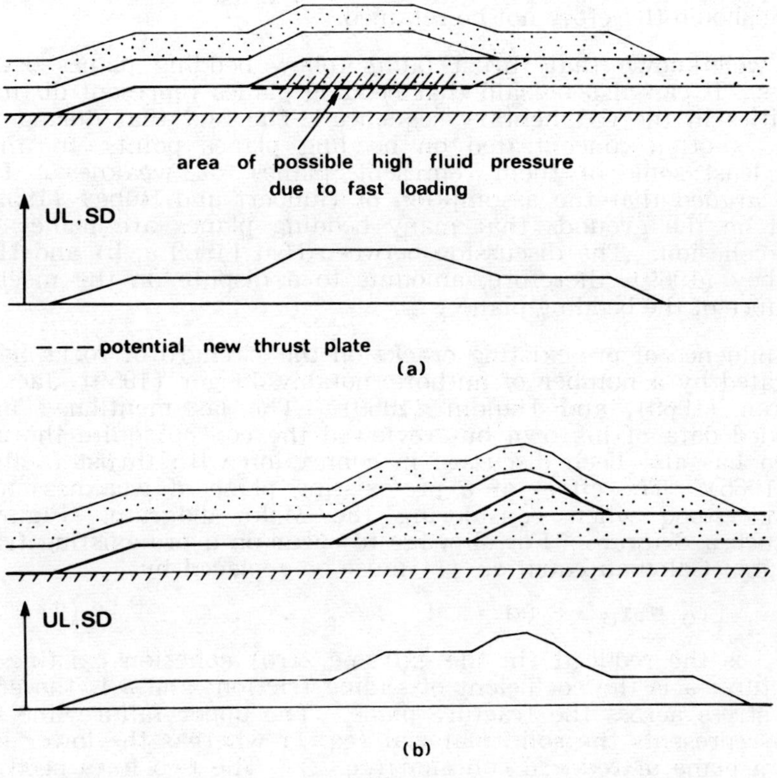

Fig. 9a. The effect of loading on the underlying sequence.

Fig. 9b. Activation of a lower thrust plane. The old thrust is carried piggyback and folded over the new step.

The concept may best be illustrated with the Mohr diagram. Figure 10a shows what happens in a porous medium under stress when the fluid pressure is raised by an amount Δp (Hubbert and Rubey, 1960). The Mohr circle describing the initial condition of stress is shifted to the left and approaches the failure envelope. As the pore pressure reaches a critical value, the Mohr circle will touch the failure envelope, and fracture of the material may be induced in this manner.

The significance of the Hubbert and Rubey papers (1959) is demonstrated by the intensive discussion that followed. Most recently Hsü

(1969a, b) criticized the authors' model of overthrust faulting. He claimed that the model works not so much because of the increased pore pressure — reduced normal stress across the failure plane — but rather since the authors chose to neglect the term cohesion (τ_0) in the Coulomb failure criterion (eq. 1) for the fracture plane:

$$\tau = \tau_0 + n\sigma \quad . \; . \; . \; . \; . \; . \; . \; . \; . \quad (1)$$

where τ_0 is the cohesion; n the coefficient of internal friction; and σ the normal stress across the plane. Hubbert and Rubey (1959) did neglect this term, with the justification that the cohesion would be gradually broken; i.e., the fracture would process at a finite speed and the term should therefore not be retained.

It is well known that thrust faults follow bedding planes over long distances. It can also be said that bedding planes represent distinct discontinuities in the sedimentary sequence. The fact that during folding slippage is often concentrated on bedding planes points to the fact that at least some of them represent planes of weakness. It may thus be argued that the assumption of Hubbert and Rubey (1959) was justified on the grounds that many bedding planes are planes of low or zero cohesion. The discussion between Hsü (1969 a, b) and Hubbert and Rubey (1969), therefore, amounts to a dispute on the mechanical significance of the bedding plane.

The influence of pre-existing cracks on the strength of rocks has been investigated by a number of authors, notably Jaeger (1959), Jaeger and Rosengren (1969), and Handin (1969). The last-mentioned has not only added data of his own but reviewed the concept quite thoroughly. The idea has also been discussed in connection with thrust faulting by Price (1965). The effect of a pre-existing plane of weakness may be best understood when considering the Mohr diagram. Figure 10b shows such a diagram. For slippage to occur on a pre-existing fracture, the original Coloumb equation (1) must be replaced by:

$$\tau_0 = \tau_0' + \mu\sigma \quad . \; . \; . \; . \; . \; . \; . \; . \quad (2)$$

where τ_0' is the reduced (in the extreme zero) cohesion existing across the fracture; μ is the coefficient of sliding friction; and σ is the effective normal stress across the fracture plane. The upper failure line in Figure 10b represents the solid material (eq. 1) whereas the lower line applies to a plane of reduced cohesion (eq. 2). The two lines need not be parallel. The slope of the upper line is given by the coefficient of internal friction, while that of the lower line is the coefficient of sliding friction. It is quite evident from the inspection of Figure 10b that the effect of such a plane will be most pronounced for materials with high cohesion, and for the condition where the coefficient of sliding friction is smaller than the coefficient of internal friction.

In Figure 10b a Mohr circle is shown. The circle does not intersect the failure envelope for solid rock, and for an unbroken piece of rock this state of stress is stable. However, the circle does intersect the failure line for a plane of weakness in this rock. Thus, if the plane of weakness is so oriented that its angle (δ) with the σ_1-axis meets the condition:

$$\delta_{min} \leqslant \delta \leqslant \delta_{max} \quad . \; . \; . \; . \; . \; . \; . \quad (3)$$

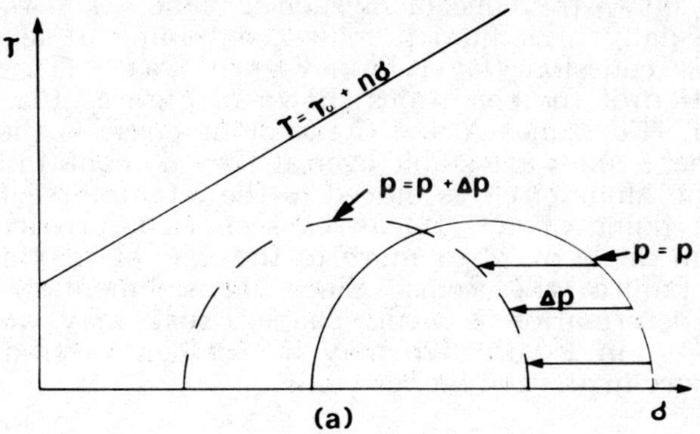

(a)

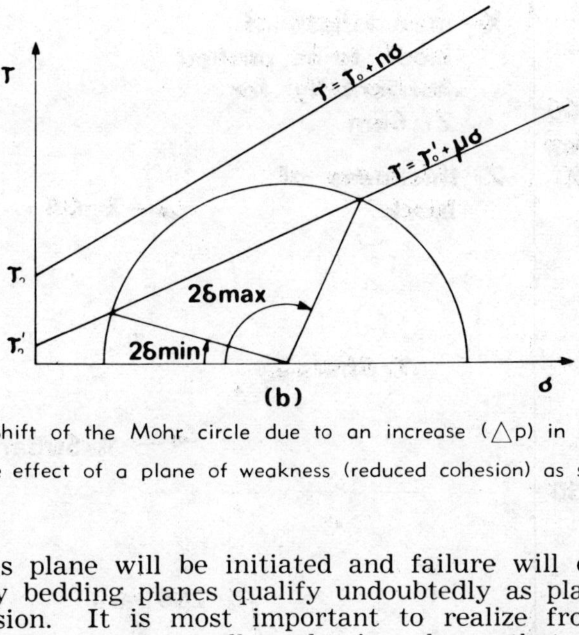

(b)

Fig. 10a. Shift of the Mohr circle due to an increase (△p) in pore pressure.

Fig. 10b. The effect of a plane of weakness (reduced cohesion) as shown in the Mohr diagram.

sliding on this plane will be initiated and failure will occur (Handin, (1969). Many bedding planes qualify undoubtedly as planes of reduced or zero cohesion. It is most important to realize from Figure 10b that δ_{min} may be a very small angle; i.e. planes that are subparallel to the σ_1-axis may be activated.

Returning to the discussion between Hsü (1969a,b) and Hubbert and Rubey (1969), the following can be said: Taking the model as set out by Hubbert and Rubey (1959), we may plot the graph shown in Figure 11. The length of the block to be pushed without destruction is plotted versus the cohesion or the pore pressure - overburden ratio (λ) respectively. The thickness of the particular block is assumed to be 6 km (3.7 mi) — not an unreasonable value according to our previous considerations. The graph clearly shows that neither a reduction of the cohesion alone nor a simple increase of the pore pressure is sufficient to increase the permissible length of the block to any appreciable extent. Evidently an increase of pore pressure across a plane of low cohesion is necessary to make the model work. The latter may be called a plane of superweakness.

Figure 12a shows the concept of such a plane. The requirement is a zone of abnormally high fluid pressure containing at least one bedding plane with low cohesion (layer B in Figure 12a). This situation leads to a combination of the conditions shown in Figures 10a and 10b, given in Figure 12b. For zones A and C the Mohr circle is the one shown to the right. These zones are stable even if they do contain bedding planes. For zone C the Mohr circle is moved to the left, intersecting the failure line for the bedding plane. It can be seen that, provided the bedding plane forms an angle of 7° or more to the axis of maximum horizontal compression, failure will occur. Since the sedimentary sequence that is subject to deformation is wedge-shaped, such may well be the case. The angle δ—— in Figure 12b may be further lowered by increasing the fluid pressure to an even higher value.

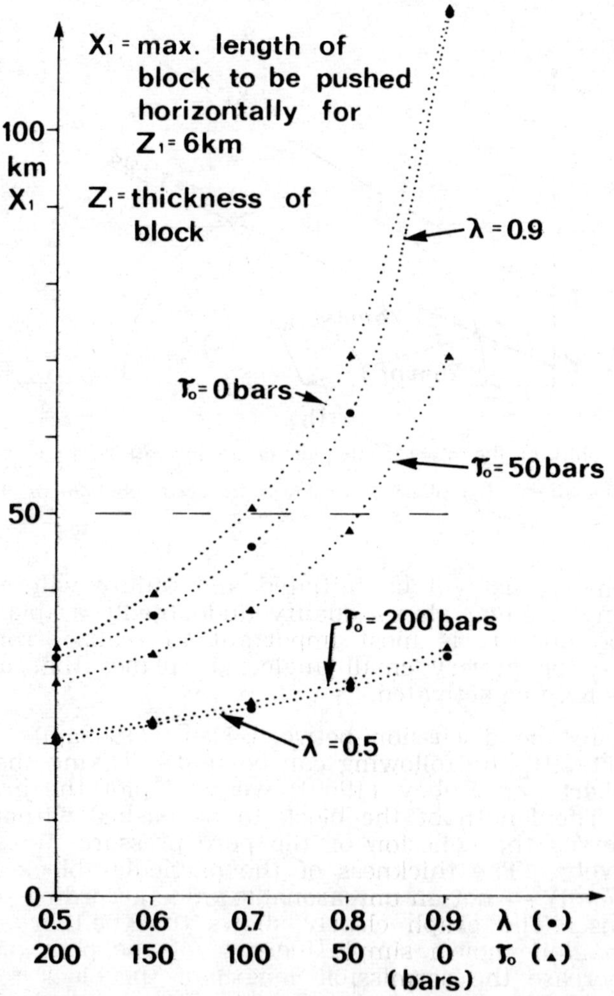

Fig. 11. The maximum length of a block 6 km (3.7 mi) thick that may be pushed as a function of the pore pressure and cohesion across the basal plane. (After Hsü, 1969a and Hubbert and Rubey, 1959).

The model of Figure 12 explains easy gliding along the links as proposed by Hubbert and Rubey (1959). It predicts the links to be concentrated in zones that are prone to high fluid pressures, such as shales, particularly when interbedded with evaporites. This is in accordance with published field observations (Douglas, 1950; Mostofi and Gansser, 1957; Dahlstrom, 1970). However, the plane of superweakness need not occur in this zone; it may well be located adjacent in the zone A or C of Figure 12a comprising carbonate or sandstones. In reality, the pore-fluid pressure in these rocks next to an overpressured shale zone will not be entirely normal but rather will drop off gradually. Thus we may well have the arbitrarily chosen conditions as shown in Figure 13. The pore pressure in the shale is 900 kp/cm², and in the dolomite adjacent to the shale 800 kp/cm². For the various mechanical parameters of importance we may assume the following numerical

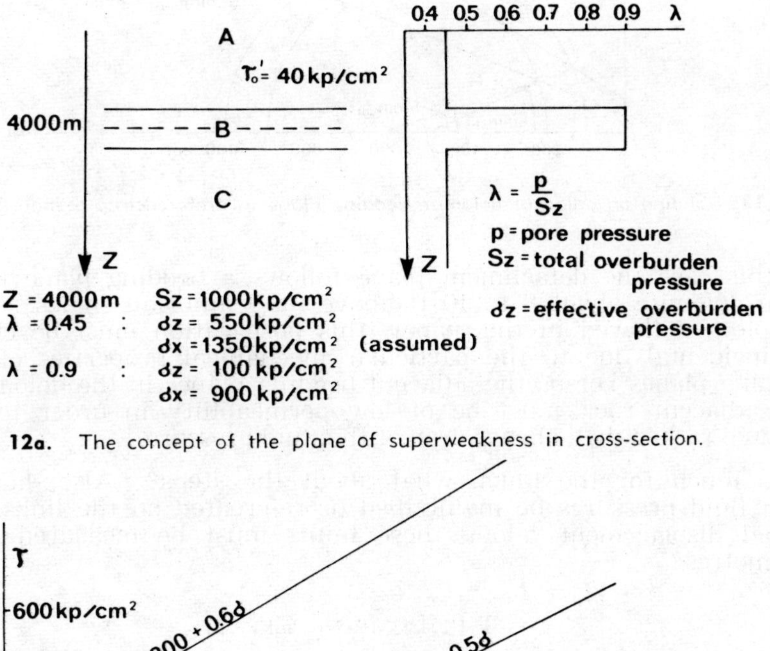

Fig. 12a. The concept of the plane of superweakness in cross-section.

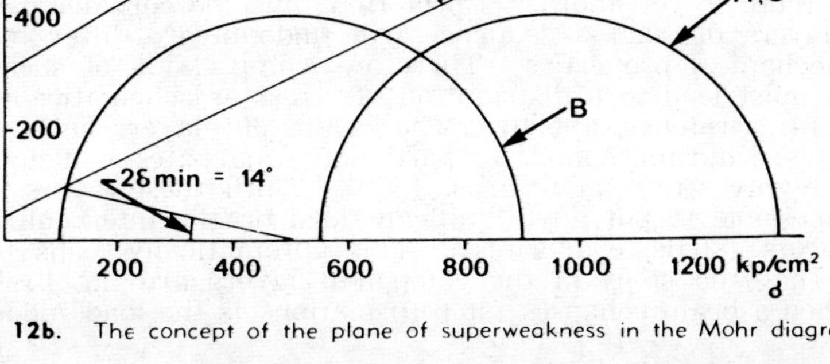

Fig. 12b. The concept of the plane of superweakness in the Mohr diagram.

values: On a bedding plane in the shale the cohesion (τ_0') is 80 kp/cm² whereas in the dolomite it is 40 kp/cm²; the coefficient of gliding friction for the shale bedding plane is 0.6 and for the dolomite bedding plane 0.4. Under these conditions, horizontal compression may well induce gliding on a bedding plane in the dolomite next to the shale rather than in the shale itself. This explanation may well apply to the observation on the Heart Mountain thrust reported by Stearns *(pers. comm.)*.

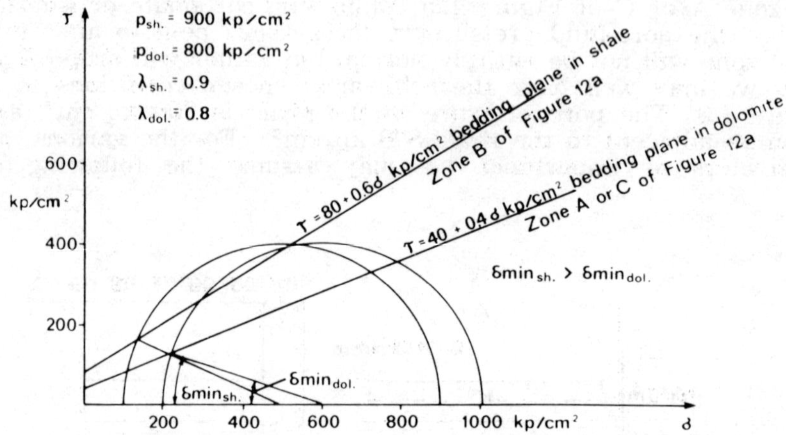

Fig. 13. Sliding on adjacent dolomite bedding plane in preference to a shale bedding plane.

In this case the detachment plane follows a bedding plane in the Bighorn dolomite about 8 to 10 ft above the Cambrian shales. In the terminology followed in this paper, this observation must be classified as an incidental due to the particular mechanical properties of the shale bedding planes versus the adjacent bedding planes in the dolomite. Also, the adjacent rock must be of low permeability in order to permit a gradual rather than abrupt drop of the fluid pressure.

So much for the links; what about the steps? Also, how can the high fluid pressures be maintained or re-created on the links, when the actual displacements along these faults must be measured in tens of kilometres?

The Pumping Effect

In order to attempt to understand thrust faulting in a layered sequence, there is yet another aspect that must be considered. The individual layers of such a sequence will undoubtedly differ markedly in their mechanical properties. Therefore, compression of such a layered medium must lead to a distribution of stress as schematically shown in Figure 14 (Gretener, 1969b). The competent layers such as massive carbonates, sandstones and, in particular, quartzites will act as beams and be highly stressed (Stearns, 1969). Until these layers are broken it is impossible to put any significant load on the intervening incompetent spacers (shale, evaporites). This automatically leads to the conclusion that the steps in the competent layers are the first to form. Only when a beam collapses (step formation) is the load suddenly transferred to the adjacent spacers and the link produced. This situation is illustrated in the Mohr diagram of Figure 15. Given are the failure envelopes for the solid rock as well as for the bedding planes. The large solid Mohr circle refers to the state of stress in the beam (e.g., dolomite) immediately before failure, whereas the small solid Mohr circle gives the state of stress in the adjacent incompetent bed (e.g., shale) at the same time. Note that the effective overburden stress (σ_z) is the same for both dolomite and shale. No allowance is made for an initially abnormal pore pressure in the shale. The effect of such an abnormal pressure would be to shift the Mohr circle for shale to the left and thereby further ease the process described below.

In Figure 15 the large solid Mohr circle (stress in dolomite) just touches the failure envelope for solid rock. A step forms, the stress in the beam is relieved (collapse of the large Mohr circle) and the load is rapidly transferred to the adjacent shale layer. Because of rapid loading, the additional load placed on the incompetent and often poorly permeable layer is carried by the fluid rather than the matrix. The rise in the fluid pressure results in a reduction of the confining (overburden) pressure. The small Mohr circle for the shale therefore grows in the manner shown in Figure 15. As it reaches a critical diameter, slippage on a bedding plane occurs and the link is formed. This is termed the pumping effect. It accounts for the initial formation and the peculiar geometry of the thrust faults.

The process may be better understood when considered in terms of a special triaxial test. A sample is in a pressure cell with confining pressure as well as axial load applied, and the pore pressure is moni-

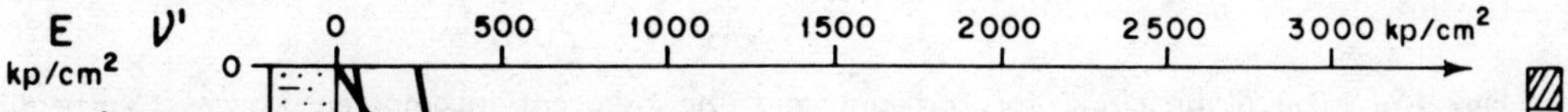

Fig. 14. Stresses in a layered sequence under horizontal compression (from Gretener, 1969b.)

tored. The axial load is slowly raised, with the result that both confining pressure and pore pressure remain constant while the effective axial load is increasing. At a given momen the rate of axial loading is drastically stepped up. Under these conditions, the fluid in the fully saturated sample is unable to move out of closing cracks and pore spaces. As a result, the fluid pressure will increase, with a corresponding drop in confining pressure, and the effective axial load will remain constant. This is the case for the shale as described above, the vertical stress component being the confining pressure and the horizontal compression being equivalent to the axial load.

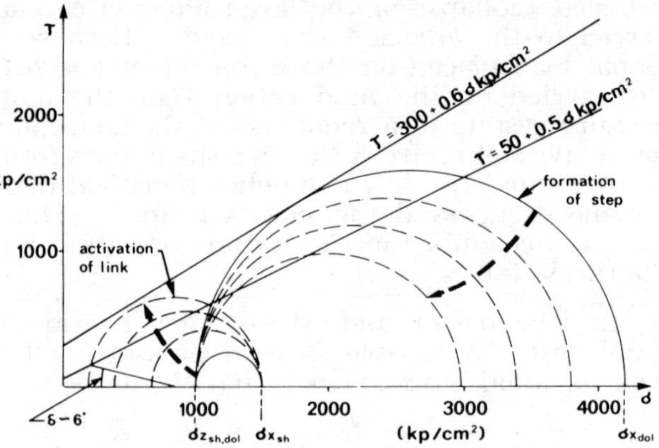

Fig. 15. The Mohr circle for a beam and the adjacent incompetent layer. Vertical stress is equal in both. Breaking the beam results in the collapse of the respective Mohr circle. The load is transferred to the spacer. Due to rapid loading only the pore pressure is increased. The result is to inflate the respective Mohr circle by decreasing the overburden pressure. A link (bedding plane) is activated as the circle reaches a critical diameter. As movement takes place, the latter Mohr circle collapses and motion on the fault is therefore limited.

This model describes what might happen during the initial formation of a thrust fault. However, we have seen that displacements along such faults are to be measured in tens of kilometres. It is not suggested that the total displacement take place at once for two reasons: First, as soon as movement occurs on the link, the respective Mohr circle must collapse and the fluid pressure decrease. Second, as is evident from Figures 3 and 4 the thrust sheet and, in particular, the beams must deform over every step. Therefore, displacement at any one time is limited, the link becoming 'stuck' because of a drop in fluid pressure, and the beam becoming locked over a step. Only segments of the fault are in motion at a given time, and the progress of the thrust sheet is much like the crawl of a caterpillar. This is the form of movement as envisaged by Barnes (1966), Hsü (1969c) and Price *(pers. comm.)*. Hsü has compared it to the stick-slip movement observed in experimental work (Brace and Byerlee, 1966). In the case of the thrust faults, sticking is related to the megaprotrusions on the fault plane — the steps — as well as to the drop in fluid pressure on the links as motion proceeds.

Thus, the process of rapid transfer of load from the beam to the incompetent layers may be repeated an unlimited number of times. High fluid pressures may be created on the links as the thrust plate moves along. A process that may further enhance the fluid pressure in the poorly permeable beds of the link is the pressurization due to shock waves. Every time the beam is unlocked, shock waves must be produced. Their effect would be to shift the Mohr circle for shale in Figure 15 even farther to the left, thereby facilitating movement on the links. The pumping effect is thus not limited to the initial condition, but repeats itself during the whole life span of the thrust fault. It represents an attempt to account for the evident ease of movement along such faults as suggested by the field observations. Clearly any acceptable mechanism must not only produce a favorable initial condition but must also account for maintenance of such a condition during the very sizable displacements.

It is interesting to note that, for this particular model, the jerky movement of the thrust plates is most important. The advance of a given point on a thrust sheet is shown in Figure 16. It is obvious that the average rate of advance is of no great importance except for the rate of loading of the underlying sequence. What is important is the form of the actual advance. Thus the old discussion of fast versus slow becomes irrelevant. The over-all movement is slow, but consists of a series of fast spurts. The average rate of progress tells only an incomplete story when the actual process consists of a series of fast, singular events (Gretener, 1967).

This explanation still leaves the positioning of the steps as an unresolved problem. Undoubtedly the step spacing will be related in some degree to the thickness of the beams. At least, in a statistical manner, one would expect the step spacing to increase with increasing beam thickness. Step spacing may also be related to gentle pre-thrust folding (Simony, *pers. comm.*), or the step formation may be entirely triggered by incidentals such as pre-existing fractures or lithological changes in the beam.

Conclusions

A number of aspects important to thrust faulting in a layered sequence have been considered. The role of the fluid pressure was introduced by Hubbert and Rubey in 1959 and the influence of the bedding plane was considered by Price in 1965. The combination of the two principles leads to the concept of the plane of superweakness: a plane with low or zero cohesion across it and situated in a layer of high fluid pressure, with the normal stress across the plane a very low value.

Field observations strongly suggest that any valid model of thrust faulting must account for a very low basal friction on the thrust plates. This suggestion appears particularly valid since Hubbert and Rubey (1959) have not considered the effect of the toe, and Raleigh and Griggs (1963) have done so only in a limited way (small toe). Extensive toes — or overthrust plates — are, however, an observed fact. Their existeice will tend to reduce the maximum lenth of plate that can be pushed without destruction, as published by Hubbert and Rubey (1959) and Raleigh and Griggs (1963). Or, as an alternative, the basal friction must be even lower than these authors assumed. Furthermore, this low

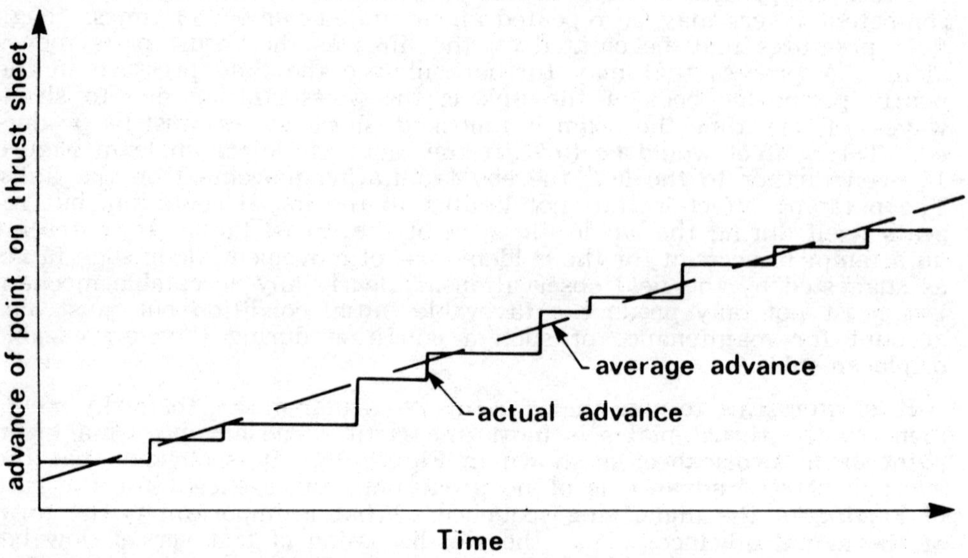

Fig. 16. Movement of a point on a thrust sheet. The actual movement is a series of jerky singular motions.

friction must be maintained in both time and space. Emplacement of a single major thrust sheet is a process that takes on the order of a million years and displaces rocks many tens of kilometres. Evidently, whatever the postulated causes of low basal friction, they must prevail during the whole time and over the total area involved.

In the present model it is suggested that the competent units called the beams carry a disproportionately high part of the horizontal load. When they break initially the load is suddenly transferred to the incompetent units, usually shales with low permeability. The effect of the sudden loading will be to raise the pore pressure in these units while reducing the effective overburden pressure, and the link is formed along a bedding plane of low or zero cohesion (Fig. 15). It is termed the pumping effect. Motion is limited because of the dissipation of the fluid pressure as movement takes place and also as the beams become locked in their movement over the steps. The process may thus be repeated many times. It is visualized that the plates move in the manner of caterpillars, with only a limited number of segments in motion at any time. Shock waves from the initial breaking of the beams or the breaking of the locks at the steps may well enhance the pore pressures in the links.

The shift of load that takes place during thrusting represents a case of fast loading, and the consequences of such action can be anticipated. High pore pressures will be induced in the impermeable layers of the overridden sequence. Thrusting is, therefore, predicted to progress outward, the lower and more forward thrusts being younger as shown in Figure 9b.

It is quite obvious that a full understanding of the mechanics of overthrust faulting is still a matter for the future. However, it is hoped that the foregoing discussion presents some ideas that may prove useful in the final analysis of these features.

REFERENCES

Armstrong, F. C. and Oriel, S. S., 1965, Tectonic development of Idaho-Wyoming thrust belt: Am. Assoc. Petroleum Geologists Bull., v. 49, no. 11, p. 1847-1866.

Bally, A. W., Gordy, P. L. and Stewart, G. A., 1966, Structure, seismic data, and orogenic evolution of southern Canadian Rocky Mountains: Bull. Can. Petroleum Geology, v. 14, no. 3, p. 337-381.

Barnes, W. C., 1966, Mechanics of overthrust faulting: Geol. Soc. America Ann. Mtg., San Francisco, [Abst.], p. 11-12.

Brace, W. F. and Byerlee, J. D., 1966, Stick-slip as a mechanism for earthquakes: Science, v. 153, no. 3739, p. 990-992.

Dahlstrom, C. D. A., 1970, Structural geology in the eastern margin of the Canadian Rocky Mountains: Bull. Can. Petroleum Geology, v. 18, no. 3, p. 332-406.

Douglas, R. J. W., 1950, Callum Creek, Langford Creek, and Gap Map-Areas, Alberta: Geol. Surv. Canada, Mem. 255, 124 p.

Gansser, A., 1964, Geology of the Himalayas: Regional Geol. Ser., L.U. de Sitter, *ed.*, New York J. Wiley and Sons Ltd., 289 p.

Goguel, J., 1962, Tectonics: San Francisco, W. H. Freeman & Co. [transl. from French ed., 1952, by H. E. Thalmann], 384 p.

Gretener, P. E., 1967, Significance of the rare event in geology: Am. Assoc. Petroleum Geologists Bull., v. 51, no. 11, p. 2197-2206.

——, 1969a, Fluid pressure in porous media — its importance in geology: a review: Bull. Can. Petroleum Geology, v. 17, no. 3, p. 255-295.

——, 1969b, On the mechanics of the intrusion of sills: Can. J. Earth Sci., v. 6, no. 6, p. 1415-1419.

Handin, J., 1969, On the Coulomb-Mohr failure criterion: J. Geophys. Res., v. 74, no. 22, p. 5343-5348.

Heim, A., 1882, Die Glarner Doppelfalte: Vierteljahrsch. Naturforschende Ges. Zurich, v. 27, no. 11.

——, 1919, Geologie der Schweiz: Cr. Herm. Tauchnitz, Leipzig, Band I & II, 1018 p.

Holmes, A., 1964, Principles of Physical Geology: London & Edinburgh, T. Nelson, 1288 p.

Hsü, K. J., 1969a, Role of cohesive strength in the mechanics of overthrust faulting and of landsliding: Geol. Soc. America Bull., v. 80, no. 6, p. 927-952.

——, 1969b, Role of cohesive strength in the mechanics of overthrust faulting and of landsliding: Reply: Geol. Soc. America Bull., v. 80, no. 6, p. 955-960.

——, 1969c, A preliminary analysis of the statics and kinetics of the Glarus overthrust: Eclogae Geol. Helv., v. 62, no. 1, p. 143-154.

Hubbert, M. K. and Rubey, W. W., 1959, Role of fluid pressure in mechanics of overthrust faulting: Geol. Soc. America Bull., v. 70, no. 2, p. 115-166.

——, 1960, Role of fluid pressure in mechanics of overthrust faulting: a reply: Geol. Soc. America Bull., v. 71, no. 5, p. 617-628.

——, 1969, Role of cohesive strength in the mechanics of overthrust faulting: Discussion: Geol. Soc. America Bull., v. 80, no. 6, p. 953-954.

Jaeger, J. C., 1959, The frictional properties of joints in rocks: Geofis, Pura e Appl., v. 43, p. 148-158.

—— and Rosengren, K. J., 1969, Friction and sliding of joints: Austr. I.M.&M. Proc., no. 229, p. 93-104.

Jones, P. B., 1971, Folded faults and sequence of thrusting in Alberta foothills: Am. Assoc. Petroleum Geologists Bull., v. 55, no. 2, p. 292-306.

Laubscher, H. P., 1961, Fernschubhypothese der Jurafaltung: Eclogae Geol. Helv., v. 54, no. 2, p. 221-282.

Longwell, C. R., 1954, Interpretation of levelling data, *in* Thomas, H. E., *ed.*, First Fourteen Years of Lake Mead: U.S. Geol. Surv., Circ. 346, p. 5.

Mostofi, B. and Gansser, A., 1957, The story behind the 5 Alborz: Oil Gas J., v. 55, no. 3, p. 78-84.

Price, R. A., 1965, Flathead Map-Area, British Columbia and Alberta: Geol. Surv. Canada, Mem. 336, 221 p.

----, 1969, The southern Canadian Rockies and the role of gravity in low-angle thrusting, foreland folding and the evolution of migrating foredeeps: Geol. Soc. America Ann. Mtg. [Abst.], pt. 7, p. 284-286.

Price, R. A. and Mountjoy, E. W., 1970, Geologic structure of the Canadian Rocky Mountains between Bow and Athabasca rivers, a progress report: Geol. Assoc. Canada, Spec. Pap. 6, p. 7-26.

Raleigh, C. B. and Griggs, D. T., 1963, Effect of the toe in the mechanics of overthrust faulting: Geol. Soc. America Bull., v. 74, no. 7, p. 819-830.

Rich, J. L., 1934, Mechanics of low-angle overthrust faulting as illustrated by Cumberland thrust block, Virgina, Kentucky, and Tennessee: Am. Assoc. Petroleum Geologists Bull., v. 18, no. 12, p. 1584-1596.

Ricour, J., 1956, Le Chevauchement de la Bordure Occidentale du Jura sur la Bresse dans la Région de Lons - Le Saunier: Bull. Ver. Schweizer Petrol. Geol. und Ing., v. 23, no. 64, p. 57-70.

Rubey, W. W. and Hubbert, M. K., 1959, Role of fluid pressure in mechanics of overthrust faulting: Geol. Soc. America Bull., v. 70, no. 2, p. 167-205.

Shaw, E. W., 1970, Vertical uplift in orogenic belts: Bull. Can. Petroeum Geology, v. 18, no. 3, p. 430-438.

Shelton, J. S., 1966, Geology illustrated: San Francisco & London, W. H. Freeman & Co., 434 p.

Skempton, A. W., 1960, Significance of Terzaghi's concept of effective stress. From theory to practice in soil mechanics: New York, J. Wiley & Sons, p. 42-53.

Staub, R., 1954, Der Bau der Glarneralpen und seine prinzipielle Bedeutung fuer die Alpengeologie: Glarus, Verlag Tschudi und Co., 187 p.

Stearns, D. W., 1969, Fracture as a mechanism of flow in naturally deformed layered rocks, *in* Baer, A. S. and Norris, D. K., *eds.*, Proceedings, Conference on Research in Tectonics: Geol. Surv. Canada, Paper 68-52, p. 79-90.

Terzaghi, K., 1950, Mechanism of landslides, *in* Application of Geology to Engineering Practice: Geol. Soc. America, Berkey Volume, p. 83-123.

Trümpy, R., 1960, Paleotectonic evolution of the central and western Alps: Geol. Soc. America Bull., v. 71, no. 6, p. 843-908.

——, 1969, Die helvetischen Decken der Ostschweiz. Versuch einer palinspastischen Korrelation und Ansätze zu einer kinetischen Analyse: Ecologae Geol. Helv., v. 62, no. 1, p. 105-142.

The American Association of Petroleum Geologists Bulletin
V. 66, No. 9 (SEPTEMBER 1982), P. 1196-1230, 34 Figs., 2 Tables

Thrust Systems[1]

STEVEN E. BOYER[2] and DAVID ELLIOTT[3]

ABSTRACT

A general geometric framework underlies the structure, evolution, and mechanical processes associated with thrust faulting. The main purpose of this paper is to review and extend this geometric framework. A certain family of lines must exist where thrust surfaces join along branch lines or end at tip lines. Starting from a description of these lines and individual thrust faults, we examine how they join into thrust systems, either as imbricate fans or duplexes. These thrust systems have distinctive map patterns commonly observed near culminations and windows. Many of the culminations have origins tied in with a particular thrust system. The order in which the fault slices form has a marked effect on the geometry of the thrust system. These systems must be identified to understand the provenance of the synorogenic sediments.

Part of a thrust belt may be dominated by one particularly large thrust sheet. In front and beneath these dominant sheets, there is a characteristic sequence of thrust systems with a regular pattern to the involvement of basement.

This overall geometric framework provides new insight into some classic areas, illustrated by a balanced cross section through the Mountain City and Grandfather Mountain windows, in the southern Appalachians and another from the Jura to the Pennines (in the western Alps).

INTRODUCTION

For almost any work in thrust belts, it is essential to establish the three-dimensional relations between faults. These geometric relations between thrusts arise, for example, in the following typical questions. Does my interpretation change if this thrust fault joins that one? Is it possible to establish the time sequence of faulting from the geometry? Is the thrust pattern associated in some way with the origin of my culmination? What does this map pattern of thrusts imply for the cross section?

In this paper we try to draw together part of the geometric principles that we find useful in deciphering three-dimensional thrust structures. This framework has reasonably wide application, and our examples come from the North American Cordillera, the Appalachians, the Caledonides, and the Alps. Although some of these geometric concepts are fairly old, they have seldom been reviewed; we have tried to establish their origin and say something about how the ideas evolved.

There are two general categories of tools useful in thrust belt analysis. The first is balanced cross sections, or cross sections that are both restorable to the undeformed state and conform to certain specified standards of admissibility (Dahlstrom, 1969; Elliott, in prep.). The same set of data may have several different solutions, all of which allow restoration and are otherwise admissible.

The second main category, and the subject of this paper, is concerned with the interconnections or relationships among faults. We shall see that the solutions of thrust belt problems fall into certain categories or types. Choice of the correct type of solution is commonly the most important single step in the interpretation. When we specify how the various parts connect to each other we may refer to the "logic" of the fault network, in the same way we speak of the logic of an electronic device. We will use a building-block approach, starting with the geometric elements and a very few thrusts, and then look at some characteristic patterns of larger numbers of thrusts.

GEOMETRIC ELEMENTS

A thrust sheet is a volume of rock bound below by a thrust fault. A useful convention is to name the thrust sheet after this underlying or leading thrust fault. This name usually continues to the trailing thrust, which joins the leading thrust along a branch line (Fig. 1). A sheet may have a distinctive stratigraphy, state of strain, or metamorphic grade, and on this basis individual sheets are often correlated long distances. There is a long tradition in the European Alps of focusing attention on the thrust sheet volumes (nappes), and we shall discuss this later in the paper.

The intersection of a thrust surface with a stratigraphic horizon is a cutoff line (Douglass, 1958, p. 132). A cutoff line could also be called an "edge" because it indicates the intersection of two surfaces. Dahlstrom's (1970, p. 352) terms "leading edge" and "trailing edge" thrust surfaces are in this sense self-contradictory, so we have shortened them to "leading and trailing thrust surfaces." J. K. Arbenz remarked (1981, personal commun.), "In

[1]Manuscript received, June 9, 1981; accepted, March 19, 1982.

[2]Sohio Petroleum Co., 633 17th St., Suite 2200, Denver, Colorado 80202.

[3]Johns Hopkins University, Department of Earth and Planetary Sciences, Baltimore, Maryland 21218. [*Editor's Note:* David Elliott passed away August 30, 1982, in Zurich, Switzerland.]

This paper is based partly on a Ph. D. thesis (Boyer 1976, 1978) and partly on a series of lectures presented at Johns Hopkins Univ., Calgary, Alberta; and Lausanne, Switzerland (Elliott 1977, 1978). This work was supported by the National Science Foundation, Earth Sciences Section, Grants 7723209 and 7926118. Field work by Boyer was also aided by The Geological Society of America, Sigma Xi, and The Johns Hopkins University Balk Fund. Elliott acknowledges the Donors of the Petroleum Research Fund, administered by the American Chemical Society, for partial support of this research. For their excellent assistance we are grateful to our reviewers, J. K. Arbenz, R. Groshong, and the students of structural geology at Johns Hopkins.

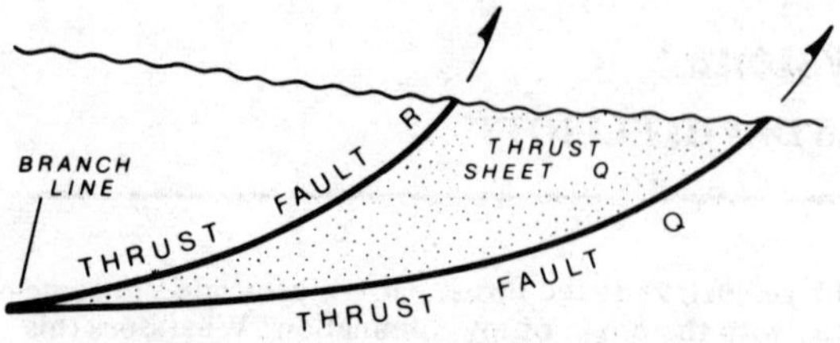

FIG. 1—Cross section through thrust sheet (Q), which is volume of rock above leading fault (Q) and below trailing fault (R).

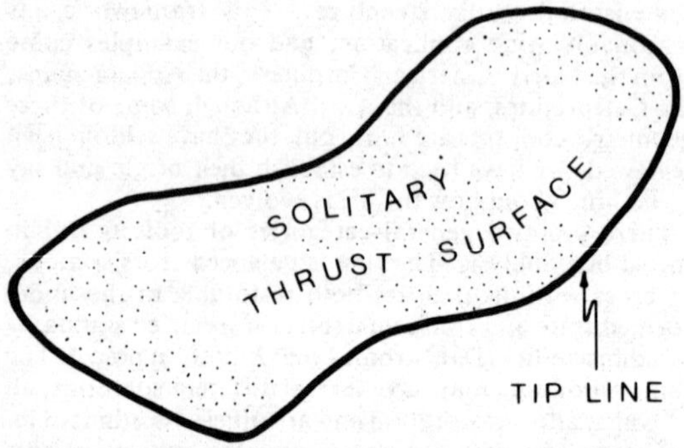

FIG. 2—Three-dimensional view of solitary thrust surface. This is buried, or blind, and its perimeter is everywhere a tip line.

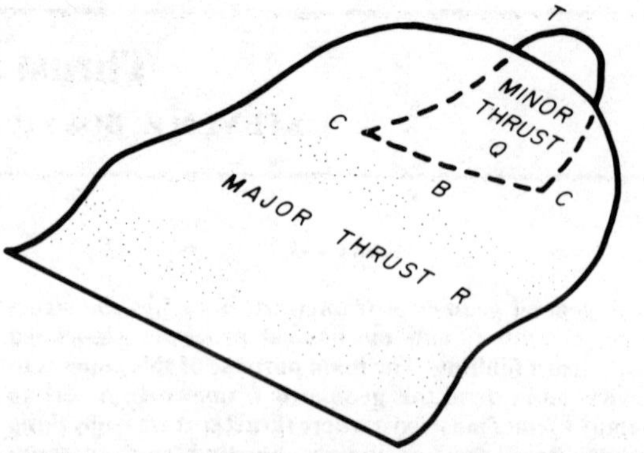

FIG. 3—Minor thrust surface (Q) branches off major thrust (R). Both thrusts meeting along branch line (B). Blind minor thrust (Q) has tip line (T) which meets branch line (B) at two corners (C).

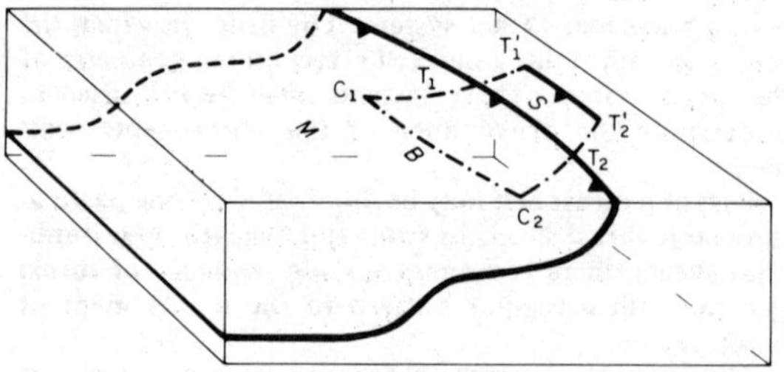

FIG. 4—Erosion cuts branching pair of thrust surfaces which might have once resembled Figure 3. Map pattern is main thrust (M) with isolated splay (S), which has one branch (B), two tip lines (T_1, T_2) that outcrop (T_1, T_2) at termination on map of fault trace.

industry usage, the term 'leading edge' refers to the line of intersection of a thrust surface with a specific stratigraphic horizon of that particular thrust sheet (e.g., the leading edge of the Madison). In fault closures the leading edge becomes the line along which the stratigraphic top seal passes to the fault seal."

So far we have concentrated on blind thrusts,[4] or those in which the tip line does not reach the ground surface (Fig. 2). These thrusts are particularly common near the frontal margin of the thrust belt. Many thrusts start blind and later turn upward to meet the synorogenic erosion surface.

Erosion through a tip or branch line produces the tip or branch points shown on maps. These are particularly important because near these special points we may observe many of the physical processes that produce thrusts (Elliott, 1976, p. 299). There are several examples illustrating the use of branch lines in the Moine thrust belt in Elliott and Johnson (1980, Fig. 17).

A splay rejoins the main fault once. Splays can crop out at the synorogenic erosion surface or arise by later erosion through an originally blind branching thrust (see Fig. 3). We can have an isolated splay (Fig. 4), where erosion cuts the tip line into two, and a diverging splay where erosion cuts the branch and tip lines once (Fig. 5). Maps show that on approaching their lateral termination, major thrusts often turn into a network of diverging splays. In a rejoining splay the tip line is fully removed and the map surface cuts the branch line twice (Fig. 6). Slightly more complex is a connecting splay, where the branch line is strung together along two different fault intersections (Fig. 7).

It is possible for an individual thrust surface to be completely surrounded by branch lines (Fig. 8), because a pair of fault surfaces may diverge after branching and then, farther up, converge to meet again at the branch line. Usually, but not necessarily, the thrust rejoins along strike as well as updip, so that the tip line is eliminated. A horse is a pod of rock completely bound by two or more such fault surfaces. This is an old and useful term (Dennis, 1967, p. 89) which we here extend beyond its original usage for pods bound by normal faults. Horses, unlike the other kinds of branching thrusts, are unlikely to meet a synorogenic erosion surface. Horses can be cut from either the hanging wall or the footwall of major thrusts and may consist entirely of inverted rocks (Fig. 9). They frequently decorate the edges of major thrusts and are particularly helpful in the field to identify thrusts that put shale upon shale. They may also provide stratigraphic information from beneath a major thrust, infor-

[4]Blind thrusts were used and described in Calgary lectures in 1977 by Elliott. We feel earlier usage exists; but were unable to locate the references.

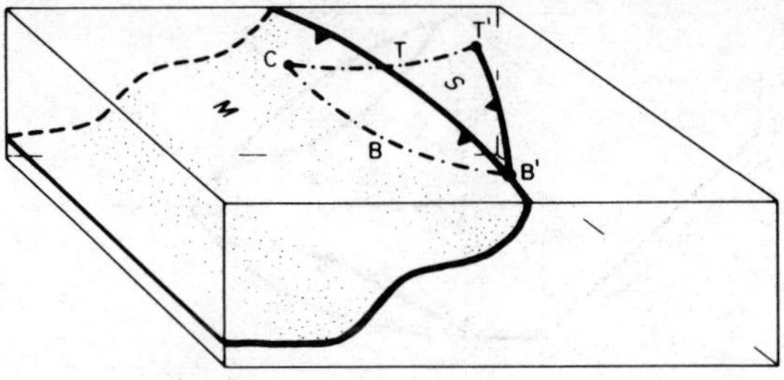

FIG. 5—Block diagram showing diverging splay (S), which has only one tip line (T) with a map termination (T′) and one branch line (B) that intersects erosion surface at B′. There is one corner C.

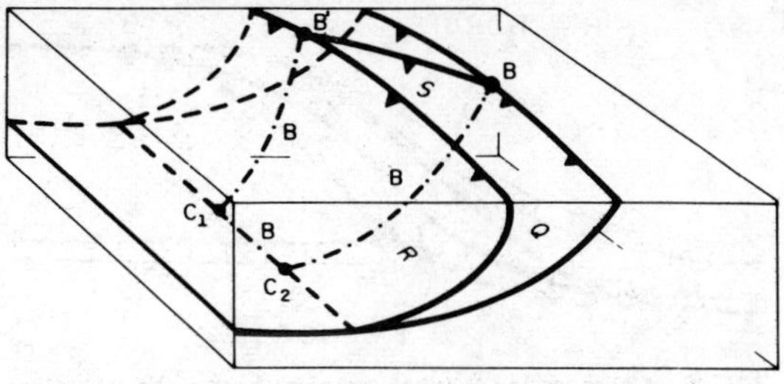

FIG. 7—Two major faults (Q, R) with connecting splay (S). Two branch lines (B) have surface terminations (B, B′) and one branch line at depth has two corners (C_1, C_2).

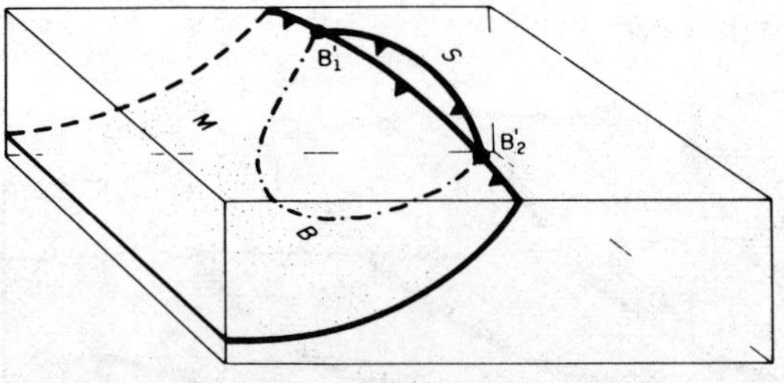

FIG. 6—Rejoining splay with one branch line (B) which intersects map at two branch points (B_1, B_2).

mation which may not be obtainable otherwise.

One must try to describe the three-dimensional geometry relative to the synorogenic erosion surface, not the current one and this can be difficult. For example, how do we distinguish a rejoining splay from an eroded horse on a map? If a rejoining splay has roughly parallel and gently plunging branch lines then cross sections can show two-dimensional pods that resemble horses (Fig. 10), how do we tell them apart?

Transfer Zones and Connectivity

A stratigraphic formation could be cut into interlocking pieces in such a way that if it were fully excavated and you started at one side of the map, you could walk around the ends of the faults on a tortuous path to the other side without having to jump across any faults (Douglas, 1958, p. 131). In this case each thrust sheet is related to its neighbor by an unfaulted envelope that acts as a transfer zone (Dahlstrom, 1970, p. 358). This description of interlocking thrust sheets is correct only for isolated and diverging splays, each of which has at least one tip line. It is not the case within rejoining and connecting splays or horses where each faulted part of the formation is surrounded by fault surfaces.

The extent to which thrust sheets are connected to each other depends on the relative length of tip and branch lines. Because all thrust surfaces are generated by tip lines (Elliott, 1976, p. 298) the degree of connectivity depends on the duration and intensity of activity in that part of the thrust belt. Other things being equal, horses are more

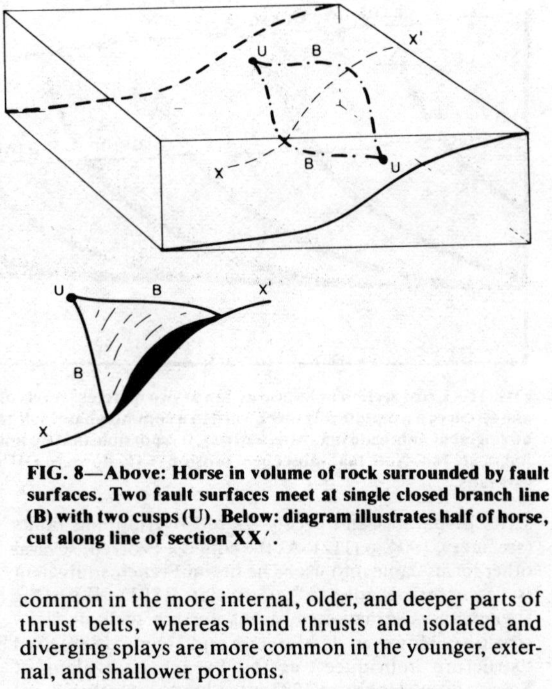

FIG. 8—Above: Horse in volume of rock surrounded by fault surfaces. Two fault surfaces meet at single closed branch line (B) with two cusps (U). Below: diagram illustrates half of horse, cut along line of section XX′.

common in the more internal, older, and deeper parts of thrust belts, whereas blind thrusts and isolated and diverging splays are more common in the younger, external, and shallower portions.

IMBRICATE THRUST SYSTEMS

So far we have concentrated on the geometric relations between two or three connected faults. Several nearby faults may join up in closely related branching array known as a thrust *system* or family (Rodgers, 1953, p. 130). Now we shall look at patterns that arise at thrust systems where a substantial number of faults are the same general shape and size.

If each thrust in a system repeats the size and shape of the neighboring thrust so that the thrust sheets overlap like roof tiles, all dipping in the same general direction, we have an imbricate system. This important special type of thrust system was recognized in 1883, when Suess defined schuppen structure as the repetition of strata by a

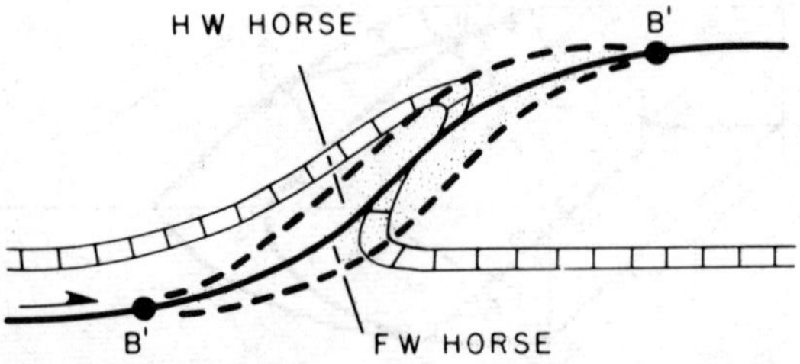

FIG. 9—Cross section through incipient horses. New fractures (dashed) may cut horses from either footwall (FW) or hanging wall (HW) of major thrust surface. Note that horses in this figure would consist entirely of inverted rocks.

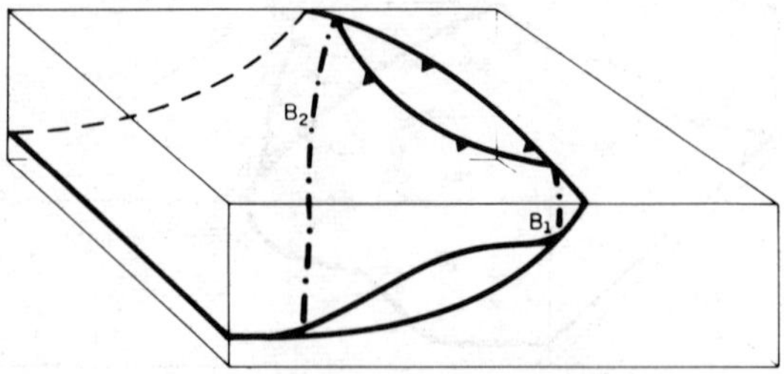

FIG. 10—Two fault surfaces meeting along two branch lines (B_1, B_2), whose map pattern resembles diverging splay and whose cross section resembles horse.

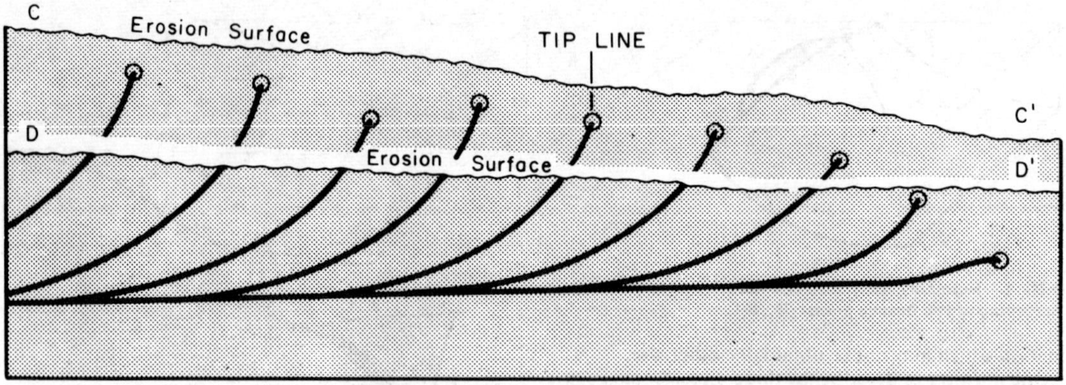

FIG. 11—Cross section of imbricate fan at two different levels of erosion. Each thrust sheet is an upward-opening crescentic slice, and all curve asymptotically downward to a common basal sole thrust. If most faults cut synorogenic erosion surface DD′ we have an emergent imbricate fan. Alternatively, it is possible that tip lines do not reach synorogenic erosion surface CC′, producing blind imbricate fan. Note that subsequent erosion (CC′ down to DD′) may obliterate any means of distinguishing two kinds of imbricate fans.

series of parallel and evenly spaced overlapping faults (see Suess, 1904, p. 112). As the concept evolved, several other terms came into use. The literal French equivalent is "structure ecailleuse" (Gosselet, 1885). Both the French and German expressions mean a scaly or flaky structure. However, de Margerie and Heim (1888) used "structure imbriquee" as the French equivalent of Suess's term. Hobbs in 1893 introduced "weatherboard structure," but later (1894) accepted a suggestion by Bernard Hobson to use imbricate structure as the English equivalent. In modern usage, schuppen zone is synonymous with imbricate zone but is less frequently employed.

Imbricate structures are an efficient means to shorten and thicken a sequence. Relative movement ". . . is trivial, as concerns adjacent members, but may in the aggregate lead to impressive telescoping of the affected zone" (Bailey, 1938, p. 607).

A sole thrust is the lower common thrust in an imbricate system (Dennis, 1967, p. 139). In an imbricate fan, a swarm of curved triangular thrust slices are asymptotically shaped downward to the sole thrust and spread out upward like an open fan (Fig. 11). Emergent imbricate fans where the faults reach the erosion surface are most common, but blind faults could produce blind imbricate fans. On approaching the synorogenic erosion surface the thrusts often increase their dip, and our impression is that they can meet the ground surface at about 60°. Imbricate fans dominate this near surface level.

An imbricate fan in which a thrust with maximum slip is at the front is a leading imbricate fan. If the thrust with maximum slip is at the back, it is a trailing imbricate fan (Fig. 12). These two terms were modified from Dahlstrom (1970, p. 352).

It is also possible to construct a thrust system with an imbricate family of subsidiary contraction faults asymptotically curving downward to a sole or floor thrust and upward to a roof thrust. This thrust system is known as a duplex (Fig. 12).

DUPLEXES

The term "duplex" first appeared in a paper by Dahlstrom (1970, p. 352), but he does not take credit for originating the duplex concept or coining the term, maintaining that these ideas were already present in the

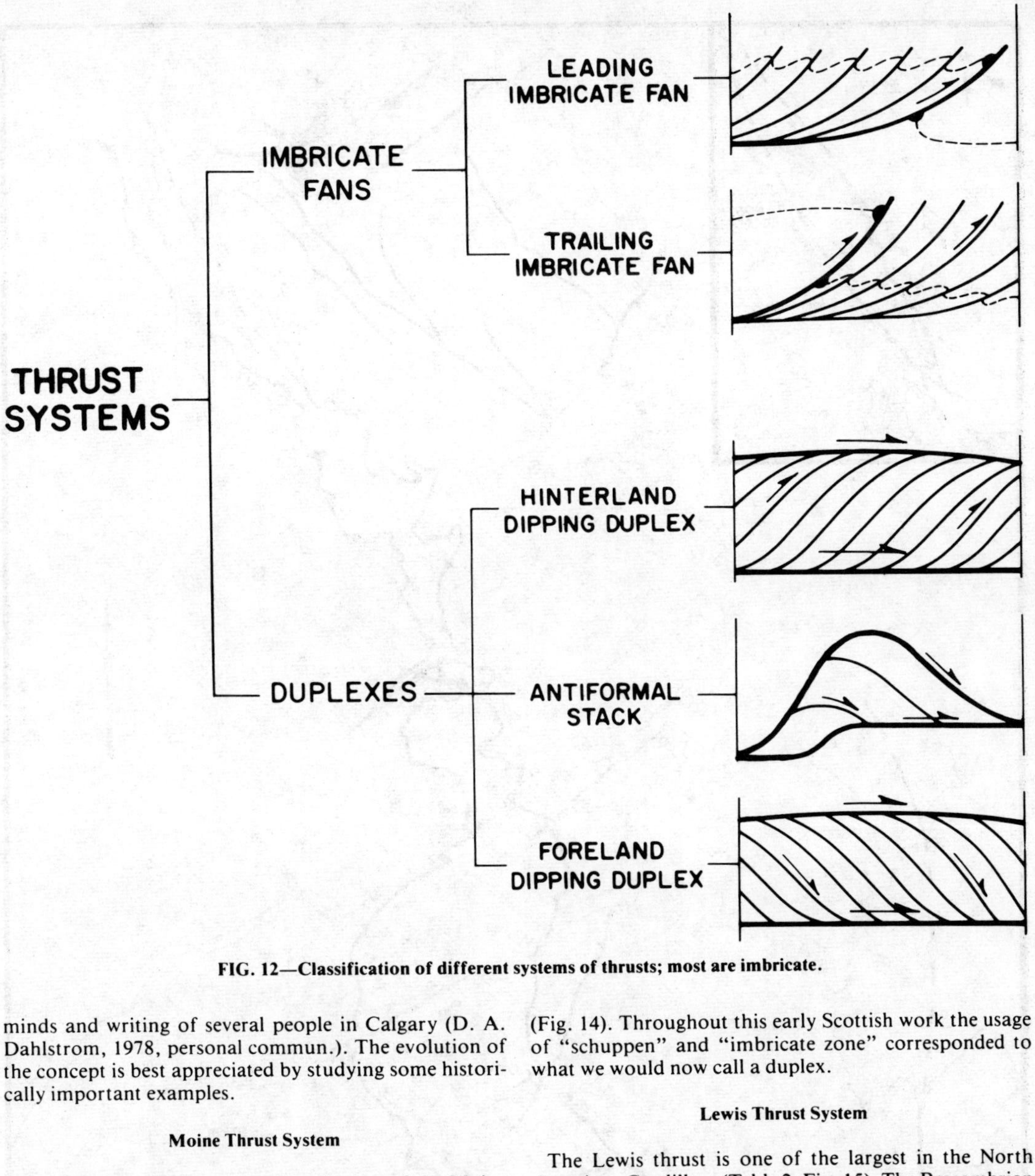

FIG. 12—Classification of different systems of thrusts; most are imbricate.

minds and writing of several people in Calgary (D. A. Dahlstrom, 1978, personal commun.). The evolution of the concept is best appreciated by studying some historically important examples.

Moine Thrust System

A 50-year controversy over the existence of the Moine thrust resulted in geologic mapping at about 1:10,000 scale, which was completed between 1883 and 1896 (Fig. 13). Early in this project, B. N. Peach interpreted the exposures on the east shore of Loch Eriboll as a duplex (McIntyre, 1954, p. 206). Shortly afterward, in an exceptionally exposed area with 0.6 mi (1 km) of local relief, a spectacular duplex zone (Foinaven duplex; Table 1) repeating a distinctive Lower Cambrian quartzite was discovered by Cadell (Peach et al, 1907, p. 491, Fig. 25) (Fig. 14). Throughout this early Scottish work the usage of "schuppen" and "imbricate zone" corresponded to what we would now call a duplex.

Lewis Thrust System

The Lewis thrust is one of the largest in the North American Cordillera (Table 2; Fig. 15). The Precambrian Belt sequence within the thrust sheet, spectacularly exposed in Glacier and Waterton Parks, shows remarkably little internal deformation, but going down through the sheet one may suddenly enter a duplex, a few hundred meters thick, whose floor is the Lewis thrust surface. This is particularly clear in the Willis (1902) illustration of the Chief Mountain klippe (Fig. 16). Douglas (1952) mapped a duplex near Mount Crandell (Fig. 17), which was cited by Dahlstrom (1970, p. 354) as his type exam-

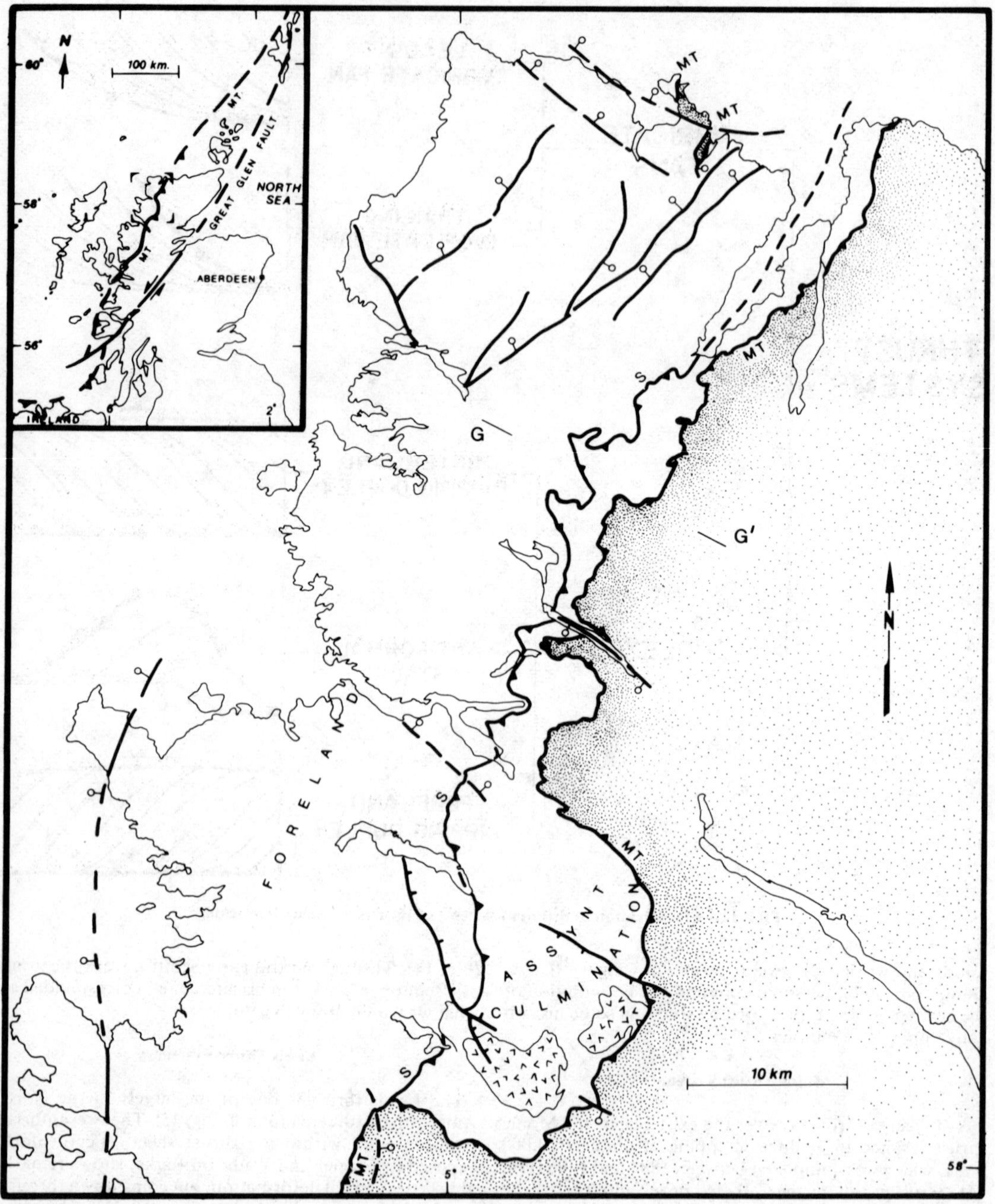

FIG. 13—Map of northern part of Moine thrust zone. MT is Moine thrust, S is Sole thrust, Moine schists stippled. Insert shows 500 mi (800 km) length of thrust zone.

ple. The Cate Creek and Haig Brook windows (Fig. 15) through the Lewis thrust sheet are particularly spectacular examples of duplexes (Dahlstrom, 1970, Fig. 24; Fermor and Price, 1976). It appears that substantial portions of the Lewis thrust are the floor of a duplex.

Klippen and reentrants demonstrate that the Lewis thrust once extended northeast of its present outcrop, possibly as much as 6 to 12 mi (10 to 20 km) (Fig. 15). Consequently, many thrusts that crop out in Cretaceous rocks, such as near the headwaters of the Carbondale River and Marias Pass, are part of a duplex whose roof was the Lewis thrust. Possibly, however, the Paleozoic rocks of the Livingstone and Sawtooth Ranges were brought up on emergent imbricate fans. Along strike, these Paleozoic rocks plunge beneath the Lewis thrust and might become part of the duplex containing the Waterton field (Gordy et al, 1977, Fig. 11b), but this is not clear, for other sections show the Waterton field as an imbricate fan (Gordy et al, 1977, Fig. 14).

A duplex is an imbricate family of horses—a "herd of horses"—but in the same way that a horse can change along strike into a splay (Fig. 10), it is possible for a duplex to change along strike into an imbricate fan.

The subject of along-strike variations in duplexes is discussed later, but it is important to emphasize here the lateral changes in the horses that make up a duplex. This change in shape of a horse, such as shown in Figure 8, is a result of oblique and lateral ramping giving the distinctive scoop shape of roof, subsidiary, or floor thrusts. The huge horses of Paleozoic carbonate rock underlying the Alberta foothills show just this sort of behavior, and are responsible for the doubly plunging culminations that affect the higher sheets. This is outstandingly illustrated by the longitudinal cross sections based on seismic and drill-hole data in Bally et al (1966, Plate 8). The Moine thrust belt also shows along-strike variation of a major duplex. In this case it is exposed at the surface for direct observation because of a regional dip of 10 to 15° which was imposed after thrusting (Elliott and Johnson, 1980, Fig. 6).

Dimensions and Internal Geometry

Duplexes have characteristic internal features (Fig. 18). Beds within a horse often trace out an elongate anticline-syncline fold pair, and bedding near the central inflection point roughly parallels the subsidiary faults. Above and below the duplex the bedding may be rela-

Table 1. Dimensions of Duplexes

Duplex	Contraction Ratio (L'/L_o)	Number of Horses (N)	Approx. Angle Subsidiary to Floor Thrust	Reference
Foinaven duplex in Moine thrust zone	0.29	34	40°	Fig. 14
Windows duplex, southern Appalachians	0.36	21	30-45°	Fig. 29
Lewis thrust (floor of duplex), North American Cordillera				
Chief Mtn.	—	2	23°	Fig. 16
Mt. Crandell	0.57	6	33°	Fig. 17
Cate Creek	0.58	2	31°	Dahlstrom (1970)
Haig Brook	0.6	12	27°	Fermor and Price (1976)
Central Appalachian Valley and Ridge	0.54	4	33°	Fig. 26
Idealized model				
Duplex constructed with kink folds	0.50	—	30°	Fig. 18

Table 2. Some Dominant Thrust Sheets

Dominant Thrust Sheet (Location)	Approximate Width Downdip (km)	Lithology
Moine (Northwest Scotland)	100	Precambrian and Cambrian siliciclastics refolded and metamorphosed to medium and high grades (Figs. 13, 14).
Semail Ophiolite (Oman)	100	Oceanic crust and upper mantle thrust over Arabian continental margin.
Austro-Alpine (European Alps)	150	Mesozoic continental margin unconformable on folded Paleozoic nonmarine sediments and quartzo-feldspathic crystalline complex of medium and high grade (Figs. 31, 34).
Lewis (North American Cordillera of U.S. and Canada)	100 to 200	Paleozoic shelf carbonates unconformable over late Precambrian shelf and slope sequence (Figs. 15, 17).
Jotun (Southwest Norway)	180	Crystalline Precambrian basement of two pyroxene granulites (Hossack et al, 1982).

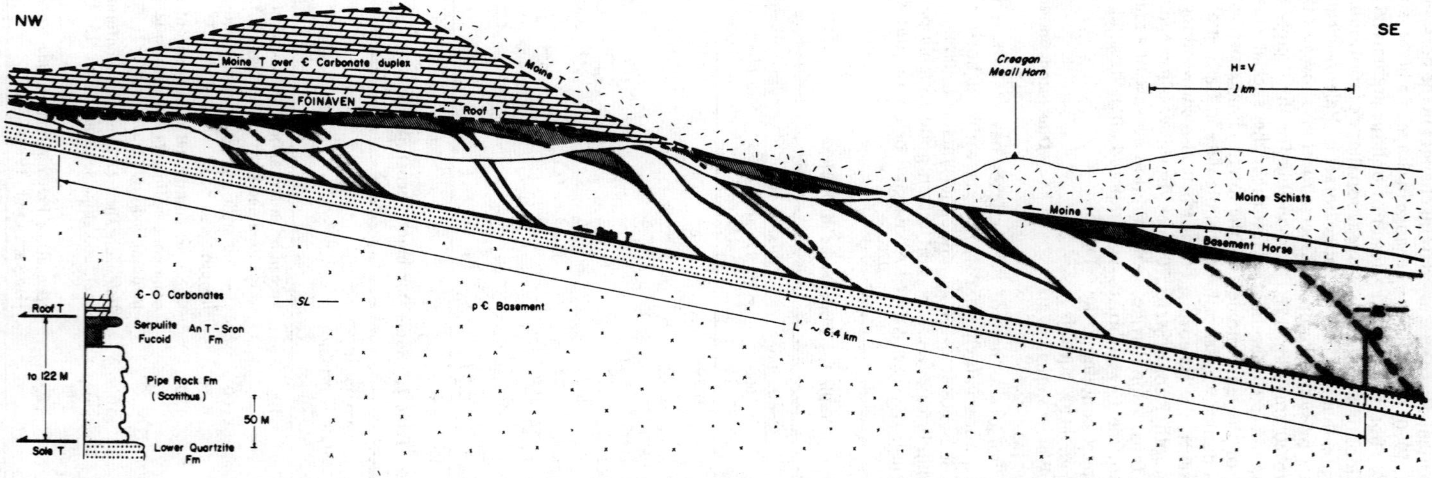

FIG. 14—Cross section of the Foinaven duplex from Moine thrust zone, line of section GG′ on Figure 13. Duplex has area PQ of 2.65 km² and is made up almost entirely of Cambrian Pipe Rock with a little An t-Sron. After Elliott and Johnson (1980, Fig. 4).

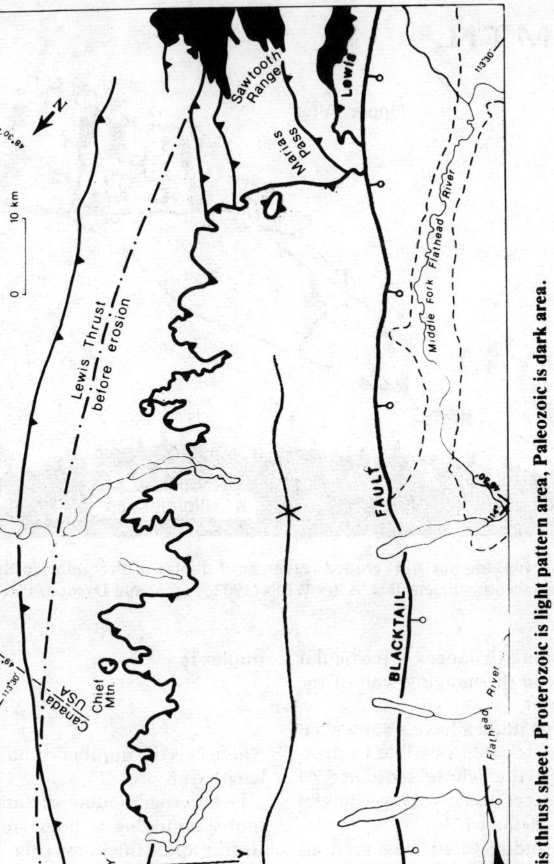

FIG. 15—Map of central portion of Lewis thrust sheet. Proterozoic is light pattern area, Paleozoic is dark area.

FIG. 16—Chief Mountain klippe, looking north, is eroded fragment of duplex in Precambrian Belt Supergroup whose floor is Lewis thrust. All lie on Upper Cretaceous siliciclastics. After Willis (1902, Fig. 5) and Dyson (*in* Nevin, 1949, Fig. 101).

tively undisturbed, and for long distances a particular stratigraphic unit may compose the hanging wall of the roof or the footwall of the floor.

The fold pairs within any one duplex have a somewhat similar shape and size, and it is often possible to draw enveloping surfaces through the whole sequence of horses. These enveloping surfaces are at very low angles to the floor thrust, seldom exceeding 15°.

While constructing current and restored cross sections through duplexes we repeatedly make a number of simple calculations (equations 1 to 7). Capitals indicate measurements over the complete thickness or length of a duplex, and small letters refer to one horse. The easiest measurements are the current duplex length (L′), structural thickness (H′), cross-section area (A), initial stratigraphic thickness of the formations making up the horses (t), and the current angle (β') between floor thrust and the central portion of the subsiding faults (Fig. 18).

We assume plane strain, with the cross-section area

$$A = H'L' = tL_o, \qquad (1)$$

where L_o is the initial length of the duplex. The overall shortening distance (S), accomplished by formation of the duplex, is

$$S = L_o - L', \qquad (2)$$

and from equation 1

$$S = A/t - L'. \qquad (3)$$

If the bed length (ℓ) within each horse is unaltered by deformation, then the total initial bed length within the duplex is

$$L_o = \Sigma\ell = N\ell_a, \qquad (4)$$

where N is the number of horses with an average bedding length of ℓ_a.

The perpendicular distance (h) between subsidiary faults bounding a horse may approximate the initial stratigraphic thickness (Fig. 18), h = t, so that the spacing between subsidiary faults, measured parallel with the floor thrust, is

$$p' = t/\sin\beta'. \qquad (5)$$

Now the total number of horses in the duplex is

$$N = L'/p', \qquad (6)$$

and the number of faults per unit length

$$N/L' = \ell/p'. \qquad (7)$$

A duplex beneath the Moine thrust provides an example of these calculations (Fig. 14). The duplex has a cross-section area (A) of 2.65 $10^6 m^2$ and a current length (L′) of 6.38 km. The average structural thickness (H′) is 415 m, although in the central and southeast part the duplex is thicker (H′ is 510 m). The duplex consists almost entirely of quartzite, although some shale is present toward the top, and the initial stratigraphic thickness (t) of the formations making up the duplex is 122 m. The initial length (L_o) is equivalent to 21.7 km from equation 1, and from equation 3 the shortening (S) is 15.3 km.

Only 20 faults, clearly observed on cliffs, were plotted

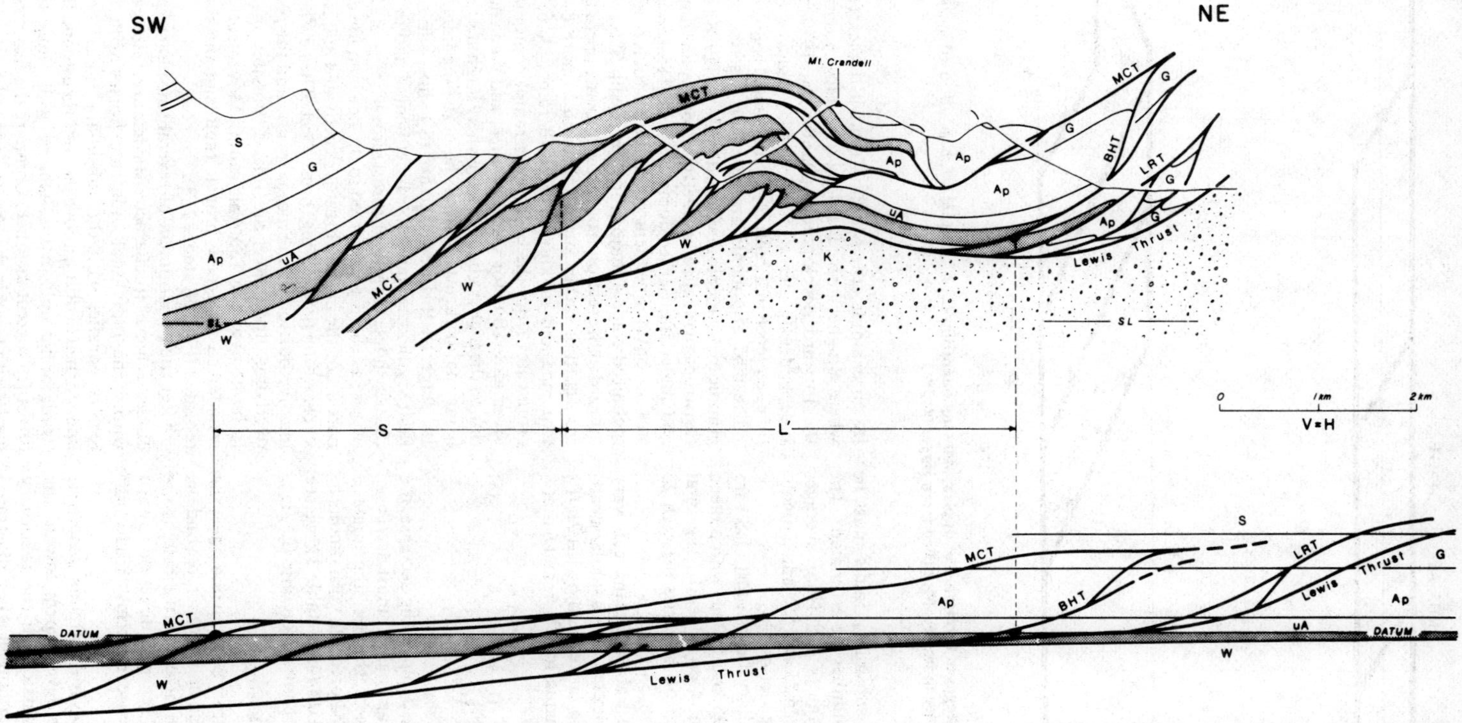

FIG. 17—Precambrian Belt Supergroup, comprising Waterton (W), lower Altyn (fine stipple), mid and upper Altyn (uA), Appekunny (Ap), Grinnell (G), and Siyeh (S), is thrust over Cretaceous siliciclastics (K, with pebble pattern) by Lewis thrust. Mount Crandell thrust (MCT) is roof and Lewis thrust is floor to duplex, and folded horse just northwest of Mount Crandell suggests that duplex developed toward foreland. Cross section is balanced (with current distance L′ between points recording a shortening of S), and is based on excellent control provided by over 2 km of local relief. Modified from Douglas (1952).

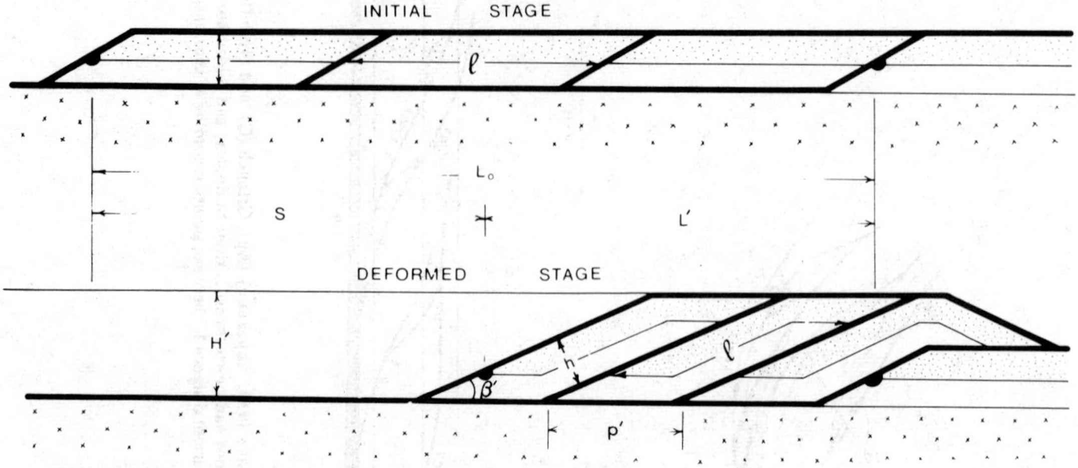

FIG. 18—Initial and deformed stages in formation of duplex, showing quantities L_o, L', S, t, H', ℓ, h, p', β' used in equations 1 through 8. Note elongate "S" folds in horses. Modified from Boyer (1978).

on the section. Many more subsidiary faults must be present, but how can this number be estimated? The angle β' between subsidiary faults and floor averages around 40°, but ranges between 25 and 55° and is smallest at the northwest end of the section.

The spacing between subsidiary faults must also vary (from equation 5) between 290 and 150 m, with a mean spacing (p') equal to 190 m. Consequently, the total number of faults (using equation 6) varies between 22 and 45, with the mean (N) being 34.

The shortening and structural thickening can vary from place to place. If the duplex involves a similar stratigraphic thickness and uniform mean bedding length (ℓ_n) in the horses, it follows from equations 4, 5, and 6 that

$$\left\{\frac{L'}{L_o}\right\} = \frac{p'}{\ell_n} = \left\{\frac{t}{\ell_n}\right\} \frac{1}{\sin \beta}. \quad (8)$$

Therefore, as the shortening and thickening decrease, the ratio (L'/L_o) increases, and the angle β' must decrease.

As the Foinaven duplex gets thinner, the angle β' becomes smaller and the local shortening must diminish. By comparing different duplexes (as in Table 1), we see that β' varies inversely with the contraction ratio (L'/L_o).

Sequence of Development

There are two sides to any thrust belt. On one side is the internal zone or hinterland, and on the other, beyond the external zone and the margin of the thrust belt, is the foreland. When we require a sense of direction toward the hinterland or foreland, we use the terms "hindward" or "forward."

In this section, we review several different geometric arguments for the successive development toward the foreland, or a forward progression, of the subsidiary faults in duplexes. The oldest argument is best illustrated by an example, the Mount Crandell duplex (Fig. 17), where a higher horse is folded over a lower one proving the forward development (Dahlstrom, 1970, p. 352). Independently, Perry (1971, p. 195) deduced the same time sequence in a central Appalachian duplex.

The next arguments are based on a series of simple graphic experiments developed by Boyer (1978). First, we construct an idealized model based on typical dimensions and angles of observed duplexes (Table 1) assuming plane strain, constant bed lengths, and kink folds (Fig. 19). In the initial stage a major thrust with slip S_o has climbed upward in the section from a lower to an upper glide horizon, cutting rather steeply through a more competent sequence and making a footwall ramp. A new crack propagates from the base of the ramp, continues in the lower glide zone for some distance, and then cuts upsection to the higher glide horizon where it rejoins the preexisting major thrust. During the next time interval (Fig. 19, stage 1), this new fracture (S_2) slips, the overlying fault segment remains fixed, yet behind and in front of the new horse the major thrust slips by $S_o + S_1$. In other words, as slip is transferred to the new and lower fault a portion of the major thrust is deactivated and rides passively within the growing thrust sheet. The new horse, the inactive portion of the major thrust, and the rest of the overlying thrust sheet are folded, possibly in kinklike fashion, over the footwall ramp. Movement is again transferred to a new branch fault and the process is repeated (Fig. 19, stages 2, 3).

After the development of several subsidiary faults the duplex achieves its characteristic features: elongate folds within the individual imbricate horses with bedding locally parallel with the subsidiary faults, relatively undisturbed bedding above the roof thrust, and the same stratigraphic unit in the hanging wall along the roof thrust for a great distance. Compare the features in Figure 19 with those in the Mount Crandell duplex (Fig. 17).

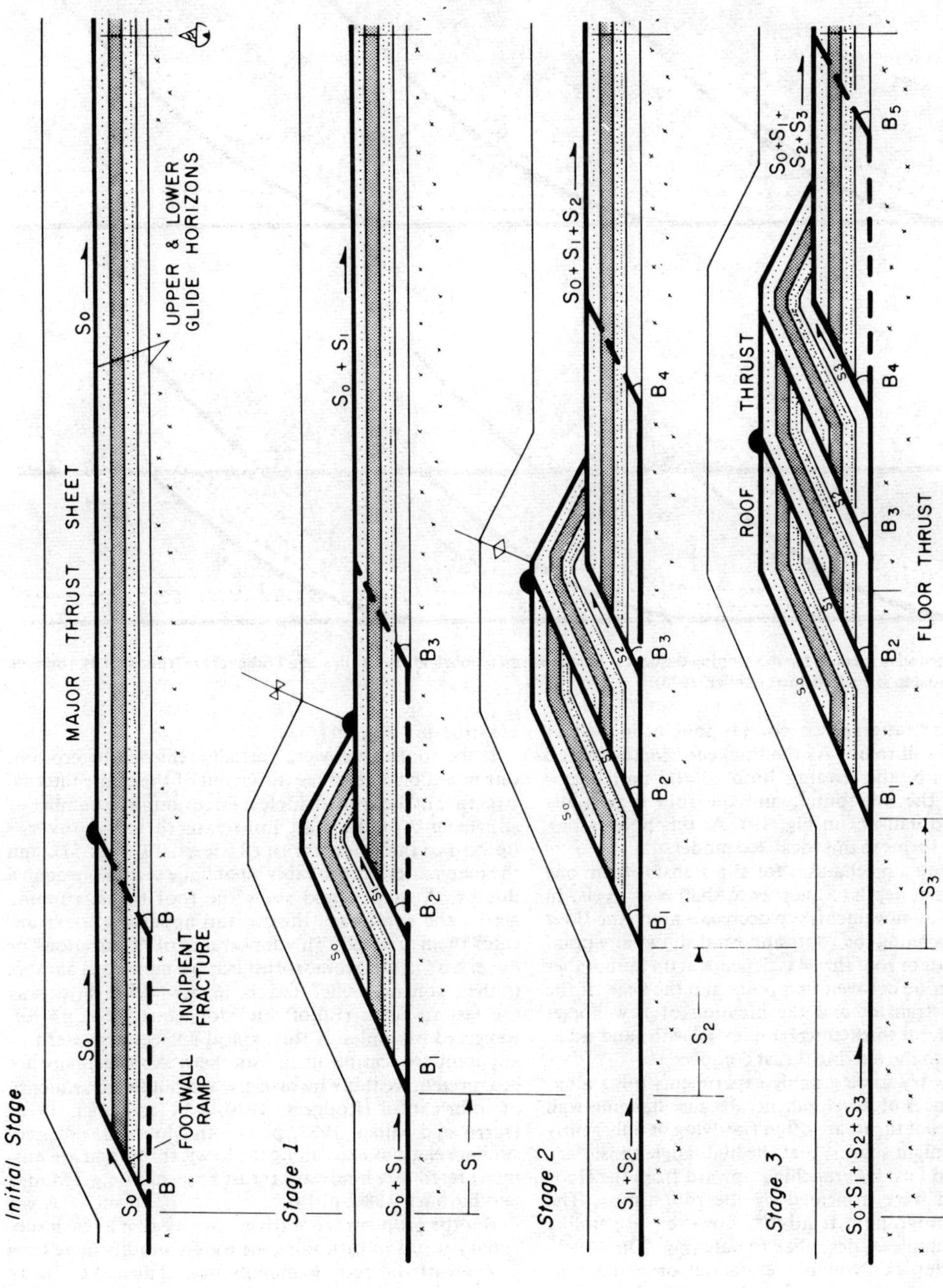

FIG. 19—Progressive collapse of footwall ramp builds up duplex. This is measured graphical experiment, assuming plane strain and kink folding, with angles and ratios of dimensions typical of natural examples (Table 1). Roof thrust sheet undergoes complex sequence of folding and unfolding, seen by following black half dot. Modified from Boyer (1978).

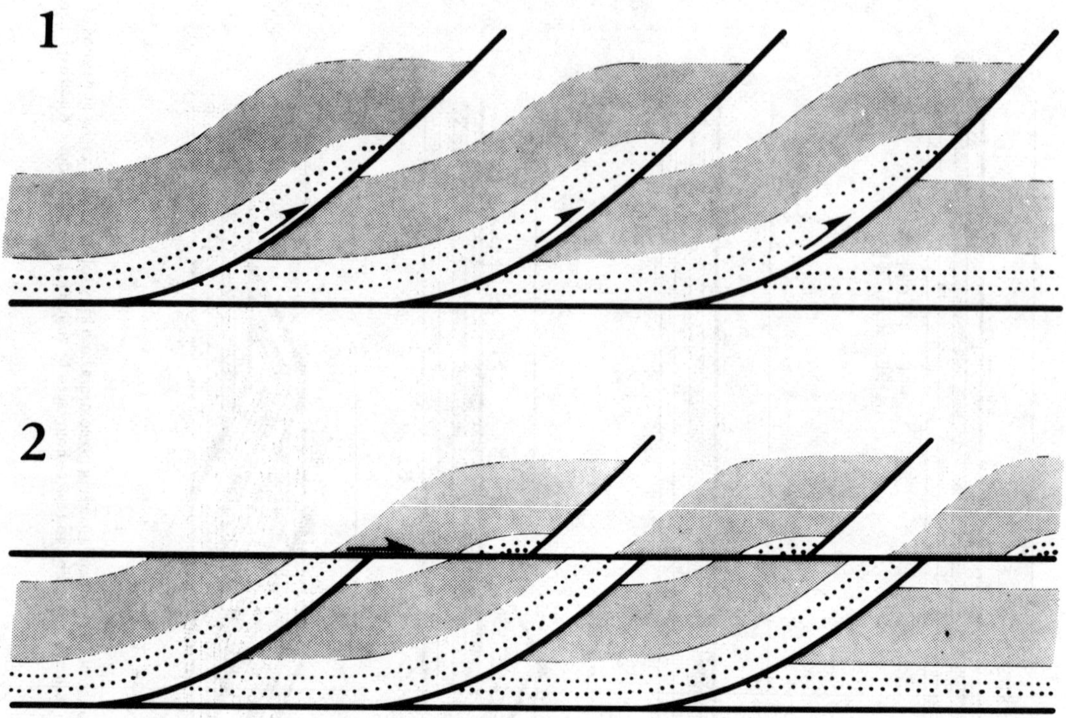

FIG. 20—Alternative method for developing duplexes. Earlier formed imbricate faults in stage 1 (above) are truncated by younger and higher thrust in stage 2 (below) (Boyer, 1978).

The faulted stratigraphic section is doubled in the area of each footwall ramp. As the thickened section enters the next ramp, the leading limb of the anticline is returned to the horizontal, and the fold is partially opened (solid half-dot in Fig. 19). At this point a new branch fault forms in this idealized model.

Duplexes are a mechanism for slip transfer from one glide horizon at depth to another at shallower levels. In the direction of movement, slip decreases along the floor and increases along the roof, and total slip at any point along the floor or roof thrust is dependent on the number of horses that lie between that point and the head of the duplex. Slip transfer and the creation of new horses causes structural thickening, duplex growth, and addition of mass to the moving thrust complex.

Let us now try some graphic experiments using alternative sequences of development. Because hanging-wall rocks of the roof thrust are often flat-lying or only gently folded, one might suggest that the high-angle subsidiary faults formed first by branching upward from the floor thrusts, then were truncated by the roof thrust. The resulting stratigraphic relations, however, are unlike those of any duplexes described to date (Fig. 20).

Could a duplex form by sequential branches that become younger toward the hinterland, a hindward progression? Here also the geometric features are not found in natural duplexes, although the imbricate fan is a possible structure (Fig. 21).

If the roof thrust were partially removed by erosion, one might be able to see the layout of the entire internal structure in the eroded duplex. For example, a number of southern Appalachian imbricate thrusts converge upward to the Rome thrust (Rodgers, 1970, p. 57), and the map pattern is probably an oblique section through a duplex. A map would show the roof thrust running across the subsidiary thrusts and appearing to "truncate" them (Fig. 22). This appearance of "truncation" or "overlap" in the Moine thrust belt led numerous authors to the erroneous belief that the higher Moine thrust was the last to form (Elliott and Johnson, 1980, p. 90). Repeated examples of this typical geometric pattern of apparent truncation in the southern Appalachians are interpreted, we think incorrectly, as a hindward sequence of imbrication (Rodgers, 1970, p. 178; Milici, 1975; Harris and Milici, 1977, p. 14). In Montana, precisely similar relations exist along the Lewis thrust and are misinterpreted as a hindward thrust sequence (e.g., Mudge and Earhart, 1980, p. 14).

But the problem is not trivial. We see that as each subsidiary thrust in turn joins the roof a steadily increasing portion of the roof is immobilized. Thus, there is an apparent paradox: a major roof thrust moved at different times along its map trace, yet everywhere it may bear the same name.

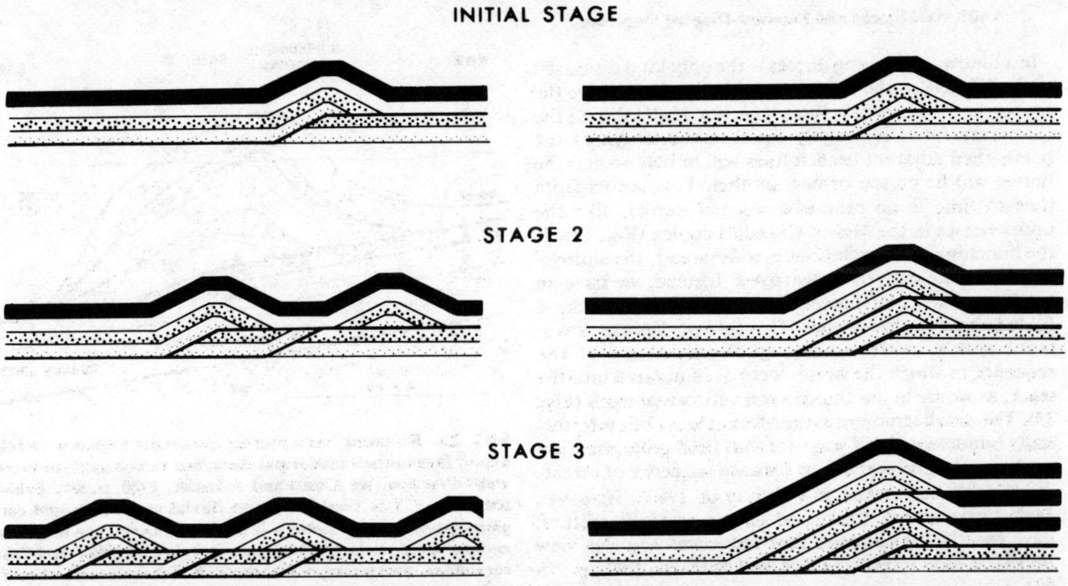

FIG. 21—Two alternative methods of hindward imbrication shown by measured graphical experiments (Boyer, 1978).

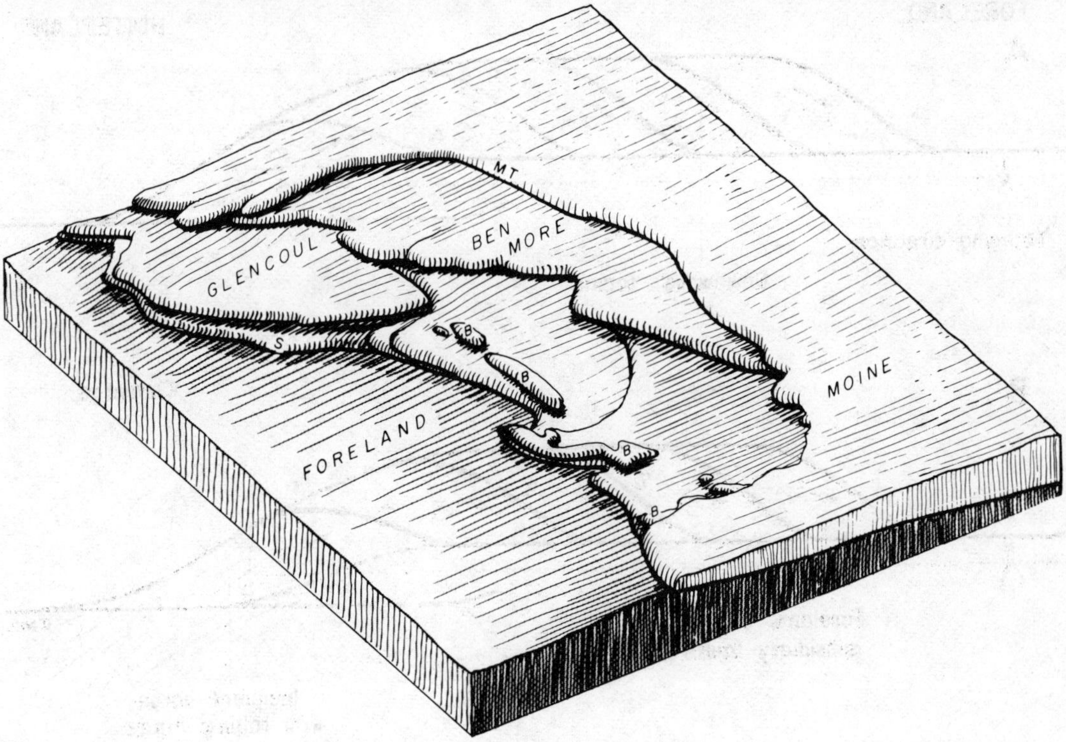

FIG. 22—Block diagram of Assynt culmination in Moine thrust zone (Fig. 13). This is oblique section through duplex. Diagram follows geologic interpretation of Elliott and Johnson (1980) and was inspired by plaster model on display at University of Edinburgh, and drawing by H. Cloos (1936, Fig. 290).

Antiformal Stacks and Forward-Dipping Duplexes

In a hindward-dipping duplex—the only kind discussed so far (Fig. 19)—the effect of shortening is to move the branch lines somewhat closer together. If the slip on the subsidiary faults is roughly equal to the length of the horse, then adjacent branch lines will bunch up and the horses will lie on top of one another. This occurs from time to time in an otherwise normal duplex, like the upper horses in the Mount Crandall duplex (Fig. 17). If the bunching of branch lines is widespread, the shingle-like imbricate pattern is destroyed. Instead, we have an antiformal stack of horses where each higher horse is folded about the lower ones. This folding dies out downward and provides unambiguous evidence of the sequence in which the horses were accumulated into the stack, as we see in the Dundonnell antiformal stack (Fig. 23). This small structure in the Moine thrust belt is historically important, for it was here that field geologists first worked out the piggyback or forward sequence of thrusting (Cadell and Horne, in Peach et al, 1907). However, fault "truncation" (discussed in the previous section) gave exactly the opposite time sequence, and this view prevailed until recently (Elliot and Johnson, 1980, p. 90-93).

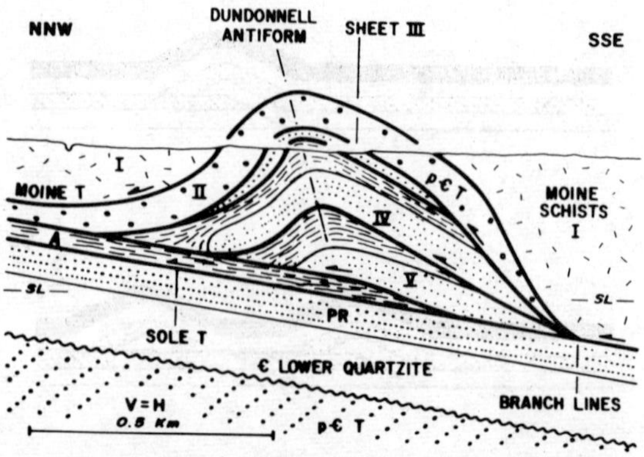

FIG. 23—Balanced cross section constructed normal to fold axis of Dundonnell antiformal stack, but at high angle to movement direction (see Elliott and Johnson, 1980, p. 90). Pebble pattern (p€T) is Torridon group fluvial sandstones and conglomerates. Stippled pattern (PR) is Cambrian Pipe Rock formation, a quartzite. Dashed pattern A, is Cambrian An t-Sron formation, with an important shale. The sheets are numbered I to V in their order of formation.

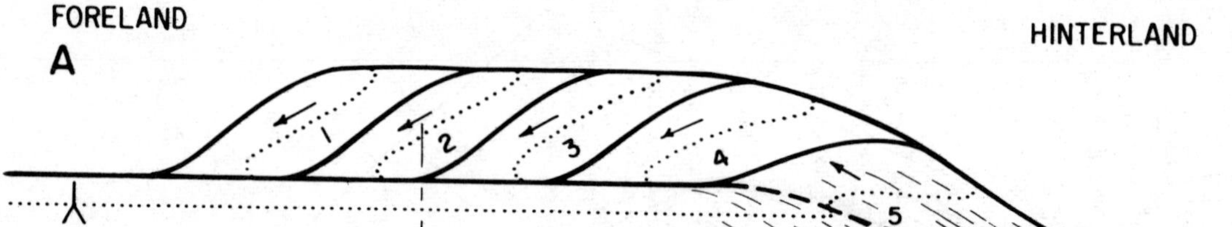

FIG. 24—Formation of forward-dipping duplexes. Horses numbered in order of formation; youngest horses (5) are not yet completely developed. Formations within horses may be predominantly right-way up (A) or upside-down (B), structures are often downward facing, but always forward facing.

In the Dundonnell structure, several of the higher and originally more internal branch lines are now over and forward of the underlying ones. This process can develop further when each branch line passes over and beyond the underlying one. We now recover once more the imbricate pattern (Fig. 24). The dip of the subsidiary faults suggests the term, forward-dipping duplexes, and the roof now contains the branch lines that started on the floor thrust. Forward-dipping duplexes were first proposed by N. Woodward (personal commun., 1981) for the southern part of the Mountain City Window.

Consider a horse just in the process of being plucked from a ramp and with cleavage developing in the surrounding rocks (Fig. 24). Stratigraphy, cross-bedding, or other primary features will usually provide the way-up or younging direction, and the cleavage/younging orientation within each horse fixes the structural facing of the horse from Shackleton's rule (Shackleton, 1958). Forward-dipping duplexes will usually consist of downward-facing horses (Fig. 24), even when the beds are upside down and originated at a footwall syncline with a rolling hinge (Fig. 24). In a similar fashion, hindward-dipping duplexes are usually composed of upward-facing horses, but both hindward and forward-dipping duplexes always have a forward structural facing. We have assumed here a particularly simple relation

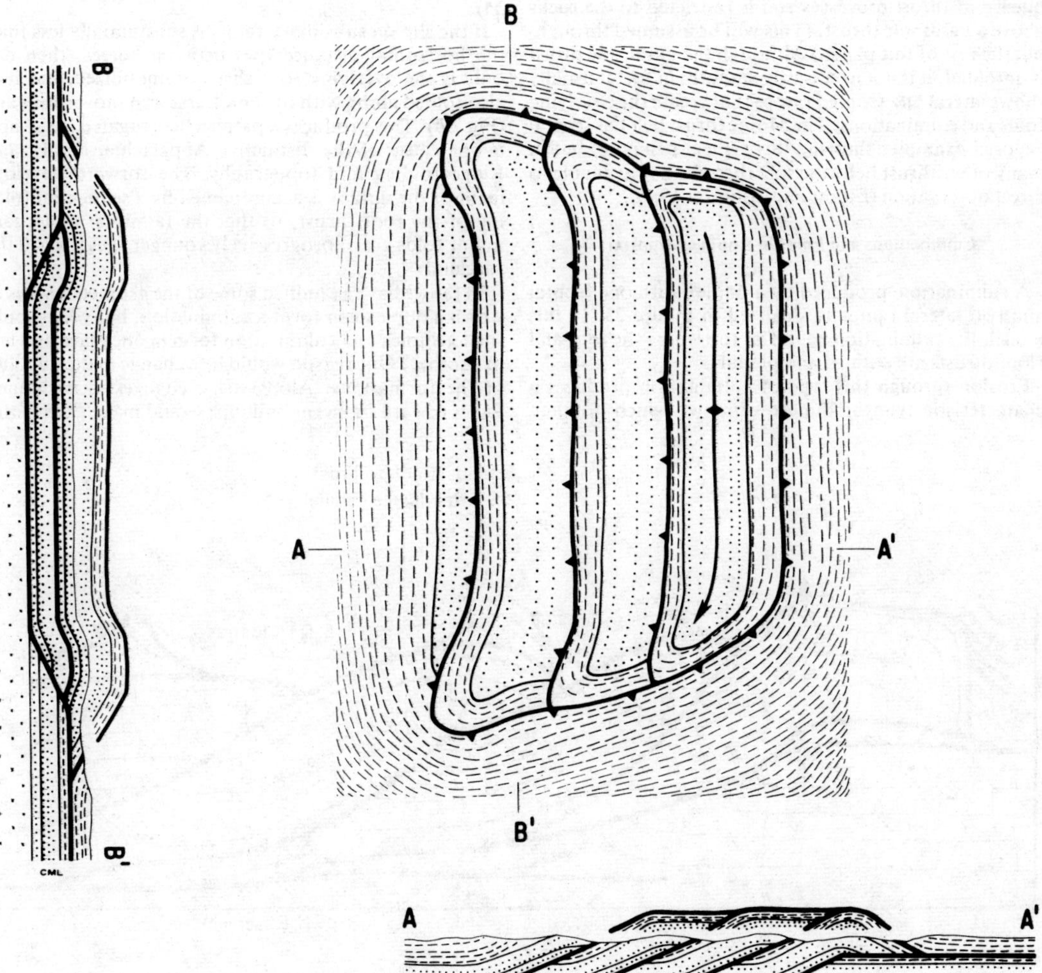

FIG. 25—One type of window into duplex. Anticline occurs at foreland side of window and overlies ramp in floor thrust. Other anticlines occur within each horse. Lateral termination of window may be lateral ramp in sole or floor thrust, with roof and floor thrusts rejoining along strike as well as across strike (Boyer, 1978).

between formation of cleavage and the duplex, and of course much more complex situations can exist.

Culminations and Windows

Windows may form by simple differential erosions into undisturbed planar faults, but usually they form by erosion through culminations. Frequently, a culmination in a thrust surface is produced after the thrust was emplaced, and this folding could, in principle, run from top to bottom through the entire cross section including basement—an interpretation widely favored in older treatments. However, balanced cross sections and seismic exploration reveal that the folding itself is a consequence of thrust processes and is restricted to the rocks above a basal sole thrust. (This will be assumed throughout the rest of this paper.) Although the basal sole thrust is unfolded, it is not necessarily smooth; in fact it usually shows lateral and fronted ramps that create the overlying folds and culminations. The Moine thrust belt has several exposed examples showing this typical behavior, but in many other thrust belts the sole thrust is not available to direct observation (Elliott and Johnson 1980).

Culminations and Folds in Roofs of Duplexes

A culmination produced by a duplex with one frontal and two lateral ramps is idealized in Figure 25. In this model, the culmination has a flat top where the roof and floor thrusts are parallel and planar.

Erosion through this type of culmination produces a characteristic type of duplex window, which Tollman (1968, p. 46) called a "window of dislodged slices" (schurflingsfenster). The distinctive features on a map are: (1) a pattern of several subsidiary faults that join the roof thrust at the window margin, (2) doubly plunging anticlines within each horse, and (3) anticline crests and subsidiary faults are roughly parallel (Boyer, 1976, 1978).

There are several other ways in which a duplex can cause a culmination, such as dissimilar initial fault trajectories, changes in stratigraphic thickness, or variable contraction in the duplex. Excellent examples occur in places in the Moine thrust belt where contoured duplex thicknesses show a series of domes, basins, and saddles, rather like peas in a pod (Elliott and Johnson, 1980, Fig. 33).

If the slip on subsidiary faults is substantially less than the horizontal distance (p′) between horses, then not only is the overlying roof thrust immobilized, but it is also folded along with the new horse as it moves forward (Fig. 26). This produces a pattern "corrugated iron" and is one cause of the distinctive Appalachian Valley and Ridge folding and topography. The forward development of the duplex is accompanied by a sequential folding of the roof thrust, so that the family of congruent major folds are of progressively younger ages toward the foreland.

So far we have examined some of the geometric ways in which a duplex can form a culmination, but why should such a duplex and culmination form in one place and not in another? One reason would be a change in the ductility of the formations. Along-strike changes in carbonate facies toward bioherm buildups would make the stratig-

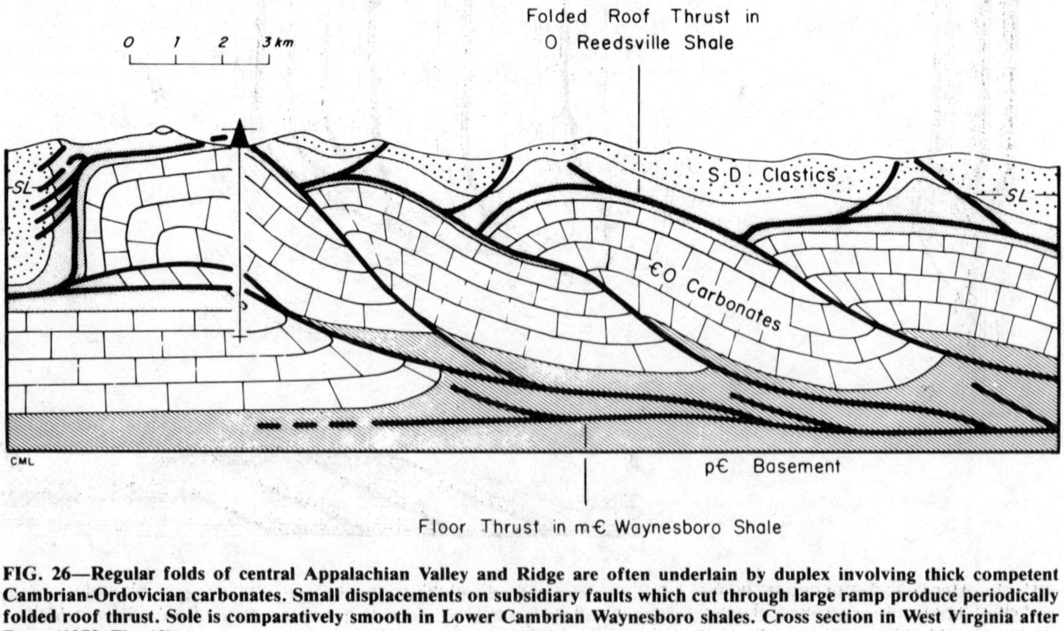

FIG. 26—Regular folds of central Appalachian Valley and Ridge are often underlain by duplex involving thick competent Cambrian-Ordovician carbonates. Small displacements on subsidiary faults which cut through large ramp produce periodically folded roof thrust. Sole is comparatively smooth in Lower Cambrian Waynesboro shales. Cross section in West Virginia after Perry (1978, Fig. 10).

raphy much less ductile and favor development of a duplex and culmination. In sandstones, small changes from carbonate to quartz cement could have the same effect. In the Moine thrust belt, local development of igneous intrusions slightly metamorphosed a shale and carbonate sequence, which during subsequent thrusting resulted in the Assynt culmination (Fig. 22).

Eyelid Windows

Now that we have looked at the possible causes of a culmination and the fault patterns within the window, let us examine the map patterns of the thrusts that frame the window.

An eyelid window is caused by the folding and erosion of an imbricate fan in such a manner that the window frame consists of several different thrust sheets (Fig. 27). The term "eyelid window" has a somewhat convoluted history. Sander (1921, p. 193, 212) originally suggested the name "scissors-window" (schernfenster), and later compared the pattern of framing thrust sheets to the overlapping blades in an iris diaphragm. Although Sander emphasized that scissors-window was a purely descriptive term, Tollman (1968, p. 47) pointed out that at one time these windows were thought to form by the opposed motion of the framing thrust sheets, like the blades in a pair of scissors. Oriel (1950, p. 46) suggested the more appropriate English name "eyelid window," and although the term was then in current use, he could find no published examples of its use.

In a discussion of the Goat Ridge eyelid window in Nevada, Gilluly (1960; Gilluly and Gates, 1965) argued that thrusts start folding as they form and are then abandoned for newer, higher, and as yet unfolded thrust surfaces (Fig. 21, right). He claimed that eyelid windows are diagnostic features of this hindward thrust progression.

Alternatively, we could have an imbricate fan that is then folded into a culmination—itself possibly owing to an underlying ramp or duplex. Erosion would produce an eyelid window, but one in which the thrusts all developed in forward progression. It seems that identical map patterns of eyelid windows could arise with either sequence of thrusting, and clearly each field example must be care fully analyzed.

The Lower Engadine eyelid window was the first one described (Sander, 1921). The Hot Springs eyelid window in the southern Appalachians is a more symmetrical example (Oriel, 1950).

Southern Appalachian Blue Ridge

Some 155 mi (250 km) southwest of our study area in the southern Appalachians, an important seismic line shows that the Blue Ridge and much of the Piedmont is a thin but southeast-thickening slab, 3 to 9 mi (5 to 15 km) thick, thrust over a Cambro-Ordovician sequence (Cook et al, 1979, 1980). We believe this basic geometry holds true over our region; and indeed some of this basic picture was anticipated from surface geologic evidence (e.g., Harris and Milici, 1977; Roeder et al, 1978). Regarding the Blue Ridge thrust complex itself, Cook et al (1979, p. 565) remarked that ". . . few seismic features are seen." Fortunately, the geologic map of the Blue Ridge provides exceptionally useful information for building a vertical cross section (Fig. 28). A key is two large windows, each on the order of 10^3 km^2—among the largest windows so far discovered in North America.

A balanced cross section (Fig. 29), in which each fault slice in the deformed state is restorable into the original stratigraphic framework, gives a picture of the fault trajectories just before movement occurred. Each formation within a fault slice has equal area in both the restored and deformed states, but in the lower grade portion of the cross section, to the northwest, where finite strains and fabric are less intense, formation thicknesses are conserved, and formation lengths remain unchanged.

What is the sequence in which the thrusting occurred? J. Rodgers feels that within most of the Valley and Ridge province the general progression is forward, but around the Grandfather Mountain and Mountain City windows he accepts the suggestion of Gilluly (1960) that the thrusts developed in hindward sequence (Rodgers, 1970, p. 170, 178). Hatcher (1978, p. 294-295) proposed a compromise in which thrusting progresses forward in two waves, so that thrusts in the first wave are cut off at the back by younger ones. We shall try and show that there are no difficulties in building up the Blue Ridge complex with a forward thrust progression.

To explain our cross section (Fig. 29), we will briefly list some conclusions in point form; these points have the same numbers on the cross section. Although we provide only partial references, it is important to note the superb work by the U. S. Geological Survey in this area (King and Ferguson, 1960; Bryant and Reed, 1970; Rankin 1970).

1. The Inner Piedmont thrust complex, composed of sillimanite-grade immature siliciclastics intruded by Paleozoic plutons, is the most internal tectonic unit and continues well to the southeast of this line of section. About 47 mi (75 km) northeast is the Smith River thrust sheet, which clearly overlies the Fries and is equivalent to part of the Inner Piedmont (Conley, 1978). This thrust complex may once have covered the Fries sheet on this line of section, but is now found only south of the Brevard zone because of deep erosion and Triassic faulting. It is the highest and largest single tectonic unit, so in a sense we can describe it as a dominant thrust sheet—a topic which we discuss later.

2. The Fries thrust sheet carries rocks that were folded several times, affected by a pervasive kyanite grade metamorphism of Ordovician age, and intruded by Silurian-Devonian granites. The Fries thrust surface juxtaposes this entire amphibolite-grade complex onto a substantially lower grade footwall (Rankin et al, 1973). We cannot yet restore the Fries or Inner Piedmont thrust complexes.

Our thrust sequence would start, then, with the Fries moving over our area of interest carrying the Inner Piedmont thrust complex on top of it.

3. The Linville Falls thrust moved next, accompanied by formation of a penetrative cleavage. In basement rocks this fabric is cataclastic and occasionally mylonite

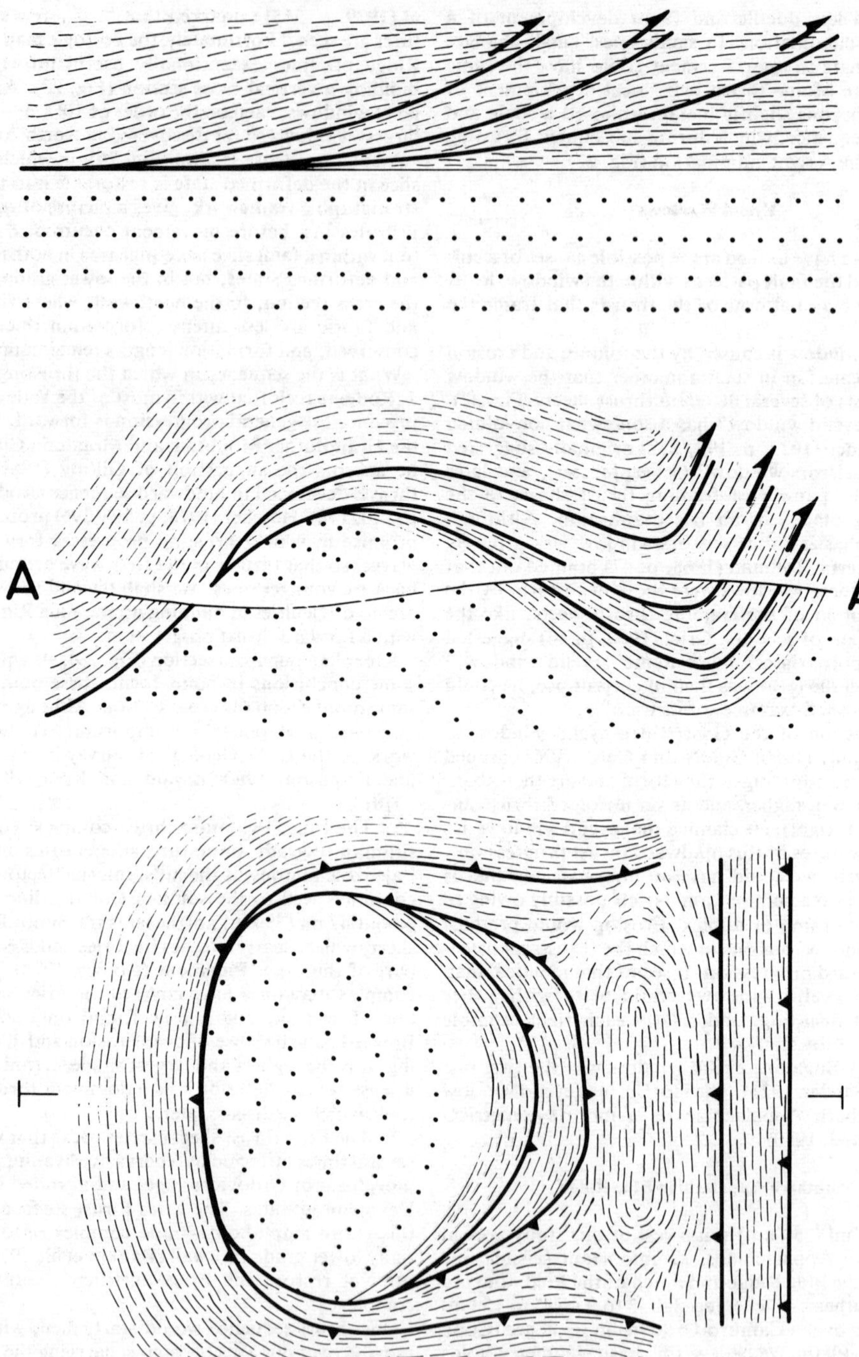

FIG. 27—Characteristic map pattern of an eyelid window (below and cross section AA′ (center). First we create an imbricate fan (center cross section), followed by sufficient but not excessive erosion (Boyer, 1978).

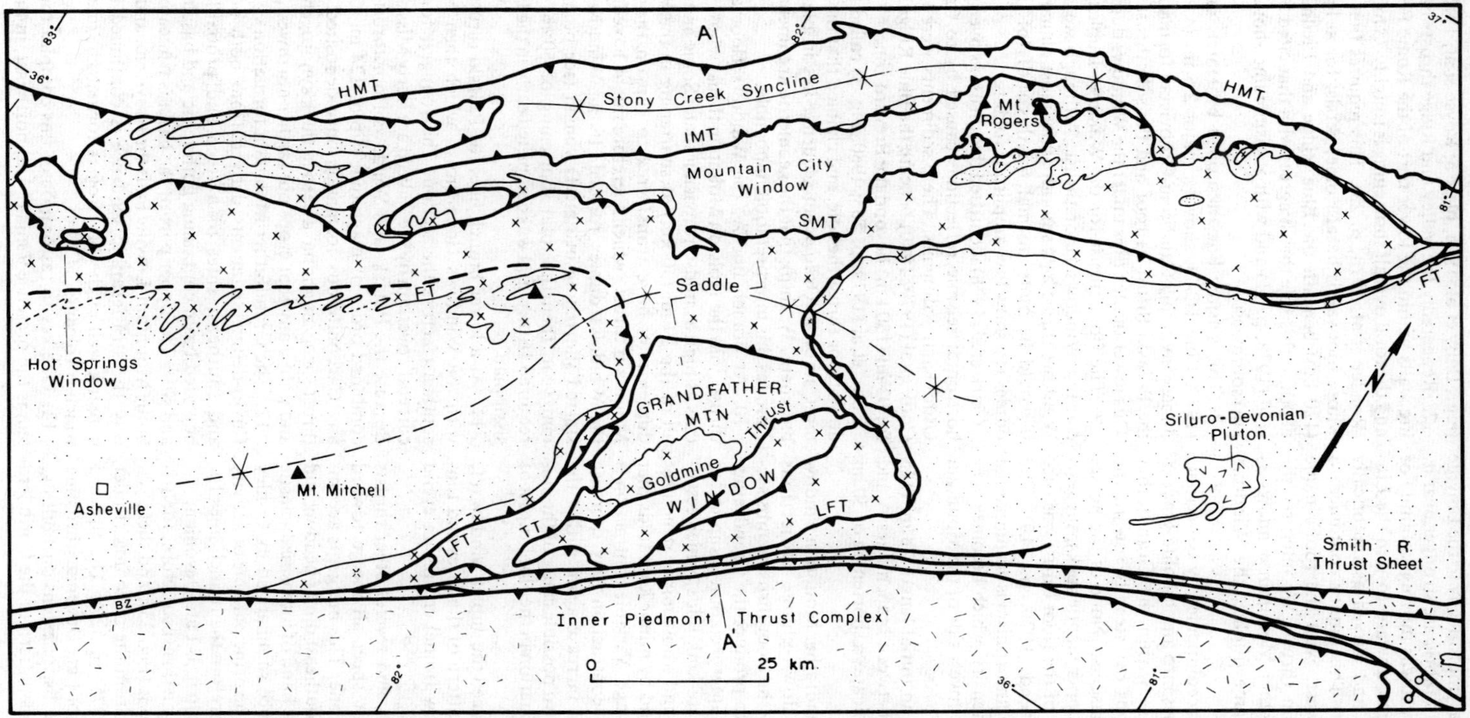

FIG. 28—Geologic map of Blue Ridge area in southern Appalachians of Tennessee, Virginia, and North Carolina. Holston Mountain thrust (HMT), Iron Mountain thrust (IMT), Stone Mountain thrust (SMT), and Linville Falls thrust (LFT) are all part of the imbricate system whose sole is intermittently exposed for over 47 mi (75 km) across strike. Precambrian basement is dashes or crosses, late Precambrian siliciclastics are stippled, Chilhowee group siliciclastics and Cambrian-Ordovician shales and carbonates are without pattern. Only major thrusts are shown, and no cross faults. Brevard fault zone is indicated by BZ. Based on Rankin et al (1973).

develops. The fabric curves asymptotically to the Linville Falls thrust and the strain and fabric are stronger near it.

Once the Linville Falls thrust was moving, a family of imbricate subsidiary faults cut upward to join the Linville Falls as a roof thrust and curved asymptotically downward to a floor thrust. This duplex continued to develop toward the northwest, with the roof's name changing to Stone Mountain thrust. We interpret the Mountain City and Grandfather Mountain windows as two culminations of the same duplex, the Windows duplex.

4. The main cleavage was accompanied by syntectonic metamorphism, which reaches into amphibolite-grade in the most southeastern horse within the Grandfather Mountain window, but drops off steadily toward the northeast and becomes greenschist. None of the horses brought up lower crustal or upper mantle rocks, suggesting that the initial trajectory of the floor thrust was subhorizontal (Bryant and Reed, 1970, p. 219). The northwestward decline in metamorphism is accompanied by a decrease of cleavage intensity, which becomes very weak and sporadic at the northwest side of the Mountain City window.

The Windows duplex had an initial length 78 mi (125 km), but its current length, between branch lines A and X, is 28 mi (45 km) which indicates a shortening of 50 mi (80 km).

5. The structure of the Grandfather Mountain window is based upon composite down-plunge projections (Boyer, 1976, 1978). Possibly the floor thrust climbs section along strike and updip to produce the Grandfather Mountain culmination. This interpretation is supported by the map pattern of large late-kink folds, which essentially define the culmination and whose minor structures are a crenulation cleavage, and by minor kink folds that deform the earlier main cleavage. Variable contraction is also a factor; for example, contraction may decrease beneath the saddle and then increase again within the Mountain City window. A third possible cause of the culmination is a still lower culmination in the floor thrust of the Mountain City window.

Erosion subsequently produced the duplex window (cf. Figs. 28 and 25). Also, the pattern of the Fries and Linville Falls thrusts,(the window frame), creates an eyelid window (cf. Figs. 28 and 27).

6. The Linville Falls, Fries, and several other thrusts beneath the Inner Piedmont sheet all merge toward branch lines in the Brevard zone. This merger is accompanied by an increase in the southeast dip of bedding and schistosity toward this narrow band of intense cataclastic deformation. Isotopic evidence summarized by Conley (1978, p. 1122) suggests that the principal time of activity of the Brevard zone, and therefore the branch lines for the major thrusts that lie within it, is Late Devonian.

The Brevard zone is not only a thrust structure, for Triassic faults can be traced into it (Fig. 28), and there is a scatter of Triassic mica ages from the mylonites (Butler and Dunn, 1968; Rankin, 1975, p. 329). The Brevard zone started as the site of major Late Devonian thrusting and was followed by Triassic listric extension faulting. There are good analogs to this pattern in the Cordillera (e.g., Bally et al, 1966, plate 5; Royse et al, 1975, p. 46).

7. Repeatedly the Valley and Ridge thrusts have hanging walls within the lower part of the Rome Formation, and the total overall shortening is roughly 75 mi (120 km) (Roeder et al, 1978, p. 23). This requires that a thrust, whose footwall is basal Rome, persist for at least 75 mi (120 km) beneath the Blue Ridge and Piedmont. The depth to this sole thrust agrees with that cited by Roeder et al (1978, p. 11) and with a seismic line shot 37 mi (60 km) southwest.

8. This seismic line, described by Harris and Milici (1977, Line TC-2), shows a major 2.5 mi (4 km) thick duplex involving Cambrian-Ordovician formations with the Pulaski thrust as roof and the sole thrust as floor. This lower duplex dips under the Windows duplex.

9. The floor thrust for the Windows Duplex we shall call the Windows floor thrust. Although we show the Windows thrust as a smooth and almost planar surface, it would be quite normal for this roof (of the lower duplex) to show low-amplitude folding.

Clark et al (1978) identified this thrust as overlying subhorizontal seismic reflections caused by to Cambrian-Ordovician sediments. These sediments were traced by Cook et al (1979, 1980) beneath the Blue Ridge and then for 93 mi (150 km) beyond the Brevard zone.

Cook et al (1979, p. 563; 1980, Fig. 3) called the Windows floor thrust the sole, and claimed that the underlying Cambrian-Ordovician sequence lies in place on its basement. We think that the geometric argument in point 7 proves this interpretation impossible.

10. At branch line A, the Windows floor and roof thrusts join the Holston Mountain thrust whose footwall cuts as high as mid-Ordovician shale. Some and possibly all of the 19 mi (30 km) of shortening in the Windows duplex may have transferred to slip on the Holston Mountain thrust, which therefore might have moved as the Windows duplex formed. If the Windows duplex started shortening in Late Devonian (see point 6) at 5 mm/year, then the total shortening of 50 mi (80 km) required 16 m.y. and was completed early in the Late Mississippian.

11. At a lower branch line, the Pulaski thrust merges with the Windows floor thrust, which then continues southeast across the section. The youngest formations currently overlain by the Pulaski are Early Mississippian. Because the Pulaski and Windows floor thrusts have a combined order of magnitude slip of 124 mi (200 km), any Carboniferous formation once deposited in those thrust sheets might have been in thick synorogenic facies and unlike any that crops out farther northwest.

12. The Grandfather Mountain formation consists of fluvio-deltaic sediments derived from both northwest and southeast sides of a steep and probably fault-bounded basin containing both felsic and basaltic flows and sills. The upper part of the basin is truncated by the Stone Mountain-Linville Falls thrust system, and in some way the stratigraphic sequence is continued into the hanging wall as the Mt. Rogers formation, one-half of which is volcanic.

13. One key to restoration of this complex thrust system is the presence 9 mi (15 km) northeast, in the Shady

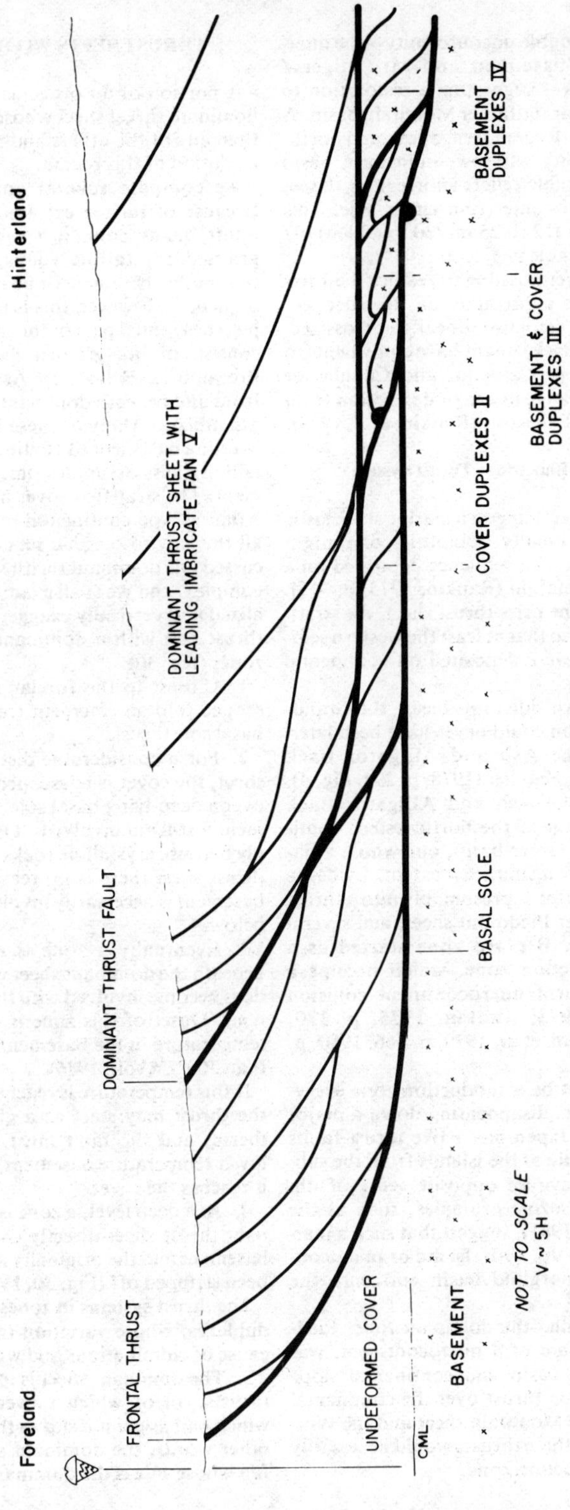

FIG. 30—Idealized sketch of thrust belt with dominant sheet shows five zones, each with characteristic thrust system and degree of basement involvement. Thrust faults at zone boundaries are bold. Vertically exaggerated and not to scale.

Valley thrust sheet, of a double unconformity—the intersection of the Unicoi/basement and Mt. Rogers/basement unconformities— suggesting a restoration to the northwest side of the Grandfather Mountain basin. A series of thrusts in the Mt. Rogers area, above and northeast of the Mountain City window, indicate a basin whose floor had considerable relief, an irregular shape, and sequences that could come from either side. This series of thrusts is too far (12 to 25 mi; 20 to 40 km) off this line of section for reliable projection.

14. Although it is not yet possible to restore the Fries thrust sheet, we do have constraints on its paleogeographic site. Formations recognized near the cross section (Ashe and Alligator Back) might have equivalents to the northeast (Lynchburg, Catoctin, and Candler or Chilhowie), indicating a range in age of deposition from late Precambrian to early Paleozoic (Rankin, et al, 1973).

Plate Tectonics and Blue Ridge Thrust System

The rocks of the Fries sheet suggest a marine siliciclastic sequence which is occasionally ophiolitic; one might infer a slope or possibly a rise sequence deposited on a newly rifted continental margin (Rankin, 1975, p. 315). In the northwest part of the Fries thrust sheet, the stratigraphic base is basement, so that at least the western sediments in the Fries sheet were deposited on continental crust.

Since we see the southeast side of its basin, the Grandfather Mountain formation could never have been laterally continuous into the Ash and Alligator Back formations, as thought by Hatcher (1978, p. 294, Fig. 4). It would appear that the Ash and Alligator Back sequences were deposited near the northwestern continental margin of a much larger basin, one whose adjacent ocean had an unknown southeast extent. Evidence of this ocean has now vanished, presumably into a thrust complex beneath the Inner Piedmont sheet, and several authors suggest that the Brevard zone started as a southeast-dipping subduction zone, which became a suture marking a continent/microcontinent collision (Odom and Fulligar, 1973; Rankin, 1975, p. 320; Hatcher, 1978, p. 295; Cook et al, 1979, p. 566, 1980, p. 213).

Is it true that there must be a subduction zone everywhere we see oceanic crust disappearing down a major thrust? The west side of Japan has active thrust faults that are on the opposite side of the islands from the subduction zone and that have an opposite sense of dip (Uyeda, 1977, p. 5). Mesozoic examples, such as the southern Andes (Dalziel, 1981), suggest that such a marginal basin can close right up, with the arc or microcontinent driven over the marginal basin and onto the continental foreland.

It would seem possible that the dominant Inner Piedmont sheet, either as an arc or a microcontinent, was thrust over the marginal basin and continental slope (Fries), and then everything thrust over the continental shelf (Linville Falls–Stone Mountain sheet and the Windows duplex), yet none of these thrusts would necessarily be traceable into the subduction zone.

THRUST BELTS WITH A DOMINANT SHEET

A portion of an orogenic belt may have an overlying dominant thrust sheet whose displacement is much larger than any of the others and whose motion dominates the evolution of this region.

We compare several dominant sheets in Table 2. Because of their great amount of slip and consequent uplift, many contain a substantial portion of coarse-grained crystalline rocks, such as an old quartzo-feldspathic basement or an oceanic cumulate and mantle sequence. However, this is not a requirement, for example, the fronted part of the far-traveled Lewis thrust sheet consists of fine-grained clastics and carbonates of the Precambrian Belt group. At least part of the rocks now in front and beneath dominant thrusts are a stratified cover assemblage. Many of these assemblages unconformably overlie a basement of continental crust, often an old crystalline gneissic complex that we refer to as foreland basement. The stratified cover in some cases may include an Atlantic-type continental-margin assemblage, complete all the way to oceanic sequences. We have already discussed one dominant thrust sheet in the Blue Ridge thrust complex, and we shall examine one in the Alps. A generalized and vertically exaggerated cross section through a thrust belt with a dominant sheet shows five different zones (Fig. 30).

1. Closest to the foreland is a zone with cover telescoped into an emergent trailing imbricate fan above a basal sole thrust.

2. For a considerable distance beneath the dominant sheet, the cover is telescoped and thickened by duplexes over a deep-lying basal sole, below which foreland basement is still not involved. It is important to note that simply because crystalline rocks are found at the surface in a thrust sheet there is no reason to assume that foreland basement is necessarily involved in the thrust belt directly below.

3. Eventually a zone is reached (Fig. 30, III) well beneath the dominant sheet where the foreland basement does become involved with the cover in duplex thrust systems. Onset of this zone is controlled by metamorphic temperature in the basement and probably is a little less than 300°C (Voll, 1976).

If this temperature is reached well within the basement, the thrust may start as a glide horizon along this isotherm, and the fault must then cut upsection though lower temperature basement, probably as a ramp, before it reaches the cover.

4. At a deep level, a zone is reached in which the dominant thrust sheet directly overlies duplexes of foreland basement and the originally intervening cover rocks have been stripped off (Fig. 30, IV).

The thrust systems in zones II, III, and IV (Fig. 30) are duplexes, whose variation in thickness is an important cause of culminations and windows.

5. The dominant sheet is itself composed of numerous thrusts, all of which moved piggyback, and none of which had as great a slip as the dominant thrust fault. In other words, the dominant sheet is a leading imbricate fan whose sole is the dominant thrust fault.

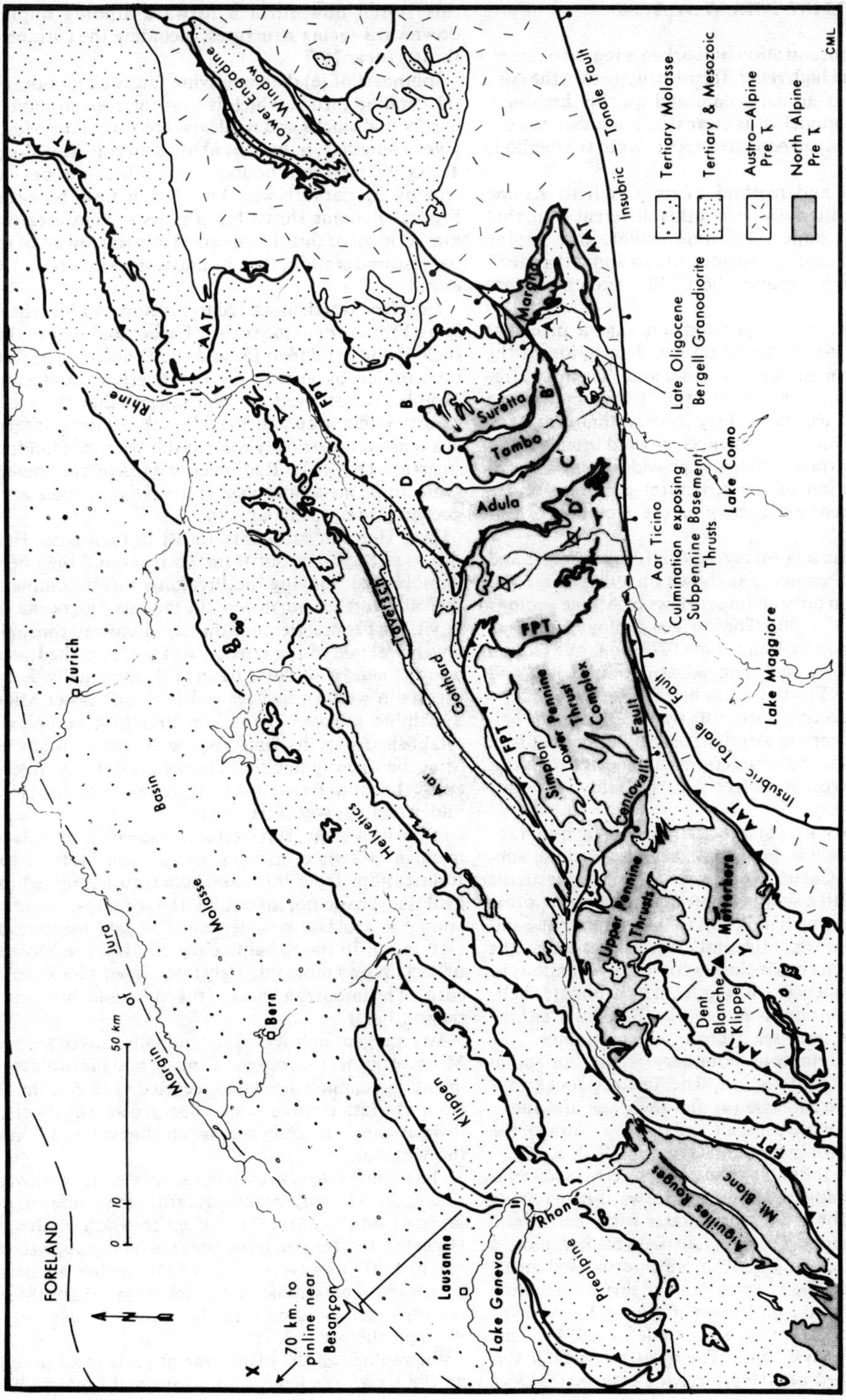

FIG. 31—Map of central Alps (modified from Spicher, 1972; Milnes, 1974).

ALPINE THRUST SYSTEMS

Helped by a concentration of work in a region of excellent exposure and high relief, thrust structures in the Alps were described at an early date and quickly became a widely accepted model. It is clear that a number of concepts developed in the Alps are precursors to the methods described here.

The emphasis and method of approach to Alpine nappes have traditionally been rather different from that outlined in this paper. To some extent, this Alpine approach is used in all the various chains about the Mediterranean, eastern Europe, the USSR, and the Himalayas.

Tollman (1968, p. 35) pointed out that a principal objective in the Alpine tradition is a description of the nappe body or thrust sheet volume, and that this is frequently accompanied by an incomplete characterization of the thrust surface. We will try to show that geometric reasoning based on thrust surfaces, tip and branch lines, and the various thrust systems, provides both an economical description of the structural geometry and a somewhat different perspective on the evolution of this classic belt.

Consider a cross section across the Jura, Prealps, and Helvetics to the Pennines, as shown on a map (Fig. 31). We rely heavily on outstanding reviews of Alpine geology by Trumpy (1973, 1980). The surface geology follows a recent cross section by Baud et al (1978), but our section is heavily modified and changed at depth. Below and northwest of the Frontal Pennine thrust these changes result in a cross section restorable by plane strain or balanced (Fig. 32). Letters identify corresponding points in the deformed and restored sections, and both letters and numbers on the cross section refer to the following points discussed in the text.

Pre-Triassic rocks (1) are principally a quartzofeldspathic crystalline basement, but also include substantial Permo-Carboniferous sediments. Mesozoic rocks (2) start with a shallow-water carbonate-evaporate Triassic sequence which indicates that the underlying siliceous basement was continental crust at that time. The Jurassic and Cretaceous include a complete Atlantic-type rifted margin sequence, varying from a carbonate shelf, below and north-northwest of the Frontal Pennine thrust, to various deep-sea deposits with ophiolites.

The Tertiary-Cretaceous boundary is the horizontal datum in the restored section. The Tertiary rocks vary from flysch (3) to molasse (4) and their age and facies reflect a growing thrust belt, advancing toward the North Alpine cratonic foreland (Trumpy, 1980, p. 50).

We will describe the balanced part of the cross section in the rough chronological order of a forward-progressing thrust system. The initial stage starts with the Frontal Pennine thrust, already carrying, in piggyback fashion, a whole sequence of originally more internal sediments and thrust sheets. The Frontal Pennine thrust, including the Prealpine sheets, glided on a planar footwall of Eocene flysch, which overlies the undeformed Helvetic carbonates (5, from e to n). The most internal of this flysch was stripped off the underlying carbonate. Does this flysch now form a forward-dipping duplex of downward-facing structures, including the Ultrahelvetic thrusts (near 20)?

Southeast of (e) the underlying faults all join the Frontal Pennine thrust, which is thus the roof of enormous antiformal stacks and duplexes. Restoring the Mesozoic gives a length (f, h, k, l, m, n) of 68 mi (110 km) along the Tertiary-Cretaceous boundary (9), which was once overlain by Tertiary flysch. This flysch now beneath the Frontal Pennine thrust has a cross section area of 170 km^2. If most of this flysch came from northwest of (n) on the restored section, then its initial thickness (t) is 1.6 km (10).

The Helvetics then started to develop and fold the overlying Frontal Pennine thrust. Earlier and higher thrusts such as the Wildhorn (6) and the Diableret (7), were in turn folded by growth of the underlying and younger Morcles thrust sheet (8). We suggest that the Helvetic thrust system is an antiformal stack, 30 times larger but in a sense geometrically comparable with the Dundonnell structure (Fig. 23). Radiometric ages of the low-grade syntectonic metamorphism in the Helvetic stack are Oligocene (Trumpy, 1980, p. 60).

Each Helvetic subsidiary thrust in turn passed on its slip to part of the roof thrust in front and then became immobilized, moving the Prealpine thrust complex farther and farther northwest. In the late Oligocene ($\sim$25 m.y.), the Frontal Pennine thrust cut upward through the lower molasse to the erosion surface, and shed pebbles into the nearby Molasse basin (13). Eventually, Helvetic thrusts in western Switzerland overrode lower Miocene Subalpine molasse. After the Prealpine complex was established over its final foot wall rocks, in late early Miocene to early middle Miocene time (13-16 m.y.), the entire Frontal Pennine thrust system and its Helvetic underpinnings were abandoned.

In the next stage, slip started at a lower level, a decollement in Triassic evaporites beneath the Molasse basin. Imbrications from this basal sole thrust emerged in the Molasse basin northwest of the Prealps, where the youngest Molasse was deposited in late Miocene time ($\sim$6 m.y.). In the Subalpine region (14), the Molasse is underlain by a more intensely imbricated Mesozoic (area cdfg). The initial thickness of the Mesozoic thins down to the southeast.

An emergent imbricate fan gradually moved across the Molasse basin (15, between a and c), and the Molasse sediment was cannabalized and recycled. The current 28 mi (45 km) width records at least 6 mi (10 km) of shortening if we assume a roughly average thickness $t \sim 3.1$ km for the Mesozoic.

The pinline (16) is chosen in the foreland, northwest of Besancon. The current distance from the pinline to point (a) is 43 mi (70 km) (17), the entire width of the Jura. However, the Jura underwent a shortening of 15 to 19 mi (25-30 km) (Chauve, 1975, p. 23), giving an original length between pinline and point (a) of 53 mi (85 km). Orogenic deformation in the Jura ended before early late Pliocene (before $\sim$2.5 m.y.).

Where does all the displacement go that accumulated on the basal sole beneath the Jura and Molasse basin?

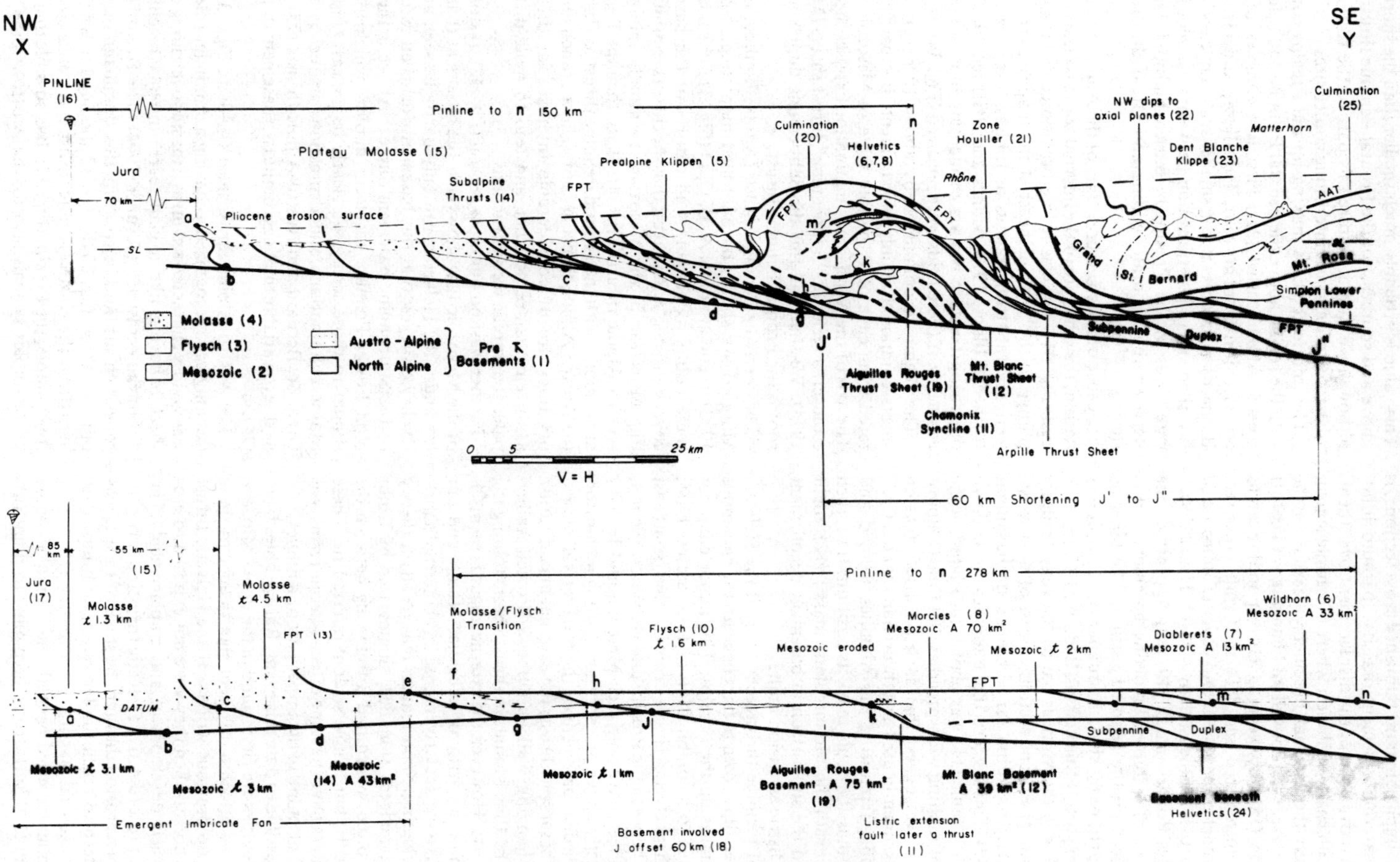

FIG. 32—Cross section through Alps along line XY′ (Fig. 31). Letters and numbers on section are location of points discussed in text.

One solution, implicit in most published cross sections, is to label the external basement massifs "parautochthonous" and to move most of the displacement over them. We favor an alternative solution where the displacement passes beneath the external basement massifs. In the undeformed state, the footwall and hanging wall cutoff must coincide (18). By keeping corresponding areas (Jheg) equal in restored and deformed state cross sections, the contact of basement and cover (J')must be 37 mi (60 km) farther southeast than is J' (cf. Hsu, 1979).

The Aiguilles Rouges massif (19) then appears as one subsidiary slice of a duplex involving basement and is most easily seen on the restored section. The tapering Mesozoic carbonate shelf sequence is completely removed from part of this massif (Badoux, 1972).

The Mesozoic in the Morcles nappe thickens abruptly (8), and the upper Eocene section consist of northwest-derived clasts, including large blocks of basement (Badoux, 1972). This suggests that a southeast side-down listric extension fault (11) was active until late Eocene time, a type of faulting widespread at that time (Trumpy, 1960, p. 882). We suggest that such a listric extension fault was subsequently used as a thrust and that the Chamonix syncline, separating the Aiguilles Rouges and Mount Blanc massifs, could mark its position. If so, then the thrust beneath the Mount Blanc massif (12) might have followed the subhorizontal basement decollement of the earlier listric extension fault.

The Frontal Pennine thrust goes over all the Helvetic sheets and then dips down to pass beneath the enormous klippe that makes up the Prealps. This major culmination (20) is due to a combination of four causes. First, horses of Ultrahelvetic flysch from farther southeast accumulated between the Prealps and Helvetics. Second, the Helvetic sheets have both their Mesozoic and Tertiary portions directly above one another, causing an antiformal stack. Third, as basement on the hanging wall moved onto the flat basal sole thrust, the external massifs arose as immense hanging wall anticlines ("Powell Valley-type," as in the Pine Mountain thrust sheet) folding the overlying thrusts. However, the external massifs themselves are a composite array of subsidiary faults and our fourth cause is a particularly intense local internal shortening and thickening. Some readers may recall the theory of "basement wedges" (Fig. 33), summarized by Umbgrove, 1950, p. 66, which had a following between the 1920s and 1950s, but is now seldom referred to. These basement wedges have the same effect—in our terms—as differential shortening in the upper part of a duplex.

It may be possible to date this event. Rare pebbles from the Helvetic nappes are first found in the upper Miocene molasse, but we attach a different significance to these pebbles than do previous writers. If the Helvetic thrust sheets are all horses, then they were not exposed to erosion during their development and emplacement. The pebbles do not record active Helvetic thrusting; instead, they mark when erosion through the stack of thrusts was sufficiently deep to reach the Helvetics, and must date growth of the external massifs.

The shortening recorded beneath the Frontal Pennine thrust is 75 mi (120 km), from the pinline to point (n). This value is probably much smaller than the shortening above and southeast of the Frontal Pennine thrust, and, although we cannot restore this part of the section, we wish to mention some interesting conjectures.

The imbricate slices making up the Prealps correlate with the region above the Frontal Pennine and below the Austro-Alpine thrust (21). First is a complex of three zones (Sion-Courmayeur, Houillere, and Great St. Bernard) each representing smaller scale duplexes. Possibly, then, there is a transition between emergent imbricate slices in the Prealps to buried multiple duplexes between the Frontal Pennine and Great St. Bernard thrusts.

Because of the structural relief of the Aar-Ticino culmination, the structure of the internal zones is clearer to the east (Fig. 31). The classic interpretation is to look down-plunge to the east-northeast at the map pattern. There were several strong phases of folding after the thrusts formed (Milnes, 1974; Milnes and Pfiffner, 1980), and our reconstruction is a very rough attempt to view the geometry, if these folds were removed (Fig. 34)

The region is dominated by about 10 major thrust sheets composed mainly of basement, but partly separated from each other by thin septa of Mesozoic cover. The axis of the Aar-Ticino culmination exposes the deepest level, a series of "Subpennine" thrust sheets (Milnes, 1974). We shall call this the Subpennine duplex and its roof is the Frontal Pennine thrust.

The overlying Pennine thrusts (Adula, Tambo, Suretta, Margna) are shown on published maps (e.g., Spicher, 1972) to end at points A, B, C, D (Fig. 31). We interpret these points as hanging wall cutoffs of the basement cover contacts, and suggest that the thrusts continue cutting upsection in the Mesozoic, eventually joining the Austro-Alpine thrust (Fig. 34). If so, then the Pennines are a duplex with the Frontal Pennine thrust as floor and the Austro-Alpine thrust as roof. These branch and cut-off lines are of considerable interest, for different arrangements are possible, and lead to quite different paleogeographic reconstructions.

Farther west on our cross section (Fig. 32), restoration of the Mesozoic (24) shows that the basement that once lay under the Helvetics can only partly be seen at map level. We suggest that this basement forms an extension of the Subpennine basement duplex. Above is the Simplon Lower Pennine basement duplex, which may be similar to those seen more clearly to the east (Fig. 34). Major culminations and depressions in the overlying Great St. Bernard and Austro-Alpine thrusts (23) could be due to variable thickness of the two basement duplexes (25).

The Austro-Alpine thrust sheet (AAT on Figs. 31, 32) has long been recognized as having a dominant role, and was vividly characterized up as a master overriding thrust sheet behaving like a giant earth-moving machine, a "Traineau ecraseur" (Termier, 1904). The five different zones beneath and in front of this dominant sheet are fairly clear (cf. Fig. 30 to Figs. 32 and 34), but we do not imply that the Austro-Alpine thrust sheet once covered, for example, the Prealps (Fig. 32). It is possible that some of the duplex systems beneath the Austro-Alpine thrust

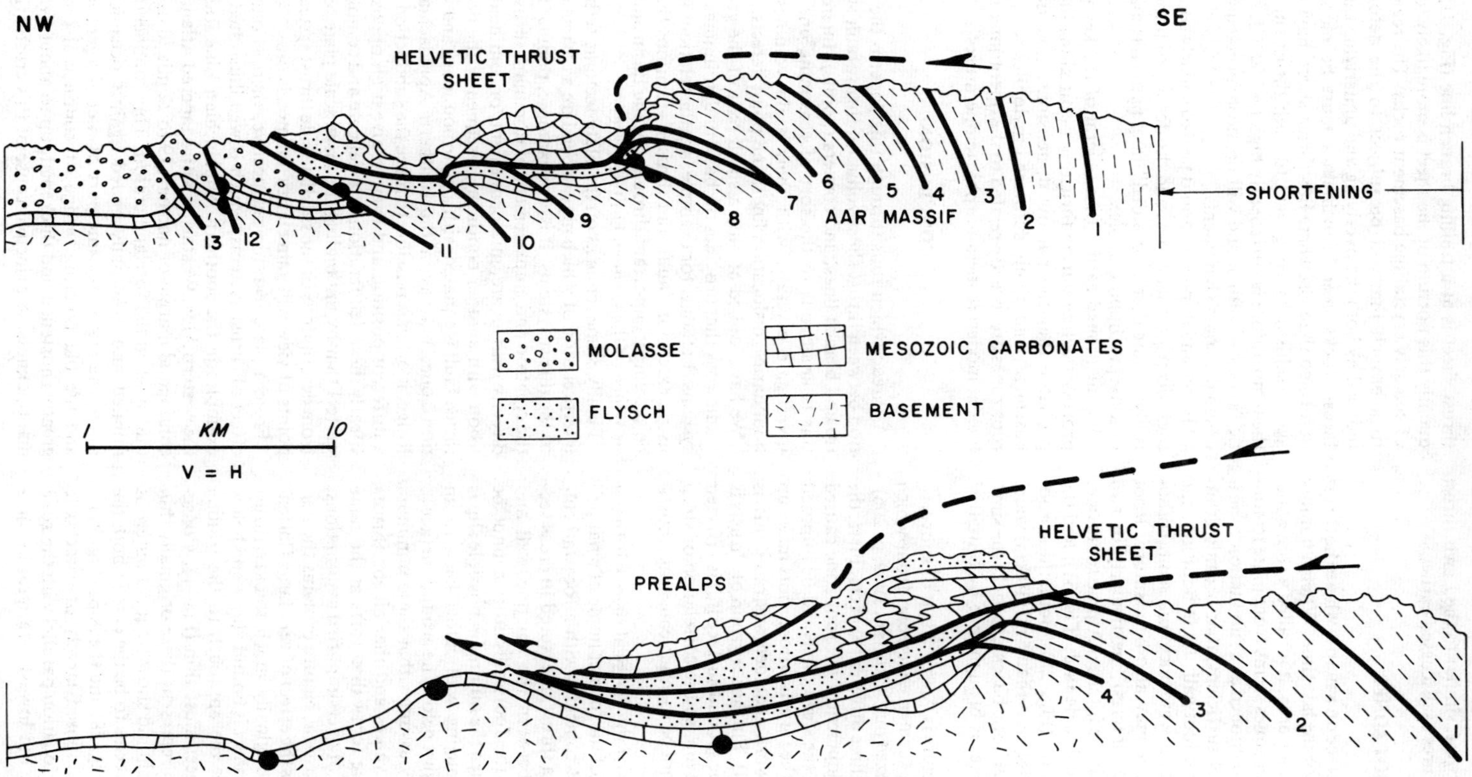

FIG. 33—Top: cross section from Molasse basin to Aar massif. Bottom: earlier stage in basin evolution (after Gunzler-Seiffert, *in* Umbgrove, 1950, Fig. 47; Holmes, 1978, Fig. 3.21). This figure illustrates that old idea of basement wedges may be equivalent to upper part of a duplex. Piggyback emplacement of Prealpine klippe and Helvetic thrust sheets by successive imbrication of underlying crystalline wedges are numbered in order of formation. Telescoping of slices steepens older wedges.

became emergent to the west, similar to the Lewis thrust and its underlying structures as discussed earlier.

ROOT ZONES

The root of a thrust surface on a map is where the thrust finally dives from view for the last time. When a number of thrusts do this in the same area, it is called a root zone. Part of the interest in root zones stems from plate tectonics, for if the root zone functioned as a subduction zone it is necessary, although not sufficient, that oceanic lithosphere once existed in the footwall. If, in addition, the two margins of the root zone were once continental lithosphere, then the root zone may also be a suture. Root zones have long been of interest, particularly in two regions already discussed: the internal Alpine root zone, just north of the Insubric-Tonale fault (Fig. 31); and the Appalachian Brevard zone (Fig. 28). Both these root zones are very narrow, often less than 6 mi (10 km) wide, and consist largely of cataclastic basement rock; the similarity of the two root zones is brought out by Burchfiel and Livingston (1967).

The classical Alpine picture is ". . . a belt of subvertical strata representing a former wide zone from which the Nappe has been squeezed out" (Rutten, 1969, p. 206). This theory of roots has three different aspects. First, the theory implies that development if the root zone caused the subvertical dips. But the steep dip could be imposed later than the motion of the thrusts whose roots make up the zone; Alpine field work that demonstrates this is reviewed by Milnes (1974). Further, a group of thrusts is more likely to intersect the ground surface if it is affected by a steeper than normal dip, so the correlation of steep dip with root zones may not be so much causal as statistical.

The second implication of the classical theory is that a root is the "inner margin of the former home of a nappe" or, in other words, a root coincides with a footwall cutoff. The slip distance of a thrust, measured in cross section, is the arc length between the hanging wall and footwall cutoffs. Often the footwall cutoff cannot be found on the map, and then the minimum possible slip is the distance from the hanging wall cutoff back to an assumed footwall cutoff just below the ground surface in the root. Classical theory assumes that these minimum slip estimates are actual values and that all the thrusts that root in the zone have footwall cutoffs at the same place—clearly an artifact of oversimplified assumptions.

The third part of the classical picture explains the narrow width of root zones as the effect of very large flattening strains imposed on originally much thicker thrust sheets. However, a thrust sheet is bound above and below by fault surfaces, and at the approach to the trailing branch line the thrust sheet must thin (Fig. 1). Consequently, root zones which represent these originally thin portions of thrust sheets should show frequent outcrops of branch lines, and this seems to be the case in both the southern Appalachians (Fig. 28) and the Alps (Fig. 31).

Why are root zones composed largely of cataclastic basement rocks, with the foliation parallel with the zone boundaries? Recall that the lowest stratigraphy in a thrust sheet is at a trailing branch line (Fig. 1). Consequently this portion of the sheet is more likely to consist of coarsely crystalline basement rocks. The region close to a branch line will be subject to the deformation induced by both the overlying and underlying thrust surfaces. Rocks near branch lines must be more highly deformed than elsewhere in a thrust sheet, having cleavage or schistosity congruent to both thrust surfaces. If the temperature is sufficiently high ($\sim 300°C$), cataclastic fabrics will form and will be most spectacular in the originally coarse basement.

Horses and duplexes should be particularly common in these deep internal zones, and the density of branch lines is high. Note that a "braided" map pattern is diagnostic of a steeply dipping duplex.

We mentioned earlier that several of the key physical processes acting during thrusting occur along lines. It is probable that the narrow dimensions, abundance of basement, and highly deformed mylonites so typical of root zones are directly related to the operation of branch lines at moderate metamorphic temperatures.

CONCLUSION

A thread which runs through this paper is the distinction between thrust sheet volumes, thrust fault surfaces, tip and branch lines, and corners (points). Interestingly, this is analogous to the sort of reasoning one uses to investigate crystals, with their grain volumes, grain boundaries, dislocation lines, and point defects.

We have looked at only a small part of the geometric framework that provides the basic underpinnings of fault systems. Left out, for example, are the relations between listric extension faults and thrusts—relations that must be extremely important throughout the Basin and Range province of the western United States.

Finally, some remarks on the plate-tectonic setting, the interpretation of which may depend on a correct reading of the thrust systems. It is, for example, popular to tie in the creation and emplacement of dominant thrust sheets directly with an arc, microcontinent, or continent in collision with another continent. Frequently the dominant thrust fault is equated with a plate boundary and subduction zone, but, as in the southern Appalachian Blue Ridge, it often is possible to place the subduction zone at a different position and even to give it an opposite dip. Clearly these deliberations are delicate and complex, and we feel that by untangling the large scale plate tectonics from descriptions of the smaller scale thrust systems both points of view will gain clarity and precision.

From a plate tectonic perspective, there is one major effect of thrust systems, which is well illustrated by our examples in the southern Appalachian Blue Ridge and the western Alps. We started with a tapered Atlantic-type continental margin of normal (19 to 22 mi; 30 to 35 km) thickness on the foreland craton. This continental crust thinned and eventually ended against oceanic crust. Later, the region became part of a convergent margin, and the old tapering wedge of continental crust was greatly shortened and thickened, up to about twice normal thickness. A principal cause of this crustal thicken-

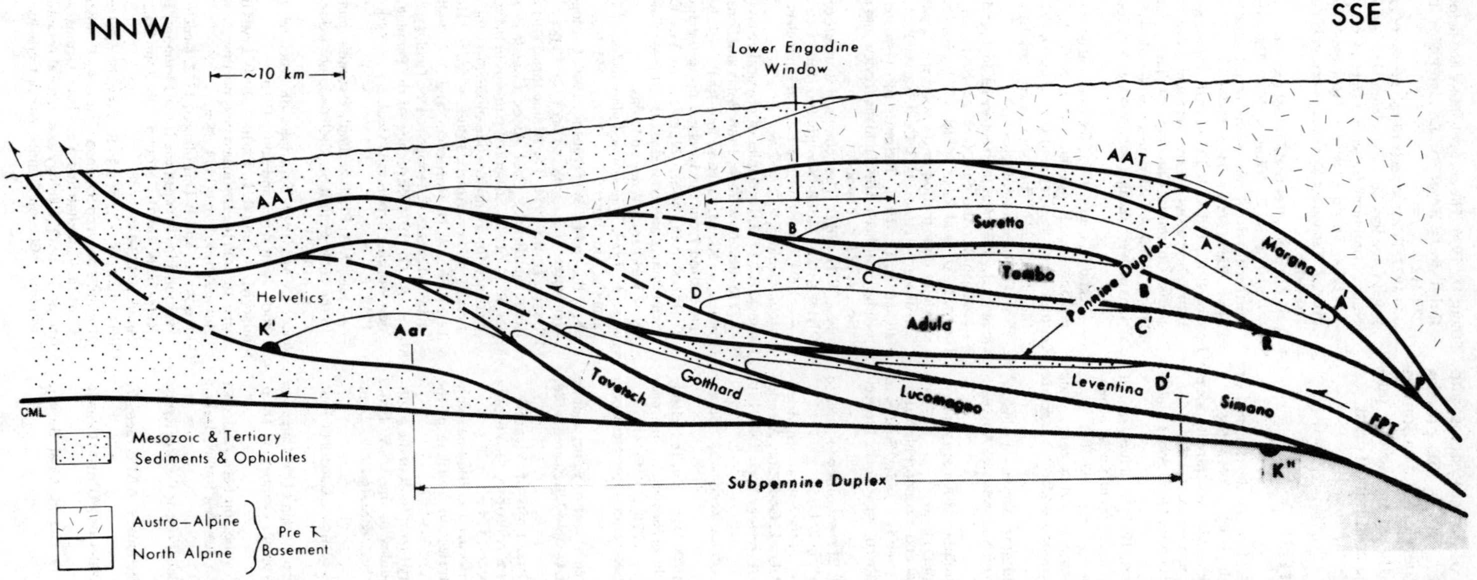

FIG. 34—Approximate down-plunge view in central Alps, west of Aar-Ticino culmination, suggesting duplex thrust systems. Only major faults which cut basement shown, but cover also shows intense but smaller scale families of duplexes. Effects of late folds removed, and scale is approximate. Letters identify points on map (Fig. 31).

ing is activity of the various kinds of thrust systems, and particularly by means of duplexes. Nor is this process restricted to shallow depths, for there are indications in the Alps and the Himalayas that thrust thickening persists down to the lower levels of the continental crust.

REFERENCES CITED

Armstrong, F. C., and S. S. Oriel, 1965, Tectonic development of Idaho-Wyoming thrust belt: AAPG Bull., v. 49, p. 1847-1866.

Badoux, H., 1972, Tectonique de la Nappe de Morcles entre Rhone et Lizerne: Materiaux Carte Geol. Suisse, new series, v. 143.

Bailey, E. B., 1938, Eddies in mountain structure: Geol. Soc. London Quart. Jour., v. 94, p. 607-625.

Bally, A. W., P. L. Gordy, and G. A. Stewart, 1966, Structure, seismic data, and orogenic evolution of southern Canadian Rocky Mountains: Bull. Canadian Petroleum Geology, v. 14, p. 337-381.

Baud, A., et al, 1978, Cross section Besancon-Ivrea zone, scale 1:200,000, Geologie Structurale, Programme 3e cycle: Pub. Lecture Notes, Inst. Geology, Univ. Lausanne.

Boyer, S. E., 1976, Formation of the Grandfather Mountain window, North Carolina, by duplex thrusting: Geol. Soc. America Abs. with Programs, v. 8, p. 788-789.

——— 1978, Structure and origin of Grandfather Mountain window, North Carolina: PhD thesis, Johns Hopkins Univ.

Bryant, B., and J. C. Reed, 1970, Structural and metamorphic history of the southern Blue Ridge, *in* G. W. Fisher et al, eds., Studies of Appalachian geology; central and southern: New York, Interscience, 460 p.

Burchfiel, B. C., and J. L. Livingston, 1967, Brevard zone compared to Alpine root zones: Am. Jour. Sci., v. 265, p. 241-256.

Butler, J. R., and D. E. Dunn, 1968, Geology of the Sauratown Mountains anticlinorium and vicinity: *in* Guidebook, Geol. Soc. America SE Section, Durham, N.C., Southeastern Geology Spec. Pub. 1, p. 19-47.

Chauve, P., 1975, Jura, *in* "Guides Geologiques Regionaux" Series: Paris, Masson and Cie, 216 p.

Clark, B. H., J. K. Costain, and L. Glover, 1978, Structural and seismic reflection studies of the Brevard ductile deformation zone near Rosman, North Carolina: Am. Jour. Sci., v. 278, p. 419-441.

Cloos, H., 1936, Einfuhrung in die Geologie: Berlin, Gebruder Borntraeger, 503 p.

Coleman, R. G., 1971, Plate tectonic emplacement of upper mantle peridotites along continental edges: Jour. Geophys. Research, v. 76, p. 1212-1222.

Conley, J. F., 1978, Geology of the Piedmont of Virginia—interpretations and problems, *in* Contributions to Virginia Geology—III: Virginia Div. Min. Resources, 154 p.

Cook, F. A., et al, 1979, Thin-skinned tectonics in the crystalline southern Appalachians; COCORP seismic-reflection profiling of the Blue Ridge and Piedmont: Geology, v. 7, p. 563-567.

——— 1980, The Brevard fault: a subsidiary thrust fault to the southern Appalachian sole thrust: Caledonides in the USA, Proc. I.G.C.P. 1979 Mtg.: Virginia Polytech. Inst. & State Univ., Dept. Geology, Mem. 2, 329 p.

Dahlstrom, C. D. A., 1969, Balanced cross-sections: Canadian Jour. Earth Sci., v. 6, p. 743-757.

——— 1970, Structural geology in the eastern margin of the Canadian Rocky Mountains: Bull. Canadian Petroleum Geology, v. 18, p. 332-406.

Dalziel, I. W. D., 1981, Back-arc extension in the Southern Andes—a review and critical reappraisal: Royal Soc. London Philos. Trans., ser. A, v. 300, p. 319-335.

de Margerie, E., and A. Heim, 1888, Les dislocations de l'ecorce terrestre: Zurich, J. Wurster, 154 p.

Dennis, J. G., 1967, International tectonic dictionary: AAPG Mem. 7, 196 p.

Douglas, R. J. W., 1952, Preliminary map, Waterton, Alberta: Canada Geol. Survey Paper 51-22.

——— 1958, Mount Head map area, Alberta: Canada Geol. Survey, Mem. 291, 241 p.

Elliott, D., 1976, The energy balance and deformation mechanisms of thrust sheets: Royal Soc. London Philos. Trans., ser. A, v. 283, p. 289-312.

——— 1977, Some aspects of the geometry and mechanics of thrust belts: Canadian Soc. Petroleum Geology, 8th Ann. Seminar Pub. Notes, Continuing Education Dept., Univ. Calgary, v. 1, 2.

——— 1978, Thrust sheets, *in* Geologie Structurale, Programme 3e cycle, 2eme semaine: Published Lecture Notes, Inst. Geology, Univ. Lausanne.

———, in prep., Restorable cross sections.

——— and M. R. W. Johnson, 1980, The structural evolution of the northern part of the Moine thrust zone: Royal Soc. Edinburgh Trans. Earth Sci., v. 71, p. 69-96.

Fermor, P. R., and R. A. Price, 1976, Imbricate structures in the Lewis thrust sheet around Cate Creek and Haig Brook windows, southeast British Columbia: Canada Geol. Survey Paper 76-1B, p. 7-10.

Gilluly, J., 1960, A folded thrust in Nevada—inferences as to time relations between folding and faulting: Am. Jour. Sci., v. 258-A, p. 68-79.

——— and O. Gates, 1965, Tectonic and igneous geology of the northern Shoshone Range, Nevada: U.S. Geol. Survey Prof. Paper 465.

Gordy, P. L., F. R. Frey, and D. K. Norris, 1977, Geological guide for the CSPG 1977 Waterton-Glacier Park field conference: Canadian Soc. Petroleum Geologists.

Gosselet, J., 1885, Sur la structure geologique de l'Ardenne d'apies M. jan Lashnlx: Ann. Soc. Geol der Nord, v. 12, p. 195-202.

Harris, L. D., and R. D. Milici, 1977, Characteristics of thin-skinned style of deformation in the southern Appalachians, and potential hydrocarbon traps: U.S. Geol. Survey Prof. Paper 1018.

Hatcher, R. D., 1978, Tectonics of the western Piedmont and Blue Ridge, southern Appalachians; review and speculation: Am. Jour. Sci., v. 278, p. 276-404.

Hobbs, W. H., 1893, The geological structure of the Housatonic Valley lying east of Mount Washington: Jour. Geology, v. 1, p. 780-802.

——— 1894, Notes on the English equivalent of Schuppenstruktur: Jour. Geology, v. 2, p. 206.

Holmes, A., 1978, Principles of physical geology, 3d ed., revised by D. L. Holmes: New York, John Wiley and Sons.

Hossack, J. R., M. Garton, and R. P. Nickelsen, 1982, The geological section from the foreland up to the Jotun thrust sheet in the Valdres area, south Norway, *in* The Caledonide orogen—Scandinavia and related areas: Uppsala Caledonides Symp., 1982, New York, John Wiley and Sons.

Hsu, K. J., 1979, Thin skinned plate tectonics during Neo-Alpine orogenesis: Am. Jour. Sci., v. 279, p. 353-366.

King, P. B., and H. W. Ferguson, 1960, Geology of northeasternmost Tennessee: U.S. Geol. Survey Prof. Paper 311, 136 p.

McIntyre, D. B., 1954, The Moine thrust—its discovery, age and tectonic significance: Geol. Assoc. Canada Proc., v. 65, p. 203-223.

Milici, R. C., 1975, Structural patterns in the southern Appalachians; evidence for a gravity slide mechanism for Alleghanian deformation: Geol. Soc. America Bull., v. 86, p. 1316-1320.

Miller, R. A., W. D. Hardeman, and D. C. Fullerton, 1966, Geologic map of Tennessee: Tennessee Div. Geology, scale 1:250,000.

Milnes, A. G., 1974, Structure of the Pennine zone (central Alps); a new working hypothesis: Geol. Soc. America Bull., v. 85, p. 1727-1732.

——— and O. A. Pfiffner, 1980, Tectonic evolution of the central Alps in the cross section St. Gallen-Como: Eclogae Geol. Helvetiae, v. 73, p. 619-633.

Nevin, C. M., 1949, Principles of structural geology, 4th ed.: New York, John Wiley and Sons, 410 p.

Mudge, M. R., and R. L. Earhart, 1980, Lewis thrust fault and related structures in the Disturbed belt, northwestern Montana: U.S. Geol. Survey Prof. Paper 1174, 18 p.

Odom, A. L., and P. D. Fullagar, 1973, Geochronologic and tectonic relationships between the Inner Piedmont, Brevard zone and Blue Ridge belts, North Carolina: Am. Jour. Sci., v. 273-A, Cooper volume, p. 133-149.

Ollerenshaw, N. C., and F. R. Frey, 1975, Structural geology of the foothills between Savanna Creek and Panther River, S. W. Alberta, Canada: Canadian Soc. Petroleum Geologists Guidebook.

Oriel, S. S., 1950, Geology and mineral resources of the Hot Springs window: Madison County, North Carolina: North Carolina Div.

Mineral Resources Bull. 60.
Peach, B. N., et al, 1907, The geological structure of the north-west Highlands of Scotland: Mem. Geol. Survey Great Britain.
Perry, W. J., 1971, Structural development of the Nittany anticlinorium in Pendleton Co., West Virginia: PhD thesis, Yale Univ., 227 p.
——— 1978, Sequential deformation in the Central Appalachians: Am. Jour. Sci., v. 278, p. 518-542.
Rankin, D. W., 1970, Stratigraphy and structure of Precambrian rocks in northwestern North Carolina, *in* G. W. Fisher et al, eds., Studies of Appalachian geology: central and southern: New York, Interscience, 460 p.
——— 1975, The continental margin of eastern North America in the southern Appalachians: the opening and closing of the Proto-Atlantic Ocean: Am. Jour. Sci., v. 275-A, p. 298-336.
——— G. H. Espanshade, and R. B. Newman, 1973, Geologic map of the west half of the Winston-Salem quadrangle, North Carolina, Virginia, and Tennessee: U. S. Geol. Survey Misc. Geol. Inv. Map I-709A, scale 1:250,000.
Rodgers, J., 1953, Geologic map of east Tennessee with explanatory text: Tennessee Div. Geology Bull. 58, pt. 2, 168 p.
——— 1970, The tectonics of the Appalachians: New York, John Wiley and Sons, 271 p.
Roeder, D., O. C. Gilbert, and W. D. Witherspoon, 1978, Evolution and macroscopic structure of Valley and Ridge thrust belt, Tennessee and Virginia: Studies in Geology 2, Univ. Tennessee Dept. Geol. Sci., 25 p.
Roper, P. J., and D. E. Dunn, 1973, Superposed deformation and polymetamorphism, Brevard zone, South Carolina: Geol. Soc. America Bull., v. 84, p. 3373-3386.
Royse, F., M. A. Warner, and D. L. Reese, 1975, Thrust belt structural geometry and related stratigraphic problems, Wyoming, Idaho, northern Utah: Rocky Mtn. Assoc. Geologists Symp., p. 41-54.
Rutten, M. G., 1969, Geology of western Europe: Amsterdam, Elsevier Pub. Co., 520 p.
Sander, B., 1921, Zur Geologie der Zentralalpen: Geol. Staatsanst: Wien Jahrb. 6, v. 71, p. 171-224.
Shackleton, R. M., 1958, Downwards-facing structures of the Highland border: Geol. Soc. London Quart. Jour., v. 113, p. 361-392.
Spicher, A., 1972, Tektonische Karte der Schweiz: Schweizer. Geol. Kommission, scale 1:500,000.
Suess, E., 1904, The face of the earth, v. 1: Trans. from 1884 ed. by H. B. C. Sollas, Oxford, Oxford University Press, 604 p.
Termier, P., 1904, Les Nappes des Alpes Orientales et la synthese des Alpes: Soc. Geol. France Bull., v. 4, no. 3, p. 711-765.
Tollman, A., 1968, Die Grundbegriffe der deckentektonischen Nomenklatur: Geotektonische Forschungen, v. 29, p. 26-59.
Trumpy, R., 1960, Paleotectonic evolution of the central and western Alps: Geol. Soc. America Bull., v. 71, p. 843-908.
——— 1973, L'evolution de l'orogenese dans les Alpes Centrales; Interpretation des donnees stratigraphiques et tectoniques: Eclogae Geol. Helvetiae, v. 66, p. 1-10.
——— 1980, An outline of the geology of Switzerland, *in* Geology of Switzerland, a guidebook: 10th Internat. Geol. Cong., Basel, Wepf and Co.
Umbgrove, J. H. F., 1950, Symphony of the earth: The Hague, Martinus Nijhoff, 220 p.
Voll, G., 1976, Recrystallization of quartz, biotite, and feldspars from Erstfeld to the Leventina nappe, Swiss Alps, and its geological significance: Schweizer. Mineralogy u. Petrog. Mitt., v. 56, p. 641-647.
Uyeda, S., 1977, Some basic problems in the trench-arc-back arc system, *in* M. Talwani and W. C. Pitman, eds., Island arcs, deep sea trenches and back-arc basins: American Geophys. Union, Maurice Ewing Ser. 1, 470 p.
Willis, B., 1902, Stratigraphy and structure, Lewis and Livingston Ranges, Montana: Geol. Soc. America Bull., v. 13, p. 305-352.

Volume 33 Number 10

BULLETIN
of the
AMERICAN ASSOCIATION OF PETROLEUM GEOLOGISTS

OCTOBER, 1949

EVOLUTION OF THOUGHT ON STRUCTURE OF MIDDLE AND SOUTHERN APPALACHIANS[1]

JOHN RODGERS[2]
New Haven, Connecticut

ABSTRACT

The first men to comprehend the structure of the Appalachian Mountains were the brothers Henry D. and William B. Rogers, who from 1835 to 1842 studied the mountains from northern New Jersey to southwest Virginia. They worked out the Paleozoic stratigraphy of the area and by means of it deciphered the folds of the middle Appalachians and the thrust faults of the southern Appalachians. They attempted to explain the structure as the result of great explosions on the southeast; beginning with Dana, however, later geologists have generally ascribed it to lateral compression.

Since the time of the Rogers brothers, many structural ideas have emerged from the study of the Appalachians, such as the geosyncline, underthrusting, erosion thrusts, and the competence of strata. Prominent among the many investigators was the group of geologists headed by Willis and Hayes, who late in the nineteenth century worked out several of the low-angle, large-displacement thrust faults of the southern Appalachians and attempted to explain the mechanics of Appalachian structure.

In recent years two schools of thought have been developing with regard to the depth of Appalachian deformation. One school holds that all large folds and faults extend down to, and are supported by, the basement; the other holds that the deformed rocks have been stripped completely off the basement along one or more great bedding-plane thrust faults.

INTRODUCTION

The Appalachian Mountains, which extend along the eastern side of North America from Newfoundland to Alabama, are classic ground for structural geology, for here many of the major concepts of the science were developed. Moreover, Appalachian structure is still a subject of fascinating interest and of much controversy, and we may confidently expect that many additional discoveries

[1] Published by permission of the director of the United States Geological Survey. Read before the Association at Pittsburgh, October 5, 1948. Manuscript received, March 11, 1949.

[2] Department of geology, Yale University. The writer wishes to thank Chester R. Longwell of Yale University, Richard E. Sherrill of the University of Pittsburgh, Berlen C. Moneymaker of the Tennessee Valley Authority, and Philip B. King, Robert A. Laurence, and Charles B. Hunt of the United States Geological Survey for critical reading of the manuscript. King took an active interest in the preparation of the paper, and his comments have been especially valuable. Those who took part in the discussion at Pittsburgh when the paper was presented have also contributed to the paper. Francis X. Bland drew the figure.

will be made and many new and fruitful ideas will be formulated here in the future. It is therefore fitting to pause in our labors and controversies and to review the evolution of thought that has created our present concepts.

Ignoring Newfoundland, we may describe the Appalachian chain as a series of three sweeping curves, each convex on the northwest (Fig. 1). These curves extend in turn from the Gulf of Saint Lawrence to New York, from New York to Roanoke, and from Roanoke southwestward. The corresponding segments of the

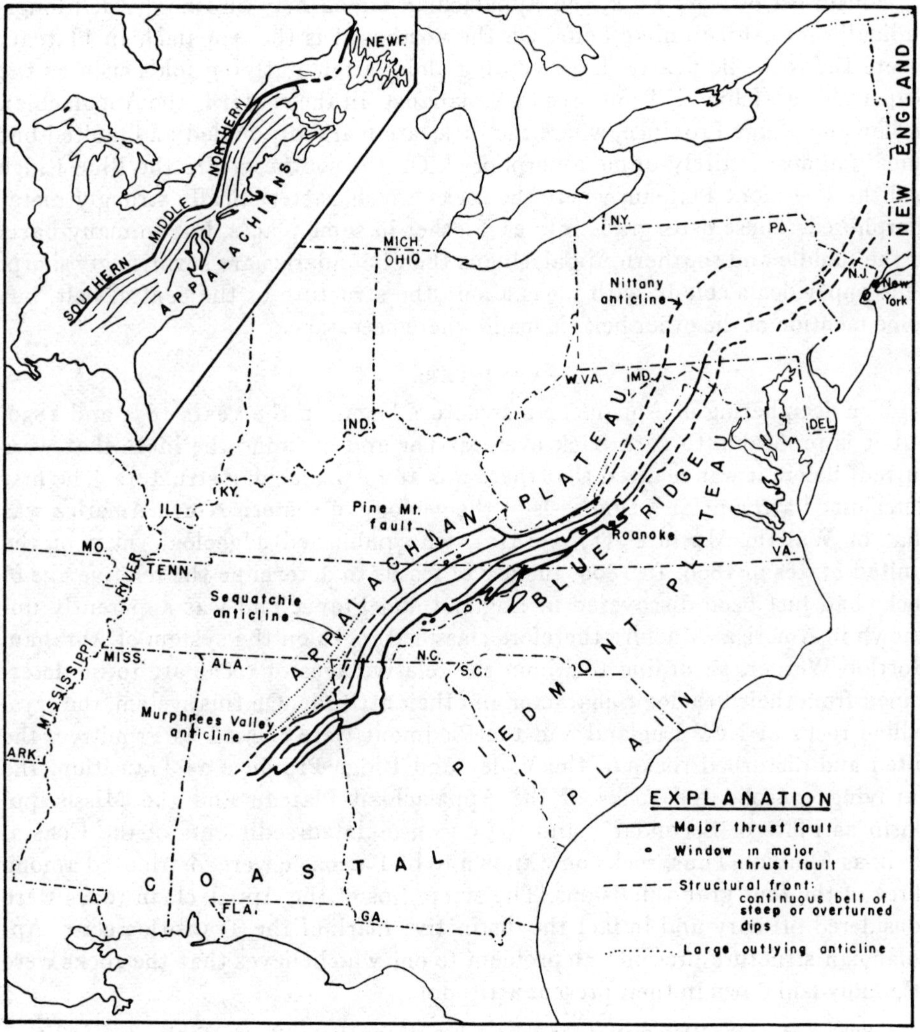

FIG. 1.—Principal structural features of Valley and Ridge province in middle and southern Appalachians.

chain, which differ considerably in structure, may be designated respectively the northern, middle, and southern Appalachians. The northern Appalachian segment was affected by at least three major orogenic episodes during the Paleozoic; it is therefore more complex than the other segments, and much of its geology is still obscure. The chief structural features of the middle and southern Appalachians, however, at least of the unmetamorphosed parts, appear to have been formed in a single period of deformation. The following discussion refers almost exclusively to these two segments.

Southwest of New York, the Appalachian region may also be divided longitudinally into three unlike belts. On the northwest is the Appalachian Plateau, where the rocks lie nearly flat excepting along a few outlying folds such as the Sequatchie anticline in Tennessee and Alabama. In the center is the Appalachian Valley and Ridge Province, where the rocks are markedly folded and faulted but remain almost entirely unmetamorphosed. On the southeast are the Blue Ridge and the Piedmont Plateau, where the rocks are characteristically strongly metamorphosed. These belts grade into each other in some places, but in many parts of the middle and southern Appalachians their boundaries are surprisingly sharp. This paper deals chiefly with ideas about the structure of the central belt, but some mention of the other belts is made where necessary.

PIONEERS

The deciphering of Appalachian structure began in the years 1835 and 1836, but it is profitable to turn back even farther and examine the ideas that were current before it was realized that there was any Appalachian structure. The first determined attempt at a synthesis of the geology of eastern North America was that of William Maclure[3] (1763–1840), who published a geologic map of the United States in 1809. In 1809, the use of fossils to determine the relative age of rocks had just been discovered in England and France and was apparently unknown in America. Maclure therefore classified rocks on the system of Abraham Gottlob Werner, according to whom the relative ages of rocks are to be determined from their lithologic character and their attitude. On this system, the crystalline rocks of New England and the Piedmont were classed as Primitive, the tilted and disturbed rocks of the Valley and Ridge Province as Transition, the flat-lying consolidated rocks of the Appalachian Plateau and the Mississippi Basin as Flötz or Secondary, and the unconsolidated sediments of the Coastal Plain as Alluvial. Thus, rocks now known to be Paleozoic were distributed among three of the four grand divisions. The steep dips of the Appalachian rocks were considered primary and in fact the distinctive mark of the Transition class. Appalachian structure presents no problem to one who believes that the rocks were originally laid down in their present attitude!

[3] William Maclure, "Observations on the Geology of the United States, Explanatory of a Geological Map," *Trans. Amer. Philos. Soc.*, Vol. 6 (1809), pp. 411–28, map. Same, revised and enlarged, *Trans. Amer. Philos. Soc.*, New Ser., Vol. 1 (1818), pp. 1–91; also separately published at Philadelphia (1817).

Approximately 20 years after Maclure published his map, the ideas of William Smith on dating rocks by their fossil content, which were revolutionizing European geology, became current in America, and, about 1835, they led to a great outburst of geologic activity. By joint study among the flat-lying rocks of New York state, Ebenezer Emmons (1799–1863), Lardner Vanuxem (1792–1848), and James Hall (1811–1898) established once and for all the standard Paleozoic section for North America. At the same time, the serious study of the Appalachians was begun by the two Rogers brothers: Henry Darwin Rogers (1808–1866) in New Jersey and Pennsylvania, and William Barton Rogers (1804–1882) in Virginia, which then included West Virginia. Though employed by different states, the brothers worked closely together; shortly they succeeded in establishing a stratigraphic column for the folded Appalachians and in unravelling the structure. In 1835 inklings of the true explanation of the tilted and disturbed strata of the area came to the brothers; during the summer of 1836 H. D. Rogers and his assistants in Pennsylvania found the stratigraphic key, and the nature, amplitude, and extent of the great Appalachian folds became apparent. During the next year or two, W. B. Rogers and his assistants worked out several of the thrust faults of southwest Virginia and showed their relations to the folds.

In 1842 the two brothers presented a summary of their findings to a meeting of the Association of American Geologists and Naturalists, the organization that later became the American Association for the Advancement of Science, in an address[4] which is a milestone in American geology. In the first part of this address the phenomena are described with a clarity that is enviable to-day. Particularly emphasized are the marked asymmetry of the Appalachian folds and the common occurrence of overturned dips, the general northwestward decrease in the degree of asymmetry and overturning, the great length and remarkable parallelism of fold axes and fault traces throughout the chain, the constant relation of thrust faults to anticlinal axes, and the contrast between the predominantly folded middle Appalachians northeast of Roanoke and the predominantly faulted southern Appalachians farther southwest.[5] Ever since the publication of this paper, the middle and southern Appalachians have been recognized as the typical folded mountain range.

In the second part of the address, the brothers attempted to explain the phenomena so well described; this part of the address is as curious and quaint as the other is straightforward and modern. On the one hand, they rejected the ver-

[4] W. B. Rogers and H. D. Rogers, "On the Physical Structure of the Appalachian Chain, as Exemplifying the Laws Which Have Regulated the Elevation of Great Mountain Chains Generally," *Assoc. Amer. Geologists and Naturalists Repts.* (1843), pp. 474–531. Reprinted in W. B. Rogers, "A Reprint of Annual Reports and Other Papers on the Geology of the Virginias," New York (1884), pp. 601–42. See also the brothers' annual reports to the State legislatures of Pennsylvania and Virginia, 1836–1842, and the final reports by H. D. Rogers on New Jersey (1840) and Pennsylvania (1858).

[5] Only a little later (*Assoc. Amer. Geologists and Naturalists, Proc. 6th Ann. Meeting* (1845), pp. 49–50), the brothers announced the additional discovery that slaty cleavage in the Appalachians and elsewhere normally lies parallel with the axial planes of folds.

tical forces previously in vogue to explain all disturbed strata as unable to account for the observed asymmetry of the folds; on the other, they rejected tangential pressure working alone as unable to explain the extraordinary parallelism. Instead they proposed to combine vertical and tangential forces, setting both in operation by a great catastrophe. They pictured the sedimentary crust as thin and flexible, lying on an indefinite thickness of fluid lava that contained gases under high pressure. Shortly after the deposition of the Coal Measures, the gases escaped in a series of sudden explosions from fractures just southeast of the present folded belt. Each explosion set up waves at the surface of the lava, causing undulations of the overlying crust to travel northwestward as in a rug shaken at one end. The explosions also produced great tangential forces which gave the waves in the lava and the undulations in the crust their pronounced asymmetry. Finally, as the lava congealed at the site of the explosions and near-by, forming the igneous rock bodies of the Blue Ridge and Piedmont, the undulations were frozen into place as the folds we now see.

This bizarre theory was apparently not accepted by the Rogers' contemporaries, but for many years geologists argued the relative merits of vertical and horizontal forces as causes of the observed deformation. Gradually the tangential theory, advocated especially by James Dwight Dana[6] (1813–1895) and impressed on the minds of generations of students through his textbooks, gained supremacy. To-day probably no geologist doubts that the folds and faults of the Appalachians are the result of lateral compression. Similarly Dana's view that the deformation took place in a single episode that marked the end of the Paleozoic has become traditional.

Somewhat later, another important generalization was drawn from the facts of Appalachian geology. In 1857, in a presidential address before the American Association for the Advancement of Science, James Hall[7] pointed out that in the Appalachian Mountains the Paleozoic sediments of the eastern United States reach their maximum thickness, and he suggested that such a relation between mountains and thick shallow-water sediments was a general rule. To Hall, the sediments were especially thick in the Appalachian region because they were localized there by a great ocean current from the northeast, and because the crust sagged under their weight, making room for more. He sought to explain the mountains solely as the result of uplift and erosion of the thick sedimentary sequence, for he believed that the folding and faulting of the Appalachian rocks was a minor effect of the subsidence of the sediments during their deposition and had nothing to do with the presence of the mountain chain. As had happened with

[6] James D. Dana, "On the Origin of Continents," *Amer. Jour. Sci.*, 2d Ser., Vol. 3 (1847), pp. 94–100.

———, "Geological Results of the Earth's Contraction in Consequence of Cooling," *ibid.*, pp. 176–88.

[7] James Hall, "Contributions to the Geological History of the North American Continent," *Proc. Amer. Assoc. Adv. Sci.*, Vol. 31 (1883), pp. 31–69. See also *Natural History of New York, Part VI, Paleontology*, Vol. 3 (1859), pp. 1–96.

the Rogers brothers 15 years earlier, Hall's induction from the facts was soon recognized as valid, but his explanation for it was not accepted. In 1873 Dana,[8] proposing the term geosynclinal for the trough of thick sediments, suggested that tangential forces best explain both the trough and the subsequent orogeny. The further vicissitudes of the idea of the geosyncline do not relate to our topic, but it is noteworthy that the idea began in the Appalachians and that the Appalachian geosyncline is the type geosyncline.

Following the Rogers brothers, many geologists carried on the study of Appalachian structure. Outstanding was the work of James Merrill Safford[9] (1822–1907) in east Tennessee, where the thrust faults of the southern Appalachians reach their greatest development. He was apparently the first to describe the Sequatchie anticline and the Pine Mountain fault, which lie somewhat northwest of the main belt of Appalachian folding and faulting. Safford was an early champion of the theory that the folds and faults were caused by tangential forces from the southeast. Later the structure of northern Alabama was studied by Eugene Allen Smith[10] (1841–1927) and his associates, one of whom, Alexander M. Gibson,[11] discovered the faulted anticline of Murphrees Valley, which in contrast to practically all other Appalachian folds is overturned toward the southeast. As they accepted the view that the Appalachians were deformed by tangential forces acting from the southeast, Smith and Gibson[12] introduced the concept of underthrusting to account for this fold.

In the middle Appalachians, the Second Geological Survey of Pennsylvania under Peter Lesley (1819–1903) restudied the folds of Pennsylvania and delineated them on fairly detailed geologic maps. Lesley and John James Stevenson (1841–1924) also studied the thrust faults of southwest Virginia.[13] Stevenson believed that these thrust faults were considerably later than the folding, perhaps of the same age as the faults cutting the Triassic rocks of New England and the Piedmont. Both Lesley and Stevenson, like their predecessors, thought of the

[8] James D. Dana, "On Some Results of the Earth's Contraction from Cooling, including a Discussion of the Origin of Mountains, and the Nature of the Earth's Interior," *Amer. Jour. Sci.*, 3d Ser., Vol. 5 (1873), pp. 423–43; Vol. 6, pp. 6–14, 104–15, 161–72.

[9] James M. Safford, *Geology of Tennessee.* Nashville (1869). Much of this material had appeared in Safford's first biennial report in 1856.

[10] See *Reports of Progress* of the Alabama Geological Survey, beginning in 1873.

[11] Alexander M. Gibson, "Report on the Geological Structure of Murphrees Valley," *Alabama Geol. Survey Spec. Rept. 4.* Montgomery (1893). Gibson made his discovery in the late eighties.

[12] Alexander M. Gibson, *op. cit.*
Eugene A. Smith, "Underthrust Folds and Faults," *Amer. Jour. Sci.*, 3d Ser., Vol. 45 (1893), pp. 305–06.

[13] J. Peter Lesley, "On the Coal Formation of Southern Virginia," *Proc. Amer. Philos. Soc.*, Vol. 9 (1862), pp. 30–33.
J. J. Stevenson, "The Faults of Southwest Virginia," *Amer. Jour. Sci.*, 3d Ser., Vol. 33 (1887), pp. 262–70.
Details in papers by Lesley in *Proc. Amer. Philos. Soc.*, Vol. 12, and by Stevenson in Vols. 19, 22, and 24.

Appalachian faults as high-angle breaks on which the movement was largely vertical upthrust.

WORK OF UNITED STATES GEOLOGICAL SURVEY

In 1887, the United States Geological Survey, formed only 8 years earlier, sent a group of young men into the southern Appalachians. Bailey Willis, the oldest of the group (he was barely 30), was in general charge of the investigation; under him Marius R. Campbell (1858–1940) worked chiefly in southwest Virginia, C. Willard Hayes (1859–1916) in northern Georgia and Alabama, and Arthur Keith (1864–1944) in east Tennessee. Within 20 years, these men published geologic maps of twenty-five 30-minute quadrangles in the Survey's folio series and worked on the geology of several more. Out of this mapping, which was more detailed than any previously done in the Appalachians south of Pennsylvania, several new ideas about Appalachian structure developed.

Probably the most fruitful general result of this work was the discovery by Hayes[14] of thrust faults exhibiting low dips (with the fault surface commonly warped) and great lateral displacement, a type not recognized before in the Appalachians. Such faults have now been found in many parts of the southern Appalachians and also in the northern Appalachians, though there they are of earlier date and argument still rages over many specific examples. In the folded middle Appalachians (omitting the controversial southeast margin), the structure is apparently simpler and large low-angle faults have nowhere been recognized. Nevertheless, the faulted area on the Nittany anticline at Birmingham, Pennsylvania, as described by Butts,[15] strongly suggests low-angle faulting to a geologist who has worked among the low-angle thrust faults of east Tennessee.

Hayes believed that the low-angle thrust faults he described were formed later than the folds and lesser thrust faults of the area, and one of them he conceived to have ridden forward over the deeply eroded crest of an anticline. For this fault he coined the term erosion thrust; the concept seems to have been incorrect for the fault to which he originally applied it, but it has been used with success in areas other than the Appalachians.

During this same period of mapping, Campbell[16] discovered in Virginia what he thought was evidence of striking unconformities in the Paleozoic series and hence of strong early and middle Paleozoic folding. Not until many years later did he recognize that he had been studying one of the largest of the low-angle

[14] C. Willard Hayes, "The Overthrust Faults of the Southern Appalachians," *Bull. Geol. Soc. America*, Vol. 2 (1891), pp. 141–52.
———, "Geology of a Portion of the Coosa Valley in Georgia and Alabama," *ibid.*, Vol. 5 (1894), pp. 465–80.

[15] Charles Butts, "Geology and Mineral Resources, Tyrone Quadrangle," *Pennsylvania Geol. Survey*, 4th Ser., A96 (1939).

[16] Marius R. Campbell, "Paleozoic Overlaps in Montgomery and Pulaski Counties, Virginia," *Bull. Geol. Soc. America*, Vol. 5 (1894), pp. 171–90. Retracted in *Virginia Geol. Survey Bull. 25* (1925), pp. 56–96.

thrust faults. Keith followed Campbell at first, mistaking faults for unconformities and as a result mixing much of Safford's excellent stratigraphy, but later he reversed himself and announced[17] the discovery of low-angle thrust faults as large as those of Hayes, on one of which the pre-Cambrian basement has been carried for miles out over younger rocks. This discovery appeared to prove that the low-angle faults were different in kind as well as in degree from the high-angle thrust faults previously recognized, many of which, as the Rogers brothers had noted, seem to have developed out of asymmetrical anticlines by breaking across the steeper limb. In contrast to Hayes, Keith believed that at least this fault was the first product of the main Appalachian deformation and that it was itself folded and faulted in a later phase of the deformation.

In the meantime, Willis attacked the theoretical implications of Appalachian structure.[18] On the basis of experimental models, he developed the concept of competence of sedimentary strata, an idea of general application in structural geology. His experiments also suggested to him that the contrast between the folded middle and faulted southern Appalachians resulted from a difference in thickness of overburden on the most competent stratum at the time of deformation, and that the individual folds and faults were localized by initial dips produced by unequal sedimentation in the geosyncline. These two ideas are stimulating and plausible but they remain unproved, for the experiments on which they were based were inadequate because the strength of the materials used was many times too great to serve as a scale model. The second of the two ideas seems inapplicable to the Appalachians, however well it may work elsewhere, because it fails to explain the remarkable length, linearity, and parallelism of most of the Appalachian structural trends. Willis and the others believed that all the deformation was the result of tangential forces, but Willis held for many years that the forces acted from the northwest, hence that all Appalachian folds and faults, excepting presumably the Murphrees Valley anticline, are underthrust.[19]

On the basis of the unconformities postulated by Campbell and Keith, Willis and his co-workers held for some years that folding recurred several times during the Paleozoic, though strongest at the end. This idea was further elaborated by E. O. Ulrich[20] (1857–1944), who maintained that deformation progressed very

[17] Arthur Keith, "Roan Mountain, Tennessee-North Carolina," *U. S. Geol. Survey Geol. Atlas Folio 151* (1907). First announced in *Science*, New Ser., Vol. 15 (1902), pp. 822–23.

[18] Bailey Willis, "The Mechanics of Appalachian Structure," *U. S. Geol. Survey 13th Ann. Rept.*, Pt. 2 (1893), pp. 211–81.

Bailey Willis and C. Willard Hayes, "Conditions of Appalachian Faulting," *Amer. Jour. Sci.*, 3d Ser., Vol. 46 (1893), pp. 257–68.

[19] Bailey Willis, *op. cit.;* retracted in *Science*, New Ser., Vol. 25 (1907), p. 867. For a general summary of the conclusions of the group, see Arthur Keith, "Outlines of Appalachian Structure," *Bull. Geol. Soc. America*, Vol. 34 (1923), pp. 309–80.

[20] E. O. Ulrich, "Revision of the Paleozoic Systems," *Bull. Geol. Soc. America*, Vol. 22 (1911), pp. 281–680. The trough and barrier idea was first set forth by E. O. Ulrich and Charles Schuchert, "Paleozoic Seas and Barriers in Eastern North America," *New York State Mus. Bull. 52* (1902), pp. 633–63, but it should be recorded here that in his later years Schuchert abandoned the idea as applied to the middle and southern Appalachians (personal communication, 1940).

slowly across the Appalachian region from southeast to northwest. Thus in early Paleozoic time the severe deformation was confined mostly to areas southeast of the present folded and faulted belt, in which its only result was gentle folding of the floor of the geosyncline into narrow subparallel troughs and barriers. Late in the Paleozoic the deformation reached and elevated the western part of the Piedmont, giving rise to the thick clastic sediments of the Pennsylvanian. Not until the middle Mesozoic did folding and faulting affect the present Valley and Ridge Province, while on the southeast the block faulting of the Triassic basins took place. The final low-angle thrust faulting Ulrich assigned to the Cenozoic. Few geologists to-day would thus extend the Appalachian orogeny over most of post-Cambrian time, but others have questioned its Permian date. Though the main deformation was almost certainly later Paleozoic, the possibility remains that it took place in a series of pulses affecting different parts of the chain at different times.

Since the folio work under Willis' direction, much geologic work has been carried on in the Appalachians, especially in the coal fields of the Appalachian Plateau. Probably the outstanding field geologist of this period was Charles Butts (1863–1946), who devoted virtually his entire life to Appalachian geology and whose work will form a basis for future investigators to build on from Pennsylvania to Alabama.

RECENT TRENDS

During the present century an important difference of opinion has slowly been developing among students of the Appalachians, though often the protagonists have hardly seemed aware of it. The difference of opinion concerns the depth of the deformation; does it or does it not involve the basement? Until approximately 40 years ago, no one thought it necessary to argue the point; it seemed self-evident that the folds and faults continued downward through the sedimentary sequence into the pre-Cambrian. Apparently the first conscious argument in favor of this view was by George H. Ashley,[21] who, discussing Willis' idea that a competent stratum in an anticline supports the overlying incompetent beds, described a low arch in the Appalachian Plateau of Pennsylvania in which no bed could have supported even itself without far surpassing the crushing strength of any rock known. According to Ashley, therefore, all the beds must have been supported from below. More recently Sherrill,[22] in a careful discussion of the folds of the plateau region, amplified Ashley's arguments and added others, tending to show that all those folds must have been continually supported from the basement. Neither Ashley nor Sherrill dealt with the possible application of this idea to the much steeper folds of the Valley and Ridge Province.

[21] George H. Ashley, "Studies in Mechanics of Allegheny Structure," *Science*, New Ser., Vol. 27 (1908), pp. 924–25.

[22] Richard E. Sherrill, "Symmetry of Northern Appalachian Foreland Folds," *Jour. Geol.*, Vol. 42 (1934), pp. 225–47.

This application has been made, however, by Nevin and especially by Ver Wiebe,[23] who argued from the limited strength of rocks, from the outpost position of such large folds as the Nittany and Sequatchie anticlines, and from certain hypotheses about the development of the geosyncline that the core of each major anticline and the upthrown side of each major thrust fault in the Appalachians is supported by an upwarp or upthrow in the basement. In other words, the major deformation took place in the basement, and the sedimentary strata are merely draped over it.

The opposite view also took form slowly. In 1921, Wentworth,[24] working on a suggestion from M. R. Campbell, recognized that a peculiar belt of faulting and crushing in Dickenson and Buchanan counties in southwest Virginia completes the outline of a great quadrilateral block (first recognized by Safford), the northwest side of which is formed by the Pine Mountain thrust fault, and the whole of which has been pushed several miles northwestward. Wentworth realized that the Pine Mountain fault has a low dip but apparently assumed that it continues indefinitely downward across the strata. In 1927, Butts[25] announced the discovery of several windows in the thrust block more than 15 miles southeast of the Pine Mountain fault, proving that, far from descending to the basement, the fault remains at a shallow depth for many miles southeastward.

In 1934, John L. Rich[26] presented a plausible mechanical interpretation of this fault, based on the concept that it developed largely along the bedding of the weakest formations in its path, and he suggested that the other Appalachian thrust faults may likewise follow bedding planes at shallow depths for many miles southeast of their present traces and that the basement beneath the faulted belt may be quite undisturbed. Faults of this kind would not be simply broken anticlines; on the contrary, Rich showed that many of the folds may rather be the result of the faulting. Moreover, the distinction between high-angle and low-angle thrust faults falls away, for the same fault may be high-angle and low-angle at different points down the dip. In recent years, increasing study of the low-angle thrust faults of the southern Appalachians has converted most of us now working in that area to views much like those of Rich. Indeed Ralph Miller,[27] in a recent abstract, has suggested as one possibility that all the thrust faults in

[23] Charles M. Nevin, *Structural Geology*, 1st ed. (1931), pp. 54, 62; 2d ed. (1936), pp. 55–57, 65–66; 3d ed. (1942), pp. 50–53, 59–60. John Wiley and Sons, New York.
Walter A. Ver Wiebe, "Geosynclinal Boundary Faults," *Bull. Amer. Assoc. Petrol. Geol.*, Vol. 20 (1936), pp. 910–38.

[24] Chester K. Wentworth, "Russell Fork Fault of Southwest Virginia," *Jour. Geol.*, Vol. 29 (1921), pp. 351–69; *Virginia Geol. Survey Bull. 21* (1921), pp. 53–66.

[25] Charles Butts, "Fensters in the Cumberland Overthrust Block in Southwestern Virginia," *Virginia Geol. Survey Bull.* 28 (1927).

[26] John L. Rich, "Mechanics of Low-Angle Overthrust Faulting as Illustrated by Cumberland Thrust Block, Virginia, Kentucky, and Tennessee," *Bull. Amer. Assoc. Petrol. Geol.*, Vol. 18 (1934), pp. 1584–96.

[27] Ralph L. Miller, "Relation of Overthrust Faults in Southwest Virginia," *Bull. Geol. Soc. America*, Vol. 56 (1945), pp. 1182–83.

southwest Virginia and east Tennessee may be branches of a single nearly flat sole fault.

Another current controversy, which in some respects runs parallel with the one just discussed, concerns the width and eastward extent of the belt that developed into a geosyncline during the Paleozoic and was deformed in the Appalachian orogeny. According to what is perhaps the traditional view, widely held among those who believe that folds continue to the basement and championed in an extreme form by Ver Wiebe,[28] the eastern margin of the geosyncline lay close to the present northwest edge of the crystalline rocks of the Blue Ridge and Piedmont. Thus, the geosyncline coincided almost exactly with what is now the belt of folded and faulted but unmetamorphosed rocks, and this belt was the principal locus of the late Paleozoic deformation. The opposite view[29] holds, on the contrary, that large bodies of the schist in the Piedmont are Paleozoic, as much of the corresponding schist in New England is now known to be. On this view, the present folded and faulted belt represents not the deepest part of the Appalachian geosyncline but merely a northwestern and lesser (*mio*geosynclinal) part. Similarly, the structure of the Valley and Ridge Province, mammoth as it is, is only marginal deformation (the term is Bucher's) with respect to the main Appalachian deformation recorded in the metamorphic and igneous rocks on the southeast. The decision between these two views must await long years of detailed and painstaking work among the crystalline schists of the Piedmont.

Thus, the natural development of ideas has led at length to two very different concepts of Appalachian structure, which may perhaps be characterized as the "thick-skinned" and the "thin-skinned" schools of thought.[30] The thick-skinned geologists, most of whom have worked chiefly in the middle Appalachians and the flat-lying rocks of the Appalachian Plateau or farther west, have reasoned from the broad low folds of the plateau to the hypothesis that the folds and faults of the Valley and Ridge Province are the main manifestation of the Appalachian orogeny and that each continues separately to the basement. The thin-skinned geologists, most of whom have worked chiefly in the highly faulted southern Appalachians, have reasoned from the great low-angle thrust faults of that area,

[28] Walter A. Ver Wiebe, *op. cit.*

[29] This view has been suggested by several writers, for example:
Walter H. Bucher, *The Deformation of the Earth's Crust*, pp. 149–61. Princeton (1933).
Marshall Kay, "Geosynclines in Continental Development," *Science*, New Ser., Vol. 99 (1944), pp. 461–62.
But the first detailed application known to the present writer is that by Ernst Cloos in his contribution to the present symposium:
Ernst Cloos, "Structures of Basement Rocks of Pennsylvania and Maryland and Their Effect on Overlying Structures" (abstract), *Bull. Amer. Assoc. Petrol. Geol.*, Vol. 32, No. 11 (November, 1948), p. 2162.

[30] These terms recall the distinction drawn by Rollin T. Chamberlin in "The Building of the Colorado Rockies," *Jour. Geology*, Vol. 27 (1919), pp. 145–64, 225–51 (esp. pp. 248–51; *cf.* also *ibid.*, Vol. 29 (1921), pp. 166–72) between thick-shelled and thin-shelled mountain ranges, but although Chamberlin classed the Appalachians as a thin-shelled range, he believed that the deformation affected the crust to a depth of 6–30 miles (see *Jour. Geol.*, Vol. 18 (1910), pp. 228–51), so that he may definitely be classed in the "thick-skinned" school.

especially the Pine Mountain fault, to the hypothesis that the Paleozoic rocks of the unmetamorphosed part of the Appalachians, caught on the margin of intense deformation on the southeast, were stripped completely off the basement and piled up against the unyielding plateau in great rootless folds and imbricate thrust sheets underlain by a few bedding-plane faults of immense displacement. The two schools of thought have gone almost as far as they can at present in setting forth the rival hypotheses and deducing their implications. It is time now to confront these hypotheses with facts.

VINTON E. GWINN *McMaster University, Hamilton, Ontario, Canada*

Thin-Skinned Tectonics in the Plateau and Northwestern Valley and Ridge Provinces of the Central Appalachians

Abstract: Folds in the Appalachian Plateau and northwestern Valley and Ridge provinces of the Central Appalachians involve allochthonous Paleozoic rocks which have been transported westward along nonoutcropping low-angle detachment thrust faults. Deep wells in the Plateau and along the Appalachian Structural Front southeast of the Plateau penetrate major overthrusts which form part of the sole fault system. The great folds along the arcuate Structural Front were formed in passive response to steplike upward shearing of the detachment thrusts from a mid-Cambrian shale glide zone in the Valley and Ridge province up through competent carbonate rocks into glide zones in the Upper Ordovician or Upper Silurian beneath the Plateau. The folds evidence shortening of the stratified sequence superincumbent on the sole thrusts.

Anticlines in the Plateau and in local parts of the Valley and Ridge provinces terminate or change trend abruptly in groups along west-to-northwest-trending lineaments. Some of the lineaments appear to reflect tear-faulting along the margins of semi-independently advancing thrust blocks, whereas others, along which no faulting is noted, seem to be zones at which the sole thrusts change stratigraphic position along the structural strike. The termination of sets of anticlines along the lineaments and the absence of uniformly offset counterparts across the lineaments in adjacent blocks indicate that a large part of the net subhorizontal translation of the thrust sheets occurred prior to most of the folding, very early in the deformation.

Both limbs of most of the anticlines in the northwestern two thirds of the folded Plateau region are broken by synclineward-dipping thrust faults at the Lower Devonian level. These thrust faults, which uplift both anticlinal flanks above a relatively depressed former axial zone, appear to flatten into bedding in evaporites of the Upper Silurian Salina group. The faults facilitate movement of material up and out of the tightening synclines onto the anticlines and are thus simply an extension of the flexural-slip folding process. Structural relief of the anticlines fault-folded in this manner is significantly reduced beneath the Salina décollement zone.

The tectonic style and mode of deformation of the "folded" Central Appalachians is, thus, practically identical with that observed in the "thrust-faulted" Southern Appalachians. The only important difference between the two regions is that the upward-shearing segments, or "toes," of thrust sheets in the Southern Appalachians have been exposed by erosion, whereas the toes of the major Central Appalachian thrust sheets are stil covered by a mile or more of stratified rocks.

CONTENTS

Geological Society of America Bulletin, v. 75, p. 863–900, 20 figs., 2 pls., September 1964

INTRODUCTION

Statement of Problem

The mechanics of deformation in the Appalachian Valley and Ridge and Plateau region can be interpreted according to either the "thick-skinned" or "thin-skinned" concepts (Rodgers, 1949). According to Rodgers (p. 1653–1654):

"The thick-skinned geologists, most of whom have worked chiefly in the middle Appalachians [Fig. 1] and the flat-lying rocks of the Appalachian Plateau or farther west, have reasoned from the broad low folds of the plateau to the hypothesis that folds and faults of the Valley and Ridge Province are the main manifestation of the Appalachian orogeny and that each continues separately to the basement [*see* Cooper, 1961, p. 100–118]. The thin-skinned geologists, most of whom have worked chiefly in the highly faulted southern Appalachians [Fig. 1], have reasoned from the great low-angle thrust faults of that area, especially the Pine Mountain fault, to the hypothesis that the Paleozoic rocks of the unmetamorphosed part of the Appalachians, caught on the margin of intense deformation on the southeast, were stripped completely off the basement and piled up against the unyielding plateau in great rootless folds and imbricate thrust sheets underlain by a few bedding-plane faults of immense displacement." (Bracketed words are author's.)

This paper is a study of the tectonics of the "unfaulted" central Appalachian Plateau and adjacent parts of the Appalachian Valley and Ridge province (Fig. 2). New data, illustrating complex subsurface structure as revealed by recent exploration for natural gas in the Appalachian basin, are combined with surface geological data: (1) to present evidence that folds of the Plateau and western Valley and Ridge provinces are surficial, compressional wrinkles developed above deep nonoutcropping bedding-plane thrust faults of large displacement, which continue northwestward beneath the Plateau from the western margin of the Valley and Ridge province; (2) to illustrate that folding at intermediate and upper levels of the sedimentary cover of the Plateau is controlled by a décollement surface within Upper Silurian Salina group evaporites and shales; and (3) to demonstrate the presence and mechanical importance of northwest-trending lineaments in the Plateau, which are apparently zones along which either the nonoutcropping detachment thrusts change stratigraphic position or along which tear-faulting occurs. Surface and subsurface anticlines terminate or change trend markedly along the lineaments.

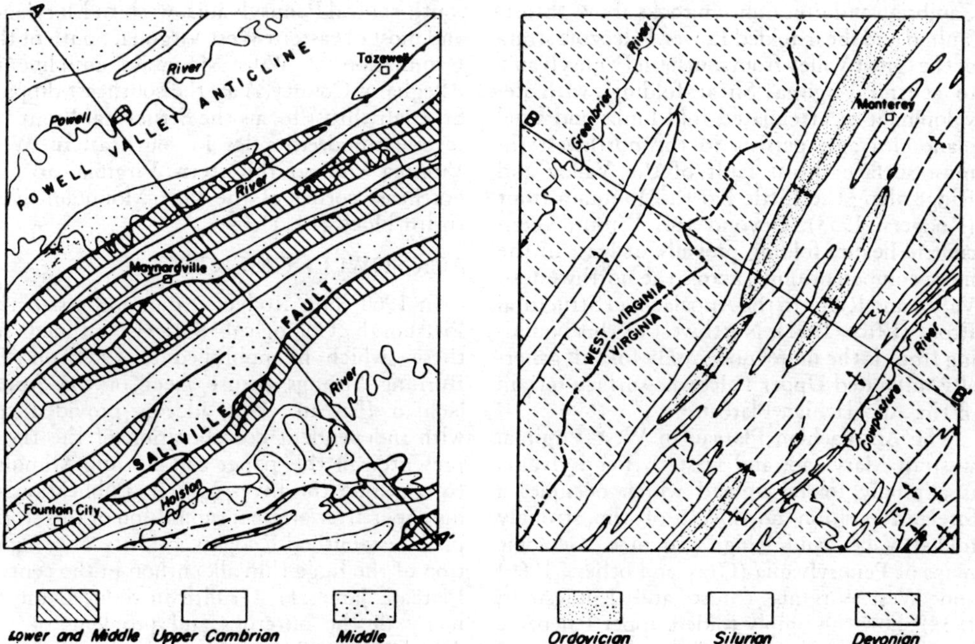

Figure 1. Apparent contrasts in tectonic style of Southern and Central Appalachians. Left, prevalent imbricate thrust faulting of the Southern Appalachians, eastern Tennessee. Right, apparent absence of faulting in the "folded" Central Appalachians, Monterey quadrangle, Virginia-West Virginia. (*After* Fig. 31, King, 1951, by permission of Princeton Univ. Press)

General Structural Setting

The Central Appalachians consist of a great arcuate salient, which extends 450 miles from northeastern New Jersey to southwestern Virginia where it intersects the arcuate salient of the Southern Appalachians. The apex of the Central Appalachian salient is in Pennsylvania.

Cross sections and maps both show that the Central Appalachians have a regular topographic and structural plan along their entire length (Fig. 2). Four parallel structural provinces extend longitudinally along the deformed belt, each with its own topographic and structural characteristics. From southeast to northwest the provinces are the metamorphic Piedmont terrain, the Blue Ridge crystalline uplift, the folded Valley and Ridge province, and the broad, less intensely deformed Appalachian Plateau or so-called Allegheny synclinorium (Kay, 1942).

King (1950; 1954) and others have shown that at the surface of the northwestern margin of the Blue Ridge anticlinorium or Front is a zone of steep flexures which raise Lower Cambrian and Precambrian rocks above thrust-faulted and shear-folded Lower Paleozoic strata of the Great Valley from southern Pennsylvania to central Virginia. Shear folding, with development of attenuated fold limbs and axial plane cleavage, persists to the northwestern-most surface thrust fault of the Valley and Ridge along the Little North Mountain front (Rodgers, 1953). Beyond this "front," concentric flexure folding prevails, at least in the more competent units, across the northwestern Valley and Ridge to the Appalachian Structural Front (Price, 1931). Northwest of this bounding front is the more moderately folded, generally unfaulted Upper Paleozoic surface terrain of the Appalachian Plateau.[1]

The Appalachian Plateau in West Virginia, western Maryland, and western Pennsylvania is a mildly folded upland which occupies a foreland position northwest of the strongly folded Valley and Ridge province. Geologic maps of Pennsylvania (Gray and others, 1960) and West Virginia (Stose and Ljungstedt, 1932) show the simply folded upper Paleozoic terrain of the Plateau and the diminution of deformation northwestward from the Valley and Ridge province. Viewed on a regional basis, the surface structure of the Plateau is a broad asymmetrical synclinorium, steeper on the southeast—the Allegheny synclinorium of Kay (1942). The surface axis of the synclinorium plunges southwestward in a smooth arc from Binghamton, New York, through Pittsburgh, Pennsylvania, and south-southwestward toward Parkersburg and Huntington, West Virginia. Because of general southeastward or eastward depositional thickening of all stratigraphic units, the structural axis of the basin as measured by depth to basement rocks is some imprecisely known distance southeast of the Appalachian Structural Front beneath the Valley and Ridge province.

The Appalachian Structural Front (Price, 1931) is the southeastern margin of the Plateau and the northwestern margin of the Valley and Ridge province. The arcuate Front is a fundamental structural declivity, formed in the area of this study by the steep or overturned west limbs of the frontal anticlines of the Nittany anticlinorium in central Pennsylvania and by Hyndman and Wills Mountain anticlines in south-central Pennsylvania, western Maryland, and most of eastern West Virginia. South of the termination of Wills Mountain anticline in Alleghany County, Va., the southeast-dipping St. Clair thrust forms the Structural Front for several hundred miles in southeastern West Virginia and southwestern Virginia, to the northeast corner of the Pine Mountain overthrust block (Pl. 1).[2]

ACKNOWLEDGMENTS

In 1960 John Rodgers in an address to the Pittsburgh Geological Society outlined his thesis, which he published in 1963, on the Burning Springs thrust sheet in the Appalachian Plateau. This address provided me with independent corroboration of the thesis presented in this paper and a strong stimulus to continue the investigation. Publication of his paper after original submission of this study justly established Rodgers' priority of recognition of the huge thin allochthon in the central Plateau. James H. Griffith introduced me to many of the outcrops and problems of the Appalachians. James H. Griffith, Glenn J. Poulter, Richard E. Jones, and Henry P. Schwarcz provided helpful checks and balances in the progress of research; John Dennison, Wallace McCord, William Overby, and R. C. Tucker of the West Virginia Geological Survey, Addison Cate, Dick Wagner, William Lytle, and William Edmonds of the Pennsylvania Geological Survey, Earl Tarr, Tidewater Oil, M. A. Vaughn, Phillips Petroleum, J. H. Miller, Manufacturer's Light and Heat Com-

[1] For a more complete summary of regional structure of the Valley and Ridge province of the Central Appalachians, the reader is referred to King (1951).

[2] This definition of the Front concurs with that proposed by Rodgers (1953; 1963), but differs from that of the originator of the term, Price (1931), and from that of Ludlum (1958). Price and Ludlum chose to include several high folds in central-eastern West Virginia in the Valley and Ridge province. If these outlying folds are considered a part of the Plateau, the Front becomes a smooth arcuate feature from Pennsylvania to Tennessee. Moreover, significant structural relationships outlined further in the paper justify the exclusion of the outlying folds and redefinition of the Front in the manner suggested by Rodgers (1963).

pany, W. A. Anderson, Cities Service Oil Company, John Rouse, Mobil Oil Company, Robert Bayles, Hope Natural Gas Company, and W. J. Perry, United Fuel Gas Company assisted in collection of critical structural data. I assume responsibility for all errors of fact or interpretation presented, but I thank John Rodgers, Philip B. King, Donald U. Wise, and Paul M. Clifford for critical reading of the manuscript submitted for publication.

The study was initiated while I was employed by Mobil Oil Company. Mrs. J. Pluhar and Donald M. Bain assisted in drafting several figures. Financial support was provided by the National Research Council of Canada and the Pennsylvania Geological Survey to complete the study.

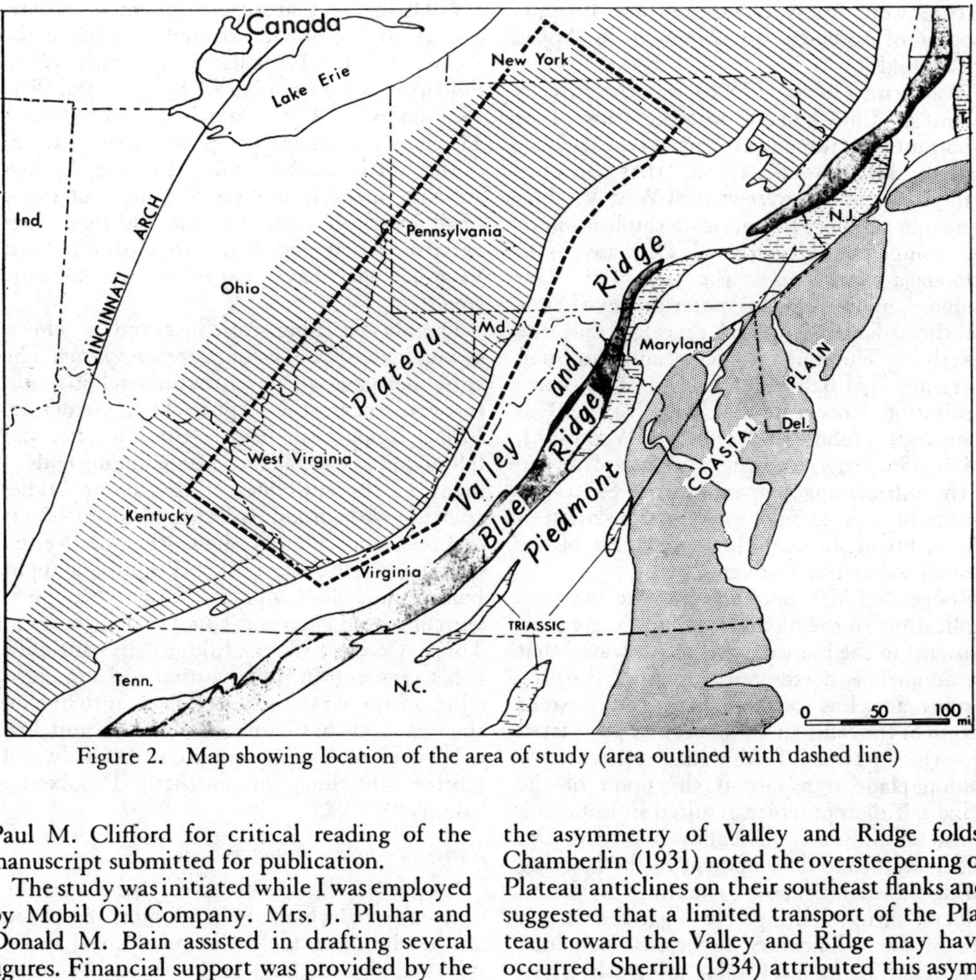

Figure 2. Map showing location of the area of study (area outlined with dashed line)

PREVIOUS INVESTIGATIONS

Early geologic investigators of the central Appalachian region (reviewed by Sherrill, 1934; Rodgers, 1949) recognized the decrease in intensity of deformation northwestward from the Appalachian Valley and Ridge into the Plateau on the northwest. They reasoned that deforming forces had been applied to the southeastern side of the folded Appalachians and that relative tectonic transport had taken place from southeast to northwest because of the northwestward decrease in size of anticlines and the asymmetry of Valley and Ridge folds. Chamberlin (1931) noted the oversteepening of Plateau anticlines on their southeast flanks and suggested that a limited transport of the Plateau toward the Valley and Ridge may have occurred. Sherrill (1934) attributed this asymmetry, however, to a southeastward thickening of the sedimentary sequence.

Ashley (1908), Sherrill (1934), and Cathcart and Myers (1934) set forth the argument in favor of basement involvement in the broad, low folds of the Appalachian Plateau, arguing that the strength of the thin cover of sedimentary rocks in the Plateau and Valley and

Ridge was not great enough to have transmitted the forces necessary to cause the folding, and concluded that the anticlines must be supported continuously at depth by uplifted basement. Fettke (1933) and Finn (1949) suggested that the Upper Silurian evaporites and Devonian shales might have undergone limited plastic or viscous flow in the cores of the anticlines. Until the late 1950's, most geologists working in the Central Appalachian Plateau were firmly committed to the philosophy of basement involvement in Appalachian folds.

Woodward (1959a; 1959b) was the first to present a "thin-skinned" interpretation of an important structure in the central Appalachian Plateau. He showed (1959a) that Burning Springs anticline in west-central West Virginia formed in response to imbricate duplication of the competent Helderberg, Oriskany, and Onondaga formations at the base of the Devonian. Unpublished dip-meter surveys below the thrust-faulted interval revealed that the underlying Silurian, Ordovician, and Cambrian strata are essentially undeformed (Oral communication, Robert Bayles, Hope Natural Gas Company). John Rodgers (*in* Woodward, 1959b; 1963) suggested that bedding-plane slip in the salt-bearing Upper Silurian Salina sequence brought about westward translation of a long, broad sheet of the upper part of the Plateau sedimentary cover.

Rodgers (1963) pointed out the regional implications of the discovery of low-angle fault transport in the Plateau. He demonstrated that the anomalous north–south-trending Burning Springs anticline occurs along the western margin of the Silurian salt facies and postulated that the increase of frictional resistance to bedding-plane transport at the point of the Salina salt disappearance resulted in imbricate thrust faulting of immediately overlying Devonian strata and development of the Burning Springs anticline above the line of facies-induced thrust faulting. He suggested that a slice of the sedimentary cover that extended from central Pennsylvania to southeastern West Virginia had moved westward a short distance along bedding thrust faults. He also suggested that the termination of folds, deviations of fold axes, and presence of irregular domes and basins in a linear zonetrending south-southeast from the southern end of Burning Springs anticline in Wirt County, W. Va., toward Mercer County, Ohio, probably reflects tear-faulting along the southwest margin of the thrust block (*see* Pl. 1).

Cate (1962) presented revisions of generalized small-scale maps of Oriskany structure in western Pennsylvania submitted in 1954 by Fettke.

SURFACE STRUCTURE

General Character of Surface Folds

Folds in the Plateau are generally arcuate, convex to the northwest, and parallel to the arc of the Central Appalachians (Keith, 1925). Fold axes swing from N. 30° E. in central West Virginia to N. 70° E. in northern Pennsylvania. The only prominent exception to this pattern is the north–south-trending Burning Springs anticline in western West Virginia, but many small anticlines with less than 100 feet of reversal on their east flanks in southern West Virginia also have irregular, unsystematic trends.

Doubly plunging anticlines and synclines, as much as 100 miles long, are *en echelon.* The anticlines have greater structural relief southeastward near the Appalachian Structural Front. Along their length, the folds rise and fall in domes and saddles. The plunging ends of many of the anticlines or the major saddles coincide with prominent changes in or offsets of fold trend. Plate 1 shows that in Preston County, northeastern West Virginia, a slight but definite offset separates Briery Mountain anticline from Etam anticline; that in southern Tucker County, W. Va., Elkins Valley anticline is between and in juxtaposition to Etam anticline on the west and Deer Park anticline on the east; and that Brown's Mountain anticline is between and in juxtaposition to Glady and Horton anticlines in northern Pocohontas County, W. Va.

Rarity of Surface Faulting

Fettke (1954, p. 7) observed that where the Central Appalachian arc bends most sharply in southeastern Clearfield County, central Pennsylvania, the foreland folds are cut by small transverse faults. Apparently, ***this is the only area where faults have been recognized as a conspicuous part of the surface geologic pattern in the central Appalachian Plateau.*** Nevertheless, small strike faults occur in the northeastern Plateau area (Finn, 1949), on Chestnut Ridge anticline in Westmoreland County, Pa. (Shaffner, 1958), on Chestnut Ridge in Monongalia County, W.

Va. (Eugene Simms, 1961, M.Sc. thesis, West Virginia Univ.), and transcurrent faults occur on Deer Park anticline in southern Tucker County, W. Va. (W. K. Overby, personal communication, June 1963).

Intensity of Folding and the Intra-Plateau Structural Front

Structural relief on anticlines decreases in step-like fashion northwestward in all parts of the Plateau. A zone of high, well-defined folds is present in the southeastern side of the Plateau from northern Pennsylvania to southern West Virginia (Pl. 1), in which anticlines rise 800–3500 feet above adjacent synclines. Farther northwest, local structural relief is less; folds are very faint in northwesternmost Pennsylvania, the northern West Virginia panhandle, Ohio, and southwestern West Virginia.

The southeastern high fold zone is bounded on the northwest (proceeding from northeast to southwest) by the Sabinsville-Tioga anticline from New York State into northern Elk County, Pa.; by Chestnut Ridge anticline from southern Indiana County, Pa., to southern Monongalia County, W. Va.; Etam anticline in Tucker County, W. Va.; Elkins Valley anticline in Randolph County, W. Va.; Brown's Mountain anticline in Pocahontas and Greenbrier counties, W. Va.; Williamsburg anticline in Greenbrier County; and Abbs Valley anticline in Monroe and Mercer counties, W. Va.

Sabinsville-Tioga anticline is the northwesternmost high fold in the outlying block of the southeastern high fold zone on the northeastern end of the Plateau. Anticlines in this part of the zone rise only 1000–1500 feet above adjacent synclines and are somewhat lower than those in southern Pennsylvania.

The changes of structural level across the frontal folds of the high southeastern Plateau zone, although of a lower magnitude, are entirely analogous to those incurred on crossing the main Appalachian Structural Front from the Valley and Ridge into the Plateau. By analogy, the high northwestern margin of the southeastern Plateau is herein designated as the Intra-Plateau Structural Front.

The continuity of the southeastern high fold zone of the Plateau is disrupted in central Pennsylvania where prominent folds plunge abruptly from the northeast and southwest into a wedgelike area, which tapers southeastward toward the Valley and Ridge area. Within it anticlines are only 300–900 feet higher than the synclines and are irregular in trend. Prominent northwest-striking high-angle surface faults (*see* Fettke, 1954), unobservable elsewhere in the Plateau, border the southeastern part of the wedge. The first folds in the Plateau which persist through this disrupted zone are Sabinsville anticline (and Caledonia syncline on its southeastern flank) in northwestern Clearfield County, Pa., and adjacent portions of Elk, Jefferson, and Indiana counties, Pa. Chestnut Ridge and Laurel Hill anticlines are not continuous through this area, despite a report to the contrary (Fettke, 1954).

West-northwest-trending lineaments, drawn through the ends of anticlines forming segments of the Intra-Plateau Structural Front or at points of prominent trend change on the frontal anticlines, coincide remarkably with terminations of anticlines or trend changes along anticlines, both northwest and southeast of the Front. Conspicuous rectilinear blocks are defined within the Plateau by the lineaments and by groups of high anticlines along the Intra-Plateau Front bounded by the lineaments. (*See* Pl. 1.)

Opposed Asymmetry of Southeastern and Northwestern Plateau Anticlines

Despite the pattern of northwest decrease in intensity of folding, it is surprising that almost all Plateau anticlines north and northwest of the Elkins Valley-Etam-Briery Mountain-Negro Mountain anticlinal trend are oversteepened on their southeastern limbs (Sherrill, 1934, p. 243), in a sense, exactly opposite to both the asymmetry of anticlines in the Valley and Ridge province and the apparent prevailing northwesterly direction of relative tectonic transport indicated by thrust faulting. Large and small anticlines in both northwestern and southeastern parts of the Plateau are oversteepened on their southeastern limbs. Thus, there is no correlation between intensity of folding and sense or degree of asymmetry.

SUBSURFACE STRUCTURE

General Statement

Deformation of competent rocks at the base of the Devonian in the low Plateau anticlines is much more complex than that at higher levels.

The few wells drilled into or below the Upper Ordovician in the Plateau reveal that below the competent Lower Devonian the

structure becomes more simple again, apparently in several stages. The writer believes that these vertical contrasts in deformational intensity and style reflect bedding-plane detachment thrust faults and décollement surfaces in zones of the Upper Silurian, the Upper Ordovician, and the top of the Lower Cambrian. Table 1 is a brief description of the Central Appalachian stratigraphic sequence.

TABLE 1. GENERALIZED STRATIGRAPHY OF THE CENTRAL APPALACHIANS

System	Lithic Unit	Approximate thickness (feet) West	Approximate thickness (feet) East
Permian	Cyclic sandstone, limestone, mudstone, minor coal	0–800	0 (Eroded)
Pennsylvanian	Cyclic sandstone, limestone, mudstone, and coal	1400	2000+–3000+
Mississippian	Mauch Chunk redbeds, Greenbrier limestone, Pocono sandstone	500–700	1500–5000
Devonian	Catskill redbeds	0–200	1200–2000
	Devonian "shale," in descending order, Chemung sandstone-shale "facies," Portage shale "facies," Tully limestone, Hamilton shale	2000–3250	4500–6500
	Onondaga limestone (NW.), Huntersville chert (SE.) Oriskany sandstone Helderberg limestone and shale	100–600	750–1000
Silurian	Salina group (NW.). Tonoloway limestone, Wills Creek shale, and Bloomsburg redbed facies (SE.): salt, gypsum, shale and carbonates in Salina; evaporites wedge out east and west	1000–1700	900–1300
	Lockport dolomite (NW.), McKenzie-Rochester shale (SE.)	200–300	300–500
	Clinton group (NW.), Rose Hill formation (SE.) Tuscarora sandstone (SE.), Medina sandstone (NW.)	450–600	1200–1500
Ordovician	Upper: Queenston shale (N. and NW.), Juniata red shale and sandstone (SE.) Oswego sandstone (eastern Plateau only)	450–1050	1000–1800
	Martinsburg shale	700–1000	1000–2000
	Middle: carbonate rocks, including Trenton, Black River, and Chazy	900–1200	1200–1600
	Lower: Beekmantown dolomite	1000	3000–4500
Cambrian	Middle and Upper Cambrian carbonates (including Conococheague and Elbrook to the east)	1100–1600	4000–5500?
	Waynesboro Formation Tomstown-Shady dolomite Chilhowee group	Thickness in Plateau not determined by drilling; units outcrop only in Blue Ridge and Great Valley to the southeast; only Upper Cambrian rocks are present in western West Virginia and western Pennsylvania because of Cambrian onlap from the southeast.	
Precambrian	Catoctin greenstone		

The following discussion of subsurface structure is divided so as to emphasize areal and vertical variations in structural configuration: (1) intermediate level Oriskany-Onondaga structure in the Plateau north and west of the Elkins Valley and Deer Park anticlines; (2) structure at deeper levels in the same area; (3) structure along the Appalachian Structural Front; and (4) structure in the intervening southeastern High Plateau zone of West Virginia, Maryland, and south-central Pennsylvania.

Intermediate-Level Structure North and West of Elkins Valley and Deer Park Anticlines

Anticlines which are simple and unfaulted at the surface are highly faulted at the level of the Onondaga and Oriskany. Contrasts between the surface and subsurface structural patterns

are readily evident in a series of small-scale regional maps of the Plateau in Pennsylvania by Fettke (1954) and Cate (1962). Figures 3 through 12 illustrate Oriskany or Onondaga structure as revealed by drilling in Pennsylvania and West Virginia.

CRESTAL DEPRESSIONS AND SYMMETRICAL THRUST FAULTING OF ANTICLINES: The most striking Lower Devonian structures are relatively depressed fault blocks ⅛–3 miles wide along or near the anticlinal axes. Several of these occur along the crest of Sabinsville-Tioga anticline in the Tioga Field (Fig. 3) and on Briery Mountain anticline in the Terra Alta Field (Fig. 4). Typically, however, there appears to be only one relatively down-faulted depression, as on Chestnut Ridge and Laurel Hill anticlines (Fig. 5).

The faults bounding the axial depressed zones are reverse faults, reflecting a thrust of the anticlinal limbs toward the anticlinal crests, and thus they can hardly be called grabens. The faults all thicken the section, repeat key beds and formation contacts, and offset subsurface structural contours in a manner consistent with reverse faulting. Deviations from the normal stratigraphic thickness of the Oriskany, the Onondaga, the Hamilton Group, and the distinctive Tully Limestone are easily discernible. Key horizons, such as the unconformity between the Oriskany and Onondaga, the glauconite zones in the Onondaga, brown "metabentonites" at or immediately above the top of the Onondaga, and the black Marcellus Shale at the base of the Hamilton Group provide precise controls on the nature of the faulting. Tectonic thickening of 50–150 per cent is common along the margin of the depressed zone near the lip of the upthrust flank blocks. Thickening of approximately 25 per cent occurs in the Onondaga-Oriskany interval of the flanks, probably due largely to intersection of dipping beds by the near vertical bore hole, but in part to crumpling and small bedding-plane faults.

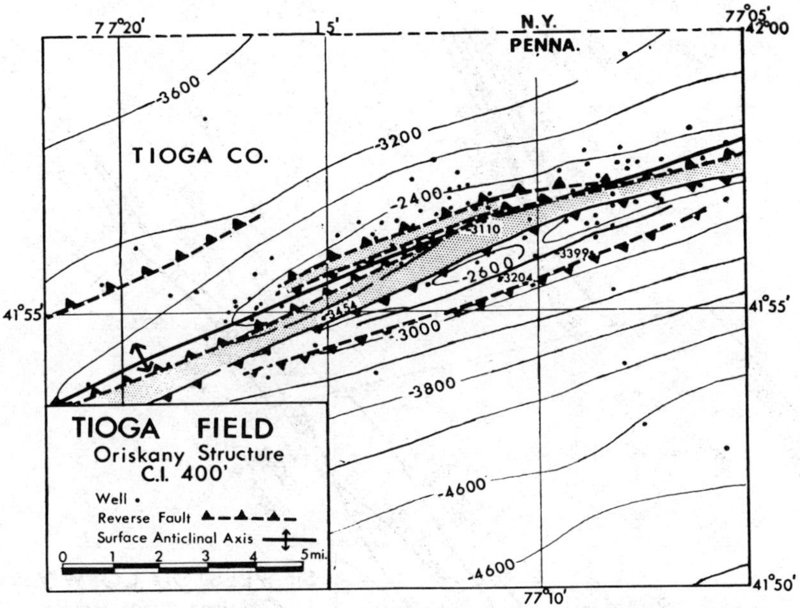

Figure 3. Depressed axial zone (stippled) created by synclineward-dipping imbricate thrust faults at the Oriskany Sandstone level, Tioga anticline, Pennsylvania

The depressed axial zone and thrusting of the limbs onto the axial zone occurs for 70 miles along the closely drilled Chestnut Ridge anticline from southern Indiana County, Pa., at least as far south as Marion County, W. Va. A structural map of the Onondaga Formation in Chestnut Ridge and Laurel Hill anticlines in southeastern Westmoreland County, Pa. (Fig. 5), demonstrates thrust of the anticlinal limbs over the depressed crestal zones. The demonstrable persistence of the faulting along the crests of these and other closely drilled folds of the Plateau provides a strong basis for lateral extrapolation of faults bounding crestal depressions on other structures which have fewer subsurface control points.

Vertical stratigraphic separation of the Oriskany along the faults is as much as 1700 feet, as on the west flank of Deer Park anticline in central Tucker County, W. Va. (Fig. 4); the separation is more typically 300–700 feet. Fault displacement decreases on anticlines northwest of the Intra-Plateau Front, in the same manner as the structural relief decreases.

In most localities where good data are available for both flanks, throw is greatest on the northwest-dipping faults in the less steeply inclined *northwest* limbs of the anticlines that are steeper on their southeast limbs. Displacement on the faults also decreases on approaching plunging parts of the anticlines. According to Finn (1949, p. 315) the faulting ". . . is associated with doming and may be part of the mechanics of doming, since displacement seems to be greatest opposite high points on the domes."

Undoubtedly, more than a single thrust fault forms the boundary of the axial depressions.

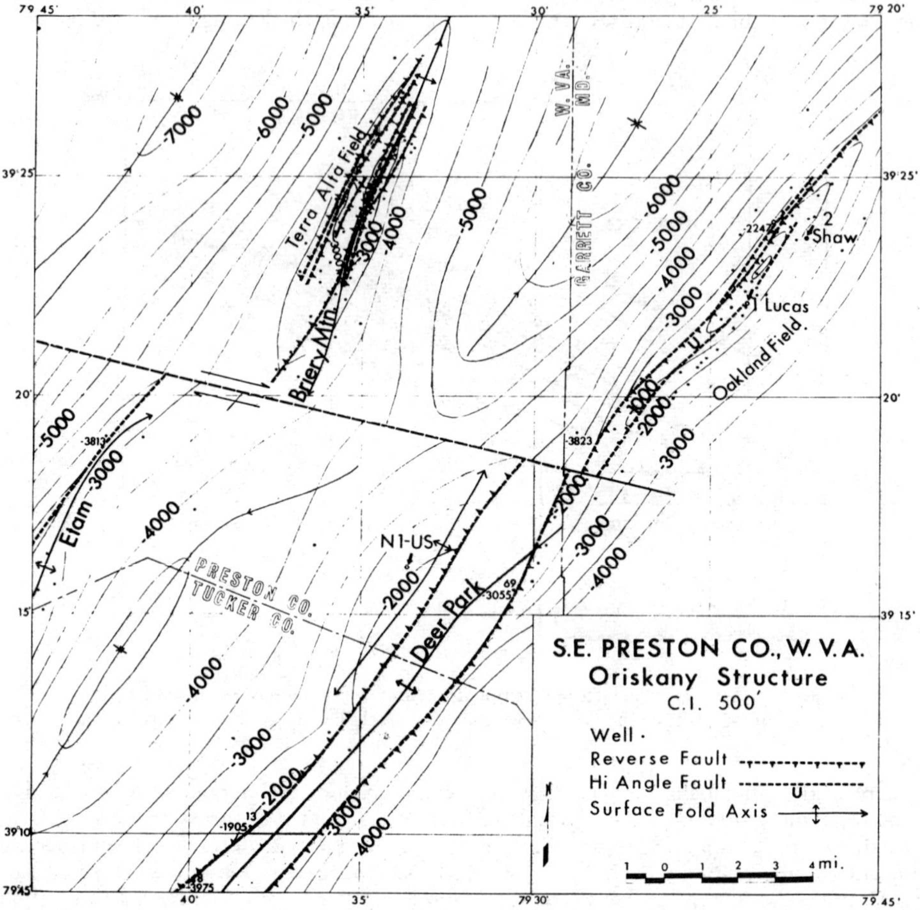

Figure 4. Offset of anticlines along a subsurface fault, Preston County, West Virginia; the Oriskany level structure of supposedly continuous Deer Park anticline undergoes marked changes in character on opposite sides of the fault, which is possibly a tear fault bounding the north margin of the Burning Springs thrust sheet or a part of the thrust where it is changing stratigraphic level along the northeast-southwest structural strike. The depressed axial zones of parts of Deer Park and Briery Mountain anticlines are stippled.

Several faults occur in the quarter-mile wide boundary fault zone in the Tioga and Leidy fields (Figs. 3, 6); faults in the boundary zones are cumulative in their effect of uplifting the flanks over the axial zone. Spacing of wells and map scale prohibit definition of more than one fault on maps of most areas.

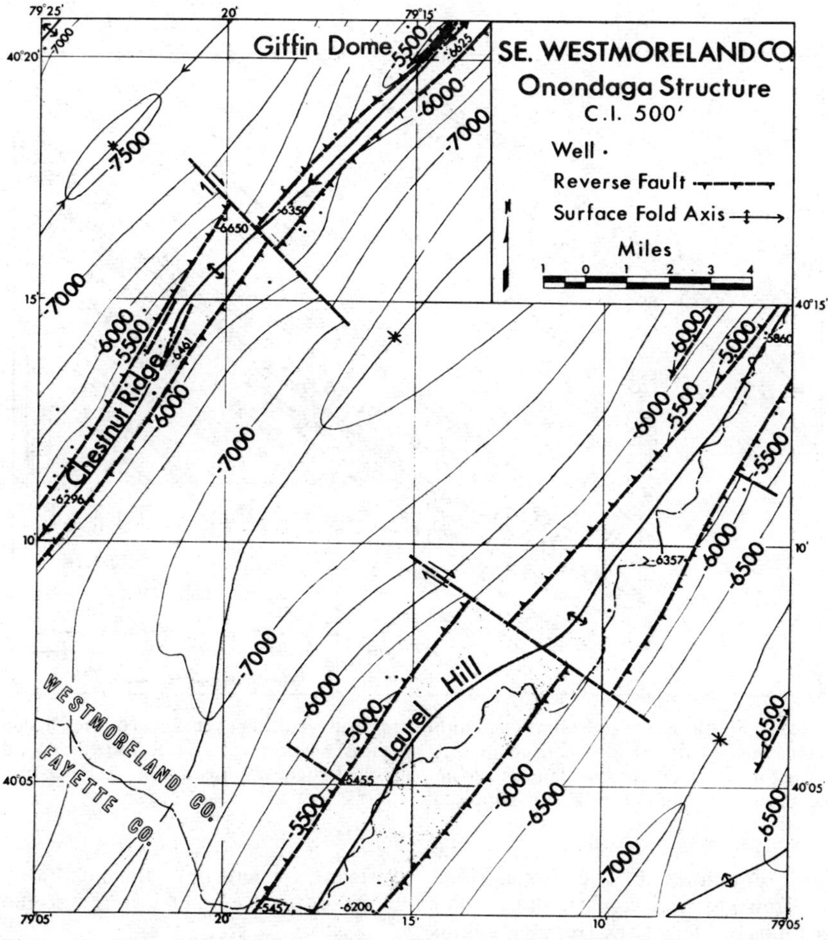

Figure 5. Depressed axial zones (stippled) and possible tear faults in anticlines of Westmoreland County, Pennsylvania; the depressed zone along the axis of Chestnut Ridge anticline is apparently continuous for more than 70 miles. Note the steep southeast limb of Laurel Hill anticline.

The thrust faulting appears to persist far down the flanks of the anticlines. Shortening at the Oriskany level has occurred along numerous thrust faults far down into Caledonia syncline in Clearfield County, Pa. (*See* Pl. 2 by Cate, *in* Lytle and others, 1961.)

Many faults, clearly discerned at the Oriskany-Onondaga level, are not reflected in structure maps of the Tully Limestone, which is separated from the Onondaga by 300 feet of shale in western Pennsylvania and by as much as 1300 feet of shale in the southeastern Plateau; *see* maps of the State Line field in northern Potter County, Pa. (Finn, 1949, p. 316–317, Figs. 3, 5). Faulting decreases upward into the Devonian shale section and is absent at the surface. Major faults at depth probably fade upward into bedding slip planes and a complex of minor folds and faults in the incompetent Upper Devonian strata.

FOLD SYMMETRY AND EXCEPTIONS TO THE

SYMMETRICAL FAULT PATTERN: As at the surface, many subsurface anticlines northwest of Deer Park anticline, such as Wellsboro, Marshlands, Tioga, Sabinsville, and portions of Chestnut Ridge and Briery Mountain (Figs. 3–6) have steeper, shorter southeast limbs at the Oriskany horizon. This asymmetry contrasts with the steeper overturned northwestern limbs in the East Fork-Wharton Field (Fig. 6). Significant tectonic thickening, such as the 200 per cent thickening of the Tully-Oriskany interval in well #51, Gaines quadrangle,[3] along the fault in Sabinsville Field, illustrates that these single northwest-dipping faults are reverse faults.

Along strike from the atypical portions of the

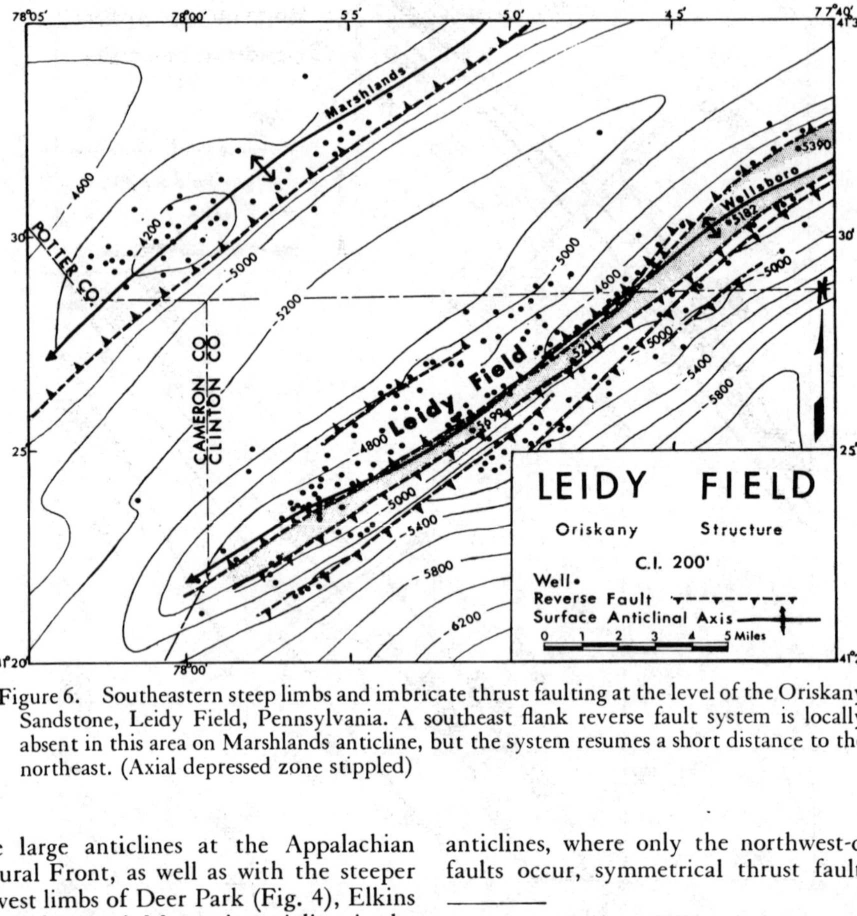

Figure 6. Southeastern steep limbs and imbricate thrust faulting at the level of the Oriskany Sandstone, Leidy Field, Pennsylvania. A southeast flank reverse fault system is locally absent in this area on Marshlands anticline, but the system resumes a short distance to the northeast. (Axial depressed zone stippled)

of the large anticlines at the Appalachian Structural Front, as well as with the steeper northwest limbs of Deer Park (Fig. 4), Elkins Valley, and Brown's Mountain anticlines in the southeastern High Plateau subprovince of West Virginia.

In addition to this ubiquitous steepening of the southeastern limbs of the anticlines in the northeast, the southeast flank thrust fault is apparently absent along portions of some folds. Only northwest-dipping reverse faults have been observed on Harrison anticline near Ulysses Field, Potter County, Pa., on Sabinsville anticline immediately to the east along the Sabinsville Field, and on Marshlands anticline anticlines, where only the northwest-dipping faults occur, symmetrical thrust faulting of

[3] Important wells are referred to by (1) topographic quadrangle permit numbers or county permit numbers assigned by the respective state geological surveys and/or (2) by operating company name followed by the land owner's name, with a number indicating the chronological position of this well in a series of wells on the owner's land (*e.g.*, Pennsylvania Oil and Gas Company #3 Jones, Latrobe quadrangle #14; West Virginia Oil and Gas Company #1 Brown, Preston County #313; Cumberland and Allegheny Gas Company #1 Lucas, Garrett County, Maryland). Data on these wells are available in the offices of the respective state agencies or in publications, except where indicated in the text.

both flanks resumes. In many localities where both flanks are faulted, displacement on the northwest flank fault is considerably greater than that on the southeast faults.

Altitudes of the Oriskany and shallower strata in successive synclinal troughs from northwest to southeast in the Plateau form a smooth curve of regional dip to the southeast to the zone where the beds turn upward and rise portant factor to consider in the determination of the ultimate nature of tectonic agents in the Plateau.

Sub-Oriskany Structure North and West of Elkins Valley and Deer Park Anticlines

Data for determination of structure below the Lower Devonian Oriskany Sandstone is provided by deep wells along Chestnut Ridge

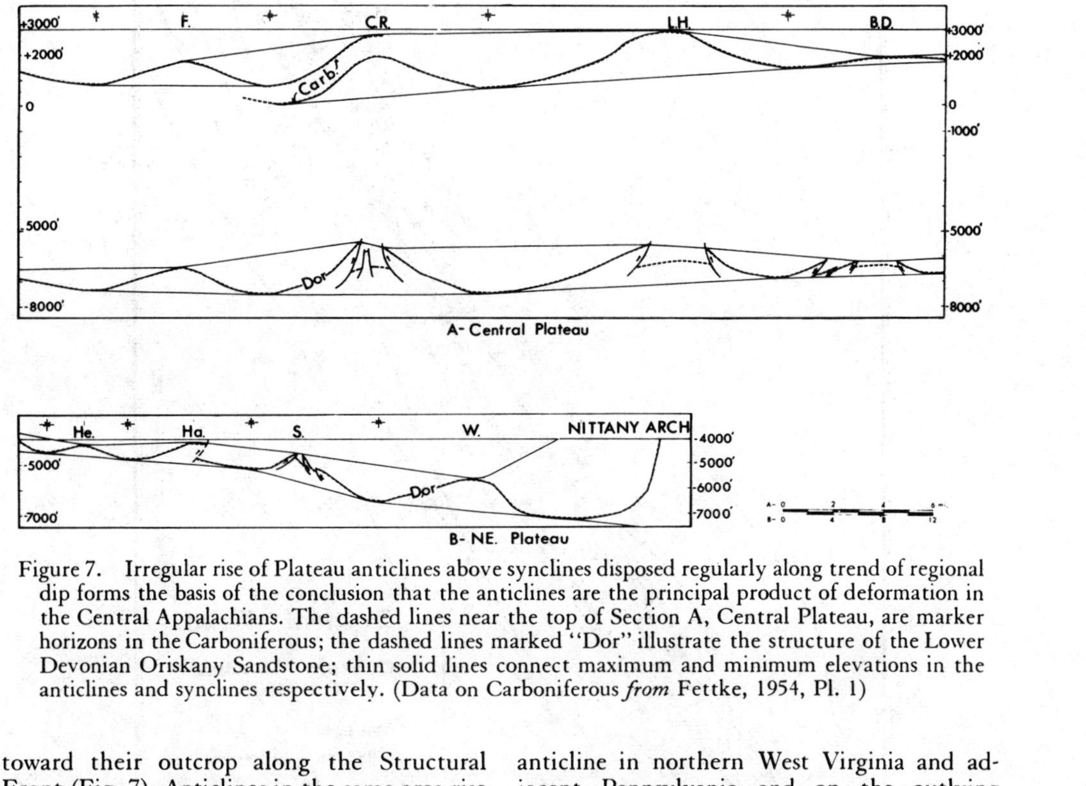

Figure 7. Irregular rise of Plateau anticlines above synclines disposed regularly along trend of regional dip forms the basis of the conclusion that the anticlines are the principal product of deformation in the Central Appalachians. The dashed lines near the top of Section A, Central Plateau, are marker horizons in the Carboniferous; the dashed lines marked "Dor" illustrate the structure of the Lower Devonian Oriskany Sandstone; thin solid lines connect maximum and minimum elevations in the anticlines and synclines respectively. (Data on Carboniferous *from* Fettke, 1954, Pl. 1)

toward their outcrop along the Structural Front (Fig. 7). Anticlines in the same area rise to variable levels.

The irregular rise of the anticlines above the regularly disposed synclines is strong evidence that the growth of the anticlines is the principal product of deformation of the Plateau. This contraverts Cooper's hypothesis (1961, p. 111) that ". . . the anticlines are purely incidental features resulting from unequal subsidence of the geosynclinal floor."

The anticlines northwest of Wellsboro anticline have nearly flat northwest limbs which are little more than terraces that interrupt the southeastward regional dip. The southeastward regional dip could be, at least in part, the result of the postdeformational tilting and is an im-

anticline in northern West Virginia and adjacent Pennsylvania and on the outlying Burning Springs anticline in Wood County and Wirt County, W. Va.

SOUTHERN CHESTNUT RIDGE ANTICLINE: Deep drilling in the Summit Field (Figs. 8, 9) immediately north of the West Virginia border indicates that the high angle reverse faults observed in the Oriskany-Onondaga units flatten down dip and pass into bedding in the tectonically weak Upper Silurian strata near the upper salt of the Salina group or in a 100-foot shale zone near the base of the Helderberg.

Well #31 (Uniontown quadrangle, Manufacturer's Light and Heat Company #8, Barton Estate) in the South Summit pool on the northwest flank (Fig. 9) penetrated 1985 feet of beds between the top of the Oriskany and the top of the upper salt of the Salina group. The normal thickness of this interval in four nearby, unfaulted wells on Chestnut Ridge is approximately 800–900 feet, with the distinctive dark entered the uppermost Salina salt at −6762 feet. As shown in Figure 9, the fault above the second dark shale must shear abruptly upward to intersect well #29 above the Onondaga; the second, deeper fault, which produces the double

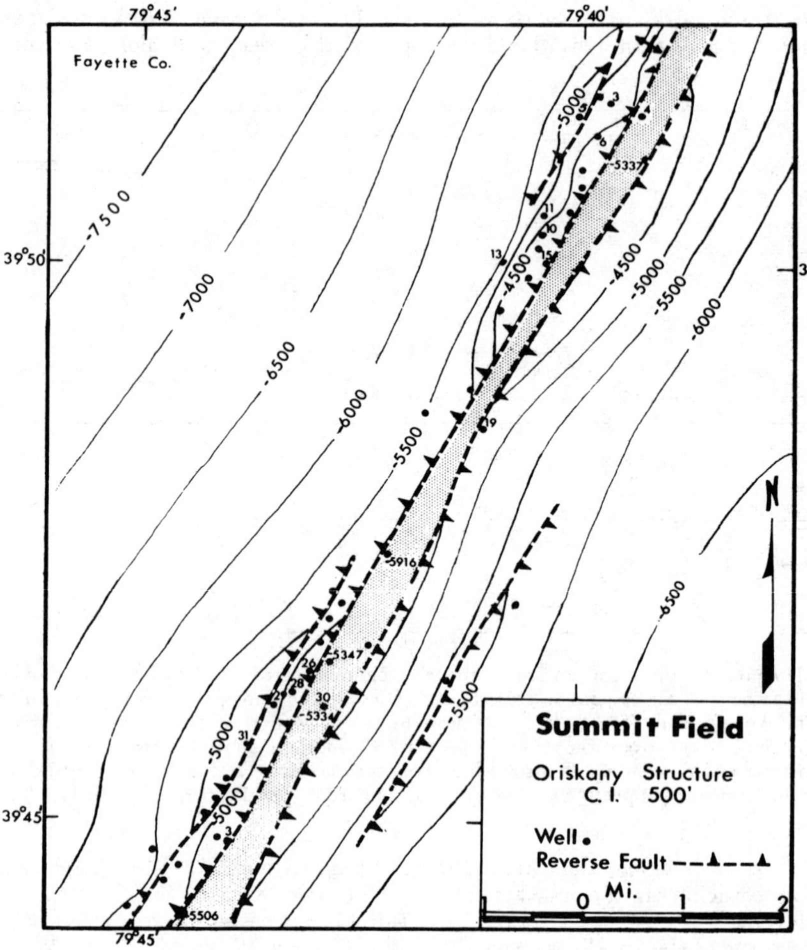

Figure 8. Depressed axial zone, Oriskany Sandstone level, Summit Field, Chestnut Ridge anticline, Fayette County, Pennsylvania; South Summit pool is on the northwest flank of the anticline near the lower left corner and North Summit Pool is on the upper right.

shale of the Helderberg 300–400 feet below the top of the Oriskany. Well #31 penetrated an overturned fold immediately below the first of two occurrences of the Lower Helderberg shale, crossed a major fault to enter the second upright repeated section of the shale, continued into a twice normal-faulted thickness of Lower Helderberg-Upper Salina beds, and finally thickness of Lower Helderberg-Uppermost Silurian strata, shears upward and eastward to form the boundary fault at the Oriskany level west of well #26. Lack of disturbance of the deep-lying Upper Silurian-Lower Devonian strata from the crestal zone to the flank is suggested by relationships between wells #30 and #31; extrapolation to the salt in well #30 places it slightly higher than the elevation at which the salt was penetrated in well #31 down the flank, suggesting that the salt and older strata dip shallowly westward in an unfaulted west limb.

Similar cross sections, in which faults must flatten at depth to satisfy reasonable geometry of the folding and faulting and known subsea elevations, can be worked out in the North Summit pool (Uniontown quadrangle wells #13, #11, #12, and #15 in one section and #5, #6, #4, and #7 in another). Well #11 included a second, thrust-repeated Oriskany section 950 feet below the first. Well #6 (Potter Development Company #1 Heyn) intersected the salt at −5971 feet after penetrating an overthickened sub-Oriskany sequence, including a thrust-faulted repetition of the Helderberg dark shale and the postsalt Upper Salina strata. Up-dip transport by bedding faults in the top of the Salina or in the Helderberg shale is again the only possible interpretation.

From the high elongate dome of Chestnut Ridge anticline on which Summit Field is situated, the anticline plunges southwestward into West Virginia to a point near Hope Natural Gas Company #1 Gribble (Harrison County well #69). Overthrusting of the limbs of the fold persists southwest as far as Marion County, W. Va., southwest of which the rate of plunge of surface beds and the Oriskany diminishes rapidly. Projection of well data from the flank uplift and the depressed axial zone onto a longitudinal section of Chestnut Ridge from the Gribble well to Summit Field (Fig. 10) demonstrates rapid convergence of the plunging Oriskany with the salt, which is almost flat for much of the distance. Elevations of the top of the upper Salina salt deviate only 200 feet from the Summit Field #8 Barton Estate to the Gribble well, in 55 miles, whereas the Oriskany surface descends nearly 1400 feet.

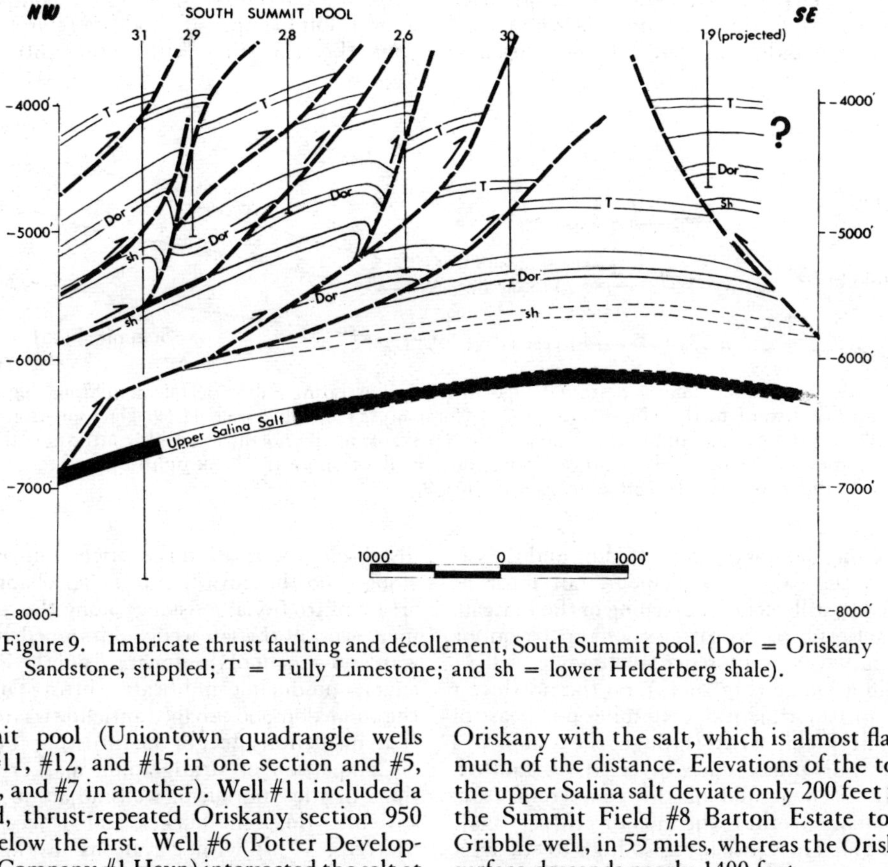

Figure 9. Imbricate thrust faulting and décollement, South Summit pool. (Dor = Oriskany Sandstone, stippled; T = Tully Limestone; and sh = lower Helderberg shale).

Significant faulting and anticlinal uplift occur above the salt and appear to be facilitated by the presence of highly incompetent strata in the Upper Silurian. Both longitudinal and transverse sections of Chestnut Ridge illustrate the folding and faulting above the incompetent rocks in the Salina and Helderberg, producing a décollement.

Uplift of the Upper Salina salt to −5971 feet in well #6 in the North Summit pool (Fig. 8, 10) perhaps suggests that locally the salt is directly involved in differential uplift and folding or faulting along the anticline. The nearly flat-lying Salina in Chestnut Ridge to the southwest suggests, however, that the thin upper salt was usually a passive, lubricating medium for thrusting of the competent overlying strata toward the anticlinal crest.

The Salina and older rocks are warped into a fold of lower relief below the décollement. Geophysical data indicate that closure at the same position on deeper horizons may be absent locally altogether. Persistence of deformation below the Salina décollement surface on Chestnut Ridge indicates that all the anticlinal role for the salt, as a simple lubricant for décollement sliding, is inferred by the data cited.

BURNING SPRINGS ANTICLINE: Burning Springs, the high, isolated, anomalously north-south trending anticline in west-central West Virginia, was produced by imbricate thrust faulting above the Salina glide zone (Fig. 12). Woodward (1959a; 1959b) described the imbricate thrust fault quadruplication of the Onondaga-Helderberg interval in Wood County well #351 (Hope Natural Gas Company #1 Power Oil Company). Rodgers (1963) attributes the thrusting within the anticline to

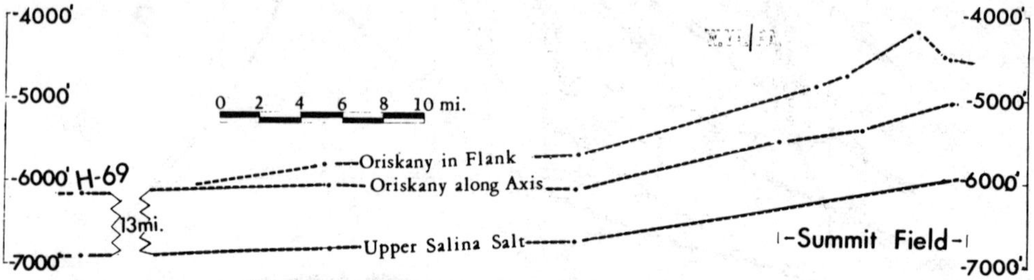

Figure 10. Section along Chestnut Ridge anticline illustrating Salina décollement from Summit southwestward to the Hope Natural Gas Company #1 Gribble well, H-69, Harrison County, West Virginia. The uplifted position of the Oriskany in the northwest flank contrasts with its position in the relatively depressed axial zone. Faults causing the flank uplift appear to flatten into the Salina décollement zone, as in Figure 9.

shortening did not occur by folding and thrusting on the Salina décollement, but must be related to still deeper shortening in the Plateau.

GIFFIN DOME, NORTHERN CHESTNUT RIDGE ANTICLINE: New Florence quadrangle well #1 in Giffin Dome (Fig. 5, 11), on the northwest flank of Chestnut Ridge 40 miles northeast of Summit Field, went through the Onondaga chert four times, once inverted. A faulted, nearly isoclinal fold of the competent basal Devonian overlies the major thrust fault (Fig. 11). The Oriskany is so high to the west in well L-1 (Latrobe quadrangle well #1) that considerable sub-Oriskany tectonic thickening must have occurred in some fashion in the intervening area, either by additional faulting or salt flowage. Another large fault may shear upward from beneath L-1, intersecting the Oriskany between L-1 and 1, as shown in Figure 11. Flowage of salt into the anticlinal flank from the depressed axial zone is unlikely because in this area of the Plateau only 150 feet of salt occurs in the 1500-foot thick Salina Group and is distributed through it in many beds in the southeastern side of the Plateau. A more passive disappearance of salt in the Upper Salina group along a north–south line of facies change. Increase of frictional resistance along the décollement west of the salt wedge-out impeded westward movement of the upper part of the Plateau cover, producing imbricate thrust faulting; the anomalous, box-shaped anticline was formed as an incidental effect of the thrusting at depth.

Comparison of stratigraphic marker elevations in the Silurian in wells adjacent to the anticline (logs in Tucker and Price, 1943; Martens, 1945) with elevations in the deep Power Oil Company test well (Woodward, 1959a) reveals no anticline in Silurian and older rocks below Burning Springs anticline. Furthermore, Burning Springs anticline merely interrupts prevailing eastward regional dip of the Onondaga in western West Virginia (*see* Fig. 12). Wirt County well #296, 12 miles south of Wood #351, was also drilled through three east-dipping, thrust-repeated Lower Devonian sections on the western lip of the Burning Springs thrust sheet.

LINEAMENTS IN THE PLATEAU: Rodgers (1963) shows that the 45-mile-long Burning

Springs anticline marks the western edge of a thin block of post-Salina strata which has glided westward ¼–½ mile along the top of the Salina. The margins of the block are clearly defined by lineaments trending east-southeast from the ends of the anticline back toward the opposite the northeast part of the St. Clair thrust. A much more prominent lineament trends S. 75° E. from the curving southern terminus of Burning Springs anticline; the lineament marks the southern terminus of Arches Fork, Wolf Summit, and other low folds to the north and south. Rodgers proposed that the southern marginal lineament, trending south-southeast, may be a line of minor slip, but probably does not bound the Burning Springs thrust block proper (*see* Pl. 1).

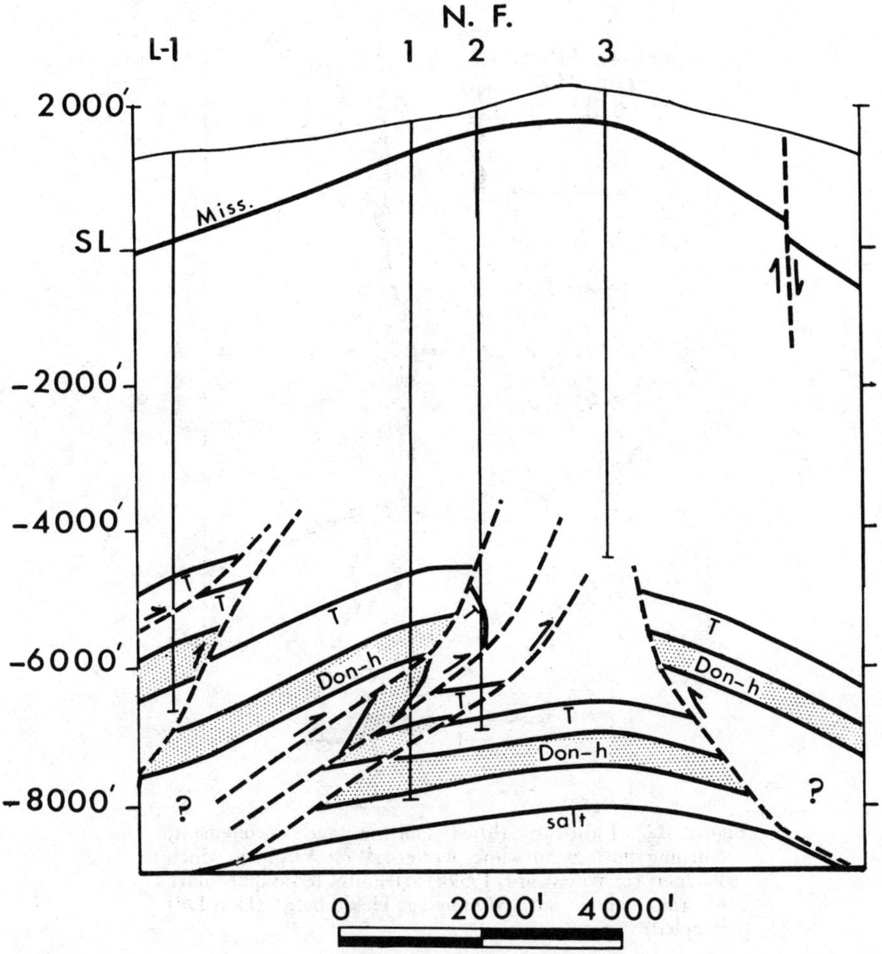

Figure 11. Imbricate thrust faulting, Giffin Dome, Chestnut Ridge anticline, Westmoreland County, Pennsylvania; at the level of the lower Devonian, the northwest flank of the anticline is an imbricately faulted, overturned fold; Gray (*in* Shaffner, 1958) interpreted the fault on the steep southeastern limb near the surface as an extension of the northwest flank fault bounding the axial depression. (Don-h = Devonian Onondaga-Oriskany-Helderberg sequence (stippled); T = Tully limestone; solid line near the surface is the base of the Mississippian Pocono Sandstone.)

High Plateau (Pl. 1). Rodgers projected the southern boundary of the thrust sheet S. 20° E. through several small, anomalously trending anticlines to a point of steep plunge on the Abbs Valley anticline in Monroe County, W. Va.,

The Burning Springs block is paralleled on the north by two lineaments, along both of which prominent offsets of anticlinal trend occur all the way back to the southeastern High Plateau and the Appalachian Structural Front. The northern and southern lineaments terminate the high Elkins Valley-Etam anticlinal trend in the northwest margin of the High Plateau. Only a low plunging portion of the Elkins Valley fold persists to the south.

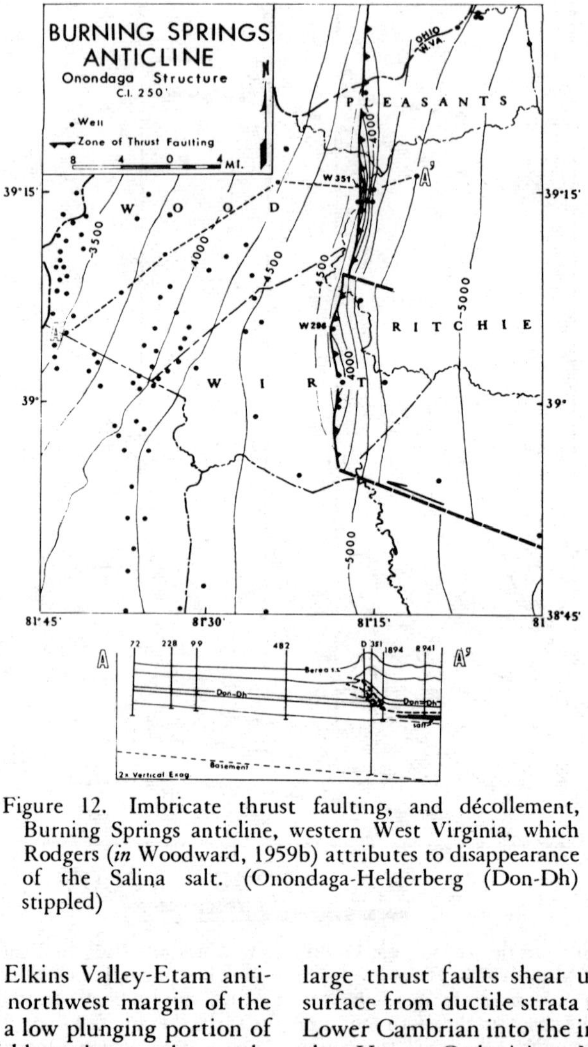

Figure 12. Imbricate thrust faulting, and décollement, Burning Springs anticline, western West Virginia, which Rodgers (*in* Woodward, 1959b) attributes to disappearance of the Salina salt. (Onondaga-Helderberg (Don-Dh) stippled)

Deep-Level Structure in Anticlines along the Appalachian Structural Front

Three deep wells and an erosional window through low-angle thrust faults in high, asymmetric anticlines along the Appalachian Structural Front indicate that this bold, arcuate zone of uplift results from changes in stratigraphic position of major, generally unexposed bedding-plane thrust faults that underlie the western part of the Valley and Ridge province and the adjacent parts of the southeastern High Plateau. The asymmetric anticlines of the Front were formed along the zone along which large thrust faults shear upward in the subsurface from ductile strata near the top of the Lower Cambrian into the incompetent beds in the Upper Ordovician Martinsburg-Juniata strata. The deep thrust faults continue westward beneath most areas of the adjacent High Plateau, apparently as sole faults from which minor branching thrusts shear upward, producing the large anticlines of the southeastern High Plateau.

BIRMINGHAM WINDOW: The only surface exposure of one of these major, low-angle overthrusts is in northeastern Blair County, Pa., on the Sinking Valley anticline, the frontal fold of the Nittany anticlinorium (Moebs and Hoy, 1959; Butts and others, 1939). In the vicinity of Birmingham village, overthrust allochthonous Cambrian rocks surround two patches of complexly deformed Ordovician and Lower Silurian strata. Shallow core drilling in the Cambrian and Lower Ordovician carbonate rocks west and south of the windows revealed that the Cambrian overlies red and gray Upper Ordovician shales along a thrust fault dipping westward toward the Plateau (Moebs and Hoy, 1959). As the only outcrop of this thrust is around the window (Pl. 1), it must dip westward from the Nittany anticlinorium. Presumably it is in the incompetent Upper Ordovician sequence beneath the east side of Plateau as shown in Figure 13B, but it could be at a somewhat higher level in the Silurian. Minimum horizontal displacement on the outcropping fault in the Tyrone area, as determined from separation of the Lower Silurian strata in the window from those on the Front several miles to the west, is 2–2¼ miles. As will become clearer after discussion of geophysical data, a deeper stepped low-angle thrust which has caused the upwarping and folding of the original subhorizontal stepped thrust to its outcropping position around the window is likely.

THRUST FAULTS IN DEEP WELLS: On the west limb of Sinking Valley anticline, 17 miles southwest of the Birmingham window near Hollidaysburg, Upper Ordovician Juniata redbeds are thrust over Salina-equivalent Upper Silurian Tonoloway strata at a drilling depth of only 2400 feet in the Hollidaysburg Oil and Gas Company #1 Rankey well (Hollidaysburg quadrangle well #2). The writer would suggest that detachment of the hanging wall sequence occurred a short distance eastward in the argillaceous Upper Ordovician glide zone, with the fault climbing upward to the evaporitic Salina glide zone where it was penetrated by the Rankey well. At least 3500 feet of west-northwestward horizontal movement has occurred along this fault, as deduced from the amount of separation of Tonoloway strata, but it does not outcrop and must be nearly parallel to the bedding of the Upper Silurian or Devonian rocks a short distance to the west of the well.

Displacement on this shallow fault is not sufficient to produce all the anticlinal shortening on this part of Sinking Valley anticline, so the deeper thrusts observed in and inferred to be present below the Birmingham window can reasonably be suspected to be present in Sinking Valley anticline beneath the bottom of the Rankey well. The prominent southwestward plunge and the southward bend of the anticlinal axis from Birmingham to Hollidaysburg may reflect a decrease in the amount of displacement on the deep thrust fault(s). Such a decrease would simultaneously produce the plunge and swing in map trend ofthe anticline (Fig. 14); southwestward from the high part of the anticline near Birmingham the allochthonous upper sheets would appear not to have advanced as far westward on the higher glide zones; in such a model, the fold above the fault would not be tightened as much as to the north and would thus plunge southwestward and assume a more southerly trend. A similar interpretation can be applied to the concomitant plunges and arcuatic trend changes of other large anticlines plunging into the Anthracite basin and the Broadtop synclinorium.

Southwestward in Pendleton County, W. Va United Fuel Gas Company #1 Sponaugle (Pendleton County well #6) was spudded in the base of the Upper Ordovician Martinsburg shale on the southeast limb of Wills Mountain anticline (Pl. 2, east end of section A-A′) and drilled 10,035 feet of Ordovician and Upper and Middle Cambrian carbonates, at which depth it crossed a major low angle thrust into Upper Ordovician shale and Middle Ordovician carbonate rocks, first inverted and then right side up and flat in the autochthonous lower block. The minimum horizontal displacement is 2¾ miles and may be much more. The altitudes of the thrust and of the Ordovician strata in the lower block are nearly the same as the elevation of the Ordovician strata in the Plateau synclines to the west and in the deep well penetrating Middle Ordovician rocks on the Elkins Valley anticline. The sequence and thickness[4] of the carbonate rocks in the allochthonous upper block match those of units between the Martinsburg and the shaley basal Elbrook-Waynesboro Cambrian(Perry, 1964). The latter formations contain the only appreciable shale in the Cambrian sequence of the Valley and Ridge province and are presumed to be the most likely position of bedding plane thrusts which have accommodated gliding of the Valley and Ridge province sedimentary

[4] After adjustment for dip revealed by dipmeter surveys (Information Release, United Fuel Gas Co., February, 1961; Perry, 1964).

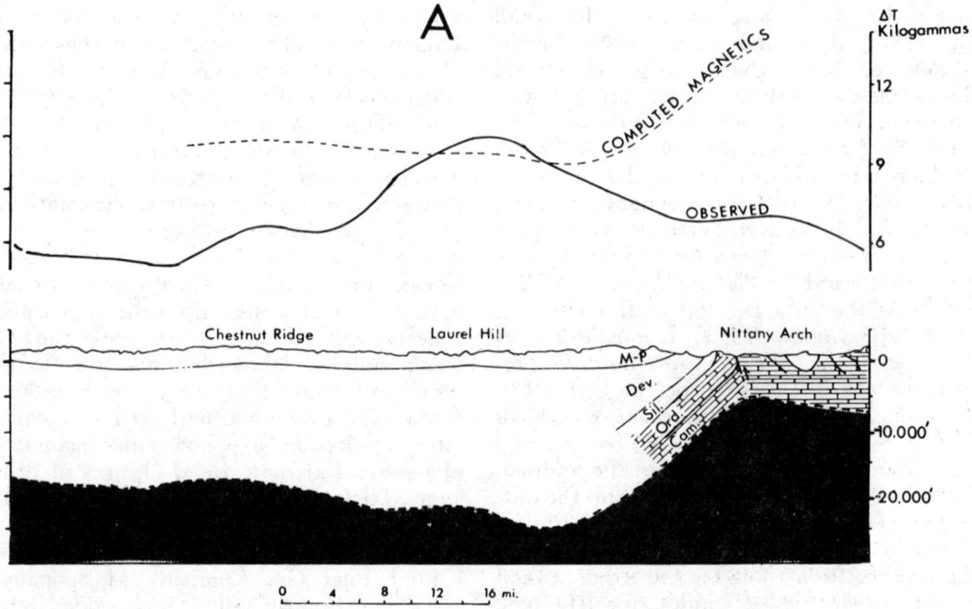

Figure 13. Thick-skinned and thin-skinned interpretations of the Appalachian Structural Front, Nittany Arch area near the Birmingham window, Blair County, Pennsylvania; A, a thick-skinned interpretation showing a comparison of a magnetic profile computed from data in cross section (constructed assuming involvement of basement in Nittany Arch uplift) with observed magnetic data (*after* Fig. 5, Joesting and others, 1949); B, thin-skinned interpretation based on shallow drill holes in the vicinity of the Birmingham window (Moebs and Hoy, 1959), presence of low-angle thrust in deep wells to the southwest along the Structural Front, and magnetic, gravity, and seismic data. (P = Pennsylvanian; M = Mississippian; Dc = Catskill Formation; Dsh = Chemung, Portage, and Hamilton groups; Dl-S = Oriskany, Helderberg, and Silurian; Ou = Upper Ordovician; Oml = Middle and Lower Ordovician; C = Cambrian)

cover (Rodgers, 1963) westward across the deep Broadtop synclinorium southeast of Wills Mountain.[5]

Another well, the Kerr-McGee #1 Martin, (Hyndman quadrangle well #1) in Bedford County, Pa., on the southeast limb of Hyndman anticline between the Rankey and Sponaugle wells, penetrated the deep thrust. It passed from Lower Ordovician Beekmantown strata of the upper block into an inverted sequence of Martinsburg, Oswego, and Juniata below a low-angle thrust. However, the well was not drilled deeply enough to establish the point at which the strata became right side up. This would have enabled determination of the amount of tectonic thickening, the minimum displacement, and data for comparison with the elevation of stratigraphic markers in deep wells on Deer Park anticline, 10–15 miles farther west on the Plateau, to be discussed later.

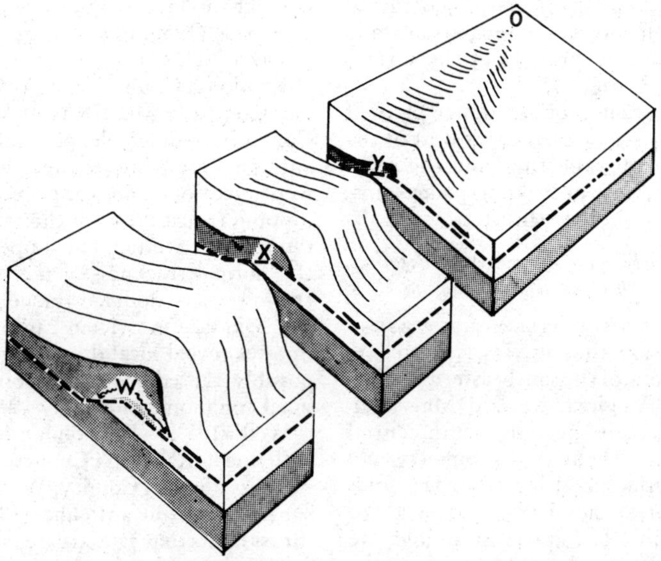

Figure 14. A simplified model to explain concomitant plunge and arcuate trend change of southern Sinking Valley anticline; plunge and trend change of the anticline are produced by a southwestward decrease in the amount of "up-and-over" movement of the hanging wall sheet onto the higher décollement glide zone; displacement $W > X > Y$, which results in a decrease in apparent shortening in the direction of decreased displacement and leads to ultimate disappearance of the fold. An *en echelon* fold pattern such as that seen in the Valley and Ridge province of the Central Appalachians is easily explained by extension of this model to include a number of noncontinuous stepped thrusts branching off the master sole thrust; each upward shear produces a flexurally folded anticline whose axis of curvature is near the point at which the thrust shears out of the sole glide zone.

NECESSITY FOR PERSISTENCE OF DEEP BEDDING-THRUST MOVEMENT BENEATH THE SOUTHEASTERN HIGH PLATEAU: Structure in the deep wells and windows indicates that the Appalachian Structural Front is caused by the upward shearing of thrust faults from Cambrian shales to incompetent zones in the Upper Ordovician or Silurian. Rocks above the detachment fault

[5] The lower Elbrook, in normal sequence immediately overlies the shales and forms the hanging wall along 90 per cent of the lengths of thrusts in the southeastern part of the Valley and Ridge province of the Central Appalachian region. Shales correlative to the Waynesboro and lower Elbrook in the Rome and Conasauga formations form the most prominent zone of bedding-plane thrust in the southern Appalachians (Rodgers, 1953).

west of the Front had to be folded (shortened) because the shortening created by movement of the older rocks up onto and westward along the Upper Ordovician thrust zone exceeds that which is created by the formation of the superficial Wills Mountain, Hyndman, and Sinking Valley anticlines. This is illustrated in cross sections of the bedding thrust zones constructed utilizing minimum amount of separation along the fault and the configuration of the surface anticline (*see* Pl. 2; Fig. 13b).

Additional shortening of the upper plate of post-Martinsburg strata must occur west of the Front in the Plateau, affecting only the rocks above the deepest zone of bedding-plane translation, *i.e.*, above the sole thrust.

Southeastern High Plateau Zone—West Virginia, Maryland, and South-Central Pennsylvania

CONTRASTS OF SOUTHEASTERN HIGH PLATEAU WITH NORTHWESTERN HIGH PLATEAU: Anticlines in the high fold zone of the southeastern Plateau in eastern West Virginia, western Maryland, and Deer Park anticline in south-central Pennsylvania differ slightly in symmetry and pattern of subsurface faulting from the folds in the High Plateau northwest and north of Deer Park anticline. The southeastern folds are steep on their northwest limbs like those in the Valley and Ridge, in contrast to the southeast ward oversteepening of the anticlines to the north and west, and many appear to lack northwest-flank reverse faults at the Oriskany level. Also, only this area's Elkins Valley and Brown's Mountain anticlines have been uplifted sufficiently to expose Middle Devonian or older strata.

More deep wells have been drilled in this area than in any other part of the Central Appalachians (Pl. 1). Data from the wells reveal that the mode and style of subsurface deformation is similar to that along the Structural Front and in the Valley and Ridge province in general to the southeast, inasmuch as it is characterized by concentric folding at the surface and low-angle detachment thrusting at depth.

ELKINS VALLEY, GLADY, AND HORTON ANTICLINES: Elkins Valley, Glady, and Horton anticlines lie in a rectilinear block of the Plateau immediately northwest of the Structural Front, set off sharply by east-southeast trending surface lineaments against which the folds plunge and terminate.

Important subsurface thrust faulting has been recognized in Elkins Valley anticline (Pl. 2, Sec. A-A′). A relatively large east-dipping thrust that duplicates Salina and Lower Devonian strata, has been penetrated by three wells (Randolph County wells #11, #7, and #102) on the southeastern limb of Elkins Valley anticline in central Randolph County, W. Va. (Similar thrusting of the southeastern flank at this level has been described by Reeves and Price (1950, Fig. 5) near the northern end of the anticline.)

Randolph County well #103, (Hope Natural Gas Company #10,228 West Virginia Board of Control), a much deeper well drilled on the anticlinal crest immediately west of the three wells mentioned above, penetrated three east-dipping thrust faults in the Upper and Middle Ordovician strata. The upper fault doubled the normal thickness of the Martinsburg and the lower two doubled the normal thickness of the Middle Ordovician carbonates. Dipmeter surveys reveal local dips as high as 77 degrees, possibly caused by drag (Robert Bayles, personal communication, July 1963).

Well #101-D (Randolph County) on Glady anticline in Randolph County, W. Va., and well #18 (Pocahontas County, W. Va.,) on the west flank of Horton anticline (Pl. 2) penetrated thrust faults that caused extensive thickening of Salina strata. In the base of the latter well, moreover, the Juniata formation was much thicker than is normal for the area, presumably because of thrust faulting. Higher in the well, a thrust fault repeated the Oriskany Sandstone.

Section A-A′ of Plate 2 extends from the Structural Front west to Elkins Valley anticline. The thrusts described in the Plateau anticlines are thought to branch off a sole thrust in the Upper Ordovician, which is the level that the principal thrust is presumed to occupy back to the Structural Front. As shown in Fig. 7 in the northern Plateau, synclinal datums of Plate 2 fall on a smooth curve and, in addition, the altitudes of the top of the Middle Ordovician limestone in the two sub-Martinsburg wells also fall on the gently west dipping plane defined by extrapolations to the top of the Middle Ordovician in synclines across the area; this conformity must indicate the detachment thrusts are in bedding in the synclines and that strata below the Upper Ordovician are essentially autochthonous. Thrust faulting of the Middle Ordovician rocks in Randolph #103 may indicate that some westward shift has occurred on a deeper décollement (dashed in as a fault within the Cambrian on Pl. 2).

The relatively great height of Elkins Valley anticline and the localization of bedding slip in the Salina décollement zone west of Elkins Valley demonstrate that the southeastern Plateau sole fault must shear upward beneath the Elkins valley fold to enter the Salina décollement zone, then continue westward 80 miles beneath the wide, thin Burning Springs thrust sheet to Burning Springs anticline. Structural detail is unknown in the small anticlines west of Elkins Valley because of the absence of deep wells.

BROWN'S MOUNTAIN ANTICLINE: Brown's Mountain anticline, the largest in the Central Appalachian Plateau, extends south-southwestward 85 miles from the lineament bounding the southern edge of the Burning Springs thrust sheet to northern Monroe County, W. Va. Lower Devonian and older rocks have their only Plateau outcrop for 50 miles along the central part of the anticlines. At both ends, the fold is a low symmetrical arch in Mississippian and Devonian rocks, but in the highly uplifted central part the western flank is overturned and broken at the surface by at least one thrust fault, the Burr fault of Price and Heck (1939). Although similar to the anticlines of the Valley and Ridge province to the southeast in size and shape, Brown's Mountain is separated from them by an 8–10 mile wide syncline which continues northeastward along the Front into Maryland and south-central Pennsylvania to the Schellsburg dome in Bedford County, Pa.

Ohio Oil Company #1 Seneca State Forest (Pocahontas County #1), immediately west of the anticlinal axis, was drilled from Lower Silurian strata through a low-angle thrust fault into the Middle Devonian Onondaga chert at a depth of only 950 feet (Pl. 2, Sec. B-B′). This fault does not crop out, but must, like the thrust fault in the Rankey well in Bedford County, pass into bedding slip at or near the base of the Devonian shale to west. The meandering Burr fault may be the trace of this thrust, or a similar one, in northernmost Greenbrier County and southern Pocahontas County, W. Va. Only 900 feet of erosion would be necessary near the Ohio Oil Company well to create an erosional fault trace east of the well like the Burr fault in all details. The Burr fault would thus form the southeast margin of a window. The trace of the bounding thrust on the western margin of the proposed window would be easily obscured by its parallelism to bedding at the base of the Devonian shale. Stratigraphic separation decreases toward either end of the Burr fault as the attitude of the fault approaches that of the bedding. Attempts to define such a structure by surface mapping were thwarted by heavy forest cover.

Section B-B′ of Plate 2 presents a generalized interpretation of Brown's Mountain anticline at the latitude of the Ohio well. The extreme tightening and uplift of the anticline preclude its formation by the upper fault alone, which has only ¾–1 mile displacement. More likely it was produced in large part by a deep-lying thrust of Valley and Ridge type that ascends from the Waynesboro to the Martinsburg level. Moreover it is probable that one or several thrust faults are shearing upward from both the Cambrian and the Martinsburg décollement. A deep well, now being drilled on the central high part of the fold by Tidewater Oil Company and associates will provide data on the deeper internal architecture of the anticline.

Brown's Mountain anticline plunges north-northeastward and terminates against the southern Burning Springs lineament. In the opposite direction, Williamsburg and Sinks Grove anticlines rise southwest of a lineament which intersects Bronw's Mountain anticline at the point in Greenbrier County, W. Va., where the anticline is offset slightly to the northwest. Immediately south of this lineament, the anticline plunges more steeply southwest for a short distance, carrying the Silurian and Lower Devonian strata below the surface. The anticlinal offset and profound structural disjuncture along the two west-northwest trending lineaments may infer the presence of zones of significant transcurrent movement.

ABSENCE OF FOLDS IN THE LOW PLATEAU NORTHWEST OF BROWN'S MOUNTAIN: The Low Plateau in eastern Kentucky and southwestern West Virginia contrasts strongly with the Plateau to the northeast because of the low relief and irregular trends of its anticlines and because of the narrowness and high amplitude of folds in the High Plateau zone on the southeast. Between the southern Burning Springs lineament on the northeast, the Brown's Mountain-Williamsburg-Abbs Valley-Dry Fork anticlinal trend on the southeast, and the Pine Mountain thrust sheet on the southwest (*see* Pl. 1), the Low Plateau lacks anticlines with more than 300 feet of structural relief for 65 miles or more northwestward to the Warfield anticline. The steeper and more continuous Warfield anticline in the western Plateau may overlie a flexure or fault in the basement that is an extension of the Kentucky River-Rough Creek fault zone, which emerges farther west in Kentucky (Price, 1931); alternately, it may be entirely analogous to Burning Springs anticline, marking the northwest limit of translation along the Salina décollement. The short, low anticlines of the Low Plateau trend anomalously north or northwest and seldom have more than 100 feet of reversal on their southeastern or eastern limbs. They are thus only minor interruptions of the prevailing west-northwestward surface regional dip.

DEER PARK ANTICLINE: The Deer Park anticline is the highest anticline north of the lineament bounding the northern margin of the Burning Springs thrust sheet. The lineament separates a long northern segment of this anticline which is apparently broken only by southeast-dipping thrusts in the Silurian and Devonian, from a slightly offset southern segment in which both limbs are thrust over a depressed crestal zone. These two segments may be two separate structures, only apparently continuous. The lineament apparently reflects a deeper zone of right-lateral transcurrent faulting (Fig. 4), along which the Etam and Briery Mountain anticlines terminate and along which the two anticlinal elements commonly thought to be continuous along the Deer Park trend have been brought into close proximity by the tear-faulting.

In the southern anticlinal segment of the Deer Park anticlinal trend in Tucker County, W. Va., shallow wells reveal symmetrical overthrusting of the limbs at the Oriskany level (Fig. 4). In addition, a critical deep well drilled by Columbian Carbon–Cities Service Oil Company on Eglon anticline, a minor fold on the larger northwest flank of Deer Park, penetrated a proportionally overthickened Silurian sequence, as if the drill penetrated beds dipping 20–25 degrees. However, the Upper Ordovician Juniata redbeds at the bottom of the well were 4200 feet thick, 3.5–4.5 times thicker than the normal 900–1200 feet, with the bottom of the formation not penetrated. This thickening must have been produced in the core of a detached décollement fold near the décollement or by drag and thrust duplication near a sole thrust. All of the structural relief of the Deer Park anticline above the adjacent synclines is accounted for by the thickening introduced in the Juniata section alone interval. Any additional structural duplication present at greater depth would indicate a syncline at depth below the anticline!

In Deer Park anticline north of the northern Burning Springs lineament, many unexposed thrusts occur in the Silurian-Devonian sequence. In Oakland Field, a major zone of thrusting rises from beneath the Tuscarora in the #2 Shaw well and splits into many imbricate splay thrusts in the Cumberland-Allegheny Gas Company #1 Lucas well and in shallower wells to the west (Fig. 15). J. H. Miller, subsurface-well geologist stated (Written communication, August 1963) that at least 26 thrust faults, were observed in the Lower Devonian-through-Middle Silurian section near the bottom of the well.

Forty-seven miles northeast of Oakland, in Somerset County, Pa., Peoples Natural Gas Company #1 Mowry (Berlin quadrangle well #1) was drilled through a twice-normal, 3438-foot thickness of the Salina group. The Tuscarora was reached at −7181 feet, whereas in the Appalachian Basin Associates #1 Miller 14 miles farther northeast on the Schellsburg dome, a high uplift on trend with the Deer Park anticline, the Tuscarora was reached at −1114 feet. This part of Deer Park anticline lies just northwest of the #1 Martin well on Hyndman anticline at the Front (*see* subsection THRUST FAULTS IN DEEP WELLS). The thrust faults within the northern part of Deer Park anticline may be splay thrust faults which branch upward from an Upper Ordovician (or mid-Cambrian décollement), creating some of the upper-plate shortening that is required by the lower-plate shortening introduced in the rise of the sole décollement zone beneath the anticlines at the Appalachian Front.

SOLE THRUST AND BRANCHING SPLAY THRUSTS IN ANTICLINES OF THE NORTHWESTERN HIGH PLATEAU: Anticlines of the High Plateau north and northwest of Deer Park anticline, such as Chestnut Ridge, Laurel Hill, Wellsboro, and Sabinsville, resemble the anticlines in the southeastern High Plateau just described, which are thought to be formed by splay faults branching from a deep detachment sole thrust or thrusts. They differ from the southeastern folds only in their slightly lower structural relief and in the general or local oversteepening of their southeastern limbs. Folding and faulting above the Salina décollement, spacing of folds, and relationship of fold height and location to the two structural fronts and lineaments are identical.

It is thus probable that anticlines of the northwestern part of the High Plateau also are not rooted; that is, instead they are situated above splay thrusts branching from a detachment sole thrust. East-dipping thrust faults of large displacement like those of the southeastern part of the High Plateau do not rise as high as the Onondaga-Oriskany level and, if present, they must flatten into bedding in the incompetent Silurian sequence.

Detailed information on the sub-Helderberg portion of Phillips Petroleum Company #1 well, drilled on the anticlinal crest several miles west in Marion County, W. Va.

GEOPHYSICAL DATA

General Statement

Published results of geophysical studies and unpublished geophysical data in oil company

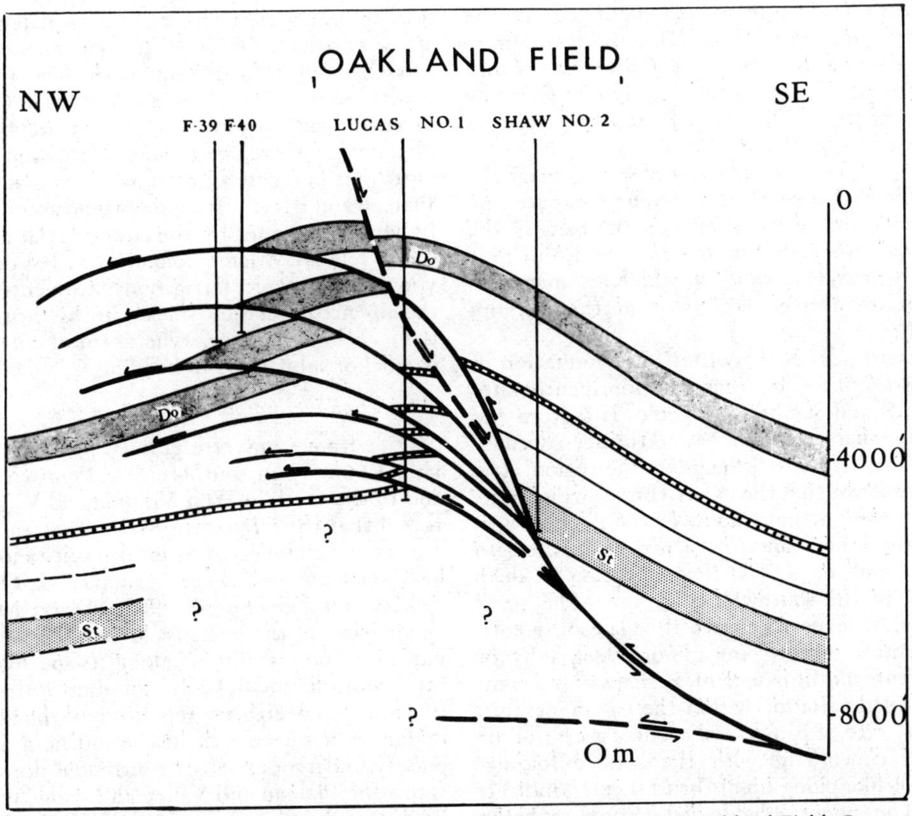

Figure 15. Imbricate thrust faulting in the core of Deer Park anticline, Oakland Field, Garrett County, Maryland; faults assumed to branch off a sole thrust. (Data on faulting in #1 Lucas, written communication from Robert Miller, Manufacturers Light and Heat Company, August, 1963.) (Do = Oriskany Sandstone, St = Tuscarora Sandstone, Om = Martinsburg Shale; dotted unit is the Williamsport Sandstone)

Walls (Preston County well #79) on the southeast flank of Chestnut Ridge anticline in western Preston County, W. Va., and New Penn Development Company #1 Heyn (Uniontown quadrangle well #6) northeastward along strike in the Summit Field will provide important control to test this assumption. Comparison of the #1 Walls, a southeast-flank well, can be made directly with Phillips #1 Finch files indicate that basement rocks of the Valley and Ridge and Plateau provinces are not deformed concordantly with the overlying sedimentary cover.

Nittany Arch-Northern Plateau Area, Pennsylvania

Magnetic surveys by Joesting and others (1949) indicate that basement structural trends, as defined by anomalies, diverge from the surface structures in the northern and central Plateau of Pennsylvania and the Nittany Arch area, as do the gravity studies by Howell and Vozoff (1953) and Ackermann and Howell (1962).

Magnetic and seismic data obtained by many oil companies indicate that the basement is much deeper beneath the highly uplifted Nittany Arch than it is in the Plateau to the northwest. Depth to basement below the Cambrian exposed on the Gatesburg and Sinking Valley anticlines near the Structural Front is estimated to be 19,000–22,000 feet (Joesting and others, 1949); comparable depths are indicated by seismic methods.

Figure 13A compares an observed magnetic profile with a theoretical profile computed by Joesting and others (1949) on the basis of the figured structure section across the Front near the Birmingham window, which assumes basement involvement in uplift of the Nittany Arch.

Figure 13B is the writer's interpretation of the structure. The interpretation incorporates surface and shallow drill hole data from the Birmingham window area and the deep décollements that occur throughout the region. The figure shows that the upper thrust, which outcrops around the window in the Arch area, is greatly folded and upwarped, compared to its −7000 to −9000 foot elevation in deep wells to the southwest and west. The small magnetic reversal beneath the Gatesburg anticline may reflect some minor topography or structure on the basement, perhaps only a compositional variation within the basement; but, at any rate, it can scarcely reflect a crystalline uplift concordant with the surface folding. Shinglelike piling-up of thrust sheets is indicated. The upper. now-folded thrust probably occurred first. This is reasonably established by application of the "law of least work." That is, the work expended to move the upper sheet up and across a pre-existing fold is orders of magnitude greater than that used in simple reinitiation of slip on the deep master glide zone followed by formation of a new step-up thrust a short distance westward. Subsequent development of a flexure in the new hanging wall, which includes the older Birmingham thrust, resulted in uplift and folding of the thrust to its present position.

Gravity surveys indicate divergence of Bougher and isostatic anomalies from surface structures in the same area. In general, however, the studies outline a gravity minimum along the southeastern edge of the Plateau, a zone of high values along the Nittany Arch, and another low zone in the position of the Broadtop synclinorium (Ackermann and Howell, 1962). The positive anomaly along the Arch is thought to be caused by density contrasts within the upper 5500 feet of the cover. The best approximation of the body causing the residual anomaly according to Ackerman and Howell (p. 222) is a wedge of material with a density difference of 0.1 g/cc which tapers to the southeast. This model conforms in all detail with the observed repetition of relatively dense Cambrian-Ordovician dolomites by low-angle thrust faults in the Nittany Arch area and along Wills Mountain anticline. The gravity minima of the synclinal Plateau and Broadtop regions would accordingly be accounted for by the lateral replacement of the relatively dense dolomites by less dense, younger Paleozoic terrigenous sediments and the absence of tectonic thickening by thrusts in the synclinal regions, where the faults are parallel or subparallel to bedding.

West Virginia-Virginia Region

Data from transverse gravity profiles along U.S. Route 30 in southernmost Pennsylvania and to the south in West Virginia and Virginia reveal that the large surface structural features are not as well reflected by local gravity anomalies (Nettleton, 1941; Hammer and Heck, 1941). Ackermann and Howell (1962) observed that the maxima of the Nittany Arch area becomes narrower and less intense along its southwestward continuation in Wills Mountain anticline, in which less overthrust repetition of the dense dolomites has occurred. In the southern area, gravity undergoes a rather continuous decrease across the Plateau and Valley and Ridge provinces to a minimum at the foot of the Blue Ridge. Only minor maxima occur along the major structures.

Magnetic surveys by Straley (1938) and Yost (1949, M.Sc. thesis, West Virginia Univ.) further document the discordance of magne ic anomalies with regional and local structural trends. The low magnitude and frequent discordance of the anomalies to surface structural features suggest that the more highly magnetic rocks of the basement were not significantly involved in the deformation.

Sears (1964) made extensive gravity and magnetic surveys from the Plateau to the Blue Ridge across the Valley and Ridge near Roanoke and Blacksburg, Virginia. Basement trends generally parallel surface structural trends, but are locally transverse to the surficial folds and faults. Sears (p. 257) stated "Geophysical data show no evidence that the basement rocks are involved in the broad overthrust faults. Regional gravity maps indicate a broad southwest-plunging synclinal basement structure which is not in keeping with known surface structures."

Salina salt did likewise in part of the western Low Plateau. Geologic and geophysical data indicate that the faults or folds do not significantly affect the basement of the region. Non-orogenic irregularities on the basement surface prior to deformation may have controlled patterns of Paleozoic sedimentary facies and thickness. These pre-orogenic irregularities may

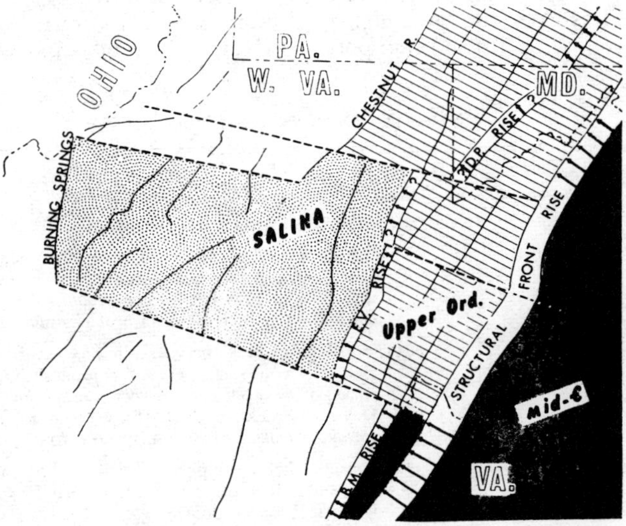

Figure 16. Areas of dominance of specific sole thrust-décollement zones in the Central Appalachians; the Structural Front Rise marks the zone along which the sole thrusts shear upward from the mid-Cambrian glide zone (cross-hatched) to the Upper Ordovician glide zone (diagonal lines); significant translation on the Cambrian décollement occurs in the Plateau only near Brown's Mountain (B.M.) anticline; E.V. = Elkins Valley anticline; D.P. = Deer Park anticline. The rise to the Salina décollement (stippled) took place beneath and produced most of shortening in the Elkins Valley anticline and probably in Deer Park anticline.

INTERPRETATION

Westward Translation of the Skin of the Central Appalachians

Surface and subsurface data from many parts of the Plateau and the western Valley and Ridge province demonstrate that stepped bedding-plane faults within incompetent stratigraphic units accommodated a westward shift of sheets of the sedimentary rocks of the folded Central Appalachian region (Fig. 16). Shales in the Waynesboro-basal Elbrook interval formed the zone of décollement in the Valley and Ridge province; the Martinsburg-Juniata argillaceous rocks served as the principal décollement in the southeastern High Plateau, and the have in turn controlled details of distribution, trend, and size of resulting faults and folds in the sedimentary cover. But the principal structures appear to be restricted to the surficial layers of the region.

Table 2 presents descriptions of the factors controlling the deformation.

Anticlines as the Products of Change of Stratigraphic Position of Bedding-Plane Thrust Faults

Anticlines are the principal products of the Appalachian deformation. These rise irregularly above the adjacent synclines, in the axial zones of which altitudes of correlative horizons lie on regular planes of regional dip (Pl. 2; Fig. 7). Drilling shows that in the synclines formations appear to have a consistent, presumably original thickness, whereas on the anticlines it shows that significant fault thickening has occurred in many units and that this has created the anticlinal uplifts (Pl. 2; Figs. 3–12).

Plateau and Valley and Ridge anticlines appear to be of three general types:

(1) Anticlines, relatively high, developed along linear zones where bedding plane thrust faults shear upward as they rise to a higher décollement-glide zone (Fig. 16) and transport significant volumes of older rocks along the higher surface; *e.g.*, Wills Mountain, Elkins Valley, Brown's Mountain, Deer Park, and Burning Springs anticlines;

(2) Anticlines of intermediate height on the high southeastern side of the Plateau, in which folds paired symmetrical thrust faults occur immediately above the Upper Silurian. These anticlines and those immediately southeast of Wills Mountain are probably produced by "splay" thrust faults branching from a major sole thrust; displacement on the splay thrusts is generally of lower magnitude, thus no considerable volumes of older rocks were transported "up and out" on a higher décollement, as in class (1) anticlines;

(3) Small, low anticlines northwest of Chestnut Ridge, Elkins Valley, and Brown's Mountain anticline, which may be faulted or unfaulted wrinkles above décollements in the Salina or the Devonian shales.

Plate 2 illustrates interpretative sections of the thrust faults moving blocks of the Plateau westward along the Cambrian, Upper Ordovician, and Upper Silurian glide zones. Middle Devonian and older rocks are exposed only on those anticlines in which significant northwestward transport of the upper allochthonous plate up to and along a higher décollement has occurred, *e.g.*, Wills Mountain, Elkins Valley, and Brown's Mountain anticlines.

Evidence from dip-meter surveys and stratigraphic sequence in a few wells suggests that considerable drag occurs in the lower block near the points where the thrusts shear upward. A minimum of drag is shown in Plate 2 in order to present a simplified picture. The inverted block of Ordovician strata in the Sponaugle well at the major thrust fault below Wills Mountain, the inverted strata below the fault northeast along strike in the Martin well in Beford County, Pa., and steep dips in the Elkins Valley deep well illustrate the possible complexity along the ascending, shear portions of the overthrust faults.

Table 2. Deformation Model

Controlling factors	Description of factors
Basin fill configuration	Southeastward-thickening wedge of sediments
Pre-orogenic intrabasinal structure	Southeastward descent of basement—between 30 and 60 feet to the mile; conformable sedimentary cover; basement irregularities unknown at present; no conspicuous hinge lines of more rapid thickening or increase of rate of basement slope
Potential zones of detachment	A, argillaceous rocks near middle of Cambrian carbonate sequence, B, argillaceous rocks between competent Ordovician and Silurian rocks, C, evaporites between competent Silurian and Lower Devonian rocks, and D, thick Devonian shale sequence below competent basal Mississippian sandstones
Wedge-outs of potential translation zones	Cambrian shales disappear westward in central or southeastern Plateau; salt appears in southeastern Plateau and disappears in western West Virginia
Style of deformation	Stepped low-angle detachment thrust faulting, with subsequent concentric folding, branching splay thrusting, and symmetrical thrust faulting of the allochthonous sheets
Tectonic agents	Gravity gliding assisted by directed subhorizontal forces

Transcurrent Faults and Changes of Stratigraphic Level along Strike of Thrust Sheets in the Plateau

More than a dozen lineaments have been defined on the surface of the Plateau. Some of the lineaments appear to mark zones of transcurrent faulting on the margins of differentially advancing dermal thrust sheets in the Plateau, as suggested by Rodgers (1963) for lineaments around the Burnings Springs anticline. However, no faulting occurs along several of the lineaments in Mississippian, Pennsylvanian, and Permian rocks in the coal fields, whose structure is well known. In these areas, some folds terminate along the lineaments, others only undergo rapid plunge and minor trend change. The best illustration of a lineament along which occurs such a combination of abrupt termination of anticlines and continous but suddenly plunging anticlines not transcurrently faulted at the surface is the lineament in northern Cambria County and central Indiana County, Pa. Chestnut Ridge and Laurel Hill anticline plunge, but continue a short distance across the lineament; other smaller anticlines terminate abruptly against the lineament from both the north and south.

It is suggested that lineaments with no associated surface faulting, such as the described, are zones along which the local sole thrust or a higher branching stepped thrust is changing stratigraphic position *along strike*. The thickness of strata being folded above the controlling glide zone would change across the lineament, leading to a change in amplitude of the fold. The controlling factor is the radius of curvature of the fold which is approximately equal to the thickness of the section involved in the folding. The upward shearing inclined or vertical non-outcropping fault connecting two glide levels along strike, in fact, is simply another kind of tear fault.

Additional detailed mapping along these lineaments is in progress and will, in combination with subsurface data, indicate the specific character of each lineament.

These lineaments are thought to be zones of changing sole thrust position in the subsurface or subvertical tear faults because:

(1) High thrust-faulted anticlines at the western margins of the rectilinear blocks indicate westward translation of the blocks;

(2) Folding and faulting is concentrated above décollement surfaces within or at the base of the thrust sheet blocks;

(3) Folds are terminated, plunge, and/or are deflected at the lineaments;

(4) Tear faulting and high-angle reverse faulting has been demonstrated in places along the marginal lineaments;

(5) Irregular domes and basins occur along the lineaments; these features are best explained by warping along the zones where the thrusts change stratigraphic position.

WESTWARD EXTENT OF THRUSTING: The westward limit of movement of the Plateau blocks is indicated by the location of the most westerly anticline bounded at one or both ends by lineaments in which east-dipping thrust faults or décollement folding occurs. Westward movement along the Salina décollement persisted farthest westward in the Burning Springs block. The low, almost east–west-trending anticline in Tyler County, W. Va., between the two northern Burning Springs lineaments would appear to reflect a clockwise, draglike rotation reflecting greater westward advance of this block to the south (*cf.*, transverse anticlines in the Pine Mountain thrust sheet, Rich, 1934).

The next most northwesterly movement was that of the block between the northern Burning Springs lineament and the central Pennsylvania lineament. This is bounded on the northwest by the Chestnut Ridge anticline or the faulted, lower anticlines immediately northwest of it. The Williamsburg anticline bounds the western edge of a block which appears to be wider than the Brown's Mountain block immediately to the north-northeast.

There is a relative absence of folding west of high Brown's Mountain anticline and the other high anticlines southwestward toward the Pine Mountain thrust sheet (Pl. 1). Folding and décollement movement appear to have been unnecessary in this area because of the uptake of all shortening during growth of the very high Brown's Mountain anticline. Shortening was resolved farther south and southwest by shearing to the surface of the St. Clair, Boissevain, and Pine Mountain thrusts obviating the uptake of shortening in the Plateau to the west.

Folds developed within a single block and were not formed prior to definition of the blocks by the transcurrent faults or transverse steps. This is suggested by the general lack of concordance of folds in adjacent blocks; the sense and amount of movement inferred by apparent separation of anticlines in adjacent blocks are contradictory along the lineaments. As the folds were not transected by any of the possible transcurrent faults, the movement along the décollements was initiated at an early stage in the deformation, apparently prior to most folding.

Mechanics of Formation of Symmetrical Thrust Faults above the Salina Décollement

Regional prevalence of symmetrical overthrusting of the limbs of anticlines at the

Oriskany level is an important feature of Appalachian Plateau tectonics. Comparably faulted anticlines occur locally in the Kuznets Basin (Rumyantsev, 1928, *in* Beloussov, 1962, Fig. 284, p. 595), but only in the Appalachian Plateau is the phenomenon a key part of the tectonic style of a large region.

The symmetrical thrust faulting is a logical outgrowth of the concentric flexural folding in a special case: that is, in the folding of an anisotropic sequence of stratified rocks of strongly varying competency. In the early stage of folding of such an anisotropic sequence, internal movement is confined to flexural bedding-plane slip, with each bed shifting anticlineward relative to the underlying bed. At the same time, the trajectory of any unit volume of rock in the flanks of a growing fold is initially vertical, becoming suddenly horizontal (Fig. 17, *after* De Sitter, 1956, p. 197, Fig. 135). Thus, besides the pervasive anticlineward shift of an upper bed relative to a lower bed, there are also two flank blocks moving upward and inward onto the crest of the anticline.

The stratigraphic succession in the Plateau includes the incompetent Salina Group which contains an upper salt unit over much of the region, capped by a competent carbonate-quartzite-chert sequence in the basal Devonian, and overlain in turn by 3000–7000 feet of argillaceous Devonian rocks. As concentric folding occurs, the upward-and-inward motion of the flanks, as well as anticlineward flexural slip of beds overlying a plane of easy slippage, would produce thrusting of the strutlike Lower Devonian competent rocks of the limbs over the passive anticlinal crest, where flexural slip is inhibited (Fig. 18B). Bedding plane slip is probably accentuated along and within the salt for considerable distances down the flank as a decrease of volume in deeper parts of the tightening synclines forces material upward and outward. Cloos (1964) has illustrated that wedgelike shortening of this character, entirely attributable to flexural slip, can amount to 1.6 times the thickness of the competent unit.

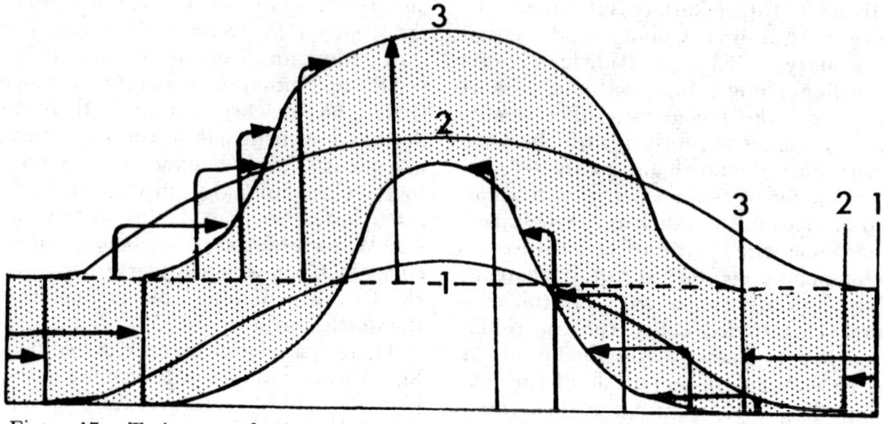

Figure 17. Trajectory of points in the flanks of a growing concentric fold, illustrating the abrupt change from a vertical to a horizontal path during folding (*after* De Sitter, 1956, p. 197, Fig. 135, by permission of McGraw-Hill Book Co.).

As the thrust faults on both flanks steepen during progressive uplift of the fold, normal stress across the thrust plane increases until the thrust locks. Further tightening of the fold produces another shear fault and further imbrication of the flanks.

The common oversteepening on the southeastern flank and the greater displacement on the northwestern flank thrust in several anticlines at the Oriskany level is in opposition to the northwestward steepening of the high anticlines of the Valley and Ridge province and southeastern High Plateau and the observed relative westward sense of translation on deep thrust faults. Sherrill (1934) attributed the opposed asymmetry of Plateau folds to southeastward regional dip at the time of deformation and to the southeastward thickening of the sequence. Regardless of regional dip direction, the steep limb of a fold will be in the direction of stratigraphic thinning, inasmuch as the radius of curvature of the fold will be shorter in that direction and the limb necessarily steeper (De Sitter, 1956, p. 239, Fig. 181). The northwest limb of Plateau folds would then be steeper, but such is not always the case. Cathcart and Myers (1934) suggested that the asymmetry developed from basement-driven deep-seated underthrusting of northwest limbs by the southeast limbs.

I tentatively suggest that the steepening of the southeast limbs is restricted to post-Ordovician strata and may have been produced by the flexural-slip folding-thrust process. The resulting greater motion on the northwest flank fault leads to development of the steeper flank on the southeast, in spite of the relative westward tectonic transport at depth.

Nature of Tectonic Agents

Hubbert and Rubey (1959) described the mechanical paradox of large overthrust faults, illustrating, as have many others, that the crushing strengths of rocks in a thrust sheet of

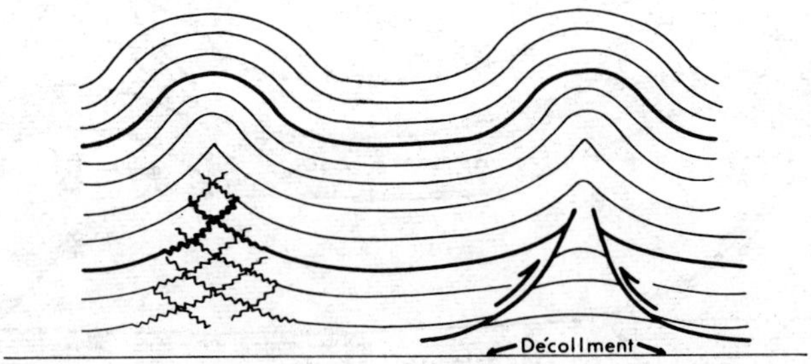

Figure 18. Symmetrical overthrusting as a consequence of room problem in the core of a concentric fold; A, Room problem in the core of a concentric fold; B, Resolution by thrust of limbs over axial depression in the tightened core zone of an anticline, with persistence of concentric folding above.

by underthrusting of the relatively stationary rocks of western anticlinal flanks by a wedge of Lower and Middle Silurian rocks moving up and to the west along a splay thrust fault which shears off the deeper décollement (Fig. 19). The movement of the inter-décollement blocks labelled B and B′ to the west onto relatively stationary C and C′ is in phase with the synclineward flexural-slip of the Silurian rocks below the upper décollement zone relative to the flexure-induced overthrusting of the Lower Devonian strata in the western flanks of the anticlines labelled Aw and A′w. The westward motion of the supra-Ordovician-sub-Devonian wedges B and B′ along the deep décollement slip surface thus accentuates relative displacement along the west flank symmetrical thrust above the Salina, inasmuch as it is in phase with the flexual slip. On the other flanks (A_E and A'_E), the deep translational motion of blocks B or B′ is out of phase with the sense of east flank flexural slip and may partially or completely cancel the relative movement induced moderate width are greatly exceeded by the external forces necessary to move the sheet on a dry, horizontal surface. This is true in the Central Appalachians, where westward translation of overthrust sheets has occurred across a 180-mile-wide zone.

Two principal means of propulsion of an upper plate over a lower autochthonous plate are possible: (1) application of externally applied surface forces or (2) the action of internally generated body forces. In the former case, Hubbert and Rubey stated (1959, p. 122), "The simplest case of propulsion by means of a surface force would involve the pushing of a block along a horizontal surface by means of surface forces applied along the rearward edge of the block. The maximum length[6] of a block that can be moved in this manner depends

[6] *Author's note:* Width across the strike of the faulted mass, as used by Hubbert and Rubey. Length and width are herein used to denote distance parallel and perpendicular to strike, respectively.

jointly upon the coefficient of friction of rock on rock and upon the strength of the rock." Hubbert's calculations suggest that the strength of rocks to externally applied forces places a maximum width of 15–18 miles on thrust sheets slipping up along a dry inclined or horizontal surface. Connate fluids and high pore fluid pressures are thought to increase the possible width. Internal body forces generated by gravity serve also to increase the permissible width of thrust sheets. During regional tilting (relative uplift in part of the thrust sheet), potential energy which can be converted to a pervasive internally generated body force is generated within the "potential" thrust sheet (Rubey and Hubbert, 1959, p. 199–200). These body forces act on each particle of the thrust sheet and tend to transport the stratified sequence without changing relative positions of its component parts.

Appalachian geologists have been painfully aware of the mechanical difficulties inherent in moving large slices of incompetent Paleozoic rocks northwestward toward the Plateau. It is this point on which the "thick-skinned" deformationists have most forcefully built their case for the necessity of involvement of basement rocks in Appalachian faults and folds. (*See* Cooper 1961; ver Wiebe, 1936; Lees, 1952.)

However, *westward translation of* thin, wide overthrust sheets of the sedimentary cover *has indeed occurred* throughout the Central Appalachians. The motive forces were: (1) northwestward upthrust of the wedge of Blue Ridge and western Piedmont rocks (Rodgers, 1953; Bucher, 1955; and others); and (2) gravity potential created within the sedimentary cover by uplift in the active internal zone of the Appalachian orogen, resulting in gravity sliding of thin thrust slices or nappes northwestward onto the passive cover of the marginal southeastern Valley and Ridge, with resulting wrinkling and thrusting in the Valley and Ridge and Plateau much like Bucher's "peel thrusting" (1955).

Uplift of the Blue Ridge massif brought

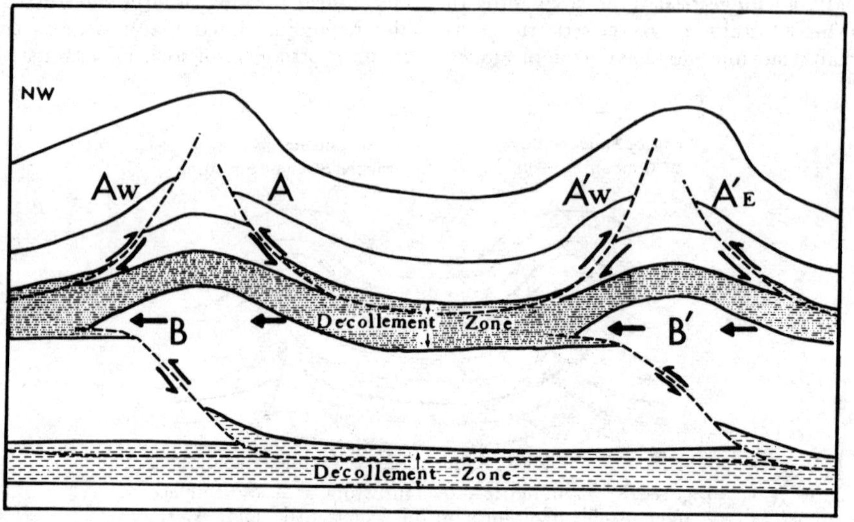

Figure 19. Origin of steep southeast limbs of Plateau anticlines; movement on the detachment sole thrust leads to further tightening of the older folds and initiation of new folds above branching splay thrusts; the wedge of competent rocks between the Upper Ordovician and Salina décollements is thrust under the northwest limb of the secondary folds, adding on that limb to displacement on the sub-Oriskany thrusts. The latter were generated primarily as a consequence of the concentric folding process, allowing movement of material up and out of the tightening syncline. Net result is relatively greater displacement on the northwest flank fault and steepening of the southeastern flank. (Aw and A'w indicate west limbs of the figured anticlines on the left and right respectively; AE and A'E, the east limbs; B and B' are the wedges of pre-Devonian, post-Middle Ordovician rocks being thrust beneath the west limbs.)

crystalline rocks to heights exceeding its present topographic crests at nearly 4000 feet above sea level. Unpublished geophysical studies in the Broadtop synclinorium near Little North Mountain thrust in West Virginia and nearby western Virginia indicate that the basement is nearly 30,000 feet deep at this point or nearly 6½ miles lower than in the Blue Ridge, only 30 miles east. The western flank of the Blue Ridge anticlinorium north of central Virginia is generally steeply inclined to the northwest or overturned and broken by discontinuous high-angle thrust faults of small displacement (King, 1950; 1954). This contrasts with the low-angle thrust faults at the contact of the crystalline rocks and sediments southwest in southwestern Virginia and Tennessee (King, 1950; King and Ferguson, 1960; Rodgers, 1953). Geophysical studies by Sears (1964) in the Virginia Valley and Ridge have shown that the basement of the Great Valley is not involved in the dermal thrust faults seen at the surface, but is warped into a gentle syncline marked only by minor geophysical anomalies which are probably only minor basement topographic irregularities or internal compositional variations.

The basement may descend from the Blue Ridge to the Valley and Ridge province in a relatively steep s-shaped flexure, as shown in schematic Figure 20. A less conservative speculation is that the crystalline rocks are themselves part of a large thrust sheet, thrust an unknown distance out over the edge of the sedimentary cover, as in the Blue Ridge of North Carolina (King, 1951; King and Ferguson, 1960). The gravity minima along the Blue Ridge uplift led Woolard to such a speculation as long ago as 1936 and 1939. (King raised the same possibility in the Geological Society of America Southeastern Section Symposium on Appalachian structure in April 1963 at Roanoke.) The displacement or shortening affected by such a thrust would have to be taken up to the northwest on outcropping or buried thrust faults in the Valley and Ridge and Plateau, inasmuch as no such major thrust fault is present along the crystalline-sedimentary contact.

Uplift of the Blue Ridge wedge, either in a broad, reverse-faulted flexure or along a deep major thrust fault, necessarily would have exerted great tectonic forces on the easterly edge of the sedimentary cover of the subsequently folded Appalachians. In addition, thrust sheets or nappes derived from the roof-like sedimentary cover of the uplifted Blue Ridge were probably emplaced on the easterly edge of the Valley and Ridge cover with the active assistance of gravity. It is proposed that the cascading nappes piled up on the east, creating an unstable mass distribution above the potential glide zones in the sedimentary sequence. Gravity thrusting to the west resulted in a more stable mass distribution by lowering the center of gravity of the load superincumbent on the glide zones. The work done in westward translation of the sheets was accomplished at least in part by utilization of the potential energy of the tilted sheets. The uplift of the Blue Ridge and the pile-up of thrust sheets of its cover on the eastern Great Valley cover produced an effect like adding a rider to a sled poised ready to slide down a gentle sloping snowy slope. The sled, portions of the Valley cover, then slid down the snowy slope, the glide zones.

The use of a "slope" in the preceding analogy implies the existence of a syndeformational westward slope on the glide zones utilized in the Valley and Ridge and in at least a part of the Plateau. This assumption is made in spite of the fact that the Waynesboro-lower Elbrook zone can be deduced, through extrapolation of stratigraphic thicknesses, to now slope *eastward* toward the Blue Ridge across most of the Valley and Ridge province at rates between 0.5 to 2.0 degrees, depending on estimations of pre-Devonian stratigraphic thicknesses. (The Upper Ordovician and Salina décollements appear to be nearly horizontal or slope slightly *westward* over portions of the Plateau in the Virginias, Maryland, and southern Pennsylvania, where drilling provides direct information. *See* Pl. 2.) It is contended that the very presence of the wide thrust sheets and wide (up to 180 miles) belt of thrusting and the mechanical properties of sedimentary rocks defy postulation of propagation of thrusting across the belt by the application of only surface forces on the edges of the thin weak sheets. One is compelled to conclude that *only internal body forces generated by the force of gravity, in combination with directed tectonic forces exerted by the upthrust Blue Ridge wedge* can begin to explain the regional thin-skinned deformation of the folded Central Appalachians.

The principal objection to this conclusion is the cited apparent eastward inclination of the principal zone of Valley and Ridge thrusting in the Cambrian Waynesboro zone. Such an attitude across the Valley and Ridge at the time of deformation, necessitating transport of

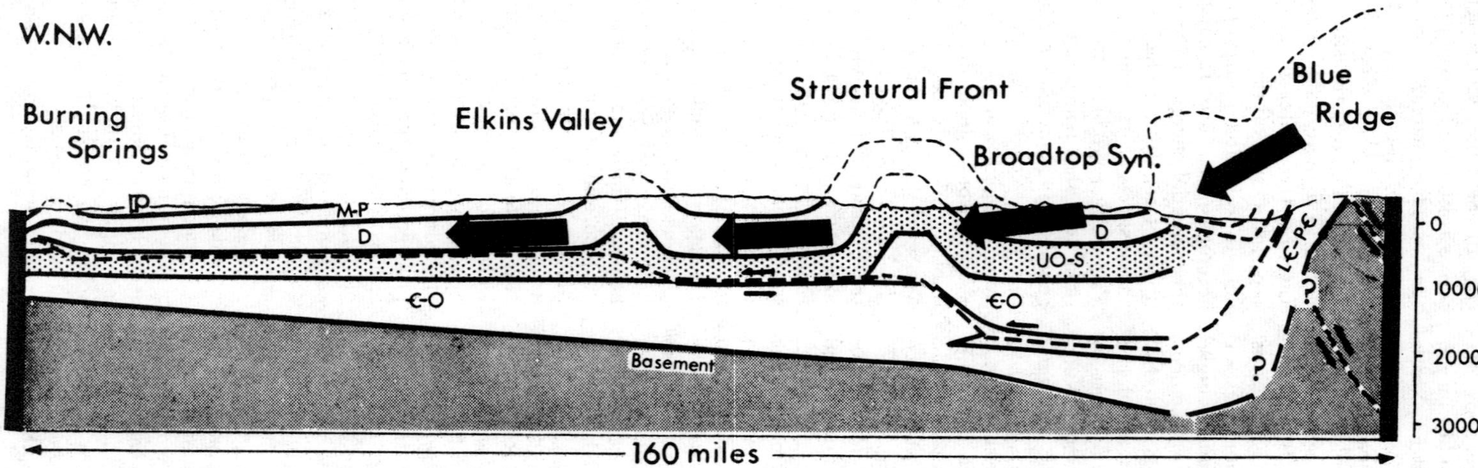

Figure 20. Schematic cross section of the Central Appalachians, Virginia-West Virginia, demonstrating the near-horizontal present-day attitudes of the décollement glide zones in the northwestern Valley and Ridge and Plateau provinces. A conservative interpretation of the rapid descent of the crystalline rocks of the Blue Ridge along a steep, slightly faulted flexure is shown here. In this interpretation, the major bedding thrust or thrusts must emerge as a bedding thrust in the eastern Great Valley or Blue Ridge area. It is conceivable, however, that the Blue Ridge is an unrooted mass, moved northwestward along a great thrust fault to overlie Paleozoic sedimentary rocks (Woolard, 1936; 1939; King, 1964). IP = Permian; M–P = Mississippian-Pennsylvanian; D = Devonian; UO–S = Upper Ordovician-Silurian; and Є–O = Cambrian and Lower and Middle Ordovician

the thrust sheets *up a slope* would appear impossible in light of the present experimental and theoretical knowledge of the behavior and strength of rocks in deformation. A definitive answer to the question of when and how some of the décollements assumed their present eastward inclination can not be given. But it is not outrageous to suppose that primary forces of tectonic compression generated the uplift in the Blue Ridge-Piedmont region and that some mechanisms operated there after deformation to cause subsidence, such as metamorphism of sediments to denser phases of smaller volume or thermal contraction, which could have caused an easterly tilting of the flanking Valley and Ridge province toward the Blue Ridge.

As with most analyses of regional overthrusting, the credibility of this thesis suggesting that forces of gravity and horizontal compression or directed tectonic stress acted on very low slopes over a distance of 180 miles in a sequence whose total thickness is 2–5 miles is open to attack on several points of interpretation. The great subsurface thrusting and the surficial nature of the deformation is documented. Rubey and Hulbert (1959, p. 200) stated "On the basis of available evidence, [the foregoing interpretation] may be the most likely explanation of the forces that have produced large scale, low angle overthrusts in the [Central Appalachian] region." (Bracketed words are author's.)

SUMMARY AND CONCLUSIONS

(1) Deformation of the folded Central Appalachians does not persist to the basement; bore holes penetrate major sole thrusts and branching splay thrusts in complexly deformed subsurface portions of Appalachian anticlines which are unfaulted concentric folds at the surface; geophysical data further substantiate these conclusions derived from consideration of subsurface and surface geologic evidence.

(2) The sole thrust-décollement zones occur largely within incompetent strata in the Middle-Cambrian, Upper Ordovician, Upper Silurian, and Middle and Upper Devonian.

(3) Northwestward movement of thin thrust sheets occurred throughout the Valley and Ridge province and over much of the Plateau. The zone of translation reaches a maximum width of nearly 180 miles in western Virginia and West Virginia between the Blue Ridge and the Burning Springs anticline near the Ohio River.

(4) Some of the northwest-trending lineaments in the sedimentary cover of the Central Appalachian Plateau are tear faults produced by differential advance of rectilinear thrust blocks; others appear to mark zones of change of stratigraphic level of the thrusts.

(5) The tectonic style of the Central Appalachians is entirely analogous to that of the Southern Appalachians (Rich, 1934; King, 1950; 1951, King and Ferguson 1960; Rodgers, 1953; and Stearns, 1954; 1955). Only one minor difference has heretofore masked the structural similarity between the contiguous regions: erosion has progressed far enough to expose the forward toes of the thrust faults of the Southern Appalachians, whereas the toes of the Central Appalachian low angle thrust sheets are still covered by an unfaulted blanket of younger sediments draped across the toes. Raleigh and Griggs' conclusion (1963, p. 829) that ". . . a thrust of large displacement must ultimately come to the surface" is placed in serious question by the nonoutcropping Central Appalachian thrust faults.

REFERENCES CITED

Ackermann, H. D., and Howell, B. F., Jr., 1962, Gravity investigations in northeastern and central Pennsylvania: Pa. Acad. Sci. Proc., v. XXXVI, p. 218–224

Ashley, G. H., 1908, Studies in mechanics of Allegheny structure (Abstract): Science, new ser., v. 27, p. 924–925

Bucher, W. H., 1955, Deformation in orogenic belts, p. 343–368 *in* Poldervaart, A., *Editor*, Crust of the Earth: Geol. Soc. America Special Paper 62, 762 p.

Butts, C., Swartz, F. M., and Willard, B., 1939, Geology and mineral resources of the Tyrone Quadrangle: Atlas 496, topographic and geological atlas of Pennsylvania: Pa. Topog. and Geol. Survey, 4th ser., 116 p.

Cate, A. S., 1962, Subsurface structure of the Plateau region of north-central and western Pennsylvania on top of the Oriskany formation: Pa. Geol. Survey, 4th ser.,

Cathcart, S. H., and Myers, T. H., 1934, Gas in Tioga County, Pennsylvania: Pa. Topog. and Geol. Survey, Bull. 107, 42 p.

Chamberlin, R. T., 1931, Isostasy from the geological point of view: Jour. Geology, v. 39, p. 19–22

Cloos, Ernst, 1964, Wedging, bedding slips, and gravity tectonics in the Appalachians, p. 237 *in* The Geological Society of America, Abstracts for 1963: Geol. Soc. America Special Paper 76, 342 p.

Cooper, B. N., 1961, Grand Appalachian field excursion; Geological Guidebook no. 1, Annual meeting of the Geol. Soc. of America: Va. Polytech. Inst. Extension Ser., p. 1–119

De Sitter, L. U., 1956, Structural geology: New York, McGraw-Hill Book Co., 552 p.

Fettke, C. R., 1933, Subsurface Devonian and Silurian sections across northern Pennsylvania and southern New York: Geol. Soc. America Bull., v. 44, p. 601–660

—— 1954, Structure contour maps of the Plateau Region of north-central and western Pennsylvania: Pa. Topog. and Geol. Survey, 4th ser., Bull. G-27, 27 p.

Finn, Fenton H., 1949, Geology and occurrence of natural gas in the Oriskany sandstone in Pennsylvania and New York: Am. Assoc. Petroleum Geologists Bull., v. 33, p. 303–335

Gray, C., Shepps, V. C., and others, 1960, Geologic map of Pennsylvania: Pa. Topog. and Geologic Survey

Hammer, S. F., and Heck, E. T., 1941, A gravity profile across the Central Appalachians: Am. Geophys. Union Trans., v. 22, p. 353–362

Howell, B. F., Jr., and Vozoff, K., 1953, Gravity investigations in north-central Pennsylvania: Am. Geophys. Union Trans., v. 34, p. 357–359

Hubbert, M. K., and Rubey, W. W., 1959, Role of fluid pressure in mechanics of overthrust faulting; Part I, Mechanics of fluid-filled porous solids and its application to overthrust faulting: Geol. Soc. America Bull., v. 70, p. 115–166

Joesting, H. R., Keller, Fred, and King, Elizabeth, 1949, Geologic implications of aeromagnetic survey of Clearfield-Philipsburg area, Pennsylvania: Am. Assoc. Petroleum Geologists Bull., v. 33, p. 1747–1766

Kay, Marshall, 1942, Development of the northern Allegheny synclinorium and adjoining regions: Geol. Soc. America Bull., v. 53, p. 1601–1658

Keith, Arthur, 1925, Outlines of Appalachian structure: Geol. Soc. America Bull., v. 34, p. 309–380

King, P. B., 1950, Tectonic framework of the southeastern United States: Am. Assoc. Petroleum Geologists Bull., v. 34, p. 635–671

—— 1951, Tectonics of Middle North America: Princeton, N. J., Princeton Univ. Press, 203 p.

—— 1954, Geology of the Elkton area, Virginia: U. S. Geol. Survey Prof. Paper 230, 82 p.

King, P. B., and Ferguson, H. W., 1960, Geology of northeasternmost Tennessee: U. S. Geol. Survey Prof. Paper 311, 136 p.

Lees, G. M., 1952, Foreland folding: Quart. Jour. Geol. Soc. London, v. CVIII, p. 1–34

Ludlum, J. C., 1958, Relation of surface and subsurface structures in West Virginia to future drilling for natural gas: W. Va. Engineer, v. 20, no. 7

Lytle, W. S., Bergsten, J. M., Cate, A. S., Fairall, V. M., Hieren, L. A., and Wagner, W. R., 1961, A summary of oil and gas developments in Pennsylvania, 1955 to 1959: Pa. Topog. and Geol. Survey, 4th ser., Bull. M45, 133 p.

Martens, J. H. C., 1945, Well-sample records: W. Va. Geol. Survey Rept., v. 17, 889 p.

Moebs, N. N., and Hoy, R. B., 1959, Thrust faulting in Sinking Valley, Blair and Huntingdon counties, Pennsylvania: Geol. Soc. America Bull., v. 70, p. 1079–1088

Nettleton, L. L., 1941, Relation of gravity to structure in the northern Appalachian area: Geophysics, v. 6, p. 270–286

Perry, W. J., 1964, Geology of the Ray Sponaugle Well, Pendleton County, West Virginia, p. 253–254 *in* The Geological Society of America, Abstracts for 1963: Geol. Soc. America Special Paper 76, 342 p.

Price, Paul H., 1931, The Appalachian structural front: Jour. Geology, v. 39, p. 24–44

Price, Paul N., and Heck, E. T., 1939, Greenbrier County: W. Va. Geol. Survey, Detailed County Rept. ser.

Raleigh, C. B., and Griggs, D. T., 1963, Effect of the toe in the mechanics of overthrust faulting: Geol. Soc. America Bull., v. 74, p. 819–830

Reeves, Frank, and Price, Paul H., 1950, Early Devonian gas in northern West Virginia and pre-Devonian oil prospects: Am. Assoc. Petroleum Geologists Bull., v. 34, p. 2095–2132

Rich, J. L., 1934, Mechanics of low-angle overthrusting faulting as illustrated by the Cumberland overthrust block: Am. Assoc. Petroleum Geologists Bull., v. 18, p. 1584–1596

Rodgers, John, 1949, Evolution of thought on structure of the middle and southern Appalachians: Am. Assoc. Petroleum Geologists Bull., v. 33, p. 1643–1654

—— 1953, Folds and faults of the Appalachian Valley and Ridge Province: Ky. Geol. Survey, Special Pub. no. 1, p. 150–166

—— 1963, Mechanics of Appalachian foreland folding in Pennsylvania and West Virginia: Am. Assoc Petroleum Geologists Bull., v. 47, p. 1527–1536

Rubey, W. W., and Hubbert, M. K., 1959, Role of fluid pressure in overthrust faulting, Part II, Overthrust belt in geosynclinal area of western Wyoming in light of fluid pressure hypothesis: Geol. Soc. America Bull., v. 70, p. 167–206

Rumyantsev, S., 1928, Tectonic dislocations observed along the northwestern margin of the Kuznets Coal Basin, and their explanations: Gorniy Zhurn., nos. 10, 11 (in Russian)

Sears, C. E., 1964, Geophysics and Appalachian structure, p. 257 *in* The Geological Society of America, Abstracts for 1963: Geol. Soc. America Special Paper 76, 342 p.

Shaffner, M. N., 1958, Geology and Mineral Resources of the New Florence Quadrangle, Pennsylvania: Pa. Geol. Survey Atlas A-57, 185 p.

Sherrill, R., 1934, Symmetry of Appalachian foreland folding: Jour. Geology, v. 42, p. 225–247

Stearns, R. G., 1954, The Cumberland Plateau overthrust and geology of the Crab Orchard Mountains area, Tennessee: Tenn. Geol. Survey Bull. 60, 47 p.

—— 1955, Low-angle overthrusting in the Central Cumberland Plateau, Tennessee: Geol. Soc. America Bull., v. 66, p. 615–628

Stose, G. W., and Ljungstedt, O. A., 1932, Geologic map of West Virginia: W. Va. Geol. Survey

Straley, H. W., 1938, Aeromagnetic reconnaissance in the Appalachian Mountains (Abstract): Geol. Soc. America Bull., v. 49, p. 1962

Tucker, R. C., and Price, P. H., 1943, Summarized records of deep wells: W. Va. Geol. Survey Rept., v. 16, 938 p.

Ver Wiebe, W. A., 1936, Geosynclinal boundary faults: Am. Assoc. Petroleum Geologists Bull., v. 20, p. 910–938

Woodward, H. P., 1959a, Structual interpretations of the Burning Springs Anticline, p. 159–168 *in* Woodward, H. P., *Editor,* A symposium on the Sandhill deep well, Wood County, West Virginia: W. Va. Geol. Survey Inv. Rept., no. 18, 182 p.

—— 1959b, The Appalachian region: New York, 5th World Petroleum Congress Proc., sec. 1, Paper 59, p. 1061–1079

Woollard, G. P. 1936, Interpretation of gravity anomalies in terms of local and regional geologic structures: Am. Geophys. Union Trans., 17th Ann. Mtg., pt. 1, p. 63–74

—— 1939, Geologic significance of gravity-investigations in Virginia: Am. Geophys. Union Trans., 20th Ann. Mtg., pt. 3, p. 317–323

Manuscript Received by the Society August 27, 1963

THE AMERICAN ASSOCIATION OF PETROLEUM GEOLOGISTS BULLETIN

MAY 1977 VOLUME 61, NUMBER 5

Flow of Weak Rocks in Appalachian Plateau Folds[1]

D. V. WILTSCHKO[2] and W. M. CHAPPLE[2]

Abstract A thin-skinned fold model of Appalachian Plateau folds in which layers do not thin in synclines is not possible geometrically. However, a two-layer model consisting of a gently folded competent layer overlying a deformed zone in which material flows from the synclines to the anticlines is both reasonable geometrically and supported by geologic evidence from the Plateau province of Pennsylvania and south-central New York. The deformed zone is composed of weak rocks—evaporites and shale—which have been deformed much more than the competent rocks above; the decollement marks the lower boundary of this deformed zone. Care should be taken in reconstructing the depocenters of rocks composing the deformed zone, because the original thicknesses have been altered by later deformation.

The Appalachian Plateau of Pennsylvania and New York may be divided into two regions on the basis of fold mechanics. In the northwestern and far western Plateau of Pennsylvania and south-central New York, Region I, the salt-bearing part of the Silurian Salina Group is the deformed zone; no other part of the stratigraphic section has flowed significantly. Synclines are broader than anticlines in the higher relief folds of Region I because much material has been removed from beneath the synclines and flanks to fill anticlines. In southwestern and central Pennsylvania, Region II, the anhydrite-rich Tonoloway formation and/or the Upper Ordovician Reedsville Shale flowed as well as salt.

Complex thrust faulting and disharmonic folding at the level of the Oriskany Sandstone contribute significantly to the structural relief, but it cannot provide the major part of the relief. Balancing the shortening present at the Oriskany level requires either stretching at this level in the synclines or significant layer-parallel shortening at shallower stratigraphic levels.

INTRODUCTION

In the past 15 years several studies have shown that folding on the Appalachian Plateau (Fig. 1; Tables 1-3) is thin skinned (Woodward, 1959; Rodgers, 1963, 1970; Gwinn, 1964, 1970; Prucha, 1968; Chute, 1972; Frey, 1973). Accepting a thin-skinned origin for the folds places strong constraints on the nature of material transport within the folds—constraints which have not been addressed by these earlier workers. This paper shows how these constraints and the available geologic information can be used to construct and test a model of material transport in Plateau folds.

Previous Work

The first well-documented model of Appalachian foreland folds was developed by Gwinn (1964), who demonstrated that the folded Appalachian Plateau province of Pennsylvania and West Virginia has been transported from southeast to northwest along a nonoutcropping detachment zone. In addition, he showed that, although the geometry is simple, their anticlinal cores are complexly faulted at depth. Curved thrust faults, representing several hundred feet of shortening, are present at the level of the Lower Devonian Oriskany Sandstone. Gwinn interpreted these faults as splays which merge with the sole thrust in the Silurian Salina Group on the flanks and again become parallel with bedding in the Middle Devonian shales in the core of the anticline (Figs. 2, 3a). According to Gwinn, these splay thrusts resulted from constriction and consequent faulting in the anticlinal cores because of a concentric-folding room problem (Gwinn, 1964, Figs. 17, 18, 20, Pl. 2).

[1]Manuscript received, July 13, 1976; accepted, October 5, 1976.

[2]Department of Geological Sciences, Brown University, Providence, Rhode Island 02912.

This work was supported by a Grant-in-Aid from Sigma Xi, The Scientific Research Society of North America, and by Grant No. GA-4182 from the National Science Foundation. We wish to thank Jo-Ann Sherwin for her kind permission to use Figure 11.

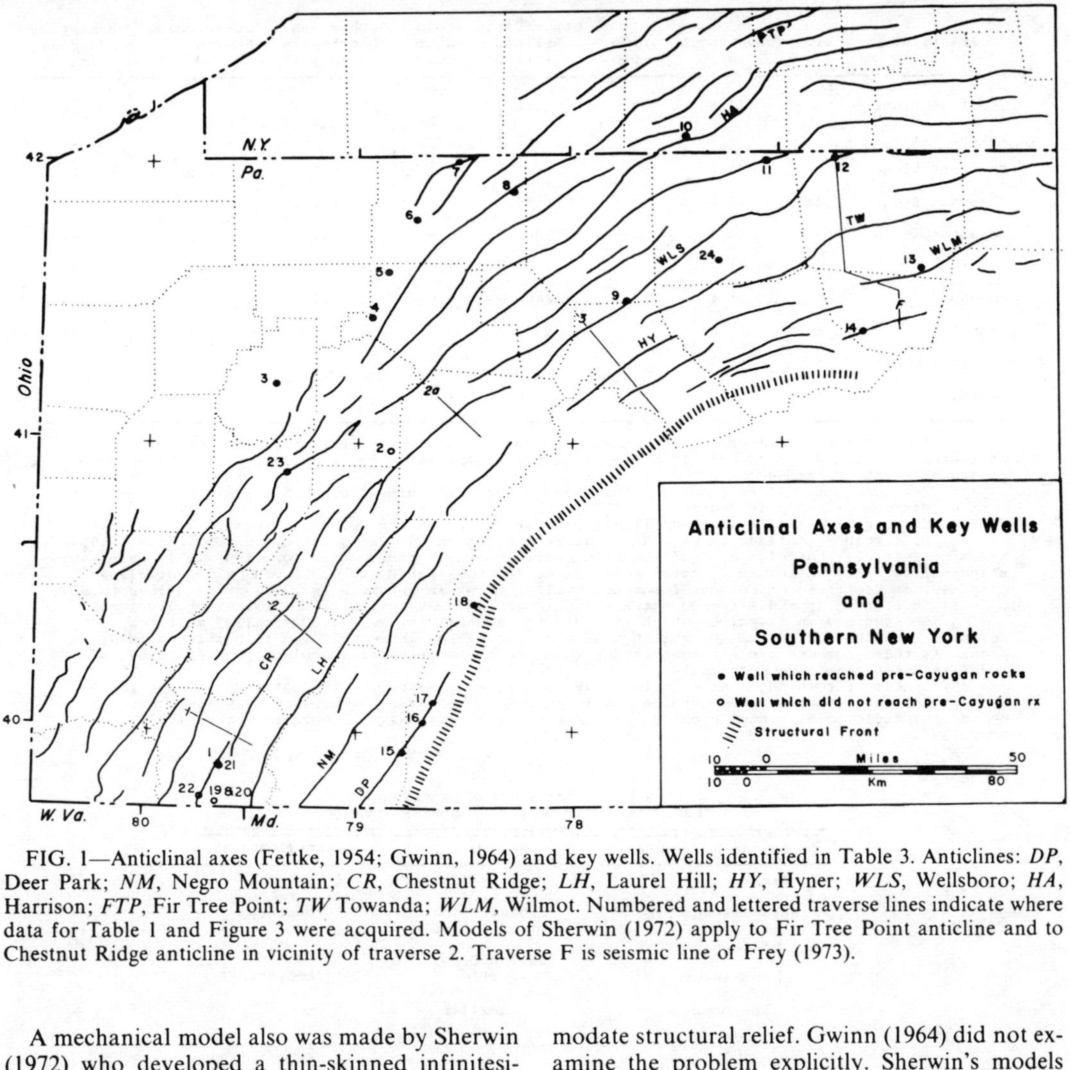

FIG. 1—Anticlinal axes (Fettke, 1954; Gwinn, 1964) and key wells. Wells identified in Table 3. Anticlines: *DP*, Deer Park; *NM*, Negro Mountain; *CR*, Chestnut Ridge; *LH*, Laurel Hill; *HY*, Hyner; *WLS*, Wellsboro; *HA*, Harrison; *FTP*, Fir Tree Point; *TW* Towanda; *WLM*, Wilmot. Numbered and lettered traverse lines indicate where data for Table 1 and Figure 3 were acquired. Models of Sherwin (1972) apply to Fir Tree Point anticline and to Chestnut Ridge anticline in vicinity of traverse 2. Traverse F is seismic line of Frey (1973).

A mechanical model also was made by Sherwin (1972) who developed a thin-skinned infinitesimal-amplitude buckling theory and tested it against data from two localities on the Plateau of Pennsylvania and New York (see Figs. 1, 2). Although Sherwin recognized that the shortening due to folding implies that the post-Silurian rocks of the Plateau have been transported northwestward, she reasoned that, because a buckling model with thinning of material under the synclines and thickening under the anticlines could explain the observed fold wavelengths, it was more reasonable than Gwinn's model. In her view, splay thrusts are a manifestation of flow due to buckling, rather than the cause of folding.

Statement of Problem

Neither Gwinn nor Sherwin dealt quantitatively with the problem of moving material to accommodate structural relief. Gwinn (1964) did not examine the problem explicitly. Sherwin's models are based on infinitesimal-amplitude theory and therefore apply only to those stages before appreciable structural relief is developed. However, there is an important qualitative difference between the two. The conventional view of Plateau folds implicitly assumes that material does not move from synclines—the prefolding and postfolding stratigraphic thicknesses in the synclinal troughs are equal, for the synclines merely slide northwestward on a lubricating layer (see De Sitter, 1964, Chap. 14; Gwinn, 1964, Fig. 17). On the other hand, Sherwin (1972) showed that a model that allows for flow from synclines is theoretically reasonable.

The no-thinning hypothesis leads to geometric difficulties when applied to the Appalachian Plateau folds. The concentric-folding room problem

TABLE 1. DATA ON PLATEAU FOLDS

Anticline	a ID	b Wavelength (mi)	c Limb Dip	d Structural Relief (ft)	e Fold Volume (mi^3)	f Observed Shortening (mi)	g "No-thinning" Shortening (mi)	h Center of Folding (mi)
Chestnut Ridge	1	11.5	9	1940	5.4±.2	0.2	28.5	16
Laurel Hill	1	9.7	9	1730				22
Chestnut Ridge	2	8.1	9	2480	1.6	0.1	8.5	11
Chestnut Ridge	2a	15.8	3	1180	2.8	0.2	14.8	112
Wellsboro	3	10.0	3	1040	0.9	0.1	4.8	42
Hyner	3	11.2	5	1180	1.0	0.1	5.3	36
Towanda	F	13.4	7	2600	1.6	0.1	8.5	5
Wilmot	F	11.6	2	1480	0.8	0.1	4.2	87
Fir Tree Point	*	5.9	2	250	-	-	-	-
Fayette	1	5.7	3	650	0.3	0.1	1.6	31

a. Identification numbers of traverses on Figure 1 from which the data in columns b - h were collected or calculated; *, data from Prucha (1968), information to calculate entries in columns e through h not available.

b. Wavelength in miles, from syncline hinge to syncline hinge; error ± 0.1 mi.

c. Maximum limb dip in degrees.

d. Structural relief (twice amplitude) in feet, adjusted for regional dip.

e. Fold volume (defined in Fig. 4), measured on convenient Carboniferous datum (Fettke, 1954) assuming that Plateau folds are sufficiently gentle that all horizons in competent layer have same shape; first entry in column is total fold volume for Chestnut Ridge and Laurel Hill anticlines combined, as are those of columns f through h also; error in column e is ± 0.1 cu mi unless noted.

f. Shortening (the difference between arc length and wavelength) in miles; error ± 0.1 mi.

g. Shortening, in miles, by which it would be necessary to move 1,000-ft (305 m) thick deformed zone into anticline to explain observed fold volume in no-thinning thin-skinned fold model. As these numbers are all much larger than their counterparts in column f, a no-thinning model is unreasonable geometrically.

h. Center of folding in miles below level of Oriskany Sandstone in anticlines. These large numbers indicate that, because decollement is only a few thousand feet below Oriskany, there can be no concentric folding room problem. See text for full discussion. Sources: Fettke (1954), Frey (1973).

TABLE 2. COMMON USAGE OF CAYUGAN TERMINOLOGY IN NEW YORK AND PENNSYLVANIA

Michigan-S.W. Ontario	New York	Pennsylvania
Absent	Manlius-Rondout	Keyser
Bass Islands Grp.	Cobleskill	Tonoloway
H { Raisin River, Put-in-Bay }	Salina Grp. Bertie	(anhydrite)
Salina Grp.	Camillus	
G		
F { 3, 2, 1 } ↑	Syracuse (salt)	(salt)
E s		
D a		Wills Creek
l		
C t	Vernon	
B ↓		
A		Bloomsburg
Guelph-Lockport	Guelph-Lockport	McKenzie-Bloomsburg

From Alling and Briggs (1961). Terminology developed in Michigan (Landes, 1945) recently has been applied to subsurface in New York and Pennsylvania (Ferguson and Prather, 1968). We break evaporite-bearing part of Upper Silurian of Pennsylvanian and south-central New York into two zones, anhydrite and salt. Lowermost one, salt zone, consists of rocks between bottom of oldest salt bed (usually in unit B) and top of youngest salt bed, in unit F. In Pennsylvania and south-central New York, unit C contains no salt and proportion of salt never exceeds 40% of total unit B through unit F section. Uppermost zone, anhydrite zone, consists of rocks between bottom of oldest anhydrite above unit F and top of youngest anhydrite. Therefore, although anhydrite commonly is interbedded with salt in salt zone, anhydrite zone contains no salt.

TABLE 3. KEY WELLS

	Well Name	Operator	Salt Zone* Thickness
1.	Leo F. Heyn	William Snee et al	1155
2.	W. H. Irwin	F. C. Deemer	156**
3.	J. A. Mays	Paul Schaffer	720
4.	Pa. Game Lands Wt. #3653	Pa. Gas Co.	520
5.	Wt. #3788 Well #3737	United Nat'l Gas	337
6.	L. D. Wilson	South Penn Oil Co.	360
7.	Delo #1	Bradford Deep Well	493+
8.	Charles Nunn et al #2	Allegheny Gas	422
9.	Pa. F & W Tr. #129	Consolidated Gas Sup.	1705
10.	Robert Olin N-650	N.Y.S. Nat'l Gas	950
11.	L. E. Shoemaker	Lycoming Gas	1705
12.	Anna Carver	Parson Bros.	1324
13.	Charles Blemle	Pure Oil Co.	1195
14.	Arthur Bennet	California Co.	0
15.	Nellie R. Mowry	Peoples Nat'l Gas et al	70
16.	Hillegas	South Penn Oil	(0)
17.	Harry Miller	Kerr-McGee Oil	(0)
18.	Edward Bigley	Hollidaysburg Gas	(0)
19.	H. M. Ryan	Sun Oil	--
20.	Ryan #2	Zenith	--
21.	E. C. Ricks	Snee & Eberle et al	1015
22.	W. H. Barton Est. #8	Manufacturers H & L	922**
23.	N. C. Martin	P.N.G.	925
24.	S. & S. Coal	Amoco	0++

* The salt zone is the salt-bearing portion of the Salina Group or from the top of the youngest to the bottom of the oldest salt.
** Did not completely penetrate the Salina Group.
\+ Approximated from Fergusson and Prather (1968).
\+\+ Drilled in a syncline.

which is called on to explain complex faulting in anticlines does not exist. This is because Plateau folds are gentle, broad features and consequently the center of folding (center of curvature) is very deep (Fig. 3, Table 1, column h). A concentric-folding room problem arises in a thin-skinned fold only if the center of folding is above the decollement zone. Because the center of folding in the tightest Plateau anticline is 5 mi (8 km) below the base of the decollement (Table 1, column h), one cannot explain the complex anticlinal faulting by a concentric-folding room problem.

To find a more acceptable model, we first will discuss briefly what is known geometrically about Plateau folds, propose a model, and then explore the geologic consequences of this model.

GEOMETRY OF THIN-SKINNED FOLDS OF PENNSYLVANIA PLATEAU

Geometry of Typical Folds

The folds of western Pennsylvania and south-central New York are broad, gentle structures (Fig. 3). A typical fold has a wavelength of about 7.5 mi (11 km), ranging between 5 and 20 mi (8 and 32 km), and a structural relief from a few hundred to more than 2,500 ft (760 m). The form of the near-surface layers is nearly sinusoidal in far western Pennsylvania, but close to the structural front, synclines are relatively broad and flat bottomed whereas anticlines are relatively narrow, especially in northeastern Pennsylvania. The anticlinal hinge commonly is closer to one of the two adjacent synclinal hinges than to the other.

There are few exposed faults. They are either small-displacement high-angle faults oriented perpendicular to fold axes (near 78°25′W long., 45°50′N lat.; Glover, 1970, p. 42) or thrust faults on the southeast limb of larger folds (Finn, 1949; Fettke, 1954, Pl. 1; Shaffner, 1958, p. 75).

In contrast to the simple surface expression of Plateau folds, the subsurface is marked by a zone of complex faulting (Cate, 1962, revising Fettke, 1954; Gwinn, 1964) and regional subsurface maps show this difference clearly. The faults terminate in the Devonian shales above the Oriskany Sandstone; the Tully is faulted only locally (Gwinn, 1964, p. 873). The lower limit, though not well defined by drilling data, is probably within the Silurian Salina Group. Additionally, two local studies showed detailed information on the complexity at depth in the northern Plateau (Prucha, 1968; Frey, 1973) and that there is a level below which folding dies out in northern Pennsylvania and southern New York.

The three general features of Plateau folds—gentle folding at the surface, complex faulting and folding in the subsurface, and a decollement—may be used to construct a thin-skinned fold model.

Thin-Skinned Fold-Model Geometry

A Plateau thin-skinned fold can be divided into two zones. The upper zone, which we call the *competent layer*, consists of those rocks that are folded nearly concentrically. Below this is a *deformed zone*, where complicated faulting and flow take place. The upper boundary of the deformed zone is the stratigraphic horizon above which there is no faulting; the lower boundary is the decollement horizon, below which the rocks are not folded. If the density change is negligible, shortening in the deformed zone must equal that in the competent layer. In a train of folds, each complete fold wavelength—from synclinal axial surface to synclinal axial surface—is a closed system where volume and shortening are conserved.

As an anticline grows, a space develops in the anticlinal core between the base of the competent layer and a line joining the base of the competent layer in the synclines (Fig. 4). Of course an actual void never is present because material comes from regions exterior to the anticlinal core to fill it. However, this conceptual fold volume has an important geometric significance. The fold volume must be filled by material coming from the zone in which we know material readjustment takes place—the deformed zone.

If synclines are not thinned, the deformed layer must be pushed into the anticline by an amount equal to the shortening to produce the structural

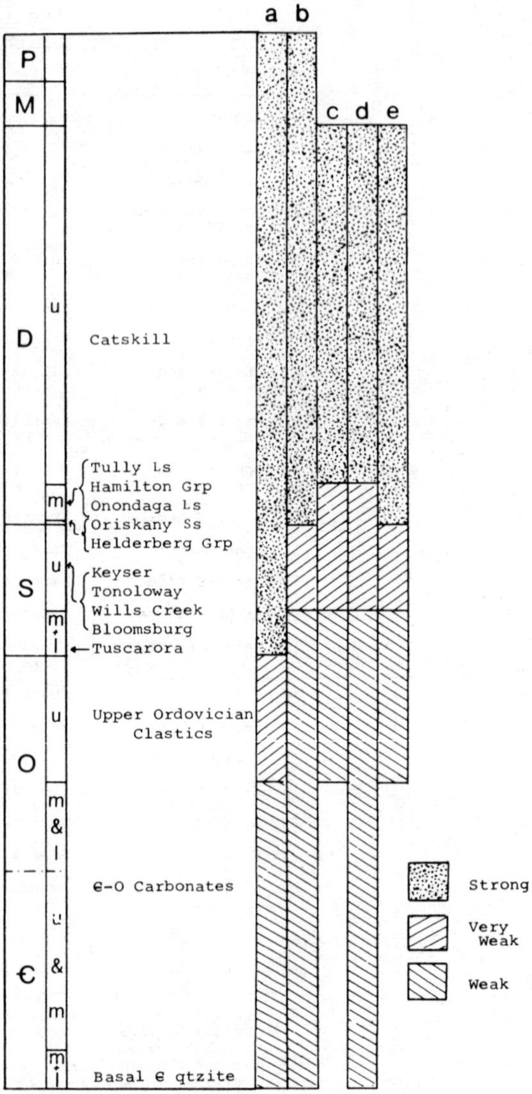

FIG. 2—Schematic stratigraphic column and mechanical models of Sherwin (1972) which predict reasonable viscosity ratios among various layers and have zone of highest velocity flow in reasonable stratigraphic position; *a* and *b* are for central Pennsylvania; other three fit Fir Tree Point anticline. Four of five successful models include Upper Silurian in their very weak layer. Approximate total thickness of section for central Pennsylvania is 21,000 ft (6,400 m).

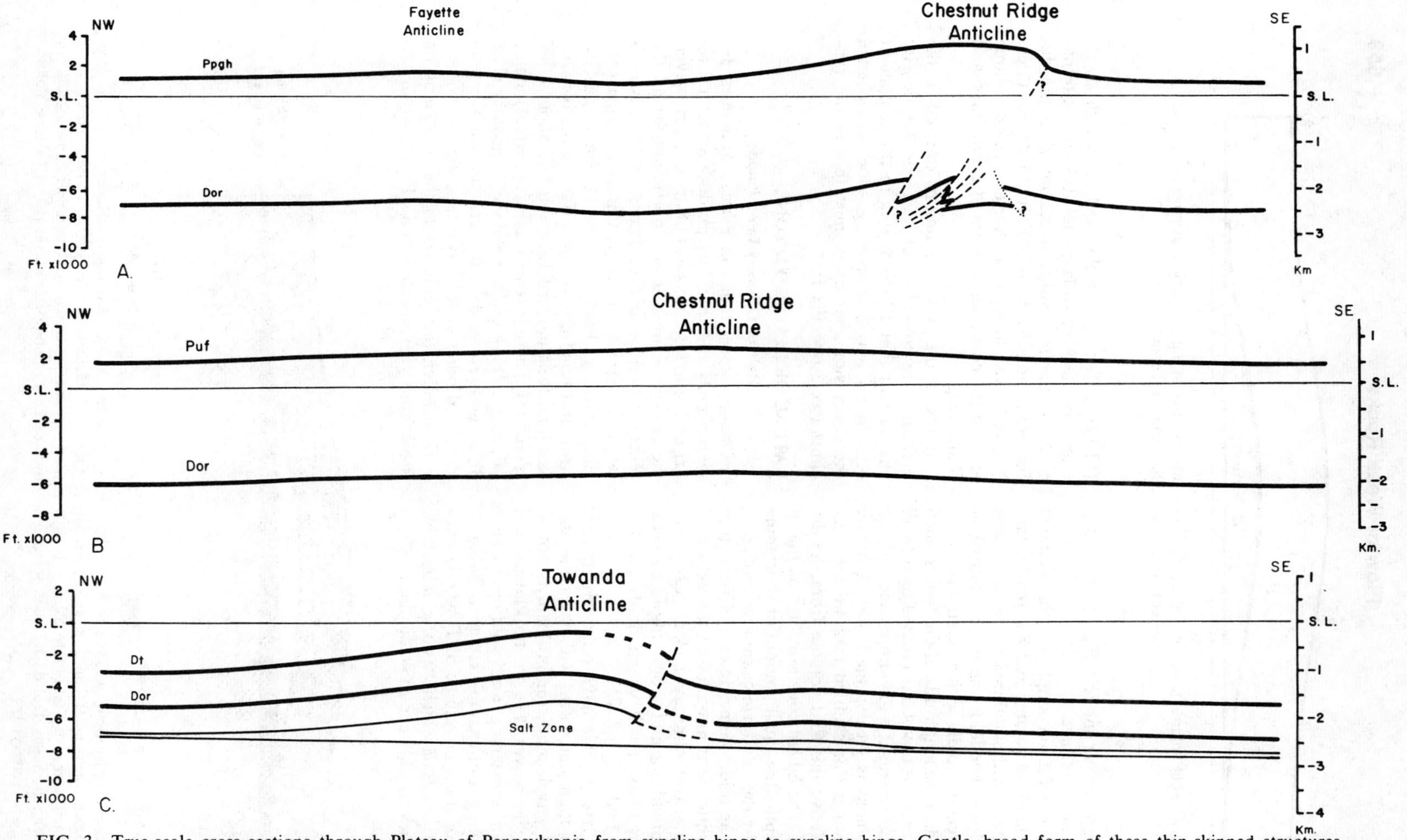

FIG. 3—True-scale cross sections through Plateau of Pennsylvania from syncline hinge to syncline hinge. Gentle, broad form of these thin-skinned structures emphasizes (1) lack of concentric-folding room problem and (2) geometric necessity of thinning units from beneath synclines. They are located on Figure 1 as traverses 2, 2a, and Towanda anticline part of *F*. Symbols: *Ppgh*, Pittsburgh coal; *Dor*, Oriskany Sandstone; *Puf*, upper Freeport coal; *Dt*, Tully Limestone. Sources: **A** uses *Dor* structure in flanks and synclines from Cate (1962) and in anticline from Gwinn (1964, Fig. 11). *Ppgh* data in **A** is from Fettke (1954) in flanks and synclines and from Shaffner (1958) in anticline. Very steep southeastern flank of Chestnut Ridge anticline in **A** is unique on Plateau of Pennsylvania and may reflect continuation of one of Oriskany-level faults. **B** uses *Dor* structure from Cate (1962) and *Puf* data from Fettke (1954). Cate (1962) constructed contours on *Dor* in synclines and flanks by projecting thickness found for Carboniferous to Oriskany interval in anticlines to synclines and flanks. **C** is after Frey (1973, Fig. 3). In **C** also, salt zone corresponds approximately to units B through F within Salina Group, that is, from the youngest to oldest salt bed (see Table 2).

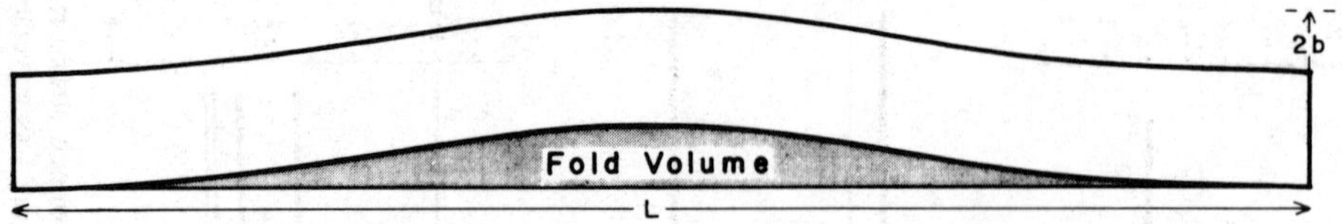

FIG. 4—Fold volume is volume of material per unit distance along fold axis that must be moved into anticline to produce structural relief.

relief. In the thinning model (Fig. 5) it is displaced from the synclines.

The geometry of a low-amplitude thin-skinned fold is such that fold volume and shortening cannot be balanced without thinning the deformed zone in synclines. The fold volume of a thin-skinned fold increases rapidly with a small increase in shortening when the amplitude of the fold is small (Fig. 6); the amount of volume created by a small amount of shortening at a low-amplitude stage of fold development is greater than that at higher amplitude. For a thin-skinned fold we are constrained to derive this fold volume from the deformed zone. The added constraint means that if the fold is gentle, the small amount of shortening represented by the fold form cannot be enough to account for the fold volume. In fact, if flow from syncline to anticline is not allowed the deformed zone must be nearly infinitely thick at nearly zero shortening for volume and shortening to balance. This is a geometric result which holds for any thin-skinned fold.

Because Plateau folds have relatively smaller fold volumes than sine waves with the same amplitudes and wavelengths, the problem is not so acute as for sine waves. Yet, for no Plateau fold geometry can shortening and volume balance without thinning. For example (refer to Table 1), to account for the fold volume of Towanda anticline, a deformed zone which was originally 1,000 ft (305 m) thick would have to be displaced 85 mi (137 km). The largest observed arc-length shortening on the Plateau is 0.2 mi (0.3 km). Alternatively, if the shortening is 0.2 mi, then the deformed zone must be 42,000 ft (13,000 m) thick. The only way to balance shortening and volume is to displace material from the synclines and flanks (Fig. 5).

Relaxing the two assumptions implicit in the preceding argument—that there is no density change and that all shortening in the competent layer is arc-length shortening—does not change the conclusion that units must thin in the synclines (see Appendix I).

NATURE OF DEFORMED ZONE

Role of Oriskany-Level Faulting

Oriskany-level faults in Plateau anticlines duplicate beds and therefore produce structural relief and represent transport of material. However, the amount is insufficient to account for all the relief or fold volume of Plateau folds. Gwinn's two cross sections through Chestnut Ridge anticline (Gwinn, 1964, Figs. 9, 11) show that the structural relief due to subsurface faults is much less than the total structural relief of about 2,500 ft (760 m). There is relief on the uppermost Salina salt and arching of the Oriskany Sandstone (Fig. 3). Thus faulting does not account for the total structural relief at these two localities.

How much thickening does the shortening on these faults represent? Gwinn's Figure 9 (our tra-

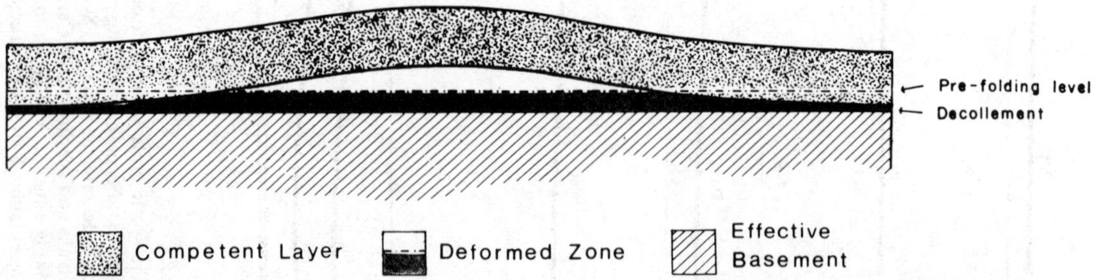

FIG. 5—Flow-model geometry. Thin-skinned Plateau fold model consists of competent layer, deformed zone which thins in synclines and thickens in anticlines, and decollement. Heavily shaded region of deformed zone is location deformed zone occupied before folding. Lightly shaded region is volume of rock that was transported from flanks and synclinal troughs. Dot-dash line is position occupied by base of competent layer before folding. Volume and shortening are conserved.

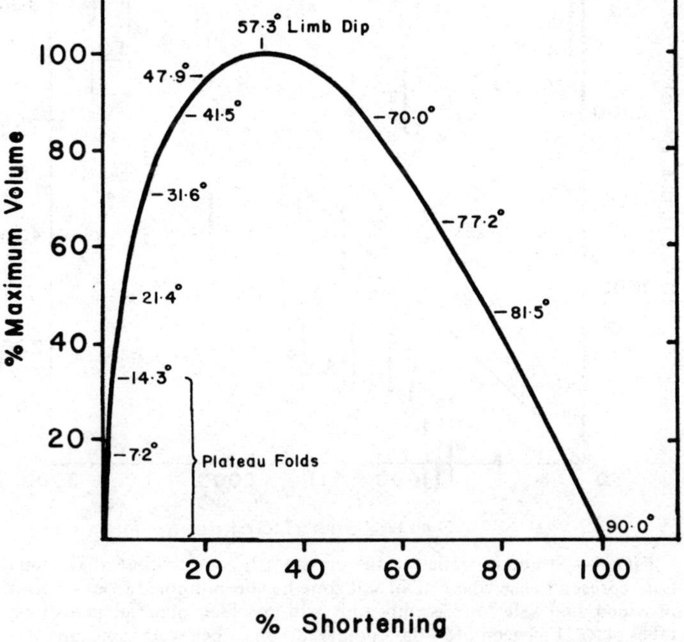

FIG. 6—Volume versus shortening for sine-wave fold form. Steep slope below about 20% shortening (slope is infinite at zero shortening) shows that relatively large fold volume is created for small amount of shortening. Fold volume cannot equal product of shortening and thickness of deformed zone during early stages of fold development unless deformed zone is unreasonably thick. Plateau folds have maximum limb dip of less than 15°. Volume is plotted as percent of maximum volume fold attains. Volume and shortening for fixed arc length are calculated using equations A2 and A6 in Appendix II.

verse 1, Fig. 1) shows 2,100 ft (640 m) of shortening on the Oriskany Sandstone over a linear distance of 5,500 ft (1,700 m). If there is no volume change, the relative thickening corresponding to this shortening is equal to the initial horizontal length divided by the final horizontal length, (5,500 + 2,100)/5,500 = 1.38. Consequently, a unit originally 1,000 ft (305 m) thick (Tully to Keyser interval) will be 1,380 ft (420 m) thick after deformation. Similarly, farther south on Chestnut Ridge anticline (Gwinn, 1964, Fig. 11; our Fig. 3a and traverse 2, Fig. 1), the relative thickening would be 1.52 or 520 ft (160 m) of structural relief per 1,000 ft (305 m) of original thickness. (Shaffner, 1958, Pl. 8, interpreted this locality as displaying a relative thickening of only 1.27). Therefore, although contributing on the order of several hundred feet (a few hundred meters) to the structural relief, the thickening associated with Oriskany-level faults cannot explain the total structural relief.

The relief on the Salina salt suggests that material adjustment takes place below the top of the Silurian. There is local geologic evidence which suggests that salt has flowed from the synclines toward the anticlines.

Thinning of Deformed Zone Under Synclines

Prucha (1968) studied the structural complexities in a salt mine in Fir Tree Point anticline (Fig. 1). Although this anticline is only a gentle warp at the surface (Table 1), at the level of the mine there has been extensive disharmonic folding, brecciation, and transport of intercalated carbonate layers and attendant flow of salt. The shortening of only a few of the higher order folds in the mine is many times that of the competent layer implied by the gross geometry of the anticline.

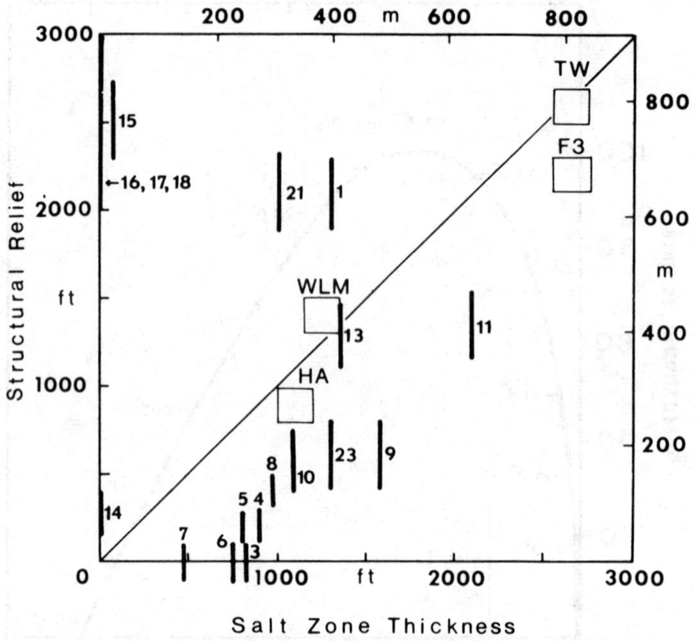

FIG. 7—Structural relief versus crestal salt-zone thickness. Diagonal line represents case where all of salt zone has been thinned from syncline, provided that salt zone is only unit within which material adjustment takes place. Fourteen of 21 data points fall on or below 45° line, and thus are consistent with simple model consisting of one deformed zone which is equivalent to salt zone. Six data points which fall above line are discussed in text. Structural relief is adjusted for regional dip. Bars are wells identified and located in Table 3 and Figure 1, and boxes are from Frey (1973): *TW*, Towanda anticline; *F3*, Frey (1973, Fig. 5); *WLM*, Wilmot anticline; *HA*, Harrison anticline. Error bars on wells are estimates of accuracy of sources from which data are derived. Error boxes indicate estimated error in measuring thickness and structural relief from seismic profiles.

The strong deformation of salt-bearing rock and the unequal shortening imply that the salt-bearing strata are thinned in the synclines.

Frey (1973) presented unequivocal evidence for large-scale thinning of salt-bearing rocks from the synclines of four folds, one in south-central New York and three from northeastern Pennsylvania (Fig. 1, traverse F plus Harrison anticline near well 10). All the structural relief on the Middle Devonian Tully Limestone (Fig. 2) is due to thinning and thickening in the salt zone, or salt-bearing part of the Silurian Salina Group, in Towanda, Wilmot, and the other two anticlines (Frey, 1973, Figs. 2-5; see Table 2). The synclinal thicknesses of the salt zone are 170, 140, and 160 ft (52, 43, and 49 m) respectively for Towanda, Wilmot, and the third anticline in Pennsylvania. (This last anticline is not identified in Frey, 1973.) By contrast, the same salt zone is about 2,600, 1,300, and 2,600 ft (790, 400, and 790 m) thick in the corresponding anticlines, respectively. Clearly, flow of salt-bearing rock is fundamental to the development of these folds.

Salt thinning in synclines also is indicated farther southwest along strike in well 24 (Fig. 1). This well, which was drilled in a syncline, found no salt. By contrast, well 9, 26 mi (42 km) southwest on Wellsboro anticline, found 1,630 ft (500 m) of salt-bearing rock. The structure indicated in well 24 is not simple. Of the three faults that have been recognized, the upper two are thrusts which repeat the Lower Devonian section twice. Their origin is unknown. The lowest fault marks the position usually occupied by the salt-bearing part of the Salina Group. Therefore, although a complete interpretation of this well is not possible, the lowest fault indicates that salt has been thinned completely from this syncline.

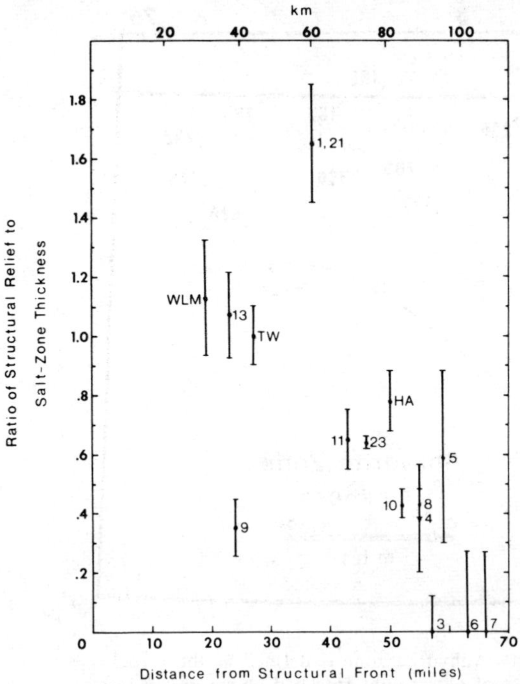

FIG. 8—Ratio of structural relief to salt-zone thickness versus distance from Allegheny structural front. With exception of wells 1 and 21, there is regular decrease in ratio of structural relief to salt-zone thickness in moving across strike away from structural front; less and less of salt zone is thinned from synclines in folds successively further from front. Wells 1 and 21 are in southwestern Pennsylvania, others are in northeastern and northwestern Pennsylvania.

Units Composing Deformed Zone

A common characteristic of the folds studied by Prucha (1968) and Frey (1973), and which is supported by well 24, is that significant material readjustment takes place within salt-bearing layers. In Fir Tree Point anticline the upper part of the Salina Group (upper part of the Syracuse Salt or unit F of Landes, 1945) has been disturbed (Prucha, 1968, p. 276, Fig. 15). Frey (1973) showed that all of the salt-bearing units are involved in his three folds. In both cases the deformed zone is confined to rocks which contain salt.

The fact that salt-bearing rocks alone constitute the deformed zone for these folds in northeastern Pennsylvania and south-central New York suggests a simple method to determine whether this is true for other areas of the Plateau. Specifically, we wish to test: (1) is flow important elsewhere; (2) if so, is there just one deformed zone; and (3) does the deformed zone consist of just salt-bearing rocks? The procedure is to assume that the flow model of Figure 5 is an appropriate one for Plateau folds. Then, any well drilled through an anticlinal crest should find a salt-zone thickness equal to or greater than the structural relief on that fold. If the relief and salt-zone thickness are nearly the same, this means that nearly all of the salt has been thinned from the adjoining synclines. If the relief is much less than the crestal salt-zone thickness, then the salt zone in the synclines has not been thinned completely; the model predicts that the salt zone will be thinned only by an amount equal to one-half the structural relief. However, if the relief is larger than the salt-zone thickness, either (1) the deformed zone is thicker than the salt zone, though there might still be only one zone of redistribution, or (2) there are several zones in which material transport has taken place.

Figure 7 shows the relation between salt-zone thickness and structural relief for all anticlinal wells (1) that penetrate the Upper Silurian Cayugan series, and (2) for which the structural relief is known; data taken from the sections of Frey (1973) also are plotted. Fourteen of the 21 data points fall on or below the 45° line which represents structural relief equal to salt-zone thickness. Wells 1, 14 through 18, and 21, which plot above the line, are discussed in the following. Thus 14 data points are consistent with the simple model of one deformed zone which is equivalent to the salt zone.

Structural relief is related both to salt-zone thickness and to distance from the boundary between the Plateau and Valley and Ridge provinces (the Allegheny structural front; see Fig. 8). All the wells drilled through anticlines more than 40 mi (64 km) from the structural front have structural relief less than salt-zone thickness. As these anticlines presumably did not experience compressive stress so high as those nearer the front, they may not have grown to the limiting structural relief imposed by thinning salt completely from their synclines. The three anticlines which have a structural relief equal to salt-zone thickness are less than 30 mi (48 km) from the front. We interpret these folds as having reached the limiting structural relief which corresponds to complete removal of salt from their synclines; this clearly is true in those folds for which Frey (1973) has published seismic sections (Wilmot and Towanda anticlines).

The structural relief on one anticline in northeastern Pennsylvania (well 14), on Chestnut Ridge anticline in southwestern Pennsylvania

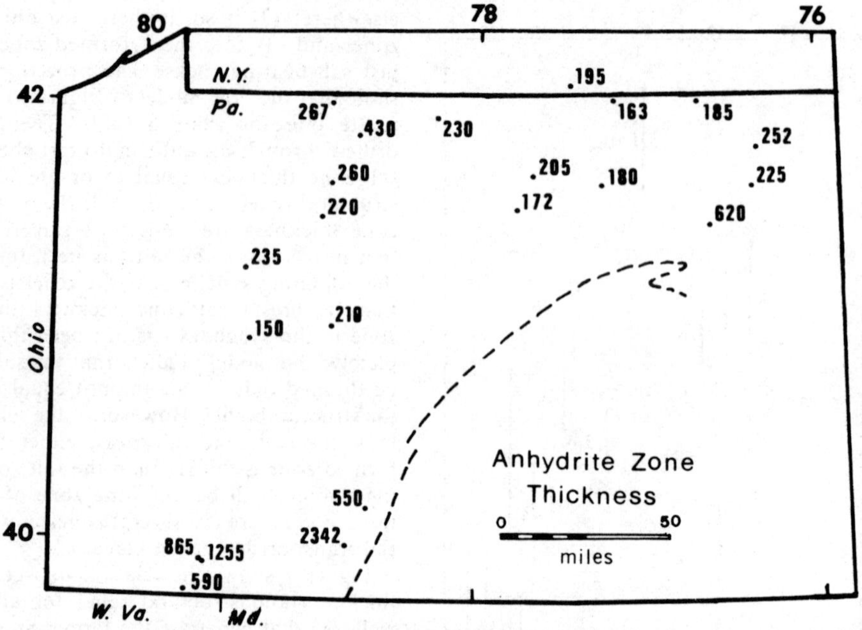

FIG. 9—Thickness of anhydrite zone (in feet). Anhydrite zone is defined as those rocks between top of youngest anhydrite bed and base of oldest anhydrite bed above unit F (see explanation of Table 2). This is roughly correlative with Tonoloway Formation. Thickness of anhydrite zone in well 15 (2,342 ft or 720 m) accounts for all but 200 ft (60 m) of structural relief in vicinity of that well. Also, 620 ft (190 m) of anhydrite-bearing rock in well 14 is twice structural relief of anticline on which 14 was drilled.

(wells 1, 21), and on Deer Park anticline near the front in south-central Pennsylvania (wells 15 through 18) cannot be explained by salt movement alone. In each of these three areas the thickness of anhydrite-bearing rocks is large (Fig. 9); therefore it is possible that gypsum dehydrating to anhydrite (Heard and Rubey, 1966) provided the weak unit which could thin in the synclines and thicken in the anticlines. An additional zone of flow in the Upper Ordovician Reedsville Shale below the competent Upper Ordovician and Lower Silurian clastic rocks is a strong possibility for the two southern areas. Shellsburg dome at the north end of Deer Park anticline shows still more complication (Gwinn, 1970, Fig. 3)—the Cambrian is thrust onto the Reedsville Shale. If this additional complication is present farther south on Deer Park anticline, then this fold is related more closely to a Valley and Ridge fold than to a Plateau fold in structural style and in the cause of its structural relief.

In addition, Gwinn's (1970, Fig. 3) cross section showed that Deer Park and its adjacent synclines have not acted as a closed system in which volume and shortening are conserved. Therefore, wells 15 through 18 on Deer Park anticline are not expected to follow the pattern of the other Plateau folds.

All of the anticlines for which the 21 comparisons between salt-zone thickness and structural relief are possible have broad gentle profiles and therefore must show thinning of the synclines if they are thin skinned. Fourteen of the data points fit a model of salt-zone flow only, two (wells 1, 21) have substantial salt-zone flow but also must exhibit thinning of other weak units in the synclines and one (well 14) must exhibit thinning of other weak units because it does not have an appreciable salt zone. The remaining four data points (wells 15 through 18) are on Deer Park anticline which is quite different in character from the rest of the Plateau folds and therefore is not expected to fit the model.

REGIONAL VARIATION IN STRUCTURAL STYLE

Variation in composition and thickness of the deformed zone is responsible for changes in structural style on the Plateau.

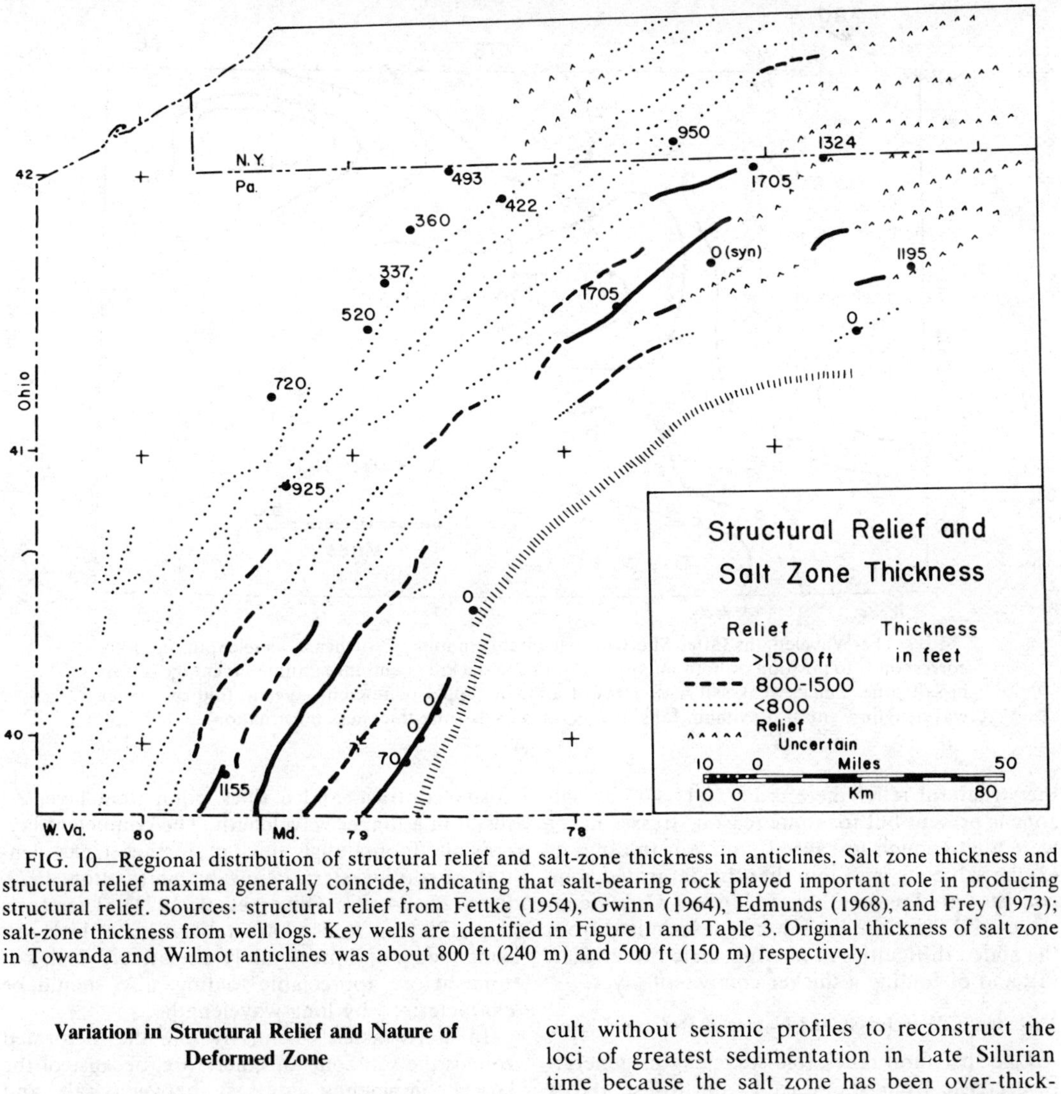

FIG. 10—Regional distribution of structural relief and salt-zone thickness in anticlines. Salt zone thickness and structural relief maxima generally coincide, indicating that salt-bearing rock played important role in producing structural relief. Sources: structural relief from Fettke (1954), Gwinn (1964), Edmunds (1968), and Frey (1973); salt-zone thickness from well logs. Key wells are identified in Figure 1 and Table 3. Original thickness of salt zone in Towanda and Wilmot anticlines was about 800 ft (240 m) and 500 ft (150 m) respectively.

Variation in Structural Relief and Nature of Deformed Zone

We already have discussed briefly the relations between the deformed zone and structural relief for northeastern, south-central, and southwestern Pennsylvania in reference to Figure 7. The salt zone is thickest in northeastern Pennsylvania (Fig. 10). Here also structural relief reaches a maximum which generally coincides with the salt-zone-thickness maximum. Structural relief decreases toward the front because the salt zone thins in that direction. Sufficient evaporite and shale were present to allow development of large structural relief in southwest and south-central Pennsylvania.

Great caution must be used in interpreting the paleoenvironmental meaning of present salt-zone thickness within anticlinal cores in northeastern Pennsylvania and elsewhere (Fig. 10); it is difficult without seismic profiles to reconstruct the loci of greatest sedimentation in Late Silurian time because the salt zone has been over-thickened in anticlinal cores. For example, if the fold volume of the fold in Figure 3c were reconstructed to a layer of length equal to that fold's arc length, the salt zone thickness before folding would be 800 ± 200 ft (240 ± 60 m). At present it is 2,600 ft (790 m) at the anticlinal crest. It is surprising that the stratigraphy is traceable at all in the subsurface in view of the large amount of over-thickening and undoubted structural complication in anticlines.

Similar caution should be used to a certain extent in reconstructing other Late Silurian (anhydrite-rich Tonoloway) and Late Ordovician (Reedsville) environments.

Nothing is known of Silurian or older rocks in the central Pennsylvania Plateau and therefore we must admit at least two possibilities as to why

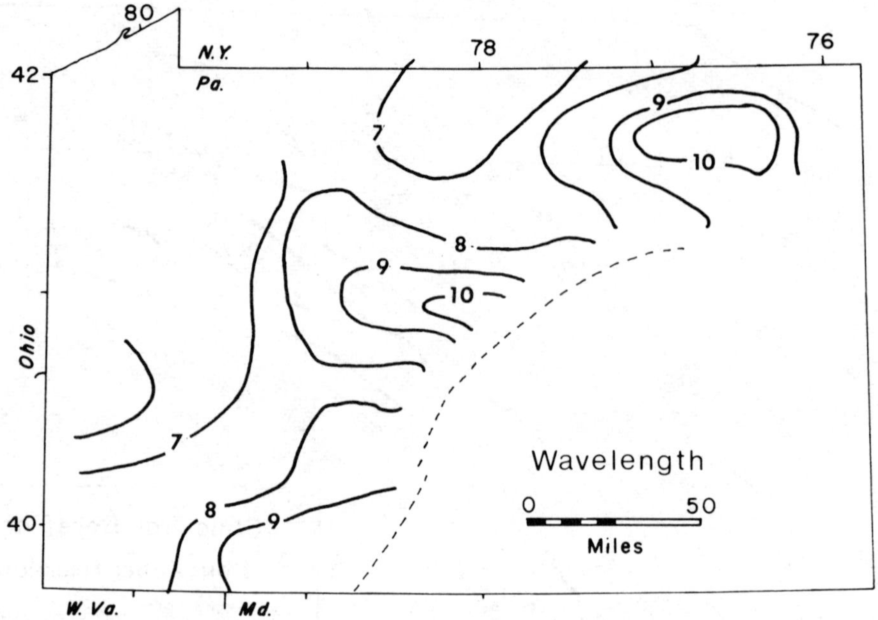

FIG. 11—Wavelengths (after Sherwin, personal commun.). Northeast wavelength anomaly corresponds to position of both maximum salt zone thickness and maximum percentage of salt in salt zone. Thick, weak salt had effect of allowing upper competent layer to fold on longer wavelength. Central maximum falls in area of no salt-zone thickness information.

the structural relief there is low: (1) a thick salt zone is present but for some reason stresses never were high enough to cause significant folding, or (2) the salt zone is so thin that the deformed zone consists predominantly of Upper Ordovician shales and therefore the lower relief resulted from the added difficulties of moving shale rather than salt and of folding a thicker competent layer.

Variation in Wavelength and Nature of Deformed Zone

Like structural relief, fold wavelengths generally decrease from southeast to northwest across strike and reach a maximum in northeastern Pennsylvania (Figs. 1, 11). Unlike structural relief, however, there is another even broader area of long wavelengths in central Pennsylvania and only a moderate departure from the Plateau-wide average of 7.5 mi (12 km) in southwest Pennsylvania. Consideration of folding theory and of the nature of the deformed zone provides insight into the nature of these trends.

Sherwin (1972) showed that, like a layer imbedded in an infinite medium (Biot, 1961), viscous thin-skinned fold models display a dominant wavelength. She showed further that this dominant wavelength for a two-layer model consisting of a competent layer overlying a weak layer depends on the ratios of the thickness and viscosities of the layers; a thick weak layer, a high viscosity contrast, and a thick competent layer all result in a longer wavelength. The competent layer is not anomalously thick in northeast, east-central, or southwestern Pennsylvania (Colton, 1970, Figs. 22, 24, 27; Oliver et al, 1971). Therefore, those Plateau folds characterized by a thick salt zone today, and presumably a relatively thick salt zone before appreciable folding, also should be characterized by long wavelengths.

In northeastern Pennsylvania the deformed zone is the salt zone and therefore, because of the large competency contrast between salt and younger-than-Silurian clastic rocks, wavelengths are long. Salt composed part of the deformed zone in southwest Pennsylvania. Consequently, the wavelengths are moderately long.

Because no wells penetrate the Salina Group in the east-central Plateau of Pennsylvania, the deformed zone's role in determining wavelengths is not known. We do not know whether the long wavelengths there are due to, (a) an abnormally thick salt zone or, (b) a thick competent layer. Sherwin (1972) found that two of her models, one corresponding to each of these two possibilities, successfully model these folds (columns a, b, Fig. 2). The deformed zone of model (a) is salt and anhydrite-bearing rock, the thickness of which has been estimated from Colton (1970). The deformed zone of the second model is the Upper

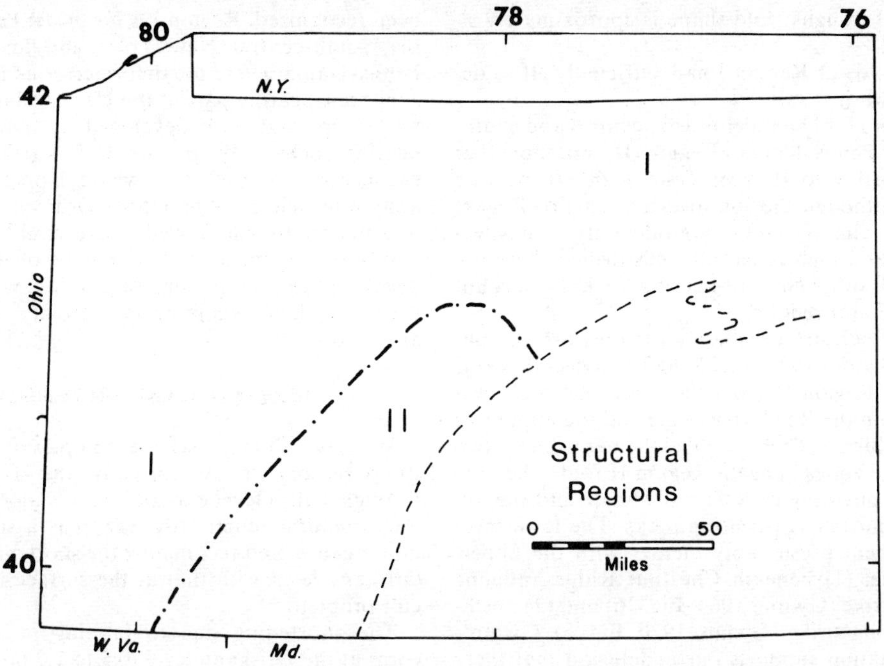

FIG. 12—Structural regions of Plateau of Pennsylvania and southern New York. Folds in Region I developed by moving only salt-bearing rock whereas those of Region II produced their structural relief by moving material in more than one deformed zone. East-central Pennsylvania is included in Region II although no subsurface information on salt-zone thickness is available from that area. Far western and far southwestern Pennsylvania are included in Region I, though no wells there completely penetrate Silurian section. In both cases surface geometric and geologic characteristics are basis for regional assignment.

Ordovician Reedsville Shale; all rocks above are the competent layer. According to model (a), the long wavelengths of east-central Pennsylvania are explained by a thick salt zone. Model (b) predicts that wavelengths are long because the competent layer is thicker whereas the ratio of thicknesses between the competent layer and deformed zone remains the same.

Structural Regions and Their Mechanics

On the basis of mechanics, the Plateau of Pennsylvania and south-central New York may be divided into two regions (Fig. 12).

Region I—Within northeastern and far western Pennsylvania as well as south-central New York, the deformed zone corresponds to the salt-bearing part of the Salina Group. The presence of thick salt explains the high relief and long wavelengths in northeastern Pennsylvania.

Broad synclines and narrow anticlines (for example, Fig. 3c) also are a result of originally thick salt. It becomes increasingly difficult to move weak material from between two relatively more competent layers as the thickness of the weak material decreases. For the simple case of flow from between two parallel plates, the stress required for a fixed rate of thinning is proportional to the reciprocal of the thickness cubed (Jaeger, 1969, p. 142). Therefore, for Plateau folds the first few feet of salt-bearing rock are the least difficult to displace from the synclines and flanks, but that difficulty increases as the bottom of the competent layer in the synclinal trough approaches the base of the deformed zone. The synclinal hinges will deepen more slowly because of this increased resistance. At the same time the synclinal flanks will continue to deepen until they in turn approach the base of the deformed zone. Thus the fold does not grow by deepening the synclinal hinges but rather by successively broadening the flat part of the synclines. Though the precise timing is unknown, the end result is clear—flat, broad synclinal troughs within the major folds of Region I.

In far western Region I, stresses were not high enough long enough to produce folds with flat synclinal troughs; fold shape is approximately sinusoidal.

The folds of Region I had sufficient salt to develop with just salt.

Region II—The relief of east-central and southwestern Pennsylvania, Region II, on the other hand, is due to flow of more than salt-bearing rocks. Although the southwestern part of Region II also is characterized by moderately long wavelengths and high relief, all wells drilled there did not find sufficient salt-bearing rock to account for structural relief.

In the subsurface, these structures are complicated. Gwinn (1970, Fig. 3) had two decollements entering Region II from the southeast, the lower one within the Reedsville Shale and the upper one in the Silurian. These two decollements imply two deformed zones beneath Region II folds, the lowermost consisting of Reedsville Shale and the upper of Silurian salt-bearing rocks. The lowermost decollement presumably merges with the uppermost one, (1) beneath Chestnut Ridge anticline across strike (Gwinn, 1964, Fig. 16) and (2) northward along strike (Gwinn, 1970, Fig. 5). Gwinn's interpretation supports our conclusion that there must be more than one horizon of readjustment in Region II folds to explain their structural relief.

Mechanics of Regions I and II—The structural development of Regions I and II probably was similar to the point where it was very difficult to move salt-bearing rock from the synclines of Region II folds. Region I folds continued to shorten by moving salt-bearing rock and consequently their synclinal troughs grew broader and anticlines narrower as the salt layers became thinner under the synclines and flanks. However, Region II folds continued to develop by moving the next weakest layers—other evaporite-rich rock (Tonoloway) and shale. The distinction between these stages in Regions I and II is not absolute; there probably was a continuous transition from the predominance of one mode of material transport to another.

SUMMARY AND DISCUSSION

1. We have shown that a thin-skinned fold model where the stratigraphic thicknesses of all units under synclinal troughs are the same before and after folding is geometrically untenable. A two-layer model where material in a deformed zone is transported from the synclines and flanks to the anticlinal core as the upper competent layer folds is geometrically valid and supported by detailed local geologic data.

2. On the basis of the nature of the deformed zone and mechanics, two structural regions have been recognized. Region I is northeast Pennsylvania, south-central New York, and far western Pennsylvania where the deformed zone is the salt zone (salt-bearing part of the Upper Silurian Salina Group) and folds developed by moving salt-bearing rocks only. Region II is southeast and east-central Pennsylvania where Upper Silurian anhydrite-rich units or Upper Ordovician shales in addition to salt flowed; there could be more than one deformed zone. The folds of Region II developed first by moving salt, if any, with anhydrite and shale becoming increasingly important as the folds grew.

Origin of Oriskany-Level Faulting

We have shown that the complex faulting at the Oriskany (more correctly the Tonoloway through Tully) level cannot be the major cause of the structural relief. However, it is a significant minor cause, and reconciling the shortening at the Oriskany level with that at the surface is a difficult problem.

The shortening due to faulting in anticlinal cores at the Oriskany level (0.5 to 1.0 mi or 0.9 to 1.6 km; Gwinn, 1964, Figs. 9, 11) is considerably greater than the arc-length shortening in the Carboniferous—one or two tenths of a mile or of a km. (Arc-length shortening at the Oriskany level other than that in the anticlinal cores is very small [Gwinn, 1964; Fig. 7]; shortening at the Oriskany level apparently is limited to the anticlines.) However, if we take a generous estimate for the uniform layer parallel shortening in the Carboniferous, then the Oriskany shortening might be equal to the sum of arc-length shortening and layer-parallel shortening in the Carboniferous. Reconciling the shortening in this fashion implies: (1) nearly complete bedding-shear decoupling at the level of the Devonian black shales, and (2) shortening in the younger formations which is largely synchronous with faulting at the Oriskany level.

An alternative explanation for the shortening is to attribute it to bending strains associated with the major Plateau folds. This explanation is geometrically adequate using the observed radii of curvature if we assume that the whole section above the Silurian salt behaved as one mechanical unit, and we take the neutral surface at the top of the unit. Thus this explanation implies no bedding-shear decoupling. In addition, it implies stretching parallel with bedding at the Oriskany level in the synclines—although the strains associated with this stretching can be less than those associated with the shortening in the anticlines because the synclines have larger radii of curvature.

Because the formations adjacent to the Oriskany are thin bedded with strong contrasts in competency, the complex structural style with disharmony and numerous faults is mechanically reasonable. The problem is to reconcile the shortening at different levels, and it seems that any model will require either stretching at the Oriskany level in the synclines or more layer-parallel shortening at the upper structural levels than has been observed.

Mechanical Comparison Between Southern and Central Appalachian Plateau Provinces

There are important similarities and differences between the central and southern Appalachian Plateau. In both areas, a sole thrust that changes stratigraphic position at the Allegheny structural front as well as farther out on the Plateau (Gwinn, 1964, Fig. 16; Harris, 1970, Fig. 1) has allowed wholesale translation of the upper sheet north and northwestward. An average of about 7 mi (11 km) has been calculated for the Pine Mountain block (Harris, 1970). There has been at least 1 mi (0.6 km) of northwestward translation in Pennsylvania. (This is the folding shortening; if the layer-parallel shortening found by Nickelsen, 1966, occurs in other parts of the Plateau, the translation could be a few times this value.) However, the deformed or decollement zones of each region are fundamentally different. The deformed zone in the southern Appalachian Plateau is a thin fault zone, the disturbance in which is confined to within a few hundred feet of the decollement (Miller and Fuller, 1954, p. 221). The deformed zone in the central Appalachians is much thicker and therefore has acted as the layer in which material adjustments necessary for fold development have taken place. The fact that the stratigraphic section in the south lacks such a thick weak layer sufficiently high in the section made development of folds difficult.

APPENDIX I. IMPORTANCE OF COMPACTION AND LAYER-PARALLEL SHORTENING

Our simple model does not take into account all the physical processes possible during the folding process. Two of these could resolve the conflict between fold volume and shortening for a no-thinning model. These are (1) compaction of rock perpendicular to bedding which is greater in synclines than anticlines, and (2) layer-parallel shortening by processes such as compaction, uniform thickening, wedging, minor folding, etc. Layer-parallel shortening in excess of arc-length shortening has the effect of moving more of the deformed zone into the anticlinal core if the material of the deformed zone does not participate to the same degree as the competent layer in this process. The added volume is equal to the layer-parallel shortening times the thickness of the deformed zone. Compaction of the synclines causes us to overestimate the fold volume by giving rise to a larger structural relief than that due to movement of material into the anticline.

There is some evidence of layer-perpendicular compaction in southwest Pennsylvania. Wells 19 and 20 (Fig. 1) were drilled side by side in Ligonier syncline. Well 1 was drilled directly north of wells 19 and 20 on the crest of Chestnut Ridge anticline and well 21 is a few miles east of well 1. Table 4 compares the key clastic units. Except for the abnormal Catskill thickness in well 21 which most likely is due to local faulting (Cate, 1962; Gwinn, 1964), there has been less than 100 ft (31 m) of thinning of shales in this syncline. Likewise, a comparison of wells 24 and 9 drilled in northeast Pennsylvania (Table 4) shows that for the only unit where comparison is possible, the Middle Devonian Hamilton shale, there could have been 150 ft (46 m) of compaction. An unknown part of these two compaction estimates is original stratigraphic variation.

There is also some evidence of layer-parallel shortening, though not more than about 10%. Nickelsen (1966) found 10% shortening in a direction perpendicular to fold axes in Pennsylvanian rocks in central Pennsylvania (near the southeast end of traverse 2, Fig. 1). He ascribes this to compaction at an early stage of uniform shortening before appreciable folding.

Nevertheless, using the maximum amount of observed layer-parallel shortening and 100 ft (31 m) of compaction in synclines relative to anticlines, shortening and volume cannot be balanced. The fold volume of a sinusoidal fold with a 7.5-mi (12 km) wavelength and a 1,600-ft (2,160 km) structural relief is 1.1 cu mi per mi (2.8 cu km per km) and its arc-length shortening is 0.03 mi (0.08 km). The volume of material that a 0.75-mi (1.2 km or 10%) layer-parallel shortening will displace from an originally 1,000-ft (300 m) thick deformed zone is 0.1 cu mi per mi (0.3 cu km per km). The arc-length shortening accounts for 0.1 cu mi per mi more. Their sum, about 0.2 cu mi per mi (0.5 cu km per km), falls short of the 1.1 cu mi per mi (2.8 cu km per km) volume of this fold form. Similarly, if 100 ft (30 m) of the structural relief were due to compaction and not movement of material, then the actual volume would be 1.0 cu mi per mi (2.6 cu km per km) not the observed volume of 1.1. The fold form still represents much more volume than can be explained by a no-thinning thin-skinned fold model.

APPENDIX II. CALCULATION OF FOLD VOLUME AND SHORTENING FOR SINUSOIDAL FOLD

Let the shape of the folded layer be:

$$y = b\left(1 + \cos\frac{2\pi x}{L}\right), \qquad \text{(A1)}$$

where b and L are constants defined in Figure 4. Then the fold volume is:

$$\int_{-\frac{L}{2}}^{\frac{L}{2}} b\left(1 + \cos\frac{2\pi x}{L}\right) dx = bL\,. \qquad \text{(A2)}$$

TABLE 4. COMPARISON OF THICKNESS OF TWO DEVONIAN SHALE SECTIONS

Formation / Well No.	1	22	20	21	24	9
Catskill	5624	5530	5510	6443	6253+	5550+
Hamilton	462	480	455	435	887	737

To calculate the shortening we must find the arc length of a sine wave of x-wave length, L. The arc length, S, may be calculated by numerical integration of the differential of arc length:

$$S = \int ds = \int \sqrt{1 + \left(\frac{dy}{dx}\right)^2}\, dx \, . \quad \text{(A3)}$$

Alternatively, the arc length of a quarter wavelength of a sine wave,

$$S = \int_0^{\frac{L}{4}} \left[1 + \left(\frac{2\pi b}{L} \cos \frac{2\pi x}{L} \right)^2 \right]^{\frac{1}{2}} dx \, , \quad \text{(A4)}$$

can be expressed as an elliptic integral by substitutions:

$$z = \frac{2\pi x}{L} \, , \quad \text{(A5a)}$$

$$\tan \Theta_0 = \frac{2\pi b}{L} \, , \quad \text{(A5b)}$$

where z is a variable of integration and Θ_0 is the dip of the layer at the symmetry point on the limb. The result of these substitutions is that the arc length of a full wavelength is:

$$S = \frac{2L}{\pi \cos \Theta_0} \int_0^{\frac{\pi}{2}} \left[1 - \sin^2 \Theta_0 \sin^2 z \right]^{\frac{1}{2}} dz \, . \quad \text{(A6)}$$

This integral is the tabulated elliptic integral of the second kind.

REFERENCES CITED

Alling, H. L., and L. I. Briggs, 1961, Stratigraphy of Upper Silurian Cayugan evaporites: AAPG Bull., v. 45, p. 515-547.

Biot, M. A., 1961, Theory of folding of stratified viscoelastic media and its implications in tectonics and orogenesis: Geol. Soc. America Bull., v. 72, p. 1595-1620.

Cate, A. S., 1962, Subsurface structure of the plateau region of north-central and western Pennsylvania on top of the Oriskany Formation: Pennsylvania Geol. Survey, 4th ser., Map 9, scale 1:250,000.

Chute, N. E., 1972, Subsurface stratigraphy and structure near the Morton Salt Company's mine on Seneca Lake, New York (abs.): Geol. Soc. America Abs. with Programs, v. 4, p. 9-10.

Colton, G. W., 1970, The Valley and Ridge and Appalachian Plateau; stratigraphy and sedimentation; the Appalachian basin; its depositional sequences and their geologic relationships, *in* G. W. Fisher et al, eds., Studies in Appalachian geology: central and southern: New York, Interscience Pub., p. 5-47.

De Sitter, L. U., 1964, Structural geology: New York, McGraw-Hill, 551 p.

Edmunds, W. E., 1968, Geology and mineral resources of the northern half of the Houtzdale 15-minute quadrangle, Pennsylvania: Pennsylvania Geol. Survey, 4th ser., Atlas A-85ab, 150 p.

Fergusson, W. B., and B. A. Prather, 1968, Salt deposits in the Salina Group in Pennsylvania: Pennsylvania Geol. Survey, 4th ser., Bull. M58 (Mineral Resources Rept.), 41 p.

Fettke, C. R., 1954, Structure contour maps of the plateau region of north-central and western Pennsylvania: Pennsylvania Geol. Survey Bull. G27, 50 p.

Finn, F. H., 1949, Geology and occurrence of natural gas in the Oriskany Sandstone in Pennsylvania and New York: AAPG Bull., v. 33, p. 303-335.

Frey, M. G., 1973, Influence of Salina Salt on structure in New York–Pennsylvania part of the Appalachian Plateau: AAPG Bull., v. 57, p. 1027-1037.

Glover, A. D., 1970, Geology and mineral resources of the southern half of the Clearfield 15-minute quadrangle, Pennsylvania: Pennsylvania Geol. Survey, 4th ser., Atlas A-85cd, 139 p.

Gwinn, V. E., 1964, Thin-skinned tectonics in the Plateau and northwestern Valley and Ridge provinces of the central Appalachians: Geol. Soc. America Bull., v. 69, p. 863-900.

——— 1970, The Valley and Ridge and Appalachian Plateau; structure and tectonics; kinematic patterns and estimates of lateral shortening, Valley and Ridge and Great Valley provinces, central Appalachians, south-central Pennsylvania, *in* G. W. Fisher, et al, eds., Studies of Appalachian geology: central and southern: New York, Interscience Pub., p. 127-146.

Harris, L. D., 1970, The Valley and Ridge and Appalachian Plateau; structure and tectonics; details of thin-skinned tectonics in parts of Valley and Ridge and Cumberland Plateau provinces of the southern Appalachians, *in* G. W. Fisher et al, eds., Studies of Appalachian geology; central and southern: New York, Interscience Pub., p. 161-173.

Heard, H. C., and W. W. Rubey, 1966, Tectonic implications of gypsum dewatering: Geol. Soc. America Bull., v. 77, p. 741-760.

Jaeger, J. C., 1969, Elasticity, fracture and flow: London, Methuen, 268 p.

Landes, K. K., 1945, The Salina and Bass Islands rocks in the Michigan basin: U.S. Geol. Survey Oil and Gas Inv. Ser. Prelim. Map 40, scale approx. 1:1,600,000.

Miller, R. L., and J. O. Fuller, 1954, Geology and oil resources of the Rose Hill district—the fenster area of the Cumberland overthrust block—Lee County, Virginia: Virginia Geol. Survey Bull., 71, 383 p.

Nickelsen, R. P., 1966, Fossil distortion and penetrative rock deformation in the Appalachian Plateau, Pennsylvania: Jour. Geology, v. 74, p. 924-931.

Oliver, W. A., Jr., et al, 1971, Isopach and lithofacies maps of the Devonian in the Appalachian basin:

Pennsylvania Geol. Survey, 4th ser., Progress Rept. 182, maps.

Prucha, J. J., 1968, Salt deformation and decollement in the Fir Tree Point anticline of central New York: Tectonophysics, v. 6, p. 273-299.

Rodgers, J., 1963, Mechanics of Appalachian foreland folding in Pennsylvania and West Virginia: AAPG Bull., v. 47, p. 1527-1536.

——— 1970, The tectonics of the Appalachians: New York, Wiley-Interscience Pub., 271 p.

Shaffner, M. N., 1958, Geology and mineral resources of the New Florence quadrangle, Pennsylvania: Pennsylvania Geol. Survey, 4th ser., Atlas A-57, 165 p.

Sherwin, J., 1972, Decollement folding: PhD thesis, Brown Univ., 150 p.

Woodward, H. P., 1959, Structural interpretations of the Burning Springs anticline, *in* A symposium on the Sandhill deep well, Wood County, West Virginia: West Virginia Geol. Survey Rept. Inv. 18, p. 159-168.

The American Association of Petroleum Geologists Bulletin
V. 63, No. 4 (April 1979), P. 647-675, 16 Figs., 2 Tables

GEOLOGIC NOTES

Conodont-Based Reinterpretation of Bane Dome—Structural Reevaluation of Allegheny Frontal Zone[1]

WILLIAM J. PERRY,[2] ANITA G. HARRIS,[3] and LEONARD D. HARRIS[2]

Abstract The F. P. Strader 1 well drilled on the crest of the Bane dome, in Giles County, Virginia, has been cited as direct evidence for basement uplift under the dome. This concept was based on the assumption that the well contained a normal stratigraphic sequence of Lower Cambrian rocks from Rome Formation at the surface to Shady Dolomite in the subsurface. Conodonts, recently recovered from cuttings of the Strader well show that the dolomites, thought to be Lower Cambrian, are actually Lower Ordovician and are part of the Knox Group. Therefore, the base of the Rome at 28 ft (8.5 m) in the Strader well is marked by a major thrust fault. We have interpreted this thrust as part of the Saint Clair fault system and attribute the formation of the Bane dome to imbrication at the trailing edge of the Saint Clair footwall.

The subsurface data refute "active-basement" tectonics and support thin-skinned tectonics in this area. Our reinterpretation adds more than 10,000 ft (3,300 m) of gas-prospective section and several attractive structural traps that remain to be drilled.

INTRODUCTION

The hypothesis of "active-basement" tectonics in the southern Appalachians of Virginia (Cooper, 1961, 1964, 1968) was put forward as an alternative to "thin-skin" tectonics (Rodgers, 1949, 1953, 1963). This hypothesis raised a controversy concerning the relation of basement to Appalachian Valley and Ridge structure. Surprisingly, a single dry hole, the California Co. F. P. Strader 1 (Fig. 1) drilled in 1948 to a total depth of only 1,443 ft (440 m) on the Bane dome in Giles County, Virginia, sparked this controversy. The Strader test began in the Rome Formation (Lower Cambrian) on the crest of the dome. According to Cooper (1964, p. 97):

> It was reasoned that since every other belt of Rome Shale in southwest Virginia was bordered on one side by a thrust fault, it was likely that if the Rome were drilled, the well would penetrate at some depth a flat thrust below which would lie an overridden younger sequence. I was then a "no-basement" adherent like most other people. The Strader test proved the Rome to be only 350 feet thick and to be underlain by beds that would be classed by anyone as Shady Dolomite. The Strader test was actually the start of my becoming disenchanted with thin-skinned interpretations of Appalachian structure, which were then deeply entrenched and quite popular. Following the thin-skin theory of deformation the most likely place for a buried thrust to occur would have been on the Bane anticline just below the Rome inlier, but it was not there.

This view of the Bane dome has been illustrated, first by Cooper (1968, Fig. 3) and later by Sears and Robinson (1971, Fig. 7).

Although several workers have challenged Cooper's view of the Strader well and Bane dome (Rodgers, 1970; Milici, 1970, 1973; Perry, 1977), no negative proof has been offered. However, cuttings from the well have been available for many years. Very detailed lithologic descriptions of Strader well cuttings and their insoluble residues by Eloise T. Jacobsen (unpub. data on file at Virginia Div. Mineral Resources, 1950) are nearly identical with her descriptions of Knox Dolomite (Upper Cambrian and Lower Ordovician) in other wells in southwestern Virginia (E. T. Jacobsen, *in* Huddle et al, 1955). After reexamining the Strader cuttings, we have found (Fig. 2) many points of similarity to the Knox: (1) dominance of very finely crystalline, slightly sandy to silty, pale-gray dolomite containing traces of chert interstratified with dark-gray algal dolomite from a depth of 18 to 228 ft (5.5 to 69.5 m), rocks lithologically identical with the Mascot Dolomite (upper Lower Ordovician); (2) very cherty, coarsely crystalline dolomite from a depth of 228 ft (69.5 m) to 450 ft (137 m), lithologically identical with the underlying Kingsport Dolomite (Lower Ordovician); (3) interstratified sandstone and oolitic chert-bearing dolomite similar to the Chepultepec Dolomite and Copper Ridge Dolomite (Lower Ordovician and Upper Cambrian) from a depth of 450 to 1,263 ft (137 to 385 m); and (4) oolitic limestone interstratified with dolomite similar to that in the western belts of the Conococheague Limestone (Upper Cambrian, equivalent to Copper Ridge Dolomite) northeast of Roanoke, below a depth of 1,263 ft (385 m).

[1]Manuscript received, May 22, 1978; accepted, October 5, 1978.

[2]U.S. Geological Survey, Reston, Va. 22092.

[3]U.S. Geological Survey, U.S. National Museum of Natural History, Washington, D.C. 20560.

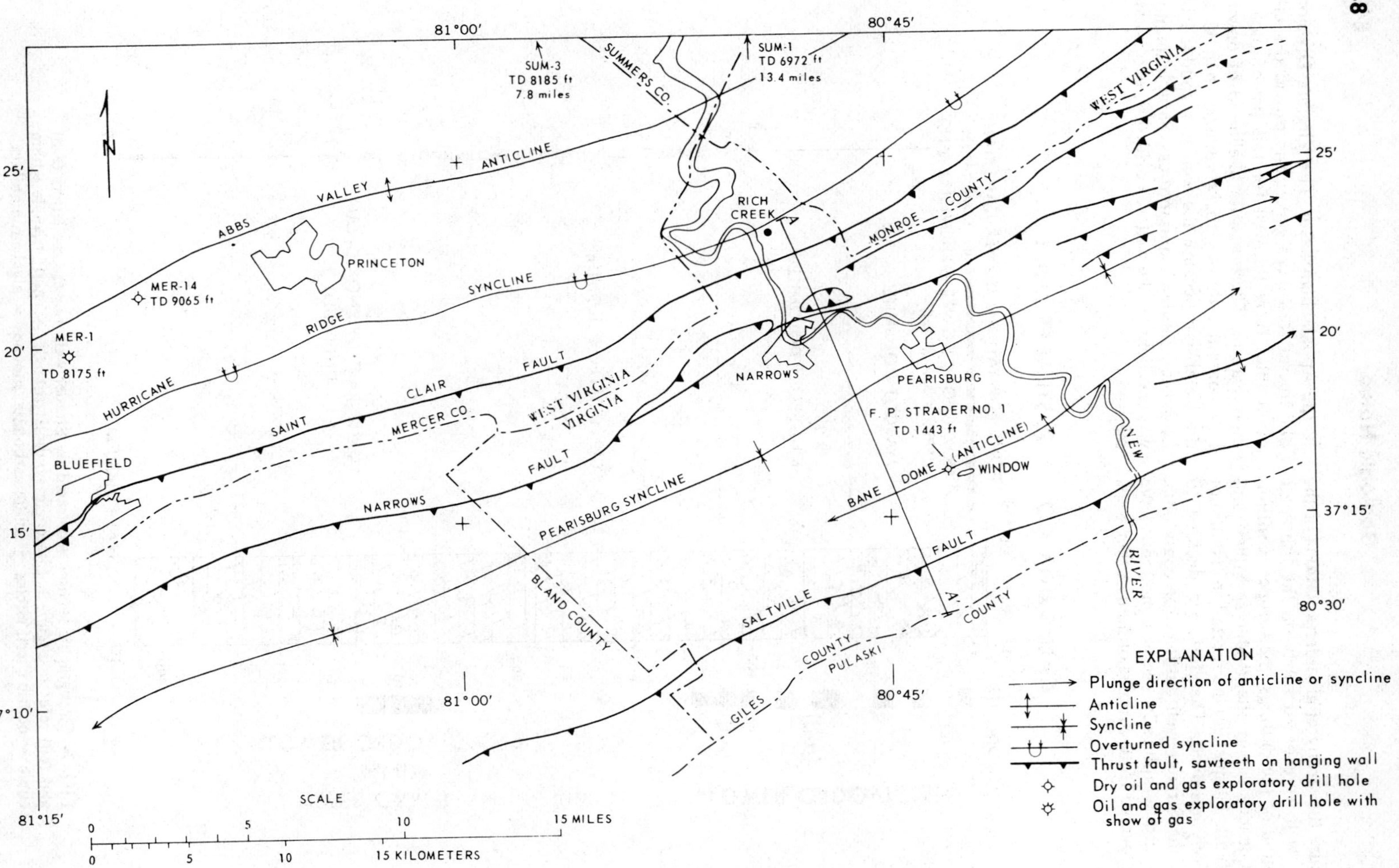

FIG. 1—Generalized tectonic map of area showing position of F. P. Strader 1 well.

Because the stratigraphic interpretations of the Strader well were based solely on the lithologic description of the cuttings, the Cooper (continuous sequence) versus the Rodgers, Milici, and Perry (faulted sequence) interpretations could not be entirely resolved. Therefore, we decided to process the Strader well cuttings for conodonts, because stratigraphically useful conodonts have been recovered from cuttings of many other wells in the Appalachian basin. The Virginia Division of Mineral Resources generously provided a cut of samples from the well for this paleontologic investigation. There is little likelihood of downhole contamination by rocks younger than Middle Cambrian because the well was spudded in Lower to Middle Cambrian rocks. Sample quality was excellent, as the well was drilled by cable-tool methods.

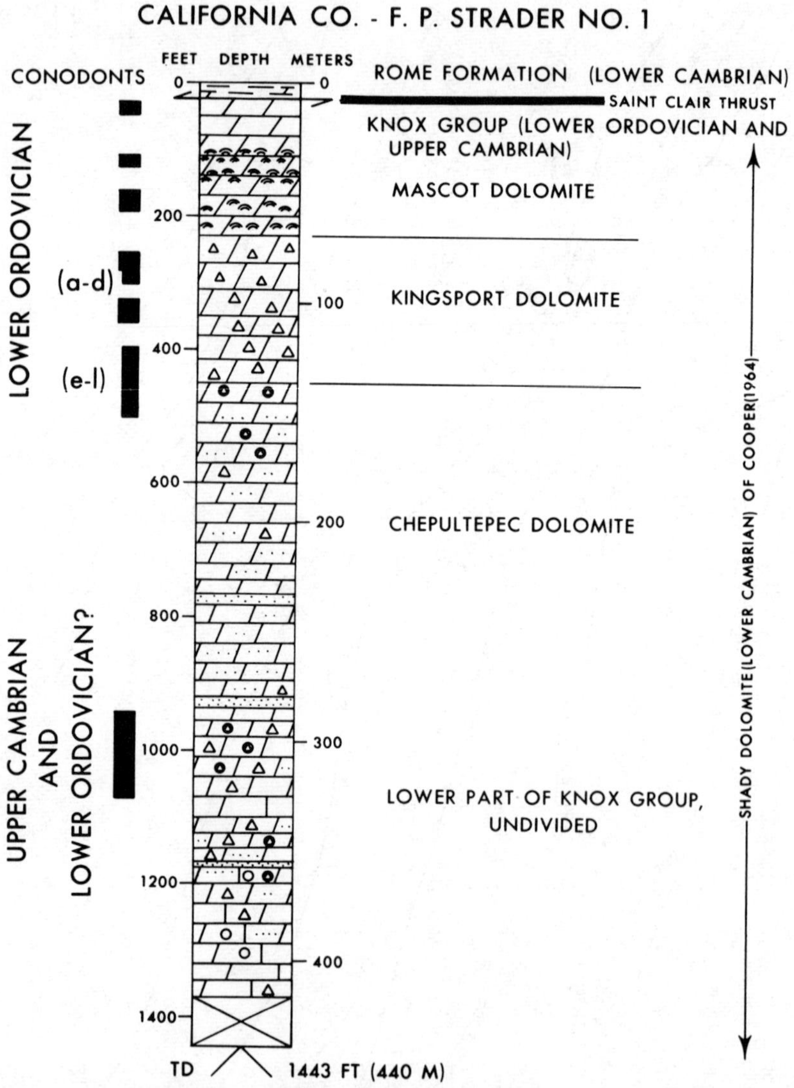

FIG. 2—Stratigraphic section penetrated by Strader well. Lower case letters (a-d) and (e-l) refer to Figure 3. Lithologic symbols: triangle = chert; triangle within circle = oolitic chert; circles = oolitic carbonate; mounds = algal stromatolites.

Early Ordovician conodonts now recovered from the well cuttings verify the thin-skinned interpretations suggested by Rodgers (1970), Milici (1970, 1973), and Perry (1977). This reinterpretation adds many thousands of meters of identified and inferred undrilled sedimentary rocks on top of basement beneath the Bane dome and significantly affects our knowledge of natural gas possibilities in the area.

GEOLOGIC, PALEONTOLOGIC, AND GEOPHYSICAL DATA

The area of this report is bounded on the northwest by the Allegheny structural front, which in Giles County, Virginia, and Mercer and Monroe Counties, West Virginia, is defined by the Saint Clair fault (Fig. 1). Between the Saint Clair fault and the Saltville fault is a locally complex belt of Cambrian to Mississippian rocks. This belt, which is 10 to 12.5 mi (16 to 20 km) wide in the Giles County area, includes the Narrows thrust on the northwest, succeeded on the southeast by the Pearisburg syncline, which in turn is succeeded by the Bane dome, "the structurally highest anticline in southwestern Virginia" (Cooper, 1961, p. 91). The regional structural setting is shown in Figure 1.

Conodonts recovered from cuttings of the California Co. F. P. Strader 1 (Figs. 1-3) prove the existence of Ordovician carbonate rocks beneath the Bane dome, as suggested by Milici (1973) and Perry (1977). Generically and specifically identifiable conodonts have been found from 110 to 502 ft (34 to 153 m) in the well (Fig. 3); generically indeterminate conodont fragments were recovered from as high as the 28 to 48-ft (8.5 to 15 m) interval to as low as the 943 to 1,073-ft (287 to 327 m) interval. All conodonts recovered from the well cuttings are Early Ordovician forms, although some range as high as the lower Middle Ordovician. The most biostratigraphically useful elements are: (1) *Scolopodus filosus* Ethington and Clark, s.f. (from 110 to 127 ft; 34 to 39 m), which ranges from high conodont Fauna C into the lower Middle Ordovician; (2) *Acodus oneotensis* Furnish s.f. (from 282 to 502 ft; 86 to 153 m), which thus far is known only from Fauna C (Ethington and Clark, 1971, Fig. 2); and (3) *Ulrichodina* sp. (from 325 to 360 ft; 99 to 110 m), a genus which is thus far unknown from rocks older than Fauna D and supposedly restricted to Fauna D (Ethington and Clark, 1971, Fig. 2), but which has recently been found in our collections from younger Canadian and oldest Middle Ordovician rocks in Tennessee, Virginia, and Maryland. The combined ranges and succession of conodonts recovered from the well indicate a Fauna C or Fau-

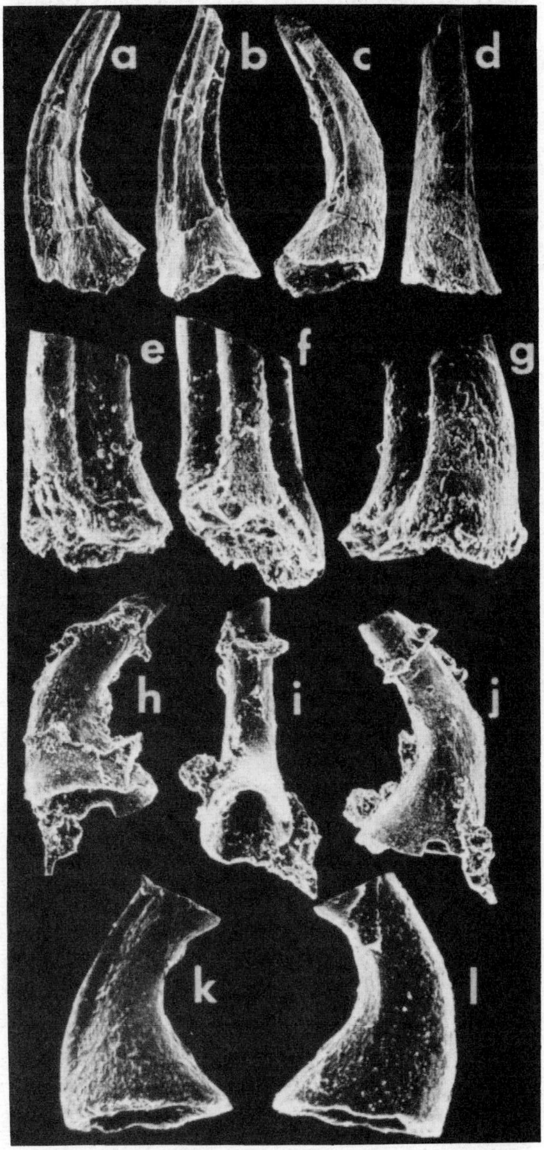

FIG. 3—Early Ordovician conodonts from California Co. 1 F. P. Strader. **a-d**, Inner lateral, oblique posterior, outer lateral, and anterior views of *Acodus oneotensis* Furnish s.f., USNM 252512, ×85; from 282 to 304 ft (86 to 93 m), USGS colln. 8798-CO. **e-g**, Outer lateral, posterior, and inner lateral views of incomplete *Acodus oneotensis?* s.f., USNM 252067, ×150; from 399 to 502 ft (122 to 153 m), USGS colln. 8781-CO. **h-j**, Opposite lateral and posterior views of oneotodiform element, USNM 252068, ×85; from 399 to 502 ft (122 to 153 m), USGS colln. 8781-CO. **k, l**, Inner and outer lateral views of *Drepanodus parallelus* (Furnish) s.f., USNM 252069, ×150; from 399 to 502 ft (122 to 153 m), USGS colln. 8781-CO.

na D age, probably late Fauna C to early Fauna D, for the conodont-bearing part of the section. On the basis of conodont data, an extensive section of carbonate of the Knox Group lies below a thrust fault under the Bane dome. Intensely sheared rocks of the upper part of the Rome Formation are present immediately above this thrust.

Recent field work reveals the presence of a small window through this thrust fault on the north side of Walker Creek 1,300 ft (0.4 km) to at least 2,700 ft (0.8 km) east of the Strader well (Fig. 1). On the basis of its conodont fauna, the massive dolostone of the Knox Group exposed in this window is of Early Ordovician age (J. E. Repetski, oral commun., 1978).

Figure 4 is a cross-sectional model of Giles County based on our revised stratigraphic interpretation. The data base used to construct the section includes a series of drill holes in the plateau north of the Saint Clair fault (Fig. 1), which provided control relative to Mississippian, Devonian, and Silurian stratigraphic thicknesses and depths to the Middle Devonian Huntersville Chert of Price (1929). Width of outcrop and bedding orientations from W. A. Moon's unpublished map of Giles County (see Fig. 4) show the thickness of Cambrian through Silurian units to be relatively uniform along the line of section (Fig. 4). The single exception is the Martinsburg Shale (Ordovician) which appears to range from about 200 m in thickness on the hanging wall of the Saltville fault to about 460 m on the flanks of Pearis Mountain to nearly 500 m on Peters Mountain. Thickness changes within the Ordovician limestones reported by Cooper (1961, 1971) are not evaluated in this report. The Middle Ordovician limestone sequence appears to thicken slightly from south to north on the basis of Moon's compilation.

The cross-sectional model (Fig. 4) is based on the following assumptions: (1) nearly flat basement; (2) constancy of formation thicknesses, with above-mentioned exceptions and except where beds are overturned resulting in tectonic thinning (Perry, 1971, 1975); and (3) material balance, to the extent that when the various fault blocks are restored to a predeformational configuration, no serious gaps or overlaps occur. Thicknesses of the Rome Formation and underlying Lower Cambrian formations under block "c" are unknown. Therefore, the Precambrian basement surface is not shown.

Sears and Robinson (1971, p. 2635) have shown a residual Bouguer gravity high in the region of the Bane dome. In support of the active-basement hypothesis, they modeled a basement ridge under the dome. Milici (1973) suggested that this positive gravity anomaly is due to a fault repetition of the dense Cambrian and Ordovician carbonate rocks in the vicinity of the dome. The gravity data are consistent with either model. Our model (Fig. 4), based on the reinterpreted Strader well, is likewise consistent with the gravity data and is similar to the Milici alternative. The residual Bouguer anomaly is 2 mgal lower on the north side of the dome, consistent with the structural model (Fig. 4) which shows chiefly clastic deposits with a bulk density probably less than 2.65 in the frontal part of fault block "a."

Edsall (1974) conducted a seismic reflection study on the northern flank of the Bane dome, more than 1.5 mi (2.4 km) east-northeast of the Strader well. The seismic study was based on 11 shotpoints along a 3,500-ft (1,067 m) traverse parallel with the axis of the dome. He detected an acoustic event between 0.6 and 0.8 sec on six of his RMS velocity scans and on four velocity-filtered records. We suggest that this corresponds to the two-way travel time expected along the shortest ray path to the northern face of fault slice "b" inferred to underlie the Bane dome (Fig. 4).

Edsall (1974, p. 47) detected an acoustic event between 1.0 and 1.1 sec on every RMS scan and on four velocity-filtered records which he termed the "most persistent event to be found on any of the records." Because of the high reflection coefficient of this event, he suggested that it originated from a major fault zone. On the basis of our estimate of interval velocity (20,000 ft/sec; 6.1 km/sec), this event may correspond to the top of the autochthonous footwall of the Saint Clair thrust fault under the northern flank of the Bane dome (block "c," Fig. 4).

STRUCTURAL INTERPRETATION

Conodont data from the Strader well and from a window on the Bane dome show that carbonate rocks of the Knox Group underlie a fault beneath intensely sheared beds of the Lower Cambrian Rome Formation. It is unlikely that the Knox section in the Strader well was derived from the Narrows thrust, because it is geometrically impossible for the Narrows thrust with less than 2-mi (3 km) displacement to have left a hanging-wall slice of Knox (block "a," Fig. 4) 5.6 to 6.3 mi (9 to 10 km) behind the base of the Knox along the Narrows leading edge (Fig. 4). The Narrows fault and associated smaller thrusts (Fig. 1) appear to be back-limb imbricates of the Saint Clair thrust.

Only one major thrust occurs north of the Bane dome to which the shallow major thrust in the Strader well can be related—the Saint Clair fault. The structurally high position of the thrust in the

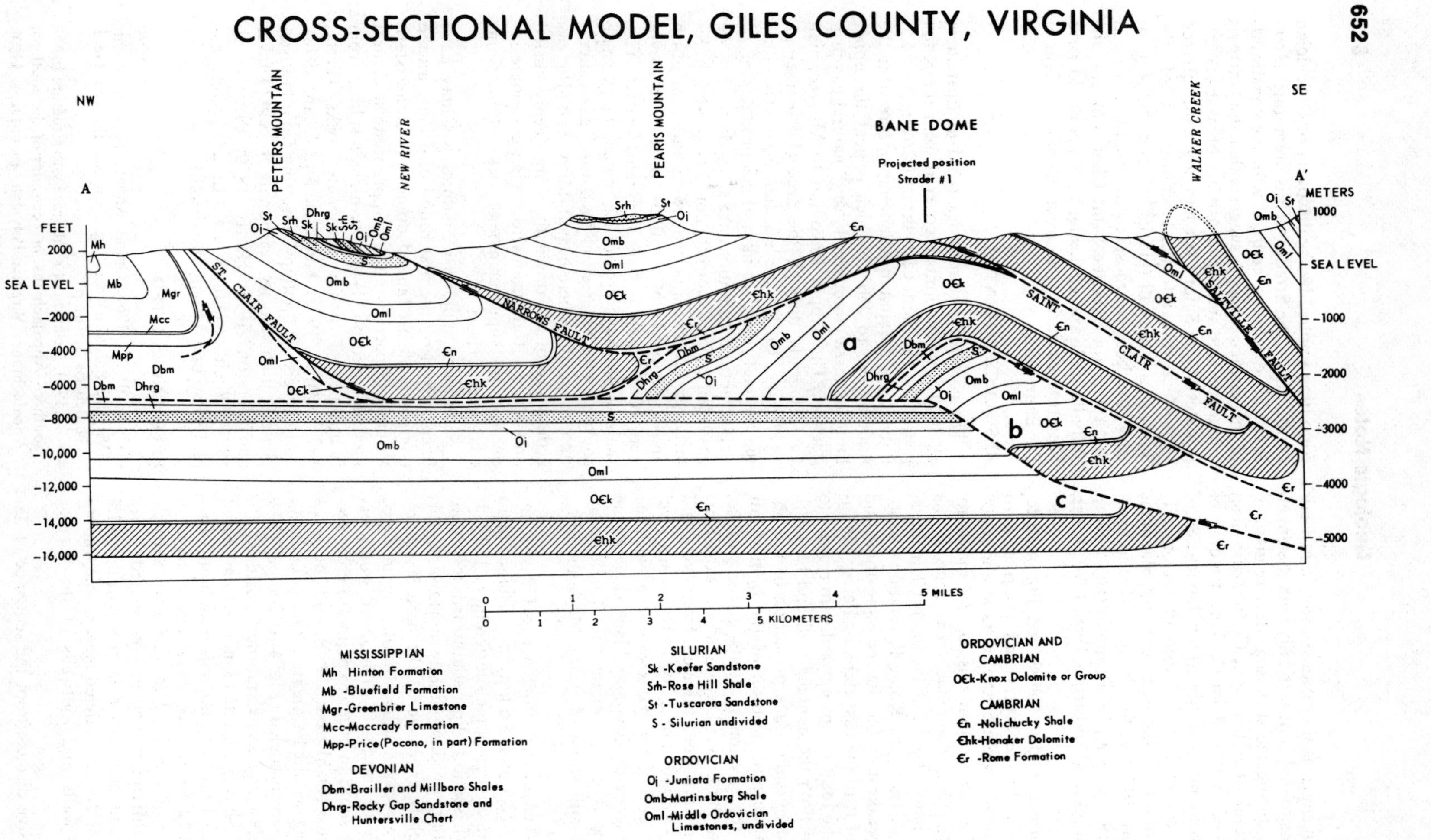

FIG. 4—Cross-sectional model of Giles County, Virginia, showing inferred subsurface configuration of Saint Clair fault system. *a*, Major fault block identified in Strader well; *b*, second fault slice, inferred to be present beneath Bane dome on basis of constraints of model; *c*, autochthonous footwall of Saint Clair thrust fault. Use of map on which cross section is based was generously permitted by W. A. Moon and W. D. Lowry.

Strader well suggests that if this is the Saint Clair thrust, it was folded concentrically with the enveloping beds during the structural development of the Bane dome after the Saint Clair thrust sheet had advanced at least 8 mi (13 km). Folding of thrust faults by massive subsurface duplication of rocks is a common mechanism in tectonically similar terranes in the southern Appalachians (Harris, 1967, 1970), in the Wyoming-Idaho overthrust belt (Royse et al, 1975), and in the southern Canadian Rockies (Dahlstrom, 1969, 1970; Jones, 1971).

Major thrusts tend to follow bedding in structurally weak, ductile layers and ramp upward through stronger, more brittle strata at about 30° or less to the compression axis, as first shown by Rich (1934). Thus the detachment zone at the base of the hanging wall of the Saint Clair thrust (the upper shales of the Rome Formation), probably is the weakest unit under the Bane dome and northwest of it to the east flank of Peters Mountain. Here the fault ramps upward with respect to the rocks of both the hanging wall and footwall, as indicated by subsurface projection of the observed attitude and position of the Saint Clair thrust and the enveloping strata at the surface. The depth to the Lower Devonian under the Hurricane Ridge syncline, inferred from well and seismic data (7,550 ft or 2,300 m below sea level at the north end of Fig. 4), is consistent with the inferred presence of Devonian black shale as the glide surface at the top of the St. Clair footwall plate below and south of Pearis Mountain (Fig. 4).

Estimated cover shortening across Giles County (Fig. 4) is 22.3 mi (35.6 km) in a distance of 14 mi (22.4 km), a megascopic shortening of 159%. If the base of section dips southward less than shown, is flat, or dips slightly northward, this value is slightly excessive. If the base of section dips more steeply southeastward, the modeled trailing edge of the St. Clair footwall would extend farther south in order to maintain cross-sectional balance (Dahlstrom, 1969), and the value of shortening would be greater. The cumulative fault separation modeled (Fig. 4) is 20.6 mi (32.9 km) at the Nolichucky (Cambrian) position: 1.6 mi (2.5 km), Narrows fault; 12 mi (19.1 km), Saint Clair fault; 5.3 mi (8.5 km), sub-Bane faults (below the folded Saint Clair thrust encountered in the Strader well); and 1.7 mi (2.8 km), Saltville fault. The difference between estimated cover shortening and cumulative fault separation (1.7 mi; 2.7 km) is due to modeled megascopic folding of hanging walls and footwalls of the various faults.

An extensive discussion of the sequence of deformation implied by the discovery of Ordovician conodonts in the Strader well is premature. However, this discovery, coupled with the constraints of the surface geology, suggests that the Narrows and associated smaller thrusts are break-back imbricates of the Saint Clair hanging wall. Inferred subsurface break-forward imbrication of the Saint Clair footwall (blocks "a" and "b," Fig. 4) may have been the most recent structural event in the area. Not only the Saint Clair, but also the Saltville fault and enclosing beds appear to have been rotated during the structural development of the Bane dome.

RESOURCE POTENTIAL

The color alteration index (CAI of Epstein et al, 1977) of conodonts recovered from the Knox just below the Saint Clair thrust (CAI = 4 to 4.5) indicates a maximum paleotemperature near the fault of 190 to 230°C, which is near the upper limit for most commercial "dry" natural gas production. The Upper Devonian to Mississippian CAI isograds through the area (Epstein et al, 1977, Figs. 16, 17) are less than 3.5—well within the thermal range for natural gas production.

If structural conditions are as modeled (Fig. 4), gas-filled fracture porosity could be present in Silurian and Devonian clastic rocks under the north and south flanks of the Bane dome. The Devonian black shale is rich in organic matter. This shale would have provided a mature black shale source rock with respect to natural gas generation, especially where heated during and after overthrusting of the originally more deeply buried and warmer Cambrian-Ordovician carbonate sequence and superincumbent rocks of the Saint Clair thrust sheet.

Significant supplies of natural gas may have been missed by the first round of drilling on the complexly thrust faulted domes of southwestern Virginia in part owing to inappropriate structural models based on (1) gravity highs resulting from fault-repeated dense carbonate rocks (Perry, 1971; Milici, 1973; Kulander and Dean, 1978), and (2) seismic "pullups" resulting from fault-repeated, high-velocity carbonate rocks (Tucker and Yorston, 1973; Harris, 1976; Perry et al, 1977). Such structures are not adequately explored by one 440-m crestal well!

REFERENCES CITED

Cooper, B. N., 1961, Grand Appalachian excursion: Virginia Polytech. Inst. Eng. Expt. Sta. Ext. Ser. Geol. Guidebook 1, 187 p.

——— 1964, Relation of stratigraphy to structure in the southern Appalachians, *in* Tectonics of the southern Appalachians: Virginia Polytech. Inst. Dept. Geol.

Sci. Mem. 1, p. 81-114.
——— 1968, Profile of the folded Appalachians of western Virginia, *in* A coast to coast tectonic study of the United States: UMR Jour., no. 1 (V. H. McNutt–Geology Dept. Colloquium Ser. 1), p. 27-64.
——— 1971, The Appalachian structural and topographic front between Narrows and Beckley, Virginia and West Virginia, *in* Guidebook to Appalachian tectonics and sulfide mineralization of southwestern Virginia: Virginia Polytech. Inst. Dept. Geol. Sci., Guidebook 5, p. 87-142.
Dahlstrom, C. D. A., 1969, Balanced cross sections: Canadian Jour. Earth Sci., v. 6, p. 743-757.
——— 1970, Structural geology in the eastern margin of the Canadian Rocky Mountains: Bull. Canadian Petroleum Geology, v. 18, p. 332-406.
Edsall, R. W., Jr., 1974, A seismic reflection study over the Bane anticline in Giles County, Virginia: Master's thesis, Virginia Polytech. Inst., 111 p.
Epstein, A. G., J. B. Epstein, and L. D. Harris, 1977, Conodont color alteration—an index to organic metamorphism: U.S. Geol. Survey Prof. Paper 995, 27 p.
Ethington, R. C., and D. C. Clark, 1971, Lower Ordovician conodonts of North America, *in* Symposium on conodont biostratigraphy: Geol. Soc. America Mem. 127, p. 63-82.
Harris, L. D., 1967, Geology of the L. S. Bales well, Lee County, Virginia—a Cambrian and Ordovician test, *in* Kentucky Oil and Gas Assoc. 29th Ann. Mtg., 1965 Proc. Tech. Sess.: Kentucky Geol. Survey, Ser. 10, Spec. Pub. 14, p. 50-55.
——— 1970, Details of thin-skinned tectonics in parts of Valley and Ridge and Cumberland Plateau provinces of the southern Appalachians, *in* G. W. Fisher, et al, eds., Studies of Appalachian geology—central and southern: New York, Interscience, p. 161-173.
——— 1976, Thin-skinned tectonics and potential hydrocarbon traps—illustrated by a seismic profile in the Valley and Ridge province of Tennessee: U.S. Geol. Surv. Jour. Research, v. 4, p. 379-386.
Huddle, J. W., E. T. Jacobsen, and A. D. Williamson, 1955, Detailed logs of 28 wildcat wells in Lee, Wise, Buchanan, Montgomery, and Dickinson Counties, Virginia: U.S. Geol. Survey Open-File Rept. 348, 1689 p.
Jones, P. B., 1971, Folded faults and sequences of thrusting in Alberta foothills: AAPG Bull., v. 55, p. 292-306.
Kulander, B. R., and S. L. Dean, 1978, Gravity, magnetics, and structure of the Allegheny Plateau/western Valley and Ridge in West Virginia and adjacent states: West Virginia Geol. and Econ. Survey Rept. Inv. 27, 91 p.
Milici, R. C., 1970, The Allegheny structural front in Tennessee and its regional tectonic implications: Am. Jour. Sci., v. 268, p. 127-141.
——— 1973, Tectonics of the southern Appalachian Cumberland Plateau and adjacent parts of the Valley and Ridge—a review (abs.): Geol. Soc. America Abs. with Programs, v. 5, p. 197.
Perry, W. J., Jr., 1971, Structural development of the Nittany anticlinorium, Pendleton County, West Virginia: PhD thesis, Yale Univ., 227 p.
——— 1975, Tectonics of the western Valley and Ridge foldbelt, Pendleton County, West Virginia—a summary report: U.S. Geol. Survey Jour. Research, v. 3, p. 583-588.
——— 1977, The structural geology of Giles County, Virginia, a preliminary reinterpretation: U.S. Geol. Survey Open-File Rept. 77-831, 7 p.
——— G. Smith, and M. W. Lee, 1977, Seismic modelling of complexly deformed fold-and-thrust-belt rocks, central Appalachians (abs.): Morgantown Energy Research Center, 1st Eastern Gas Shales Symposium Preprints, p. 327.
Price, P. H., 1929, Pocahontas County: West Virginia Geol. Survey County Repts., 531 p.
Rich, J. L., 1934, Mechanics of low-angle overthrust faulting as illustrated by Cumberland thrust block, Virginia, Kentucky, and Tennessee: AAPG Bull., v. 18, p. 1584-1596.
Rodgers, J., 1949, Evolution of thought on structures of middle and southern Appalachians: AAPG Bull., v. 33, p. 1643-1655.
——— 1953, The folds and faults of the Appalachian Valley and Ridge province: Kentucky Geol. Survey, Ser. 9, Spec. Pub. 1, p. 150-166.
——— 1963, Mechanics of Appalachian foreland folding in Pennsylvania and West Virginia: AAPG Bull., v. 47, p. 1527-1536.
——— 1970, The tectonics of the Appalachians: New York, John Wiley, 271 p.
Royse, F., Jr., M. A. Warner, and D. L. Reese, 1975, Thrust belt structural geometry and related stratigraphic problems, Wyoming-Idaho–northern Utah, *in* D. W. Bolyard, ed., Symposium on deep drilling in the central Rocky Mountains: Rocky Mtn. Assoc. Geologists, p. 41-54.
Sears, C. E., and E. S. Robinson, 1971, Relations of Bouguer gravity anomalies to geologic structure in the New River district of Virginia: Geol. Soc. America Bull., v. 82, p. 2631-2638.
Tucker, P. M., and H. J. Yorston, 1973, Pitfalls in seismic interpretation: Soc. Exploration Geophysicists Mon. 2, 50 p.

The American Association of Petroleum Geologists Bulletin
V. 59, No. 2 (February 1975), P. 193-216, 23 Figs.

Subsurface Positive Elements within Ouachita Foldbelt in Texas and Their Relation to Paleozoic Cratonic Margin[1]

RICHARD L. NICHOLAS[2] and ROGER A. ROZENDAL[2]
New Orleans, Louisiana 70160, and Houston, Texas 77001

Abstract Two major subsurface positive elements within and marginal to the Ouachita foldbelt in Texas, south and east of the foredeep basins, have been defined by recent drilling and seismic data.

Uplifted foreland carbonate rocks of early Paleozoic age were drilled on the Devils River uplift in southwest Texas in Shell 1 Stewart. This structure is interpreted to be a differentially uplifted autochthonous basement block partly overridden by the interior zone of the Ouachita foldbelt.

The presence of a second major linear positive element has been established in east-central Texas by seismic records and by Shell 1 Barrett. A faulted anticlinal structure, herein named the "Waco uplift," is at least 25 mi behind the leading edge of overthrust Ouachita facies rocks. A massive section of intensely deformed foreland carbonate rocks of early and middle(?) Paleozoic age, overlain by allochthonous low-grade metaclastic rocks of Ouachita facies, was drilled in the Barrett well. The uplift appears to be similar in structural position and history to the Devils River uplift, but the Waco uplift is probably a parautochthonous block which was partly involved in late Paleozoic orogenic thrusting.

The Waco and Devils River uplifts are subregional elements superimposed on a broad elongate gravity maximum which can be traced from the Ouachita Mountains to near the Marathon area. This maximum, thought to be the result of uplift of the dense mafic crustal layer, may have marked the edge of the early and middle Paleozoic (preorogenic) continent. A well-defined, persistent, depositional and tectonic boundary between the constructional carbonate miogeocline and thinner contemporaneous pelitic deposits of the geosyncline was also apparently localized along or near the periphery of this inferred crustal margin. Therefore, the two uplifts are interpreted to be external massifs in structural and stratigraphic positions similar to those of the Green Mountains-Blue Ridge chain of basement uplifts in the Appalachians.

Isotopic dates from both Ouachita facies and foreland rocks of the cratonic margin record multiple Paleozoic thermal events and probable accompanying deformation along the tectonic belt. The data also confirm the ancestral middle, and possibly early, Paleozoic origin of the Devils River and Waco uplifts. The presence of a post-Precambrian, pre-Late Cambrian volcanic and metasedimentary province in the area of the Devils River uplift is suggested by Rb-Sr dating of rocks drilled in the 1 Stewart well.

Foreland strata in the 1 Stewart are essentially undeformed, but the rocks in the 1 Barrett are metamorphosed to greenschist facies and highly contorted and sheared. Some formation fluids and minor hydrocarbon shows were recovered from these tests, but the quality of the reservoirs was generally poor.

Introduction

The Ouachita system is a Paleozoic geosyncline and orogenic belt which borders the southern edge of the central United States craton. The foldbelt extends more than 1,300 mi (2,100 km) from near the southern terminus of the Appalachians to western Texas, where it passes into Mexico (Fig. 1). The foldbelt is covered unconformably over most of its extent by Cretaceous and younger deposits of the Gulf coastal plain and is exposed only in the Ouachita Mountains of Oklahoma and Arkansas and in the Marathon region of western Texas. The geology of the concealed areas is known from borehole and geophysical data.

The Paleozoic sedimentary rocks and style of deformation of the foldbelt are markedly different from time-equivalent rock sequences and structures of the adjacent craton. The boundary between the two provinces is not clearly defined everywhere, but along most of its length rocks of the geosyncline have been thrust north and west over autochthonous, or nearly autochthonous, cratonic rocks. For most of its length, too, the foldbelt is bordered by a chain of asymmetric foredeep basins containing thick upper Paleozoic synorogenic clastic rocks.

The contrast in facies between the two provinces is most obvious in the preorogenic rocks of early and middle Paleozoic age. The margin or "hinge" between the two facies was a fundamental tectonic and stratigraphic boundary which separated the cratonic platform from the marginal Ouachita geosyncline. The boundary is not exposed and its original position and form only can be inferred from indirect surface and subsurface data.

In recent years, Shell Oil Company and others have explored extensively along the frontal zone of the western Ouachita system. This exploration has confirmed the existence of two major subsurface positive elements within and adjacent to the oro-

[1]Manuscript received, September 24, 1973; accepted, July 12, 1974.

[2]Shell Oil Company.

Published with permission of Shell Oil Co. This study is based on the work of many geologists and geophysicists within Shell Oil Co. A general acknowledgment is, therefore, due all those who have helped. We particularly thank John Tsiaperas for assistance in making the gravity interpretations. J. K. Arbenz and A. W. Bally contributed many ideas on the regional geology of this area. Most of the isotopic-age determinations were performed by Shell Development Co. W. A. McLaughlin and R. E. Scarborough of Shell Development Co. and R. F. Story, formerly with Shell Development Co., made significant interpretations of the isotopic dates. A. W. Bally, R. E. Denison, and Jules Braunstein reviewed the manuscript.

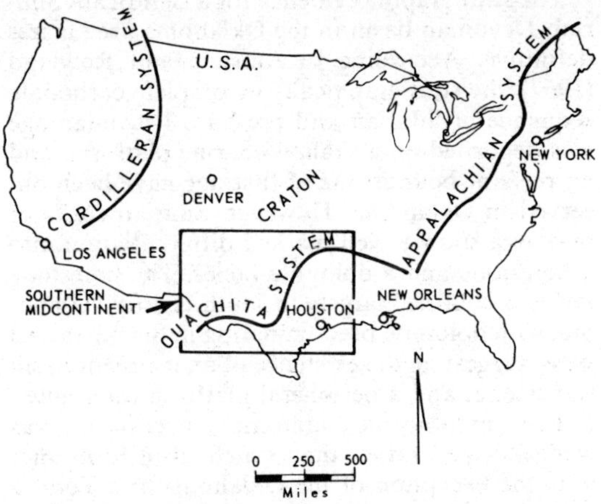

FIG. 1—Paleozoic geosynclines of United States and outline of southern Mid-Continent.

genic belt, and has provided significant new geologic and geophysical evidence relating to the history of this area. An earlier paper by Rozendal and Erskine (1971) reported data from one of the key deep tests drilled in central Texas during this latest phase of exploration. The purposes of this paper are to describe the two major structures and to speculate on the nature and location of the boundary between the early and middle Paleozoic (preorogenic) craton and Ouachita geosyncline.

The regional geology of the foldbelt and adjacent craton has been described in comprehensive regional summaries by Flawn *et al.* (1961) and Ham and Wilson (1967). In addition to the pioneer work by King (1937), detailed field studies of preorogenic rocks in the Marathon area by McBride and his students (McBride, 1969) and Thomson and McBride (1964), Wilson (1954), and others applying contemporary sedimentologic concepts have provided critical new information on the early Paleozoic development of the western part of the Ouachita system. Comparable detailed studies of the preorogenic succession in the Ouachita Mountains are lacking. The best modern reference to that area is Ham's (1959) summary of pre-Stanley strata in the Arbuckle-Ouachita Mountain region. The following stratigraphic and structural summaries are taken mainly from these and other publications on the geology of the region.

Stratigraphic Summary

The stratigraphic section of the region is summarized in Figure 2. The history of the Ouachita-Marathon foldbelt conforms in many respects to King's (1969) geosynclinal cycle of preorogenic, orogenic, and postorogenic phases. We have adopted this terminology and, in addition, have

TIME m.y.	SYSTEM	CRATON WEST TEXAS	GEOSYNCLINE MARATHON	CRATON S.OKLAHOMA	GEOSYNCLINE OUACHITAS
230	PERMIAN	OCHOA GUADALUPE LEONARD WOLFCAMP	"WOLFCAMP"	"PERMIAN"	
280	PENNSYLVANIAN	VIRGIL MISSOURI DES MOINES ATOKA MORROW	GAPTANK HAYMOND DIMPLE	VIRGIL MISSOURI DES MOINES ATOKA MORROW SPRINGER	ATOKA JOHNS VALLEY
310	MISSISSIPPIAN	CHESTER PRE-CHESTER	TESNUS	CANEY SYCAMORE	JACKFORK STANLEY
345	DEVONIAN	WOODFORD "DEVONIAN"	CABALLOS NOVACULITE	WOODFORD	ARKANSAS NOVACULITE
405	SILURIAN	"UPPER SILURIAN" FUSSELMAN		HUNTON	MISSOURI MT. BLAYLOCK
425	ORDOVICIAN	MONTOYA	MARAVILLAS	SYLVAN	POLK CREEK
				FERNVALE VIOLA	BIGFORK
		SIMPSON	WOODS HOLLOW FORT PEÑA	SIMPSON	WOMBLE
		ELLENBURGER	MARATHON	ARBUCKLE	BLAKELY MAZARN CRYSTAL MT. COLLIER LUKFATA
500	UPPER CAMBRIAN	WILBERNS RILEY	DAGGER FLAT	TIMBERED HILLS	not exposed

FIG. 2—Generalized correlation chart for Ouachita geosyncline and adjacent cratonic provinces of West Texas and southern Oklahoma.

attempted to relate the stratigraphy and structure of the cratonic margin to the structural stages of the geosyncline. This classification is in basic agreement with the concept of the development of the Marathon geosyncline in an early and middle Paleozoic "forming" stage and a late Paleozoic "filling" stage, as described by Thomson and McBride in 1964. The sequence of events in the Ouachita system also can be interpreted within the concepts of global plate tectonics (Keller and Cebull, 1973).

The basic paleogeographic and depositional provinces described are the craton and the geosyncline. The sedimentologic, lithologic, faunal, and thickness distinctions between the two gross facies are most obvious in preorogenic Late Cambrian through Devonian and earliest Mississippian sediments (Figs. 2-4). Cratonic rocks of this age range in thickness from about 5,000 ft in the West Texas basin to more than 11,000 ft in southern Oklahoma. The rocks are mainly constructional shallow-marine shelf carbonate with smaller amounts of clean sandstone and shale, and contain a characteristic shelly fauna. The sequence is interrupted by many regional and subregional unconformities which reflect extensive and frequent epeirogenic movements of the craton.

Such terms as "foreland," "platform," "Arbuckle," and "miogeosynclinal" facies also have been used by many workers to designate this lithologic

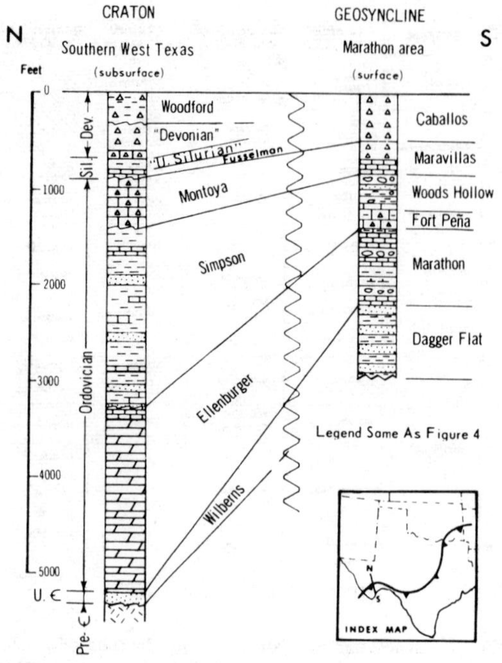

FIG. 3—Preorogenic stratigraphy of craton in West Texas and geosyncline in Marathon area.

assemblage. Most of the strata are remarkably uniform in lithology and fauna over vast areas of the craton and, with the exception of several Upper Ordovician and Silurian units in the ancestral Tobosa and Oklahoma basins (Fig. 5), are characterized by gradual facies changes. These basins were the sites of persistent downwarping during most of early and middle Paleozoic time. Pronounced subsidence began during the Cambrian in the Oklahoma basin and during the Early Ordovician in the Tobosa basin. Subsidence continued in both basins through Devonian time. Most Ordovician rocks thicken gradually into these depocenters and contain relatively greater amounts of limestone and fine terrigenous clastic rocks than do those in the flanking shelf platforms. It is clear from a study of the rocks, however, that shallow-marine conditions prevailed in both the basin and platform environments.

Post-Early Silurian patterns of sedimentation differed in the two basins. A well-defined deep-water embayment fringed by an organic reef developed mainly in Late Silurian time in West Texas (McGlasson, 1967). This sediment-starved cratonic basin apparently opened to the south into the even deeper geosyncline. Early and Middle Devonian deep-water chert and dark siliceous limestone subsequently filled the residual Silurian basin.

The stratigraphic evidence for a significant Silurian-Devonian basin in the Oklahoma area is less definitive. According to Amsden and Rowland (1967), the stratigraphically incomplete carbonate sequence of Silurian and pre-Late Devonian age was deposited on a shallow-marine platform, and no reefs or boundstone of that age have been observed in Oklahoma. However, Silurian rocks of that area show a well-marked differentiation into a limestone and a dolomite facies. The limestone facies is confined largely to south-central Oklahoma, with dolomite predominating in the north and west, suggesting the existence of an incipient basin (limestone) and a peripheral platform (dolomite).

The preorogenic cratonic succession was wedge-shaped rather than synclinal in form and, with the exception of the Oklahoma and Tobosa embayments, was deposited on a relatively stable platform which sloped gently toward the geosyncline. The deposits apparently attained a maximum thickness at the edge of the craton or continental margin and thinned toward the sea and the deeper water, sediment-starved geosynclinal basin.

The section at this tectonic-stratigraphic margin is marked in the Marathon area by an allochthonous Cambrian-Ordovician transitional slope facies containing proximal slide breccias derived from mixed terrigenous and shallow-water sedimentary, igneous, and metamorphic sources. The larger boulders of the breccia are mostly shallow-water, intertidal, and supratidal carbonate rocks of Cambrian-Ordovician age containing a benthonic shelly fauna. The lithic and faunal facies of the exotics closely resemble the correlative subjacent autochthonous cratonic rocks. The breccia is emplaced in a starved, deeper water geosynclinal facies consisting mostly of graptolitic shales, dark carbonate mudstones, and fine-grained siliceous rocks. The wedging out of the foreland debris from northwest to southeast across the Marathon outcrops suggested to King (1937), Wilson (1954), and McBride (1969) that the source area was just northwest of the depositional basin. On the basis of this stratigraphic evidence, we conclude that a cratonic platform and geosynclinal trough with a well-developed shelf-slope-basin geometry existed during early Paleozoic time somewhere south of the present Marathon area. The seaward edge of this cratonic platform or miogeocline (Dietz and Holden, 1967) appears to have been a persistent, well-defined, but intermittently unstable depositional and structural margin during most of the early and middle Paleozoic.

Although no early Paleozoic proximal slide breccias have been reported from the Ouachitas, similar platform-basin depositional relations may have existed in the eastern part of the Ouachita

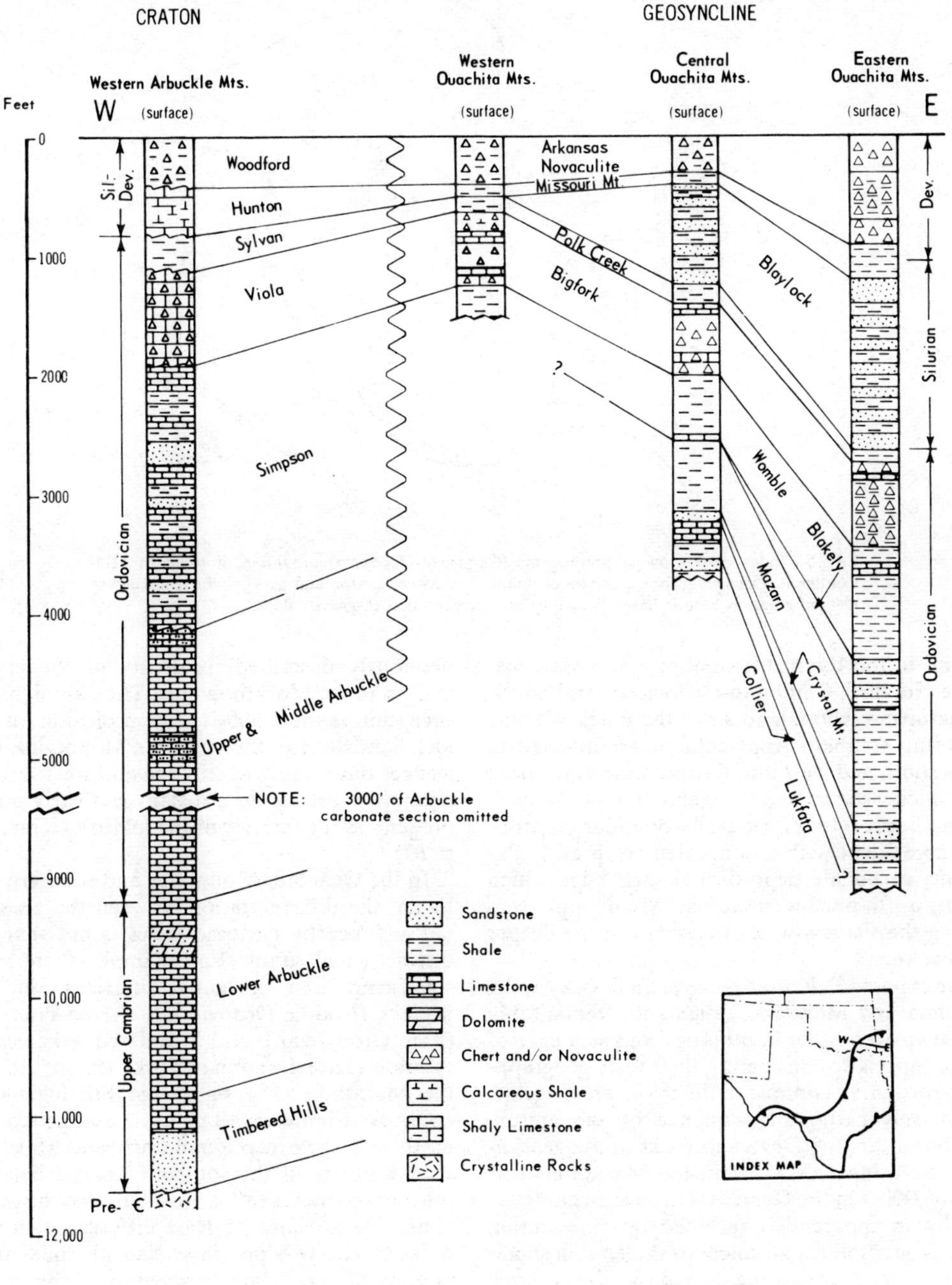

FIG. 4—Preorogenic stratigraphy of craton in southern Oklahoma and geosyncline in Ouachita Mountains.

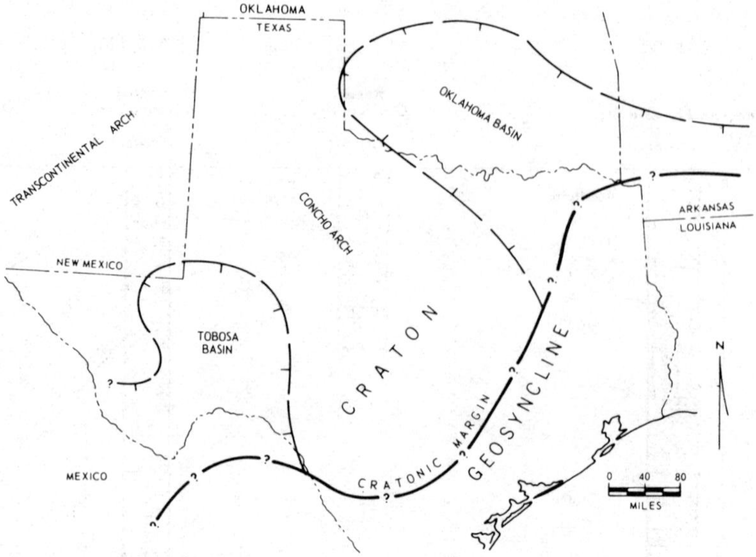

FIG. 5—Index map showing preorogenic (Cambrian-Devonian) provinces of southern Mid-Continent. Paleogeographic position of boundary between craton and geosyncline is unknown; its configuration here is based primarily on interpretation of gravity data.

system in northern Mississippi and Alabama. There, the thick Cambrian-Ordovician and Silurian platform-carbonate rocks of the Black Warrior basin and southern Appalachians are inferred to grade south and west into thinner time-equivalent dark siliciclastic rocks of Ouachita facies. According to Thomas (1972), the facies boundary is probably coincident with a concealed steep early Paleozoic, carbonate depositional shelf edge which trends north-northwest across Mississippi, separating the shallow-water craton from the deeper geosyncline.

The exposed Paleozoic geosynclinal rocks in the Ouachita and Marathon salients are remarkably similar to each other in lithology and sequence of facies, especially considering their wide geographic separation. In contrast to the thick, preorogenic, dominantly carbonate sequence of the craton, Cambrian through Devonian rocks in the geosyncline are thinner (3,000 ft in the Marathons and about 6,000 ft in the Ouachitas) and are characterized by an apparently continuous marine section of black graptolitic and siliceous shales with smaller amounts of sandstone and dark cherty limestone; the section is capped by a distinctive middle Paleozoic white chert and novaculite. Subsidence of the geosyncline greatly exceeded sedimentation during this time span and resulted in a starved basin or leptogeosyncline (Thomson and McBride, 1964). This basin is devoid of stratigraphic evidence of tectonism or orogeny other than the previously described instability of the cratonic margin in the Marathon area. The only significant exception is the 1,500-ft section of Silurian Blaylock Sandstone of the Ouachita Mountains, which wedges out rapidly northward and westward, and possibly is related to a distant and early pulse of orogeny in the interior of the foldbelt (King, 1969, p. 63).

In the Ouachita Mountains and southern Oklahoma, the differentiation between the geosynclinal and nearby cratonic facies is not sharp and distinct in all units. For example, there are no significant lithic or faunal differences in Viola-Bigfork (Middle Ordovician), Sylvan-Polk Creek (Late Ordovician), and Woodford-Arkansas Novaculite (Late Devonian) rocks (Ham, 1959). In the Marathon area significant lithologic differences exist between all cratonic and geosynclinal units, with the exception of Early and Middle Devonian cherts of the southern Tobosa basin and the correlative Caballos Novaculite of the geosyncline. These times of least differences in facies between the two provinces can be equated with periods of significant downwarp of the cratonic margin and concurrent major marine transgressions. Apparently during maximum transgression, the cratonic shelf edge ceased to act as a significant facies barrier between the Ouachita geosyncline and the adjacent craton.

Folk (1973) has suggested that the Caballos Novaculite was deposited in a peritidal environment.

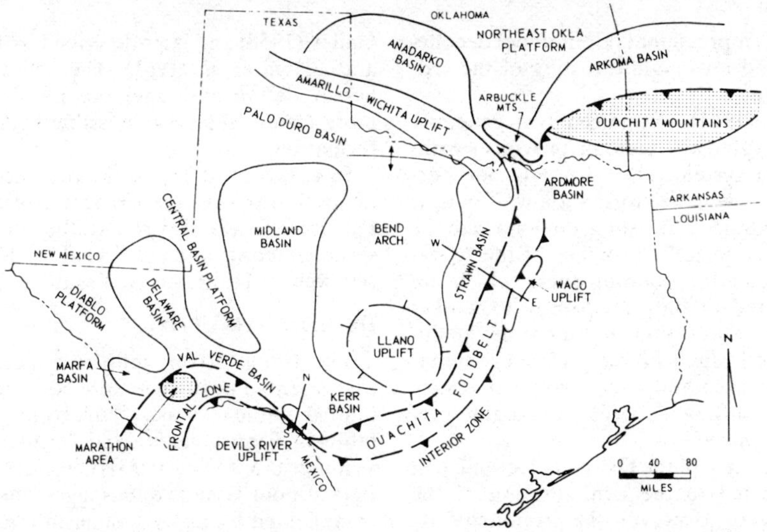

FIG. 6—Index map showing major tectonic elements of southern Mid-Continent during Mississippian, Pennsylvanian, and Permian, and location of Devils River uplift and Waco uplift. Tectonic provinces in Ouachita foldbelt from Flawn (1961). North-south and west-east lines refer to locations of cross sections of Figure 23.

However, McBride and Thomson (1970) presented convincing evidence of deep-marine deposition of that formation. This interpretation is supported by McGlasson (1967), who showed that the regional setting in this area during the Silurian-Devonian was that of a relatively deep cratonic embayment (Tobosa basin) opening into the even deeper Caballos geosyncline.

The initial major orogenic deformation within the internal zone of the geosyncline began in Mississippian time. As a result, a nearly empty extracratonic geosyncline, which formed during the preorogenic stage, was essentially filled to capacity with thick foredeep deposits of synorogenic siliciclastic detritus (Stanley-Jackfork; Tesnus-Haymond). This filling marked the end of deposition of separate cratonic and geosynclinal facies, and the younger Pennsylvanian and Permian clastic wedges were deposited for the most part on the collapsed cratonic platform in foredeep troughs (exogeosynclines). Elsewhere on the craton, the orogenic stage is characterized by an overall transgressive Pennsylvanian cycle of deposition followed by a regressive Permian cycle. Complex patterns of abrupt carbonate (platform)-terrigenous (basin) facies changes reflect the late Paleozoic tectonism of the craton (Fig. 6).

The postorogenic phase was marked by epeirogenic uplift of the cratonic margin and orogenic belt. This resulted in restricted conditions on the craton and the eventual filling of the intracratonic basins with fine siliciclastic rocks, basinal limestones, and evaporites, and, finally, the termination of marine conditions in the Late Permian.

STRUCTURAL SUMMARY

With the exception of the pre-Late Cambrian southern Oklahoma geosyncline, the preorogenic phase of the southwestern craton was characterized by relative structural stability and mild epeirogenic movements. The principal structural elements of the craton formed at this time were two wide, relatively shallow, marginal cratonic basins separated by the broad, southeast-plunging, low-relief Concho arch (also called "Texas arch" or "Texas peninsula" by some writers; Fig. 5). The basins may have been more like embayments which were open ended to the geosyncline at various times, rather than structural closures. The structural evidence for these ancestral provinces largely has been removed by more severe late Paleozoic orogenic deformation.

Figure 6 shows the principal structural provinces of the southern Mid-Continent which developed during the late Paleozoic (Mississippian, Pennsylvanian, and Permian) orogenies. The craton is characterized by large, block-faulted (germanotype) uplifts involving the crystalline basement and the overlying sedimentary cover, although some of the major faulted uplifts (*e.g.*, Arbuckle-Wichita axes) display significant compression. Deformation within the geosyncline is dominated by compressional (alpinotype) decollement folding and low-angle thrusting of the sedimentary section.

Flawn *et al.* (1961) recognized two generally parallel petrographic and tectonic provinces within the Ouachita system: (1) a frontal zone bordering the craton, where deformation was by flexure and "thin-skinned" thrusting, analogous to the Valley and Ridge tectonic province of the Appalachian system, and containing rocks which are unmetamorphosed or only incipiently to weakly metamorphosed, and (2) an interior zone, analogous to the Blue Ridge tectonic province, consisting of highly sheared and contorted sedimentary rocks that have undergone weak to low-grade regional metamorphism.

The sinuous course of the Ouachita tectonic belt probably is related to the configuration of the preorogenic craton. However, the prominent late Paleozoic Marathon and Ouachita structural salients undoubtedly were accentuated by the funneling effect of the preorogenic Tobosa and Oklahoma embayments. The geometric relations of these two Paleozoic embayments striking at high angles into the orogenic belt also suggest the "failed arms" or aulacogen structures generated at triple-plate junctions (Burke and Dewey, 1973).

The amount of crustal shortening along the cratonic margin is unknown but has been estimated to be at least 50 mi in the Ouachita Mountains (Hendricks, 1959), and a minimum of 15 mi in the Marathon area (King, 1937).

Bordering the Ouachita tectonic system is a series of late Paleozoic foredeep basins which are marginal to the tectonic belt. These are the Strawn, Kerr, and Val Verde basins in Texas, and the Arkoma basin in Oklahoma and Arkansas (Fig. 6). The recent drilling and seismic work by Shell Oil Company in this trend have confirmed the presence of two major subsurface positive elements south and east of these foreland basins: the Devils River uplift and the Waco uplift. The Central Mississippi ridge on the south flank of the Black Warrior basin is in a similar tectonic position (Vernon, 1971). These structures, behind or near the frontal edge of the thrust belt, are peripheral to the cratonic margin. The nature and histories of these linear positive elements are not entirely understood, but all three uplifts are bordered by major late Paleozoic faults associated with the Ouachita orogeny.

Following extensive geophysical surveys, two wells were located by Shell Oil Company, one on the crest of the Devils River uplift and one on the Waco uplift. Shell 1 Stewart in Val Verde County was drilled on the feature south of the Val Verde basin which was named the Devils River uplift by Galley (1958) and later described by Flawn (1959), and Flawn *et al.* (1961). The new data reported herein define and date the Devils River uplift more clearly than was possible with previous information.

In east-central Texas, the presence of a second major linear positive element covered by lower Paleozoic foreland rocks has been established by seismic records and confirmed by the Shell 1 Barrett well in Hill County, Texas.

Devils River Uplift

The Devils River uplift is a northwest-southeast-oriented positive feature 50-60 mi long that is limited on the northeast flank by major high-angle faults of Pennsylvanian and Permian(?) ages with as much as 5,000 to 10,000 ft displacement (Fig. 7). Fair to poor seismic data suggest that the uplift is complicated by many smaller internal faults both parallel with and oblique to the main axis. There is insufficient control to delimit the structurally complex south flank of the uplift, which is overridden, in part, by an allochthonous undifferentiated sequence of highly sheared, low-grade, metamorphic rocks of the interior zone. There does not appear to be an extensive folded frontal zone north of the more highly deformed interior zone.

In the Shell 1 Stewart, Cretaceous strata are present from the surface to approximately 2,000 ft, lying directly on 1,500 ft of normal foreland facies Ellenburger dolomite (Fig. 9). Below the Ellenburger is 500 ft of dolomitic quartzite with interbeds of sandy dolomite of probable Late Cambrian age. From 4,000 to 6,800 ft there is a metasedimentary sequence with some interbedded metavolcanic rocks. From 6,800 ft to total depth there is another 2,900 ft of mostly metaigneous rocks with some metasedimentary layers. The radiometric ages of these metamorphic rocks will be discussed later. The three principal seismic reflections in this area are derived from the base of the Cretaceous, the top of the Paleozoic carbonate section, and an intrabasement basaltic rock (Fig. 8). At the Stewart well, the base-of-Cretaceous and top-of-Paleozoic reflectors are coincident.

Much of the Devils River uplift apparently is underlain by a Precambrian intrabasement basaltic rock which has been mapped from seismic reflections and was penetrated by the Shell 1 Stewart. The reflection has been correlated with a metamorphosed dense epidote-tremolite schist at 9,070 ft in the well. This rock was probably originally basaltic. The structure mapped on the basis of this intrabasement reflector is a faulted domal complex essentially conformable with the configuration of the structure of the lower Paleozoic carbonate section (Fig. 10).

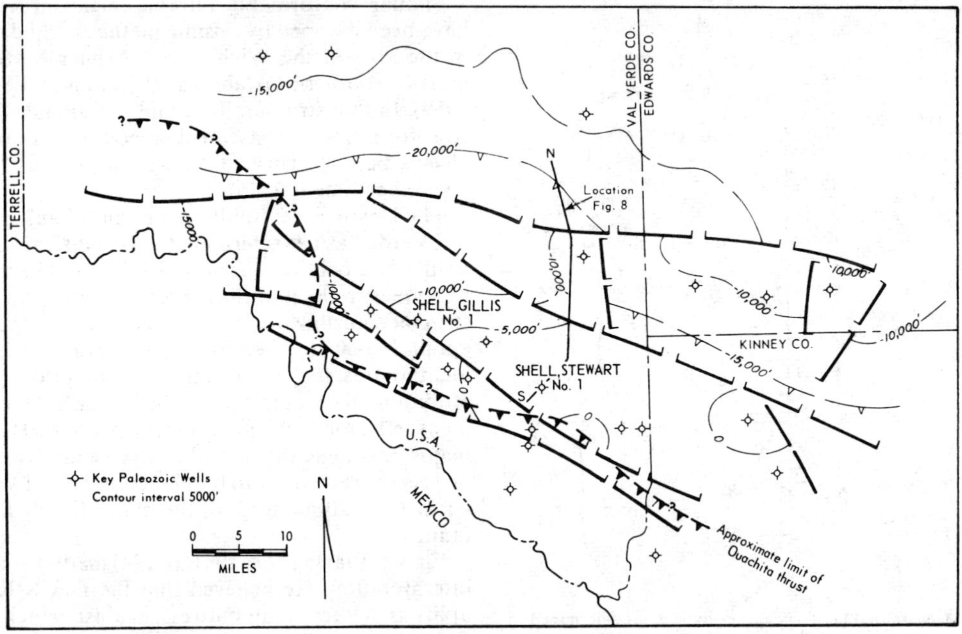

FIG. 7—Generalized structure map of Devils River uplift from seismic and subsurface data. Datum is top of Ellenburger. Contour interval 5,000 ft. Control wells shown penetrated Ellenburger or older formations or possible metamorphosed equivalents.

FIG. 8—Seismic line with geologic interpretation across Devils River uplift. For line location see Figure 7.

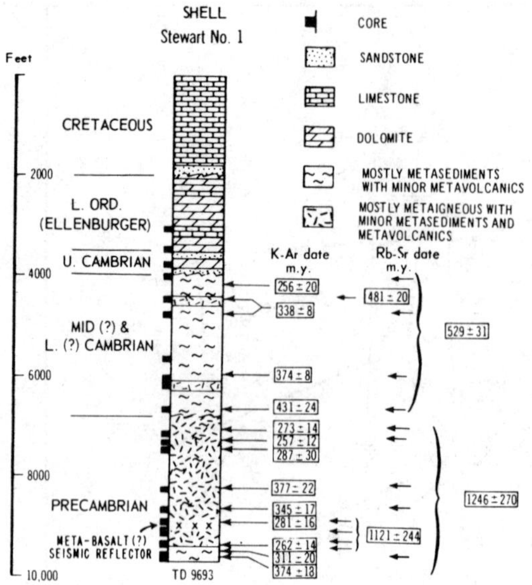

FIG. 9—Generalized lithology and isotopic dates determined in cores from Shell 1 Stewart, Val Verde County, Texas.

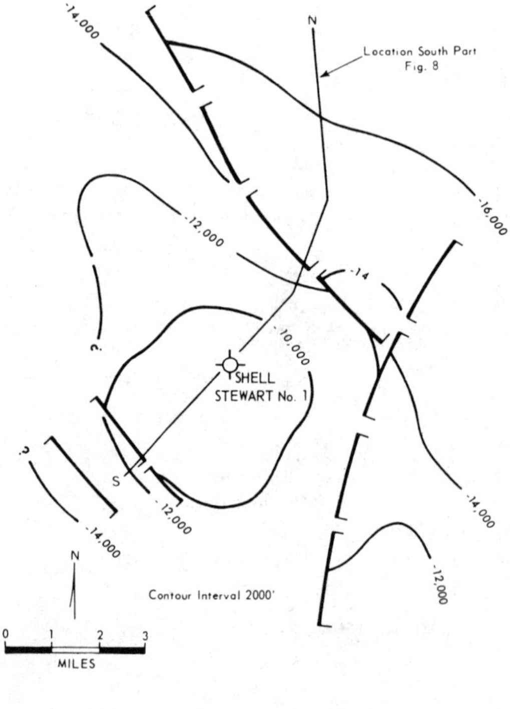

FIG. 10—Generalized structure map of Precambrian intrabasement seismic reflection on Devils River uplift. Contour interval 2,000 ft.

Similar conformable intrabasement structures have been mapped by seismic methods and drilled in the area of the Wichita and Arbuckle Mountains of southern Oklahoma (Widess and Taylor, 1959). In that structurally complex region, layered stratiform flows of basement igneous rocks cannot always be distinguished on seismic records from layered sedimentary rocks.

The Devils River uplift on the south flank of the Val Verde basin is interpreted to be a differentially uplifted autochthonous or rooted block. The intrabasement reflection has been mapped on the structure northward from the crest to a depth of about 16,000 ft. Therefore, if the structure is allochthonous, the level of decollement at the northern boundary would have to be beneath the basement reflector at a depth greater than 16,000 ft. No major low-angle thrust faults have been identified on seismic records or in boreholes in the Val Verde basin 15 to 20 mi north of the major Devils River fault.

Flawn (Flawn *et al.,* 1961, p. 144) made a similar interpretation. He believed that the Devils River uplift is a "resistant buttress against which the Ouachita system was thrust." The effect of this craton-margin positive element was to shield the Pennsylvanian and Permian siliciclastic rocks of the adjacent Val Verde basin from the characteristic compressive tectonics so common in the other marginal basins.

Present information does not permit a unique structural interpretation of the Devils River uplift. It is possible that the element was uplifted synchronously with the Ouachita thrusting, and even significant postorogenic uplift cannot be excluded.

Waco Uplift

In east-central Texas, an 80-mi-long and 10-mi-wide northeast-trending, doubly plunging anticlinal complex cresting at a depth of about 13,000 ft has been defined by seismic mapping and confirmed by the drilling of the Shell 1 Barrett well in Hill County, Texas. This uplift, about 25 mi behind the leading edge of the Ouachita facies rocks, is here named the "Waco uplift" after the nearby central Texas city. Details of lithology of 1 Barrett and the local structure were discussed in a previous paper by Rozendal and Erskine (1971) and will be reviewed only briefly here.

Interpretation of the seismic data shows the structure to be an asymmetric anticlinal block bounded by steep dip and normal(?) faults on the southeast flank, by large faults on the northwest flank, and further complicated by a sinuous northeast-southwest-trending medial fault that divides the structure longitudinally into separate culminations which have as much as 7,000 ft of simple and fault closure (Fig. 11). Although the dip of the faults cannot be documented from seismic records, we interpret the western boundary fault and the medial fault to be thrusts.

In the Shell 1 Barrett, Cretaceous and thin, possibly Jurassic strata are present from the surface to 3,904 ft, and unidentified phyllite, slate, and quartzite are present in the interval 3,904-13,595 ft (Fig. 12). A massive 6,060-ft-thick marble section is present from 13,595 to 19,655 ft, the top of which is believed to be associated with the primary seismic reflector on this structure. Cores of the marble show intense deformation characterized by compressional shear and flowage structures. Underlying the marble unit down to a depth of 19,780 ft is a medium-gray, biotite-rich quartzite 125 ft thick. This quartzite in turn overlies a medium-to-coarse crystalline quartz diorite, which is present to the total depth of 20,310 ft, a 530-ft penetration of the diorite.

Seismic reflections on this structure from the massive carbonate section are remarkably coherent (Fig. 13). Apparently the marble interval has retained its stratiform integrity in spite of the severe structural disturbance of individual layers seen in cores.

The marble with ptygmatic folding in this well differs markedly from the undeformed carbonate rocks in the 1 Stewart on the Devils River uplift. However, as noted by Flawn *et al.* (1961) and Rozendal and Erskine (1971), several wells drilled previously on the Devils River uplift penetrated intensely deformed carbonate rocks of unknown age which are strikingly similar to the marble in the 1 Barrett.

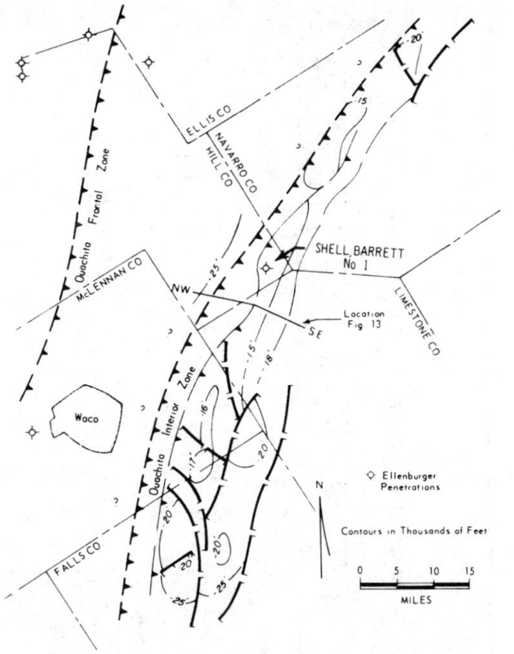

FIG. 11—Generalized structure map of Waco uplift based on seismic data. Contour interval is variable as shown. Datum is top of lower or middle(?) Paleozoic marble section. Frontal zone and interior zone (Flawn *et al.*, 1961) or overlying Ouachita thrust sheets are shown.

Isotopic Age Data

Paleontologic data are not sufficient to define the ages of any of the pre-Mesozoic units in either of the wells. All that is left of the fossil record are some unidentified inarticulate brachiopods in the phyllite section of the Barrett well and crinoid columnals in the carbonate section in both wells. It is assumed that the formations penetrated below the base of the Mesozoic are Paleozoic in age. The marble section in the Barrett well may represent a Cambrian-through-Devonian carbonate section similar to that in West Texas and southern Oklahoma. As paleontologic data are lacking, an attempt to date the pre-Mesozoic formations and time of deformation in these and other wells and outcrops along the foldbelt and cratonic margin was made by Shell Development Company using Rb-Sr and K-Ar methods.

The K-Ar method of geochronology is a particularly useful tool for determining the postmetamorphic cooling history of an orogen. In the opinion of Hurley and Rand (1968), most of the measured dates probably reflect the uplift, erosion, and cooling of the rocks through a critical isotherm of about 100-200°C, rather than time of highest metamorphic temperatures or climax of orogenesis.

The isotopic dates in the Shell 1 Barrett fall into three basic groups suggesting three cooling phases, possibly related to major orogenic events, as shown in Figure 12. The oldest date is Devonian. Some of the structure at the Barrett well could have been formed at this time by vertical uplift or by initial thrusting in the geosyncline. The Mississippian-Pennsylvanian ages probably record cooling following the emplacement of the thrust sheets of the frontal and interior zones of the Ouachita structural belt. The Permian ages may be associated with the cooling of the section during the last stages of deformation or by postorogenic regional uplift of the entire geosynclinal area. The latter dates possibly are comparable to the Late Pennsylvanian-Early Permian ages recorded from adularias in fractured but unfolded quartz veins in the Ouachita Mountains proper (Bass and Ferrara, 1969). The Early Devonian age recorded at the top of the phyllite section appears to be from detrital

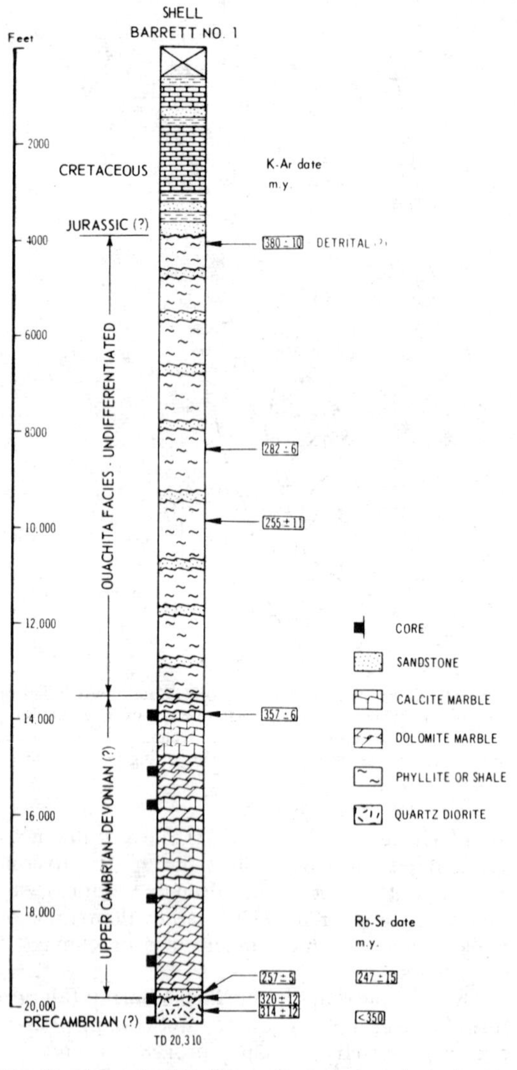

FIG. 12—Generalized lithology and isotopic dates determined in cores from Shell 1 Barrett, Hill County, Texas.

material and related to provenance rather than a time of local degasification. The concordant K-Ar and Rb-Sr 250-m.y. date at 19,700 ft is determined from biotite in a quartzite. The concordance of dates suggests that the biotite recrystallized at this time and possibly is related to hydrothermal activity.

In the Shell Stewart 1, there were 19 K-Ar determinations and 4 Rb-Sr isochrons constructed by the use of 17 separate determinations from rocks drilled between 3,800 and 9,624 ft (Fig. 9). All the analyses from this interval are from low-grade (greenschist facies) metamorphosed volcanic, sedimentary, and intrusive rocks. The most reasonable interpretation of the Rb-Sr whole-rock isochron values from this metamorphic facies appears to be, therefore, that they are related to time of deposition or emplacement. On this basis, the deeper Rb-Sr dates below 7,000 ft are definitely Precambrian in age (1.1 to 1.2 b.y.). However, the low-grade metasedimentary and metavolcanic rocks penetrated between depths of about 4,000 and 7,000 ft are much younger and probably were deposited about 500 m.y. ago. This indicates the presence of an undefined post-Precambrian, pre-Late Cambrian volcanic and metasedimentary province in the area of the Devils River uplift. The section is similar in lithology and age to the thousands of feet of pre-Late Cambrian metamorphic, volcanic, and intrusive rocks of the Wichita province of southern Oklahoma described from both the subsurface and surface by Ham (1963).

As in the Barrett well, the K-Ar age determinations in 1 Stewart indicate three basic cooling events: (1) Devonian (377-345 m.y.), (2) Mississippian-Pennsylvanian (329-281 m.y.), and (3) Permian (273-256 m.y.).

The Devonian episode may represent the initial uplift and cooling of the Devils River structure. The Mississippian-Pennsylvanian dates probably are related to the time of major orogenic thrusting of the frontal and interior allochthons of the Ouachita foldbelt. The Permian dates most likely record hydrothermal effusions of fluids related to the postorogenic uplift of the orogen, and not penetrative tectonism, although corroborative structural and stratigraphic data are lacking.

A single 431-m.y. K-Ar date from a metasediment at 6,680 ft and a 481 ± 20 m.y. Rb-Sr measurement in the 4,350-4,400 ft interval from a slightly metamorphosed volcanic rock may represent an even earlier Silurian-Ordovician thermal episode in this area.

Three K-Ar isotopic determinations were made on low-grade metamorphics from Shell 1 Gillis, 15 mi northwest of the 1 Stewart (Figs. 7, 14). This well is on the northwest flank of the Devils River uplift, and the 274-300-m.y. range of dates indicates a Permian-Pennsylvanian thermal history comparable to some of the events recorded in the Stewart well. The three samples from this well may be interpreted as showing a normal cooling history of a rock system after regional metamorphism. It appears to be a straightforward case of the top of the sequence cooling first (gives the older age) and the bottom of the sequence cooling last (gives the younger age). Conversely, the inverse isotopic-age relations, *i.e.*, younger dates over older dates, in both the 1 Barrett and 1 Stewart may reflect a more complex structural history than simple differential uplift and cooling of the metamorphic

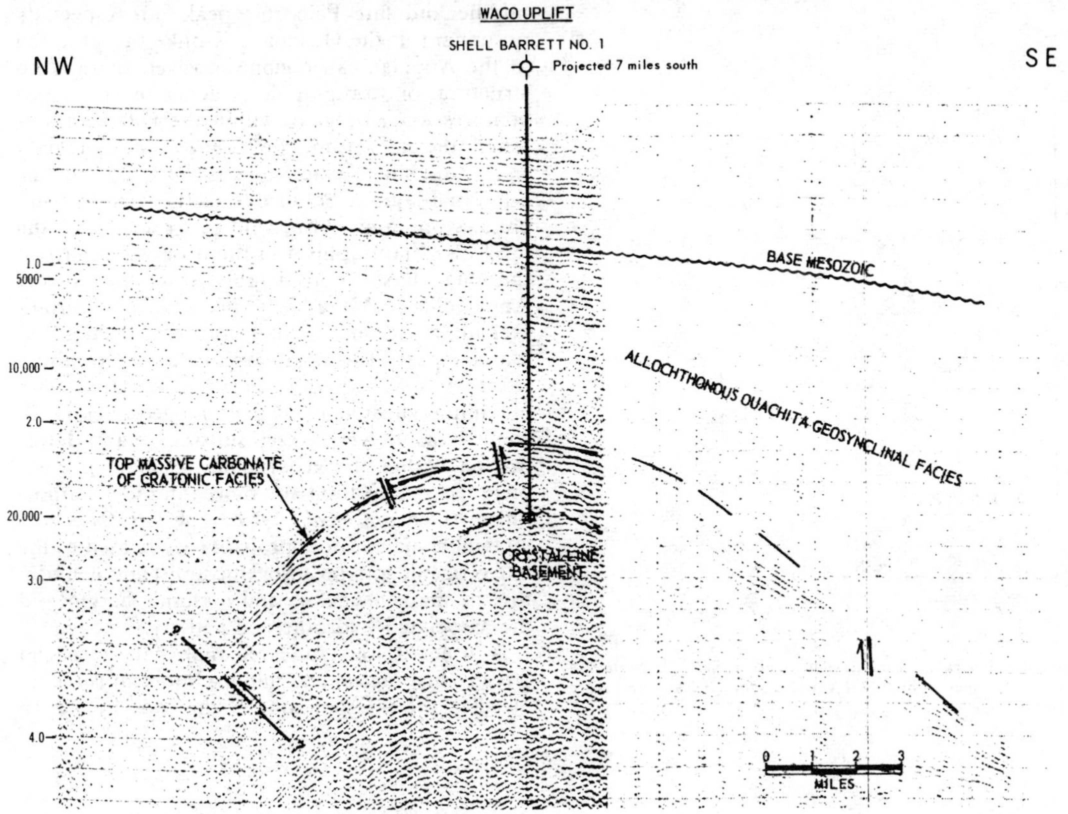

FIG. 13—Seismic line with geologic interpretation across Waco uplift. For line location see Figure 11.

system. The younger-over-older isotopic sequences may be the result of repetition caused by thrusting (and cooling) of as many as three fault slices within the 2,000-7,000-ft interval. Alternatively, the shallower rocks with the younger dates possibly cooled elsewhere and subsequently have been thrust over deeper rocks which cooled at an earlier time.

Summary of Isotopic Dates

Figure 15 is a compilation of Rb-Sr and K-Ar isotopic dates for this region. Although there is some sampling bias resulting from the concentration of data at a few localities, there is sufficiently wide geographic sampling to indicate a long and complex history of thermal activity in and adjacent to the geosyncline from Arkansas to northern Mexico (Fig. 16).

The billion-year-old Precambrian basement terrane shows significant thermal and tectonic overprinting in both the middle and late Paleozoic. The most common and widely distributed overprints are in the 380 to 300-m.y. (Devonian-Pennsylvanian) range, which includes both the preorogenic and orogenic phases of the geosyncline. The cluster of late Paleozoic K-Ar dates probably reflects both the final orogenic tectonism and the postorogenic regional uplift and cooling of the entire Ouachita system and cratonic margin.

The long history of low-grade metamorphism in the Ouachita system had been postulated by Flawn (Flawn *et al.*, 1961, p. 124), who suggested, on the basis of petrography, that at least two and possibly three periods of metamorphism had occurred in the interior zone of the fold belt prior to the final Permian-Pennsylvanian orogeny.

The pattern of Paleozoic isotopic ages in the Ouachita-Marathon tectonic belt and adjacent craton is broadly synchronous with that of the Appalachians (Harper, 1968; Kulp and Eckelmann, 1961; Carpenter, 1970; Hatcher, 1972). Thus the Taconian (Ordovician), Acadian (Devonian), and Appalachian or Alleghenian (Permian-Pennsylvanian) orogenies of the Appalachian region apparently all have radiometrically documented counterparts in the Ouachita system. The

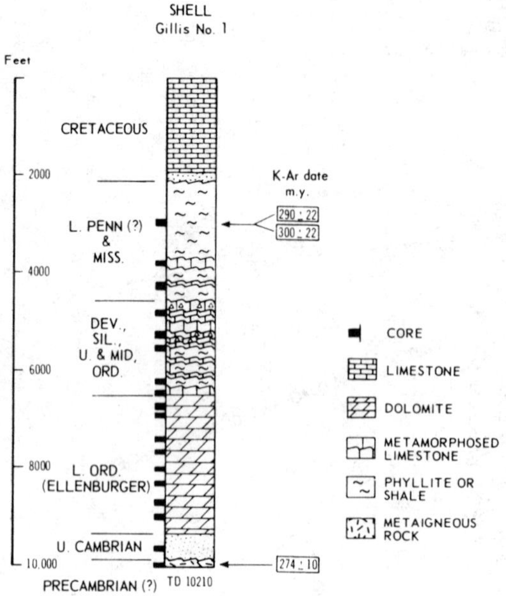

FIG. 14—Generalized lithology and isotopic dates determined in cores from Shell 1 Gillis, Val Verde County, Texas. See Figure 7 for location.

middle and late Paleozoic peaks are especially prominent in the Ouachitas. Unlike the situation in the Appalachian region, however, there is no structural or stratigraphic evidence of any major compressional orogeny within the Ouachita system in the early or middle Paleozoic, and the only significant siliciclastic wedge is the 1,500 ft of the Silurian Blaylock Sandstone in the eastern Ouachita Mountains. Also, with few exceptions, the level of metamorphism in the observable part of the Ouachitas is considerably lower than in the Appalachians. More likely, many of the Silurian-Devonian radiometric dates in the Ouachita system are part of a metamorphic "front" resulting from low-level thermal events associated with maximum subsidence of the leptogeosyncline. In some areas, however, deep subsidence and deformation within the geosyncline resulted in complete recrystallization and more intense metamorphism (Denison *et al.,* 1969). The late Paleozoic metamorphic episodes probably are related to the late orogenic and postorogenic uplift and denudation of the geosyncline and cratonic margin and the associated hydrothermal activity.

A zone of low-grade regional metamorphism (greenschist facies) and minor plutonism is recognized in the internal zone of the geosyncline. In

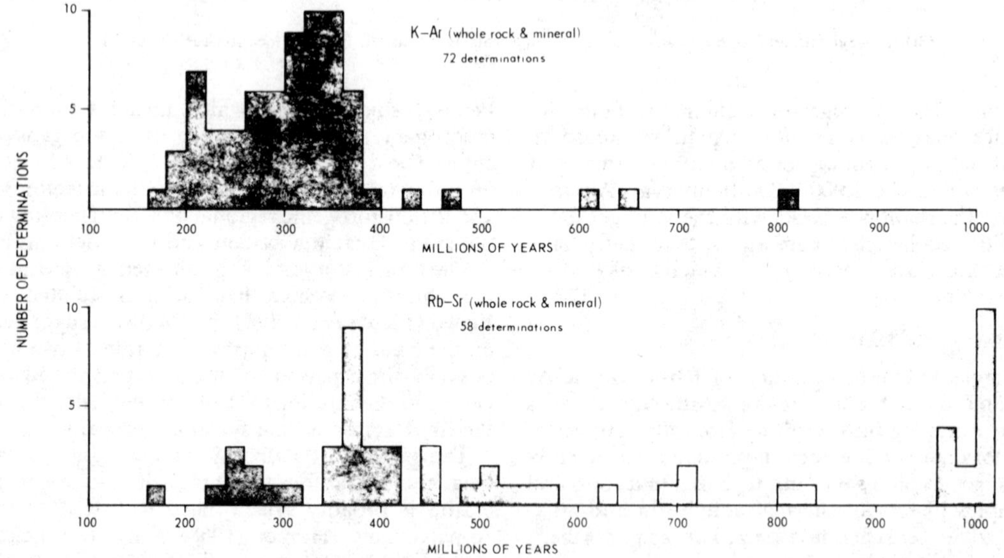

FIG. 15—Histogram of Rb-Sr and K-Ar dates from Ouachita foldbelt and cratonic margin in southern Mid-Continent. Data are from Shell Oil Company and published sources. See Figure 16 for significant data localities.

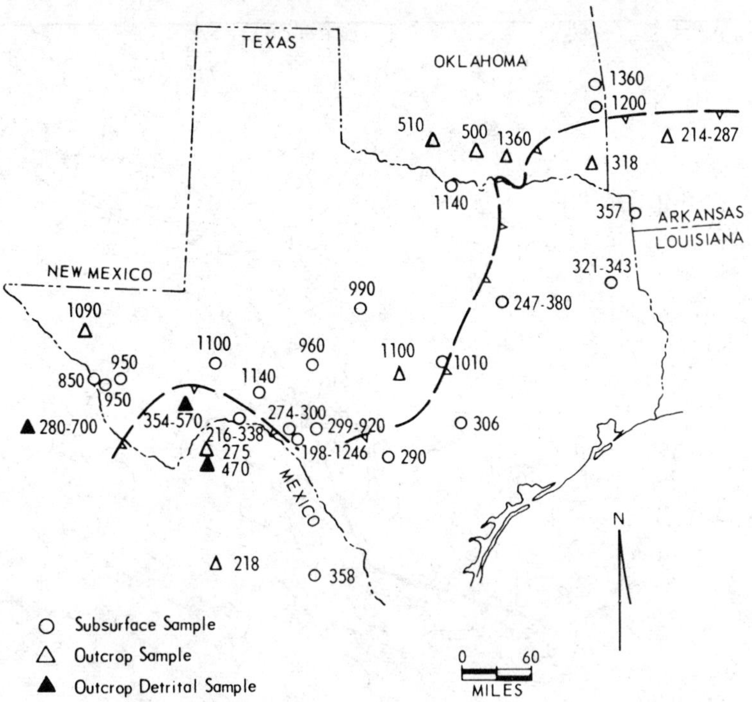

FIG. 16—Significant isotopic dates in Ouachita foldbelt and adjacent cratonic margin. Data are from Shell Oil Company and published sources.

some places, such as the Broken Bow-Benton uplift or "core" area of the southern Ouachita Mountains, the metamorphic maxima are coincident with anticlinoria and clearly are related to orogenic structure.

The perplexing condition in the Stewart well, where a metamorphosed basement yielding isotopic ages ranging from Ordovician through Permian is overlain by a relatively undisturbed Cambrian-Ordovician carbonate sequence, is not explained easily. As stated previously, several other wells in the general vicinity of the Stewart well, including the Shell 1 Gillis, penetrated low-grade metamorphosed cratonic carbonate and siliciclastic rocks of Paleozoic age. Obviously, the Devils River uplift is a very complex element and many variations in the degree of metamorphism of the sedimentary section can be demonstrated from one well to the next on this structure. The more competent dolomitic carbonate rocks apparently were more resistive to deformation than the impure carbonate and siliciclastic ones.

The radiometric data obtained from the Barrett and Stewart wells comprise the most comprehensive set of subsurface data of this type known to the writers. This geochronologic evidence is very significant in interpreting the evolution of the two uplifts and surrounding areas. However, the age distributions are wide ranging and complex, indicating considerable thermal overprinting, and it seems clear that there are still many unanswered questions concerning the geologic history of the Waco and Devils River uplifts.

GRAVITY

The structural grain of the southern Mid-Continent is well expressed on the regional isostatic gravity map (Fig. 17). All of the major structural uplifts are reflected by gravity maxima, whereas the deeper basins coincide, for the most part, with gravity minima.

Although the overall gravity pattern is broadly coincidental with basement structure, some broad-wavelength anomalies, one of which is also of high amplitude, obviously result from density variations deep within the basement rather than from structural relief. The most prominent of these anomalies is a 200-mi-long gravity minimum which is aligned east-west across central Texas from near the New Mexico border to near the Ouachita tectonic belt.

The structural grain of the Marathon-Ouachita orogenic belt is represented by a series of disconnected gravity minima and a paralleling, nearly

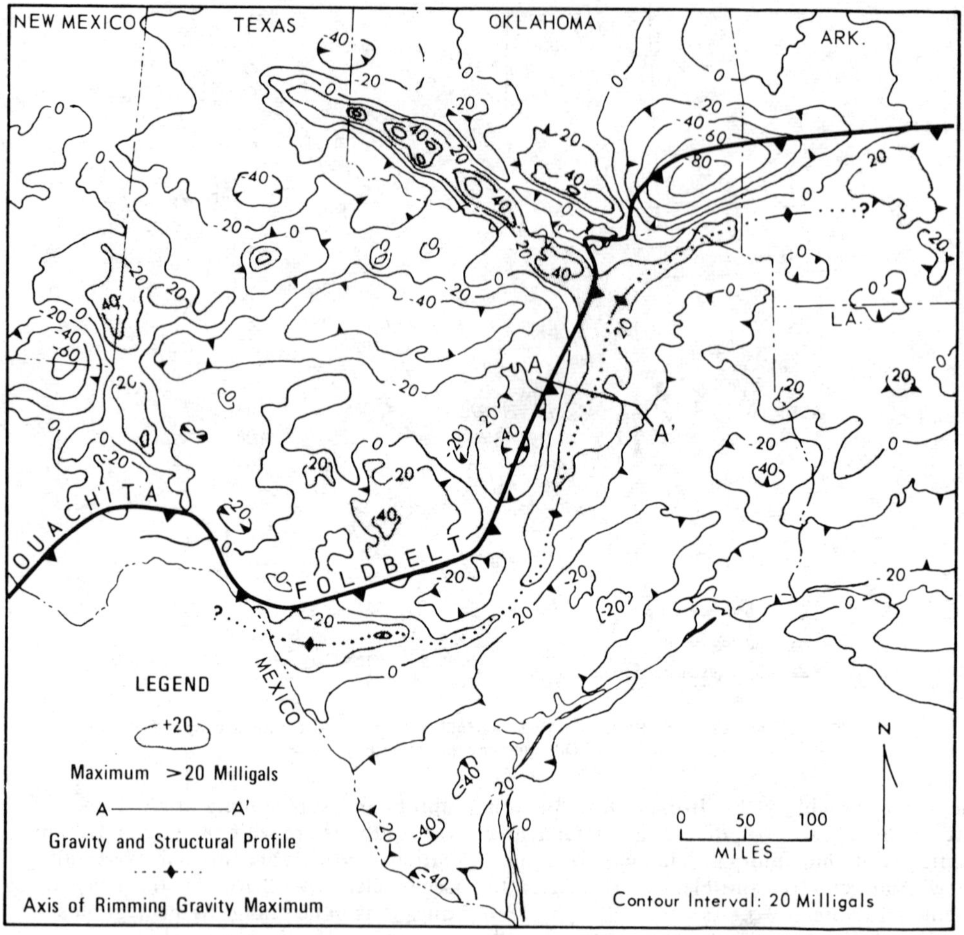

FIG. 17—Isostatic gravity map of southern Mid-Continent. Data are from Shell Oil Company.

continuous, linear gravity maximum. The gravity minima correspond in general to the cratonic marginal basins (or exogeosynclines) and presumably reflect the thick wedges of synorogenic and post-orogenic siliciclastic rocks of less than normal densities deposited in structurally depressed basement troughs adjacent to the frontal zone of the geosyncline.

One of the most significant gravity minima in North America is centered beneath the Ouachita Mountains. This minimum, covering an area of about 12,000 sq mi, including the adjacent Arkoma basin, is a 90-mgal Hayford-Bowie isostatic gravity anomaly. Lyons (1961, p. 96-97) believed that the density difference to account for this anomaly must extend into the crust and "only a downwarp of the crust into the mantle" will explain the minimum satisfactorily. He further suggested that a Moho low or closure, 59 km (37 mi) deep, coincides with this gravity minimum.

In the south and east, and generally paralleling the belt of gravity minima, is a complementary elongate gravity maximum, with many swells and saddles, which can be traced for 500 mi as an almost continuous feature from the Ouachita Mountains to near the Marathon area. The anomaly is asymmetric in many places, with a steeper gradient on the north and west toward the continent. The maximum locally reaches a positive 40-mgal value and corresponds fairly closely with the axis of greatest metamorphism in the interior zone of the geosyncline as defined by Flawn *et al.* (1961).

The Broken Bow-Benton anticlinorium or core area of older Palezoic low-grade metamorphics in the southern Ouachita Mountains coincides with this gravity maximum. Lyons (1961, p. 102) speculated that this 30 to 40-mi-wide structure co-

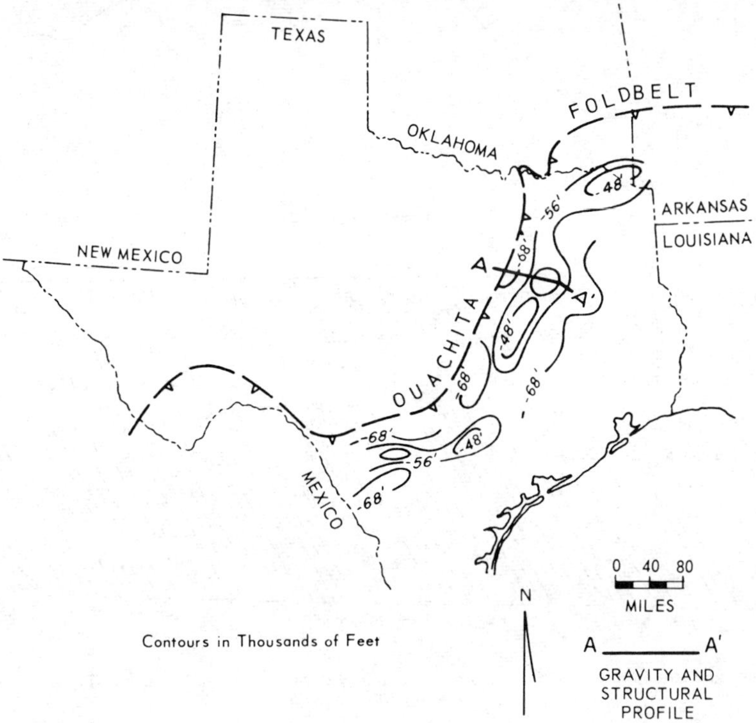

FIG. 18—Structural configuration of 3.0 g/cc (6.8-7.2 km/sec) "gabbroic" crustal layer. From interpretation of isostatic-gravity data by J. Tsiaperas.

incides with a basement ridge as well as an elevation of the Moho. Tsiaperas (Shell Oil Co., personal commun.) also interprets this feature to be a basement uplift but believes that the main gravity anomaly is the result of thickening of the basic intermediate crustal layer. Undoubtedly, a part of the anomaly also results from the presence of the relatively dense low-grade metamorphic rocks of the interior zone which are juxtaposed above the basement high and less dense clastic rocks in the foreland basins.

The smaller Waco and Devils River uplifts are superimposed on the broader rimming gravity maximum, suggesting a genetic relation. However, the major isostatic anomaly along this trend probably originates principally from deformation of the basic intermediate crustal layer at a depth ranging from 40,000 to 60,000 ft (Fig. 18).

Figure 19 is a model of the earth's crust that reproduces the observed gravity profile; assumed densities are shown and the assumption has been made that any deformation of the Moho level subsequently has been eliminated. The section also shows that the smaller and more local Waco uplift (15,000 to 20,000 ft) is superimposed on the steep west flank of the major intracrustal feature. A similar relation between near-surface basement structure and an intermediate layer crustal upwarp probably exists at the Devils River uplift on the south flank of the Val Verde basin.

Geologic Significance of Rimming Gravity Maximum

The principal isostatic gravity maximum paralleling the Ouachita-Marathon geosyncline is thought to result principally from the thickening of the dense (3.0 g/cc) basic intermediate crustal layer. A similar gravity anomaly has been described from the Green Mountain anticlinorium in the northern Appalachians by Bean (1953) and Diment (1968; Fig. 20). The Green Mountain structure is one of a series of linear, Precambrian-cored, faulted anticlinoria that extend from Newfoundland to the southern Appalachians, *i.e.*, the Long Range of Newfoundland, the Green Mountains and Berkshire Hills of New England, the Hudson-New Jersey Highlands, and the Blue Ridge (Fig. 21), and which lie just east or southeast of the Appalachian Valley and Ridge province or its equivalent. In many places this chain of 900 to

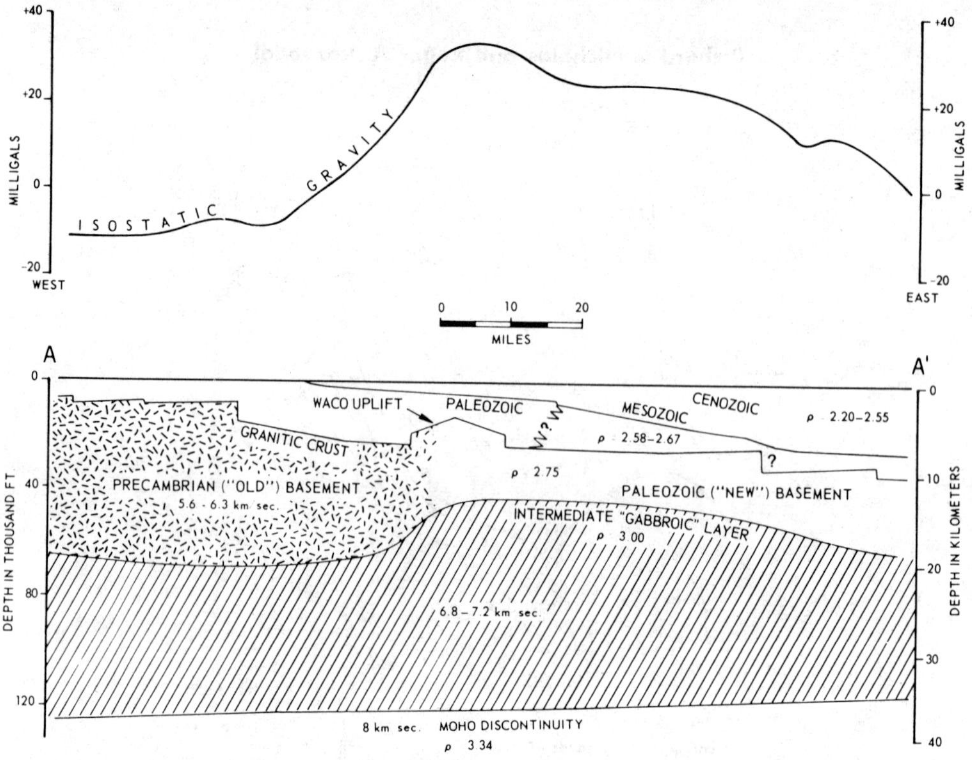

FIG. 19—Gravity and structural profile *A-A'* across Waco uplift. Interpretation by J. Tsiaperas. For profile location, see Figures 17 and 18.

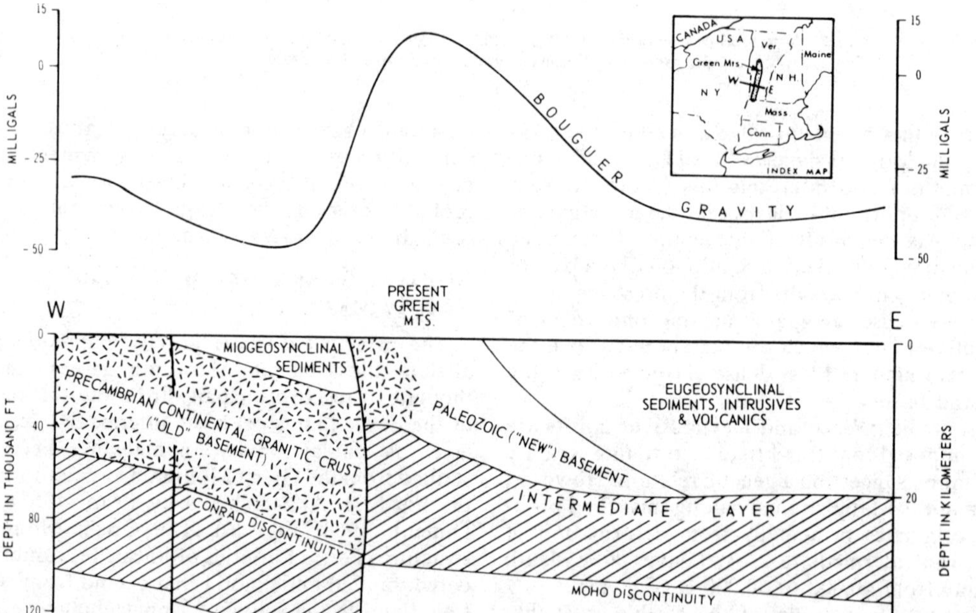

FIG. 20—Bouguer gravity and schematic crustal configuration across Green Mountains in Vermont and Massachusetts (modified from Diment, 1968). Main gravity maximum is interpreted to be result of Middle Ordovician uplift of dense sub-Conrad rocks coincident with subsidence of miogeosyncline and eugeosyncline.

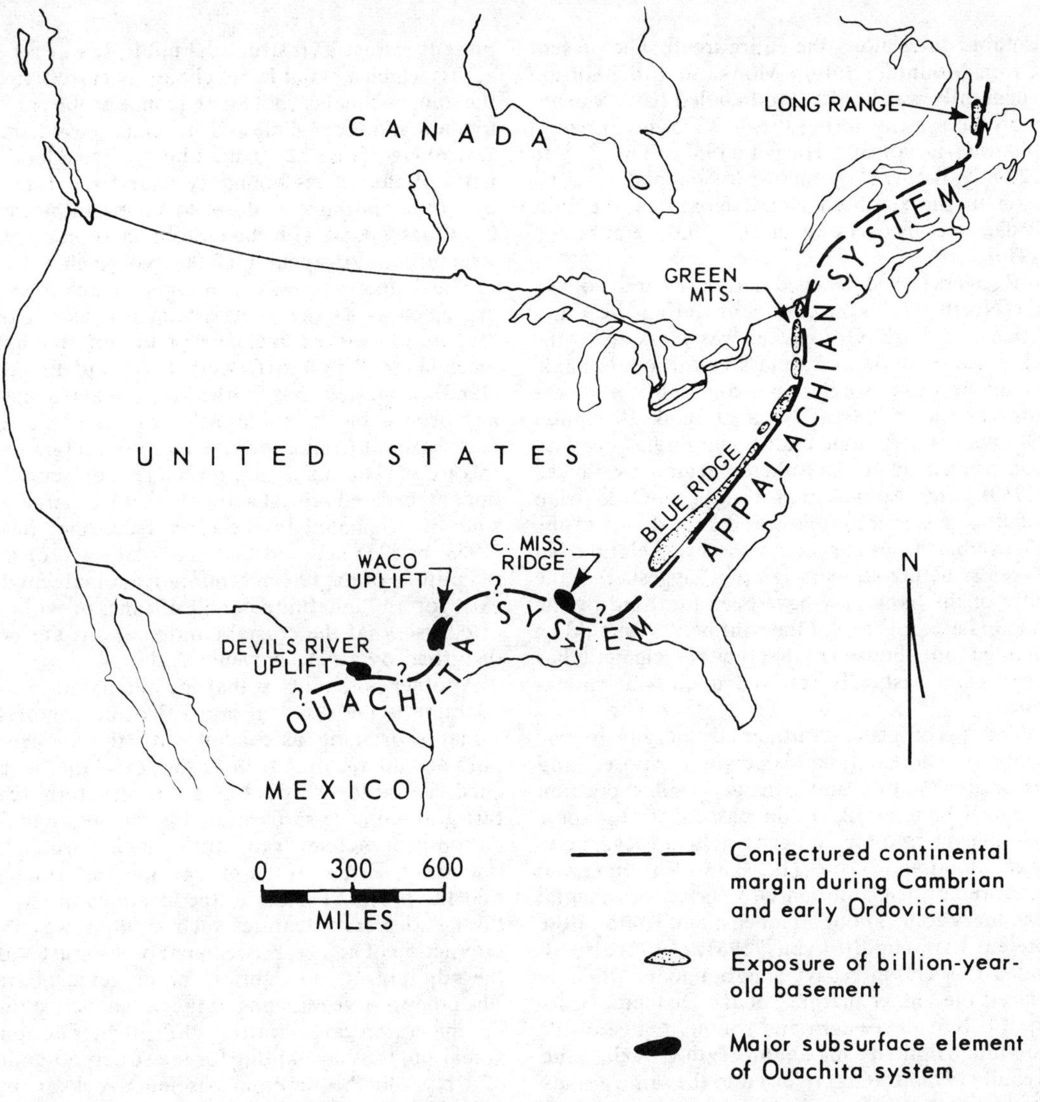

FIG. 21—Eastern and southern margins of North America during Cambrian and Early Ordovician. Continental border in Appalachians from Rodgers (1968). In Ouachita system, billion-year-old basement is documented only on Devils River uplift but is believed to be present on both Waco uplift and Central Mississippi ridge.

1,100 m.y.-old Precambrian basement ridges approximately marks the boundary between the internal and external zones of the orogen and separates the "eugeosynclinal" rocks on the east from the "miogeosynclinal" facies which generally lie west or northwest. Diment (1968, p. 401) believed that the linear positive gravity anomalies, which are nearly coincident with the Precambrian ridges in the central and northern Appalachians, mark a fundamental crustal boundary for the length of the Appalachian system. Early Paleozoic continental crust may have terminated just east of this boundary. In the southern Appalachians the major positive gravity anomaly centers many tens of miles southeast of the Blue Ridge Precambrian outcrops. This suggests considerably more thrusting of the Blue Ridge in the south than farther north, where the basement anticlinoria and gravity maxima coincide.

The ancestral (early and middle) Paleozoic origin of these uplifts is implied clearly by stratigraphic and structural relations and confirmed by isotopic dates along the entire trend. The present Green Mountain-Sutton Mountain anticlinorium apparently was uplifted and cooled from a metamorphic system during the 425-375-m.y. interval (Silurian-Devonian; Harper, 1968). The 375 to 320-m.y. dating (Devonian-Mississippian) is common in the southern Appalachians in the Blue Ridge province (Long *et al.,* 1959; Carpenter, 1970).

Rodgers (1968) believed that the eastern edge of the North American continent during the Cambrian and Early Ordovician was marked by the abrupt termination of a massive carbonate bank, or miogeocline, similar in many respects to the edge of today's Great Bahama Bank. A similar Ordovician carbonate craton margin has been reconstructed in the Cordilleran region by Lowell (1960). The Appalachian Cambrian-Ordovician bank edge seems to follow and lie just east of the Precambrian anticlinoria; this close relation between structure and stratigraphy suggests that the edge of the bank may have been localized by the original eastern edge of the continent. It should be pointed out, however, that neither element has been palinspastically restored in this interpretation.

The preorogenic continental margin in the southern Mid-Continent is nowhere exposed, and its original nature and paleogeographic position can only be inferred. On the basis of stratigraphic relations in cratonic and geosynclinal rocks, however, Cloud and Barnes (1957) and Wilson (1954) interpreted the Cambrian-Ordovician continental boundary to be about 50 mi east and south of the present Llano uplift. Lyons (1961, p. 102) also alluded to a craton-geosyncline boundary when he placed the "inner margin" of the Ouachita structural belt at the eastern and southern edge of the continuous gravity maximum feature. Keller and Cebull (1973, p. 1662) referred to the same gravity maximum anomaly and believed the anomaly is caused by the "now-inactive core or root" of a late Paleozoic cordilleran-type plate-tectonic-subduction system.

We place the preorogenic Paleozoic cratonic margin of the southern Mid-Continent farther south and east than the boundary shown by these authors, and with a somewhat different model (Figs. 5, 19). The rimming intracrustal gravity maximum appears to be analogous to the Green Mountain anomaly in the northern Appalachians in many respects, although the Green Mounains were mobilized more highly during major Taconian and Acadian orogenies. On this basis, it seems reasonable to infer that the seaward termination of the cratonic preorogenic carbonate miogeocline was localized along or near the outer margin of the present intracrustal structural uplift. It was a belt across which a major facies change occurred, from the miogeoclinal carbonate sequence of the craton to a much thinner shale and siliceous geosynclinal assemblage (Fig. 22). Undoubtedly considerable parts of this facies boundary were fixed further by contemporaneous down-to-the-basin normal faults associated with the cratonic margin crustal structure and deepening of the geosyncline.

The Paleozoic cratonic margin in this area is expressed as a gravity maximum because of the "fossilized" excess thickness of the intermediate basic layer. Bird and Dewey (1970) and Diment (1968) suggested that in the Appalachians there has been some lateral transfer of dense mantle materials out from under the miogeosyncline into this ridge. The maximum probably represents an uncompensated crustal anomaly that has persisted since it originated in the early Paleozoic. Bean (1953, p. 523) believed that a crustal warp of the magnitude of the Green Mountain anomaly might exist for an "indefinite period of time, providing, of course, that the crustal configurations are not disturbed by a new orogenic cycle."

Another possibility is that the anomalous crustal ridge is the result of late Paleozoic compressional shortening associated with the orogenic phase of the tectonic belt. In that case, the thickened crustal layer would be a late structural feature and not necessarily related to the preorogenic craton-geosyncline stratigraphic or the structural boundary. Zen (1972), for example, considered that the process leading to the locations of structural and gravity features such as these was the interaction of an oceanic segment of the crust with the adjoining craton during the orogenic phase. The compressive plunging of the oceanic crust under the craton caused rafting of the lighter cratonal margin, thus accounting for the structural uplift of 1-b.y.-old Precambrian basement rocks. Continued compression then led to later thrusting of the uplifted cratonal margin.

Structural Interpretation

It appears that the Waco and Devils River uplifts are in similar structural positions and have similar structural histories. However, the data are insufficient to determine definitely the structural style of either. The Devils River uplift on the south flank of the Val Verde basin may be a differentially uplifted autochthonous or rooted block (Fig. 23). However, in the Shell 1 Barrett on the Waco uplift, the section of cratonic carbonate rocks with extensive shear and flowage structures beneath allochthonous, weakly metamorphosed clastic rocks ("Ouachita facies") of undetermined age derived from the internal zone of the geosyncline indicates

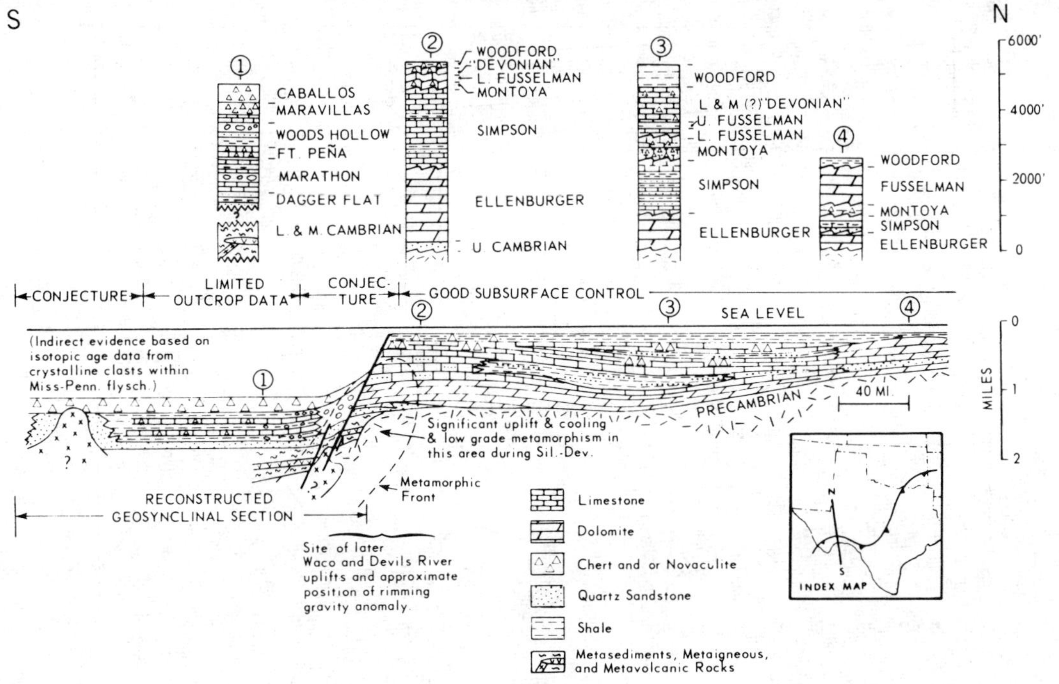

FIG. 22—Restored section of preorogenic (Upper Cambrian-Devonian) rocks from southern craton to Marathon area, Texas.

that the structure is a type transitional between the clearly autochthonous brittle basement blocks of the craton and the definite allochthonous overthrusts of the Ouachita orogenic belt. The general asymmetry of the anticlinal structure toward the continent and its linear continuity paralleling the orogenic belt suggest that the structure is at least partly involved in the compressional system. However, because (1) cratonic facies and the underlying Precambrian crystalline basement (which has subsequently been remobilized and thermally overprinted) are involved in the deformation, and (2) much of the Waco uplift appears to have been accentuated by the insertion of a buried basement block under older, overlying thrusts, the uplift is interpreted to be parautochthonous or semirooted, and probably has moved laterally only a minor distance.

In some orogens, the boundary region between the external or frontal and the internal zone is occupied by a belt along which old basement and its cover are brought up by late orogenic thrusting and folding (such as in the Blue Ridge of the Appalachians). In Alpine terminology, the structures of this zone are called "external massifs" and show basement ages that still can be clearly preorogenic (cratonic basement) or that are transitional to younger metamorphic ages of the internal zone (Armstrong, 1966). The Devils River and Waco uplifts possibly may be interpreted as fragmentary orogenic external massifs, localized along or near the cratonic margin in a belt across which there was a major preorogenic facies change from the miogeoclinal carbonates of the craton to a much thinner shale and siliceous geosynclinal assemblage. The Central Mississippi ridge on the south flank of the Black Warrior basin may be a similar type of structure (Vernon, 1971).

Hydrocarbon Potential of Cratonic Margin

No hydrocarbon shows were noted in the Shell 1 Stewart. A drill-stem test from 2,384 to 2,410 ft recovered 26 bbl of fresh water with a chloride content of 200 ppm from the Ellenburger (Lower Ordovician). From a drilling break between 5,646 and 5,648 ft, there was a salt water flow for 30 minutes with a chloride content of 47,900 ppm. Coarsely crystalline calcite in the samples from this interval may indicate the presence of a fault.

In the Shell 1 Barrett, methane shows on the hot-wire mud logger were prevalent in the interval 13,400 to 13,800 ft, and a 20-ft flare burned for about 30 minutes during a trip at 13,911 ft. We assume most of this gas came from near the top of

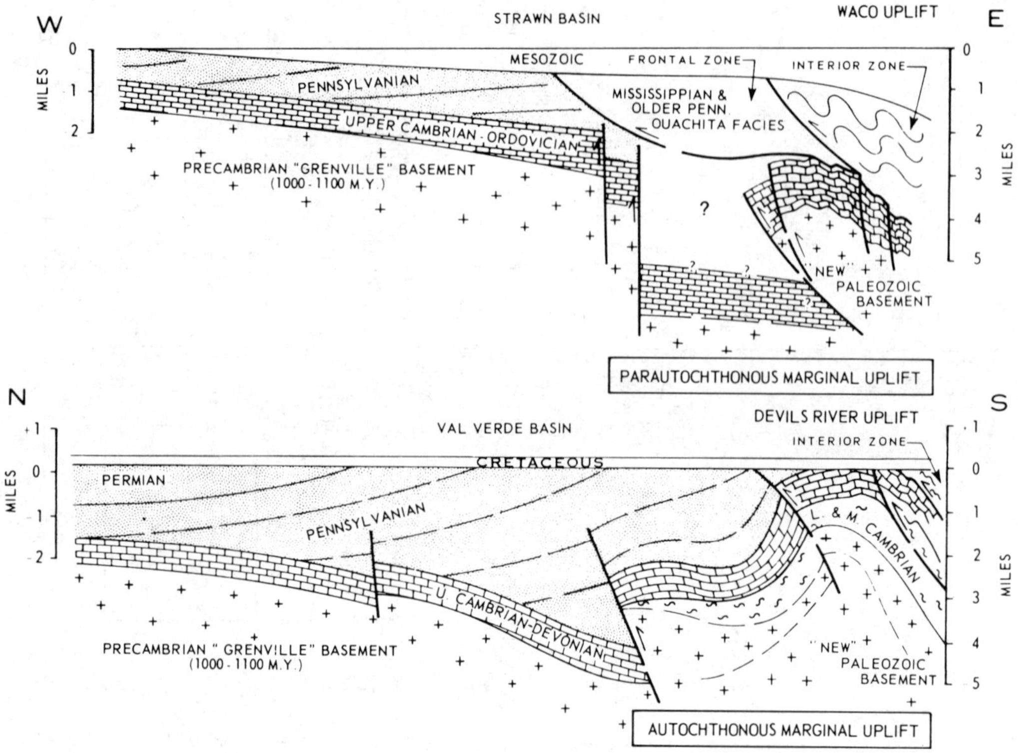

Fig. 23—Schematic cross-sections showing interpretive structural styles of Waco and Devils River uplifts. Refer to Figure 6 for general area of sections.

the marble section. Hot-wire methane shows were recorded between 19,080 and 19,440 ft, but at the same time an increase in salinity occurred. The well began flowing salt water during a trip at 19,701 ft, and chlorides increased to 42,000 ppm. An open-hole drill-stem test between 15,027 and 18,662 ft yielded 53 bbl of drilling fluid with 34,500 ppm chloride, and no trace of hydrocarbons.

The complexity of structure, extensive low-grade thermal and dynamic metamorphism, and subsequent increase in diagenetic effects result in poor reservoirs and make a mobile cratonic-margin province and adjacent geosyncline a habitat generally unfavorable for hydrocarbons. In addition, the Val Verde basin in the foreland of the western part of the Ouachita system contains large quantities of carbon dioxide. However, (1) major structural traps and (2) the possibility of some thermally bypassed areas suggest that potential still exists for significant accumulations of gas. Opportunities for discovering such accumulations are probably best on early formed structures in the "cool" areas where the paleothermal history is apparently much more favorable for the preservation of hydrocarbons and reservoirs. Other prospective areas are the poorly defined internal "successor" basins or platforms on the south side of the Ouachita Mountains (Vernon, 1971). These basins were formed behind or within the geosyncline after the main phase of tectonism and plutonism had finished and could be very prospective hydrocarbon provinces.

Summary

The recent seismic and drilling activity by Shell Oil Company in the Ouachita structural belt in Texas has confirmed the presence of at least two major positive elements marginal to the Ouachita tectonic belt, south and east of the foreland basins. The ancestral nature of these uplifts is demonstrated by isotopic dating. These basement structures developed along or near the site of the preorogenic cratonic margin and, in the case of the Waco uplift, 25 mi or more south and east of the leading edge of the frontal thrusts. The exact nature of these cratonic platform-edge structures is not known. Subsurface data are so limited that we cannot tell whether the structures existed during the preorogenic phase as low-relief marginal welts or as simple, relatively passive, structural and depositional terraces coincident with the cratonic margin. The time of formation and the type of structure are important in determining the possibilities for migration and trapping of hydrocarbons in this area.

Problems

Some fundamental geologic problems need to be resolved before the history of the Ouachita geosyncline is clearly understood.

1. The nature of the junction with the Appalachian system is still unresolved. Furthermore, the extent and direction of the orogen south and west of the Marathon area into northeast Mexico essentially are undefined. However, geologic and isotopic data from widely spaced and isolated Paleozoic outcrops in northeastern and eastern Mexico suggest that the Ouachita System extended southward in the Huastecan structural belt of eastern Mexico, possibly connecting with the Northern Central American Paleozoic foldbelt. De Cserna (1971) recognized both Taconian and late Paleozoic deformation along the Huastecan belt. Denison (1970) also studied rocks in northeastern Mexico which are petrographically "interior zone" and which give radiometric dates ranging from Precambrian to late Paleozoic.

Meyerhoff (1967, p. 224) contended that the Northern Central American orogen cannot be a simple extension of the Ouachita foldbelt because of the polarity reversal, *i.e.,* the foreland of the Northern Central American orogen is apparently on the north and east whereas in the Ouachita system the foreland is on the north and west. However, Paleozoic outcrops in Mexico and adjacent Central America are limited and widely scattered, and the foreland-geosyncline relations are questionable. Furthermore, the Appalachians have been interpreted as a two-sided symmetrical system (Williams, 1964; King, 1951), and Aubouin (1965) reported that bilateral symmetry is common in other geosynclines. In addition, the complex and divergent sedimentary and structural polarities in some plate-tectonic models described by Dewey and Bird (1970) suggest that the apparent polarity reversal between the Ouachita and the Northern Central American orogen does not necessarily preclude a direct connection between these two foldbelts.

2. Only a small part of what is apparently a much wider orogenic belt is known. Therefore, the concept of a single depositional basin evolving through long and simple "forming and filling" stages (Thomson and McBride, 1964) may be more apparent than real. The exposed rocks of the Ouachita and Marathon salients and the rocks drilled to date may not be typical axial sediments of the geosyncline.

3. In terms of global plate tectonics, the preorogenic (Late Cambrian-Devonian) phase of the Ouachita system differs mainly in timing from the model proposed by Bird and Dewey (1970) for the early and middle Paleozoic evolution of the northern Appalachians. Tectonism and accompanying plutonism were much less severe in the known part of the Ouachita system during the early and middle Paleozoic than in the Appalachians, and no exogeosynclines older than Mississippian developed in the Ouachitas. Furthermore, the open-sea tectonic setting of the leptogeosynclinal Silurian and Devonian siliceous deposits in the Ouachita system obviously differed from the supposed Devonian continental collision and clastic wedge of the northern Appalachians. Whereas Taconian and Acadian compressional tectonics, clastic wedges, and associated plutonism and metamorphism are common to many parts of the Appalachians, all evidence indicates that the Ouachita depositional basin developed primarily from earliest Paleozoic through Devonian time as an extensional feature. We believe that much of the low-grade metamorphism and thermal activity along the cratonic margin were associated with the maximum deepening of the basin during the Silurian-Devonian leptogeosynclinal phase. Possibly the Ouachita embayment was a simple extension of the Appalachian system but was too far removed from the eastern North American subduction zone to be affected by compressional deformation until the late Paleozoic. If so, then the plate-tectonic history of the two orogens differs primarily in the timing of events rather than in the absence of continent-continent collision in the Ouachitas, which was suggested by Keller and Cebull (1973) and Cebull and Keller (1974).

The pronounced reentrants in the cratonic margin in the Ouachita and Marathon embayments during the extensional preorogenic phase of the geosyncline may reflect the existence of triple-plate junctions in those areas, of the type described by Burke and Dewey (1973). These writers named the southern Oklahoma-Ouachita indentation the "Dallas junction." The "failed arms," striking at high angles to the Ouachita system, were reactivated during the late Paleozoic compressional orogeny. Major continental structural elements developed in these older rift basins in the form of the Central Basin Platform complex in West Texas and the Amarillo-Wichita and Arbuckle elements in southern Oklahoma.

4. Any reconstruction of the Ouachita system also must take into account the presence of relatively undeformed late Paleozoic (Pennsylvanian-Permian) continental and shallow-marine sediments in the subsurface of the Gulf coastal plain south of the Ouachita Mountains (Vernon, 1971).

Within the constraints of available data, the post-Desmoinesian Ouachita foldbelt in this area must be confined to a relatively narrow corridor, less than 75 mi across, between the Arkoma basin of eastern Oklahoma and western Arkansas and the late Paleozoic platform or successor basin in northeastern Texas, southern Arkansas, and northern Louisiana.

Speculatively, the Ouachita system and the associated external massifs may be manifestations of a secondary arcuate foldbelt related to an Appalachian subduction zone and flanked on the south by a relatively stable late Paleozoic intermontane or successor basin or basins. The primary Appalachian orogenic zone might extend in a more or less straight line southwestward and form the southeastern margin of this intermontane area (H. Widmer, Shell Oil Co., personal commun.). If these assumptions are valid, deep-water marine conditions probably were present over large parts of the present Gulf of Mexico during early and middle Paleozoic time. Most of these lower and middle Paleozoic sediments may have been "erased" or highly deformed by pre-late Paleozoic movements associated with the closing of a proto-Atlantic. Possibly a few stable areas, analogous to the Ordovician and Silurian-Devonian subcrops of northern Florida, escaped destruction (Woods and Addington, 1973). Subsequently, a thick section of pre-Desmoinesian clastic sediments was deposited, deformed, and then covered by shallow marine Desmoinesian shelf sediments. The Desmoinesian section apparently is overlain unconformably by mostly nonmarine Late Pennsylvanian and Permian siliciclastic rocks (Vernon, 1971; Meyerhoff, 1973).

The relatively undeformed Desmoinesian shallow-marine shelf sediments in this area are similar to the Gaptank Formation (Desmoinesian-Missourian-Virgilian) of the Marathon area. The Gaptank Formation is a postorogenic molasse-like facies that records the final stage of basin filling in the Marathons.

Few data are available to support these conjectures, and information concerning the nature and regional distribution of a stable Paleozoic platform and successor basin on the apparent seaward side of the Ouachita foldbelt must await additional deep drilling in the Gulf basin.

References Cited

Adams, J. E., 1954, Mid-Paleozoic paleogeography of central Texas: Shale Shaker, v. 4, no. 6, p. 238-241.

Amsden, T. W., and T. L. Rowland, 1967, Silurian-Devonian relationships in Oklahoma, *in* International Symposium on the Devonian System, v. 2: Alberta Soc. Petroleum Geologists, p. 949-960.

Armstrong, R. L., 1966, K-Ar dating of plutonic and volcanic rocks in orogenic belts, *in* Potassium-argon dating: New York, Springer-Verlag, p. 117-133.

Aubouin, J., 1965, Geosynclines: New York, Elsevier, 335 p.

Bass, M. N., and G. Ferrara, 1969, Age of adularia and metamorphism, Ouachita Mountains, Arkansas: Am. Jour. Sci., v. 267, p. 491-498.

Bean, R. J., 1953, Relation of gravity anomalies to the geology of central Vermont and New Hampshire: Geol. Soc. America Bull., v. 64, p. 509-537.

Bird, J. M., and J. F. Dewey, 1970, Lithosphere plate-continental margin tectonics and the evolution of the Appalachian orogen: Geol. Soc. America Bull., v. 81, p. 1031-1060.

Burke, K., and J. F. Dewey, 1973, Plume-generated triple junctions: key indicators in applying plate tectonics to old rocks: Jour. Geology, v. 81, p. 406-433.

Carpenter, R. H., 1970, Metamorphic history of the Blue Ridge province of Tennessee and North Carolina: Geol. Soc. America Bull., v. 81, p. 749-762.

Cebull, S. E., and G. R. Keller, 1974, Plate tectonics and the Ouachita system in Texas, Oklahoma, and Arkansas: reply: Geol. Soc. America Bull., v. 85, p. 147-148.

Cloud, P. E., and V. E. Barnes, 1957, Early Ordovician sea in central Texas, *in* Paleoecology: Geol. Soc. America Mem. 67, p. 163-214.

Cserna, Z. de, 1971, Taconian (early Caledonian) deformation in the Huasteca structural belt of eastern Mexico: Am. Jour. Sci., v. 271, p. 544-550.

Denison, R. E., 1970, Basement rock framework of parts of Texas, southern New Mexico and northern Mexico (summ.), *in* Geological framework of the Chihuahua tectonic belt—Symposium in honor of Ronald K. DeFord: West Texas Geol. Soc., p. 4-6.

——— G. E. Kenny, W. H. Burke, Jr., and E. A. Hetherington, Jr., 1969, Isotopic ages of igneous and metamorphic boulders from the Haymond Formation (Pennsylvanian), Marathon basin, Texas, and their significance: Geol. Soc. America Bull., v. 80, p. 245-256.

Dewey, J. F., and J. M. Bird, 1970, Mountain belts and the new global tectonics: Jour. Geophys. Research, v. 75, p. 2625-2647.

Dietz, R. S., and J. C. Holden, 1967, Miogeoclines in space and time: Jour. Geology, v. 74, p. 566-583.

Diment, W. H., 1968, Gravity anomalies in northwestern New England, Chap. 30, *in* Studies of Appalachian geology: Northern and maritime: New York, Interscience, p. 399-413.

Emery, K. O., and E. Uchupi, 1972, Western North Atlantic Ocean: topography, rocks, structure, water, life, and sediments: AAPG Mem. 17, 532 p.

Fish, J. E., 1972, Crustal structure of the Texas Gulf Coastal Plain: Corpus Christi Geol. Soc. Bull., v. 11, no. 8, p. 1-29.

Flawn, P. T., 1959, Devils River uplift, *in* Guidebook of the geology of the Val Verde basin: West Texas Geol. Soc., p. 74-78.

——— A. Goldstein, Jr., P. B. King, and C. E. Weaver, 1961, The Ouachita system: Texas Univ. Pub. 6120, 401 p.

Folk, R. L., 1973, Evidence for peritidal deposition of Devonian Caballos Novaculite, Marathon basin, Texas: AAPG Bull., v. 57, p. 702-725.

Galley, J. E., 1958, Oil and geology in the Permian basin of Texas and New Mexico, *in* Habitat of oil: AAPG, p. 395-446.

Ham, W. E., 1959, Correlation of pre-Stanley strata in the Arbuckle-Ouachita Mountain regions, *in* The geology of the Ouachita Mountains—a symposium: Dallas Geol. Soc. and Ardmore Geol. Soc., p. 71-86.

——— 1963, Basement rocks and structural evolution of southern Oklahoma—a summary, *in* Field Conf. Guidebook: Ardmore Geol. Soc., p. 11-18.

——— and J. L. Wilson, 1967, Paleozoic epeirogeny and orogeny in the Central United States: Am. Jour. Sci., v. 265, p. 332-407.

——— R. E. Denison, and C. A. Merritt, 1964, Basement rocks

and structural evolution of southern Oklahoma: Oklahoma Geol. Survey Bull. 93, 302 p.

Harper, C. T., 1968, Isotopic ages from the Appalachians and their tectonic significance: Canadian Jour. Earth Sci., v. 5, no. 1, p. 49-59.

Hatcher, R. D., Jr., 1972, Developmental model for the southern Appalachians: Geol. Soc. America Bull., v. 83, p. 2735-2760.

Hendricks, T. A., 1959, Structure of the frontal belt of the Ouachita Mountains, *in* The geology of the Ouachita Mountains—a symposium: Dallas Geol. Soc. and Ardmore Geol. Soc., p. 44-56.

Hurley, P. M., and J. R. Rand, 1968, Continental radiometric ages, *in* The sea, v. 4, New concepts of sea floor evolution: New York, Wiley Interscience, p. 583.

Keller, G. R., and S. E. Cebull, 1973, Plate tectonics and the Ouachita system in Texas, Oklahoma, and Arkansas: Geol. Soc. America Bull., v. 84, p. 1659-1666.

King, P. B., 1937, Geology of the Marathon region, Texas: U.S. Geol. Survey Prof. Paper 187, 148 p.

——— 1951, The tectonics of middle North America: Princeton, New Jersey, Princeton Univ. Press, 203 p.

——— 1969, The tectonics of North America—a discussion to accompany the tectonic map of North America, scale 1:5,000, 000: U.S. Geol. Survey Prof. Paper 628, 95 p.

Kulp, J. L., and F. D. Eckelmann, 1961, Potassium-argon isotopic ages on micas from the southern Appalachians, *in* Geochronology of rock systems: New York Acad. Sci. Annals, v. 91, p. 408-416.

Long, L. E., J. L. Kulp, and F. D. Eckelmann, 1959, Chronology of major metamorphic events in the southeastern United States: Am. Jour. Sci., v. 257, no. 8, p. 585-603.

Lowell, J. D., 1960, Ordovician miogeosynclinal margin in central Nevada: 21st Internat. Geol. Cong., Copenhagen, Rept., pt. 7, p. 7-17.

Lyons, P. L., 1961, Geophysical background of Arkoma basin tectonics, *in* Symposium on the Arkoma basin: Tulsa Geol. Soc. Digest, v. 29, p. 94-104.

McBride, E. F., ed., 1969, A guidebook to the stratigraphy, sedimentary structures and origin of flysch and pre-flysch rocks of the Marathon basin, Texas: Dallas Geol. Soc., 104 p.

——— and A. Thomson, 1970, The Caballos Novaculite, Marathon region, Texas: Geol. Soc. America Spec. Paper 122, 129 p.

McGlasson, E. H., 1967, The Siluro-Devonian of west Texas and southeastern New Mexico, *in* International Symposium on the Devonian System, v. 2: Alberta Soc. Petroleum Geologists, p. 937-948.

Meyerhoff, A. A., 1967, Future hydrocarbon provinces of Gulf of Mexico-Caribbean region: Gulf Coast Assoc. Geol. Socs. Trans., v. 17, p. 217-260.

——— 1973, Late Paleozoic of western Gulf Coastal plain, *in* A study of Paleozoic rocks in Arbuckle and western Ouachita Mountains of southern Oklahoma: Shreveport Geol. Soc. Guidebook, p. 31-37.

Paine, W. L., and A. A. Meyerhoff, 1970, Gulf of Mexico basin: interactions among tectonics, sedimentation, and hydrocarbon accumulation: Gulf Coast Assoc. Geol. Socs. Trans., v. 20, p. 5-44.

Rodgers, J., 1968, The eastern edge of the North American continent during the Cambrian and Early Ordovician, Chap. 10, *in* Studies of Appalachian geology: northern and maritime: New York, Interscience, p. 141-149.

Rozendal, R. A., and W. S. Erskine, 1971, Deep test in the Ouachita structural belt of central Texas: AAPG Bull., v. 55, p. 2008-2017.

Thomas, W. A., 1972, Regional Paleozoic stratigraphy in Mississippi between Ouachita and Appalachian Mountains: AAPG Bull., v. 56, p. 81-106.

Thomson, A., and E. F. McBride, 1964, Summary of the geologic history of the Marathon geosyncline, *in* The filling of the Marathon geosyncline—Symposium and guidebook: Soc. Econ. Paleontologists and Mineralogists Permian Basin Sec., p. 52-60.

Vernon, R. C., 1971, Possible future petroleum potential of pre-Jurassic, western Gulf basin, *in* Future petroleum provinces of the United States: AAPG Mem. 15, p. 954-979.

Widess, M. B., and G. L. Taylor, 1959, Seismic reflections from layering within the Precambrian basement complex, Oklahoma: Geophysics, v. 24, p. 417-425.

Williams, H., 1964, The Appalachians in northeastern Newfoundland—a two-sided symmetrical system: Am. Jour. Sci., v. 262, p. 1137-1158.

Wilson, J. L., 1954, Late Cambrian and Early Ordovician trilobites from the Marathon uplift, Texas: Jour. Paleontology, v. 28, p. 349-385.

——— and O. P. Majewske, 1960, Conjectured middle Paleozoic history of central and west Texas: Texas Univ. Pub. 6017, p. 65-86.

Woods, R. D., and J. W. Addington, 1973, The Northern Gulf Basin Paleozoic frontier, *in* A study of Paleozoic rocks in Arbuckle and western Ouachita mountains of southern Oklahoma: Shreveport Geol. Soc. Guidebook, p. 38-41.

Zen, E-A., 1972, The Taconide zone and the Taconic orogeny in the western part of the northern Appalachian orogen: Geol. Soc. America Spec. Paper 135, 72 p.

THE AMERICAN ASSOCIATION OF PETROLEUM GEOLOGISTS BULLETIN
Vol. 62, No. 9, September, 1978

GEOLOGIC NOTES

Tectonics of Brooks Range Ophiolites, Alaska[1]

DIETRICH ROEDER[2] and CHARLES G. MULL[3]

Abstract Ophiolites in the Alaska Brooks Range of probable Paleozoic crustal age are erosional remnants of stacked thrust sheets and are composed of two complexes formed at different sites and tectonically assembled during orogeny. An upper complex is an incomplete ophiolite suite with mafic and ultramafic cumulates and small remnants of a sheeted-dike complex. A lower thrust sheet consists of basaltic flows which probably were extruded in the lower plate during the Jurassic to Neocomian obduction of the ophiolites over the Arctic Alaskan craton and its late Paleozoic south margin composed of cratonic shelf edges and rifts or basins with thinner crust.

INTRODUCTION

Ophiolite bodies in the Brooks Range of northern Alaska (Tailleur, 1970) form the highest unit of a stack of unmetamorphosed thrust sheets which was transported to the north, folded, and locally intruded by felsic plutons, possibly produced by anatexis along an intracrustal shear surface. These ophiolite bodies are significant to the hydrocarbon explorationist in the two following respects.

1. Their setting within the Brooks Range is clearly that of an obduction or collision orogene (Dewey and Bird, 1970), but the location is in the Eastern Cordillera, an area believed almost unanimously to contain only intracratonic or foreland thrust belts (Price and Mountjoy, 1970; Roeder, 1973; Burchfiel and Davis, 1975). Although a correlation between plate-tectonic setting and hydrocarbon potential is not established for all types of orogenes, this exception should be significant.

2. The Brooks Range ophiolites comprise widespread sheets of basalt in flow contact with foreland rocks, apparently extruded just prior to obduction. Although major aspects of these mafic flows are not yet known, they contain significant data about the architecture of the cratonic margin covered by the obduction.

The purpose of this paper is to introduce the second of the two aspects because it appears to have widespread application in other collision belts.

REGIONAL SETTING

In the Brooks Range (Fig. 1), about eight large (200 to 700 sq km) and 10 small (1 to 10 sq km) mafic and ultramafic igneous bodies collectively may be interpreted as an ophiolite complex. These bodies, arranged in two diffuse strike belts, are erosional remnants of the highest unit within a stack of northward-transported thrust plates. Similar in cross section and evolution to other collision belts such as the Alps, the imbricate stack is warped into broad anticlinoria and synclinoria. It contains, from south to north, a steeply south-dipping root zone, an anticlinal belt with the foreland emerging in windows (Dutro et al, 1976; Mull et al, 1976), a major synclinorium, and a frontal belt of smaller detached folds. The ophiolites crop out discontinuously along the root zone and within synclinal and axial depressions of the regional synclinorium.

The dish-shaped internal structure, the regional setting in a synclinal-axial depression, the total absence of metamorphic halos, the lack of gravimetric and magnetic expression except in the immediate vicinity, and a growing number of precise analogies worldwide suggest that the ophiolites are shallow-soled klippen. Once continuous with the root zone, the ophiolites constitute a thrust plate of oceanic crust and upper mantle

[1]Manuscript received, July 30, 1976; accepted, December 28, 1977.

[2]Department of Geological Sciences, University of Tennessee, Knoxville, Tennessee 37916.

[3]U.S. Geological Survey, Menlo Park, California 94025.

This note has benefited from discussions, field checks, and reviews with W. Patton and I. Tailleur, although they do not necessarily agree with all of the conclusions reached here. Thanks are due Exxon Company, U.S.A., Houston, for permission to publish the results of a field study of Siniktanneyak Mountain made by the writers in 1970.

Author Identification Number
0149-1423/78/B009-0007$03.00/0

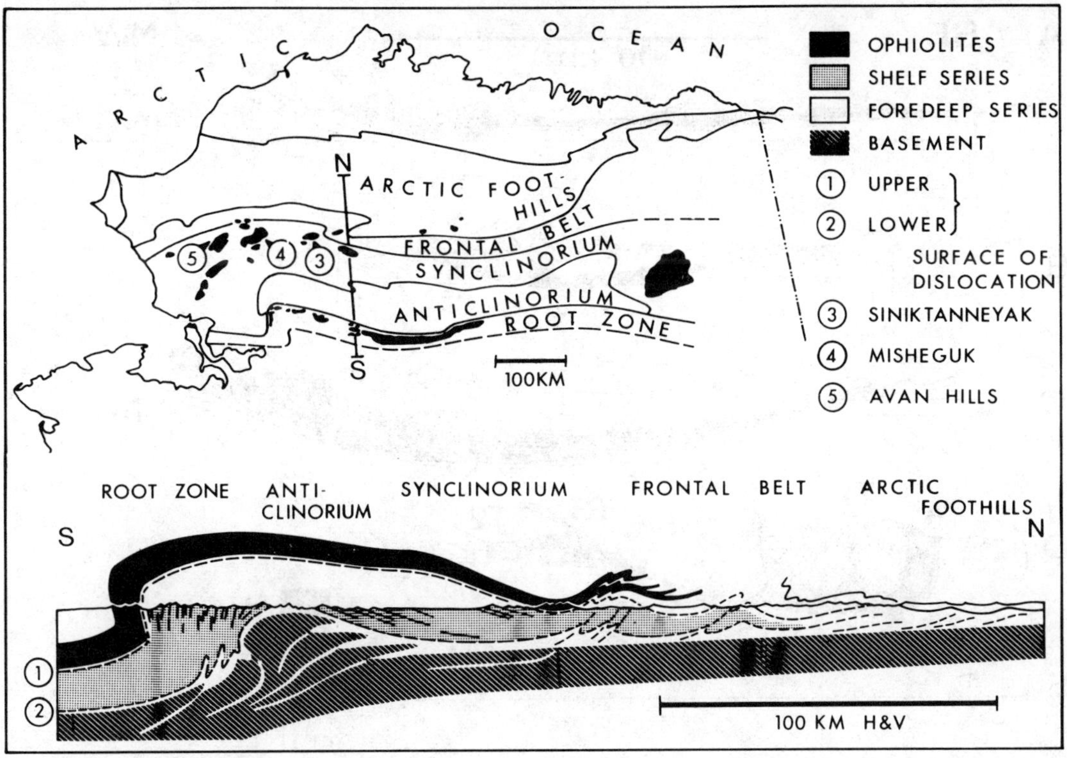

FIG. 1—*Top*: Index map of northern Alaska showing gross zonation of Brooks Range and location of section S-N. *Bottom*: Schematic interpretive structure section S-N of Brooks Range showing, by dashes, two principal surfaces of dislocation between ophiolite slab and shelf series, and between shelf series and foredeep and/or basement.

emplaced on a shelf series by northward thrusting or obduction over a minimum distance of 120 km.

A lower group of dislocation surfaces between the shelf wedge and the parautochthonous complex within the windows is probably younger than the obduction and belongs to a post-collision phase, when tectonism consisted of intraplate compression and suture progradation (Bird et al, 1975; Roeder, 1977). Dutro et al (1976) and Mull et al (1976) discussed the geology of one of the lower surfaces of dislocation; the present note is concerned with the uppermost thrust sheet and its surface of dislocation.

OPHIOLITE BODIES

Three of the larger bodies, Siniktanneyak Mountain, Misheguk Mountain, and Avan Hills (3, 4, and 5 on Figs. 1, 2) have similar structure and composition and may be representative of some or all of the other ophiolite bodies. They contain three major rock types: basaltic flows, cumulate gabbros, and cumulate chromite-bearing peridotites to dunites. Minor basalt in dikes and intermediate to granitoid plutonic masses are present at Siniktanneyak Mountain. The structural arrangement of all rock types is substantially different from the oceanic layer sequence (e.g., Moores and Jackson, 1974); it can be attributed in part to thrust imbrication.

Fine-grained mafite makes up over half of the ophiolite volume, forms a 0.5 to 1 km thick layer at the base of the larger klippen and makes up all of the smaller klippen. Flows 1 to 30 m thick, with and without pillow structure, are locally separated by red and black cherts (radiolarites) in 5 to 20 cm beds and laminated ribbons. However, there are no known sediment bodies suggestive of a continuous oceanic sediment layer above the mafite.

In the larger klippen, the mafite is overlain by gabbros and ultramafites and/or serpentinite

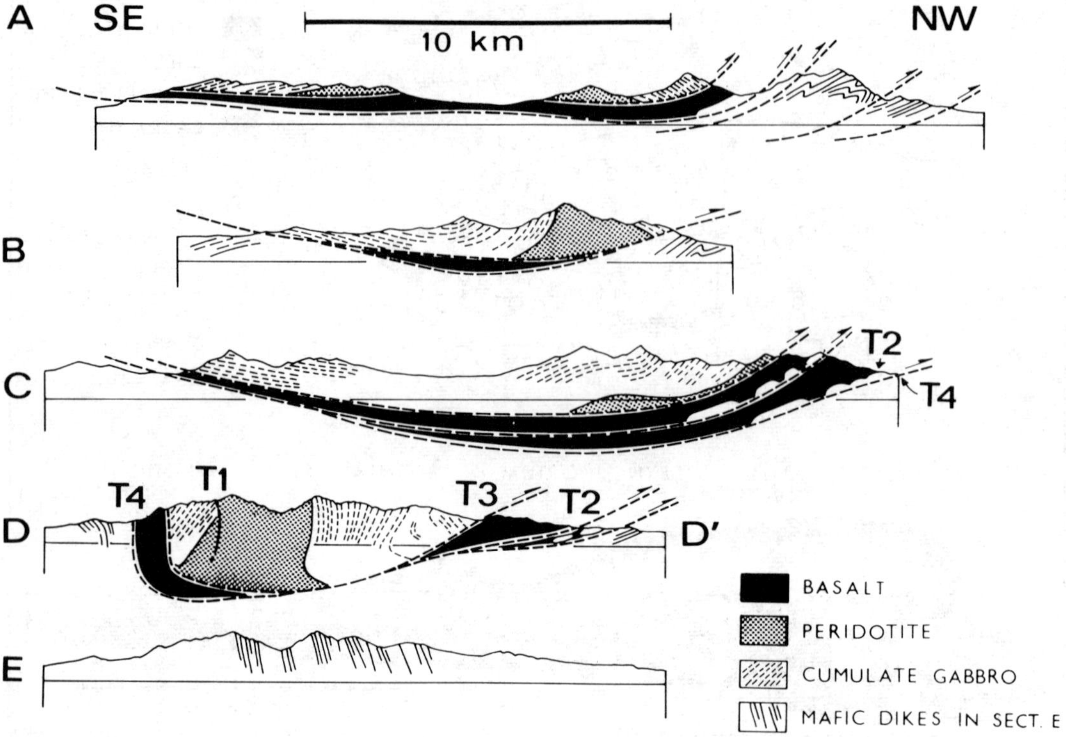

FIG. 2—Structure sections across ophiolite klippen of western Brooks Range. **A,** hills near Avan River. **B,** west part of Misheguk Mountain. **C,** east part of Misheguk Mountain. **D,** Siniktanneyak Mountain (see Fig. 3). **E,** same as **D,** showing location and attitude of basaltic dikes. *T1,* fault at boundary between gabbro and peridotite. *T2,* contact between basalt and Devonian carbonate fragments. *T3,* thrust contact between cumulates and basal mafic unit. *T4,* thrust at base of ophiolite complex. Numbers express inferred age sequence; 1 is oldest.

with cumulate layering which is locally but significantly discordant to the top and base of the underlying basalts. The contact between the basalts and the overlying cumulates is a synclinally folded thrust fault almost parallel with the flow surfaces. The base of the basal mafic layer is a thrust fault and overlies nonmetamorphic Mesozoic sedimentary rocks. Large slivers of Devonian carbonate rock, possibly thrust wedges or melange clasts, occur locally between the basal thrust and the basalt, with extrusive or flow contact between the carbonate beds and basalt. Flow contacts between Mesozoic rocks and basalt have not been observed. At least one of the thrusts within the basalt layer also is lined with such clasts and/or slivers, suggesting a complex sequence of flow and thrusting.

Within the coarse crystalline complexes, the order of crustal layers is normal. The gabbroic rocks generally overlie the ultramafic rocks, with the contact presumably representing an ancient fragment of the Moho. The largest volume is occupied by cumulate olivine gabbros with strong color contrast between inch-thick feldspar-rich and pyroxene-rich layers. They grade downward into olivine-rich peridotites with chromite in continuous layers 1 to 50 mm thick and in discontinuous streaks, flasers, and pods parallel with the cumulate layering in the gabbro. A 100-m thick zone near the contact contains interbedded gabbro and peridotite; commonly it is faulted oblique to the layering.

In some exposures, the cumulate layering is grossly conformable to the overall synclinal structure, but major structuring, of vaguely anticlinal shape, is discordant to the basal thrust in all three of the complexes studied.

In analogy to other ophiolite complexes (E. M. Moores, personal commun.), this structuring probably was caused by deformation of the edges of the magma chamber during seafloor spreading. Discontinuous steep faults discordant to the contact between gabbro and peridotite (Fig. 3) also infer that these structures are not likely to be part of the transport fabric.

Other rock types in Siniktanneyak Mountain include serpentinite, diorite, and granitoid rocks in dikes and small plutons intruding the cumulates. Minor rock types also include basaltic dikes 2 to 50 cm thick that are subvertical and occur in parallel swarms within the gabbro. Their orientation varies much less than that of the cumulate layering, creating areas in which the dikes and the layers are parallel and likely to be misinterpreted in aerial reconnaissance. These dikes can be interpreted as the basal part of a sheeted-dike complex and allow estimation of the amount of erosion down to the present outcrop level. On the basis of data compiled by Moores and Jackson (1974), the missing section should have been 2.5 to 3.5 km thick. This figure provides some constraint to the amount of uplift that the synclinorium and the frontal belt of the Brooks Range experienced after collision and before the end of thrusting.

GEODYNAMIC INTERPRETATION (Roeder)

From the foregoing, it follows that the Brooks Range ophiolite complex is composed of two major independent thrust masses—a basaltic unit below and a gabbroic-ultramafic unit above. This complex was tectonically assembled during a Mesozoic or Tertiary event of Brooks Range thrusting. By their composition, both units could be restacked to form a normal oceanic crustal profile from peridotite through gabbro and sheeted dikes into pillow basalt and chert. However, only the upper thrust mass constitutes a true ophiolite suite, whereas the basalts of the lower thrust mass appear to have been extruded, in post-Devonian time, onto Devonian shelf sediments north of the oceanic basement manifest in the ophiolite suite.

Although not entirely conclusive, the setting and structure of the complex suggest that the northward obduction of oceanic crust and upper mantle downbent, fractured, and overran an Atlantic-type margin that was differentiated into sialic-cored platforms and basins or rift zones with thinner crust. This tentative conclusion is derived from (1) considerations of the possible mechanism of basalt extrusion, (2) the timing of the events, and (3) the known mechanisms of arc-continent collisons.

1. The flows of the basal unit were emplaced on an offshore shelf during an event of rapid subsidence, which probably was not associated with initial rifting on an extensional continental mar-

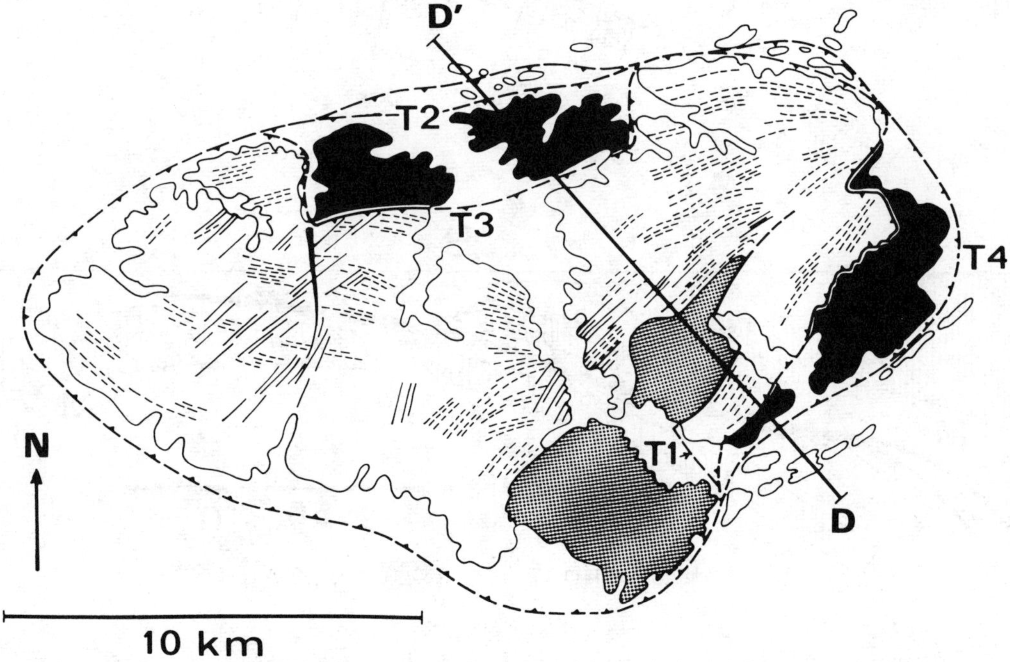

FIG. 3—Photogeologic sketch map of Siniktanneyak Mountain, showing location of section D-D′ of Figure 2. Pattern legend and indexing of contacts as in Figure 2.

gin but rather during the downbending of a lower plate arriving in an oceanic trench (Fig. 4). This interpretation is based on an estimate of the subsidence rate indicated by carbonate beds below the basalt and radiolarites above it. If the basaltic magma had been generated in the upper plate, such as in the magmatic arc, it would have traveled across the entire width of the arc-trench gap and spilled onto the foreland (Fig. 4B). Alternatively, the magma could have been generated and extruded within the downbent and stretched lower plate in front of the colliding arc (Fig. 4A).

The second interpretation is consistent with evidence provided by the informal Anisak formation (Charles G. Mull, unpub. data) within the Ipnavik thrust sequence (Tailleur et al, 1966). Widely distributed in the central and western Brooks Range, this unit contains abundant diabase sills and dikes intruded into a series of Mississippian dark carbonate rocks, cherts, and shales. It also contains thin slivers of Devonian limestone at the basal thrust. Mafic volcanism in a similar stratigraphic position thus extends far into the foreland of the ophiolites.

Fracturing during subductive downbending is well supported from recent arc-trench systems. Tensional first motions of earthquakes and recent normal faults (Spence, 1977) are common in trench outer walls worldwide, and some of the basalt flow above acoustic basement in JOIDES boreholes near trenches may have been produced at such tensional fissures.

Also in older obduction systems, basal mafic units are known, such as in the Eohellenic system of Greece (Jacobshagen et al, 1976), in the Bay of Islands complex of Newfoundland (Williams and Smyth, 1973), in the Loralai ranges of Pakistan (Jones, 1961), and in New Caledonia (Brothers and Blake, 1973). Some of these mafites are similar to the Brooks Range mafites and likely extru-

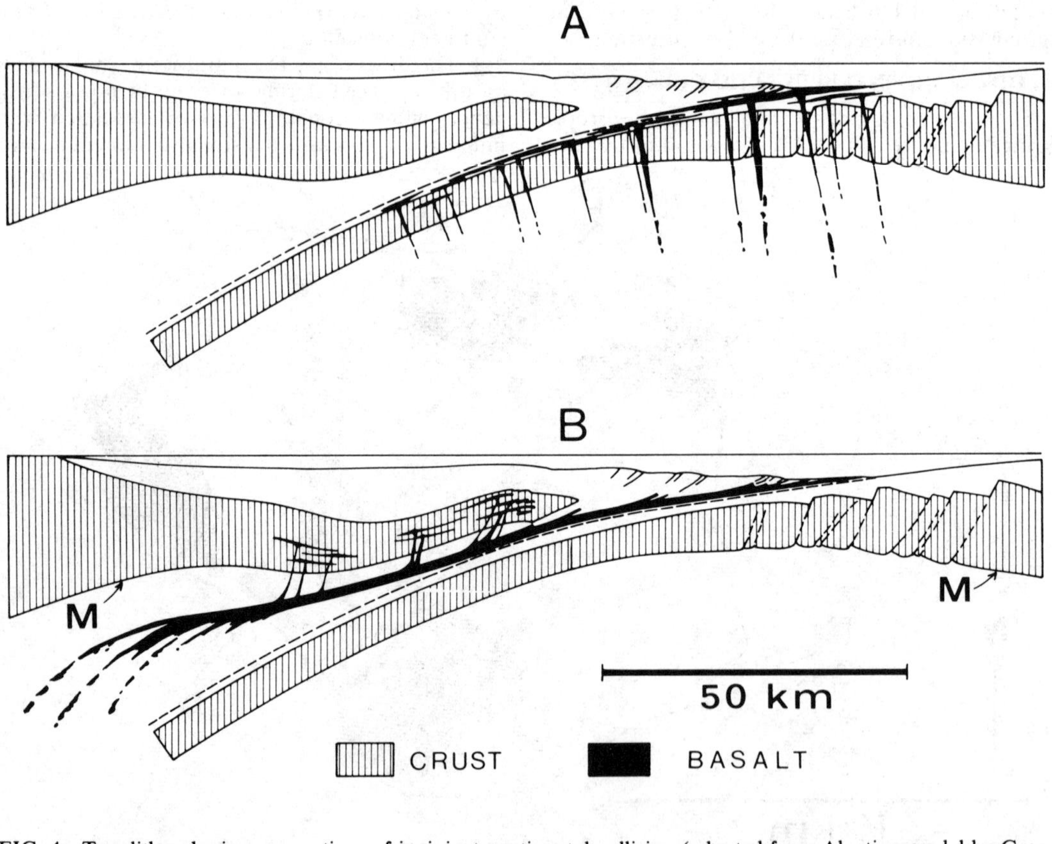

FIG. 4—Two lithospheric cross sections of incipient continental collision (adapted from Aleutian model by Grow, 1973). Arc system in upper plate (left) and Atlantic-type margin with rotated crustal fault blocks in lower plate (right) are separated by suture (dashed line). In **A**, basalt is generated within and extruded on lower plate. In **B**, basalt is mobilized in upper plate and extruded in foreland on lower plate. *M*, = Mohorovičić discontinuity.

sive basalts in flow contact with shelf rocks. Some may represent boulder material as part of ophiolite melanges produced and emplaced during obduction. Still other occurrences may be fragments of the ophiolitic top layer of oceanic crust rearranged in the stacking order by repeated and cross-cutting low-angle thrusting.

2. The crustal ages of neither the ophiolite nor the basal mafite are evident in the wide range of radiometric dates (Tailleur, 1973), but the ophiolite most probably is coeval with Permian ophiolitic basalts in the root zone on the south side of the Brooks Range (Patton, 1973). Based on the proposed correlation with the informal Anisak formation, the crustal age of the basal mafite is post-Mississippian. Based on the argument of rapid subsidence during extrusion, the mafite should be coeval with the arrival of the trench in the vicinity of the south-facing shelf of the Alaskan Arctic plate. This arrival is dated by the pebbly shales of the Jurassic Kingak Formation. Possibly related is a Jurassic K:Ar date from a basalt within foredeep sediments (Brosgé and Tailleur, 1971).

After the collisional contact, tectonism persisted in the Brooks Range for about 100 m.y. through the Paleogene, while the location of the main surface of movement prograded into the foreland, as can be inferred from the stratigraphic record.

Early Cretaceous mudstones, deposited on a diachronous and possibly unconformable base (Mull et al, 1976) in the foreland of the obducted ophiolites and now exposed in the front ranges of the Brooks Range, contain olistostromes with exotic boulders and ophiolitic material and fit plausibly as the accretionary wedge (Karig and Sharman, 1975) of the north-facing arc-trench system. The presence of clasts of Mississippian and Devonian limestone suggests that part of the Alaskan Arctic plate had been incorporated into the upper plate. This Neocomian-age shear surface, therefore, must have been located in the foreland, perhaps at or just below the basement top.

Regional mapping (Charles G. Mull, unpub. data) shows that during Late Cretaceous and Paleogene times, the orogeny and presumably the plate convergence continued with diminished strain rates. There is evidence of erosive truncation of thrust plates and the accretionary wedge, of cannibalization of depocenters, of thick-skinned open folding, and of steep rethrusting. At this stage, the shear surface was certainly located within the crust. This is indicated in the Central Brooks Range anticlinorium by greenschist-grade metamorphism and small plutons, most probably anatexites released by strain heat owing to friction along an intracrustal shear surface similar to the low-velocity layer observed in the Alps (Giese, 1968; Roeder, 1977) and to the basement thrusts in the Himalayas (Bird et al, 1975).

3. The age span of orogeny in the Brooks Range is far too large for a collision over an undifferentiated cratonic shelf at a normal rate of plate convergence greater than 2 cm/year (LePichon et al, 1973), although the tectonic style suggests normal rates. A more realistic model, therefore, would assume at least one trend of cratonic microcontinents south of the Alaskan Arctic craton separated from the craton by rift systems or small marginal basins with oceanic or mixed crust. Although not yet reconstructed in detail, this paleogeography is suggested by the thrust sequences and their complex facies changes (Martin, 1970; Brosgé and Tailleur, 1971; Mull, unpub.). The assumed Jurassic flow contacts of submarine basalt imply a submarine topography exposing Devonian rocks at an erosion surface. The absence of detritus from the Innuitian orogeny affecting the Alaskan Arctic craton (Brosgé and Tailleur, 1971) in the Late Devonian may suggest that the carbonate rocks represent a zone palinspastically remote from the craton.

The subductive removal of this complex lower plate may have occurred either on one subduction zone with a prograding suture or on a succession of subduction zones that were separate and in part active at the same time. It is likely, but not known, that the basin evolution of the Arctic Alaskan shelf was influenced by the subduction mechanism. In other collision orogenes, such as the Alps, both prograding subduction and multiple subduction are possible models (Roeder and Bögel, 1978).

REFERENCES CITED

Bird, P., M. N. Toksöz, and N. H. Sleep, 1975, Thermal and mechanical models of continent-continent convergence zones: Jour. Geophys. Research, v. 80, p. 4405-4416.

Brosgé, W. P., and I. L. Tailleur, 1971, Northern Alaskan petroleum province, *in* Future petroleum provinces of the United States—their geology and potential: AAPG Mem. 15, p. 69-99.

Brothers, R. N., and M. C. Blake, Jr., 1973, Tertiary plate tectonics and high-pressure metamorphism in New Caledonia: Tectonophysics, v. 17, p. 337-358.

Burchfiel, B. C., and G. A. Davis, 1975, Nature and controls of Cordilleran orogenesis, western United States—extensions of an earlier synthesis: Am. Jour. Sci., v. 275-A, p. 363-396.

Dewey, J. F., and J. M. Bird, 1970, Mountain belts and the new global tectonics: Jour. Geophys. Research, v. 75, p. 2625-2647.

Dutro, J. T., et al, 1976, Geologic significance of

Doonerak structural high, central Brooks Range, Alaska: AAPG Bull., v. 60, p. 952-961.

Giese, P., 1968, Die Struktur der Erdkruste im Bereich der Ivrea-Zone: Schweizer. Mineralog. Petrog. Mitt., v. 48, p. 261-284.

Grow, J. A., 1973, Crustal and upper mantle structure of the central Aleutian arc: Geol. Soc. America Bull., v. 84, p. 2169-2191.

Jacobshagen, V., H. Risch, and D. Roeder, 1976, Die eohellenische Phase, Definition und Interpretation: Deutsch. Geol. Gesell. Zeitschr., v. 127, p. 133-145.

Jones, A. G., ed., 1961, Reconnaissance geology of part of western Pakistan; a Colombo Plan cooperative project (a report published for the government of Pakistan by the government of Canada): Toronto, Hunting Survey Corp., 550 p.

Karig, D. E., and G. F. Sharman, 1975, Subduction and accretion in trenches: Geol. Soc. America Bull., v. 86, p. 377-389.

LePichon, X., J. Francheteau, and J. Bonnin, 1973, Plate tectonics: Amsterdam, Elsevier, 300 p.

Martin, A. J., 1970, Structure and tectonic history of the western Brooks Range, De Long Mountains and Lisburne Hills, Northern Alaska: Geol. Soc. America Bull., v. 81, p. 3605-3622.

Moores, E. M., and E. D. Jackson, 1974, Ophiolites and oceanic crust: Nature, v. 250, p. 136-139.

Mull, C. G., et al, 1976, New structural and stratigraphic interpretations, central and western Brooks Range and Arctic Slope: U.S. Geol. Survey Circ. 733, p. 24-26.

Patton, W. W., 1973, Reconnaissance geology of the northern Yukon–Koyukuk province, Alaska: U.S. Geol. Survey Prof. Paper 774-A, 17 p.

Price, R. A., and E. W. Mountjoy, 1970, Geologic structure of the Canadian Rocky Mountains between Bow and Athabasca Rivers—a progress report, *in* Structure of the southern Canadian Cordillera: Geol. Assoc. Canada Spec. Paper 6, p. 7-25.

Roeder, D. H., 1973, Subduction and orogeny: Jour. Geophys. Research, v. 78, p. 5005-5024.

——— 1977, Continental convergence in the Alps: Tectonophysics, v. 40, p. 339-350.

——— and H. Bögel, 1978, Geodynamic interpretation of the Alps, *in* H. Closs, D. Roeder, and C. K. Schmidt, eds., Alps, Apennines, Hellenides: Internat. Geodynamics Project Report, no. 38, Stuttgart, Schweizerbart (in press).

Spence, W., 1977, The Aleutian arc—tectonic blocks, episodic subduction, strain diffusion, and magma generation: Jour. Geophys. Research, v. 82, p. 213-230.

Tailleur, I. L., 1970, Structure and stratigraphy of western Arctic Alaska (abs.): AAPG Bull., v. 54, p. 2508.

——— 1973, Possible mantle-derived rocks in western Brooks Range, *in* Geological Survey research 1973: U.S. Geol. Survey Prof. Paper 850, p. 64-65.

——— B. H. Kent, and H. N. Reiser, 1966, Outcrop geologic maps of the Nuka-Etivluk region, northern Alaska: U.S. Geol. Survey Open-File Rept.

Williams, H., and W. R. Smyth, 1973, Metamorphic aureoles beneath ophiolite suites and alpine peridotites—tectonic implications with west Newfoundland examples: Am. Jour. Sci., v. 273, p. 594-621.

Volume 33 Number 9

BULLETIN
of the
AMERICAN ASSOCIATION OF
PETROLEUM GEOLOGISTS

SEPTEMBER, 1949

INTERPRETATIONS OF FOOTHILLS STRUCTURES, ALBERTA, CANADA[1]

THEODORE A. LINK[2]
Toronto, Ontario, and Calgary, Alberta

ABSTRACT

During the last decade geophysical exploration in the Foothills area of Alberta has aided considerably in arriving at rough approximations, and in some cases fairly reliable data, with respect to the nature of the structural conditions above and beneath major sole faults. The compilation of structural cross sections on the basis of such surveys, plus knowledge derived from previous subsurface drilling and surface geology, are the main theme of this contribution.

Several examples of structures mapped and drilled are given, together with a hypothetical case wherein the difficulties of interpretations are discussed. The structures described are Turner Valley, Highwood uplift, Jumping Pound, Brazeau uplift, Stolberg, Coalspur, Savannah Creek, and the hypothetical Alpha structure.

INTRODUCTION

The tectonic forces which gave rise to the building of the Rocky Mountains and Foothills of western Canada were operative from west to east. There appear to be contradictions to this statement, but when the data are fully studied and analyzed, they generally fit into the picture and thus substantiate the original contention. Furthermore, the intensity of the resultant deformation decreases gradually from west to east, so that the less complicated structures are at the eastern edge of the Foothills. A logical corollary to this is the observed fact that folded faults and folded overthrust sheets are more common in the interior or western part of the Foothills nearer the mountains.

The construction of idealized cross sections through the Foothills is, in some respects, an enjoyable pastime, but detailed cross sections which indicate ultimate depths to prospective oil zones is another matter, and these must be attempted

[1] Read before the Association at Los Angeles, March 26, 1947, through the courtesy of O. B. Hopkins, vice-president of the Imperial Oil Limited. The writer is indebted to the various members of the geological and geophysical staff of the Imperial Oil Limited at Calgary, Alberta, as indicated on the figures and in the text, for their respective contributions. Manuscript received, March 14, 1949.

[2] Geologist, Room 3100, 25 King Street West, Toronto, Ontario, and 810A First Street West, Calgary, Alberta, Canada.

by the petroleum geologist. The object of this contribution is to present illustrations showing results of some of the recent findings in the Foothills on the basis of surface geology, seismic-reflection surveys, and subsequent deep drilling.

EXPLORATION IN FOOTHILLS

Since the discovery of large volumes of gas and distillate in 1924 in the Turner Valley field, exploration in the Foothills belt of Alberta has varied in intensity from year to year, dependent on geologic thinking and results obtained. The

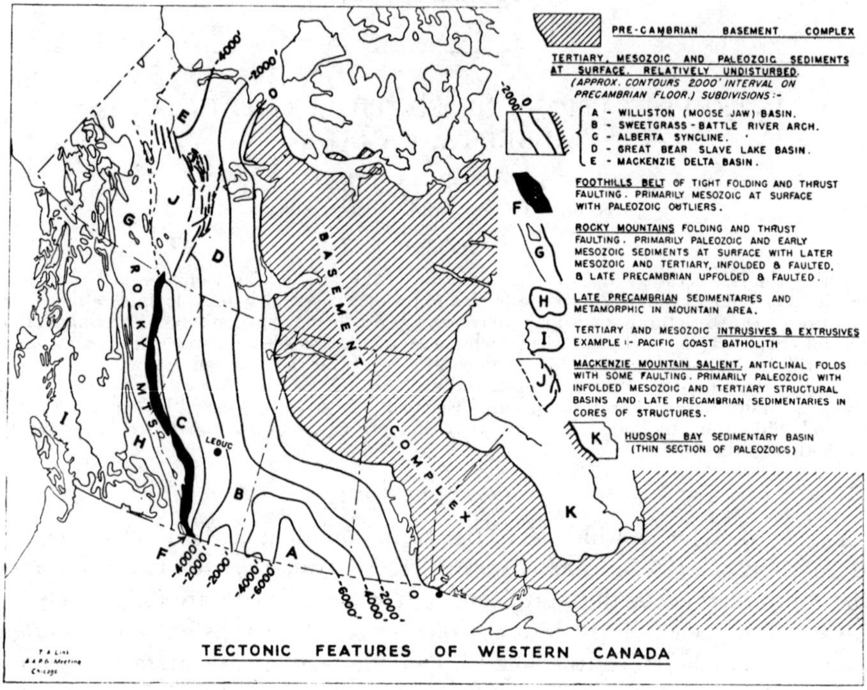

FIG. 1

discovery of wet gas and distillate in the Paleozoic limestone below the surface thrust sheet of the Jumping Pound structure by the Shell Oil Company of Canada, Ltd., in 1944 gave rise to the recent serious exploration activity in the Foothills which is now on the wane, in spite of the subsequent discovery, by the Canadian Gulf Oil Company, of what may be a distillate field near Pincher Creek. The high cost of seismic surveys in the rugged Foothills terrane, the necessity for costly road construction to well sites, and the excessive drilling costs, because of the great depths of the prospective oil zones, call for large expenditures of money; consequently, some of the more recent wildcats drilled in the Foothills belt of Alberta were joint affairs wherein two, and in some instances as many as five, oil companies combined forces and resources to drill such tests. Since 1942, 18 holes have been drilled in the Foothills, based on geologic and seismic-reflection work, but to date only the Jumping Pound and Pincher Creek structures appear to justify hope that these discoveries may lead to something commercial, like the west-flank oil accumulation at Turner Valley. The total amount of money expended by the industry on the Jumping Pound structure alone, for surface geologic, seismic, and gravity-meter surveys, leases, and drilling is probably more than $2,000,000, the largest part of which was expended by the Shell Oil Company of Canada. The Pincher Creek project has promise of similar expendi-

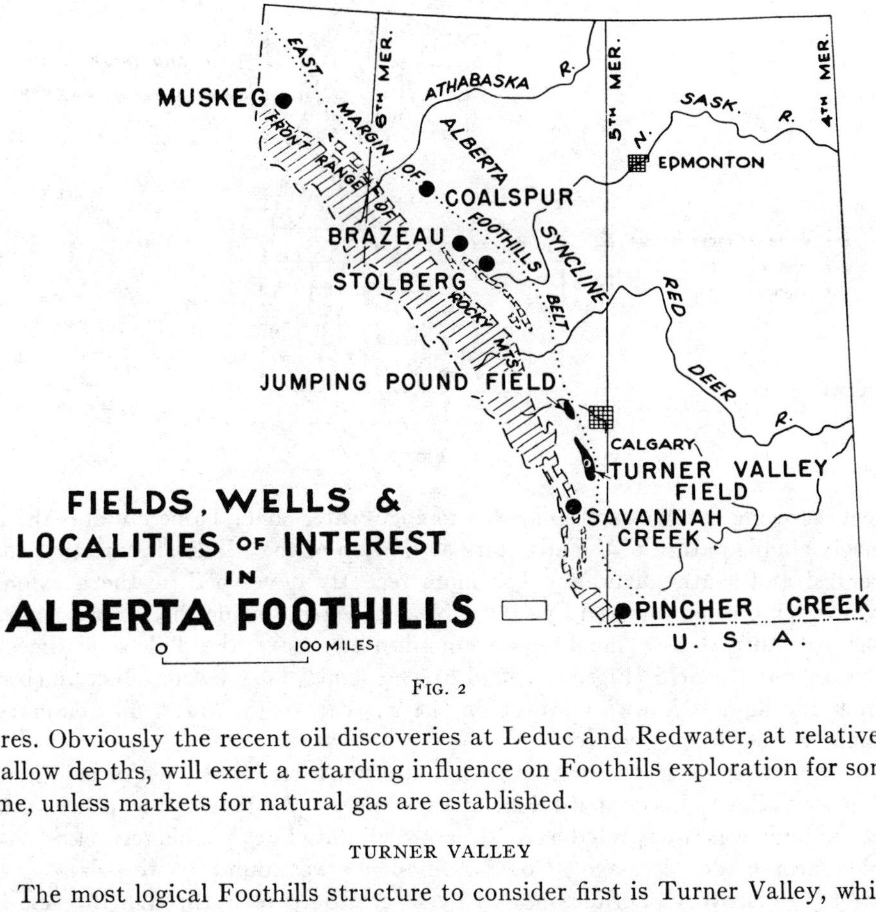

FIG. 2

tures. Obviously the recent oil discoveries at Leduc and Redwater, at relatively shallow depths, will exert a retarding influence on Foothills exploration for some time, unless markets for natural gas are established.

TURNER VALLEY

The most logical Foothills structure to consider first is Turner Valley, which has now been so thoroughly drilled that the nature of the structure and the oil and gas accumulation is fairly well established. On the index map (Fig. 2) are shown the relative positions of the Turner Valley, Jumping Pound, Stolberg, Brazeau, Coalspur, Savannah Creek, Muskeg, and Pincher Creek structures.

Figure 3 is a structure-contour map of Turner Valley (contour interval 1,000

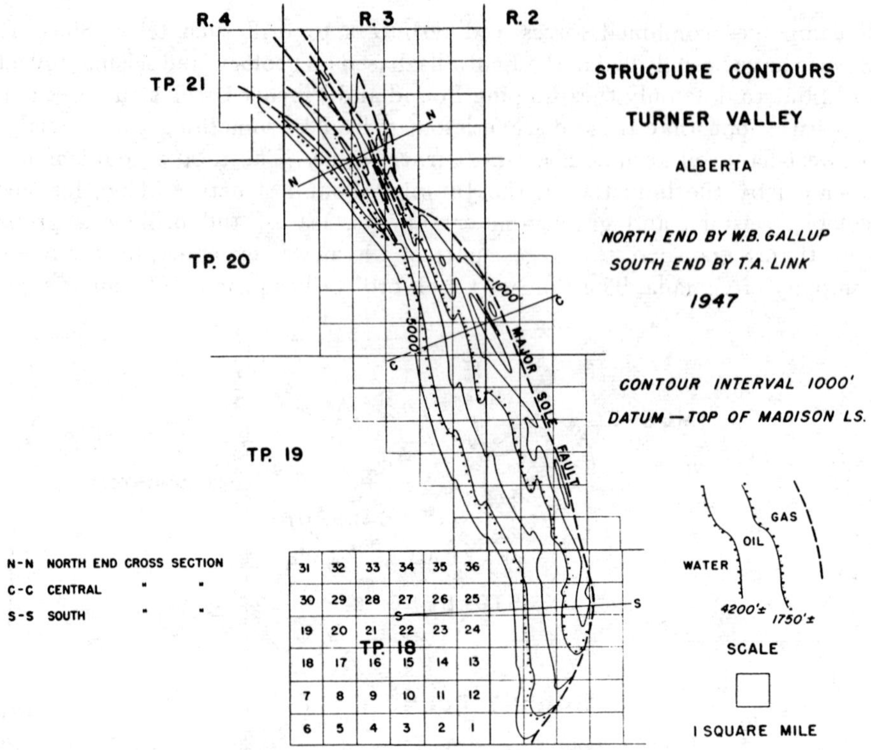

FIG. 3

feet) showing the gas cap, oil area, and edge-water zone. To be noted is the relatively simple nature of the structure at the top of the Madison limestone in the central and south parts, and the more recently developed northern extension which is more complicated by faults. Near the crest, inside the surface rim rock, the structure is also much more complicated above the Paleozoic limestone throughout the field (Figs. 4, 5, and 6).The structural relief or effective closure, from the highest known contour on the gas cap to the water-oil boundary, is slightly more than 5,000 feet.

The writer desires to withdraw his interpretation of the major sole fault of Turner Valley in his contributions of 1934 and 1935[3] wherein the upthrown mass of the fault was interpreted as a "decapitated anticline." (This retraction should have been made years ago.) Contrary evidence was found by west-flank drilling after the discovery of crude oil, and by several deep tests through the sole fault near the center of the field. However, there is still the possibility, adhered to by

[3] Theodore A. Link and P. D. Moore, "Structure of Turner Valley Gas and Oil Field, Alberta," *Bull. Amer. Assoc. Petrol. Geol.*, Vol. 18, No. 11 (November, 1934), pp. 1417–53.
Theodore A. Link, "Types of Foothills Structures of Alberta, Canada," *ibid.*, Vol. 19, No. 10 (October, 1935), pp. 1427–71.

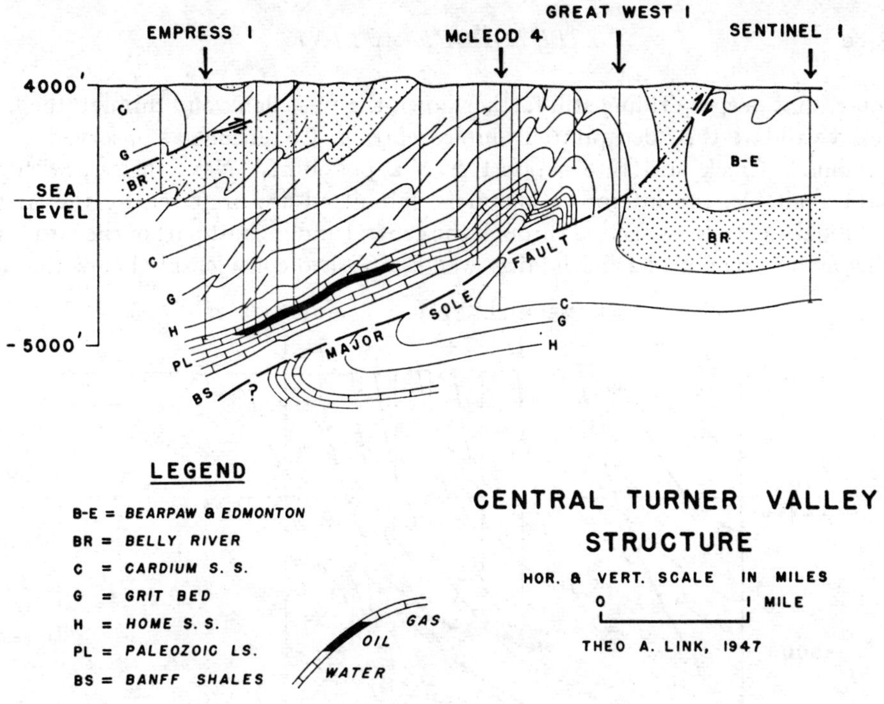
EMPRESS I
McLEOD 4
GREAT WEST I
SENTINEL I
4000'
SEA LEVEL
-5000'
C
G
BR
B-E
H
PL
BS
MAJOR SOLE FAULT
LEGEND
B-E = BEARPAW & EDMONTON
BR = BELLY RIVER
C = CARDIUM S. S.
G = GRIT BED
H = HOME S. S.
PL = PALEOZOIC LS.
BS = BANFF SHALES
GAS
OIL
WATER
CENTRAL TURNER VALLEY
STRUCTURE
HOR. & VERT. SCALE IN MILES
0 1 MILE
THEO A. LINK, 1947

FIG. 4

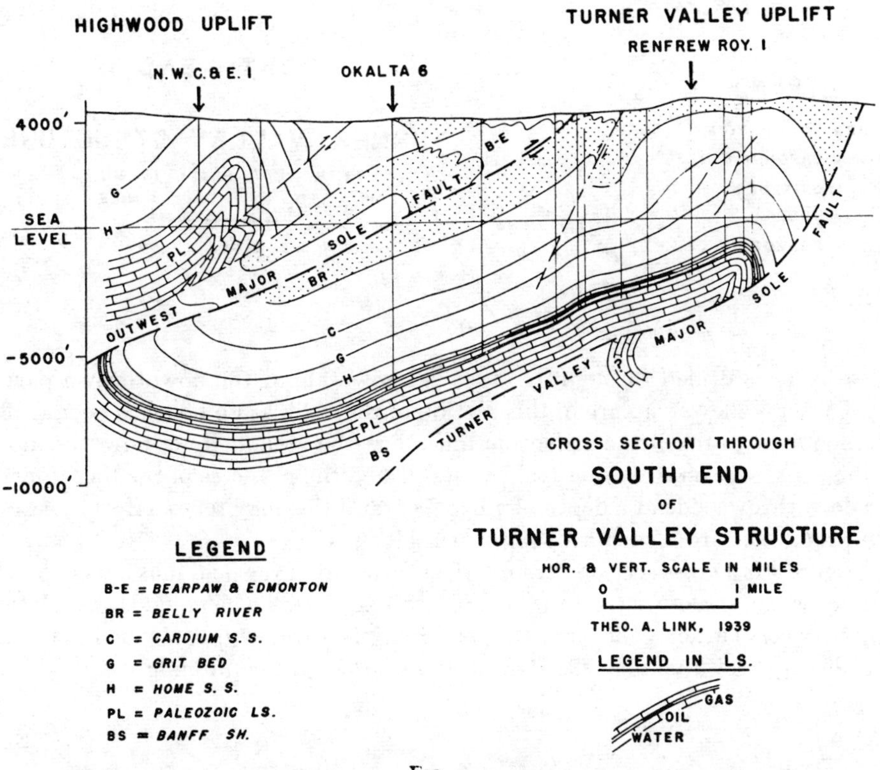
HIGHWOOD UPLIFT
TURNER VALLEY UPLIFT
RENFREW ROY. I
N. W. C. & E. I
OKALTA 6
4000'
SEA LEVEL
-5000'
-10000'
G
H
PL
B-E
BR
C
BS
MAJOR SOLE FAULT
OUTWEST
TURNER VALLEY MAJOR SOLE FAULT
CROSS SECTION THROUGH
SOUTH END
OF
TURNER VALLEY STRUCTURE
HOR. & VERT. SCALE IN MILES
0 1 MILE
THEO. A. LINK, 1939
LEGEND
B-E = BEARPAW & EDMONTON
BR = BELLY RIVER
C = CARDIUM S. S.
G = GRIT BED
H = HOME S. S.
PL = PALEOZOIC LS.
BS = BANFF SH.
LEGEND IN LS.
GAS
OIL
WATER

FIG. 5

some, that deeper drilling still farther down the west flank and through the fault could validate the "decapitated" interpretation. Conditions as observed at the Savannah Creek structure support such a possibility. Nevertheless, with the data available, the interpretations here submitted are preferred by the writer.

Figure 4 is a cross section through the central and highest part of the structure. The major sole fault has been fairly well established considerably below the west

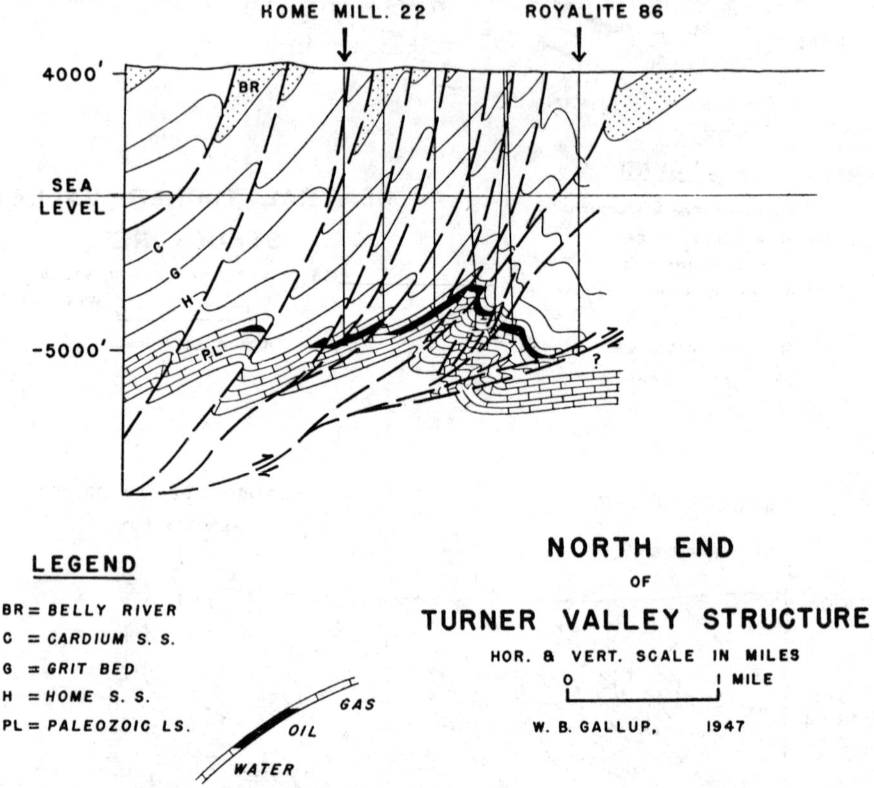

FIG. 6

flank by tests drilled through it. Little is known about the downthrown part of the Turner Valley structure in this section of the field, as no hole has been drilled deep enough to reach the Paleozoic limestone the second time; however, north of this cross section, Royalite No. 73 established the presence of the limestone on the downthrown side at a depth of 7,630 feet, and this may renew efforts to establish production from such a downthrown block.

Figure 5 is a cross section through the south end of the field illustrating a more simple type of structure than the central cross section. To be noted is the relationship between Turner Valley and the west-lying Highwood uplift, which is another overthrust sheet involving the Paleozoic limestone but containing water and dry gas near the top, though this mass is much higher structurally than Turner Valley. The sole fault underlying this thrust sheet is termed the "Out-West fault."

Figure 6 is a typical cross section through the north end of Turner Valley which shows clearly the numerous faults cutting the Paleozoic limestone and the resultant much more complicated nature of the structure. To be noted in particular is the presence of limestone fault-blocks on the east flank. These contain oil at a level lower than the water-oil contact on the west flank. This, coupled with

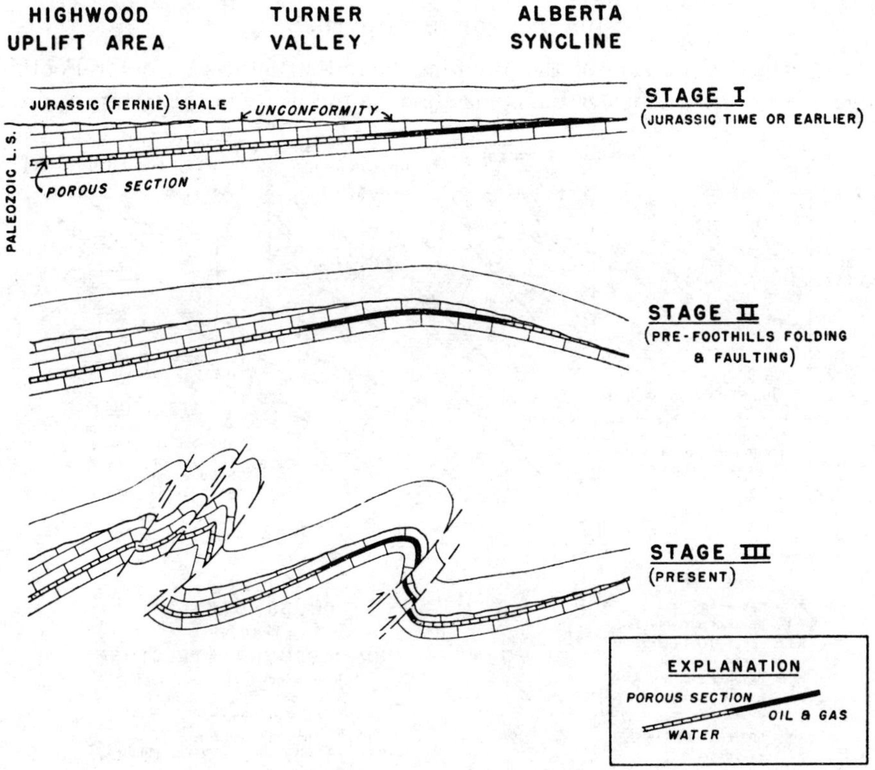

FIG. 7

the water in the west-lying Highwood uplift at a much higher level, as shown in Figure 5, suggests that the oil and gas accumulations and water segregation took place previous to the development of the thrust fault, as suggested in Figure 7, in which are shown the probable stages of the growth of the Turner Valley structure and the related Highwood uplift. It is not necessary to postulate an anticlinal uplift to effect the segregation of the oil, gas, and water as shown in Stage II. The same results could have been obtained under the assumption that the segregation took place in a west-dip stratigraphic-trap accumulation, such as a coral reef, or against the unconformity on the east, without going through the anticlinal stage. However, the history as depicted in Figure 7 is regarded as more probable. It is now believed that the porosity of the producing zones in Turner Valley is in reef limestones of the biostrome rather than the bioherm type.

With this brief outline of Turner Valley in mind, some of the more recently drilled Alberta Foothills structures are described. On only two of these has oil been found, but knowledge of these structures may be helpful if the exploration pendulum should swing back to the Foothills.

JUMPING POUND STRUCTURE

The surface expression of the Jumping Pound structure, an outer Foothills fold 20 miles north of Turner Valley, is almost a duplicate of the latter. Approxi-

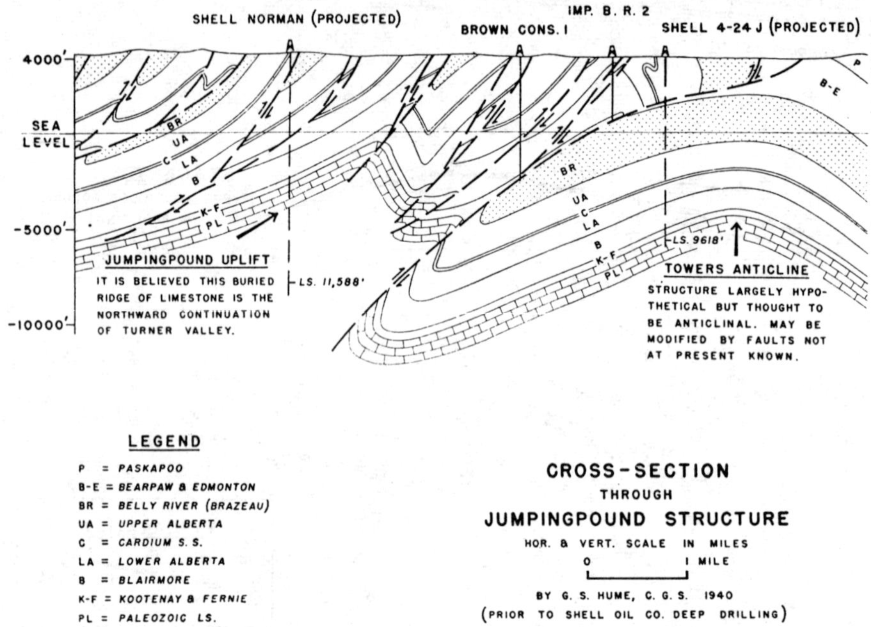

FIG. 8

mately 20 years ago the Imperial Oil Limited drilled two tests through a shallow-depth thrust fault, and abandoned the prospect because the probable depth to the Paleozoic limestone was estimated at 10,000 feet—somewhat too deep for the cable tools which were then being used. Figure 8 is a cross section of the Jumping Pound structure as published by G. S. Hume of the Canadian Geological Survey in 1940, showing the position of one of these Imperial-drilled locations and that of the Brown Oil Consolidated. A seismic profile was run by the Heiland Research Corporation for the Brown Oil Company over this structure during 1940, and on the basis of further seismic work, and possibly Hume's cross section,[4]

[4] G. S. Hume, "The Structure and Oil Prospects of the Foothills of Alberta between Highwood and Bow Rivers," *Canada Geol. Survey Paper 40–8* (1940).

the Shell Oil Company of Canada drilled two deep tests during the years 1942–1944. The first of these, the Shell's Norman 1, superimposed on this illustration (Fig. 8), and shown at the extreme left, encountered water in the porous part of the Paleozoic limestone which was reached at a depth of 11,588 feet, much lower than expected on this cross section. The second test, the Shell's 4-24-J, projected and superimposed at the right on the cross section, reached the top of the Madison limestone at a depth of 9,618 feet which, if this projection is reasonable, is very close to that anticipated by Hume on the basis of his published 1940 cross section. A large flow of gas and distillate was encountered in this hole. A third hole, drilled structurally 1,000 feet downdip only $\frac{3}{4}$ mile from the second

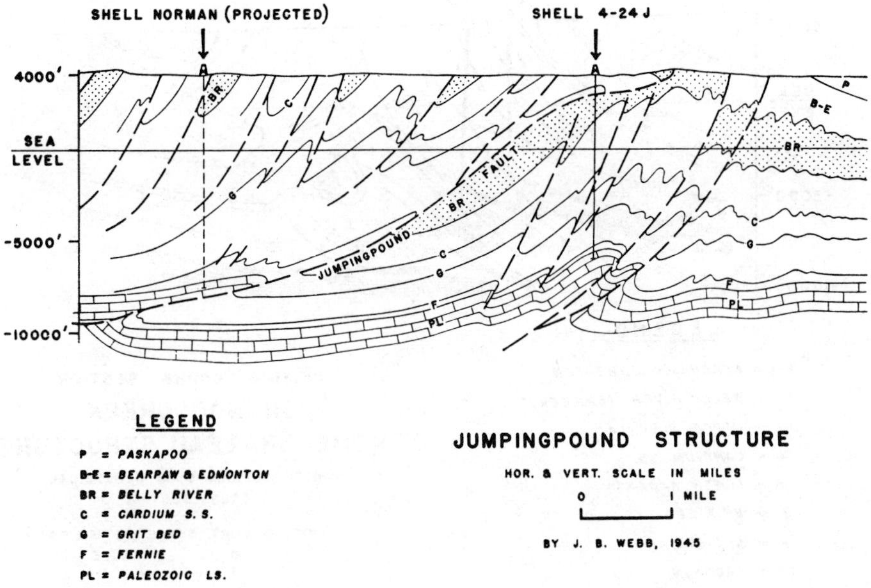

FIG. 9

hole, encountered water; a fourth, slightly up the dip from the third, encountered water, gas, and distillate; and a fifth was completed as a gas producer in the Madison at 9,793 feet, after being drilled to the total depth of 14,443 feet and after passing through a fault and encountering the Madison limestone the second time at 10,735 feet, in which block drilling was continued into the Cambrian.

Figure 9 is a cross section prepared by J. B. Webb, prior to completion of this last well, in which he postulated a branching major fault cutting the Paleozoic limestone. This cross section is indeed very close to the true picture as revealed by the drilling of the last well in which a repetition by one fault was disclosed.

To be noted is Hume's suggestion that the fold postulated by him is probably faulted and more tightly folded, which indeed it was. Different versions of this lower structure could be constructed, but the main point in this figure is an example of a prospective oil and/or gas field below a thin surface overthrust sheet which appears to be unrelated to the one above, and still another block at greater depth.

BRAZEAU STRUCTURE

Six companies joined in a venture to drill the Brazeau structure, of which an interpretation as prepared by Imperial geologists is presented herewith. This is an inner Foothills fold of great magnitude.

Figure 10 is a cross section through the Brazeau structure where one hole

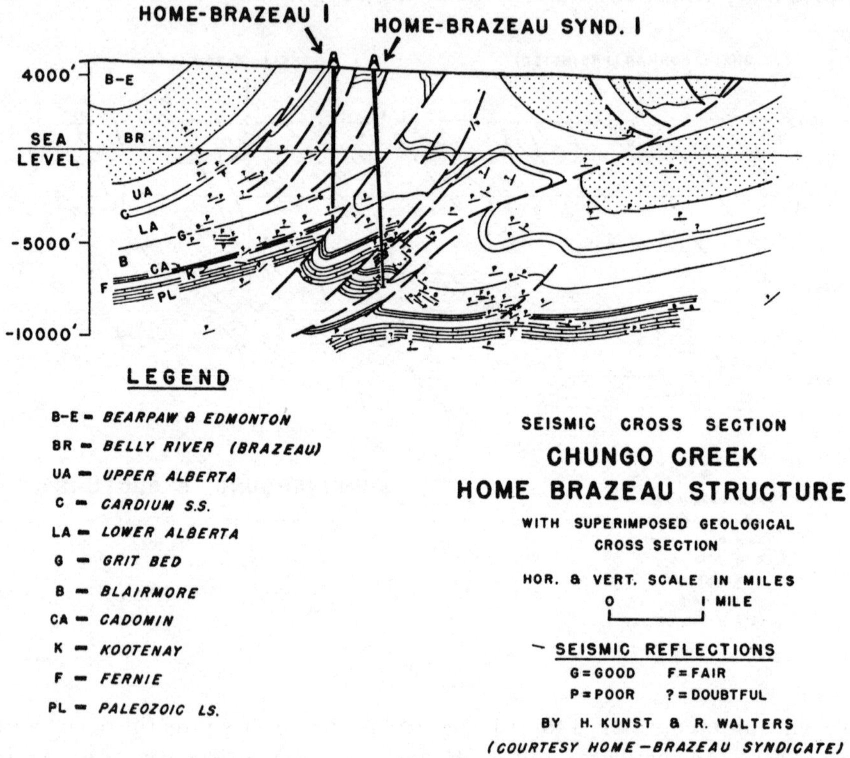

FIG. 10

(Home-Brazeau 1), which had to be abandoned because of mechanical difficulties, failed to reach the objective. The second, or joint venture, reached the top of the Paleozoic limestone at a depth of 9,498 feet, only to encounter another fault, and then enter the limestone for the second time at a depth of 11,351 feet. On this cross section are shown the seismic reflections obtained by running a profile across the structure. This cross section is an attempt to reconcile the seismic data with the surface geology and the logs of the holes drilled.

Figure 11[5] is a cross section along a line slightly north, prepared before the drilling of the wells and without the use of seismic data—in other words, a section based on surface geology only. The Paleozoic limestone was expected at a depth of 4,500 feet, but the drilling of the first and second hole did not confirm this expectation and the top of the limestone was encountered at a depth of 9,498 feet, which was within an error of 50 feet predicted by the seismologists who, however, did not anticipate the faulting encountered below.

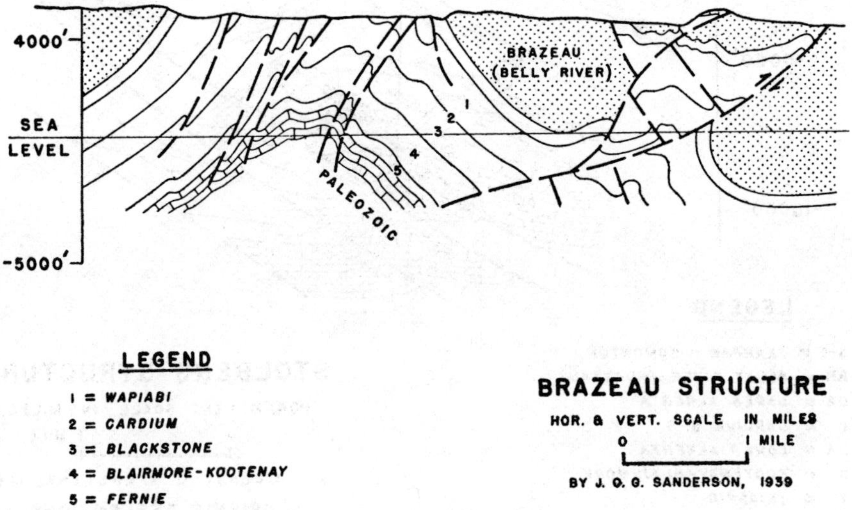

FIG. 11

STOLBERG STRUCTURE

The Stolberg structure (Fig. 12) was drilled jointly by the Shell and Imperial to the total depth of 13,747 feet—at that time the deepest hole in western Canada. On the basis of seismic data, indicated on the cross section, the Paleozoic limestone was expected at a depth of approximately 9,000 feet, but a depth of 12,218 feet was drilled before the limestone was encountered, and from that depth to the bottom of the hole the limestone was repeated by faulting three times (the lowest at the depth of 13,118 feet). There was no faulting of major significance in the formations above the limestone. The misjudgment in depth was the result of miscorrelation of beds in the complicated fault zone where Blairmore reflections were mistaken for limestone reflections. As in previous cross sections, the seismic data, surface geology, and the well logs are reconciled as closely as possible. Obviously, numerous other interpretations could be made, but the general picture would probably be the same. It is superfluous to suggest that no two

[5] J. O. G. Sanderson, "Geology of the Brazeau Area," *Canadian Inst. Min. Met. Eng.*, Vol. 42 (1939), pp. 429–42.

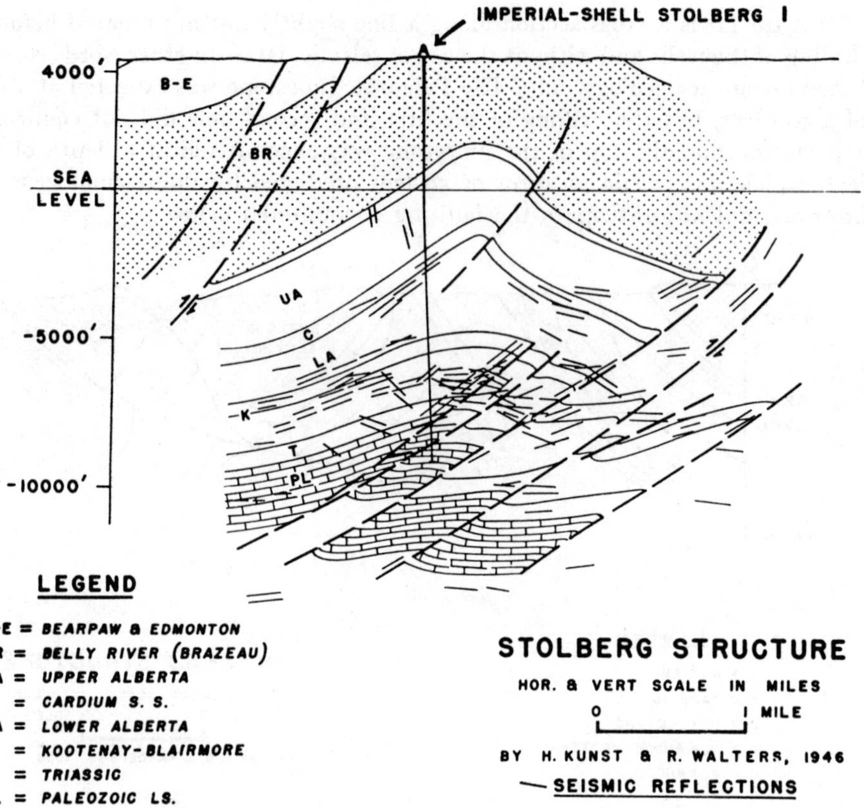

FIG. 12

geologists and/or geophysicists would produce exactly the same picture even though they are supplied with the same data, and even though they had the same training. The Stolberg structure is an outer Foothills fold, and major faulting is absent from the surface down to the top of the Paleozoic limestone.

COALSPUR STRUCTURE

A most interesting structure (as far as giving leeway to the imagination and variety of interpretation) is the Coalspur structure drilled by the Imperial Oil Limited in 1945, where the top of the limestone was encountered at 12,756 feet. One interpretation of it is illustrated in Figure 13. In this part of the Foothills belt some of the outer folds exhibit, at the surface, steeper dips on the west side, as shown in this illustration. Several geologists have suggested an active push from the east to account for this apparent reversal of conditions, well knowing that the entire Rocky Mountain and Foothills orogeny indicates active forces from the west. The deep test and the seismic records seem to indicate clearly that the forces came from the west, and that the underthrusting is surficial.

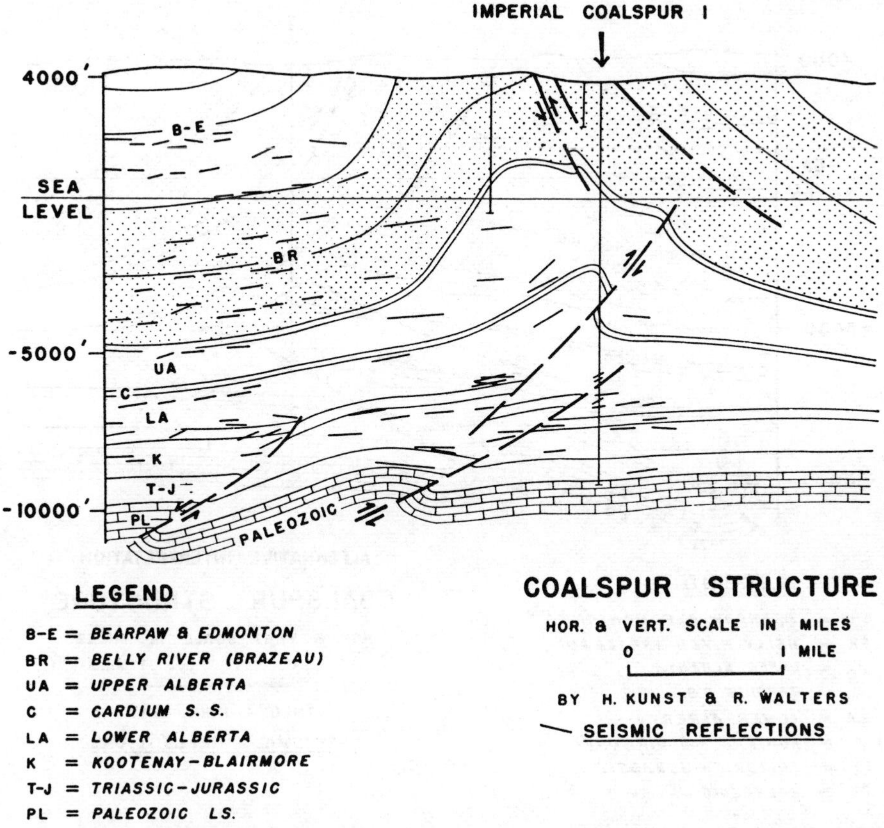

FIG. 13

The interpretation shown here postulates limited near-surface underthrusting and the typical overthrusting at depths, but limited to such an extent that essentially no structure exists. The low seismic dips at depth on the west flank are not used in this interpretation, and in Figure 14 is shown an alternate version indicating more pronounced underthrusting (nevertheless surficial) which uses the low-dip seismic-reflection records encountered at considerable depths on the west flank and the very high dips observed and plotted at the surface.

SAVANNAH CREEK STRUCTURE

The presence of large-scale folded faults and/or folded thrust sheets in the Alberta Foothills has been recognized for some time, and much has been published regarding that subject during the last decade. This contribution could not be regarded as adequate without at least one clean-cut and undeniable example. Figure 15 shows the geological conditions as observed in the area of the north

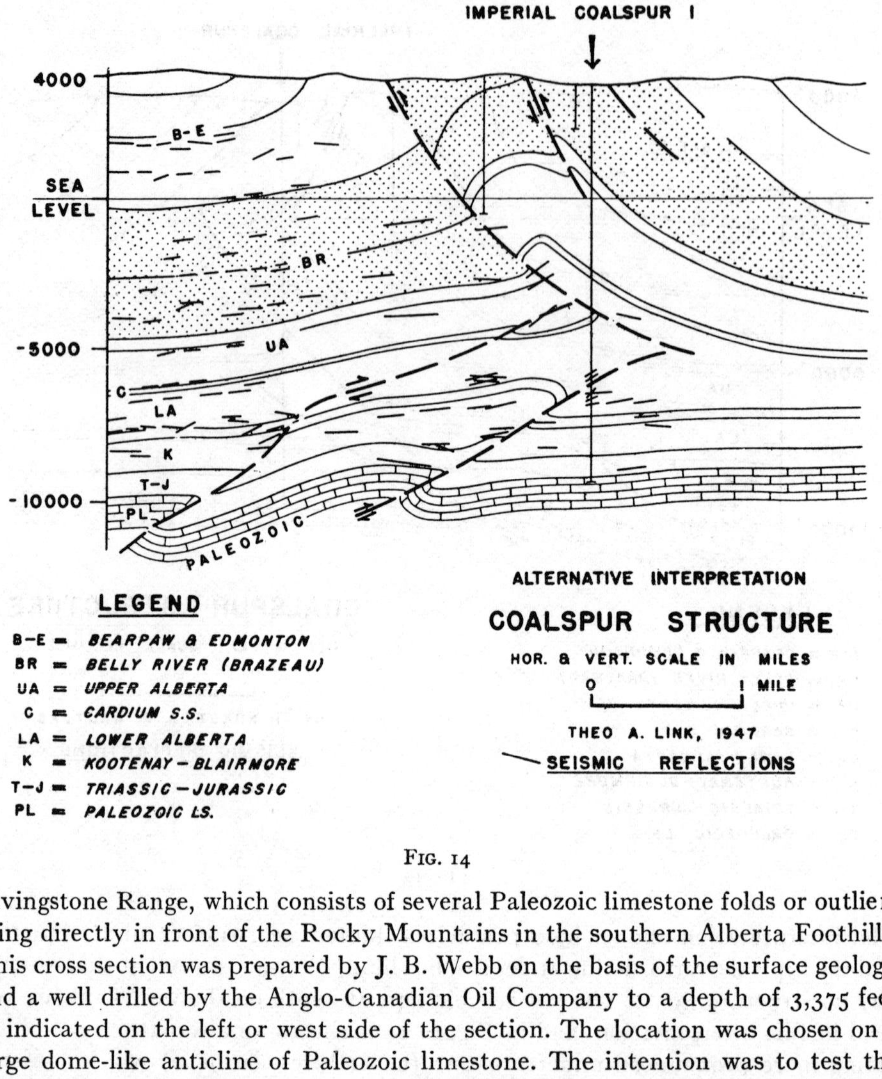

FIG. 14

Livingstone Range, which consists of several Paleozoic limestone folds or outliers lying directly in front of the Rocky Mountains in the southern Alberta Foothills. This cross section was prepared by J. B. Webb on the basis of the surface geology and a well drilled by the Anglo-Canadian Oil Company to a depth of 3,375 feet as indicated on the left or west side of the section. The location was chosen on a large dome-like anticline of Paleozoic limestone. The intention was to test the Devonian and older Paleozoic but, after drilling hardly more than 40 feet, the drill passed through a major sole fault into the Upper Cretaceous "Benton." This fault must lie at a lower elevation downstream along the creek east of the well location. The creek bottom is Paleozoic limestone, as indicated under note B. According to Webb, the fault emerges at the surface farther east after involving above it another fold which, only 2 miles south of the line of section, as indicated under note C, involves another Paleozoic limestone fold. Webb's interpretation is the most conservative possible, and there is no doubt that the sole fault is warped or folded.

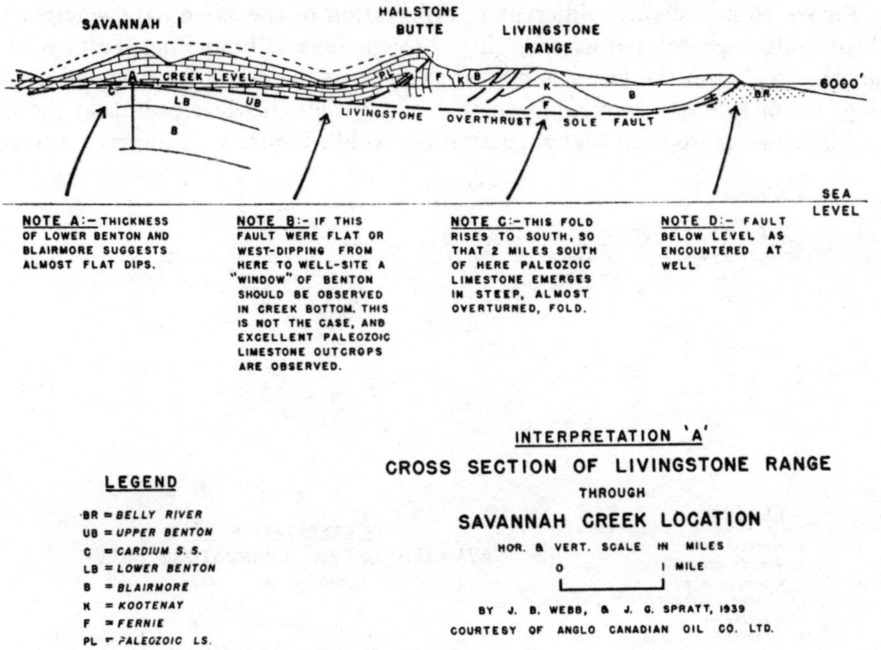
SAVANNAH I
HAILSTONE BUTTE
LIVINGSTONE RANGE
CREEK LEVEL
6000'
LIVINGSTONE OVERTHRUST
SOLE FAULT
SEA LEVEL
NOTE A:- THICKNESS OF LOWER BENTON AND BLAIRMORE SUGGESTS ALMOST FLAT DIPS.
NOTE B:- IF THIS FAULT WERE FLAT OR WEST-DIPPING FROM HERE TO WELL-SITE A "WINDOW" OF BENTON SHOULD BE OBSERVED IN CREEK BOTTOM. THIS IS NOT THE CASE, AND EXCELLENT PALEOZOIC LIMESTONE OUTCROPS ARE OBSERVED.
NOTE C:- THIS FOLD RISES TO SOUTH, SO THAT 2 MILES SOUTH OF HERE PALEOZOIC LIMESTONE EMERGES IN STEEP, ALMOST OVERTURNED, FOLD.
NOTE D:- FAULT BELOW LEVEL AS ENCOUNTERED AT WELL
LEGEND
BR = BELLY RIVER
UB = UPPER BENTON
C = CARDIUM S. S.
LB = LOWER BENTON
B = BLAIRMORE
K = KOOTENAY
F = FERNIE
PL = PALEOZOIC LS.
INTERPRETATION 'A'
CROSS SECTION OF LIVINGSTONE RANGE
THROUGH
SAVANNAH CREEK LOCATION
HOR. & VERT. SCALE IN MILES
0 1 MILE
BY J. B. WEBB, & J. G. SPRATT, 1939
COURTESY OF ANGLO CANADIAN OIL CO. LTD.

FIG. 15

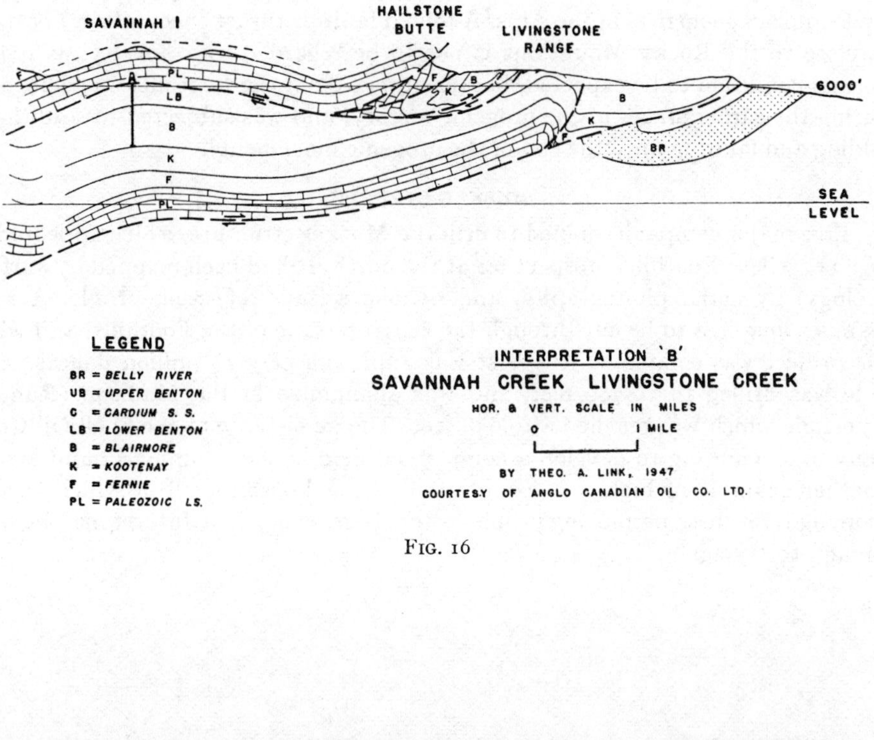
SAVANNAH I
HAILSTONE BUTTE
LIVINGSTONE RANGE
6000'
SEA LEVEL
LEGEND
BR = BELLY RIVER
UB = UPPER BENTON
C = CARDIUM S. S.
LB = LOWER BENTON
B = BLAIRMORE
K = KOOTENAY
F = FERNIE
PL = PALEOZOIC LS.
INTERPRETATION 'B'
SAVANNAH CREEK LIVINGSTONE CREEK
HOR. & VERT. SCALE IN MILES
0 1 MILE
BY THEO. A. LINK, 1947
COURTESY OF ANGLO CANADIAN OIL CO. LTD.

FIG. 16

Figure 16 is a slightly different interpretation of the same data wherein the folded fault is postulated as emerging through several branching faults west of the other fold, and the latter is indicated as being underlain by a second, lower-lying, major sole fault emerging on the east. Under this interpretation the first fault is folded more than that suggested by Webb. Figure 17 is another interpre-

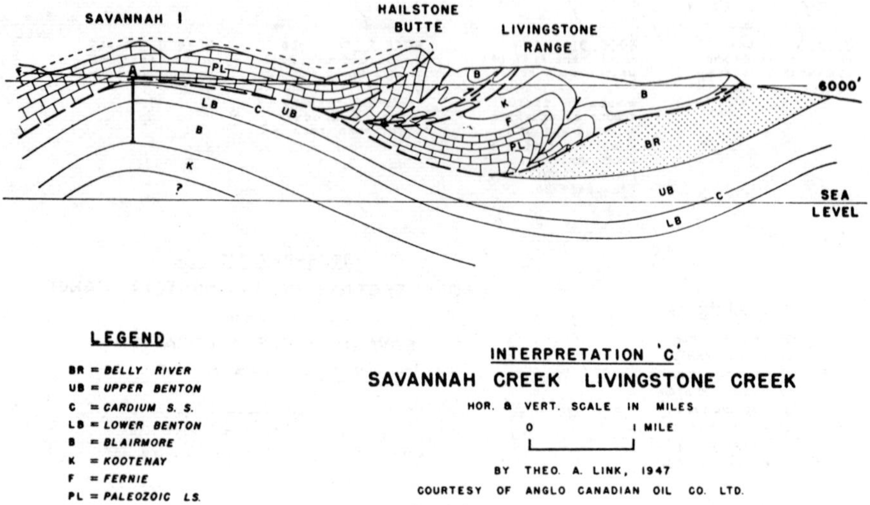

FIG. 17

tation, essentially the same as that of Webb, but with extreme folding of the major thrust fault, and more like some of the folded thrust sheets or nappes as mapped by European geologists in the Alps. A folded fault or thrust sheet in the Foothills so close to the Rocky Mountains is not to be regarded as a novelty, as it is a normal condition to be expected, because this part of the Foothills was deformed during the earlier stages of the orogenic history, and was subjected to additional folding and faulting until the end of the orogenic movements.

MUSKEG STRUCTURE

Five major companies joined to drill the Muskeg structure, which appeared to be an excellent Foothills prospect far at the north. It had been mapped by surface geology, by aerial photography, and by one seismic-reflection profile. A road 75 miles long had to be cut through the rough terrane of the Foothills, and when the project was completed, the cost was approximately 1½ million dollars. This hole was drilled to 10,709 feet, and was abandoned in the Madison (Rundle) limestone which was reached at 9,638 feet. The persistence of the Shell Oil Company in attempting to develop a commercial field in the Jumping Pound area is another example of high-cost exploration in the Foothills, all of which throws more light on these perplexing problems for those who, in the future, may be bold enough to try again.

PINCHER CREEK STRUCTURE

Since delivery of this paper at the Los Angeles meeting of the Association in 1947, an important discovery of wet gas and distillate was made by the Gulf Oil Company in the Foothills area near Pincher Creek, Alberta.

This discovery was the result of an intensive seismic-refraction survey which indicated the probable presence of a Paleozoic limestone block at a depth of approximately 12,000 feet. According to officials of the company, the drill encountered the Madison (Rundle) limestone within 400 feet of the depth predicted by the seismic surveys. Unfortunately, no cross section is available for submittal in this contribution. However, this discovery will undoubtedly be the forerunner of renewed exploration in the Foothills areas of Alberta and British Columbia, and these will add to the oil and gas possibilities of western Canada. As permission to export natural gas from Alberta to other parts of Canada and potential consuming centers in the near-by northwestern United States is being sought, discoveries of this nature may eventually play a very important part in granting this permission and in the ultimate economy of Alberta and western Canada.

GENERAL REMARKS

It is clear that with the aid of geophysics and deep drilling, considerable progress has been made in the interpretation of Alberta Foothills structures at depth. The data obtained by drilling the Turner Valley structure, additional surface geological mapping, and the drilling of one or several holes on other prospects have all contributed toward a better understanding. These data, implemented with seismic-reflection records, make it possible to predict with a more reasonable certainty what is to be expected beneath the surface. However, as no two Foothills structures are exactly alike, and as no two cross sections of the same structure are identical, correct interpretations or predictions are not now, and probably never will be, possible.

This presentation began with Turner Valley where data are sufficient to present very reasonable interpretations. The other examples presented were taken from other parts of the Alberta Foothills where considerable information is available but where speculation must still be resorted to in spite of excellent surface geological mapping, aerial photographs, and seismic-reflection data. In some places more than one deep drilling test has added more information but it is obvious that reasonably correct final interpretation of a structure can not be obtained until it has been almost completely exploited. Experience in Turner Valley has taught that each new drilling location caused a slight, and in some places a radical, change of interpretation as the field was being drilled. The last part of this contribution consists of an example of the method of attack in the interpretation of Foothills structures on the basis of surface geology and seismic-reflection records.

ALPHA STRUCTURE

In Figure 18 is indicated a typical geological traverse usually available after mapping a representative Foothills structure. The structure profile is plotted

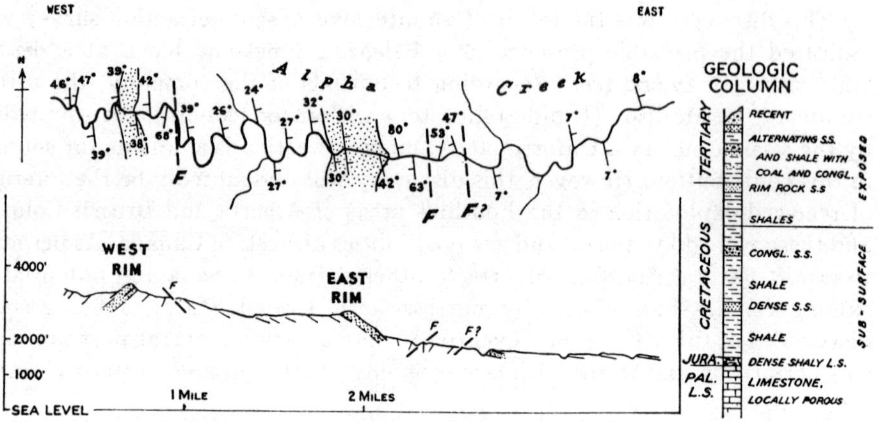

"ALPHA" STRUCTURE

SURFACE GEOLOGY & PROFILE

FIG. 18

below the traverse of Alpha Creek, where reasonably good outcrops were observed and mapped as indicated. Limited to an almost insignificant vertical range, the geologist must construct the cross sections on the basis of which a deep test, costing possibly ½ million dollars, might be drilled. The method of attacking the problem is dependent on the geologist's experience and previous training. As a result, no two geologists will submit identical interpretations, and in many cases widely divergent interpretations are submitted based on the same field data.

Figure 19 (interpretation A) is a conservative, or orthodox, cross section which interprets the fault within the rim rock to be a minor underthrust fault, and the zone of faulting east of the rim rock as a zone of major overthrust faulting. A slight amount of thickening of sediments is postulated at the crest of the structure, thus placing the prospective producing zone below bed A at a depth of approximately 4,400 feet at the drilling site indicated. In constructing this cross section the geologist also had available the measured geologic section and the stratigraphic sequence indicated in the geologic column on the east. The unexposed part of this geologic column was presumably measured west of the map and cross-section area.

Figure 20 (interpretation B) is a cross section in which the geologist interprets the faulting observed east of the rim rock as of minor significance, but regards the fault observed within the rim rock, as well as the steeper west-flank dips, as

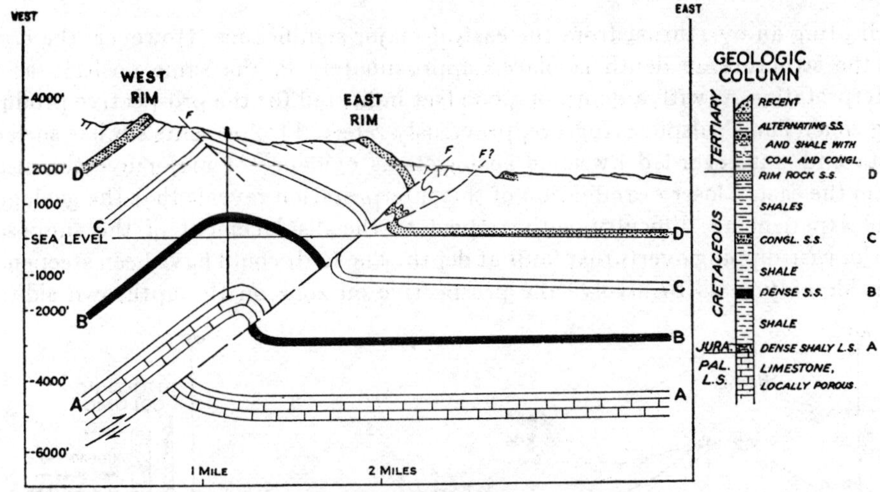

"ALPHA" STRUCTURE

INTERPRETATION A

FIG. 19

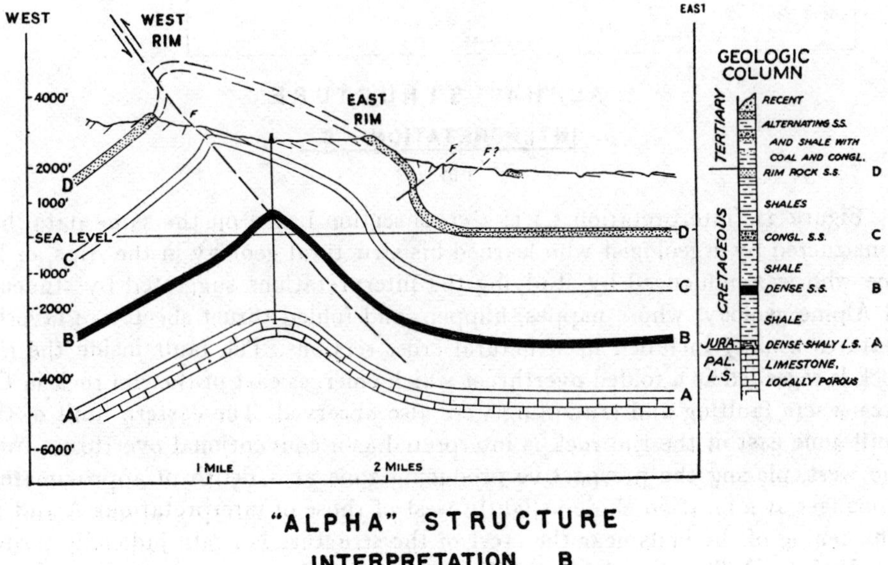

"ALPHA" STRUCTURE

INTERPRETATION B

FIG. 20

indicating an overthrust from the east of major significance. However, the crest of the structure at depth is placed approximately in the same position as in interpretation A, with a depth of 5,000 feet indicated for the prospective producing zone. The Coalspur structure, previously referred to, presents similar surface data and was regarded by some geologists as evidence of major overthrusting from the east. Closer examination of this interpretation reveals that the geologist was experiencing difficulty with respect to the displacement of the supposed major east-dipping overthrust fault at depth. The fault could have been steepened considerably so as to involve the prospective oil zone on the upthrown side.

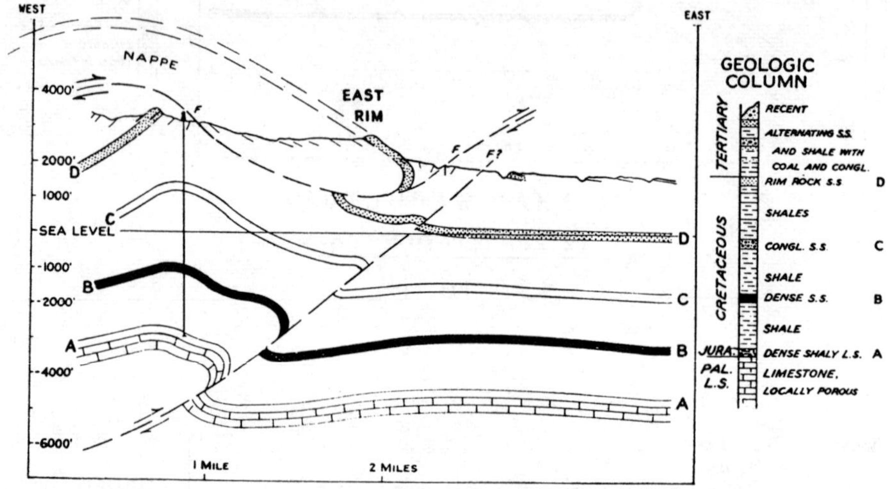

"ALPHA" STRUCTURE
INTERPRETATION C

FIG. 21

Figure 21 (interpretation C) is a cross section based on the same data, but constructed by a geologist who learned his structural geology in the Alps, or by one who was influenced by studying the interpretations suggested by students of Alpine geology, whose nappes, klippen, and folded thrust sheets are favorite features usually included in structural cross sections. The fault inside the rim rock is regarded as a folded overthrust which emerges east of the rim rock in the area where faulting and crumpling were also observed. The eastern limit of the fault zone east of the rim rock is interpreted as a conventional overthrust from the west, placing the prospective producing zone at a depth of approximately 6,000 feet at a location chosen slightly west of those of interpretations A and B. Thickening of the beds near the crest of the structure is again judicially postulated. Hake, Willis, and Addison[6] interpreted the Brazeau structure previously

[6] B. F. Hake, Robin Willis, and C. C. Addison, "Folded Thrust Faults in the Foothills of Alberta," *Bull. Geol. Soc. America*, Vol. 53 (February, 1942).

described as being composed of at least six folded overthrust sheets. Folded faults and thrust sheets are present in the Alberta Foothills, as illustrated in the Savannah Creek area (Figs. 15, 16, and 17), but their use should not be overdone.

The preceding three interpretations do not exhaust the possibilities. It is obvious that a geophysical survey is now in order. Figure 22 shows the reflections obtained from a seismic-reflection profile across the structure. This is an excellent set of reflection records rarely obtained. The reflections are graded good, fair, and questionable, and with these data it is obvious that the geologists and the geophysicists must cooperate. It looks much like a hopeless task, but with knowledge of the stratigraphic section and sequence, and of previously drilled Foothills

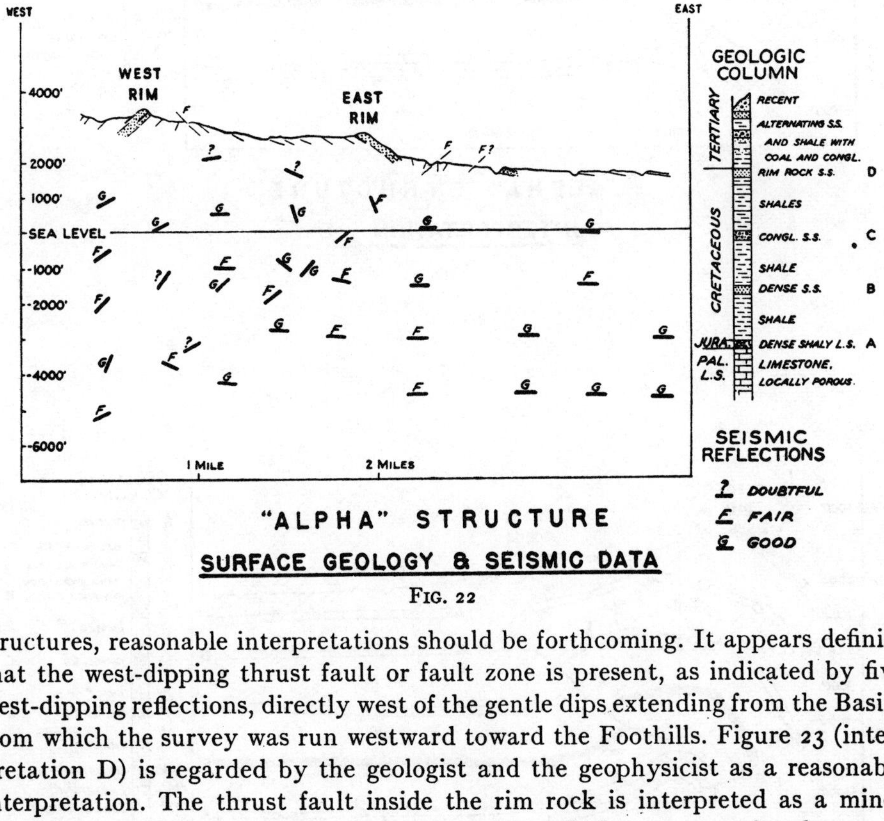

FIG. 22

structures, reasonable interpretations should be forthcoming. It appears definite that the west-dipping thrust fault or fault zone is present, as indicated by five west-dipping reflections, directly west of the gentle dips extending from the Basin, from which the survey was run westward toward the Foothills. Figure 23 (interpretation D) is regarded by the geologist and the geophysicist as a reasonable interpretation. The thrust fault inside the rim rock is interpreted as a minor underthrust adjustment fault. Two other thrust faults are postulated west of the major sole fault at depth, and considerable thickening and thinning of beds are indicated. The prospective producing zone is expected at a depth of 4,000 feet at a location chosen close to the selections made under the other interpretations. Obviously, other interpretations of similar nature could be made, but a radically different one is submitted as Figure 24 (interpretation E), wherein the geologist favors the theory of overthrusting from the east. In this section the

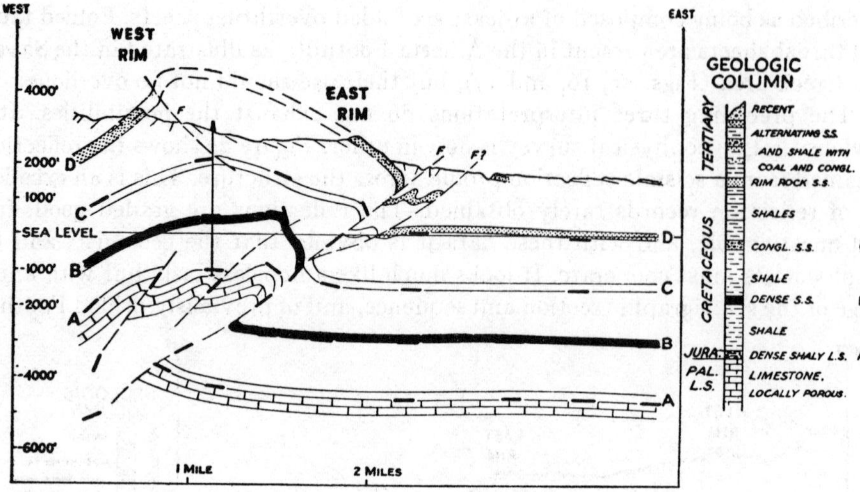

"ALPHA" STRUCTURE

INTERPRETATION D

FIG. 23

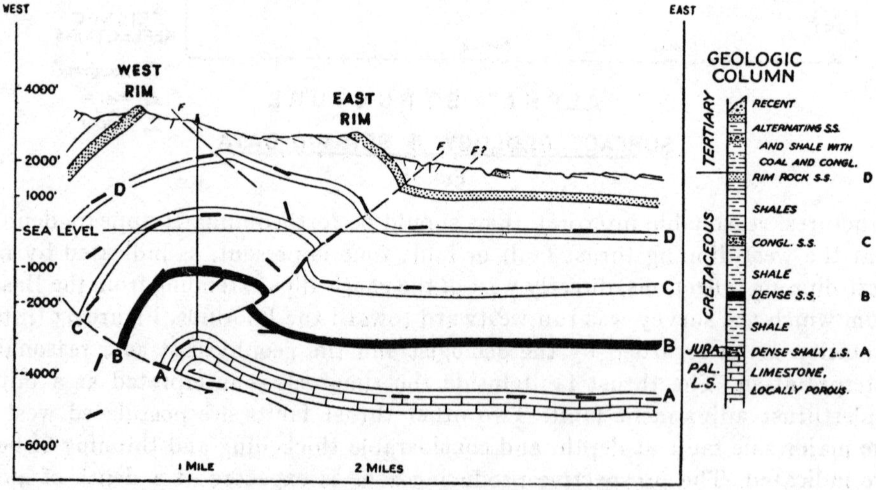

"ALPHA" STRUCTURE

INTERPRETATION E

FIG. 24

west-dipping reflections and the observed outcrop of the thrust faults east of the east rim rock are interpreted as a minor adjustment underthrust, and the steeper dip of the west rim rock is emphasized so that the postulated east-dipping sole fault is suggested as emerging somewhere west of this cross section. Trouble is encountered in the geologic section in dealing with the bed forming the rim rock. The prospective producing zone is postulated at a depth of 6,000 feet at a location close to those chosen in the previous interpretations. It is to be noted that several seismic-reflection records were disregarded or ignored, as was usually the case in many of the cross sections previously shown. This is very common practice, but it generally leads to misinterpretations.

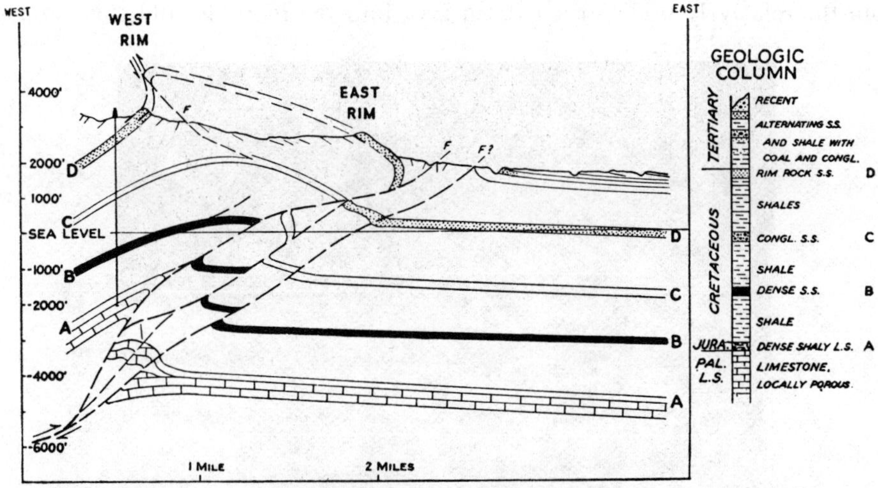

"ALPHA" STRUCTURE

FINAL INTERPRETATION

FIG. 25

The question is now "What is the correct interpretation of the surface geologic and seismic-reflection data available for the Alpha structure?" Figure 25 is a cross section of this structure correct in practically every detail. The branching major overthrust fault or fault zone is very much like the north end of Turner Valley, and also similar to the Brazeau structure previously described. Observe the east-dipping fault blocks below the main fault, and the gradual decrease in displacement of the overthrust faults toward the surface, and the decrease in displacement at depth of the underthrust fault inside the rim rock which terminates as a bedding-plane fault on bed C. Note also the upward bend of the top overthrust fault and the thickening of beds near the crest of the faulted anticlinal fold. To reach the prospective producing zone on this structure above the major sole fault it is necessary to locate the well west or outside of the rim rock, and the depth will be found at approximately 5,200 feet as indicated. In other words, all previous interpretations and well-location recommendations would have been failures, as the limestone would have been encountered on the downthrown side of the thrust sheet.

The logical question for the reader to ask now is: "How does the writer know that this is the correct interpretation?" The answer is, the writer made the structure in the laboratory, and Figure 26 is a picture of a section cut through it. Space will not permit pointing out where the mistakes in interpretation were made and the reader can do that for himself. It suffices to say that miscorrelation of beds above and below the major sole faults is a primary cause of misinterpretation of seismic data in the Foothills. Also, seismic velocities commonly change in passing from the relatively undisturbed Basin area into the foothills, and this is one of

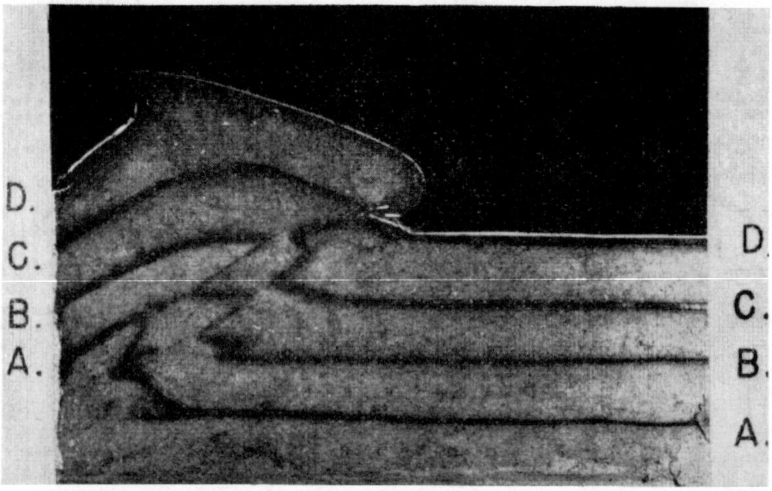

FIG. 26.—Cross section of Alpha structure as made in laboratory, and of which Figure 25 is a drawing. ("Tertiary" of Figure 25 was overburden in this experiment, and removed before cutting the section.)

the several causes of miscalculation. Difficulty in differentiating between reflections from fault planes and bedding planes is another, and rough terrane with resultant large variations of surface elevations also contributes to the confusion. The geophysicists can probably enumerate many other technical reasons. Nevertheless, it must be admitted that the first interpretation based on the combination of the surface geological and seismic-reflection data (interpretation D) was indeed very reasonable and probably much closer than is usually the case in actual practice. However, on several occasions in the Alberta Foothills the prospective Paleozoic limestone has been reached at depths very close to those predicted by the seismic surveys. The Brazeau structure is an example. Unfortunately in this case the repetition of the prospective Paleozoic limestone was not predicted and only the drilling of the deep test proved such repetitions.

FOLDED FAULTS

In Figures 25 and 26 the upper overthrust fault can be clearly traced from depth (at the left) eastward toward the surface, and it is slightly arched with concave side down. The thrust fault directly beneath is also slightly arched, but the last or lower overthrust fault is essentially straight at an angle of 45°. The manner in which this may take place is illustrated in Figure 27. In the progressive development of this structure, stage I shows the first overthrust fault, developed at an angle of approximately 45°, and fairly straight. In stage II, a second overthrust fault has developed, again at an angle of approximately 45°, but the first one shows signs of up-bowing or warping. This is because the underthrust fault, developed near the surface, caused the mass to rise higher on the west side. The first overthrust fault was bent upward correspondingly. The amount of rise is indicated by the arrows near the surface. In stage III, a third overthrust fault has

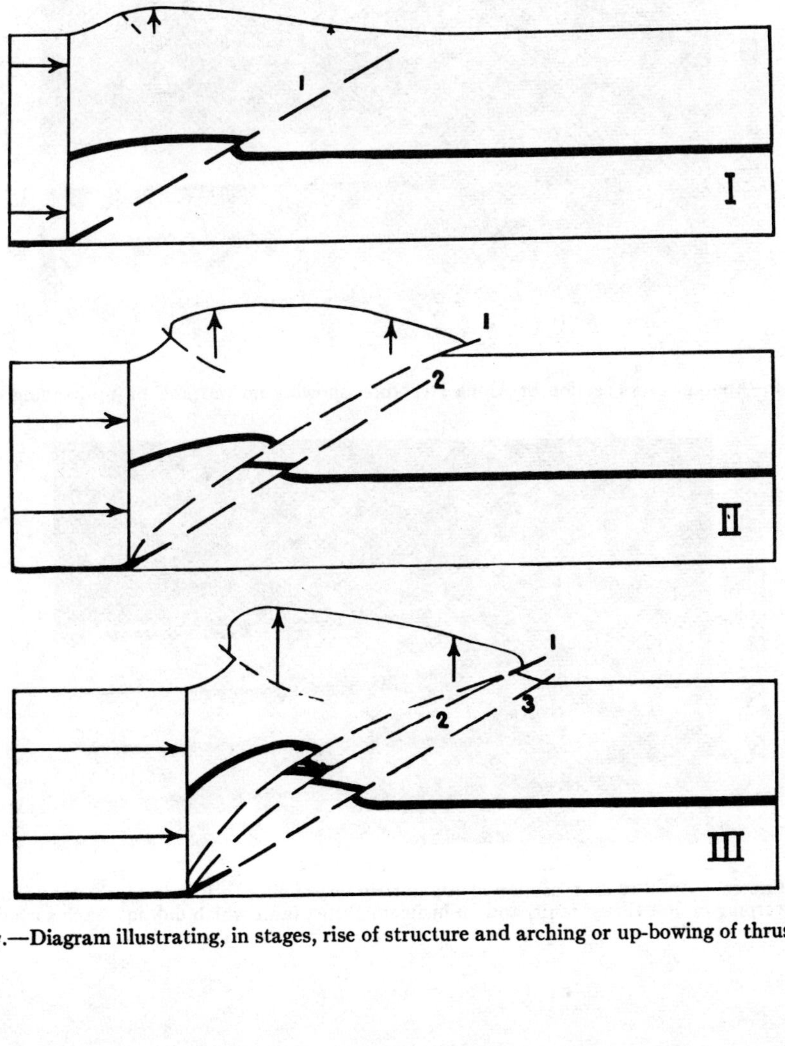

FIG. 27.—Diagram illustrating, in stages, rise of structure and arching or up-bowing of thrust faults.

developed, again in a straight line, more or less at an angle of 45°, and the two previously developed overthrust faults have been further arched or bowed upward because of the continued rise of the mass on the west side. Thus, it is apparent that folding or bending of overthrust sheets is entirely possible and highly probable.

Another cross section through the same structure is indicated in Figure 28 but in this case practically all of the uplift of the mass was on the east side where the overthrust fault emerged at the surface. Here the three major thrust faults were not arched or bent upward, but continued in their original, generally straight-line direction, at an angle of slightly less than 45°. This seems to indicate that during the same orogenic movement overthrust fault planes might and might

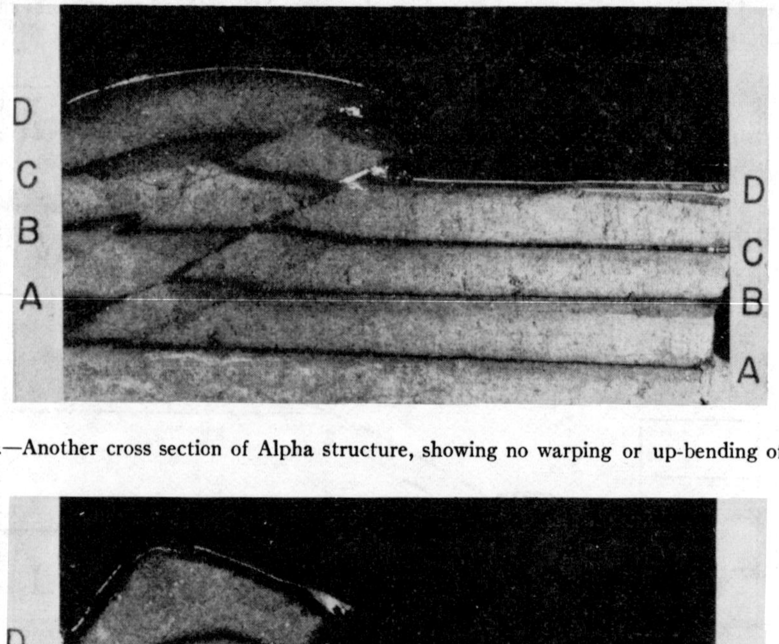

FIG. 28.—Another cross section of Alpha structure, showing no warping or up-bending of faults.

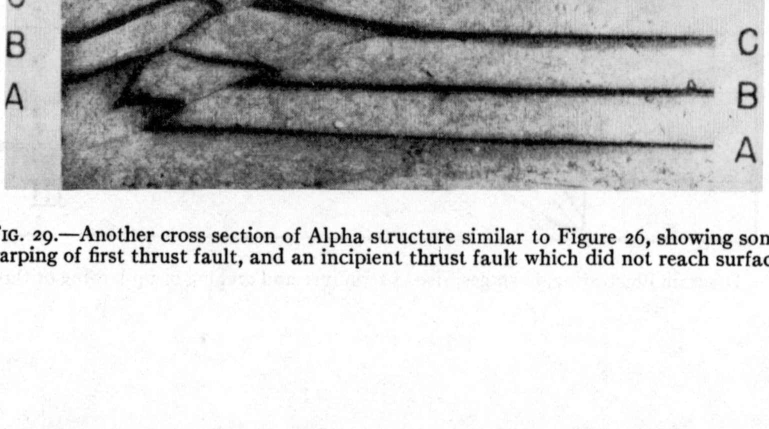

FIG. 29.—Another cross section of Alpha structure similar to Figure 26, showing some warping of first thrust fault, and an incipient thrust fault which did not reach surface.

not be warped, depending on local conditions in the same structure. Figure 29 is another cross section through the same structure which is more or less intermediate between the two illustrated in Figures 26 and 27.[7]

CONCLUSIONS

Following is a summary of the main features which the writer has attempted to convey.

1. The structural geology of the Foothills of western Canada is complicated, and far from being completely understood. Aided by recent deep drilling and the application of geophysical methods, some progress has been made toward solving the problem of choosing well sites for testing the prospective producing zones of the Paleozoic.

2. Underthrusting or apparent overthrusting from the east is a near-surface phenomenon, in some instances initiated during the incipient stages of the development of folds, or as an adjustment feature only, but never a primary or major phenomenon.

3. Warping, up-bowing, and folding of overthrust faults and sheets are recognized features of Foothills structure, less common near the eastern edge, but becoming more pronounced westward toward the Rocky Mountains.

4. Because of variations of stress intensity, direction and mode of operations, and change of sedimentary sections and other factors, no two Foothills structures are alike. This also applies to cross sections on the same structure, as illustrated in Turner Valley and in the laboratory.

[7] For a complete set of cross sections through this model see Figures 12–17, *Bull. Amer. Assoc. Petrol. Geol.*, Vol. 15, No. 4 (April, 1931), p. 395.

BULLETIN OF THE AMERICAN ASSOCIATION OF PETROLEUM GEOLOGISTS
VOL. 35, NO. 11 (NOVEMBER, 1951), PP. 2316-2347, 14 FIGS.

FOLDED FAULTS IN ROCKY MOUNTAIN FOOTHILLS OF ALBERTA, CANADA[1]

J. C. SCOTT[2]
Calgary, Alberta

ABSTRACT

A series of thin fault plates, lying in the Brazeau area of the central foothills of Alberta, are described. Developed at the surface in beds of the Upper Cretaceous Alberta group and Brazeau formation, they are folded as a group into several anticlines and synclines. This structure is attributed to two distinct periods of movement. It is shown that the disturbance causing the folding of the fault plates also formed Chungo anticline, the major structure of the Brazeau area. The folded fault system is found to extend throughout a large part of the adjoining foothills, indicating a widespread influence of the two episodes of movement. Many folded faults are found in other parts of the foothills and mountains and a similar tectonic history throughout is suggested. It is postulated that the entire foothill belt and at least the outer ranges of the Rocky Mountains were subjected to low-angle thrust faulting followed by large-scale folding with attendant thrust faulting. It is not suggested that these movements each occurred uniformly throughout the foothills and mountains; in the southern foothills there is some evidence of two periods of faulting prior to the major folding.

INTRODUCTION

The chief purpose of this paper is to demonstrate the existence of a crustal movement which was capable of producing marked folding subsequent to the large-scale thrust faulting characteristic of the foothills of Alberta and the mountains on the west. The demonstration depends simply on proving that several thrust blocks formed by the first disturbance have been folded later as a unit, in a manner which can not be attributed to folding or warping during the formation of the individual faults. These are the folded faults set forth in the title.

The proof is supplied by illustrations of the geology of the Brazeau structure in the central foothills of Alberta. There are many indications that the folding movement occurs elsewhere and was widespread. These are discussed later. However, the writer knows of no other part of the foothills or mountains where the evidence is so excellently exposed.

A description of the Brazeau structure also serves the purpose of bringing attention to the relationship of folded faults to anticlinal areas in the foothills which have been or may be considered for their oil and gas possibilities. The practical value of a correct understanding of the order of events is obvious to any petroleum geologist. The probability that the compressive forces which produced folding occurred later than most of the faulting has an important bearing upon structural interpretation. The folding, if present, would also play an important part in the accumulation of oil and gas. Recognizing it will not solve, of course, all the problems encountered in finding oil and gas in the foothills;

[1] Read before the Association at Banff, Alberta, September 6, 1950. Manuscript received, March 22, 1951.

[2] Division geologist, Husky Oil and Refining, Ltd.; formerly geologist, Shell Oil Company, Inc. The writer thanks E. P. Williams and A. E. Kliske for valuable assistance in the field. This paper is based on their work as well as on that of the writer. Thanks are also expressed to L. M. Clark and H. H. Beach for helpful suggestions and ideas; to I. D. Crawford, L. M. Clark, and E. P. Williams for critical reading and editing of the manuscript.

recognition will at least aid in understanding the structural deformations involved.

Folded faults in the foothills and mountains of Alberta have long been recognized: McConnell (16) in his "Report of the Geological Structure of a Portion of the Rocky Mountains," in 1886, described undulation of the fault plane (the McConnell fault) at the base of the Paleozoics at the "Gap" where the Bow River leaves the mountains and in the south fork of the Ghost River; Willis, in 1902 (23) from his work in Montana, considered that there was a strong probability that the Lewis thrust is folded; Daly (4) also observed warping of the Lewis thrust near the Forty-Ninth Parallel; Hume (12) arrived at the same conclusion in 1933 from his work in the Waterton area where he found a fault, subsidiary to the Lewis in pre-Cambrian strata, which had been folded; in 1939 Webb and Spratt[3] demonstrated the necessity of at least one folded or warped fault underlying the Savannah Creek structure in the northern part of the Livingstone Range.

A paper entitled "Folded Thrust Faults in the Foothills of Alberta" by Hake, Willis, and Addison (10) was published in 1942. This was the result of field work done during the summer of 1929 in the Cardinal district, a part of the area under discussion in this presentation. The work was reported at a meeting of the Geological Society of America in 1934 (9). Although a considerable amount of skepticism was held regarding the conclusions, the paper contributed, along with more detailed and more widespread mapping, to the realization that folded faults were not uncommon.

Clark (2), mapping in the front ranges of the Rocky Mountains near the Bow River, confirmed McConnell's earlier observations. Maps and other publications concerning various parts of the foothills and mountains of Alberta show that geologists of the Geological Survey of Canada have had an increasing awareness of folded faults since the early nineteen-forties. MacKay's map and notes on the Wawa Creek area (19) and his map of the central foothills belt of Alberta (18) show that he recognized folded or warped faults in 1941 in the area covered by Hake, Willis, and Addison, and later by the writer and his associates. Hume (13) added to his earlier contribution with the description of a folded fault, which he mapped in 1940, in the Pekisko area of the southern foothills. Hage (8) found evidence of similar faulting in the same region a year later. In 1943 Crombie (3), and in 1944 Henderson (11), found evidence of folded thrusts in and near the southern part of the Brazeau Range. Erdman, in 1944, continued the work started in 1943 by Crombie, and presented his findings in a memoir published by the Geological Survey of Canada in 1950 (7). Included is a detailed description of the folded thrust faults observed, as well as an excellent summary of the history of work concerning folded faults in Alberta. Also published in 1950 is a memoir by Douglas (5) on the Callum Creek, Langford Creek, and Gap map-

[3] Cross section published in Link's "Interpretations of Foothills Structures, Alberta, Canada."

areas, and a map by the same author (6) of the Mount Head area, both concerning the southern foothills. In the former the author discusses the mechanics of faulting, particularly referring to the Livingstone thrust, a folded fault of large throw with a surface trace extending along the foothills in front of the Livingstone Range for many miles. The publication is based on work done during the field seasons of 1944 to 1947. In 1947 and 1948 Douglas mapped more folded faults in the Mount Head area adjoining localities where Hume and Hage had previously described them.

Where an opinion is expressed, those writers publishing prior to Hake, Willis, and Addison have taken, more or less as a matter of course, the view that folding of the faults they describe occurred separately, after the break. It is doubtful, however, that satisfactory proof was available to them. In the Cardinal area, Hake *et al.* were able to show the presence of several thrust sheets folded as a group in anticlines and synclines over a distance of several miles. If their work is accepted it will be agreed readily that here at least is sufficient proof of two distinct pulses of movement, however closely one followed the other. However, there has been considerable reluctance on the part of many geologists to accept this evidence; partly because of doubt concerning the clearness; partly due to experience elsewhere in the foothills, where large areas show no plain evidence of the same conditions; and, perhaps, partly due to disagreement with the conclusions expressed regarding the mechanics involved.

Whether or not the evidence presented by Hake *et al.* is accepted, most later authors, with the exception of Douglas and Link, conclude that later folding holds true for folded faults they describe. Douglas considers the folding, of the Livingstone thrust at least, to be part of the faulting process. Link (14), in a recent contribution to the *Bulletin* of the Association, expresses ideas similar to those of Douglas. Based on experimental models, he states his belief that folded faults in the foothills belt are produced by warping almost synchronous with the thrusting. Obviously the faults on a single structure formed in this manner would be limited both in number and intensity of folding. As a consequence he considers that the folding is restricted to fairly gentle curves, and the faults on any one structure to two. Link therefore concludes that Hake, Willis, and Addison were mistaken in their field observations.

In this last regard Link is expressing an opinion which is also held by others. The writer mapped the Cardinal area in detail at a later date, and found himself in agreement with Hake *et al.* This appears to be sufficient reason for adding to publications on the subject at the present time. Added to this, however, the mapping on which this paper is based covered the entire Brazeau structure (of which the Cardinal district is part), and the remaining foothills on the east. The structure was mapped previously by Sanderson (21) and MacKay (18). MacKay, in his map entitled "Foothills Belt of Central Alberta," indicated the presence of folded faults in the area in a general way. With this exception, the complexities due to folded faults have not been described outside of the Cardinal district.

BRAZEAU AREA

The area encompassing the Brazeau structure is shown on the index map (Fig. 1), and its position relative to major streams and other areas discussed in this paper is indicated. It is the southwest half of a block mapped during the field seasons of 1944 and 1945. The Brazeau River, flowing from west to east, divides the area roughly into a northern third and a southern two-thirds. A generalized map of the areal geology of the southern part, extending a short distance north of the Brazeau River, is shown in Figure 3. The work of Hake, Willis, and Addison was concentrated on the southwestern side of the northern third; the Canyon Creek[4] section (Fig. 5), directly south of the Brazeau River, marks the southern limit of the area they mapped. It has an important part in their, and the present, interpretation of the geology of the region.

The area is rough and forested, with a maximum relief of approximately 2,000 feet, and local relief ranging from 500 to 1,300 feet. Long strike ridges formed on the resistant sandstones of the Cardium and Brazeau formations, broken by transverse valleys of the major streams, are characteristic of much of the area. In many localities where the geology is complicated by low-dipping fault plates, the topographic expression is likewise more complex. Single ridges give way to multiple ridges and straight ridges to arcuate ridges. In general the bedrock exposures are good along the transverse streams, and it is along these streams that most data are available. Excellent exposures are also to be seen on many ridges and are invaluable in working out the structure. As elsewhere in the foothills, it is possible to connect cross sections of the geologic structure, as viewed in parallel streams some distance apart, by their aid.

The field work was done with the aid of vertical aerial photographs, making it possible for a field party of three geologists, unencumbered by survey equipment, to cover large areas thoroughly in a comparatively short time. Data plotted on the photographs were transferred to a base prepared from maps published by the Topographic Survey of Canada. These maps were also very useful in the field, particularly in providing elevations at critical points. The work was detailed and all parts of the area were searched carefully for outcrops.

It is thought that no useful purpose is achieved by showing the whole of the area originally mapped. Although the writer's observations were at variance with Hake *et al.* concerning parts of the geology of the Cardinal area, no important difference resulted. Since the differences can hardly be shown at a scale required for this paper, no new map is submitted and the reader is referred to their publication for a general view of the northern part of the Brazeau structure. Figure 3 shows only the generalized geology of the part of the Brazeau structure with which this paper is immediately concerned: the Canyon Creek area, containing the key to the structual habit; and the culmination area showing Chungo anticline and the relationship of seemingly "normal" foothill structure to the

[4] **Known also as Elk Creek and thus named by Hake, Willis, and Addison.**

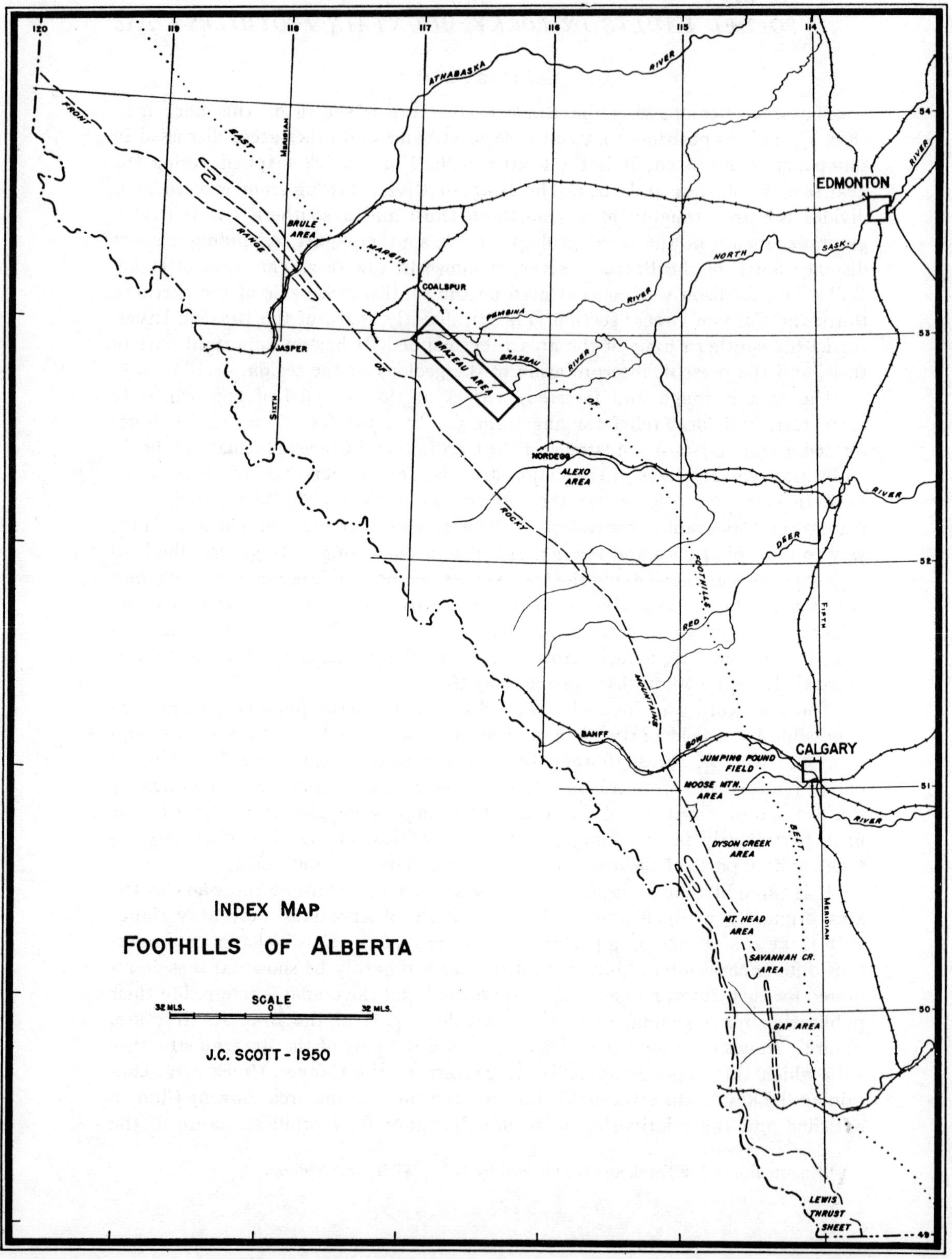

FIG. 1.—Index map showing: approximate limits of foothills of Alberta; Brazeau area; other areas of interest discussed in this paper.

Canyon Creek folding. Although generalized, the map truly represents the work of the writer's group in its principal features and interpretation.

It may be interesting to note that the presence of folded faults in this area was recognized somewhat independently from Hake, Willis, and Addison. Field work by the writer's party was started near the south end of the Brazeau structure in the vicinity of the Blackstone River. It was carried northwestward along the trend of the surface exposures of the Alberta group to Brown Creek, before folding of the thrust sheets involving these beds became evident. It became very clear upon examination of the Canyon Creek section and, as work continued

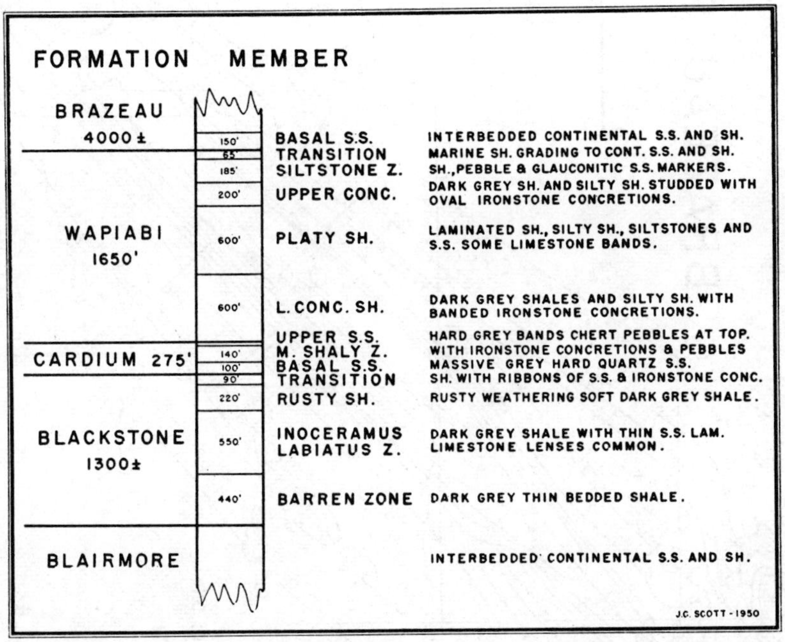

FIG. 2.—Formations and members of Alberta group, Upper Cretaceous, showing thicknesses and giving some of characteristic lithology. Blairmore is Lower Cretaceous.

northwestward, it was realized that this was not an isolated phenomenon. At this stage of the mapping, "Folded Thrust Faults in the Foothills of Alberta" was made available for study and comparison.

STRATIGRAPHY

A generalized log of the stratigraphic section lying, at or near the surface, in the folds of the Brazeau structure is shown in Figure 2. The beds of the Blackstone, Cardium (Bighorn), and Wapiabi formations form the Alberta group, and, together with the basal beds of the Brazeau formation, constitute the units involved in the faulting described in this paper. The Barren zone of the Blackstone formation is not entirely exposed anywhere in the area, and the Blairmore group makes no appearance at the surface.

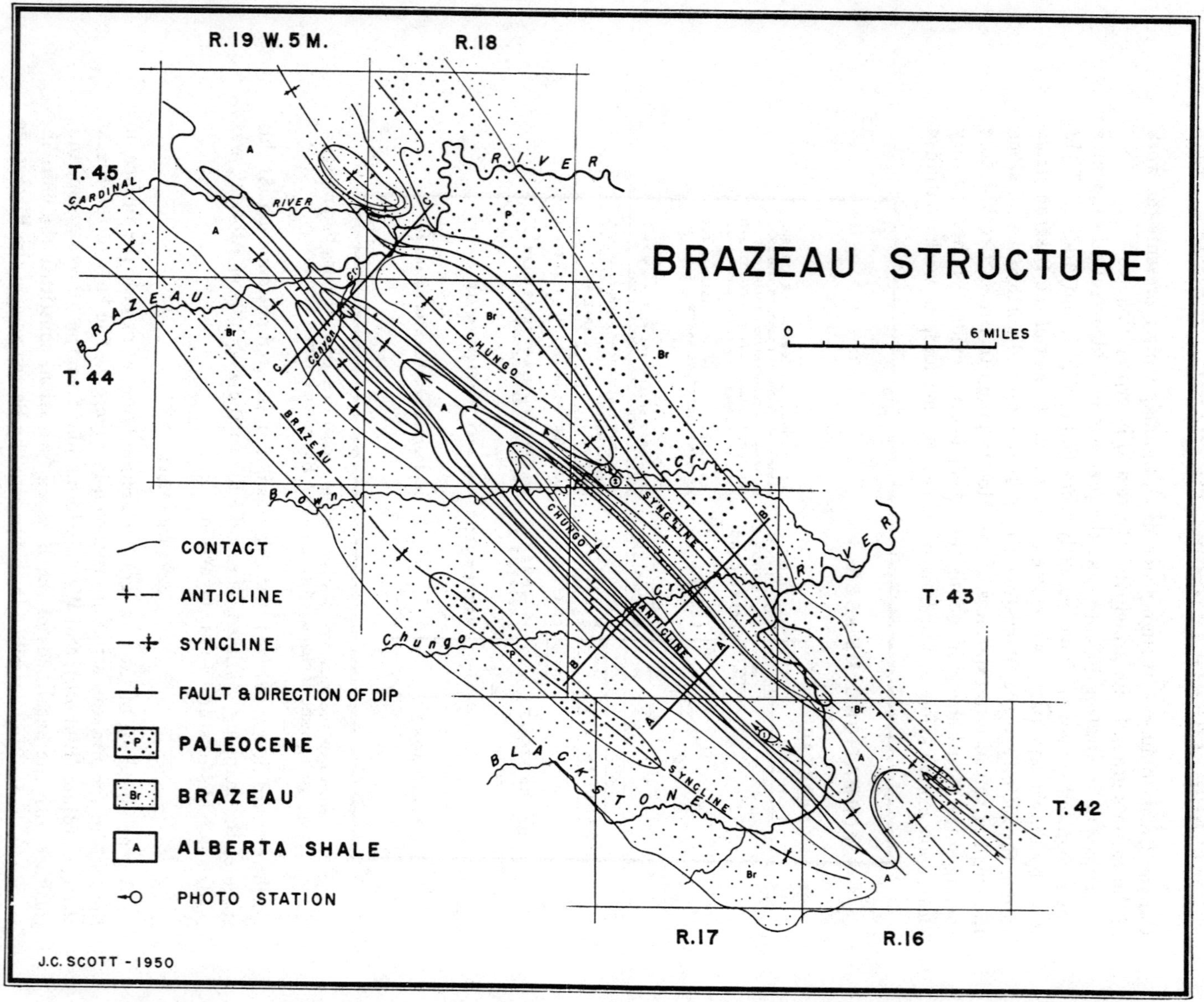

FIG. 3.—Generalized geological map of southern part of Brazeau structure, Alberta. All faults shown are thrusts from southwest. Individual formations of Alberta group not shown.

Information regarding the lower part of the Barren zone, the Blairmore group, and older formations down to Devonian, can be obtained from outcrop areas on the Brazeau Range, along strike toward the southeast. Additional stratigraphic data on formations down to the Mississippian are available from well sections. The oldest rocks penetrated by near-by wells are beds of the Mississippian Rundle formation. These are in turn overlain by the Fernie and Nikanassin formations of Jurassic age, followed by the Lower Cretaceous, Cadomin, Luscar, and Mountain Park formations of the Blairmore group. The Blairmore is overlain by the marine strata of the Blackstone, Cardium, and Wapiabi formations (Alberta group) which, with the probable exception of the basal part of the Barren zone, are Upper Cretaceous in age. An upward transition to continental deposition marks the boundary between the Wapiabi and the overlying Brazeau formation. Beds lying above the Brazeau formation were considered, at the time of mapping, equivalent to the Upper Cretaceous Edmonton formation of the plains at the east. Paleobotanical evidence showing these beds to be Paleocene in age has since been assembled by Bell (1). They are similar lithologically to the underlying Brazeau beds. Only gradual changes can be noted in the entire sequence from the base of the Brazeau upward; individual exposures, with a few exceptions, are difficult or impossible to correlate.

The opposite is true of the Alberta group. The Blackstone, Cardium (Bighorn), and Wapiabi formations can be zoned, each into several units or members. These are shown in Figure 2. Each member has lithologic characteristics, which, with paleontological evidence, make it possible to distinguish it from others in all but the smallest outcrops. The principal lithology of each zone or member is shown in the figure with a few of the characteristics and markers. Particularly distinguishable are the Cardium sandstones and the zones near the Wapiabi-Brazeau contract. These are prominently shown as markers in the accompanying cross sections. For a more detailed lithologic description, reference should be made to "Zones in Alberta Shale in Foothills of Southwestern Alberta" by Webb and Hertlein (22). The units set up in their work, with some modifications, have been used in the present paper. Hake, Willis, and Addison followed the same lithologic framework, based on stratigraphic studies by Webb. The most noticeable difference between their work and the present, in dividing the map units, concerns the Cardium formation: they use Malloch's (15) term "Bighorn" rather than Cardium and include the transition beds below in this formation; the writer follows the usage of Webb and Hertlein. Regardless of the nomenclature, the identifying features can be readily recognized. This coupled with excellent outcrops in critical areas makes possible the unravelling of the structural complexities.

STRUCTURE

GENERAL

The main features of the southern two-thirds of the Brazeau structure are shown in Figure 3. The structure is anticlinal in form, with younger Brazeau and

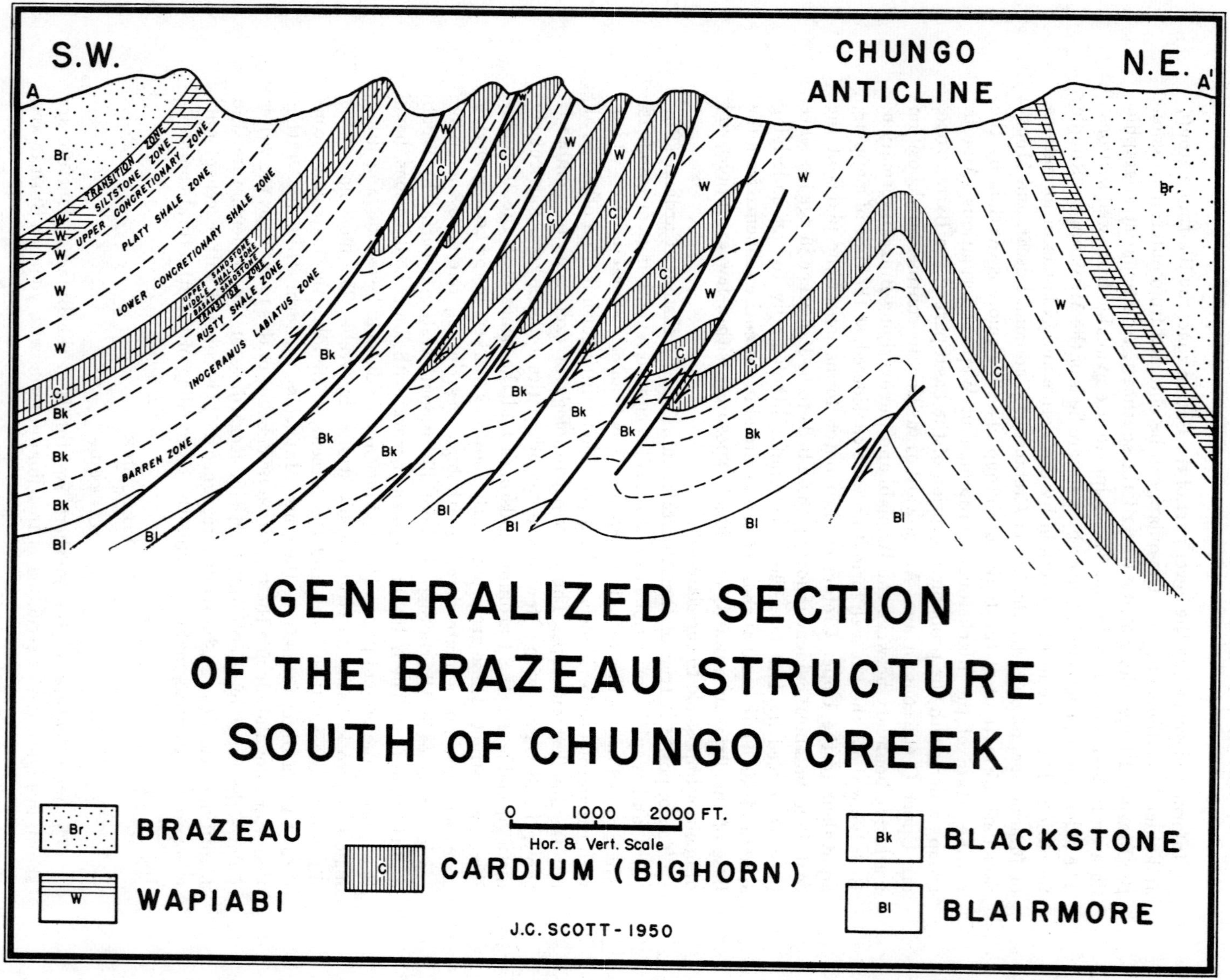

FIG. 4.—Generalized section of Brazeau structure south of Chungo Creek. Formation contacts, light solid lines; members shown by dashed lines. Faults, heavy solid lines; arrows in direction of movement.

Paleocene beds exposed on each flank, and with the older Alberta group exposed in the center. For this illustration, the Alberta group is not subdivided into its component formations, although the mapping of the faults shown are dependent on their recognition. This is particularly true of the Cardium. Were it possible to show the surface distribution of this formation, the existence of the indicated faults would be apparent. However, the scale of the drawing is such that more detail would obscure the important features. Therefore, not only is the Alberta group shown undivided, but some of the minor folds are omitted. This generalization is offset in part by information on the cross sections, which should give an understanding of the nature of the structure between.

Three main features of the Brazeau structure can be distinguished: the Brazeau syncline, the Chungo anticline, and the Chungo syncline. The Brazeau syncline, extending along the southwestern side of the structure, is developed at the surface in unbroken beds of the Brazeau and Paleocene. Along its northeast flank, the basal beds of the Brazeau formation form a resistant cap over the more easily eroded Wapiabi shales. This condition has produced a scarp which is the most prominent structural and topographic feature in the area. The ridge is unbroken except for the breaches made by major transverse streams. Rock exposures on the ridge and in the intervening stream valleys, bearing out the topographic evidence, show that the northeastern flank of the Brazeau syncline is an unbroken unit from south of the Blackstone River to north of the Pembina River, a distance of more than 60 miles. All workers in the area are in agreement on this point.

Below this escarpment, northeast, the beds of the Wapiabi and Cardium formations dip into the syncline in a section very little disturbed by faulting. It is in this belt that the Wapiabi can be measured best. Farther northeast, generally just past the first Cardium exposures, the repetition produced by the faults on the southwest flank of the Chungo anticline is encountered. For the area in the southern part of Figure 3 this is characterized topographically by many strike ridges held up by the resistant sandstones of the Cardium formation; north of Brown Creek the ridges become more complex due to the folding of the fault slices across the northwestern plunge of Chungo anticline.

On the northeast flank of Chungo anticline, erosion at the Wapiabi-Brazeau contact has formed a southwestward facing escarpment which, for its length of about 10 miles, is equally prominent with that along the Brazeau syncline. A little farther northeast, a similar ridge, formed in the same manner, extends diagonally through T. 44, R. 18, from the Brazeau River to Brown Creek. These two structural units have also been recognized by all geologists who have worked in the area.

EVIDENCE OF FOLDED FAULTS

Generalized section south of Chungo Creek (Fig. 4 and *AA′* in Fig. 3).—This section represents very closely actual conditions on this part of the Brazeau structure, surface profile as well as subsurface geology. The construction is based on data from surface geology and the well sections on Chungo Creek at the northwest. The escarpments at the Wapiabi-Brazeau contact on the southwest and northeast flanks of the Chungo anticline, previously referred to, and the topographic expression of the repetitions of the Cardium formation are apparent. Figure 11, a photograph taken from photo station No. 1 (Fig. 3), looking northwest to the line of section, presents a view of the valley along the axis of Chungo anticline with the escarpment at the northeast and some of the Cardium ridges at the southwest.

The section is structurally similar to part of *BB′*, the Chungo Creek cross section, but *AA′* has the Cardium formation involved in the imbrications, at the surface, to a greater extent. It is shown to illustrate the nature of the faulting along this part of the Brazeau structure, without involving the wider concept shown in the Chungo section. The imbrications are typical of many other foothill structures, and there has long been general agreement with the basic principle shown in the construction. Geologic studies in the foothills by many geologists have shown that the type of repetition illustrated here has resulted from thrusts from a westerly direction.

All the faults in the cross section appear as thrusts from the southwest, cutting the bedding at moderate to low angles. At each fault, all the beds in the southwest, or overthrust block, are older than those in the northeast block. Also, in all places, on both sides of the thrust, the fault cuts younger and younger beds upward and northeastward. This is regarded as typical of faults resulting from thrusting from the southwest. Exactly the opposite would be expected if thrusting from the northeast (or underthrusting from the southwest) occurred. In other words, if this section were to be reversed (turned northeast for southwest) a fault plane followed upward and southwestward would cut younger and younger beds. This difference between thrusting from the southwest and thrusting from the northeast is important to remember in the discussion of an area where northeast-dipping limbs of folded thrusts from the southwest can be, and have been, mistaken for thrusts from the northeast.

It is noteworthy that if only the part of the area represented by cross section *AA′* had been mapped, there would be no apparent evidence to suggest later folding.

Canyon Creek cross section.—The Canyon Creek section (Fig. 5 and *CC′* in Fig. 3) is representative of the type of structure seen in the northern part of the Brazeau area. It corresponds with the Elk Creek section of Hake, Willis, and Addison at the southern limit of their Cardinal district. Superficially at least, there is considerable difference between this section and section *AA′*. However, they have in common as a reference point, the escarpment of the Wapiabi-Brazeau contact on the northeastern limb of the Brazeau syncline: this appears at the left of both sections.

In section *CC′*, the general profile from the Brazeau escarpment at the southwest to the Brazeau River level at the northeast is shown as a solid line. The bottom of the Canyon Creek valley and the profile of the hill south of the line of section are shown as dotted lines, and are approximately to scale. Contacts are shown as light lines, and faults as heavy lines, both dashed where the geology is projected above the surface. The members of the formations are not shown here but it should be remembered that they were readily recognized and used in the mapping. All the thrusts shown, like those in Figure 4, are from the southwest.

The outstanding feature concerning this section is the small amount of con-

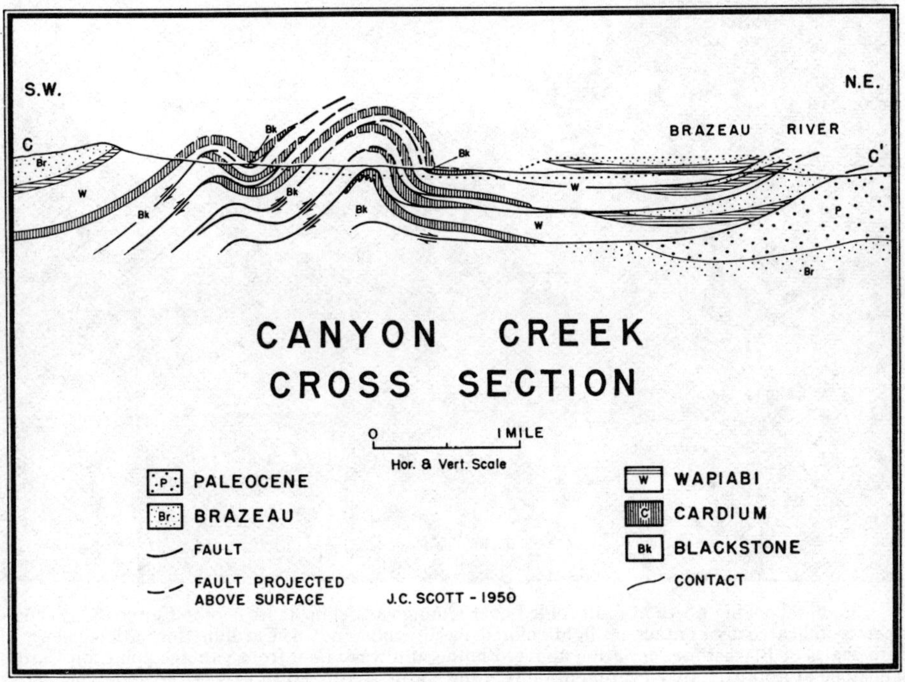

FIG. 5.—Canyon Creek cross section.

jecture needed in constructing it. Complex structures, such as those shown here, are seldom clearly exposed. In this case, critical places are completely exposed along Canyon Creek. Other exposures show that all parts fit wonderfully well, making impossible major deviations from this construction.

Following along the surface profile from the southwest, the section is uncomplicated downward to a point below the first Cardium outcrop. The second and third appearances could easily be interpreted as a continuation of the first, forming a single band folded into an anticline and syncline, were it not for the plainly seen occurrence of older Blackstone beds lying in the axis of the syncline on younger Wapiabi. Obviously a fault, such as the upper one shown, is required to explain its presence. Continuing northeastward, the second thrust is made apparent by repetition of the Cardium on the southwest flank of the Chungo anticline (between the third and fourth surface appearances). The next two faults are readily located by repetition of zones in the Blackstone formation which is exposed in good, but not continuous, southwest-dipping section.

At about this point Canyon Creek earns its name by entering a narrow steep-walled valley. From here northeastward, for more than a mile, bedrock is prac-

FIG. 6.—Looking north at fault folded over Chungo anticline at its core in Canyon Creek. Fault trace is indicated by contact of light-colored basal sandstone of Cardium formation below, and dark shales of Blackstone formation above. Photograph was taken from vantage point, on Cardium sandstone at approximate elevation of fault, a few yards southwest of axis of the fold, which strikes northwest. Fault can be traced from near stream level beyond photograph at left, upward and northeast to a point, near apex of the anticline, in canyon wall in foreground opposite observer. Beneath this part of fault, Cardium can be seen, crumpled and contorted, down to stream level. Uncrumpled slice of Cardium sandstone lies at fault plane and has been dragged along it.

At right, on east side of canyon, fault can be walked from apex, near point of observation, down to where it disappears in valley floor downstream, beyond spur capped by dark Blackstone shale. On this spur, fault dips northeast with Blackstone above and Cardium below.

Two exposures of basal Cardium sandstone can be seen in middle distance. These lie in upper two fault plates in the northeast flank of Chungo anticline (Fig. 5).

tically continuously exposed and surface formations can be seen in three dimensions. The next fault, the lowest appearing at the surface, underlain by contorted Cardium beds at the core of Chungo anticline, can be walked from the floor of the canyon at the southwest, over the crest of the anticline, and down to where it disappears into the floor of the canyon at the northeast. There is no guesswork about this: the Blackstone above and the Cardium below are exposed completely. Furthermore, the fault can be seen clearly truncating younger and younger beds of the Blackstone, as progress is made from the southwest to northeast. It is not possible to present one view showing the fault rising from the creek and descending again to it; however, Figure 6, a photograph looking north diagonally across the strike, shows dark shale of the Blackstone rising over the light sandstone of the Cardium on the southwest limb to near the apex of the anticline. The Blackstone and Cardium fault contact can be seen again at the right side of the photo-

FIG. 7.—Looking north across Canyon Creek in the fault complex of the northeast flank of Chungo anticline. The three prominent light-colored sandstone bands are repetitions of the basal member of the Cardium formation in the three upper thrust plates of Figure 5. The two at the right were visible in the background of Figure 6. The Cardium sandstone at the left lies above the Blackstone shown in Figure 6 and is part of the same fault plate; the fault between this plate and the next above dips steeply northeast separating the Cardium and the dark shale of the Blackstone to the right, which appears in the triangular exposure in the foreground. Detail of fault is shown in Figure 9. From this point the section is normal northeastwards through Blackstone, basal Cardium sandstone (center), middle Cardium shale, upper Cardium sandstone (in line of trees at right) to Wapiabi (out of sight in depression behind trees). The Cardium sandstone at the right is part of the upper fault plate of Figure 5. Photograph by E. P. Williams.

graph descending on the northeast limb. Beyond the picture, on the right, the contact can be traced back over the apex to the locality of the viewpoint.

On the northeast side of the anticline, the next fault is seen in the canyon dipping steeply northeast, separating northeast-dipping Cardium overlain by Blackstone. This appears at the left in Figure 7; a close-up of the fault is shown in Figure 9. Continuing northeastward, the observer passes upward through a normal unfaulted section from Blackstone to Cardium to Wapiabi, until the uppermost fault plane on the east side of the anticline is reached. This can be seen entering the ground with a steep northeast dip, then curving to a gentle northeast dip which carries it, in plain sight, along the canyon walls for more than ½ mile. The basal sandstone member of the Cardium formation, in the underlying fault plate, can be seen in the center of Figure 7; the same member, in the overlying plate, is at the right side of Figure 7, and is shown on both sides of the canyon in Figure 8. A view of the locality where the upper fault intersects the surface is shown in Figure 10, its position indicated by the discordance in the shale bedding at left center. The strata beneath the fault (on the southwest) are Wapiabi, and those above (on the northeast) are Blackstone transitional to the Cardium sandstones seen at the right. In general the bedding planes above and below the upper fault are folded similarly to it but dip a few degrees more steeply northeast. Throughout the distance the fault appears in the canyon wall, it can be seen to

FIG. 8.—Looking north-northeast down Canyon Creek to Brazeau River (position shown by light-colored bank in middle distance) and beyond to hills extending northwest from mouth of Cardinal River. Two fault plates in hills are excellently exposed in cliffs at right. In foreground, nearly flat beds of basal Cardium sandstone are visible on both sides of canyon. Exposure at left is same as that appearing at right of Figure 7. Upper fault in northeast flank of Chungo anticline appears in shale a few feet below sandstone and steepens toward left to about same degree. This is locality of Figure 10. Photograph by E. P. Williams.

cut younger and younger beds, from southwest to northeast, on both sides of the fault: on the upper side, from Blackstone to Cardium to Wapiabi; on the lower side, beds higher and higher in the Wapiabi. The angle of intersection is generally low, but exceptions occur as might be expected (at least due to drag). Downstream for some distance from where it makes its appearance at the locality of Figure 10, the Cardium is found lying at an angle little steeper than the fault. Farther downstream, in a short interval, beds of the formation are seen to change to a vertical attitude with tops northeast and they disappear against the fault. Beyond this point the fault plane is traced in Wapiabi shale to where it disappears, dipping northeastward, into the valley floor. The actual trace can be seen as a sharp break separating different zones of the formation. The beds are little disturbed, but distortion and minor folding due to drag occur at some places. The complications Hake *et al.* refer to in the lower part of their Elk (Canyon) Creek section (10, pp. 307, 319–20) were not observed, although the area was carefully searched. Minor fault slices dragged from the underlying block or left as "pockets" due to local deviations in the major fault break are known to occur in the area, but it is not possible that any but very small features of this type could be

FIG. 9.—Looking northwest, closer view of fault separating Cardium and Blackstone formations in foreground of Figure 7. Basal Cardium sandstone on left; Blackstone shale, containing fine sandstone beds, on right. Tops at right. Photograph by E. P. Williams.

present here. To repeat, the fault can be traced along excellent exposures to where it disappears into the creek botton, dipping northeast.

From this point, to a point in the creek on strike with the hill shown by dots on the cross section, outcrops are good but are broken by covered intervals at one or two places. Exposures are all of the Wapiabi formation in close to normal sequence but there is some evidence of a minor southwest-dipping thrust which is not shown in section *CC'*. It is here that Hake and his co-authors may have expected the gently northeast-dipping fault exposed upstream to reappear with a

FIG. 10.—Looking northwest at upper fault in northeast flank of Chungo anticline in Canyon Creek. Base of Cardium sandstone appearing in Figures 7 and 8 shown at right, with transition beds to underlying Blackstone shales beneath. Fault intersects surface at left center, where discordance appears, dipping northeast not so steeply as the overlying beds. Wapiabi shale beds at left underlie fault. Fault is hidden underneath talus, appearing in middle of photograph, for a short distance, but reappears beyond photograph at right, dipping gently northeast with Cardium above and Wapiabi below in similar attitudes. Photograph by E. P. Williams.

southwest dip. This (if possible at all in the short distance between its disappearance and the unbroken unit of the hill) would necessitate a sharp change in the angle at which the fault cuts the strata of the Wapiabi; an angle of almost 90° between the two would be necessary. It is highly improbable that this happens, as all evidence indicates the same low angle between bedding and faulting in the northeastern part of the cross section as that appearing in the part already described. This is well illustrated by two fault slices situated on a high hill directly north of the junction of the Cardinal and Brazeau rivers. Excellent exposures show the fault separating them (considered to be the same as the one under discussion) cutting upward and eastward at a low angle through Brazeau beds beneath and the transition beds of the Wapiabi above. Wapiabi strata are finally completely cut off, and the fault continues in Brazeau beds. The same relationship can also be seen at the underlying fault which closely parallels the upper.

FIG. 11.—View along Chungo anticline looking northwest toward line of cross section *AA'* from photo station 1 (Fig. 3). Valley is formed in Wapiabi shales lying along crest of anticline, and more resistant beds of overlying Brazeau formation, dipping steeply northeast, form hills at right. Beyond view, at right, escarpment at Wapiabi-Brazeau contact swings across southeast-plunging nose of Chungo anticline to vicinity of viewpoint. Cardium sandstones, dipping steeply southwest and repeated in several fault plates, form ridges along left side of valley.

The shallow syncline occupying the eastern part of the cross section is shown to be an unbroken unit from the southwest flank to the northeast flank, where upper Wapiabi and basal Brazeau beds can be followed along the north end of the hill outlined in the cross section. Along the banks of the Brazeau River, below the hill, contorted but essentially flat Wapiabi strata suggest the presence of a fault beneath. The three west-dipping faults at the northeast end of the section are well marked, good exposures showing the repetition of beds. There is also some indication of the upward truncation of section northeastward which is so well displayed in faults exposed in Canyon Creek. The underlying fault of the three is the last of the series and may be regarded as the master thrust. The section beneath it is unbroken for a considerable distance toward the east. Well exposed Paleocene, followed by Brazeau beds, can be seen down the Brazeau

River for several miles. The dip is generally 5°–15° southwest but one or two broad, gently folded, anticlines and synclines occur.

Throughout the preceding discussion, attention has been drawn to the manner in which the faults cut the beds of the strata involved. They are everywhere seen to be cutting upward in the stratigraphic column from southwest to northeast, no matter which way they are dipping. This agrees with the construction of the cross section as a whole; each fault as it is traced from southwest to northeast is shown cutting younger and younger beds. There is no strain in making the facts fit. They can not possibly point to thrusts from the northeast where faults are northeast-dipping, even if it were not possible to trace one fault of the series across the major anticline.

FURTHER EVIDENCE OF FOLDED FAULTS

The Canyon Creek cross section was chosen to illustrate the folded faults of the area because the continuous exposures at critical locations left little opportunity for guesswork. Particularly valuable in this regard are the exposures in which the fault at the core of Chungo anticline appears. However, the folds and faults at Canyon Creek can be traced northwestward and are equally well exposed at many places. In most localities several faults can be seen, in all places migrating upward in the section in a northeastern direction as they are traced up and down through folded strata. It is well stated by Hake, Willis, and Addison, that the acid field test of geologic understanding is the prediction of what is to be encountered on the other side of the hill. Thus, observations on the Brazeau River, upstream from the mouth of Canyon Creek to the Brazeau syncline, bear out those made along Canyon Creek; the Cardinal River section ties in readily with that of the Brazeau River.

In the hills north of the Cardinal River as many as five or six repetitions of the Cardium formation can be seen, involved in such a way that no other explanation but separation by folded faults can account for their presence. Excellent examples of folded fault plates can be seen along strike northwest to a point north of the Pembina River beyond which the Alberta group is confined to a narrow belt. All of this is well illustrated by Hake *et al.* Not illustrated by them are the fault plates in the upper Wapiabi and lower Brazeau beds on the hill north of the mouth of the Cardinal River and discussed in connection with the Canyon Creek section. These features lie beyond their map area and lack of detailed knowledge of them, on their part, probably explains the difference between this work and theirs toward the northeast end of the Canyon Creek section. The continuation of the fault plates on the Cardinal ridge can be mapped northwestward but exposures are rare and reliance is placed upon the excellent exposures near the Cardinal.

In general, with careful field work, there is no possibility of misconstruing the over-all surface and near-surface structure of the northern part of the area. The greatest likelihood of error involves the correlation of individual faults from stream to stream, through localities of few exposures, where thrust plates are numerous, closely spaced, and similar in attitude and stratigraphy.

South of Canyon Creek the most spectacular illustrations of folding of the imbricated beds of the Alberta group are found along the axis and southwest flank of Chungo anticline between Brown and Canyon creeks where some of the highest and most rugged hills of the area are to be found. On one of these, just north of Brown Creek, sandstones of the Cardium formation are seen repeated in a shallow syncline which deepens and plunges northwestward toward Canyon Creek. Farther along this ridge two broad "shells" of the Cardium formation lie, gently dipping, one superimposed on the other, along the plunging crest of the anticline. The upper one is presumably equivalent to the Cardium slice underlying the lowest fault exposed in Canyon Creek. This indicates the presence of still deeper faults along that part of the Canyon Creek section.

There is some indication of folding of the faults directly southeast of Brown Creek, but no further evidence is found until the south end of the structure is reached. A satisfactory interpretation of the structural geology of the area south and east of the Blackstone River in T. 42, R. 16, does not appear to be possible without accepting the presence of folded faults. Along the Blackstone River, beds of the Brazeau formation, beneath the lowest thrust, are greatly contorted suggesting the presence of the fault above. Southeast of the river, scattered exposures of the Brazeau formation can be found for a short distance only. The Cardium formation can be traced in almost continuous exposures from the southwestern flank of Chungo anticline across the plunge to the northeast flank; there it can be followed northward to where exposures end a few hundred feet south of the Brazeau outcrops; the Cardium beds strike directly toward beds of the Brazeau. For these reasons, the two formations must be separated by a fault which is also swinging around the plunging nose of the anticline. Brazeau beds appear again in the center of the township, involved in a syncline hundreds of feet in elevation above the river exposures. In this locality the beds of the Wapiabi-Brazeau contact, forming a prominent escarpment similar to those previously described, can be traced from one flank of the syncline to the other. On the southwest flank, repetition of northeast-dipping beds in the upper Wapiabi attest to the presence of the upper fault; on the northeast flank, basal Brazeau beds rest on higher Brazeau beds, indicating its presence and upward migration in the section. On the northeast, beds of the Brazeau formation above the master fault lie, with underlying Paleocene beds, in a syncline, a situation calling for a folded fault.

DRAG FOLDING

In previous pages, evidence of drag folding occurring in the structure described has not been discussed. Continuity through several folds and not the direction of movement has been the important point in showing the existence of folded faults. Direction of movement is important, of course, to the consideration of the structure as a whole. However, it is likely that the invariable upward migration of the faults through the stratigraphic section, from southwest to northeast, does more to prove thrusting from the southwest in the area than drag folds. A study of the drag fold evidence regarding the direction of movement falls more into the category of checking its dependability than providing additional proof.

Drag has been helpful however, and some of the larger drag folds, particularly those involving the Cardium formation, obviously indicate movement of the overthrust block from the southwest. A good example of this is presented by the previously described relationship of the Cardium to the upper fault plane on the northeast side of Chungo anticline in Canyon Creek. Of more interest for future diagnostic value in areas where exposures are smaller and less numerous, are the small contortions along the fault planes encompassed within a few feet. In the writer's opinion very few of these gave clear-cut evidence, which opponents would not dispute, favoring thrusting from the southwest. A surprising number of the fault breaks seen had very little sign of any disturbance whatever. However, good drag folding showing overthrusting from the southwest was available in at least one critical area where all the available evidence was welcomed. This was found in the east-dipping limbs of the faults appearing in the Chungo Creek cross section. Some evidence from the smaller drag folds was also obtainable along the Canyon Creek section. As an example, an overturned drag fold in Wapiabi shale of the overthrust slice block above the upper fault can be seen at the point where the fault disappears, dipping northeastward, into the bottom of the canyon.

EFFECT ON STRUCTURAL INTERPRETATION

It is apparent that structures such as those appearing along the line of the Canyon Creek section could be formed only by faulting, followed some time thereafter by folding. In proving the continuity of the faults as shown, the direction of movement was immaterial. It was only necessary to show that the faults had uniform characteristics throughout, whether dipping northeast or southwest. Furthermore, it was shown that one fault of the series can be traced continuously from a southwest to a northeast dip at one locality. As discussed, the common characteristics include: upward migration in the section from southwest to northeast; and drag folding indicating movement of the overthrust block in the same direction. It is certain, therefore, that the thrusts are from the southwest. The habit of the faults shown in section *AA′*, south of Chungo Creek, is also substantiated, for the two cross-section areas are correlated by the unbroken flank of the Brazeau syncline. Also, although no sign of folding of the faults in section *AA′* appears, they must be involved in the same folding as those in Canyon Creek.

On the east side of the Canyon Creek cross section it has been shown that the southwest thrusts, with their roots in the southwest side of the structure, can be traced unquestionably into fault plates on the northeast side. The prominent escarpment based upon the Wapiabi-Brazeau contact, mentioned previously as extending diagonally from southeast to northwest through T. 44, R. 18, is formed on one of these. The escarpment forms the side of a synclinal hill, upheld by Brazeau beds, which is clearly seen to be unbroken from Canyon Creek to Brown Creek. The north end is shown in the northeast half of Canyon Creek section, and an idea of the appearance of the south end is given in Figure 12, a photograph taken from a position marked by photo station 2 (Fig. 3). Since folded thrusts lie under the north end of the unit, they must lie under the southern end,

FIG. 12.—Looking north across Brown Creek from photo station 2 (Fig. 3), at south end of synclinal hill which extends northwest to Brazeau River near mouth of Canyon Creek. Exposures of thick sandstones in basal part of Brazeau formation and shape of hill depict structure. Wapiabi beds are found underlying Brazeau in normal contact at base of bluff on left side of hill. They rest in fault contact on younger Brazeau beds, lying at river level. All strata, and fault, dip northeast at this locality.

which extends into, and is part of, the system of fault plates lying along the northeast flank of Chungo anticline. From this evidence, it can only be concluded that the whole structure is involved in thrusts from the southwest which were folded following the faulting, and Chungo anticline must have been formed by this folding movement. Once that is established, many previously puzzling geologic relationships fit into place like parts of a jig-saw puzzle, and a reasonable interpretation of the Chungo anticlinal area can be made.

Sufficient work was done on the southeast, just beyond the limits of Figure 3, to show that folded faults are involved in the Brazeau Range. Paleozoic strata of these mountains overlook, from the southwest, the syncline shown in the center of T. 42, R. 16, in its southeast continuation, beyond the map area. The region extending northeast from the Brazeau structure to the eastern margin of the foothills was mapped later, and found to be complicated by folded thrusts. These, though fewer in number, are much more difficult to map because of the scarcity of marker beds in the continental Cretaceous and Tertiary sediments

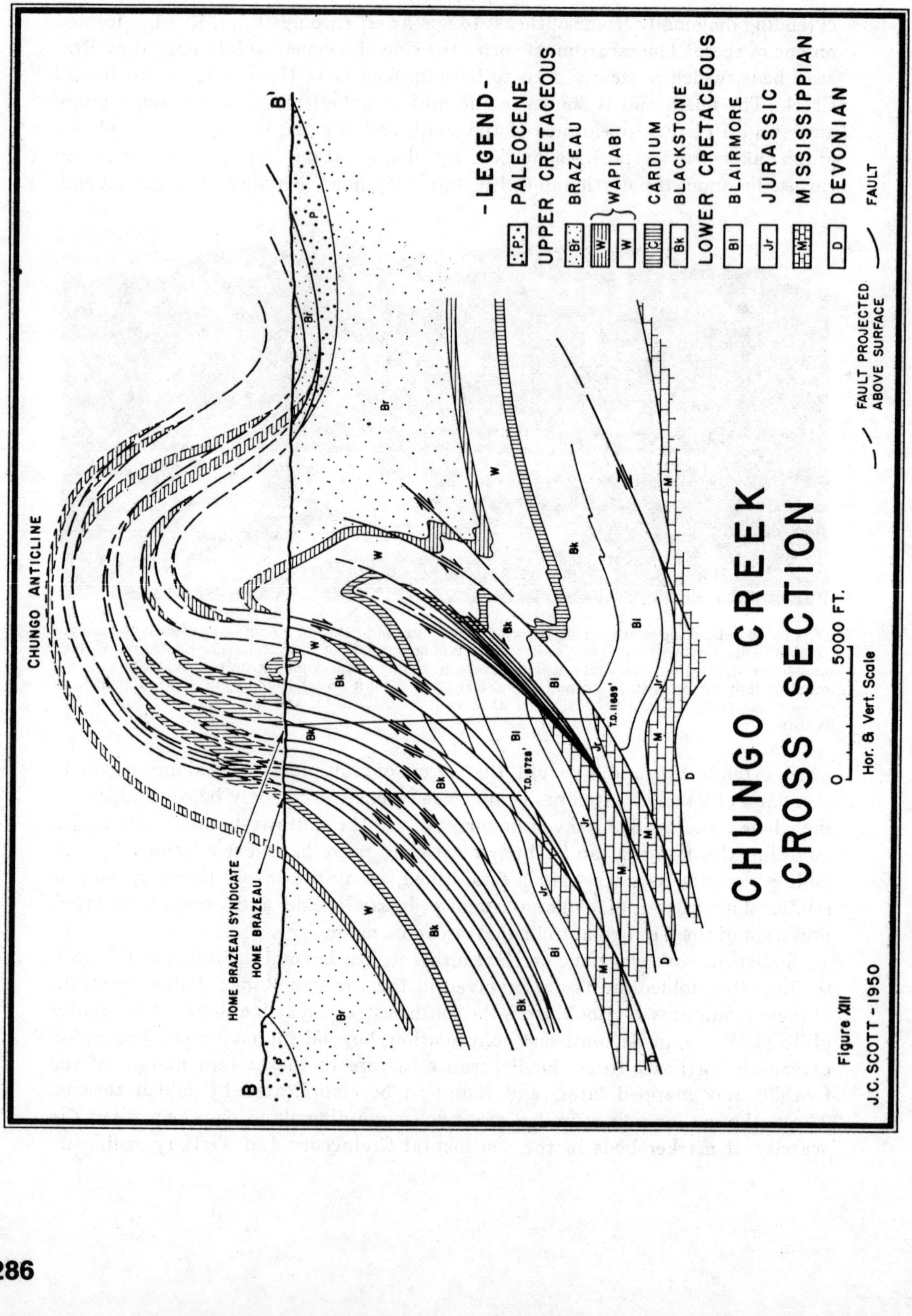

FIG. 13.—Chungo Creek cross section.

they cut. Were it not for the evidence afforded by the Brazeau structure, it is likely that these faults would have been overlooked or explained in a different way. From this mapping, it is clear that the area covered by folded faults directly correlated with those in the Cardinal district is of considerable size. The magnitude suggests that the structure of the area is of more than local importance.

CHUNGO ANTICLINE

In the present discussion an attempt has been made to show that the existence of the Chungo anticline is dependent on the same forces responsible for the folding of the faults along strike toward the northwest, and that these same faults are involved. It should now be possible to discuss features of the anticline itself with the assurance that the reader is aware of the reasons for the construction of the Chungo Creek cross section, shown in Figure 13.

The only direct evidence, in the vicinity of Chungo Creek, indicating that the northeast-dipping faults cutting Wapiabi and Brazeau beds on the northeast flank of Chungo anticline are thrusts from the southwest is the drag folding already mentioned. With only this information available, considerable reluctance to rely upon it would be held, in view of the apparent difficulties in joining the faults on the northeast side with those on the southwest. Other evidence, however, makes this solution unavoidable. The chief difficulty develops in attempting to find a root in the imbrications in the southwest flank of the anticline for the major thrust appearing in the northeast. The present section[5] shows one way this may be explained and serves as a framework for considering other ideas.

Contrasted to the map in Figure 3, all the known faults are shown. It is postulated that the lowest surfacing faults developed first, followed in turn by each succeeding higher fault. Each higher fault is considered to converge toward the northeast on the one below, finally coinciding, thus producing greater throw upward or northeastward from their junction. In the cross section, two different sources are shown for faults following these courses. A lower group is considered to spring from a single major thrust originating in Paleozoic strata, forming, in effect, a series of fault lenses reaching their maximum development in the shales of the Alberta formations. Miniature lenses of this type have been found lying along fault planes at localities in the Cardinal River area. The upper group are shown as originating as bedding-plane slippage in the Blackstone and Blairmore formations, a concept used before the drilling of Home Brazeau Syndicate No. 1, when it was believed that few of the many faults in the Mesozoic extended to the Paleozoic. This group also served to increase the depth to the top of the Mississippian to a level expected from seismic operations. Some faults are bound to originate in this way, as Hake, Willis, and Addison have pointed out. However, at the present time the writer is inclined to favor the first explanation, which in-

[5] This cross section is modified from one showing the subsurface geology by L. M. Clark at Calgary, and is based on conceptions formulated by Clark and the writer.

volves the principle of the master thrust. Both explanations have a certain amount of support from surface geology, for some of the many imbrications of the southwest flank of the Chungo anticline appear to merge or die out along strike northwest and southeast. This is particularly evident at the southeast plunge of the anticline where there may be more than the two faults shown finally carrying through to the northeast flank, but where there is definitely not room for six or more. At the northwest plunge of Chungo anticline, surface evidence indicates that at least two deeper folded faults lie beneath the lowest exposed in the core of the fold in Canyon Creek. This would mean that five faults at that location must merge to three northeastward.

No doubt other explanations are possible. It is conceivable that one of the faults appearing at the surface in the southwest flank of Chungo anticline has far more throw than indicated, and may be in reality the master thrust. The most likely one to select for this would be the fourth fault southwest of the axis. Those nearer the axis could easily be interpreted as later faults formed at the time of folding. In any event some of the faults near the axis are now thought to be due to breaking in conjunction with the folding; such faults are necessary at depth at Chungo Creek, and at one or two localities elsewhere in the Brazeau structure.

The Chungo Creek cross section is based on well information and seismic data as well as surface evidence. The second of the two wells, Home Brazeau Syndicate No. 1, was drilled after the writer's field work was completed. Although considerable faulting was expected in the tightly squeezed core of Chungo anticline, involvement of the Mississippian to the extent shown in this hole was a surprise. Since the well was drilled, Erdman (7) has published his study of the Alexo area in which he described folded faults cutting Paleozoic strata in the core of the Brazeau Range. It is now reasonable to expect that repetitions of these beds are folded in the Brazeau structure and that some of the faulting cutting Paleozoic beds preceded the formation of Chungo anticline.

Considerable differences are to be seen in cross sections of the Chungo anticline from its earlier exploration to the present. One of the first was presented by Sanderson in his "Geology of the Brazeau Area" (21, pp. 429–42); one of the more recent (as well as Sanderson's section) is given by Link in his "Interpretations of Foothills Structures, Alberta" (14, pp. 1484–85). The increase in complexity as more information became available is noticeable. The most outstanding difference between the present cross section and those presented by Sanderson and Link concerns the treatment of the northeast-dipping faults appearing in the flank of Chungo anticline. These faults have been described here as part of the system of folded thrusts from the southwest. Sanderson and Link treat them as overthrusts, or backthrusts, from the northeast. According to them, the faults are associated with major thrusting from the southwest extending from the depths of Chungo anticline to the surface at the point where the lower folded fault shown in Figure 13 finally emerges.

Regardless of where the truth lies, the opposing constructions point to entirely different concepts of the geological processes involved in the formation of Chungo anticline. They therefore illustrate remarkably well the influence of surface geology upon subsurface interpretation. No matter what other form of exploration is used upon foothill oil prospects, a detailed surface map should be made, including as much of the surrounding area as possible.

SURFACE EXPRESSION OF FOLDS

The geology of folds involving faults as well as strata is obviously more complex than that of an unfaulted section. One interesting feature, which becomes apparent as mapping progresses, is the offset between the axial planes of the

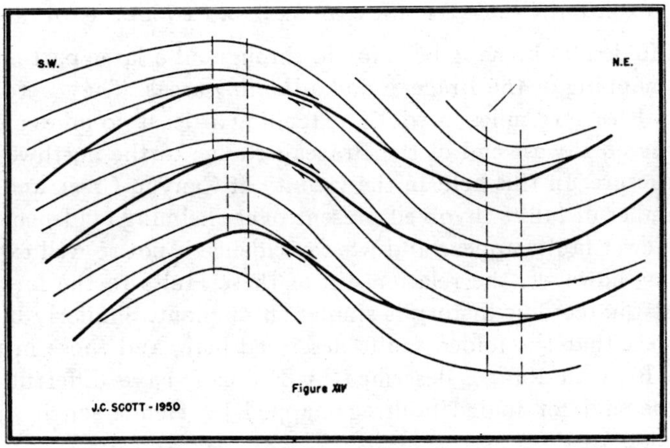

Fig. 14

faults and the bedding, due to the angle between the two. An illustration of this is shown in Figure 14 where the faults are drawn as approximately parallel heavy lines and the bedding as lighter lines at a fairly constant angle to the faults. The locations of the axes of the folds are shown by the dashed lines. The anticlinal axis of the bedding falls southwest of the axis of the anticline of the fault traces; the synclinal axis of the bedding falls northeast of the axis of the fault syncline. If the folds were asymmetric the same relationships for the same conditions would hold true but each axial plane would be inclined instead of standing vertical, as shown. Whether the folds are symmetric or not, divergence or convergence of the fault planes, and variation, within and between fault blocks, in the angle at which the faults intersect the bedding, will shift the apparent axis at different depths in the same fold. This is illustrated in the diagram; in the anticline, the axis of the bedding above the lower fault lies northeast of the axis of the bedding in the other plates. The faults cut these beds at a smaller angle.

In a normal unfaulted section, beds at different depths in an asymmetric fold exposed in hilly terrane will have offset axial lines due to the inclined axial plane. This effect, and the effect of the variations just discussed, appears to a marked degree in parts of the Brazeau area, due to sharp steps in hills and exposures, from resistant beds (Cardium sandstones) in one fault block to the same beds in another. This is well illustrated by Hake, Willis, and Addison in their sketch map of the northern end of the Cardinal district (10, p. 319). Similar offsets occur in the area shown in Figure 3 of the present paper, but are mostly ignored because of the scale of the map. Not shown, for instance, are offsets of the surface trace of the axis between different fault slices, along most of Chungo syncline. These may be rather marked. Some indication of this is shown in the southeast development of the syncline.

REGIONAL SIGNIFICANCE OF BRAZEAU FOLDED FAULTS

Related folded faults have been found throughout a large part of the foothills during the mapping of the Brazeau and adjoining areas. They were found to occupy a belt at least 15 miles in width, extending at least 50 miles along the foothills from the northwest end of the Brazeau Range to the northwest end of the Brazeau structure. In this belt, in the vicinity of Canyon Creek and other localities, the number of faults involved present overwhelming evidence of later folding. Where fewer faults appear and where evidence is not so well exposed, this is not apparent; however, the relationship of these faults to the former makes it obvious that the tectonic history is similar throughout. Similarly it is unreasonable to suggest that the folded faults described here, and those in the southern part of the Brazeau Range, described by Erdman, have different causes. The same may be said for folded faulting mapped by Henderson in the adjoining Fall Creek map-area.

Southward the writer knows of no reports pertaining to folded faults between the Fall Creek area and the vicinity of the Bow River. Most previous descriptions of isolated examples of folded faults in the Alberta foothills (and mountains) have come from the Bow River valley and south, where most intensive geological work in the Alberta disturbed belt has been performed. Several examples have been cited previously. These faults have been recognized at so many places along a belt extending from the Moose Mountain region to the Livingstone thrust in the Gap area, that it is now apparent that they form a continuous series with a common origin. The area in which they can be found is not confined to this belt alone. Warping or folding of the major thrusts of the mountain belt, extending from the ranges west of the Livingstone Range southeast along strike to the Lewis thrust, has been known for a long time. The folding of the Lewis thrust sheet is one of the earliest recorded for the Rocky Mountain front. There is also some evidence of folded faulting in the eastern foothills between the Gap area and the Bow River. As an example, one, at least, appears to be present in the Jumping Pound anticline at the edge of the foothills west of Calgary.

The folding of the McConnell thrust at the mountain front, at the Bow River and north, has already been mentioned. It appears to be part of the same folding as that in the mountains on the south and in the belt south of Moose Mountain.

L. M. Clark in his cross sections[6] prepared to illustrate the structure of the foothills and eastern Rocky Mountains has indicated that folded faults extend well into the mountains in the Bow River area.

From the foregoing, it is evident that folded faults are a widespread feature of the southern foothills and mountains of Alberta. Since most of those described are major thrusts, the large scale of the faulting makes it difficult to observe the relationship of two or more in the same structure. Therefore, where found in open folds, it is not impossible to conclude, as Link has done (14, pp. 1493–1501), that the folding or warping was contemporaneous with the faulting. In dealing with single folded faults it is also possible to apply the theory (discussed under the next heading) Douglas used to explain movement of the Livingstone thrust.

Either of these theories could apply to faults of the southern area without conflicting with the evidence of the Brazeau structure, since the two folded areas are not as yet shown to be connective. In the absence of conflicting evidence however, it is much more reasonable to conclude that the folding and faulting disturbances have been similar in each area, although not necessarily contemporaneous. Rather than conflicting, information available strengthens this conclusion. Hage (8, pp. 75–78), from work on the Barwell Mountain and Dyson Mountain folded faults in the Dyson Creek area, has concluded that folding occurred later than the faulting. In the Mount Head area, cross sections drawn by Douglas showing several faults folded together support this viewpoint. This is not suggested nor necessarily agreed to by Douglas however.

So far little evidence has been uncovered northwest of the Brazeau area to show the presence of folded faults, but there is every reason to suspect their presence. North of the Athabaska River the writer has observed a warped or folded fault overlying the west flank of the Muskeg structure; in the Brûlé area, along the Solomon Creek valley, exceptional thicknesses and lithologic repetitions in the Cardium formation have been seen in a folded area. This strongly suggests folded faulting, for the Cardium has a very uniform development throughout the foothills.

The writer is convinced that further detailed mapping in the foothills and mountains will bring to light folded faults along most of the length of the disturbed belt. It is evident that they will be very difficult to find in certain poorly exposed areas, or areas where the surface formations are entirely continental Cretaceous or Tertiary deposits.

ORIGIN OF THRUSTS AND FOLDING

The scope of the present paper is not intended to include a detailed discussion of the theory of the mechanics involved in the formation of the foothills. However, certain conclusions are suggested by the structures described, and by observations made elsewhere in the foothills.

First of all, the theory presented by Hake, Willis, and Addison (10, pp. 325–

[6] Appearing with the Field Trip Guide Book for the Banff regional meeting of the American Association of Petroleum Geologists, September, 1950.

31) to explain the folded faults in the Cardinal district now appears to be untenable (or, at least, to require modification), considering the evidence presently known. They postulate the formation of large asymmetric folds in competent Paleozoic strata; these create a "rolling up" of incompetent Mesozoic strata which finally produces low-angle thrust faults in these last sediments alone. Continuation of the major folding then is thought to produce an anticlinal buttress in the deeper beds on the side of the Mesozoic synclinal basin away from the application of force. This and other causes would then transform the rotational shearing stresses to one more uniformly distributed vertically, causing folds in the thrust slices. The presence of folded faults in the Devonian and Mississippian strata of the adjoining Brazeau Range calls for an explanation which this theory does not give. It is possible that the concepts applied to Mesozoic beds could be applied to the deeper beds, but the Brazeau Range is itself one of the asymmetric folds of Paleozoic rock, which, according to the foregoing theory, is responsible for the folded fault structure. This, and the widespread folded faulting, involving Mesozoic and Paleozoic strata in both foothills and mountains, requires a more general explanation.

Also to be considered is the theory proposed by Douglas (5, p. 79 *et seq.*) to explain the folding of a major thrust in the southern Alberta foothills. He follows Rich (20) in applying the latter's interpretation of movement along the Cumberland thrust sheet of the Appalachian Mountains to his interpretation of the movement along the Livingstone thrust. This involves the concept of the thrust following the bedding plane in definite zones of slippage, cutting diagonally across more resistant beds, to a higher zone of slippage, and so on. This produces anticlines and synclines in the overthrust mass without folding the beds beneath. The anticlinal areas above the fault, situated over the steeper diagonal cuts on the beds beneath, are then thought to be the loci of folding (involving the fault, and beds below and above) as the result of resistance to further movement among the fault.

The writer is in general agreement with ideas expressed concerning zones of slippage and the varying angles at which a thrust may cut the bedding, dependent on the resistance. Evidence of this is abundant throughout the foothills and many examples can be seen in the Brazeau area alone. Zones of slippage are considered to be, as suggested by others, the seat of imbrications associated with major faults: the extensive fault system in the Alberta group of the Brazeau structure is probably an excellent example. The writer does not favor the degree to which the theory is carried nor the conclusion expressed. In any event the proof does not lie in the correctness of Rich's ideas. The comparison Douglas presents between the Livingstone thrust sheet and the Cumberland thrust sheet is not convincing as shown, and, even as shown, depends to a considerable degree on interpretation. The variation of the angle between the fault planes and the bedding planes is almost certainly a fairly gradual change. Even in incompetent beds the angle is noticeably one of a few degrees at many places. Where the angle is greater, it is suggested that the accommodation may be as much due to adjustment of the strata to the fault, as the fault to the strata, the result being a fairly straight fault path with gentle flexures, having long radii of curvature, occurring at points of major differences in the resistance to slippage. It has been shown in the Brazeau area that it is not necessary for folding other than this mild refraction to take place, and thrusting may be repeated many times. It seems quite likely, however, that with the advent of compressive forces producing folding, the warps of the fault would have some influence upon the folding pattern. The formation of the Chungo anticline, where faulting passes from the weaker strata of the Alberta group to the stronger Brazeau beds (and cuts more steeply), may be an example of such an occurrence. Whether this is so or not, there appears to be as much stratigraphic control of the faulting and folding in Chungo anticline as there is at the Livingstone thrust; in Chungo anticline there can be little question concerning the relationship of the folding to the faulting.

In order to fit the facts of at least the area surrounding the Brazeau structure, it appears that the first event (or events) of tectonic importance occurring in or subsequent to Paleocene time, was the formation of a series of low-angle thrust faults, pushing from the west or southwest with attendant imbrications, cutting all sediments at present exposed. Highly significant in this regard is the generally uncomplicated and commonly uniformly low angular relationship between the faults and the bedding of the strata they cut. This would be expected as a result of low-angle thrusting of flat or gently west-dipping beds.

The occurrence of thrusting from the east, or underthrusting, at the eastern margin of the foothills is also accounted for by low-angle thrusting. Where these faults occur, the same low angular relationship between fault and bedding appears to be present. A fault of this type, apparently formed prior to the folding movement, was mapped near Foothills, northeast of the Brazeau area.

As far as the west thrusts alone are concerned, the comparatively small difference in angle between the faults and the bedding, generally found, could be explained by steeper thrusting of steeper west-dipping beds. However, this would necessitate, first a pronounced regional tilt toward the west, and later, a reverse tilt, prior to the subsequent folding. The faults and bedding involved in the synclines and anticlines of the folds of the next "pulse" of movement lie, in general, in attitudes closely comparable with those that might be expected from similar folding of an originally flat undisturbed section.

Folds range from fairly open to moderately tight, and, from those of the type developed in beds of the Alberta group in the Brazeau structure, to some of great size. The larger probably predominate. As expected and discussed in earlier pages, faulting has accompanied the formation of the tighter folds, which ordinarily display asymmetry with westward dipping axes. Examples of faulting of this type can be found in the Brazeau area. The Brazeau Range is almost certainly bounded on the northeast by a major thrust associated with the folding period; it is very likely that other similar anticlinal mountain structures are as well. In some places later faulting is believed to have followed the steeper parts of the folded fault planes. This seems to have happened to the folded fault in the Brazeau Range described by Erdman, and to the Livingstone thrust in the Gap area mapped by Douglas.

The foregoing is thought to apply not only to the Brazeau area and its environments, but also to the whole foothill belt and outer mountains, undoubtedly in varying degrees and intensities in different areas. An order of tectonic events has been suggested, but it is not intended to imply that these events necessarily occurred at the same time throughout the region. On the contrary, there is some indication that folding and faulting took place in the outer and middle foothills in the south prior to the formation of the Lewis and the underlying Twin Butte overthrust.

CONCLUSIONS

The evidence available from bedrock exposures in certain parts of the Brazeau area of the central foothills of Alberta is more than sufficient to prove the existence of several thrust faults, folded as a group into a series of anticlines and synclines of considerable amplitude. The only reasonable explanation lies in two distinct movements: the first, faulting, the later, folding. These faults, all the result of thrusting from the southwest, can be traced into a major anticlinal structure of the foothills (Chungo anticline) and show it to be the result of the same folding. The folded faulting is extensive and involves sediments ranging from Devonian to Paleocene, underlying a large area adjoining the Brazeau structure. It is the writer's opinion that faults found in other parts of the foothills and in the mountains on the west have the same origin and that the later folding has affected the belt as a whole.

In general, without considering causes, the order of tectonic events is thought to be as follows.

1. Low-angle thrusting from the southwest. The thrusting involves rocks at least as old as Devonian, and probably as old as the pre-Cambrian Belt series. Zones of imbrications occur along the major thrusts and are concentrated for the most part in the more incompetent sediments. Underthrusting toward the northeast, at the outer margin of the foothills, is part of this movement.

2. Folding on a regional scale, accompanied in its more intense phases by adjustment faulting, commonly taking the form of steep southwestward dipping thrusts.

BIBLIOGRAPHY

1. BELL, W. A., "Uppermost Cretaceous and Paleocene Floras of Western Alberta," *Geol. Survey Canada Bull. 13* (1949).
2. CLARK, LESLIE M., "Geology of Rocky Mountain Front Ranges near Bow River, Alberta," *Bull. Amer. Assoc. Petrol. Geol.*, Vol. 33, No. 4 (April, 1949), pp. 614–33.
3. CROMBIE, G. P., "Alexo Map-Area, Alberta," *Geol. Survey Canada Paper 44-18* (1944).
4. DALY, REGINALD ALDWORTH, "Geology of the North American Cordillera at the Forty-Ninth Parallel," *ibid., Mem. 38* (1912).
5. DOUGLAS, R. J. W., "Callum Creek, Langford Creek, and Gap Map-Areas, Alberta," *ibid., Mem. 255* (1950).

6. ———, "Mount Head, Alberta," *ibid.*, *Paper 50-8* (1950).
7. ERDMAN, O. A., "Alexo and Saunders Map-Areas, Alberta," *ibid.*, *Mem. 254* (1950).
8. HAGE, C. O., "Folded Thrust Faults in Alberta Foothills West of Turner Valley," *Trans. Roy. Soc. Canada*, 3d Ser., Sec. IV, Vol. 36 (1942), pp. 67–78.
9. HAKE, B. F., WILLIS, ROBIN, AND ADDISON, C. C., "Folded Sheet Thrusts in the Foothills of Alberta," *1934 Proc. Geol. Soc. America* (1935), p. 324.
10. ———, "Folded Thrust Faults in the Foothills of Alberta," *Bull. Geol. Soc. America*, Vol. 53 (1942), pp. 291–334.
11. HENDERSON, J. F., "Fall Creek Map-Area, Alberta," *Geol. Survey Canada Paper 45-19* (1945).
12. HUME, G. S., "Waterton Lakes-Flathead Valley Area, Alberta and British Columbia," *ibid.*, *Sum. Rept. 1932*, Pt. B (1933), pp. 1–20.
13. ———, "A Folded Fault in the Pekisko Area, Foothills of Alberta," *Trans. Roy. Soc. Canada*, 3d Ser., Sec. IV, Vol. 35 (1941), pp. 87–92.
14. LINK, THEODORE A., "Interpretations of Foothills Structures, Alberta, Canada," *Bull. Amer. Assoc. Petrol. Geol.*, Vol. 33, No. 9 (September, 1949), pp. 1475–1501.
15. MALLOCH, G. S., "Bighorn Coal Basin, Alberta," *Geol. Survey Canada Mem. 9-E* (1911).
16. MCCONNELL, R. G., "Report on the Geological Structure of a Portion of the Rocky Mountains, Accompanied by a Section Measured near the 51st Parallel," *ibid.*, *Ann. Rept.* (New Ser.), Vol. II, 1886 (1887), Pt. D, pp. 1–41.
17. MACKAY, B. R., "Folded Thrust Faults in the Brazeau River District, Alberta," *Proc. and Trans. Roy. Soc. Canada*, Abstract in Appendix C, 3d Ser., Sec. IV, Vol. 35 (1941), pp. 189–90.
18. ———, "Foothills Belt of Central Alberta, Alberta," *Geol. Survey Canada Paper 43-3* (1943).
19. ———, "Wawa Creek, Alberta," *ibid.*, *Paper 43-10* (1943).
20. RICH, J. L., "Mechanics of Low-Angle Overthrust Faulting as Illustrated by Cumberland Thrust Block, Virginia, Kentucky, and Tennessee," *Bull. Amer. Assoc. Petrol. Geol.*, Vol. 18 (1934), pp. 1584–96.
21. SANDERSON, J. O. G., "Geology of the Brazeau Area," *Trans. Canada Inst. Min. and Met.*, Vol. 42 (1939), pp. 429–42.
22. WEBB, J. B., AND HERTLEIN, J. O., "Zones in Alberta Shale ('Benton' Group) in Foothills of Southwestern Alberta," *Bull. Amer. Assoc. Petrol. Geol.*, Vol. 18, No. 11 (November, 1934), pp. 1387–1416.
23. WILLIS, BAILEY, "Stratigraphy and Structure, Lewis and Livingstone Ranges, Montana," *Bull. Geol. Soc. America 13* (1902), pp. 305–52.

BULLETIN OF THE AMERICAN ASSOCIATION OF PETROLEUM GEOLOGISTS
VOL. 43, NO. 5 (MAY, 1959), PP. 992-1025, 11 FIGS.

STRUCTURE AND ACCUMULATION OF HYDROCARBONS IN SOUTHERN FOOTHILLS, ALBERTA, CANADA[1]

F. G. FOX[2]
Calgary, Alberta, Canada

ABSTRACT

The southern foothills of Alberta are composed of much compressed, imbricately faulted Paleozoic and Mesozoic rocks in an elongate belt between the Great Plains and the Rocky Mountains. Within the belt structural shortening evidently has been achieved mainly by thrust faulting, accompanied by the development of drag folds, most of which are not large. The major thrust faults lie in well defined sliding zones, and they generally transect the bedding at relatively low angles. This is true for supposedly competent as well as incompetent beds, although there is apparently some refraction of fault planes between beds of differing competence. At and near the surface the faults commonly dip steeply, but this is a result of rotation and is not indicative of the dips the faults had when formed.

Structures that originated as folds and were later faulted may not exist. Turner Valley anticline, which is commonly regarded as a faulted fold developed from a protofold, might be an exception, but the evidence is by no means conclusive.

In this area oil and gas have been found in Cretaceous, Mississippian, and Devonian rocks. The reservoirs tapped to date are associated with thrust faults and it appears that, with the possible but unproved exception of the Turner Valley pool, all the traps are fault traps. Migration of hydrocarbons might have started very early, but was not completed before the close of the Laramide revolution. There is no reason to suppose that all hydrocarbons migrated at once, or in any one direction, or that the process of migration was uninterrupted.

INTRODUCTION

In Alberta the foothills belt lies in the western part of the southern half of the Province, adjacent to the Rocky Mountain front. It is about 250 miles long and 12–25 miles wide. It has three geographical subdivisions. The southern part extends from the International Boundary to the Bow River, the central part lies between the Bow and Athabaska rivers, and the northern part extends from the Athabaska to the British Columbia boundary (Fig. 1).

This structural belt, sometimes euphemistically referred to as the "disturbed belt," comprises an area of intensely faulted and folded sedimentary rocks. It strikes generally northwest, but there are three major strike segments. From the International Boundary to Crowsnest Pass the strike is about 135°, from Crowsnest Pass to the North Saskatchewan River it averages 170°, and from the North Saskatchewan to the British Columbia boundary it averages 135°. The boundaries are not everywhere clearly distinguished. On the east the boundary is usually drawn to include the most easterly definite and persistent dips. This *apparent* boundary very nearly coincides with the present western limit of Paleocene beds. On the west it is the frontal thrust of the Rocky Mountains. As this is not everywhere the same fault there are places where the western boundary is ill defined.

[1] Read before the Association at Los Angeles, March 13, 1958. Manuscript received, November 20, 1958.

[2] Triad Oil Co. Ltd., Calgary, Alberta.

Preparation of this paper has been encouraged by the Management of Triad Oil Co. Ltd., Calgary, and permission to publish it is gratefully acknowledged. For helpful discussion of many problems and reading of the manuscript thanks are due to J. C. Scott, R. Lakeman, P. E. Kent, A. N. Thomas, E. W. Best, F. A. McKinnon, W. J. Hennessey, J. C. Sproule, and W. R. S. Henderson.

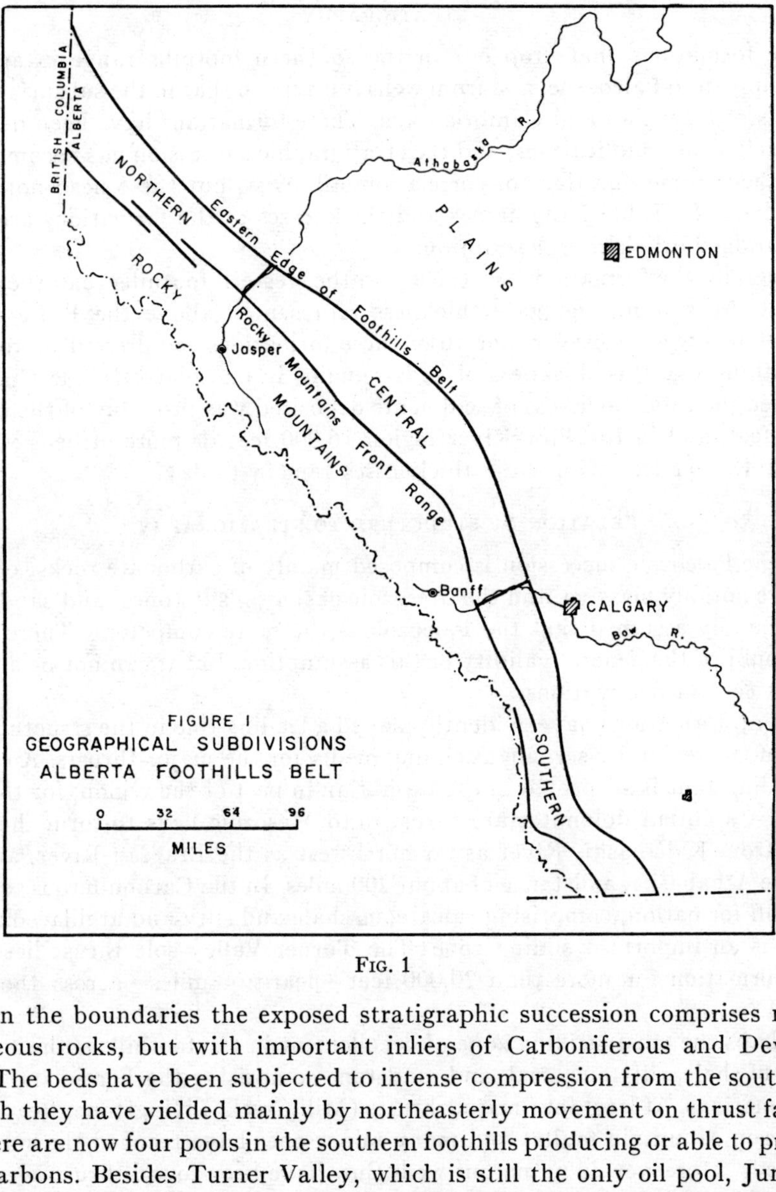

FIGURE I
GEOGRAPHICAL SUBDIVISIONS
ALBERTA FOOTHILLS BELT

FIG. 1

Between the boundaries the exposed stratigraphic succession comprises mainly Cretaceous rocks, but with important inliers of Carboniferous and Devonian rocks. The beds have been subjected to intense compression from the southwest, to which they have yielded mainly by northeasterly movement on thrust faults.

There are now four pools in the southern foothills producing or able to produce hydrocarbons. Besides Turner Valley, which is still the only oil pool, Jumping-pound and Pincher Creek produce gas, naphtha, and sulphur, and Savanna Creek is being developed as a gas pool. In recent months important discoveries of gas have been made in the Shell Waterton, Texaco Castle River, and Texaco Gladstone Creek wells. In the central foothills gas has been discovered in the Triad-B. A.-Stolberg and B. A.-Triad-Lovett River wells, which are 110 and 150 miles respectively northwest of Bow River.

STRATIGRAPHY

The formations that crop out in the southern foothills range in age from Mississippian to Paleocene, and from wells it is known that in the subsurface there are beds of Devonian and Cambrian age. These formations have been described in detail in other publications, and the stratigraphic succession has become firmly established. Some questions of correlation still exist, but these need not be discussed here. In Table I the names and thicknesses of the formations are given, with abridged lithological descriptions.

Generally the formations are thicker in the western foothills than they are in the east. Maximum aggregate thickness represented above the Proterozoic is about 30,000 feet. However, not all of these formations are present in any area, and nothing like this thickness of beds remains in one place. In the Crowsnest Pass area the total thickness of sediments deposited was probably of the order of 22,500 feet, and in the Bow River region 16,000 feet or more of beds were deposited. Rather less than these thicknesses remain to-day.

RELATION OF STRUCTURE TO STRATIGRAPHY

As the Paleozoic succession is composed mainly of carbonate rocks, of which some are notably massive, and the Mesozoic of shales, siltstones, and sandstones, it is generally assumed that the Paleozoic is the more competent. There can be little doubt of the general validity of this assumption, but it can not be accepted without certain reservations.

Several formations have evidently played a leading role in the structural evolution of the region by serving as sliding media for the major thrusts. An important sliding zone is suspected in the Cambrian in part of the region, for the same massive Cambrian dolomites are thrust onto Mesozoic beds to form the Front Range from Kananaskis River as far northwest as the Brazeau River, and perhaps the Athabaska, a distance of about 200 miles. In the Carboniferous sequence the Banff formation, comprising calcareous shale and silty and argillaceous limestones, is an important sliding zone. The Turner Valley sole thrust lies in the Banff formation for more than 20,000 feet—nearly 4 miles—across the strike (Fig. 4).

Perhaps the most striking zone of decollement in the foothills is the weak sequence of shale, siltstone, coal, and sandstone comprising the Fernie and Kootenay formations. At least four major thrusts (Mill Creek, Livingstone, Bear Creek, and Dyson Mountain faults) and many minor ones have reached the surface in these beds. Moreover, in numerous wells they have been found thrust over much younger Mesozoic beds, on faults that involve other formations at the surface.

The Blackstone formation ranges in thickness from 450 to about 1,000 feet, and comprises fissile shale with some siltstone and sandstone. Unfaulted sections are rarely found—indeed it is normal to find the formation cut by numerous thrusts, all at low angles to the bedding, and with important aggregate throw and displacement. (Throughout this paper "throw" refers to stratigraphic throw,

TABLE I

System	*Formation*	*Thickness in Feet*	*Lithologic Description*
Tertiary Paleocene	Paskapoo-Porcupine Hills	2,000	Sandstone, shale, mudstone, thin coal seams, cobble bed at base
		Disconformity	
	Willow Creek	350–2,700	Sandstone, shale, and mudstone
Cretaceous Upper	Edmonton—(Blood Reserve, St. Mary River, Lower Willow Creek)	1,000–1,500	Sandstone, shale, and coal, conglomeratic at base in places
	Bearpaw	0–600	Shale, siltstone, with thin sandstone beds
	Belly River	1,200–4,000	Sandstone, shale, some coal, basal sandstone very massive in places
	Wapiabi	1,100–1,800	Siltstone and shale
	Bighorn (Cardium)	30–450	Sandstone, arenaceous shale, lentils of chert conglomerate
	Blackstone	450–1,000	Siltstone and shale
	Crowsnest	0–1,800	Agglomerate, tuff, essentially confined to Crowsnest area
Lower	Blairmore	1,000–2,300	Sandstone, shale, some thin limestone, bentonitic, and tuffaceous beds
		Disconformity	
	Kootenay	50–700	Sandstone, carbonaceous shale, and coal
Jurassic	Fernie	100–900	Siltstone, shale, and fine-grained sandstone
		Disconformity	
Triassic	Spray River	0–50	Dolomite and limestone
		Disconformity?	
Permian?	Rocky Mountain	0–400	Arenaceous dolomite, limestone, quartzitic sandstone, siltstone, basal conglomerate in some places
		Disconformity	
Carboniferous Mississippian	Rundle group (Tunnel Mt.-Mt. Head-Turner Valley-Shunda-Pekisko)	900–2,000	Limestone, dolomite, some calcareous shale, commonly cherty, in places anhydritic
	Banff	500–900	Silty dark limestone, calcareous shale
	Exshaw	10–35	Black shale
		Disconformity?	
Devonian	Palliser	650–1,150	Limestone, dolomite, characteristically massive
	Alexo	100–200	Dolomite, silty to arenaceous
	Fairholme	1,350–1,750	Dolomite, limestone, argillaceous and silty limestone, anhydrite
	Ghost River(?)	270+	Variegated shale, dense dolomite, edgewise conglomerate
		Disconformity	
Cambrian		2,550±	Limestone, dolomite, argillaceous limestone, with sandstone or quartzite at base
		Unconformity	
Proterozoic		10,000+	Limestone, dolomite, argillite, quartzite, shale. Known only in Lewis overthrust sheet

and "displacement" to the distance between the faulted ends of beds as measured along the fault plane.) An excellent example of the response of the Blackstone beds to stress is provided by the Triad-B. A.-Stolberg well in the central foothills, 110 miles northwest of Bow River. In this well the Blackstone, which should be about 900 feet thick, was drilled through a vertical interval of 4,500 feet. This thickness must have been achieved either by great diapiric thickening or by repetition on numerous thrust faults. In either case, the incompetence of the Blackstone has permitted substantial structural shortening.

The Wapiabi formation is also composed of shale and commonly faulted, and in some places may be a major sliding zone. Its importance in this respect, however, is apparently much less than that of the Blackstone and Kootenay-Fernie formations.

Faults and their associated folds can be found in any formation in the foothills and some have important throw. It is generally true, however, that at the surface the major faults are found in the few formations discussed.

REGIONAL STRUCTURE

The southern foothills are bounded on the east by the Great Plains and on the west by the Front Ranges of the Rocky Mountains (Fig. 2). Both boundaries usually are represented on regional maps by long smooth sweeping curves. This is convenient but misleading.

On the east the thrust-faulted and closely compressed structures characteristic of the foothills seem to end just west of the west edge of Paskapoo exposures. However, there is no gradual diminution of intensity of compression from west to east—on the contrary the most easterly foothills structures are fully as complex as the most westerly ones. Moreover, in the Husky-Northern Pekisko wells, drilled 8 miles southeast of Turner Valley, faults were encountered in drilling that lie beneath unfaulted Paskapoo beds. The Consumer Co-op well, drilled 7 miles southeast of Turner Valley, penetrated at least one thrust fault that does not, as far as we know, cut the Paskapoo formation. Outcrops are relatively abundant and the Paskapoo beds dip uniformly east and at low angles; there is not the slightest suggestion that they are faulted (Fig. 3). Thus it seems obvious that the true eastern limit of foothills structure lies buried beneath the Paskapoo formation.

The western boundary is locally more distinct, but can not properly be delineated by a single unbroken line. South of the latitude of Savanna Creek it is marked by the base of the High Rock Range and the Clarke Range, which lie on the Lewis overthrust. However, from Savanna Creek northward the western edge lies farther east, at the base of the Highwood Range. This becomes part of the Fairholme Range in the vicinity of the Bow River Valley, and from the Highwood River north it lies on the McConnell fault. Thus the western boundary as envisaged by the writer must be shown by two lines *en echelon* as it is not everywhere marked by the same range or the same fault. It is a physiographic rather

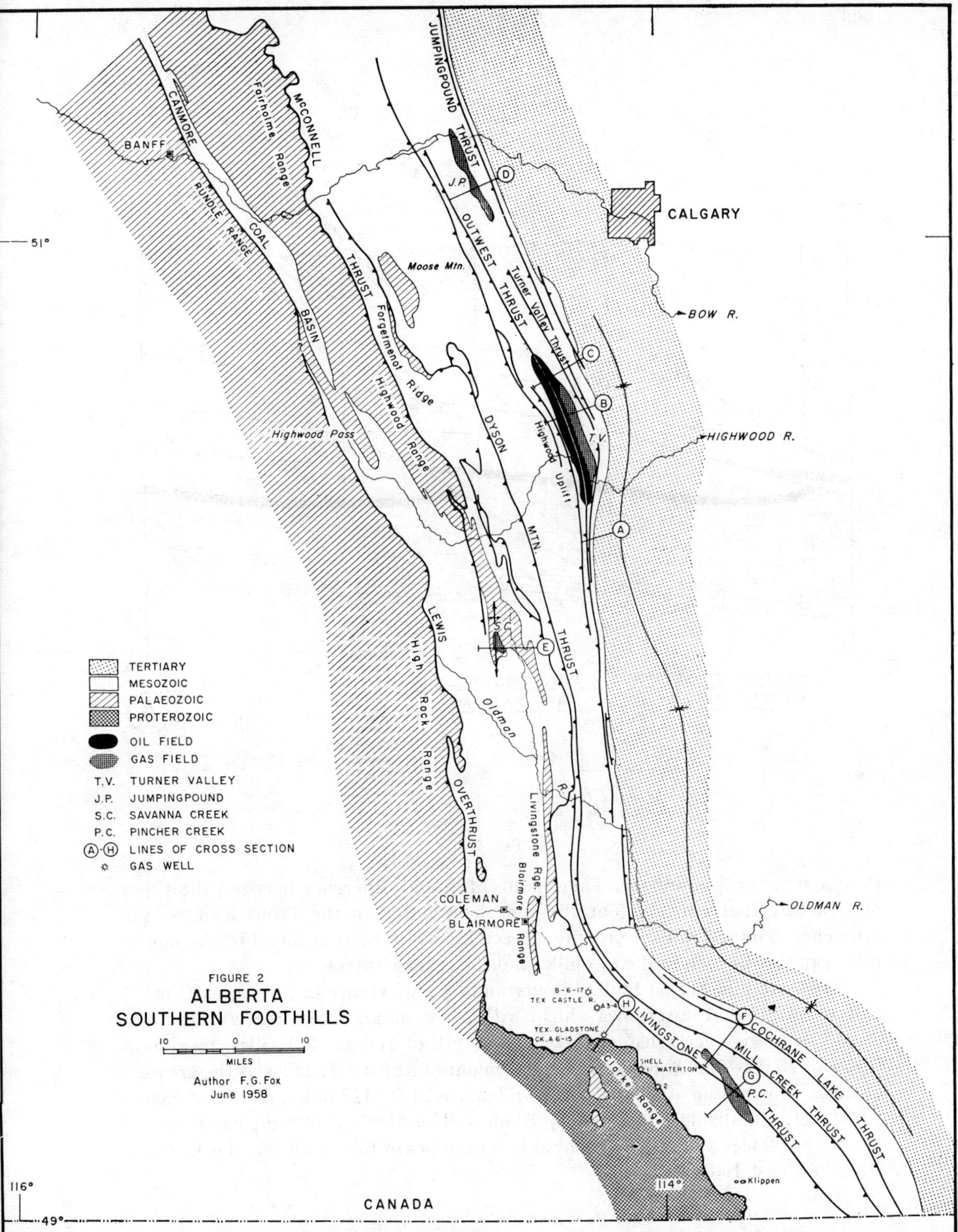
JUMPINGPOUND THRUST
McCONNELL
Fairholme Range
CANMORE
BANFF
RUNDLE RANGE
COAL
BASIN
J.P.
D
CALGARY
51°
OUTWEST THRUST
Turner Valley Thrust
Moose Mtn.
THRUST
Forgetmenot Ridge
Highwood Range
BOW R.
C
B
DYSON
Highwood Uplift
T.V.
HIGHWOOD R.
Highwood Pass
MTN.
A
THRUST
LEWIS
High Rock Range
S.C.
E
Oldman
OVERTHRUST
Livingstone Rge.
R.
Blairmore Range
COLEMAN
BLAIRMORE
OLDMAN R.
TERTIARY
MESOZOIC
PALAEOZOIC
PROTEROZOIC
OIL FIELD
GAS FIELD
T.V. TURNER VALLEY
J.P. JUMPINGPOUND
S.C. SAVANNA CREEK
P.C. PINCHER CREEK
(A)-(H) LINES OF CROSS SECTION
GAS WELL
FIGURE 2
ALBERTA
SOUTHERN FOOTHILLS
10 0 10
MILES
Author F.G. Fox
June 1958
B-6-17
TEX CASTLE R.
A3-4
H
TEX. GLADSTONE
CK. A 6-15
LIVINGSTONE
SHELL WATERTON
Clarke Range
F
COCHRANE LAKE THRUST
MILL CREEK THRUST
G
P.C.
THRUST
114°
Klippen
116°
49°
CANADA
UNITED STATES

FIG. 2

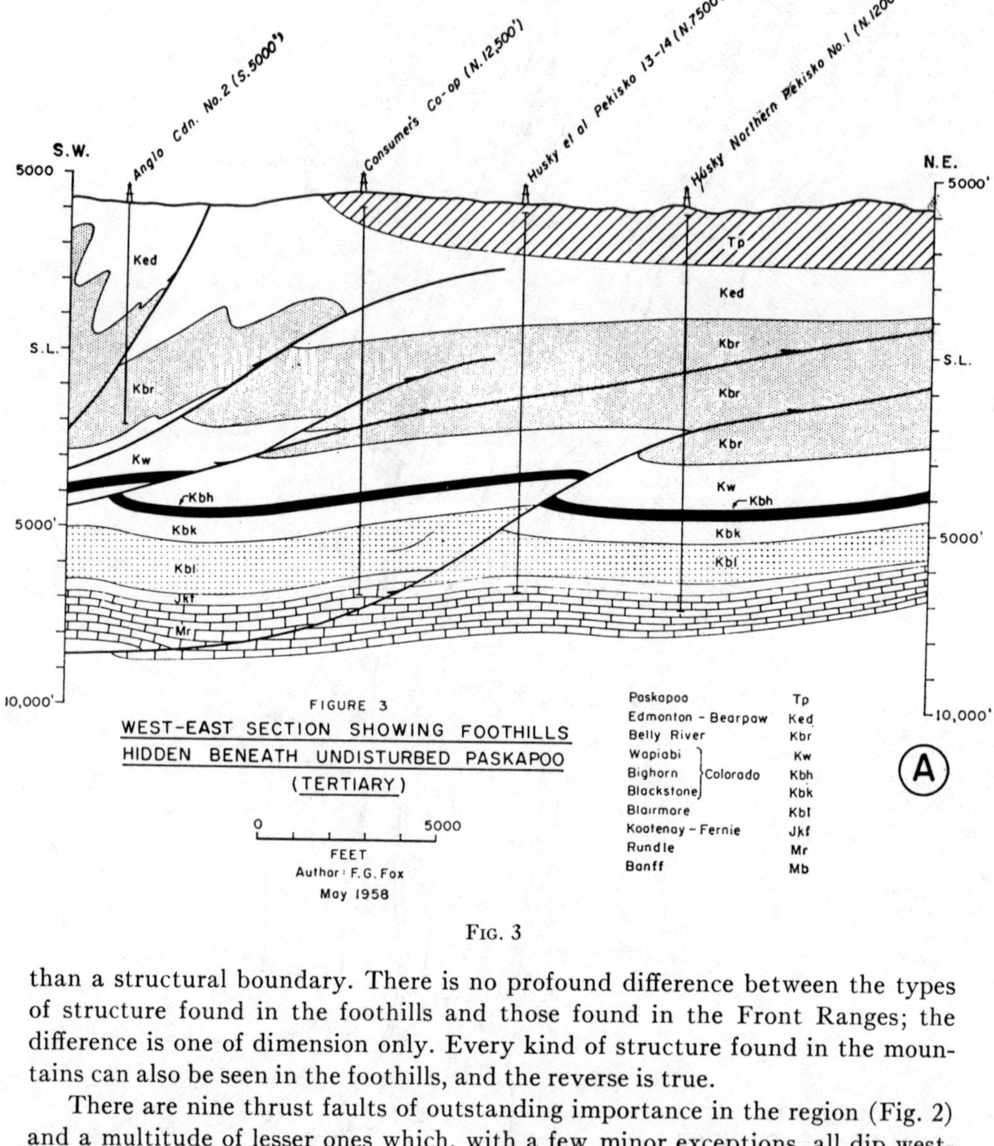

FIG. 3

than a structural boundary. There is no profound difference between the types of structure found in the foothills and those found in the Front Ranges; the difference is one of dimension only. Every kind of structure found in the mountains can also be seen in the foothills, and the reverse is true.

There are nine thrust faults of outstanding importance in the region (Fig. 2) and a multitude of lesser ones which, with a few minor exceptions, all dip westerly. The Lewis overthrust, with known length of at least 175 miles, maximum throw of 20,000–25,000 feet, and displacement of at least 25 miles, is the greatest of these. Most of the others have known lengths of 50–125 miles, throws of 5,000–8,000 feet, and displacements of $1\frac{1}{2}$–8 miles. The McConnell fault has throw of about 13,000 feet and displacement of several miles in Bow Valley, and it has been traced at least 160 miles.

Near the eastern margin of the foothills easterly dipping thrust faults are not uncommon. Such faults are thought to be characteristic frontal structures. Throw and displacement are not great and they probably do not extend to any great depth. East-dipping faults in the central and inner foothills are uncommon, but are found in some places. Probably they were once frontal faults; all that are seen now are the eroded root zones.

The lesser faults of the region require no comment except that, although individually they may have relatively small throw and displacement, they have permitted in sum a very great amount of shortening. Moreover, several minor surface thrusts can unite at depth to produce a major displacement.

In this part of the foothills four important inliers of Paleozoic rocks are present, Moose Mountain, Forgetmenot Ridge, Livingstone Range, and Blairmore Range. The oldest beds exposed in any of these are of Mississippian age. The latter three are known to be thrust blocks, and the probability is that Moose Mountain is also underlain by a major sole thrust. If it is not, it is unique in the foothills.

The best known structures are those in which bore holes have been drilled, and even they are not fully known. Even where many wells have been drilled, as in Turner Valley, major problems remain unanswered and widely divergent structural interpretations are possible.

In the ensuing discussion reference is made to the dips of various fault planes, and to the angles at which they have cut across bedding planes. The dips are important, but the angles at which faults cut bedding planes are much more important and revealing.

TURNER VALLEY

Turner Valley oil and gas pool lies near the eastern edge of the foothills about 30 miles southwest of Calgary. The first well was drilled on the structure in 1913, near gas seepages along the south fork of Sheep Creek, and was completed as a gas well in the Blairmore formation. Eleven years elasped before the Paleozoic reservoir was discovered in 1924, and this important event was followed by 12 more years of sporadic development of the gas-cap area. Crude oil was not discovered until 1936, when a well drilled on the west flank encountered the Rundle at 6,396 feet and yielded an initial flow of 860 barrels per day of 45° API oil from a depth of about 6,500 feet. This well, Turner Valley Royalties No. 1, is generally regarded as the discovery well, although as Spratt and Taylor (1936, p. 715) have pointed out the Model well No. 1 provided a hint of the presence of crude oil in 1930. Model No. 1 was a small well, however, and the significance of its "discoloured" naphtha was not generally appreciated at the time.

After the successful completion of the Turner Valley Royalties well development of the west flank proceeded rapidly and reached its culmination with development of the Millarville sector, which is really part of the west flank, during the second great war. There are now about 300 oil wells and 96 gas wells capable of production from the Turner Valley formation of the Rundle group, which is of Mississippian age. No drilling is being done at present.

The structure as seen at the surface is an anticline with a core of Blackstone, Bighorn, and Wapiabi shales, and Belly River sandstone on both flanks. The Belly River can be traced without interruption around the south end of the structure, where there is simple southerly plunge, but the northwesterly closure of the Belly River beds is interrupted by several faults and minor folds. The shales exposed in the core are cut by numerous minor faults, and the whole anticline is underlain by the Turner Valley sole thrust. It is, in fact, a drag fold, and it owes its existence to movement on the sole thrust. Other faults of considerable importance are the Millarville and Outwest thrusts (Figs. 4 and 5).

The Turner Valley thrust is a major fault at depth, the throw being about 5,000 feet maximum and displacement at least 17,500 feet in the central part of the structure. There at depth beds of the Mississippian Banff formation have been thrust onto the Cretaceous Wapiabi shale. However, both throw and displacement decrease southeast and northwest from the central area. For example, at the north end of the producing area the total throw and displacement on the Turner Valley thrust, even with the added effect of several subsidiary thrusts, are only of the order of 3,500 and 2,000 feet, respectively. Moreover, there is a truly remarkable decrease in throw and displacement updip on the fault from the leading edge of the Rundle block. At the surface the fault involves only Belly River and Edmonton beds (Figs. 4 and 5); its throw ranges from about 1,000 to 2,500 feet and displacement ranges along strike from 1,000 to 2,000 feet.

The net result of these factors is that the sole thrust, which intersects the surface about $1\frac{1}{4}$ miles east of the axis of the fold, is arcuate and convex to the east in plan. It has been mapped northwest and southeast beyond the limits of the field.

Gallup (1954, p. 399) has described the sole fault as being concave to the east in surface plan, and as having "plunge" northwest and southeast from the central sector. However, the "plunge" of the fault plane is more apparent than real; it is due to the fact that the wells that penetrated the fault plane in the central sector of the field are updip from those that passed through it on the ends of the structure. The anticline or drag fold overlying the fault has plunge in two directions, but the fault plane itself has none.

Another fault of major importance, although modest proportions, is the Millarville thrust. This fault has its known southern extremity on the west limb of the structure near the center of the field, whence it strikes northerly across the northwest-plunging end of the main Turner Valley thrust block. On the northeast flank the strike swings northwest. On this fault the northwest flank has overridden the remainder of the plunging end of the anticline, forming the Millarville fault block. This is itself broken into segments by several subsidiary faults. Gallup (1954, p. 400) has described the Turner Valley and Millarville thrust blocks as separate structures, as the former is the reservoir in the southern two thirds,

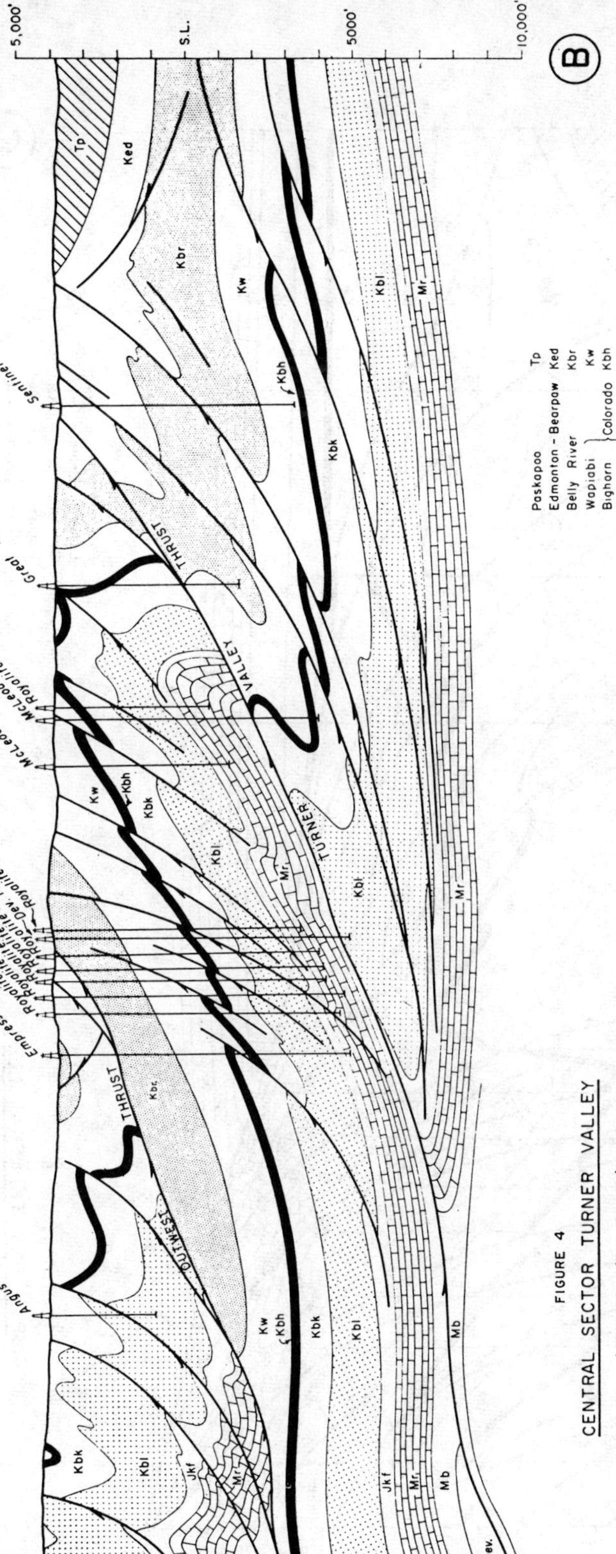
FIGURE 4
CENTRAL SECTOR TURNER VALLEY
(In part after Gallup)
0 5000
FEET
Author: F.G. Fox
May 1958
S.W.
N.E.
5000'
S.L.
5000'
10,000'
Angus I (S.1250')
Empress Synd. (N 440')
Royalite 75
Royalite 74
Royalite 67 (N 1225')
Royalite 80 (N 1350')
Brev. Test (S.3700')
Royalite 81 (S.800')
Royalite 17 (S.2841)
McLeod 5
McLeod 4
Great West I
Sentinel I (N.2460')
SOUTHWEST THRUST
TURNER VALLEY THRUST
Tp
Ked
Kbr
Kw
Kbh
Kbk
Kbl
Jkf
Mr
Mb
Dev.
Paskapoo Tp
Edmonton - Bearpaw Ked
Belly River Kbr
Wapiabi Kw
Bighorn Colorado Kbh
Blackstone Kbk
Blairmore Kbl
Kootenay - Fernie Jkf
Rundle Mr
Banff Mb
B

FIG. 4

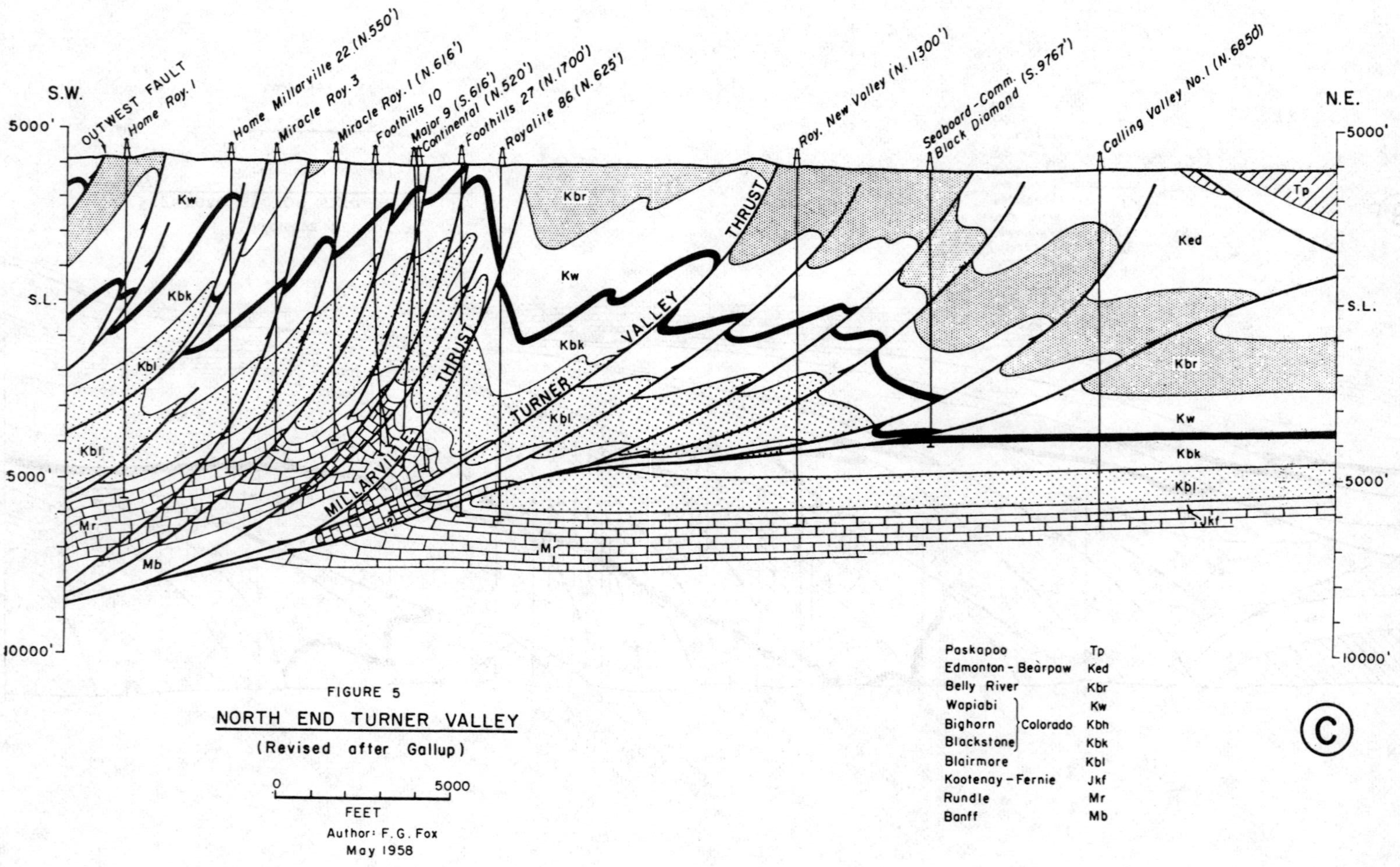
S.W.
N.E.
OUTWEST FAULT
Home Roy. 1
Home Millarville 22 (N.550')
Miracle Roy. 3
Miracle Roy. 1 (N.616')
Foothills 10
Major 9 (S.616')
Continental (N.520')
Foothills 27 (N.1700')
Royalite 86 (N.625')
Roy. New Valley (N.11300')
Seaboard-Comm. Black Diamond (S.9767')
Calling Valley No. 1 (N.6850')
5000'
S.L.
5000'
10000'
TURNER VALLEY THRUST
MILLARVILLE THRUST
Kw
Kbk
Kbl
Kbr
Ked
Tp
Mr
Mb
Jkf
FIGURE 5
NORTH END TURNER VALLEY
(Revised after Gallup)
0
5000
FEET
Author: F.G. Fox
May 1958
Paskapoo Tp
Edmonton - Bearpaw Ked
Belly River Kbr
Wapiabi Kw
Bighorn Kbh
Blackstone Kbk
Colorado
Blairmore Kbl
Kootenay - Fernie Jkf
Rundle Mr
Banff Mb
C

Fig. 5

and the latter in the northern third of the field. However, it seems apparent that the Millarville fault is subsidiary to the Turner Valley thrust, and the Millarville block is part of the northwest-plunging end of the Turner Valley block; they are merely parts of the same major thrust sheet. Reservoir data support this view.

A third major fault, the Outwest thrust, lies on and partly conceals the west flank of the anticline. It is probably the sole thrust of the Highwood uplift, which is the next large structure west of Turner Valley. At the surface, where the throw is about 3,500 feet, Wapiabi and lower Belly River beds have been thrust onto higher Belly River. The throw increases downdip so that beneath the Highwood uplift it is about 6,000–7,000 feet, and displacement is of the order of 16,000 feet. The near-surface dip of the fault varies considerably. West of the central part of the field it has been penetrated by several wells and its dip is known to be about 35°. Shortly below the surface this is reduced to about 20°. Near the north end of the field the fault is thought to dip much more steeply near the surface—in Figure 5 it is shown dipping west at 60°.

In addition to these major thrusts numerous minor faults are present, particularly in the northern half of the field. Many of these are not recognized at the surface; their presence was revealed by drilling.

Gallup (1954), Link (1949), Hume (1957), and others regard Turner Valley as the descendant of a proto or ancestral fold, which had its beginnings at the close of Paleozoic time, and endured additional folding at the close of Jurassic time, when the accumulation of oil is supposed to have taken place. Gallup says (1954, p. 410), "The Laramide orogeny merely accentuated the folding and faulted the rocks of a pool already formed." This might be so but the evidence is inconclusive.

It is well known that at the close of the Paleozoic era there was an epeirogenic uplift of much of Alberta and eastern British Columbia, which resulted in erosion and eastward truncation of the Permo-Pennsylvanian and older beds. The Nevadian orogeny occurred in latest Jurassic or earliest Cretaceous time, and it culminated in the uplift of the Cordilleran geanticline on the site of the present Selkirk Mountains. East of this lay the Rocky Mountain geosyncline, its east shelf extending far beyond the present foothills area. It is not unreasonable to suppose, as Gallup does, that these orogenic movements caused the development of folds within the bordering geosyncline.

The following principal reasons for considering Turner Valley to represent such an early fold appear to be weighty, but on close examination they fail to carry conviction.

1. The absence of oil from, and the presence of water in, the Highwood uplift, which adjoins Turner Valley on the west. The Rundle in this uplift stands about 5,500 feet structurally higher than the lowest oil-bearing Rundle of Turner Valley. Where tested it has yielded only water with traces of oil and gas. This appears to pose an awkward problem. However, if a fold was formed at Turner Valley at the close of Paleozoic time and filled with oil as a result of Nevadian movements at the end of Jurassic time, and if the Highwood uplift is a Laramide

structure, the answer appears to be clear—the Highwood uplift was formed after the oil had migrated updip into the Turner Valley protofold. But it is not so clear as it seems. According to Gallup (1954, p. 404), "It is now recognized that the porous sections of Turner Valley are, for the most part, of organic origin," and "The porous zones of Turner Valley can not be correlated with porous zones of the Paleozoic uplifts only 1–2 miles west. . . . " These two statements raise a serious doubt in the writer's mind that the apparent absence of oil from the Highwood uplift must be attributed to development of a protofold at Turner Valley. Porous zones of organic origin are not noted for their lateral continuity, but rather for their lack of it. There may never have been any direct connection between the Turner Valley reservoir and the porosity in the Highwood uplift, and if there was not, the absence of oil from the uplift need not be explained by any special sequence of tectonic events. The evidence is far from conclusive.

2. The belief that differences in thickness of as much as 80 feet in the Kootenay formation (and presence of sandstones in the Millarville area that are not found in the central area), and 50 feet in the Fernie formation must be attributed to early uplift. These thickness variations seem to be impressive, occurring as they do in a relatively small area, and there is evidence of early uplift in the general area. However, considerable apparent variations in thickness of the incompetent Kootenay and Fernie sandstones and shales are to be expected wherever they are involved in an imbricate fault system, as they are at Turner Valley. Also, the uplift known to have occurred was regional, not local, and thin Fernie sections are found over a broad area, not only at Turner Valley. Finally, the Kootenay has a regional erosional disconformity at its top, the Fernie has one at its base and one within it, and these disconformities converge eastward. Thus substantial differences in the thicknesses of the two formations, even on a local scale, do not necessarily require the postulation of a protofold.

3. Hume (1957) says: "Drilling in Turner Valley confirmed the belief that accumulation had preceded faulting, because at the north end of the field the oil-water line in one fault block is much lower than in another, although presumably at the time of accumulation they were at the same level." It is true that the different oil-water levels suggest accumulation prior to development of the north end *subsidiary* thrusts, but it does not follow that it occurred prior to development of the sole thrust. The original trap might have been a thrust sheet, the north end of which was broken up later.

It seems to the writer, therefore, that although Turner Valley anticline apparently could have been developed from a protofold, it is not definite that it did. It could just as well be nothing more than a great drag fold, formed as a result of eastward movement on the Turner Valley thrust during the latest stages of the Laramide orogeny. If this is so, the oil is contained in a fault trap, into which it migrated very little, if any, earlier than the close of the Laramide orogeny, and certainly not before the formation of the Turner Valley thrust.

The problem is of more than academic importance. If there was a Nevadian protofold the Turner Valley oil might have migrated and been trapped as early as the end of Jurassic time, and other oil was either trapped in similar folds or migrated updip into the plains area. If there was not, then the migration of oil can not have been completed before the end, or nearly the end, of the Laramide revolution, when the Turner Valley thrust block was formed, and any similar thrust block might be found to contain oil or gas.

Whatever its earlier history Turner Valley is now a pool in a compound thrust sheet.

JUMPINGPOUND

Jumpingpound gas pool lies near the eastern edge of the foothills about 20 miles northwest of Turner Valley and 22 miles west of Calgary. It is about 15 miles long and 2 miles or less wide. The pool is in typical foothills country so far as its geology is concerned, but the hills are lower and less heavily timbered than most of the foothills.

Drilling began at Jumpingpound in 1914 when four wells were spudded. Three were abandoned in 1914: one when the rig blew down, one at 600 feet, and the third at 1,255 feet. The fourth struggled along for 7 years before it was finally abandoned in 1921. During those days the Blairmore was the only producing formation in Turner Valley (and the foothills) and it seems certain that these four wells were intended to test the Blairmore. However, the Rundle reservoir was discovered in Turner Valley in 1924 and it is safe to assume that all Jumpingpound holes drilled after that date had the Rundle as their principal objective. The old Northwest Company drilled four, one each in 1926 and 1928 and two in 1936. All were failures. The two earlier ones were abandoned when it became evident that the Rundle could not be expected at less than about 10,000 feet, which was too deep for cable tools. The two holes drilled in 1936, presumably also with cable tools, both penetrated faults and were abandoned at depths of 1,652 and 2,019 feet. Of these early wells at least five were located within the present productive area, and had the tools, money, and faith been available to drill them deep enough they would have been successful. In 1939 the Brown Consolidated well was drilled on the west flank of the structure, and was abandoned at 6,885 feet when it became evident that a major thrust had been penetrated at 6,570 feet. Following this, and after a seismic survey, Shell drilled a dry hole, Shell Norman, to 12,056 feet, about 4½ miles west of the present field. This well reached the Rundle at 11,588 feet.

Finally, in 1944, Shell drilled the Shell 4-24-J well (later renamed Unit 1), which reached the Rundle at 9,618 (-5,598) feet and encountered a large flow of gas and distillate. It was completed at 9,947 feet. The Unit 2 well, drilled ¾ mile west of Unit 1, was about 1,000 feet structurally lower on the Rundle and found only water. Unit 3, drilled halfway between Nos. 1 and 2, found water, gas, and distillate, and established the water table at 6,436 feet subsea.

There are now 14 wells in the field. Of these 11 are gas wells, two are abandoned as marginal and non-commercial, and one is dry and abandoned. Two wells, Units 4 and 8, were drilled through a sole thrust and into the subthrust block, where Rundle, Banff, Devonian, and Cambrian (?) beds were penetrated. The subthrust beds are water-bearing.

Jumpingpound structure is generally anticlinal at the surface. The core of the fold exposes Blackstone shale in four small areas surrounded by Bighorn and Wapiabi beds. The Blackstone is contorted and broken by several faults. Belly River beds lie along the east flank and the northern half of the west flank, and 3 miles south of the Bow River, show some closure around the Wapiabi beds. The closure is interrupted by two west-dipping thrust faults but is reasonably distinct. Along the west flank the Belly River can be traced southeastward to about the center line of the fold (4 miles southeast of the line of section, Fig. 6), beyond which it disappears beneath a west-dipping thrust. Southwest of this fault Belly River, Wapiabi, Bighorn, and Blackstone beds appear repeatedly on a series of imbricate faults. Although the Bighorn in the crestal area of the fold is believed to close round the Blackstone, outcrops are non-existent in the critical localities and surface evidence for southeasterly closure of the structure as a whole is indeed tenuous.

The east flank is faulted, the Belly River having been thrust, on the Jumpingpound fault, onto Edmonton strata. As interpreted in Figure 6, the Jumpingpound thrust is the most easterly exposed fault. On Figure 7 it is considered to be the next known fault west.

Figure 6 was drawn through the Units 1 and 2 wells, and Unit 3 was projected about 800 feet to the line of section. It is a conservative though complicated interpretation, of which the dominating elements are the Jumpingpound thrust, the sole fault, and the intervening imbricate wedge of Mesozoic beds. The Jumpingpound fault is definitely folded. It is easily recognized in the wells; in Units 1 and 3 Blackstone beds overlie Belly River, and in Unit 2 farther west and downdip, Blairmore overlies Belly River as it should. In the vicinity of the wells there are six small flyer faults, and the main thrust emerges $1\frac{1}{2}$–2 miles east of the axis of the fold. In the field area the throw is about 6,000 feet and displacement is nearly 4 miles.

The sole fault is of more modest dimensions, its throw and displacement being 1,500 and 2,500 feet, respectively. It has been penetrated in two wells, Unit 8 north and Unit 4 south of the line of section, but it has not been seen at the surface. A flyer from this fault, penetrated in Units 1 and 3, divides the reservoir into three segments according to Martin (1956).

The imbricate wedge of Mesozoic beds, cut by at least six high-angle faults, is consistent with all available data, but it poses problems. None of the faults is known to exist above the Jumpingpound thrust, and it seems a fair assumption that they have their roots in the lower Mesozoic beds and extend upward to, but not across, the thrust. Such an imbricate system could have evolved in either of two ways. If the Jumpingpound fault formed before the sole fault, the interven-

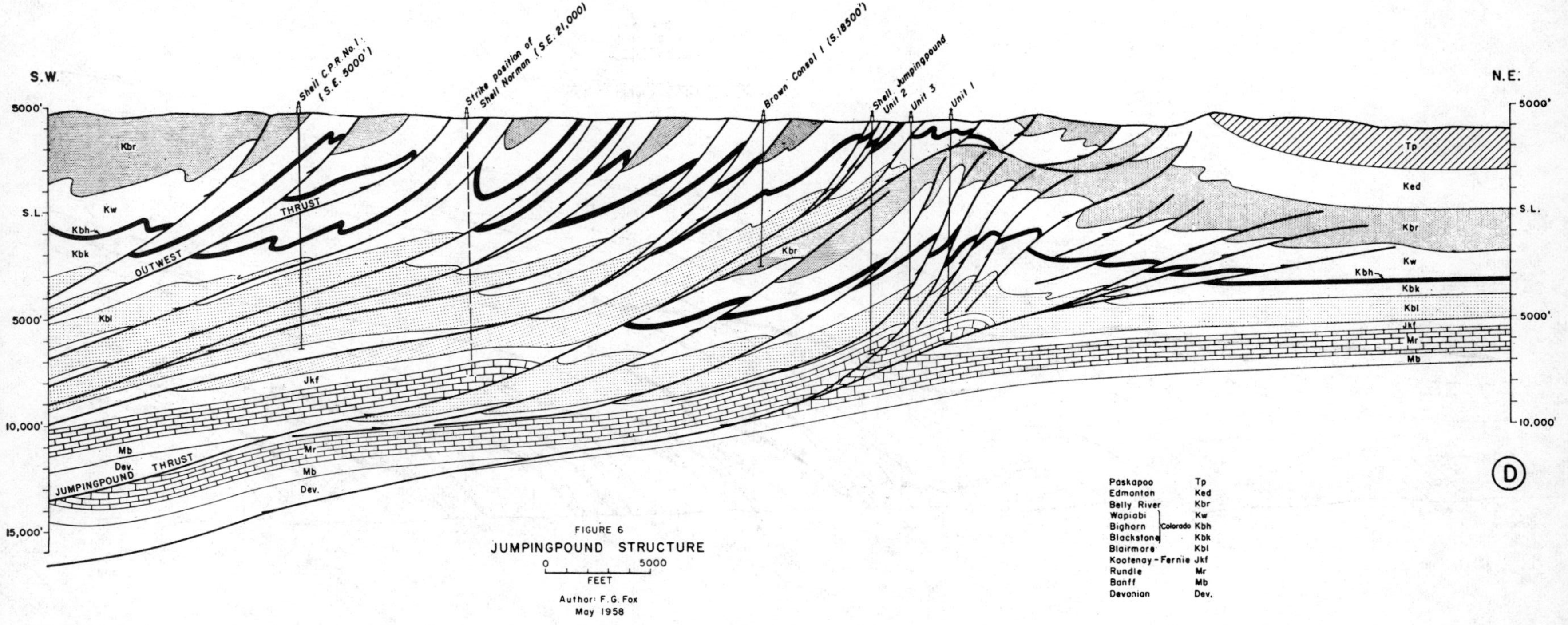
S.W.
N.E.
5000'
S.L.
5000'
10,000'
15,000'
Shell C.P.R. No. 1 (S.E. 5000')
Strike position of Shell Norman (S.E. 21,000)
Brown Consol 1 (S. 18500')
Shell Jumpingpound Unit 2
Unit 3
Unit 1
Kbr
Kw
Kbh
Kbk
Kbl
Jkf
Mr
Mb
Dev.
Tp
Ked
THRUST
OUTWEST
JUMPINGPOUND THRUST
D
FIGURE 6
JUMPINGPOUND STRUCTURE
0 5000
FEET
Author: F. G. Fox
May 1958
Paskapoo Tp
Edmonton Ked
Belly River Kbr
Wapiabi Kw
Bighorn Colorado Kbh
Blackstone Kbk
Blairmore Kbl
Kootenay-Fernie Jkf
Rundle Mr
Banff Mb
Devonian Dev.

FIG. 6

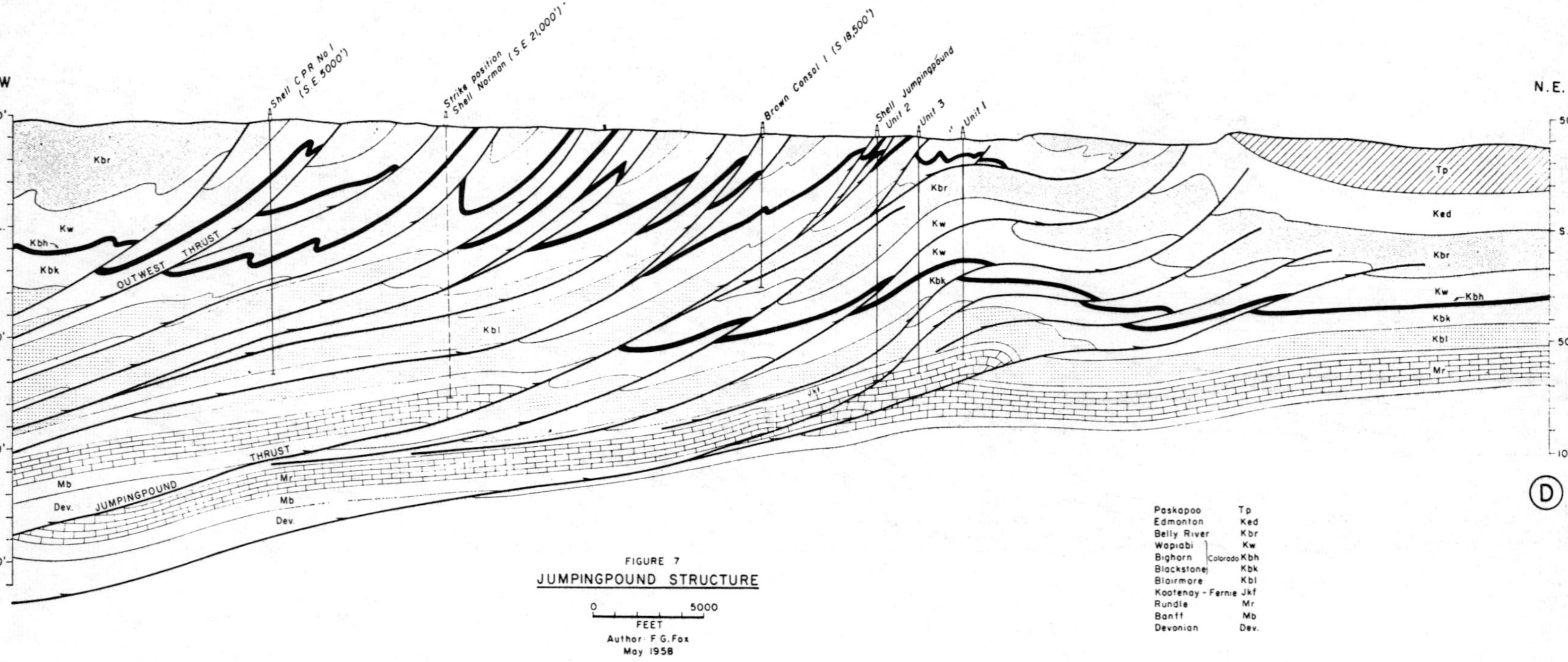
S.W
N.E.
5000'
S.L.
5000'
10,000'
15,000'
5000'
S.L.
5000'
10,000'
Shell C.P.R. No. 1 (S.E. 5000')
Strike position Shell Norman (S.E. 21,000')
Brown Consol 1 (S 18,500')
Shell Jumpingpound Unit 2
Unit 3
Unit 1
Kbr
Kw
Kbh
Kbk
OUTWEST THRUST
Kbl
Jkf
Mr
Mb
Dev.
JUMPINGPOUND THRUST
Tp
Ked
D
FIGURE 7
JUMPINGPOUND STRUCTURE
0
5000
FEET
Author: F.G. Fox
May 1958
Paskapoo Tp
Edmonton Ked
Belly River Kbr
Wapiabi Kw
Bighorn Colorado Kbh
Blackstone Kbk
Blairmore Kbl
Kootenay - Fernie Jkf
Rundle Mr
Banff Mb
Devonian Dev.

Fig. 7

ing imbricate wedge itself probably would have been developed later during the folding of the former. Whether it caused the folding of the Jumpingpound thrust, or was caused by it, can not be told. Alternatively, if the sole fault formed first and the imbricate system next the Jumpingpound thrust later could have sheared off the near-surface extensions of the imbricate faults, and carried them northeastward and upward some miles, where all trace of them has been lost due to erosion.

West of the field the Jumpingpound fault dips at 25° for about 3 miles, beyond which the dips are reduced to 10°–15°. The throw is reduced to about 3,000 feet, but displacement remains large, being about 3 miles. The fault is thought to underlie the Rundle found in the Shell Norman well, but was not penetrated.

Figure 7 illustrates a substantially different and simpler interpretation of the same data, and is preferred by the writer. Two of the main structural elements, the sole and Jumpingpound thrusts, are essentially unchanged, although the latter is shown to be less sharply folded and the former more distinctly folded than in Figure 6. The difference is to be seen in the intervening Mesozoic beds, where the excessively complicated imbricate structure has been replaced by a simple series of sub-parallel thrusts, which cut across the Mesozoic rocks at relatively small rather than large angles. They are all folded in harmony with the folding of the two main thrusts.

This structure probably developed in response to a single phase of compression —the final or nearly final pulse of the Laramide orogeny. The faults developed in succession from west to east, or from the top of the section down in non-folded beds, at low angles to the bedding planes. Formation of the Jumpingpound thrust was followed after some interval of steady pressure by development of the next fault below, and this by the next, and the process ended with the movement on the sole fault, which was accompanied by folding of the overlying beds and fault planes. The folding was caused partly by the rise of the Rundle on the sole fault and partly by the continuing compression. Compression, however, was diminishing and the whole process stopped when the Laramide orogeny ended—it simply ran down. Had compression continued long enough to move the Rundle eastward about 3 miles, and upward 3,000–4,000 feet, the resulting structure probably would have resembled the Turner Valley structure.

There is nothing in either interpretation to suggest that the Jumpingpound structure developed from a fold in the Rundle or above it. The leading edge of the Rundle block is presumed to have suffered some distortion due to drag, but the implication that the faults developed in unfolded beds is clear. Moreover, this is true of all the structures shown on Figures 6 and 7. The only folds to be seen were caused by movement on faults; that is, they are drag folds.

The gas and naphtha are, therefore, contained in a fault trap; if there were no sole fault there would be no trap and no Jumpingpound pool. Only the fault prevented the escape of the hydrocarbons updip into the plains.

SAVANNA CREEK

Savanna Creek gas pool lies about 25 miles southwest of Turner Valley and 60 miles southwest of Calgary, in the Front Range of the Rocky Mountains. It is the most spectacular field in Canada, in both its scenic and geological aspects. Development is proceeding on and around Plateau Mountain—a somewhat flat-topped anticlinal mountain 9 miles long, 3 miles wide, and standing 8,200 feet above sea-level. The surface of the plateau is about 1,000 feet above the timber line. Local relief ranges from 500 to more than 1,500 feet.

The first well (Anglo-Canadian-Savanna Creek No. 1) was drilled in 1939. It was located near the crest of Plateau anticline and near the base of the Rundle group, and was intended to test the Devonian rocks. However, it passed from Rundle to Upper Cretaceous rocks at very shallow depth (less than 50 feet) and was abandoned in the lower part of the Blairmore formation at 3,372 feet (Fig. 8). Although the well was a failure it proved the presence near the surface of a thrust fault. The next well, Phillips-Husky-Northern Target (hereafter referred to as PHNT) Savanna Creek No. 1, was drilled in 1952. In this well the Rundle was penetrated at 4,439 feet and yielded some gas, but a second major thrust was encountered at 5,182 feet. The well was, therefore, suspended in the Kootenay formation at 5,628 feet. In 1954 drilling in the same well was resumed and the Rundle was penetrated a second time at 6,401 feet. In the lower part of this block in strata thought to represent the Turner Valley formation, and also in other parts of the Rundle group, large quantities of gas were discovered. The well was drilled through the Rundle and reached bottom in the Banff formation at 9,040 feet. Development has been relatively slow but there are now six wells ready to produce gas and one is being drilled. One well has been abandoned.

Savanna Creek area has been described in some detail by Scott, Hennessey, and Lamon (1957).

Plateau Mountain anticline is a broad deceptively simple fold in the Rundle formation. West-flank dips do not exceed 25°; dips on the east flank are 10° or less. The axis of the structure strikes nearly due north and south. It plunges south from the crest maximum at about $4\frac{1}{2}$°; the north plunge is less distinct and does not exceed 2°. It is oblique to the Highwood Range, which strikes northwesterly. Thus the north-plunging end of the fold passes from the Front Range into the foothills, and the south-plunging end merges with the west flank of the Livingstone Range. On the west it is flanked by a strongly asymmetrical syncline containing Blairmore and Blackstone beds; on the east by the Livingstone River syncline and Hailstone Butte.

The structure section (Fig. 8) was drawn through the discovery well. It is similar to a recent section by Scott *et al.* (1957), and, in its uppermost part, to a section by Webb and Spratt (1939, published by Link, 1949). Link (1949) has also published a section similar to this.

Certainly the most outstanding feature of this section is the presence of three major and numerous minor thrust faults folded in harmony. Although the folded

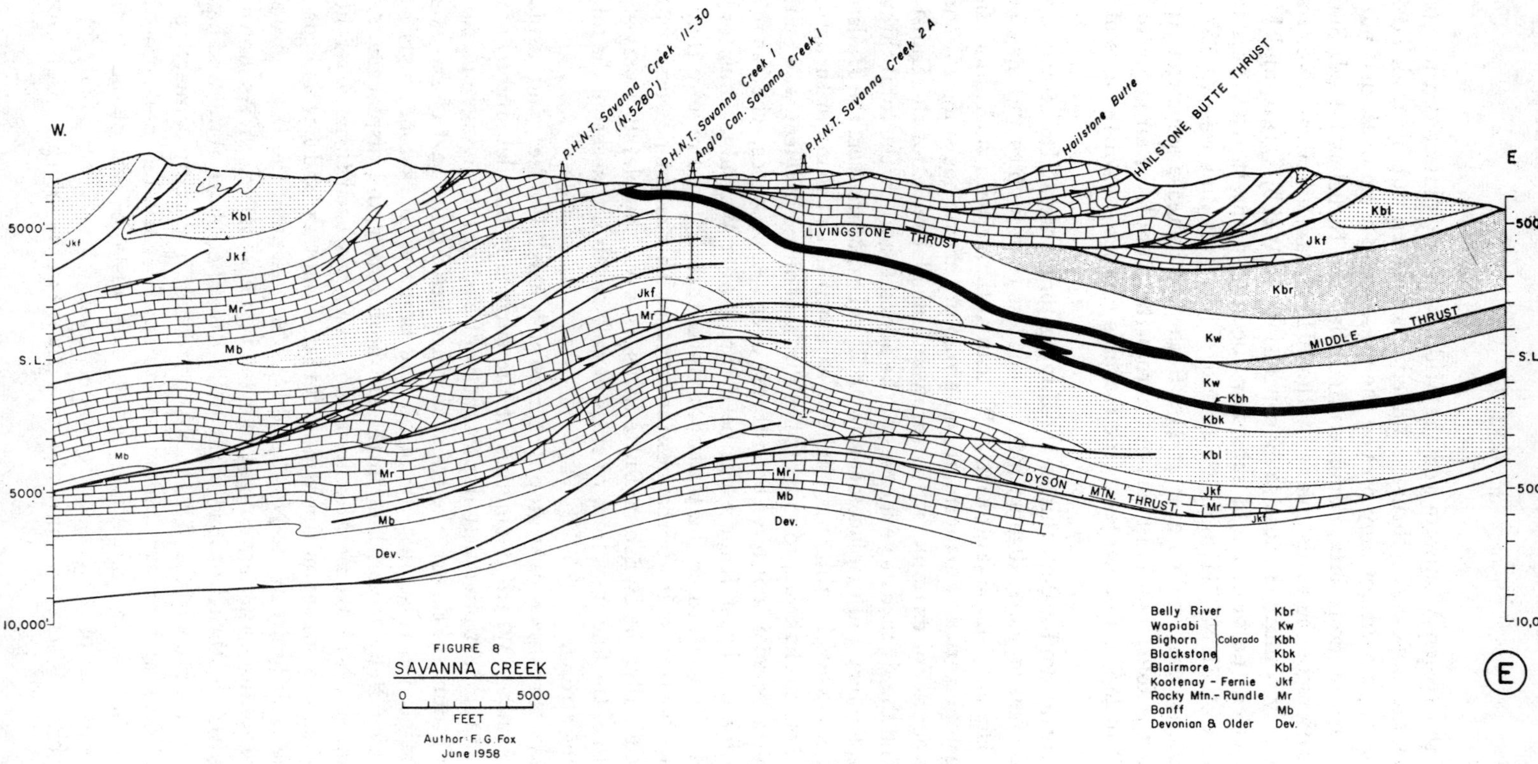
W.
E
5000'
S.L.
5000'
10,000'
P.H.N.T. Savanna Creek 11-30
(N.5280')
P.H.N.T. Savanna Creek 1
Anglo Can. Savanna Creek 1
P.H.N.T. Savanna Creek 2 A
Hailstone Butte
HAILSTONE BUTTE THRUST
LIVINGSTONE THRUST
MIDDLE THRUST
DYSON MTN. THRUST
Kbl
Jkf
Mr
Mb
Dev.
Kbr
Kw
Kbh
Kbk
FIGURE 8
SAVANNA CREEK
0
5000
FEET
Author: F.G. Fox
June 1958
Belly River Kbr
Wapiabi Kw
Bighorn Colorado Kbh
Blackstone Kbk
Blairmore Kbl
Kootenay - Fernie Jkf
Rocky Mtn.- Rundle Mr
Banff Mb
Devonian & Older Dev.
E

FIG. 8

character of the Dyson Mountain and the Middle thrust is inferred, there is no question that the uppermost thrust is folded. As pointed out by Link (1949), the Anglo Canadian well not only proved the presence of a major thrust, but, in conjunction with outcrop data, demonstrated that the thrust is folded. Subsequent penetrations of the thrust plane by the PHNT Savanna Creek No. 1 and 2A wells not only confirmed these early conclusions of Webb, Spratt, and Link, but have revealed the presence of subsidiary bedding-plane thrusts. Less than 1,000 feet west of PHNT Savanna Creek No. 1 the fault dips west at 15° and the same distance east it dips east at 11°. Both dips increase somewhat down the flanks of the fold. The fault has cut across the Rundle at an angle to the bedding of 5° or less, and emerges in the foothills 6 miles east of the axis of the fold.

There is little doubt that this is the Livingstone thrust. Its stratigraphic throw is 6,500 feet or more; displacement is not less than 9 miles and may be as much as 12. The only single fault east of Plateau anticline with such dimensions is the Livingstone thrust. Furthermore, no group of thrusts has sufficient throw and displacement. For example, if the throw and displacement of the Hailstone Butte fault and the six thrusts next east of it be added together, the total throw is only about 4,200 feet and displacement about $2\frac{1}{4}$ miles. Moreover, the Hailstone Butte thrust, a low-angle fault that might be assumed to be the major thrust, can be eliminated on mechanical grounds. It has throw of only 1,500 feet, and on it at Hailstone Butte Rundle beds have been thrust over Fernie shales, whereas in the wells on the folded thrust Rundle has been thrust over Wapiabi. Thus for the folded fault in the wells to be the Hailstone Butte thrust would require that the fault cut down section as it approaches the surface. This might be possible, but it is unlikely. Finally, the Hailstone Butte thrust dies out completely a very short distance south of the line of section, whereas in wells south of the section the folded thrust remains strong.

The Middle thrust has stratigraphic throw of 1,500–2,100 feet, and displacement of 1–$2\frac{1}{2}$ miles. Where it emerges at the surface is not definitely known. On the ground the next obviously important fault east of the Livingstone thrust is the Dyson Mountain thrust, the trace of which lies 8 miles east of the axis of Plateau anticline. On this fault, Wapiabi, and perhaps Blackstone, beds have been thrust over Belly River, but northward both throw and displacement increase, so that south of Moose Mountain Kootenay has been thrust over Belly River. In Plateau anticline, on the Middle thrust, Kootenay and Fernie beds are cut off by the fault between the No. 1 and No. 2A wells, and the Blairmore and Blackstone formations are not thought to extend far east of No. 2A. As shown by Figure 8, the Middle thrust probably has Belly River beds both above and below it where it reaches the surface. This seems to preclude any updip connection between the Middle and Dyson Mountain faults. The Middle fault probably emerges at the surface in Belly River beds a short distance east of the Livingstone thrust and west of the Dyson Mountain thrust.

This conclusion is tentative only. A minor change in the interpretation shown in Figure 8 to suggest a greater eastward extension of Wapiabi, Blackstone, and Blairmore beds on the Middle thrust, and an assumption that throw and displacement on the fault increase northward, as they do on the Dyson Mountain fault, would permit the joining of the two faults. However, at the moment the writer is inclined to think that the Dyson Mountain fault lies deeper in Plateau anticline, as indicated by Figure 8.

The presence of the lowermost fault shown on the section is not merely inferred; it was penetrated by the PHNT Savanna Creek No. 3A well, nearly 2 miles north of the line of section. Stratigraphic throw on the fault is at least 2,500 feet. The displacement is unknown, but here shown to be about 3 miles. If this fault rises, with slightly increasing dips, east of the line of section, it should reach the surface about 8 miles east of Plateau anticline. Moreover, it can be inferred that at the surface the beds directly overlying the fault should belong to the Wapiabi formation. Both the assumption and inference are justified by surface data. There is an important thrust 8 miles east of Plateau anticline, and on it Wapiabi shale has been thrust onto Belly River beds; it is the Dyson Mountain fault.

The lowest Rundle penetrated to date, that below the Dyson Mountain fault, is probably the trailing edge of the Rundle of the Highwood uplift. At some depth below this Rundle there ought to be at least one, and perhaps two, more faults, the Outwest and Turner Valley thrusts. They may be merged or not. In either case the fault, or faults, should be present and should be nearly parallel with the bedding planes in Cambrian or, perhaps, Proterozoic beds.

The numerous subsidiary faults that cut the Rundle are also of great interest because of the uncommonly low angles of their intersection with it. This is especially true of the uppermost sheet. The Hailstone Butte fault has been mapped (Scott *et al.*, 1957) and was penetrated in the No. 2A well. There can be no doubt that it transects the bedding at an exceedingly small angle—less than 5°. A smaller fault, above the Hailstone Butte thrust, can be seen in the cirque wall east of the No. 3A well, 2 miles north of the line of section, cutting the beds at about 10°. The subsidiary faults penetrated in the No. 1 and No. 11–30 wells are *inferred* to cut the Rundle at angles of 5°–15°. They are known to cut the Blairmore at angles of this order.

It might be argued that these faults should be shown to cut at greater angles across the Rundle and older Paleozoic than they do across Mesozoic beds. This would be consistent with the long held conviction that the Paleozoic is much more competent than the Mesozoic. It would also suggest application of the principle of step faulting as proposed for the Appalachians by Rich (1934) and persuasively applied to the southern foothills by Douglas (1950) and Hume (1957). However, it would not be consistent with what is seen in the uppermost thrust sheet, where the sole thrust (Livingstone thrust) and all the subsidiary thrusts are very low-angle (to the bedding) faults. Moreover, there is no suggestion whatever that any of the faults are step faults.

Some explanation for the folding of the faults is required. Douglas (1950) and Hume (1957) have suggested that some folded faults have been developed from step faults. Douglas (1950, pp. 84–85) says "Folding of the fault plane is, in part, only the result of the accentuation of its initial anticlinal and synclinal bends. An initial anticlinal bend is present in the thrust plane where it changes from a diagonal, crosscutting shear to a bedding plane thrust . . . , and an initial synclinal bend where it changes again to a diagonal shear. . . . " As applied to the Livingstone thrust farther south, in the vicinity of the Oldman River, Douglas' explanation appears to fit the observed facts very well. However, at Savanna Creek it does not apply (Scott *et al.*, 1957). There is no evidence that the faults cut across the Rundle at notably greater angles than they do across the Mesozoic formations. There is no suggestion that the present folded nature of the thrust planes is attributable to peculiarities in their original shapes. The Livingstone thrust, which may be a step fault at Oldman River, is a very low-angle thrust at Savanna Creek.

To the writer it seems likely that at Savanna Creek the faults cut at low angles across Paleozoic and Mesozoic beds alike, that folding did not begin until the faults were all formed, and that when folding began movement on the faults ceased. The thrust blocks then were folded together, perhaps as the result of drag over a deep-seated (Turner Valley-Outwest?) thrust, or perhaps as a result of their own inertia.

If there is a change in the character of the Livingstone thrust between Savanna Creek and Oldman River, it probably can be related in large part to the stratigraphy of the Rundle group. The presence locally of evaporites, shaly beds, or even thinly bedded rather than massive limestone and dolomite might greatly reduce its competence. For example, in the Triad-B. A.-Stolberg well the Rundle was penetrated three times. The upper two Rundle bodies are both thin slices lying on low-angle thrusts in evaporitic Rundle beds. They were simply peeled off the top of the formation (Thomas, 1958). Evaporites have not been observed in outcrop at Savanna Creek but they have been logged in the wells, and furthermore, there are several thinly bedded shaly zones.

It seems clear that the Paleozoic, and particularly the Rundle, in places is less competent than we have supposed, and is capable of forming structures similar in nearly every respect to structures in Mesozoic beds. Comparison of Figure 8 with Figure 7 reveals a remarkable general similarity between the Savanna Creek and Jumpingpound structures. Different beds are involved in the two structures, and the former is broader and higher, but they are otherwise much alike and they probably formed in the same way.

The entrapment of gas here was possible because the migration of hydrocarbons was not completed before the faults were formed. The trap was made by the development of these faults, and the folds we now see are merely distortions of the original fault trap. Once this was formed, folding of the thrust blocks localized the trapped hydrocarbons, but it was the faults that prevented their escape updip in the first place.

PINCHER CREEK

The Pincher Creek gas-condensate pool is 18 miles north of the International Boundary and 100 miles due south of Calgary, in about the middle of the foothills. It is not in typical foothills country (except structurally) as the characteristic elongate ridge and valley topography is missing. Most of the country is of low relief, and the plains extend to the mountain front.

The earliest drilling in Alberta was done in the general Pincher Creek area. Oil seeps along Cameron Brook in Waterton Park had been known to the Indians, presumably for generations, and by 1886 at least one white man, the famous John G. (Kootenai) Brown, had learned of them and put the oil to use. A survey of the plains on the east was made in 1884 by a banking firm, but drilling did not commence until 1891, when a well was started 2 miles southwest of the present field. It was abandoned at 190 feet and its successor, drilled 12 miles southwest in about 1893, was rather less successful.

In 1901 a well was started on Cameron Brook and drilled to 1,020 feet, where oil was found in the Cretaceous below the Lewis overthrust. This strike gave rise to many ill founded rumors of spectacular production, which were denied by its backers. It undoubtedly did produce some oil, although apparently the quantity was small, and it encouraged further drilling. At least 11 more wells were drilled nearby in the next 6 years, without success.

Drilling in the foothills had its real beginning in 1907 when a well was drilled to 1,800 feet near Pincher Creek. Exploration was slow, however, and during the next 40 years only 16 wells were drilled in the general area. All were dry and the deepest one was abandoned at 7,189 feet. Twelve of them were drilled after 1920. At least six of the early wells were drilled within 5 miles of the boundaries of the present field.

In 1942 Canadian Gulf Oil Company commenced geophysical work in the southern foothills. Gravity-meter surveys were made first and these were followed by extensive geophysical surveys during the next 4 years. In 1947 Canadian Gulf Pincher Creek No. 1 was drilled to 12,516 feet and completed in Rundle beds. On production tests it indicated an open-flow potential of 45 million cubic feet of gas and 1,670 barrels of condensate per day. Subsequent developments have been slow, mainly for economic reasons. The field, about 16 miles long and $2\frac{1}{2}$ miles wide, now has 10 wells testing or on production. Two others have been abandoned.

In 1929 and 1930 one company drilled 17 core holes in the vicinity, presumably seeking structural data. These ranged in depth from 225 to 1,754 feet.

The surface geology in the area reveals nothing of the presence of either a fold or fault slice at depth. The reservoir underlies a series of rather thin thrust sheets of Belly River and Wapiabi beds, and even the strike of the surface beds is at variance with the strike of the reservoir. There is no suggestion of anticlinal structure—on the contrary the distribution of the beds indicates that the series of thrust sheets might represent the severely faulted east limb of a southeasterly plunging syncline lying on the Harland Lake thrust. West of the north end of the field there is a suggestion that Belly River, Wapiabi, Bighorn, Blackstone, and Blairmore beds are closing westward around such a syncline. The west limb of the syncline, if it is a syncline, lies hidden beneath a fault.

West of the field area and the zone of imbricate faults there is a belt 6–9 miles wide, characterized by gentle folds and a few faults involving Colorado and Belly River beds, and extending to the mountain front. The Belly River is at the west against the mountains with the Colorado below it at the east.

Northeast from the field for 3 miles at the north end and 5 miles at the south end the exposures are mostly Belly River in thin steeply inclined fault sheets. Farther northeast to the foothills front younger beds are found with the Belly River in the fault slices. East of Cochrane Lake thrust in the frontal crumpled zone, uppermost Cretaceous and even Paleocene beds are found.

The area is one in which mapping and correlation of structures are perplexing as most of the outcrop is confined to stream channels. Also, the lack of good topographic expression of structure makes the tracing of beds and contacts most difficult.

Figures 9 and 10 are structure sections drawn through the northern and central parts of the field respectively (Fig. 2). Both illustrate the absence of anything resembling an anticline at the surface.

Both sections are characterized by the remarkably low angles at which the beds are presumed to lie at depth, and by the extremely low dips of the fault planes at depth. It is possible to make quite different sections, but there is strong evidence to support the interpretations offered here.

Near the western end of Figure 9, two wells, Alberta Gas and Fuel Yarrow, and Alberta Gas and Fuel Drywood, provide an important clue to the structure. Both wells penetrated a regular section (excepting a minor repetition of Blackstone beds at Drywood) until they were part way through the Kootenay formation. In the Kootenay both passed through a fault, the Drywood well to Wapiabi and the Yarrow well to Blackstone beds, and in each the fault has 2,000–2,500 feet of throw. As there is no fault of this magnitude known to reach the surface between the two wells it is reasonable to suppose that both penetrated the same fault, the Twin Butte thrust, and further, that this fault reaches the surface east of the Drywood well. Thus, very low dip for at least one of the thrusts is fairly established.

Further evidence is provided by the attitude of the reservoir beds. Erdman *et al.* (1953, p. 156) point out that the Rundle beds dip southwest at 4°–8° in the field area. Also, the dips are known from geophysical data to remain at 10° or less for some distance west of the field (Erdman, pers. comm.). Moreover, the drilled thicknesses of the Blairmore and Blackstone beds directly above the Rundle indicate that they must be nearly flat.

With the establishment of the very low westerly dip of the reservoir block and of the Twin Butte thrust, it follows that the intervening structures also dip west

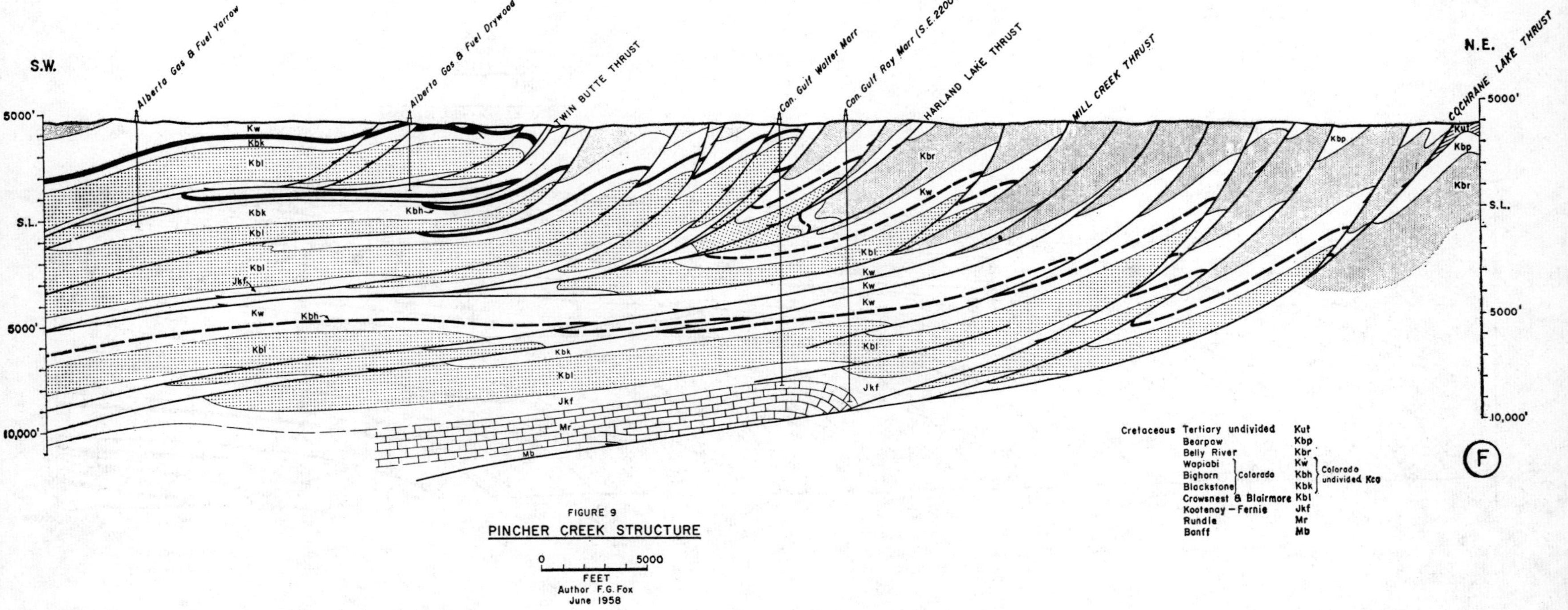
S.W.
N.E.
Alberta Gas & Fuel Yarrow
Alberta Gas & Fuel Drywood
TWIN BUTTE THRUST
Con. Gulf Walter Marr
Con. Gulf Ray Marr (S.E. 2200')
HARLAND LAKE THRUST
MILL CREEK THRUST
COCHRANE LAKE THRUST
5000'
S.L.
5000'
10,000'
Kw
Kbk
Kbl
Kbh
Jkf
Kbr
Kbp
Kut
Mr
Mb
F
Cretaceous Tertiary undivided Kut
Bearpaw Kbp
Belly River Kbr
Wapiabi Kw
Bighorn Kbh
Blackstone Kbk
Colorado
Colorado undivided Kcg
Crowsnest & Blairmore Kbl
Kootenay – Fernie Jkf
Rundle Mr
Banff Mb
FIGURE 9
PINCHER CREEK STRUCTURE
0 5000
FEET
Author F.G. Fox
June 1958

FIG. 9

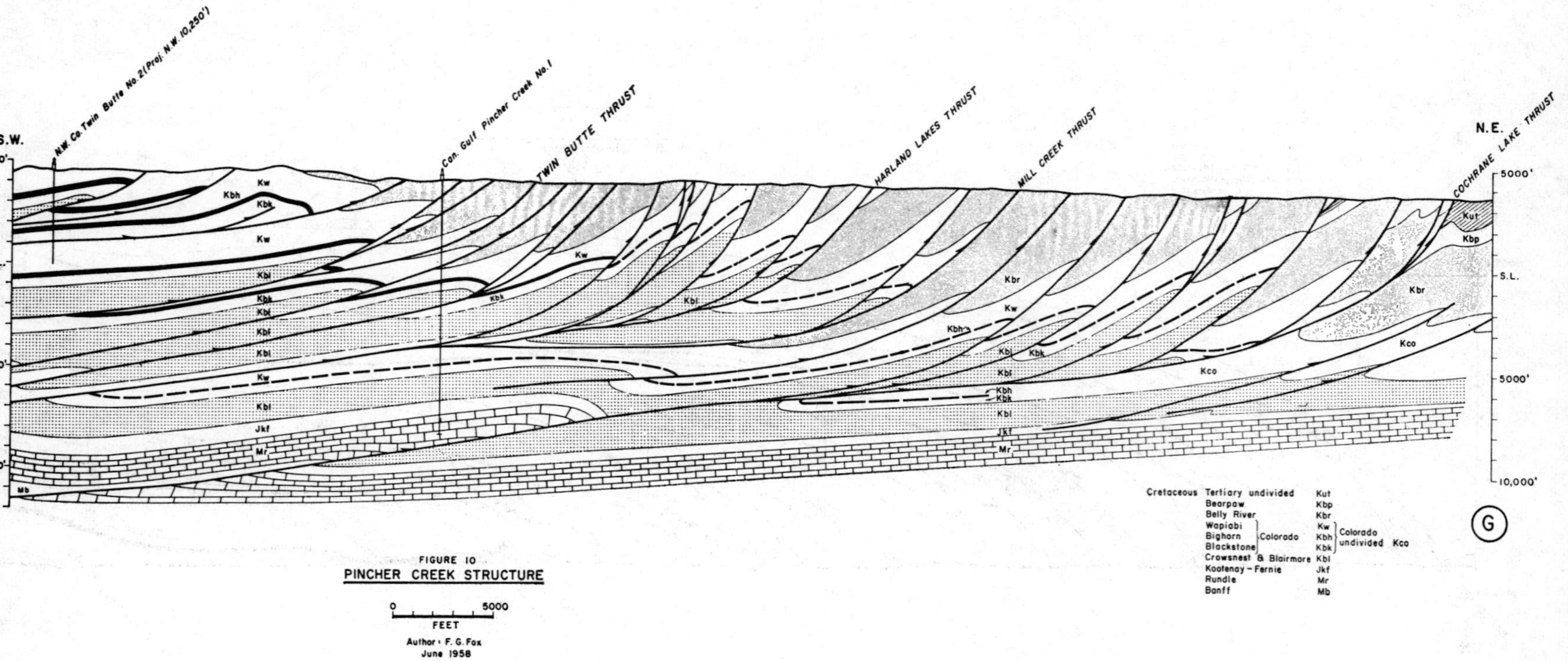

N.W. Co. Twin Butte No. 2 (Proj. N.W. 10,250')
Con. Gulf Pincher Creek No. 1
TWIN BUTTE THRUST
HARLAND LAKES THRUST
MILL CREEK THRUST
COCHRANE LAKE THRUST
S.W.
N.E.
5000'
S.L.
5000'
10,000'
Kw
Kbh
Kbk
Kbl
Kw
Jkf
Mr
Mb
Kbr
Kco
Kut
Kbp
FIGURE 10
PINCHER CREEK STRUCTURE
0
5000
FEET
Author: F. G. Fox
June 1958
Cretaceous Tertiary undivided Kut
Bearpaw Kbp
Belly River Kbr
Wapiabi Kw
Bighorn Colorado Kbh
Blackstone Kbk
Colorado undivided Kco
Crowsnest & Blairmore Kbl
Kootenay - Fernie Jkf
Rundle Mr
Banff Mb
G

FIG. 10

at very low angles—faults and bedding planes alike. Moreover, it also follows that the faults and strata east of the Gulf Marr wells (Fig. 9), however steep they may be at the surface, most probably are also nearly flat at depth. To assume that they are steep east of the Marr wells and nearly flat west of them would pose a difficult, though not insoluble, geometrical problem.

No well has yet been drilled through the Rundle and the sole thrust presumed to lie beneath. The presence of this fault is inferred from the fact that the eastern termination of the Rundle is abrupt, and the inference is supported by geophysical data and by drilling. A fold with a very steep northeast limb could produce an equally abrupt apparent termination, but experience in other areas and the presence of so many known faults in this area lead to the conclusion that the reservoir is not an anticline but a thrust sheet. The east dips at the leading edge of the Rundle block do not necessarily indicate that the structure is a faulted fold. They are mainly the result of drag against the fault, which probably formed in non-folded beds. The throw and displacement on the sole fault are unknown quantities. On Figure 10 the throw is shown to be about 3,000 feet and displacement 3 miles, but these values might well be far from the truth.

No single thrust of major proportions emerges at the surface east of the wells, and the place of emergence of the sole fault is, therefore, in doubt. The Mill Creek thrust is a considerable fault, but in the face of the evidence favoring low dips at depth it does not seem likely that it underlies the Rundle block. However, the Cochrane Lake thrust has been traced a considerable distance (Douglas, 1952) and is an important fault. It is, therefore, assumed to be the sole thrust. It is thought that it merges downward with several of the faults west of it (four in Figure 9 and seven in Figure 10), and that it is a major thrust at depth. An alternative possibility is that the sole thrust does not appear at the surface anywhere. Should this be so, the Cochrane Lake fault and its subsidiary faults might be even more nearly flat than they are shown to be.

Not far west of the field is the present eastern edge of the Lewis overthrust sheet. The eastward movement of this enormous thrust mass has evidently had a profound influence on the geology of the area. This is partly to be seen in Figures 9 and 10. The thrust sheet on which the Drywood and Yarrow wells (Fig. 9) were drilled, and the sheet on which the Twin Butte well (Fig. 10) was drilled are characterized not only by the very low dips of the thrust planes, but also by the remarkably gentle folds and low bedding dips associated with them from the surface down. These structural conditions persist westward to the mountain front. Farther east, as shown in both sections, the near-surface dips are steep. In the writer's opinion, the gentle structures on the west have low dips because they were once overlain by the Lewis overthrust sheet, the weight of which prevented any considerable folding, rotation, and steepening of the beds and fault planes below. At the latitude of Pincher Creek the thrust sheet probably extended about 9 miles east of its present front. Only two small klippen about 3 miles northeast of the mountains now remain.

The faults under the gentle westerly structures probably lie very near to the attitudes at which they were originally formed. Farther east, where they were free of excessive overload, they were rotated and steepened along with the bedding planes, especially near the surface.

The Lewis overthrust had other effects. At Pincher Creek the general strike at the surface is about N. 45° W., whereas the strike of the reservoir body (Rundle) is N. 30° W. For this reason the strike of the reservoir is commonly thought to be peculiar. The fact is there is nothing very peculiar about the strike of the Rundle; on the contrary it is reasonably consistent with the strike of the foothills as a whole. It is the strike of the surface beds that is peculiar. They have been bulged eastward in Pincher Creek area and south (Fig. 2) by the eastward movement of the Lewis thrust sheet, which apparently stripped the Colorado and younger beds from their foundations and pushed them eastward, accompanied by considerable piling and imbrication in front. The Rundle, being deeply buried and relatively remote from the thrust sheet, was less affected by its movement and, furthermore, was affected later than were the Cretaceous beds. This process accounts for the apparent great disparity in shortening between the Rundle and the Upper Cretaceous beds, and for their divergent strikes.

Erdman *et al.* (1953, p. 156) suggest that the reservoir block, with the Kootenay/Fernie and Blairmore beds immediately overlying it, is an early Laramide structure that was later overridden by the near-surface structures. This implies (as they point out) that the Lewis overthrust, which was the cause of the shallower structures, was formed later than the reservoir block and the Cochrane Lake thrust. The writer's view is that the Lewis overthrust is the earlier fault. As it pushed forward it moved a large part of the Cretaceous sequence with it, resulting in the formation below and in front of it of a series of low-angle faults. In front of the thrust sheet the subthrust Cretaceous beds were free to move, and rotation of beds and fault planes resulted. However, as rotation proceeded and the angles of dip increased, resistance to further rotation also increased. Ultimately the point was reached where relief from the stress was attained more readily by the formation of a new fault or series of faults farther in front of the thrust sheet, and these in turn suffered rotation. The end result was the formation of the Cochrane Lake thrust and rotation of its forward part, but this was only the end because the Laramide revolution was ending. Had it continued unabated more faults would have been formed east of the Cochrane Lake thrust. Thus Pincher Creek reservoir is in a very late, rather than an early, Laramide structure.

The Cochrane Lake thrust is not the only one involving the Rundle formation. The Mill Creek thrust and perhaps the Twin Butte thrust also involve the Rundle west of the line of section. The Waterton and Castle River blocks both lie structurally above and west of the Pincher Creek block. (This is not to imply that they lie on either of the faults last named).

A striking feature of both sections is the simplicity of the reservoir block compared with the structures that lie above it. Both also show how a major displacement at depth can be masked at the surface by a maze of subsidiary and relatively minor faults.

Pincher Creek is clearly an accumulation in a fault trap. There is no evidence that a fold ever existed there, and only the formation of the Cochrane Lake thrust and eastward dislocation of the Rundle prevented the escape of the hydrocarbons updip.

CASTLE RIVER-WATERTON

During the past 2 years important discoveries have been made in the inner foothills about 12 miles west of the north end of Pincher Creek field. To date, five wells have made wet-gas discoveries in Mississippian and Devonian rocks, and five more wells are being drilled.

The area being explored lies close in front of the Lewis overthrust, north of which for $3\frac{1}{2}$–$4\frac{1}{2}$ miles Belly River to Kootenay beds lie in regular succession, dipping generally southwest beneath the thrust. Farther north and east, several west-dipping faults cause repetition of parts of the Bighorn to Kootenay succession. Seven miles northeast of the mountain front is the Mill Creek anticline, on which two dry holes have been drilled. The structure in the area has exerted a strong influence on the topography, and the countryside exhibits typical foothills ridges and valleys.

This series of discoveries began with the completion of Shell Waterton No. 1 in early 1957, followed shortly by the Texaco Castle River A-3-4 well. Subsequently, Texaco has completed two wells, Gladstone Creek A-6-15 and Castle River B-6-17, and Shell has completed Waterton No. 2 and is drilling two more wells. All have found gas and naphtha in the Rundle formation, and the Texaco Castle River B-6-17 well also yielded strong gas flows from Upper Devonian rocks.

At this stage in the development of the area firm conclusions regarding the subsurface structure are not possible, but a pattern seems to be emerging. It is tentatively suggested that the Castle River A-3-4 and B-6-17 wells reached the same Rundle thrust block, and that the Gladstone Creek A-6-15 and Waterton No. 1 and No. 2 wells reached a different, higher block. In the ensuing discussion these are alluded to as the Castle River and Waterton blocks respectively.

In the Waterton block, the Rundle was penetrated in Shell Waterton No. 1 at −5,239 feet; in Waterton No. 2 at −5,224 feet; and in the Gladstone Creek A-6-15 at −6,847 feet. Thus between Waterton 2 and 1 there is 15 feet of plunge in 4 miles and between Waterton 1 and Gladstone Creek there is 1,608 feet of plunge in $7\frac{1}{2}$ miles. This indicates that the crest maximum of the thrust block is southeast of Waterton 1 and that the block plunges northwesterly.[3] What its configuration is farther southeast will be learned only by further drilling.

[3] Data from a well newly completed show that the northwesterly plunge is not simple; the Waterton block is divided into two segments by a subsidiary thrust.

No conclusions can be drawn yet about the plunge of the Castle River block. The B-6-17 well encountered the Rundle at −6,281 feet and the A-3-4 well penetrated it 521 feet lower, at −6,793 feet. The latter well, however, is downdip from the former and this accounts for some, if not all, of the difference in elevation. Whether any of the difference is attributable to plunge can not yet be told.

Figure 11 presents an interpretation of the structure based on the belief that its evolution was generally similar to that of the Pincher Creek field. A close similarity between this section and the western parts of Figures 9 and 10 should be noted.

Although there is an abundance of outcrop data in the area as well as information from three wells, one close to and two on the line of section, the real key to this interpretation is the Crowsnest formation in outcrop and in the wells. At the surface it is found at A, B, and C. In the Gladstone Creek well it was penetrated at A′, B′, and C′, and in Castle River it was found at C″. It is abundantly clear that in Gladstone Creek Crowsnest A′ can be in continuity with A at the surface (admittedly interrupted by a minor fault), B′ with B at the surface, and C′ with C″ in Castle River A-3-4 and C at the surface. This is the simplest way to fit them together. Thus it is established that the dips of the beds and fault planes need not remain steep to any considerable depth.

An obvious and acceptable alternative interpretation can be made by connecting Crowsnest B′ in Gladstone Creek with A at the surface, and C′ with B. To do so requires a re-interpretation of the fault system and the acceptance of a more steeply inclined series of thrusts, at least between the wells. It also calls for a larger and higher third thrust mass of Paleozoic rocks west of Gladstone Creek. In Figure 11 this is an inconspicuous body. The interpretation offered here is preferred because it is simpler.

Several features of the structure as portrayed in Figure 11 should be noted. Most obvious is the simplicity of the structure of the reservoirs as compared with the overlying structures. No less important are the very low angles at which the faults are shown to transect the beds, and, therefore, the near parallelism of the bedding and fault planes. Further concerning the faults, it should be remarked that they are assumed to cut the Paleozoic at slightly less acute angles than those of their intersection with the Mesozoic beds. Finally, the nearly flat attitudes of beds and fault planes from the Gladstone Creek well west are directly attributable to the fact that the Lewis thrust sheet once extended at least 3 miles northeast of its present front in this locality.

The wells now being drilled should greatly advance the solution of the outstanding structural problems, although most of them are on strike extensions of the Waterton thrust block. More wells intended to probe the Castle River block, and wildcat wells at the northeast across the strike will ultimately permit a more firm interpretation of the structure. Meanwhile the relations of the Waterton and Castle River blocks to the Pincher Creek block are unclear.

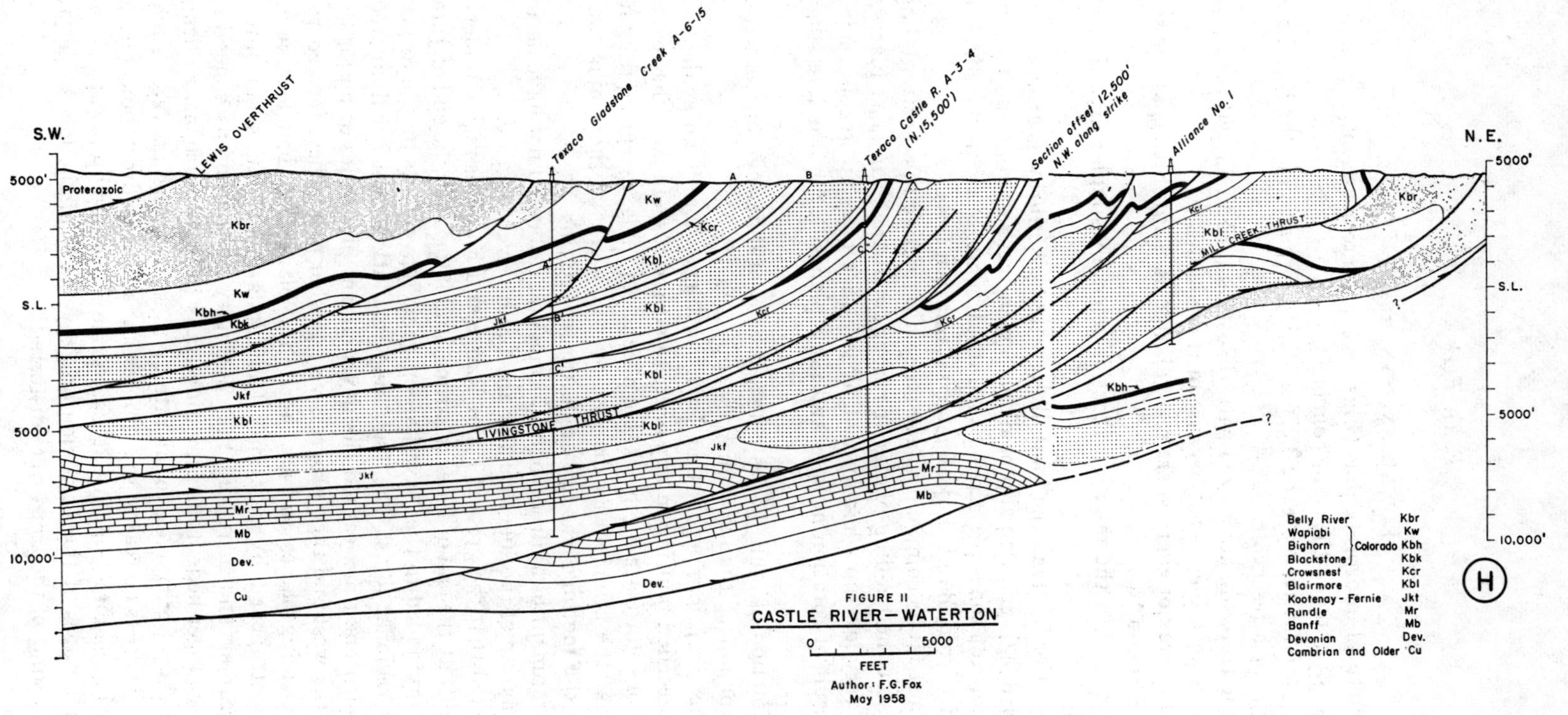
S.W.
N.E.
LEWIS OVERTHRUST
Texaco Gladstone Creek A-6-15
Texaco Castle R. A-3-4
(N.15,500')
Section offset 12,500'
N.W. along strike
Alliance No. 1
5000'
S.L.
5000'
10,000'
Proterozoic
Kbr
Kw
Kbh
Kbk
Kcr
Kbl
Jkf
Mr
Mb
Dev.
Cu
LIVINGSTONE THRUST
MILL CREEK THRUST
FIGURE II
CASTLE RIVER—WATERTON
0
5000
FEET
Author: F.G. Fox
May 1958
Belly River Kbr
Wapiabi Kw
Bighorn Colorado Kbh
Blackstone Kbk
Crowsnest Kcr
Blairmore Kbl
Kootenay-Fernie Jkf
Rundle Mr
Banff Mb
Devonian Dev.
Cambrian and Older Cu
H

FIG. 11

REVIEW AND SYNTHESIS

In the structural interpretations (Figs. 3–11), there are many common characteristics and all point to the same conclusions.

1. In every productive area the Turner Valley formation of the Rundle group is the reservoir, or a reservoir.

2. All of the traps are associated with thrust faults and all, with the possible exception of Turner Valley, exist because of these faults.

3. The faults are not everywhere recognizable as major breaks at the surface (Figs. 4, 5, 6, 7, 9 and 10).

4. The structure of the reservoir block is generally simpler than that of the overlying beds.

5. On the major faults in the foothills, stratigraphic throw is generally of the order of 5,000–8,000 feet, and slip or displacement ranges from one to 8 miles. The most important zones permitting such displacements are in the Blackstone, Kootenay and Fernie, Banff, and unnamed Cambrian beds.

6. The thrust faults commonly have steep dip relative to a horizontal plane, but they transect the bedding planes at rather small angles in most places, and probably had very low dip when formed. Where they now dip steeply their attitudes are a result of rotation of beds and fault planes during the late stages of their evolution.

7. The refraction of fault planes between beds of differing apparent competence is less pronounced than is commonly supposed.

8. With the exception, perhaps, of Turner Valley, there is no evidence that local folding preceded faulting in any of these structures—in fact, the reverse is true. The sections illustrate a remarkable absence of folds, excepting drag folds, and suggest that the principal mechanism by which crustal shortening was accomplished in the foothills was low-angle thrust faulting.

9. The entrapment of hydrocarbons, in most places, could not have occurred before the formation of the sole thrust of any productive structure.

It is suggested that the present Front Range and Foothills region underwent epeirogenic uplift at the end of Paleozoic time, and later, near the end of Jurassic time, perhaps responded to the Nevadian orogenic movements by the formation of very broad regional folds. The Laramide revolution, which culminated no earlier than Paleocene time, caused the development of a series of major thrust faults, first in the west and then progressively farther east. These formed at low angles to the bedding planes and any horizontal plane, but their leading edges were generally steepened by rotation. The fundamental faults cut into the Paleozoic succession, and disrupted the continuity of the porous zones in the Devonian and younger beds, thus providing traps for hydrocarbons. Indigenous gas and oil, as they were forced to migrate by these movements, were caught in the fault traps. There is no reason to suppose that all the hydrocarbons migrated at one time, or in one direction, but at least some moved updip toward the east. During the final stages of the Laramide orogeny many of the low-angle faults were folded and some of the traps were distorted.

It is implicit in the whole of the foregoing argument that migration of gas and oil could have begun, but could not have been completed, before the formation of the most easterly fault that cuts the reservoir beds, which also appear to have been the carrier beds. No less clear is the implication that there ought to be oil-bearing structures other than Turner Valley, and this does not depend on the question of whether there were local protofolds formed at the end of Jurassic time. In either case it is difficult to believe that only one structure in the whole foothills belt caught and held any of the migrating oil.

REFERENCES

BEACH, F. K., AND IRWIN, J. L., 1940, *The History of Alberta Oil*, Dept. Lands and Mines, Gov. Alberta, Edmonton.

BEACH, H. H., 1943, "Moose Mountain and Morley Map Areas, Alberta," *Geol. Survey Canada Memoir 236.*

CLARK, L. M., 1954, "Geology of Rocky Mountain Front Ranges near Bow River, Alberta," *Western Canada Sedimentary Basin*, Amer. Assoc. Petrol. Geol., pp. 29–46.

DE WIT, R., 1953, Devonian Stratigraphy in the Rocky Mountains South of Bow River; *Alberta Soc. Petrol. Geol., 3d Ann. Field Conf. and Sympos.*, pp. 105–07.

DOUGLAS, R. J. W., 1950, "Callum Creek, Langford Creek, and Gap Map-Areas, Alberta," *Geol. Survey Canada Memoir 255.*

———, 1953, "Carboniferous Stratigraphy in the Southern Foothills of Alberta," *Alberta Soc. Petrol. Geol., 3d Ann. Field Conf. and Sympos.*, pp. 68–88.

ERDMAN, O. A., 1950, "Alexo and Saunders Map-Areas, Alberta," *Geol. Survey, Canada Memoir 254.*

———, BELOT, R. E., AND SLEMKO, W., 1953, "Pincher Creek Area, Alberta," *Alberta Soc. Petrol. Geol. 3d Ann. Field Conf. and Sympos.*, pp. 139–57.

GALLUP, W. B., 1954, "Geology of Turner Valley Oil and Gas Field, Alberta, Canada," *Western Canada Sedimentary Basin*, Amer. Assoc. Petrol. Geol., pp. 397–414.

HUME, G. S., 1941, "A Folded Fault in the Pekisko Area, Foothills of Alberta," *Trans. Roy. Soc. Canada*, Vol. 25, Sec. IV, pp. 87–92.

———, 1957, "Fault Structures in the Foothills and Eastern Rocky Mountains of Southern Alberta," *Bull. Geol. Soc. America*, Vol. 68, pp. 395–412.

HAGE, C. O., 1942, "Folded Thrust Faults in Alberta Foothills West of Turner Valley," *Trans. Roy. Soc. Canada*, Vol. 36, Sec. IV, pp. 67–78.

LINK, T. A., 1935, "Types of Foothills Structures of Alberta, Canada," *Bull. Amer. Assoc. Petrol. Geol.*, Vol. 19, No. 10, pp. 1427–71.

———, 1949, "Interpretations of Foothills Structures, Alberta, Canada," *ibid.*, Vol. 33, No. 9, pp. 1475–1501.

———, 1953, "History of Geological Interpretation of the Turner Valley Structure and Alberta Foothills," *Alberta Soc. Petrol. Geol., 3d Ann. Field Conf. and Sympos.*, pp. 117–33.

MARTIN, R., 1956, "The Jumpingpound Gas Field," *ibid., 6th Ann. Field Conf. Guide Book*, pp. 125–40.

MCLAREN, D. J., 1953, "Summary of the Devonian Stratigraphy of the Alberta Rocky Mountains," *ibid., 3d Ann. Field Conf. and Sympos.*, pp. 89–104.

NORTH, F. K., 1953, "Cambrian and Ordovician of Southwestern Alberta," *ibid.*, pp. 108–16.

———, AND HENDERSON, G. G. L., 1954, "Summary of the Geology of the Southern Rocky Mountains of Canada," *ibid., 4th Ann. Field Conf. Guide Book*, pp. 15–81.

RICH, J. L., 1934, "Mechanics of Low-Angle Overthrust Faulting as Illustrated by Cumberland Thrust Block, Virginia, Kentucky, and Tennessee," *Bull. Amer. Assoc. Petrol. Geol.*, Vol. 18, pp. 1584–96.

RUTHERFORD, R. L., 1944, "Regional Structural Features of the Alberta Foothills and Adjacent Mountain Ranges," *Trans. Roy. Soc. Canada*, Vol. 38, Sec. 4, pp. 71–77.

SCOTT, J. C., 1951, "Folded Faults in Rocky Mountain Foothills of Alberta, Canada," *Bull. Amer. Assoc. Petrol. Geol.*, Vol. 35, No. 11, pp. 2316–47.

———, 1953, "Savanna Creek Structure," *Alberta Soc. Petrol. Geol., 3d Ann. Field Conf. and Sympos.*, pp. 134–39.

———, HENNESSEY, W. J., AND LAMON, R. S., 1957, "Savanna Creek Gas Field, Alberta," *ibid., 7th Ann. Field Conf. Guide Book*, pp. 113–31.

SPRATT, J. G., AND TAYLOR, V., 1936, "Oil Prospects along the West Flank of the Turner Valley Gas Field," *Bull. Canadian Inst. Min. and Met.*, Vol. 29, No. 10, pp. 713–22.

THOMAS, A. N., 1958, "Note on a Geological Cross Section through the Nordegg Area," *Alberta Soc. Petrol. Geol., 8th Ann. Field Conf. Guide Book*, pp. 121–27.

The American Association of Petroleum Geologists Bulletin
V. 55, No. 2 (February 1971), P. 292-306, 10 Figs., 1 Table

Folded Faults and Sequence of Thrusting in Alberta Foothills[1]

P. B. JONES[2]
Calgary 1, Alberta

Abstract Thrust faults in the Alberta Foothills were folded by development of step-thrusts and other thrust-generated structures beneath them. Because step-thrusting is a pattern of faulting found in layered rocks of varying competence, folded faults may be expected in any thrust belt where the stratigraphic sequence is inhomogeneous and higher thrusts are emplaced before underlying ones. By invoking folded faults, several structures in the Alberta Foothills can be reinterpreted without the geologic and geometric inconsistencies inherent in existing interpretations.

Although a west-to-east sequence of thrust emplacement in the Alberta Foothills is implicit in the mechanics of folding of thrusts, some structural relations indicate an east-to-west sequence. The apparent contradiction may be resolved by the proposition that the time taken for a thrust to form, spreading both across and along strike, is greater than the time interval between initiation of successive thrusts.

Introduction

The Rocky Mountain orogeny (White, 1959) in the Alberta Foothills and eastern Rocky Mountains involved Precambrian to Paleocene sedimentary rocks (Table 1). Deformation consisted largely of overthrusting from the west along zones of *décollement* at various levels overlying the crystalline basement, which was not itself involved (Bally *et al.*, 1966; Keating, 1966). Faults other than thrusts are rare in the area under discussion. In this paper the term "fault" is used to denote thrust faults.

The presence of folded faults was demonstrated by Hake *et al.* (1942) in the central Foothills (Fig. 1). Folded faults also were described by Hage (1942) in the southern Foothills. Subsequent references to folded faults have been concerned mainly with those two areas. Scott (1951), MacKay (1943), and Erdman (1950) mapped other folded faults in the central Foothills: Douglas (1950; 1958a, b) described similar faults in the southern Foothills and Front Ranges of the Rocky Mountains. One purpose of this paper is to show that folded faults may be present anywhere in the Foothills and Rockies, and to present evidence for their presence in structures previously regarded as unaffected by folded faults.

Two explanations for the folding of faults have been given. Douglas (1950, Figs. 16, 17) and Bally *et al.* (1966, p. 37) regarded the folding as an integral part of the process of thrusting. Scott (1951, p. 2347), Dahlstrom *et al.* (1962, p. 394), and others attributed the folding to a discrete compressional phase which followed the period of thrusting in the disturbed belt.

In profile, thrust faults are curved and commonly are associated with tilted and folded strata, so that it is rarely possible to determine whether the curvature is the original form of the thrust plane or is the result of folding. It is therefore appropriate, before describing folded faults, to consider thrust faults that have not been folded.

Faults in Crimson Lake Area, Alberta

Killick (1954, p. 35) suggested that linear features in the vicinity of the sharp bend in the North Saskatchewan River (Fig. 1), east of the foothills belt, might be faults. Following photogeology by K. W. Roth, field checks by R. B. Sanders and the writer showed that two northwest-trending linears are southwest-dipping thrusts, the Sylvester Creek and Brewster Creek faults (Fig. 2). At the surface the Sylvester Creek fault dips 40° southwest. The footwall strata are horizontal. About 2,000 ft of beds above the fault also dip southwest, subparallel with the fault plane. In the subsurface, a seismic profile shows the fault plane flattening at depth, passing into a bedding-plane fault in the Alberta Group. The Brewster Creek fault appears to be similar. Wells drilled west of the fault trace penetrated thickened sections in the Alberta Group (Fig. 3). Farther west, wells drilled through both thrusts penetrated normal thicknesses in the Mesozoic sequence, with no

[1] Manuscript received, February 18, 1970; accepted, June 11, 1970. Published by permission of Amerada Hess Corporation.

[2] Amerada Hess Corporation.

The writer is indebted to R. L. Zell, F. G. Fox, P. Verrall, and E. L. Fitzgerald, who detected many errors and omissions in the manuscript and suggested improvements. Figures were drawn by the Amerada Hess drafting department. The typescript was prepared by Merlyn Trigg.

Table 1. Formations in Alberta Foothills (N = Northern, C = Central, S = Southern)

System	*Formation*		*Lithology*	*Thickness (Ft)*
Tertiary	Paskapoo (N, C)	Porcupine Hills (S)	Nonmarine sandstone, shale, coal	2,000– 4,000
			Unconformity	
		Willow Creek (S)	Nonmarine sandstone, shale, mudstone	350– 2,700
Cretaceous, Upper	Brazeau (N, C)	Edmonton (S) Bearpaw (S) Belly River (S)	Mainly nonmarine sandstone, mudstone, shale, coal	2,000– 5,000
	Wapiabi		Marine shale and siltstone	1,100– 1,800
	Cardium		Marine sandstone, shale	30– 450
	Blackstone		Marine shale	400– 1,000
Cretaceous, Lower		Crowsnest (S)	Agglomerate, tuff	0– 1,800
	Mountain Park (N, C) Luscar (N, C)	Blairmore (S, C)	Nonmarine sandstone, shale	900– 6,500
	Cadomin (N, C)		Conglomerate	0– 200
			Unconformity	
Jurassic	Nikanassin (N, C)	Kootenay	Nonmarine sandstone, shale, coal	100– 4,000
	Fernie Group		Marine shale, limestone, sandstone	100– 1,300
			Unconformity	
Triassic	Spray River Group		Marine dolomite, siltstone	0–1,200
			Unconformity	
Permian	Ishbel Group		Marine sandstone, chert, sandy dolomite	0– 1,400
			Unconformity	
Pennsylvanian	Kananaskis		Marine silty dolomite	0– 300
Mississippian	Rundle Group		Marine limestone and dolomite	750– 3,000
	Banff		Marine argillaceous limestone	500– 1,100
	Exshaw		Marine black shale	0– 40
			Unconformity	
Devonian	Palliser		Marine limestone, dolomite	650– 1,200
	Alexo		Silty limestone and dolomite	20– 600
	Fairholme		Marine limestone, shale, dolomite, local basal clastic unit	950– 1,600
			Unconformity	
Cambrian			Marine limestone and dolomite, sandstone and shale	700– 2,500
			Unconformity	
Precambrian	Purcell Group (S) Mountains only		Dolomite, argillite, limestone	7,000–15,000

indication of the existence of faults in the section. The Stolberg structure, the most easterly Foothills structure in which Paleozoic strata are involved in thrusting, is present 20 mi southwest of the outcrop of the Sylvester Creek fault. Two fault slices of Mississippian carbonates overlie the autochthon. Figure 3 shows the correlation of these two faults with the Sylvester Creek and Brewster Creek faults, via bedding-plane faults, as suggested by Thomas (1958, p. 127).

The Sylvester Creek and Brewster Creek faults are not folded. The footwall strata are horizontal and the tilting of the hanging wall strata is due to fault movement only. According to Fox (1959, p. 1024), the arcuate profile of many faults in the disturbed belt is the result of folding or rotation of thrusts having a uniform low dip. If this were true of the Sylvester Creek fault, footwall and hanging wall strata would be affected equally. They are not, hence the arcuate profile must be related to the stress pattern causing the thrusting (Hafner, 1951, p. 385), modified by the vertical anisotropy of the sedimentary sequence.

The Sylvester Creek and Brewster Creek

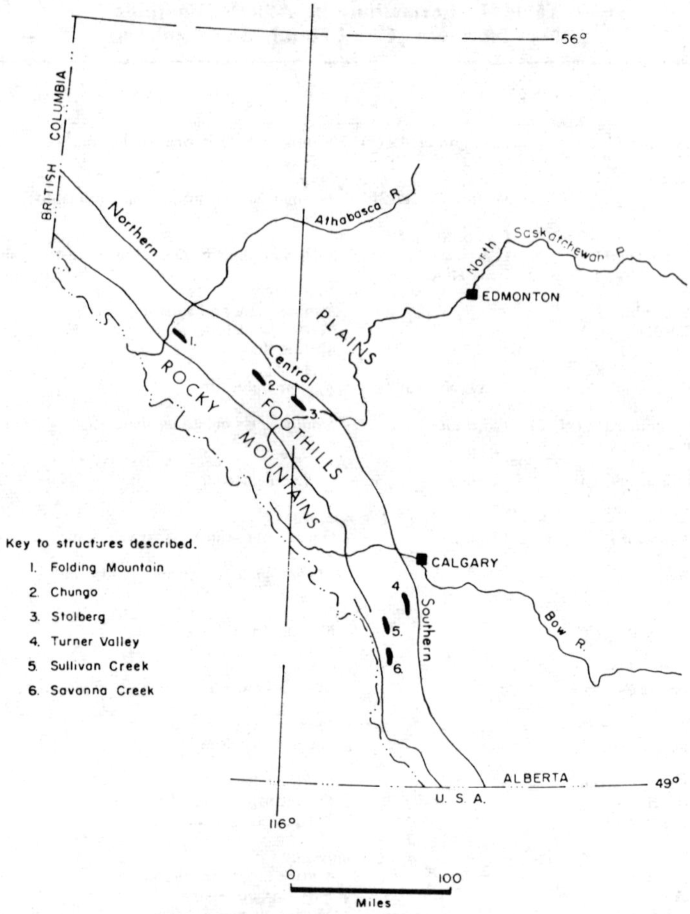

FIG. 1.—Location map, Alberta Foothills.

faults transect the Paleozoic in the Stolberg structure at low angles, parallel the bedding of the Alberta Group for 16 mi, and curve upward to the surface cutting the footwall strata at up to 40°. They are step faults, and their profiles in cross section (Fig. 3) are probably typical of a thrust fault before folding.

Fault Structures in Central Foothills of Alberta

Stolberg Structure (Fig. 3)

Three wells have been drilled on the Stolberg structure, and extensive seismic work has been conducted by several oil companies. The warping of the Ancona thrust over the subsurface structure is shown both in plan (Fig. 2) and cross section (Fig. 3). Although the Ancona thrust and adjacent strata are folded, the underlying faulted slices of Mississippian rocks are barely folded. The Ancona thrust appears to have been folded by the emplacement of the Sylvester and Brewster Creek faults beneath it, as well as by the imbrications in the intervening Upper Cretaceous strata. The lower of the two faults that displace the Mississippian is not folded, presumably because no structure was developed beneath it to cause folding. If folding were due to a discrete compressional phase following thrusting, the entire sequence would have been folded. The slight folding of the overthrust slices of Mississippian rocks can be attributed entirely to thrust movement. Apparently, the original profile of the Ancona thrust was similar to the Sylvester Creek fault, curving upward to the surface in the same way. The present anticlinal and synclinal profile is believed to have resulted from upwarping of the near-horizontal sector of the thrust plane by emplacement of the thrusts beneath it.

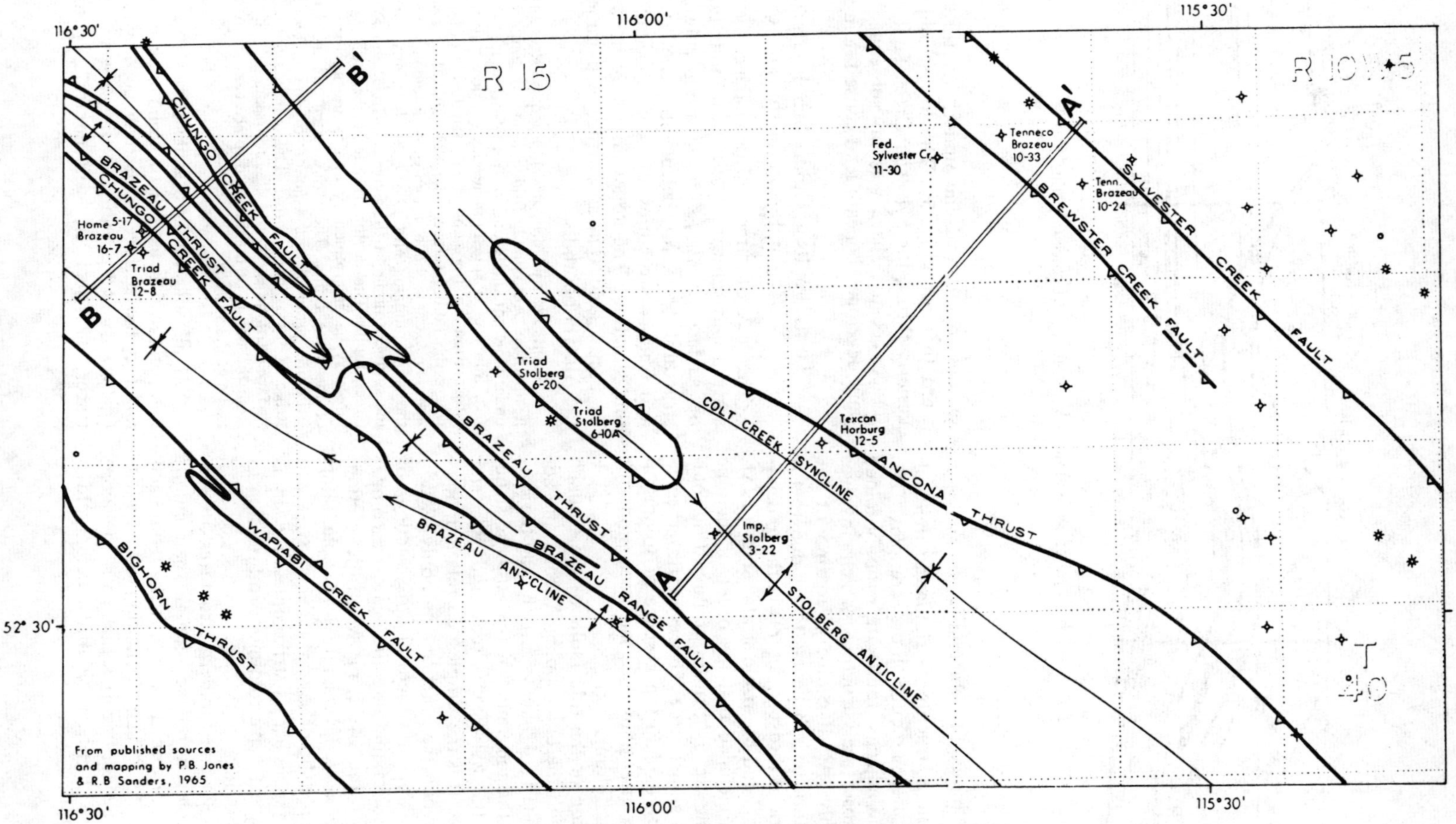

FIG. 2.—Simplified tectonic map of part of central Foothills.

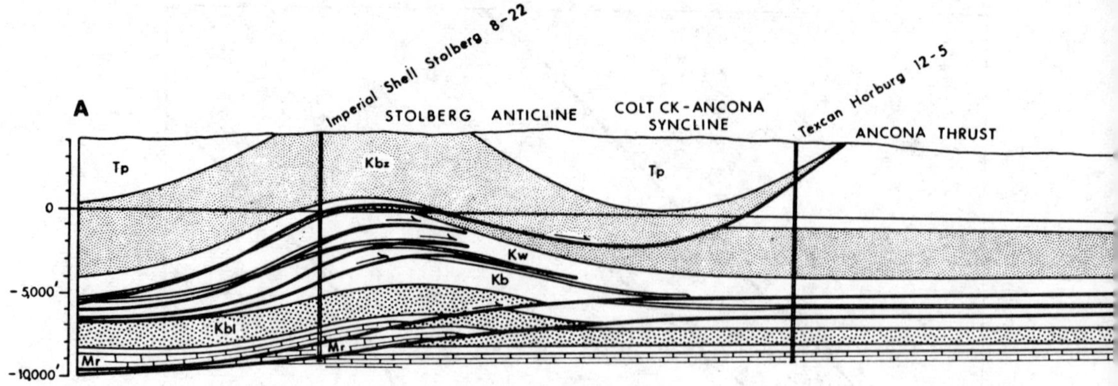

FIG. 3.—Structural cross-section A-A′, Stolberg structure to plains.

Chungo Structure (Fig. 4)

The Chungo structure is similar to the Stolberg structure, although more complex. The Brazeau and adjacent faults are strongly folded over a subsurface structure which involves faulting only. As at Stolberg, folding of the overlying thrusts is believed to be due to the development of thrust structures in the underlying Paleozoic and Mesozoic strata. Scott (1951, p. 1347) concluded that the order of tectonic events in the area is thrusting followed by folding on a regional scale. However, this mechanism fails to explain the absence of folding at depth, even after allowance has been made for disharmonic folding.

Correlation of the sole faults of the Chungo structure with the outcropping Ancona and adjacent fault (Fig. 4) is less certain than the correlation of the sole faults of the Stolberg structure with the Sylvester Creek and Brewster Creek faults. It is the simplest correlation, however, and avoids the practice commonly used in the construction of structural cross sections of projecting sole faults into the nearest incompetent part of the overlying Mesozoic section, where they are conveniently lost. Sole faults involving a net slip of several miles must extend for long distances across strike unless the slip is taken up by extensive folding or imbrication in the hanging wall.

The Chungo and Stolberg structures are adjacent and closely related, though their culminations are not directly across strike from each other (Fig. 2). Their formation can be explained by a west-to-east sequence of development of step thrusts, each thrust being folded by development of the underlying thrust toward the east (Fig. 5). The lowest and most easterly thrust, the Sylvester Creek fault, is not folded because there was no subsequent step fault developed beneath it to cause folding.

Thrust Faults as Guides in Subsurface Exploration in Foothills

The lack of direct correspondence between surface and subsurface structure limits the use of surface geology for locating subsurface structural traps for petroleum. The Stolberg and Chungo structures are typical examples of this situation. The following suggestions may be useful guides in subsurface exploration in the Alberta Foothills.

1. Faults are folded by development of structures beneath them. Anticlinally folded faults are indicators of underlying anticlines and/or thrust structures.
2. Study of a sole fault at the surface may provide information on the subsurface structure on the west where the fault intersects the Paleozoic. For example, there appears to be a rough correspondence between the south-plunging end of the Stolberg structure and the southern limits of the Sylvester Creek and Brewster Creek faults. The northern limit of the Stolberg structure may be determined by finding the northern limits of the same faults. The strike extent of the Chungo structure probably corresponds to the extent along strike of the Ancona thrust. The culminations of subsurface structures should lie approximately downdip from the points of maximum displacement of their sole faults at the surface.

Not all sole faults of subsurface structures involving overthrust Paleozoic have been recognized at the surface. This may be due to (1) Quaternary and possibly Tertiary cover of fault traces; (2) lack of well control, or misin-

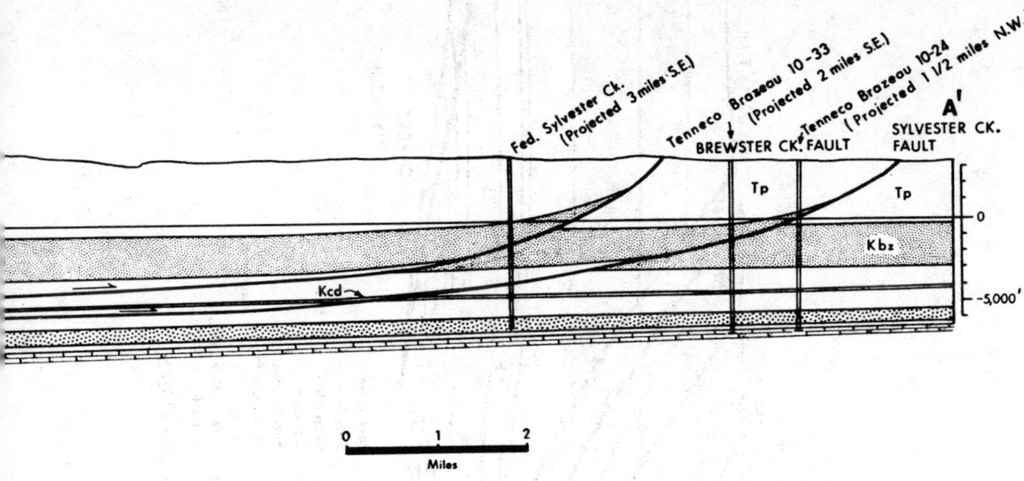

Tp	Paskapoo
Kbz	Brazeau
Kw	Wapiabi
Kcd	Cardium
Kb	Blackstone
Kbl	Mountain Park, Luscar } Blairmore
Jfn	Fernie-Nikanassin
Mr	Rundle

terpretation of well data; (3) passage of sole faults into bedding plane faults which are unrecognizable, and result in the faults reaching the surface many miles east of the structures they have generated; and (4) bifurcation of major thrusts into numerous smaller ones whose aggregate slip is equal to the net slip of the major thrusts.

Origin of Folded Faults

Hypotheses of formation of folded faults fall into two groups. First, it is supposed that the development of the Foothills belt took place by a period of thrusting, followed by a period of folding during which the thrusts were folded. Scott (1951, p. 2347) concluded that the sequence of tectonic events in the Chungo area involved a period of thrusting followed by an episode of compressional folding on a regional scale. Similarly, Dahlstrom *et al.* (1962, p. 394) attributed the synclinal form of the Lewis thrust in southern Alberta and British Columbia to a second cycle of deformation. In both cases, however, the structures underlying the folded faults are simple fault structures with little or no folding. If folding of the Lewis plate had taken place after thrusting, the footwall section would also have been involved. Bally *et al.* (1966, cross section E-E′) show the Lewis thrust folded over the Sage Creek structure in the west and the Waterton structure in the east. Both these structures are the result of faulting, and the syncline in the Lewis thrust plate is not present in the underlying strata.

Douglas (1950, Figs. 16, 17) showed that if thrusting took place along a stepped fault plane and if the compressive stress was not relieved completely by the thrusting, folding of the fault plane would take place, accentuating any existing irregularities. Hume (1957, p. 407) stated:

> It is inevitable, therefore, as shown by Douglas, that at the point where the dip of the fault changes upward from a steeper to a gentler slope, an anticline will develop and produce not only a fold in the fault itself but in the beds above and below it.

Figure 6 represents a process of folding of thrusts described by Verrall (1968, p. 114). It shows that the folding of fault planes involves neither the separate episode of folding proposed by Scott nor accentuation of the stepped profile in the manner shown by Douglas. The only essential mechanism is the emplacement of a step-thrust beneath the fault concerned. Figure 6 shows that (1) movement of the fault along a stepped thrust plane may not affect the underlying strata but inevitably causes folding of the overlying sequence, including any thrusts —thus, fault A, which is not a step-thrust, is folded by movement of fault B, a step-thrust, beneath it; (2) emplacement of successive step-thrusts B and C, if superimposed on one another, accentuates the folding higher in the sec-

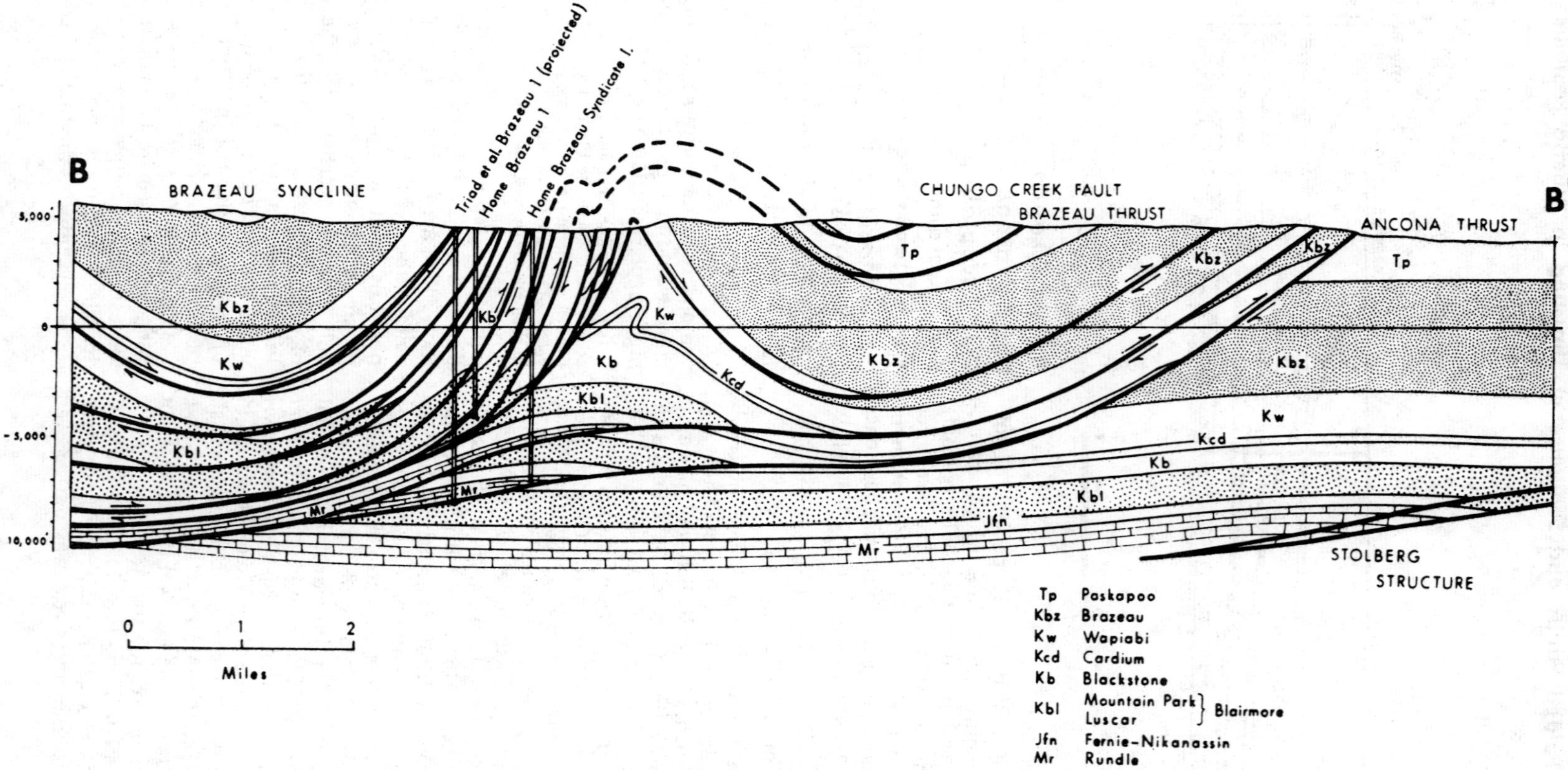

FIG. 4.—Structural cross-section B-B′, Chungo structure. Modified after Douglas (1958b).

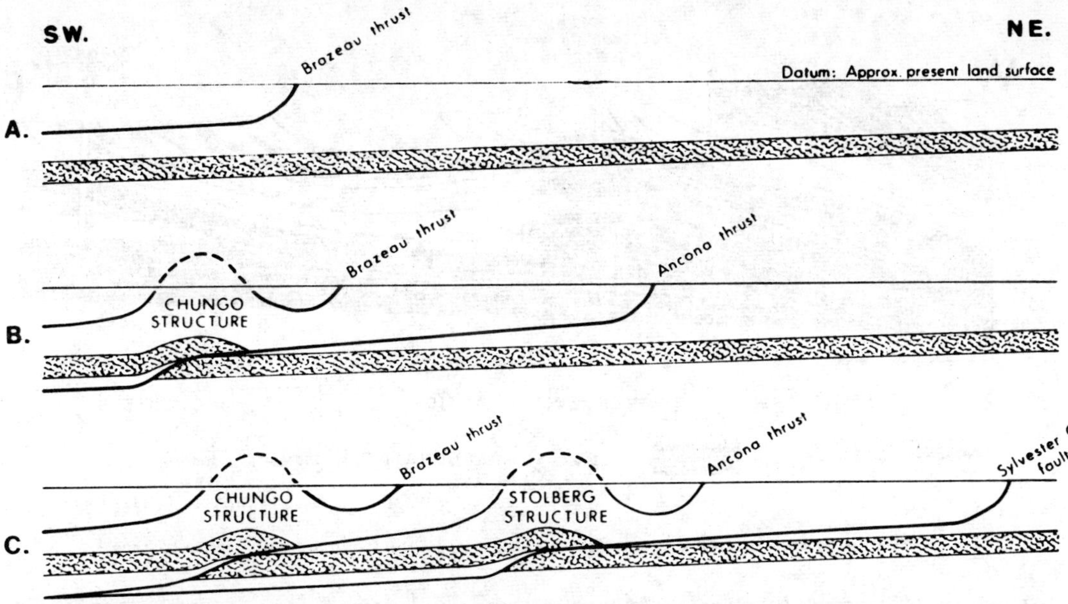

FIG. 5.—Development of Stolberg and Chungo structures, diagrammatic, not to scale. Brewster Creek and other intervening faults have been omitted for clarity. **A.** Emplacement of Brazeau thrust. **B.** Emplacement of Ancona thrust, folding of Brazeau thrust, and formation of Chungo structure. **C.** Emplacement of Sylvester Creek fault, folding of Ancona thrust, and formation of Stolberg structure

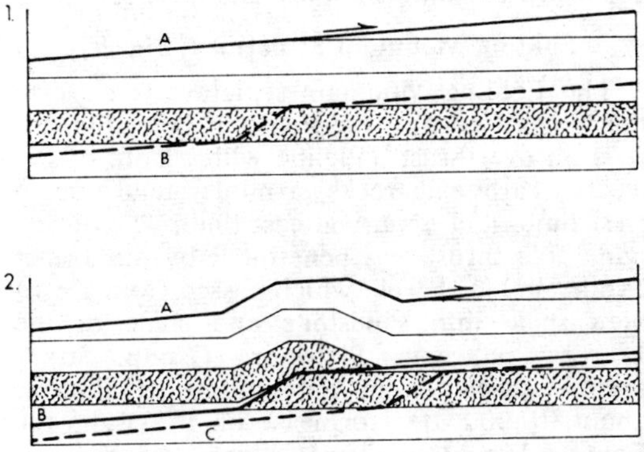

FIG. 6.—Diagram showing development of folded faults. **1.** Planar fault A, incipient step-fault B. **2.** Movement on step-fault B folds overlying sequence, including fault A. Incipient step-fault C. **3.** Movement on C folds fault B and increases folding of fault A.

tion; and (3) thrusts are concentrated in certain parts of the section, in which they appear to be of little significance, because they are bedding-plane or near bedding-plane thrusts. Only two conditions are required for this process to take place. At least one fault, preferably a lower one, need be a step-thrust, and the sequence of emplacement of thrusts is from higher to lower, *i.e.*, from west to east in the Foothills belt of Alberta.

The formation of structures such as the Savanna Creek structure in the southern Alberta Foothills can be explained simply by this mechanism. Fox (1959, p. 1013) pointed out that there is no evidence that the folded faults penetrated by wells are step faults. However, the autochthon was not penetrated and there are several major faults above it that were not penetrated by drilling, any of which may be a step fault. Similarly, the folding of the Dyson Mountain thrust in the Sullivan Creek structure (Fig. 7) can be explained in terms of a stack of stepped thrust plates.

Step faults are common in the Alberta Foothills. Geologic maps of the area and well records show that faults are much more common in some formations than others (Fox, 1959, p. 994). If faults cut all formations at the same

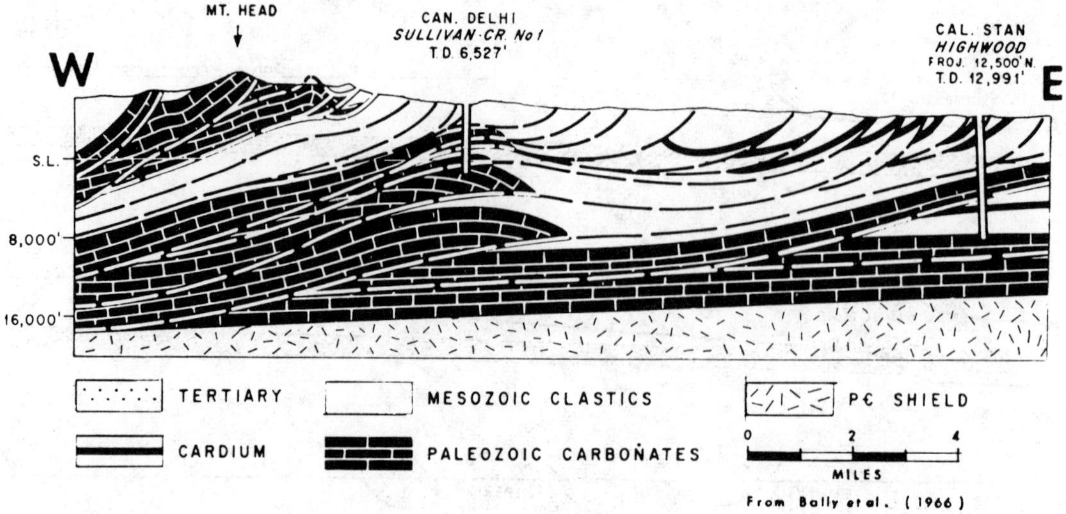

FIG. 7.—West-east cross section through Sullivan Creek structure. Compare with Figure 6, stage 3.

angle, their stratigraphic distribution would be random. Their concentration in less competent formations requires that they cut those formations at much lower angles than the others—*i.e.*, they are step faults. The step faulting is due to variations in competence through the stratigraphic column. Thus, folded faults are to be expected in the development of any thrust belt where the stratigraphic column is inhomogeneous and the order of thrusting is from higher to lower.

The mechanism described explains the formation of folded fault structures, such as Savanna Creek and Sullivan Creek, but it may not account completely for the more extreme folding of higher thrusts in the Stolberg and Chungo structures. In these structures there is considerable thickening of the incompetent shales of the Alberta Group. Over double the normal thickness of Blackstone shale was penetrated in Triad Stolberg 6–20. Fox (1959, p. 996) suggested that the thickening is due to flowage of shale or to repetition of the Blackstone by faulting. The shale-sandstone-shale sequence of the Alberta Group is an ideal sequence for the generation of step-thrusts. Given sufficient faults, even the extreme folding of the Brazeau thrust over the Chungo structure (Fig. 4) could have been produced by the process shown in Figure 6.

Reinterpretation of Foothills Structures

The formation and occurrence of folded faults may be more common in the Alberta Foothills and Rocky Mountains than are indicated by existing geologic maps and reports. Much of the early mapping of the Foothills took place before folded thrusts were recognized, or when they were regarded as exotic phenomena. There are many structures that should be reexamined. Two examples follow.

Folding Mountain Structure (Fig. 8)

The Folding Mountain structure is close to the western edge of the Foothills belt (Fig. 1). It is an overthrust anticline with a core of Paleozoic carbonate rocks, a moderately dipping west limb, and a vertical east limb. The underlying sole thrust was penetrated by the Jasper Syndicate No. 1 well, which passed from Devonian shale into sandstone and shale of the Lower Cretaceous Blairmore Group. Stratigraphic displacement (minimum net slip) is about 10,000 ft. Northeast of the well, the Folding Mountain thrust, which cuts the east flank of the structure, has been correlated with the sole thrust penetrated by the well (Webb, 1955; Mountjoy, 1959). However, net slip of the Folding Mountain thrust at outcrop near the well is not over 2,400 ft (Mountjoy, 1959, cross-sections A-B, C-D). Mapping by the writer indicates a slip of not more than 400 ft just updip from the well. If this is the sole thrust, there is a tremendous amount of "stretching" of the Blackstone beneath the thrust. The earlier cross section also implies that there is a complete reversal of dip direction only a few feet beneath the surface, for

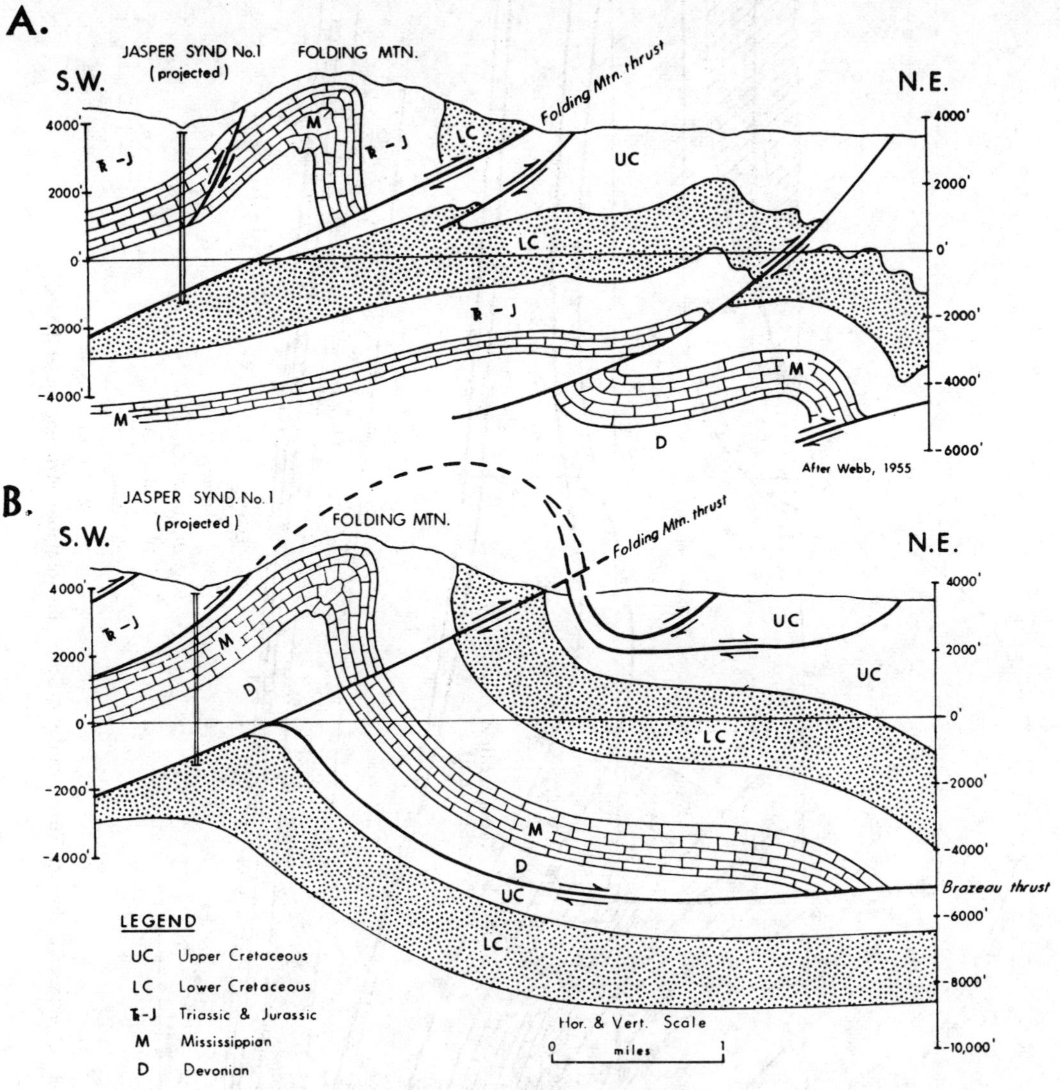

FIG. 8.—Cross sections through Folding Mountain structure. **A.** Interpretation without folded faults. **B.** Interpretation involving folded faults.

which there is no evidence. A simpler alternative explanation is offered.

In the Livingstone and Brazeau ranges of the Foothills, Douglas (1950, 1956) showed that faults cutting east limbs of similar anticlinal structures are minor imbrications of major sole faults which themselves crop out several miles farther east. This interpretation can be applied to the Folding Mountain thrust. This thrust, which cuts the east limb of the fold, is believed to be an imbrication of a folded major sole fault that was formed at or near the end of the folding. The sole fault itself is folded parallel with the overlying strata beneath the Brazeau syncline, as shown in Figure 8, emerging on the east flank of the syncline as the Brazeau thrust (Mountjoy, 1959).

Turner Valley Structure (Fig. 9)

Turner Valley is the oldest oil and gas field in Alberta. The overthrust anticline was drilled in 1913, and gas was found in the Lower Cretaceous Blairmore Group. Later drilling proved the presence of oil beneath the gas cap and of

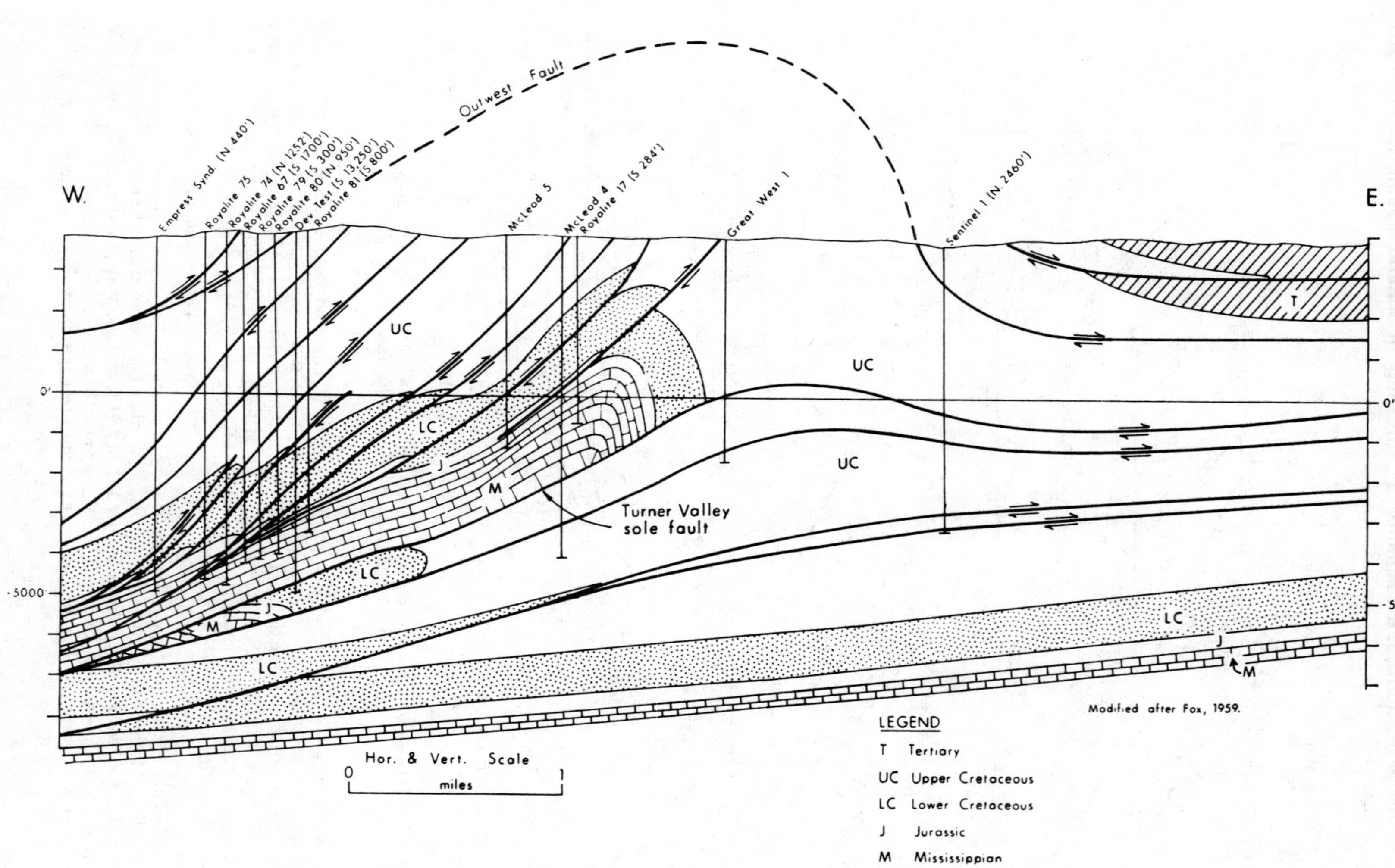

FIG. 9.—West-east cross section through Turner Valley structure, interpretation involving folded faults.

another reservoir in Mississippian carbonates, underlain by a major thrust fault. Almost 400 wells have been drilled on the structure; at least 11 penetrated the sole fault.

Geologic maps and cross sections of the structure by Hume (1931), Gallup (1951, Plate I), and Fox (1959, Fig. 4; 1969, Plate 5b) illustrate two problems that are inherent in current structural interpretations. The first is that, although the Turner Valley sole fault, which was penetrated by wells in the central part of the field, has a minimum net slip of 2 mi, the fault outcropping in the east flank of the structure, mapped as the Turner Valley sole fault (Gallup, 1951, Plate I), has little or no net slip. Hume (1939, p. 9) suggested that the sole fault bifurcates upward and is represented at the surface by numerous small faults, each having small net slip, whose aggregate slip is equal to the slip of the sole fault at depth. Fox (1969, Plate 5b) showed the sole fault dying out before it reaches the surface. The second problem is that the width of outcrop of the east-dipping Cretaceous Edmonton and Belly River Formations exposed on the east flank of the Turner Valley anticline is as much as four times greater than the width calculated from their true formation thicknesses and observed dips. Hume (1931, p. 9) suggested that this was the result of repetition by west-dipping faults. However, although thrusting causes repetition of beds, the outcrop width is increased only if the thrusts dip in the same direction as the involved beds. If they dip in the opposite direction, the result is a reduction of the outcrop width. Conventional interpretation of the Turner Valley structure fails to explain why there is no surface fault having a displacement comparable to that of the sole fault, and why the strata on the steep east flank are thickened instead of thinned.

Figure 9 shows an alternative interpretation, involving folded faults. Strata in the east flank are shown to be thickened by east-dipping thrusts, folded roughly parallel with the underlying anticline in the Paleozoic carbonates. It is probable that the numerous smaller thrusts in the west flank also extend in folded form into the east flank. The Turner Valley sole fault does not steepen and reach the surface in the east flank of the structure, but extends eastward as a low-angle fault. This interpretation explains both the excessive outcrop width and the absence of a major west-dipping thrust in the east flank. Fox (1959, p. 998) described faults in the Mesozoic section several miles southeast of Turner Valley, underlying horizontal Tertiary strata. The Turner Valley sole fault may be one of these.

The Outwest fault has a slip of 6 mi (Bally *et al.*, 1966, cross-section C-C′). Its folding over the Turner Valley structure is suggested by Gallup's (1951, Plate I) geologic map, which shows the Outwest and adjacent faults in the west flank of the Turner Valley anticline converging on a similar fault in the east flank, as they are traced toward the south-plunging anticlinal nose. It is assumed that the Outwest fault maintains its character as a near-bedding-plane thrust and follows the strata across the fold axis into the east flank. The absence of faults south of Turner Valley that are big enough to be correlated with the Outwest fault suggests that it does not continue southward.

The Folding Mountain and Turner Valley structures are not unique. Similar structures are present in other parts of the Foothills and Rocky Mountains. The examples given show how structural interpretations involving folded thrusts can reduce geologic and geometric inconsistencies. Foothills structures should be examined carefully for evidence of folded thrusts. Structural cross sections of some geologic maps of the Foothills attempt to reconcile the surface geology with well data by showing abrupt reversals of observed dip direction of strata within a few feet of the surface. Many of these dubious constructions can be eliminated by a structural interpretation involving folded faults. Such interpretations may allow reevaluation of the petroleum prospects of some areas and indicate the presence of underlying structures that are suitable for petroleum accumulation.

Sequence of Thrusting in Canadian Rockies and Foothills

Several authors have discussed the sequence of thrusting in the disturbed belt, both on a local and on a regional scale. The temporal relation between folding and thrusting—*i.e.*, whether folding preceded, followed, or was contemporaneous with thrusting—is also involved. Douglas (1950, Fig. 26) showed how back-limb thrusts form in sequence from east to west within a major overthrust sheet in the southern Alberta Foothills. Hume (1957, p. 411) follows Douglas in stating that back-limb thrusts are younger than their associated sole faults and formed from east to west, whereas the sole faults themselves formed in the reverse sequence, from west to east.

It is difficult to apply Douglas' argument because of the problem of distinguishing back-limb thrusts from major sole faults. Most major thrusts pass along strike into minor back-limb thrusts within the next major thrust plate below and on the east. An example is the McConnell thrust, underlying the eastern Rockies through much of southwest Alberta. It passes southward into a back-limb thrust within the Livingstone major overthrust sheet of the Foothills. Similarly, a small back-limb thrust within the McConnell overthrust sheet passes southward into the Lewis thrust, which is a major thrust in southern Alberta. Thus, strict application of Douglas' (1950) principles of back-limb thrusting leads to the conclusion that the Lewis thrust is later than the Livingstone thrust, the next major thrust to the east, and that the entire disturbed belt was formed in an east to west sequence.

Many geologic maps of the Alberta Foothills show major thrusts intersecting thrusts east of them. This relation led Choquette (1959) and Hunt (1956) to conclude that the order of thrusting was from east to west and that some Foothills structures existed long before they were overridden by thrusts developed west of them. However, many structures have steep to vertical flanks, and it is difficult to visualize an overthrust mass moving over such relief.

Working in the disturbed belt of Montana, Deiss (1943, p. 256) concluded that, because the Lewis thrust truncates the westernmost "high-angle" thrust of the Sawtooth Range to the east, it is the younger structure. Hurley (1959, Figs. 6, 7) also stated that the Lewis thrust is younger than the structures which it overrode. Both these authors favor an east-to-west sequence of thrusting. In the same area, Childers (1963, p. 160–162) proposed a west-to-east sequence of major thrusting, combined with a variable sequence of thrusting within each major thrust plate. He also related the folding of the Brazeau and associated thrusts in Alberta to development of thrust structures beneath them.

Fitzgerald (1962, Fig. 5) explained the warping of the McConnell thrust in the Ghost River area of Alberta by subsequent emplacement of the Panther River thrust beneath and to the east of it. The Panther River thrust itself is not shown as a folded thrust.

Bally *et al.* (1966, Figs. 6, 13) discussed the overriding and cutting of thrusts by those developed west of them. In an attempt to reconcile these relations with a west-to-east sequence of thrusting, they suggested that the relation is surficial only, and is reversed at depth. Early formed thrusts were bedding-plane slides from which imbrications cut upward to the surface. Subsequent thrusts formed below and on the east, cutting steeply upward through both the bedding-plane slides and their imbrications, and reaching the surface west of their predecessors. This explanation is unacceptable to the writer because it implies a different manner of emplacement for the earlier and later thrusts, for which there is no evidence. Moreover there is no example in the area of a low-angle thrust being displaced by a high-angle one having the same dip direction.

Hypothesis of Thrust Development

There is evidence for both east-to-west and west-to-east sequences of thrust development. A hypothesis concerning the sequence must be based on the premise that items of evidence presented are complementary and not mutually exclusive.

The structural relations between thrusts in the Alberta Foothills and Rockies are clearly shown on maps and cross sections. It is from these relations that the conflicting conclusions regarding thrust sequences have been drawn. It should be realized, however, that field relations show only the sequence of *termination* of events. They do not show the sequence in which those events started. The longer the time interval involved in the deformation of the Foothills and Rockies, the greater is the probability of lack of correspondence between sequence of initiation of thrust emplacement and sequence of termination of thrust movement. Furthermore, study of maps and cross sections may lead to a subjective impression that the relations show a sequence of discrete events. There is no reason why this should be so.

The mechanism of thrusting envisaged by the writer involves three assumptions, the first of which was stated by Douglas (1958b, p. 130). They are:

1. A thrust develops as an initial break at a given point and spreads along strike in both direction.
2. The amount of slip of a thrust is more or less proportional to the duration of its movement. A thrust that has a net slip of 5 mi moved over a much longer period than one whose net slip is 1 mi. Within one thrust, the sector where the net slip is greatest is the sector that moved longest. Although it is unlikely that thrusts moved at the same speed, the range of velocities is probably small.
3. The time interval between emplacement of successive thrusts is considerably smaller than the time required for full development of a single thrust.

If these assumptions are true, particularly

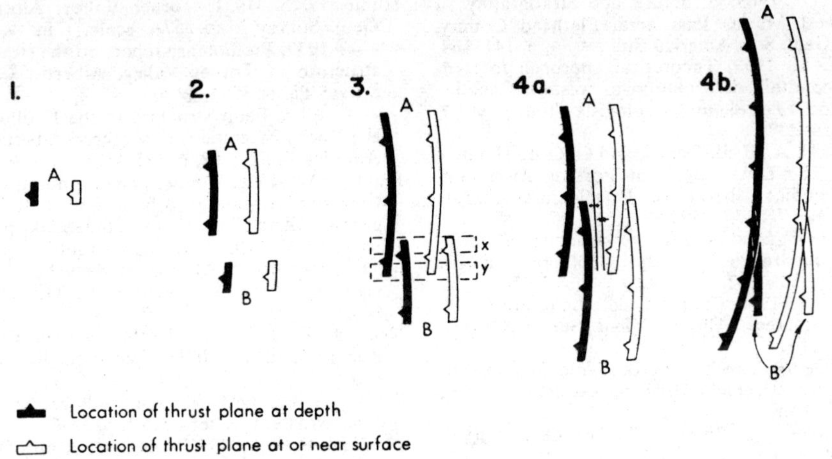

FIG. 10.—Diagram of show time and space relations between successive thrusts (plan view). **1.** Initial break of thrust A. **2.** Movement of A, initial break of thrust B. **3.** Movement of A and B. At X, thrust A is older than B. At Y, thrust A is younger than B. **4a.** Movement of thrust A ceases. Continued movement of B causes folding of thrust A. **4b.** Movement of B ceases. Continued movement of A causes it to override B.

the third, then in many places the sequence of initiation of emplacement is of no significance, and the final relations are largely independent of it. A major thrust such as the Lewis thrust, moving for a long period, would cut and override a minor one east of it, moving for a short period, regardless of which was initiated first (Childers, 1963, p. 160). The cutting of eastern by western thrusts would depend on the difference in duration of their movement or magnitude, not the sequence in which they were initiated.

Figure 10 shows how structural relations indicating an east-to-west sequence of thrusting may have been caused by a west-to-east sequence of thrust initiation. The folding of thrusts cannot be explained by an east-to-west sequence of thrust initiation. It appears, therefore, that at least all the larger thrusts were initiated in a west-to-east sequence, although locally the sequence may have been reversed.

During the Tertiary deformational history of the disturbed belt, major thrusts such as the McConnell and Lewis thrusts were active for much longer periods than lesser faults. The sequence of thrusting was predominantly from west-to-east, but because major thrusts moved for a longer period, they overrode smaller ones on the east that started to move later and stopped earlier.

Conclusions

1. Folded faults are believed to have been formed by the development of new thrusts and associated structures beneath them. In the Alberta Foothills and Rockies this means that faults were folded by subsequent emplacement of thrusts cropping out farther east.

2. Step-thrusting results in the folding of both the overlying strata and any pre-existing faults present in the hanging wall sequence.

3. Folded faults may be expected to occur in any thrust belt if the stratigraphic column is inhomogeneous and the sequence of thrusting is from higher to lower.

4. Folded faults overlie deeper structures that may contain hydrocarbons. Sole faults, where exposed, may reveal significant information on the buried structures which they generated several miles down-dip.

5. Thrusting in the Alberta Foothills and Rocky Mountains was initiated mainly in a west-to-east sequence. In many cases, however, order of termination of thrust movement was from east to west. This apparent paradox is explicable if it is accepted that (a) thrusts develop from an initial break and spread along strike; (b) amount of slip is dependent on duration of fault movement; and (c) the time interval between initiation of successive thrusts is considerably less than the time involved in the emplacement of a single thrust.

References Cited

Bally, A. W., P. L. Gordy, and G. A. Stewart, 1966, Structure, seismic data, and orogenic evolution of southern Canadian Rocky Mountains: Bull. Canadian Petroleum Geology, v. 14, p. 337–381.

Childers, M. O., 1963, Structure and stratigraphy of the southwest Marias Pass area, Flathead County, Montana: Geol. Soc. America Bull., v. 74, p. 141–164.
Choquette, A. L., 1959, Theoretical approach to foothills and mountain deformation of western Canada: Alberta Soc. Petroleum Geologists Jour., v. 7, no. 10, p. 234–237.
Dahlstrom, C. D. A., R. E. Daniel, and G. G. L. Henderson, 1962, The Lewis thrust at Fording Mountain, British Columbia: Alberta Soc. Petroleum Geologists Jour., v. 10, no. 7, p. 373–395.
Deiss, C. F., 1943, Stratigraphy and structure of southwest Saypo quadrangle, Montana: Geol. Soc. America Bull., v. 54, p. 205–262.
Douglas, R. J. W., 1950, Callum Creek, Langford Creek and Gap map-areas, Alberta: Geol. Survey Canada Mem. 255, 124 p.
——— 1956, George Creek, Alberta (geologic map with marginal notes): Canada Geol. Survey Paper 55–39, scale 1 in. = 1 mi.
——— 1956, Nordegg, Alberta: Canada Geol. Survey Paper 55–34, 31 p.
——— 1958a, Chungo Creek map-area: Canada Geol. Survey Paper 58–3, 45 p.
——— 1958b, Mount Head map-area, Alberta: Canada Geol. Survey Mem. 291, 241 p.
Erdman, O. A., 1950, Alexo and Saunders map-areas, Alberta: Canada Geol. Survey Mem. 254, 100 p.
Fitzgerald, E. L., 1962, Structure of the McConnell thrust sheet in the Ghost River area, Alberta: Alberta Soc. Petroleum Geologists Jour., v. 10, no. 10, p. 553–574.
Fox, F. G., 1959, Structures and accumulation of hydrocarbons in southern Alberta Foothills, Alberta, Canada: Am. Assoc. Petroleum Geologists Bull., v. 43, no. 5, p. 992–1025.
——— 1969, Some principles governing interpretation of structures in the Rocky Mountain orogenic belt: London Geol. Soc. Spec. Pub. 3, p. 23–41.
Gallup, W. B., 1951, Geology of Turner Valley oil and gas field, Alberta, Canada: Am. Assoc. Petroleum Geologists Bull., v. 35, no. 4, p. 797–821.
Hafner, W., 1951, Stress distributions and faulting: Geol. Soc. America Bull., v. 62, p. 373–398.
Hage, C. O., 1942, Folded thrust faults in Alberta Foothills west of Turner Valley: Royal Soc. Canada Trans., 3rd ser., sec. 4, v. 36, p. 67–78.
Hake, B. F., R. Willis, and C. C. Addison, 1942, Folded thrust faults in the Foothills of Alberta: Geol. Soc. America Bull., v. 53, p. 291–334.
Hume, G. S., 1931, Turner Valley, Alberta: Canada Geol. Survey Map 257A, scale, 1 in. = 1 mi.
——— 1939, Preliminary report on the stratigraphy and structure of Turner Valley, Alberta: Canada Geol. Survey Paper 39–4, 19 p.
——— 1957, Fault structure in the Foothills and eastern Rocky Mountains of southern Alberta: Geol. Soc. America Bull., v. 68, p. 395–412.
Hunt, C. W., 1956, Panther dome: a minor orogen of the Canadian cordillera: Alberta Soc. Petroleum Geologists 6th Ann. Field Conf. Guidebook, p. 44–55.
Hurley, G. W., 1959, Overthrust faulting and Paleozoic gas prospects in Montana's disturbed belt: Billings Geol. Soc. 10th Ann. Field Conf. Guidebook, p. 98–108.
Keating, L. F., 1966, Exploration in the Canadian Rockies and Foothills: Canadian Jour. Earth Sci., v. 3, p. 713–723.
Killick, G. H., 1954, Surface faults in the Alberta syncline: Western Miner (Feb.), p. 35–37.
MacKay, B. R., 1943 (Preliminary Map) Foothills belt of central Alberta: Canada Geol. Survey Paper 43–3.
Mountjoy, E. W., 1959, Miette, Alberta: Canada Geol. Survey Map 40–1959, scale, 1 in. = 1 mi.
North, F. K., and G. G. L. Henderson, 1954, Summary of the geology of the southern Rocky Mountains of Canada: Alberta Soc. Petroleum Geologists 4th Ann. Field Conf. Guidebook, p. 15–81.
Scott, J. C., 1951, Folded faults in Rocky Foothills of Alberta, Canada: Am. Assoc. Petroleum Geologists Bull., v. 35, no. 11, p. 2316–2347.
Thomas, A. N., 1958, Note on a geological cross-section through the Nordegg area: Alberta Soc. Petroleum Geologists 8th Ann. Field Conf. Guidebook, Aug., p. 121–127.
Triad Oil Co. Ltd., 1958, Geological map of Nordegg area, *in* Alberta Soc. Petroleum Geologists 8th Ann. Field Conf. Guidebook, Aug., scale 1 in. = 4 mi., 203 p.
Verrall, P., 1968, Observations on geological structure between the Bow and North Saskatchewan rivers: Alberta Soc. Petroleum Geologists 16th Ann. Field Conf. Guidebook, p. 106–118.
Webb, J. B., 1955, Cross-section of Foothills belt east of Jasper Park: Alberta Soc. Petroleum Geologists 5th Ann. Field Conf. Guidebook, p. 130a.
White, W. H., 1959, Cordilleran tectonics in British Columbia: Am. Assoc. Petroleum Geologists Bull., v. 43, no. 1, p. 60–100.

The Cordilleran foreland thrust and fold belt in the southern Canadian Rocky Mountains

R. A. Price

SUMMARY: The thick (~40 km) slab of Hudsonian (>1750 Ma) continental crust that extends under western Canada from the Canadian Shield can be followed westward, on the basis of its distinctive magnetic anomalies, its influence on the Bouguer gravity values, the results of deep seismic refraction experiments, and the results of geomagnetic depth sounding of the deep electrical conductivity structure, to the Kootenay Arc. The Kootenay Arc is basically a W-facing monocline of crustal dimensions, across which the change in structural level involves an aggregate stratigraphic thickness of up to 20 km. It marks the western edge of the continental craton over which the displaced supracrustal rocks have been draped.

Balanced structure sections of the thrust and fold belt, which take into consideration the deep crustal structure, as constrained by the geophysical data, show that: (i) in early Campanian time the continental crust that now lies beneath the western Rocky Mountains and the Purcell anticlinorium was covered with the platformal Palaeozoic to Upper Jurassic rocks and the exogeoclinal Mesozoic rocks that now form the northeasterly verging imbricate thrust slices of the eastern Rocky Mountains; (ii) the Cordilleran miogeocline developed outboard from the edge of the continental craton, on tectonically attenuated continental crust, or oceanic crust; and (iii) tectonic shortening of about 200 km in the supracrustal rocks in the Rocky Mountains must be balanced at a deeper level, W of the Kootenay Arc, by the shortening of the oceanic or attenuated continental crust.

The net convergence between the Cordilleran magmatic arc and the autochthonous cover on the continental craton is a type of intra-plate subduction that was antithetic to the SW verging subduction zone marking the boundary of the North American Plate. The basement of the back-arc or marginal basin, in which the miogeocline formed, was consumed; but the adjoining continental margin was not. The foreland thrust and fold belt is a shallow subduction complex that was tectonically prograded over the margin of the continental craton, as the supracrustal cover scraped off the down-going slab was piled up against the overriding mass, and spread laterally eastward under its own weight.

The foreland thrust and fold belt of the North American Cordillera is a zone of easterly verging shallow thrust faulting and décollement folding, up to about 300 km wide, that follows the boundary between the Cordilleran miogeocline and the North American craton from the Yukon Territory of northern Canada to southeastern California (King 1969; Wheeler *et al.* 1974; Burchfiel & Davis 1972). From whence it may extend across southern Arizona and New Mexico and southwestern Texas into the Sierra Madre Oriental of Mexico (Corbitt & Woodward 1973), and finally into Guatemala and Honduras (de Cserna 1971). Within this zone of 'thin-skinned' deformation, an easterly tapering wedge of supracrustal rocks, comprising parts of the miogeocline, the cover of the cratonic platform, and the overlying exogeoclinal wedge of synorogenic clastic wedge deposits, was horizontally compressed and tectonically thickened, as it was displaced eastward relative to the underlying undeformed craton. The distinctive attributes of structural style and tectonic setting of this belt are characteristic of many other foreland thrust and fold belts, including: the Helvetic and Jura Mountains of the northern Alps (Bernoulli *et al.* 1974), the Valley and Ridge Province in the western Appalachian Mountains of the United States (Rodgers 1970), the Labrador 'trough' along the boundary between the Churchill Province (Proterozoic–Aphebian) and Superior Provinces (Archean) in the eastern Canadian Shield (Dimroth 1970), and the Asiak fold and thrust belt along the boundary between the Bear Province (Proterozoic–Aphebian) and the Slave (Archean) Province in the northwestern

Canadian Shield (Hoffman 1973; Hoffman *et al.* 1978). Foreland thrust and fold belts appear to represent a geotectonic phenomenon that is widely distributed in space and time, but for which there is, as yet, no simple global explanation.

Foreland thrust and fold belts display obvious similarites in structural geometry to the zones of imbricate thrust faulting and associated folding that occur above subduction zones along the inner slope of trenches in which there is a thick sedimentary fill (Hamilton 1977; Seeley 1977); and some foreland thrust and fold belts have been interpreted as the result of a 'collision' between a passive continental margin and island arc or another continental margin, during which part of the supracrustal cover of the underriding passive margin was scraped off against the overriding plate (Dewey & Bird, 1970). However, many foreland thrust and fold belts cannot be formed in this way because they represent relatively small ($d \times 10^2$ km) displacements distributed within a single physically continuous lithospheric plate (intra-plate displacements), rather than relatively large ($d \times 10^3$ km) displacements across a suture along which one discrete plate has overridden another quite different one (inter-plate displacements).

The foreland thrust and fold belt of the North American Cordillera is a good example of an intra-plate thrust and fold belt. Although there are conspicuous changes in the details of tectonic setting, structural style and history of deformation from one segment to another of the foreland thrust and fold belt of the North American Cordillera, the segment containing the southern Canadian Rocky Mountains is as typical as any. It is a particularly appropriate place for a more detailed analysis of the nature and tectonic significance of this foreland thrust and fold belt, because its essential features are not obscured by younger sedimentary and volcanic deposits, or superimposed deformation of the Basin and Range type, and because it has been relatively thoroughly studied by deep drilling and seismic reflection exploration for hydrocarbon traps (Fox 1959; Dahlstrom 1970; Shaw 1963; Keating 1966; Bally *et al.* 1966) as well as by systematic mapping of surface exposures. The main objective of this paper is to elucidate the geological structure, tectonic evolution, and geotectonic significance of the segment of the Rocky Mountains thrust and fold belt between latitude 48°N and latitude 52°N (Fig. 1), and of the Cordilleran foreland thrust and fold belt in general.

Structure of the southern Canadian Rocky Mountains

The structure of the southern Canadian Rocky Mountains is dominated by thrust faults, almost all of which are easterly verging. Flexural-slip folds have developed in conjunction with the displacements along the thrust faults; and many of the thrust faults die out upward in the cores of anticlines or downward in the cores of synclines, where they mark the centres of curvature for strata that are concentrically folded (Price 1964*b*, 1965). The mechanical anisotropy due to the stratigraphic layering has exerted a profound influence on the style of the deformation. The thrust faults commonly follow bedding glide zones that are linked by ramps along which the faults step across the stratigraphic layering; and the thrust

FIG. 1. Geological map of the foreland thrust and fold belt of the North American Cordillera between 49° and 52°N latitude. The lines labelled W–E and SW–NE mark the locations of the structure sections presented in Fig. 2. Symbols identifying the more important faults are as follows: Bo-Bourgeau; BR-Bull River; Ca–Chatter Creek; Co–Columbia River; Fl–Flathead; Ha–Hall Lake; Ho–Hope; Jo–Johnston Creek; Le–Lewis; Li–Livingstone; Mc–McConnell; Moy–Moyie; Mo–Mons; Ne–Newport; Pi–Pipestone Pass; Pu–Purcell; RMT–Rocky Mountain Trench; Ru–Rundle; Sa–Standfast Creek; Si–Simpson Pass; Sm–Sulphur Mountain; StM–St. Mary.

Symbols identifying batholiths are as follows: Ba–Battle Mountain; Bu–Bugaboo; By–Bayonne; F–Fry Creek; H–Horsethief Creek; Ka–Kaniksu; Ku–Kuskanax; N–Nelson; W–White Creek. The Valhalla gneiss complex is identified by the symbol V. The information which is compiled and interpreted in this map is taken from maps by: Balkwill, *et al.* (in press *a, b*); Benvenuto (1978); Bielenstein, *et al.*, (1971); Campbell & Loofbourow (1957); Douglas (1951, 1952, 1958); Fyles (1964, 1967); Fyles & Hewlett (1959); Gilman (1972); Glover (1978); Harrison & Jobin (1963, 1965); Harrison & Schmidt (1971); Henderson (1954); Höy (1977, 1978); Huntting *et al.* (1961); Johns (1970); Leech (1959, 1960); Little (1960); Miller (1974 a–d); Miller *et al.* (1961); Johns (1970); Ollerenshaw (1975, 1978); Park & Cannon (1943); Price (1962); Price & Mountjoy (1978 *a–d*, 1979 *a–b*, in press); Price *et al.* (in press); Price *et al.* (in press *a, b*); Balkwill *et al.* (in press *a & b*); Read & Wheeler (1976); Reesor (1973); Ross (1959); Sears (1979); Simony & Wind (1970); Thompson (1972); Wheeler (1963); Williams (1949); Yates (1964, 1971); Zwanzig (1973).

sheets have been folded (and unfolded) as they moved along the stepped fault surfaces (Douglas 1950). Flexural-slip folding and thrust faulting have proceeded concurrently as two different manifestations of the same basic process of horizontal compression and relative northeasterly upward translation of the supracrustal rocks. Many of the thrust faults have been folded together with the beds they cut, because of displacements on other curved faults which developed later, at lower structural levels (Douglas 1950, 1952, 1958; Dahlstrom 1970). Displacements on the thrust faults range up to a maximum of at least 50 km in the case of the Lewis Thrust (Price 1965; Dahlstrom *et al.* 1962); but nevertheless all the faults die out along the strike, and the rock mass is physically continuous around the ends of each fault. On a regional scale, the faults comprise a penetrative array of discrete, discontinuous overlapping and interfingering slip surfaces; and at this scale, the deformation can be viewed as a kind of compressive plastic flow (Nye 1952) involving both large distortion (E–W compression and vertical thickening) and large easterly relative translation, without any overall disruption and loss of cohesion (Price 1973).

The tectonically foreshortened easterly tapering wedge of supracrustal rocks is firmly attached in the E to the undeformed supracrustal rocks comprising the platformal and exogeoclinal deposits that cover the western flank of the Canadian Shield. The western part is penetrated by granitic intrusions and intercalated with and overlain by calc-alkaline volcanic rocks comprising a regional calc-alkaline magmatic belt that has been attributed to the easterly subduction of oceanic lithosphere beneath the Cordilleran orogenic belt (Monger *et al.* 1972; Monger & Price 1979). There are some structures in the deformed supracrustal wedge that reflect an easterly horizontal stretching; but, as will be shown below, the magnitude, location, and timing of this horizontal stretching leave no doubt that the development of the foreland fold and thrust belt involved a large net convergence between the magmatic belt and the North American craton. Therefore, the relative easterly translation of

---

FIG 2. Structure Sections and palinspastically restored sections through the Cordilleran foreland thrust and fold belt in southern Canada. The locations of section are given in Fig. 1. Symbols identifying equivalent points on the hanging-wall side of the more important faults in each structure section and its palinspastically restored counterpart are listed: Bo–Bourgeau; BR–Bull River; BT–Burnt Timber; Bz–Brazeau; Co–Coleman; HL–Hall Lake; In–Inglismaldie; La–Lac des Arcs; Le–Lewis; Li–Livingstone; Mc–McConnell; Mi–Mill Creek; OB–Old Baldy; RMT–Rocky Mountain Trench; Ru–Rundle; Sm–Sulphur Mountain; Wa–Watson; Wh–Whaleback.

The horizontal sea-level datum for the restored sections is the boundary between marine and non-marine Upper Jurassic rocks. In the overlying synorogenic clastic wedge deposits of the exogeocline the Upper Jurassic and Lower Cretaceous Kootenay Formation is designated–JKk; the Lower Cretaceous Blairmore Group in designated–Kbl, and the Upper Cretaceous marine Alberta Group is designated Kag. The shallow structure shown in the eastern part of Section W–E is adapted, with modifications, from sections by Douglas (1950), Price (1962); that between the Bull River (BR) and Rocky Mountain Trench (RMT) faults is modified after Leech (1958) and Höy (1978); and that in the Kootenay Arc follows Fyles (1967) and Höy (1977). The eastern part of Section SW–NE, above −2400 m, is adapted with modifications from sections at a scale of 1:50 000 by Price & Mountjoy (1970*a*, 1970*b*, 1973*a*, 1973*b*) and Ollerenshaw (1972*a*, 1972*b*) and follows the basic pattern of a section by Bally *et al.* (1966, plate 1B and plate 3) that is based on seismic reflection data and deep drilling. The shallow structure of the Purcell anticlinorium is adapted with modifications from a section at 1:250 000 by Reesor (1973); and that in the Kootenay Arc follows a section by Fyles (1964).

Locations of wells drilled for oil and gas, from which logs were used in the preparation of the structure sections are shown by heavy lines and identified by number as follows: 1–Spring Point 2-4-10-29W4 (projected 460 m S); 2–Calstan C&E Cow Ck. 76-30-8-1W5 (projected 8400 m S); 3–Marjon Lundbeck No 1. (projected 4100 m S) (log by Douglas 1950); 4–Texaco Wilmont Todd Ck A-1 3-6-10-2W5 (Projected 4100 m S); 5–Union Quaich 10-3 10-3-9-3W5 (Projected 4500 m N); 6–Imperial Quaich 3-3-10-3W5 (Projected 4100 m); 7–Triad Union Quaich 10-21-9-3W5 (Projected 100 m S); 8–Texaco Livingstone East 3-20-9-3W5 (Projected 500 m N); 9–Coseka *et al.* Coleman 4-23-9-4W5 (Projected 500 mN); 10–Gulf PCP Coleman 7-33-10-4W5 (projected 460 m S); 11–California Standard-Crowsnest 6-14-8-5W5 (Projected 11500 mN); 12–Sinclair *et al.* Racehouse 15-29-9-5W5 (Projected 2100 m S); 13–Consumer Shell Cremanc 10-21-29-5W5 (projected 1200 m SW); 14–Dome Winchell 10-18-24-5W5 (Projected 1500 m S; 15–Western Imperial wildcat 6-16-18-6W5 (Projected 2400 m SW); 16–Imperial Fina Pacifac Benjamin 3-14-38-W75 (Projected 800 m SW).

The logs of the wells are given in the Schedule of Wells Drilled for Oil and Gas, Province of Alberta, published annually by the Energy Resources Conservation Board, Calgary, Canada.

the magmatic belt, and the attendant horizontal compression of the supracrustal wedge, must be balanced by an equivalent large horizontal displacement and/or compression of a basement of continental or oceanic lithosphere at a deeper crustal level. This raises a number of important questions concerning the foreland thrust and fold belt:

1. What is the minimum amount of net horizontal shortening across it?
2. How is this shortening accommodated at a deep crustal level?
3. What is the significance of the foreland fold and thrust belt in terms of the geotectonic evolution of the whole of the Cordilleran Orogen?

In attempting to answer these questions it is necessary, first, to consider the nature of the northeasterly tapering wedge of supracrustal rocks, in order to provide a framework for interpreting the structures that have developed within it.

Tectonostratigraphic framework

The northeasterly tapering wedge of supracrustal rocks, within which the structures of the foreland fold and thrust belt have developed, consists of a series of distinctive tectonostratigraphic assemblages, each comprising a suite of rock units, the overall characteristics of which imply deposition in a particular tectonic setting. These tectonostratigraphic assemblages record the existence of contrasting eugeoclinal, miogeoclinal and platformal tectonic domains, and of distinct preorogenic and synorogenic stages of tectonic evolution (Fig. 1). The preorogenic stage spans the interval from late Proterozic to Middle Jurassic, and involves a western eugeoclinal domain characterized by the widespread occurrence of basic to intermediate and acidic volcanic rocks and of immature, wacke-type clastic rocks; and eastern platformal and miogeoclinal domains characterized by mature, shallow-water clastic and carbonate rocks that become thicker and more shaly westward, from the platform to the miogeocline. The synorogenic stage spans the interval from late Jurassic to Palaeogene and is characterized by clastic deposits that are of western or else local provenance. These accumulated mainly in a fore-deep or exogeocline that migrated northeastward from late Jurassic to Palaeocene time (Bally *et al.* 1966), as the continental lithosphere subsided isostatically in reponse to the load imposed on it by tectonic thickening and relative northeasterly displacement of the overlying supracrustal rocks (Price 1973).

The Belt-Purcell Supergroup, the oldest assemblage in the supracrustal wedge, has a maximum exposed thickness of about 11 km in the Purcell anticlinorium in southern Canada (Reesor 1973), and about 20–25 km in NW Montana (Harrison 1972). The lower part, which is up to at least 6 km thick in southern Canada, consists mainly of fine-grained quartz-wacke turbidites and dark pelites that grade upward through light grey, green, purple and red shallow-water argillites and sandstones, into the carbonate rocks of the middle part. The great volume of relatively homogeneous fine-grained sediment comprising the very thick lower part of the Belt-Purcell assemblage appears to have been supplied by a very large river system of the size and character of the modern Mississippi, and to have been prograded outboard from the margin of the continent (Price 1964*a*). From the thickness and volume of the Belt-Purcell sediments, and the fact that they lie athwart the structural grain of the buried crystalline basement complex, one may infer that the basin in which they accumulated formed during the initial stages in the rifting of a former Precambrian continental mass which was eventually severed to produce the Pacific margin of the North American continent (Sears & Price 1978). The middle and upper parts of the Belt-Purcell assemblage consist of shallow-water carbonate and terrigenous clastic, tidal-flat and flood-plain deposits that accumulated over the prograded delta plain. Eastward thinning and convergence of all the units in the Belt-Purcell assemblage is accompanied by the disappearance of the turbidite facies, and by the appearance of coarse feldspathic detritus that apparently came from the nearby margin of the basin (Price 1964*a*).

The Windermere Supergroup, of late Proterozoic age, provides a sharp contrast with the Belt-Purcell assemblage, which it overlies unconformably. Coarse-grained feldspathic wacke sandstone and quartz-feldspar pebble conglomerates are interbedded with green and grey pelites, and locally with carbonate rocks. Conglomeratic mudstones, some of which have been interpreted as glacial deposits (Aalto 1971), occur on the unconformity at the base of the assemblage, and locally also at various other levels, including the top. The clasts in these conglomerates consist of distinctive rock types that occur in the underlying Belt-Purcell Supergroup. They are of local provenance, and imply high structural relief that exposed Belt-Purcell rocks, whilst Windermere strata accumulated nearby to thicknesses of up to 9 km (Lis & Price 1976). In contrast, the coarse feldspathic wackes had an external provenance and were transported southwestward (Young *et al.* 1973) over the Belt-Purcell rocks to their site of deposition. The Eo-Cambrian quartz sandstones, which form the base of both the overlying Lower Palaeozoic miogeoclinal and platformal assemblage and the eugeoclinal assemblage, unconformably overlap the thickest part of the Windermere Supergroup, and also the Upper and Middle Belt-Purcell strata, in the Purcell anticlinorium in southern Canada. Relationships beneath the unconformity show that the St Mary Fault (Fig. 1, 49°30′N, 116°15′N), which is now an important right-hand reverse fault (Rice 1941; Leech 1958, 1962) and is linked to thrust faults in the western Rocky Mountains, follows the locus of an older NE trending structure across which there were up to 13 km of stratigraphic separation with the northwest side down, during the deposition of the Windermere Supergroup (Lis & Price 1976).

The Lower Palaeozoic platformal sequence consists of about 1 km or less of shallow-water carbonate rock and interbedded shale. It thickens westward in the miogeocline, to about 5 km adjacent to a carbonate bank margin that is now located in the centre of the Rocky Mountains (Fig. 1, 49°40′N, 115°10′W to 52°0′N, 117°20′W). At this carbonate bank margin all the Cambrian and Ordovician formations, except one Upper Cambrian limestone unit and an Upper Ordovician and Lower Silurian dolomite unit undergo a relatively abrupt change in facies to light-coloured shale and argillaceous limestone (Aitken 1971). Farther W, there is a profound change across the Purcell anticlinorium between the light-coloured miogeoclinal shale facies and the coeval eugeoclinal facies. The latter consists mainly of dark graphitic pelites, but contains significant intercalations of basic volcanic rocks and feldspathic grits and wacke-type sandstones (Fyles 1964) that must have had either a western or a local provenance. The Lower Palaeozoic platformal and miogeoclinal strata appear to have accumulated as shelf and slope deposits along a continental margin (Monger *et al.* 1972) that was prograding westward into a back-arc or marginal basin. Repeated basic volcanism punctuates the record of deposition in this basin; and it includes intermittent incursions of igneous clastic detritus, that presumably were derived from a magmatic arc situated further west, (Monger & Price 1979). Abrupt thickness and facies changes, which occur in the miogeoclinal rocks of the Purcell and western Rocky Mountains, record the tilting of large blocks, perhaps caused by displacements on deep down-to-the-basin listric normal faults.

Important unconformities separate the Lower Palaeozoic and Upper Palaeozoic assemblages. Mississippian and younger eugeoclinal rocks, consisting of dark pelites with intercalated sandstones, limestones and basic volcanic rocks, overlie the Lower Palaeozoic assemblage with angular unconformity. A basal conglomerate, containing clasts among which there is random orientation of a low-grade metamorphic tectonite foliation, shows that the Lower Palaeozoic rocks in the eugeoclinal domain were deformed and metamorphosed prior to late Mississippian time (Read & Wheeler 1976). In the platformal and miogeoclinal domain, Upper Devonian shallow-water carbonate rocks unconformably overlap Silurian, Ordovician and Upper Cambrian formations toward the E (Price & Mountjoy 1970, Figs. 2, 3) across the flank of a broad epeirogenic structure, the Alberta arch (Douglas *et al.* 1970, Fig. VIII–15). However, in the Purcell Mountains and Rocky Mountains near Cranbrook (Fig. 1, 49°30′, 113°30′) relationships beneath the unconformity show that the Moyie-Dibble Creek Fault, another right-hand reverse fault like the St Mary Fault, follows the locus of an older northeasterly trending structure across which there was stratigraphic separation of more than 10 km, with the NW side down, during the deposition of the Lower Palaeozoic assemblage (Leech 1958, 1962; Norris & Price 1966; Price 1972).

The Triassic–Middle Jurassic assemblage is represented on the cratonic platform by about 100 m of Jurassic marine shale; but it is more than 1000 m thick in the miogeocline, where it includes about 600 m of Triassic shallow-water marine shale, siltstone, quartz sandstone, carbonate rock and evaporites. The coeval eugeoclinal assemblage, which is only preserved W of the Purcell anticlinorium and the Kootenay arc (Fig. 2) appears to be more than 10 km thick (Little 1960; Read & Wheeler 1976) and consists of dark pelites, volcanogenic sandstones, and mainly andesitic volcanic rocks. Some of the mid to late Jurassic granitic plutons have been deformed along with the rocks they have intruded, and they may be partially coeval with the Jurassic volcanic rocks; whereas the Cretaceous granitic plutons cut across previously developed folds and faults, and have no obvious extrusive counterparts (Gabrielse & Reesor 1974).

The dramatic changes in thickness and lithofacies that occur along, as well as across, the strike of the northeasterly tapering wedge of Proterozoic–Jurassic supracrustal rocks involve important changes in mechanical properties, and have exerted a profound influence on the nature and orientation of the structures that have developed within it. They account for the development of a series of distinct linear tectonic subprovinces in the Rocky Mountains, including the Foothills (platformal Palaeozoic and exogeoclinal Upper Cretaceous), Front Ranges (platformal to miogeoclinal Palaeozoic and exogeoclinal Jurassic–Lower Cretaceous), Eastern Main Ranges (miogeoclinal carbonate facies) Western Main Ranges (miogeoclinal Cambrian shale facies), and Western Ranges (miogeoclinal Cambro-Ordovician shale facies with Ordovician–Silurian carbonate rocks). Each of these displays distinctive characteristics of structural style and physiography, as well as stratigraphy (North & Henderson 1954; Price & Mountjoy 1970). Most change abruptly along the strike at about 49°20′N latitude because of changes in sedimentary facies and thicknesses resulting from the influence of the transverse, northeasterly trending structures that were active during Proterozoic and Palaeozoic time. For example, SE of this zone, where the thick and relatively homogenous Belt-Purcell assemblage above the Lewis thrust fault is overlain by very thin Lower Palaeozoic and Upper Palaeozoic assemblages that are in the platformal facies, broad open folds in relatively flat-lying beds are carved into castellated mountain massifs, and a structural and physiographic style that is characteristic of the eastern Main Ranges further N, extends out to the edge of the Foothills. The steeply-dipping, moderately thick thrust slices of Palaeozoic platformal to miogeoclinal carbonate rocks, that form the characteristic linear mountain ranges of the Front Ranges subprovince further N, are missing completely. The conspicuous changes in structural trend that occur across this transverse zone can be related, as will be shown below, to changes in the configuration and orientation of the margins of the basin in which the various tectonostratigraphic assemblages accumulated.

Balanced structure sections

Reliable estimates for the minimum net tectonic shortening or convergence across the thrust and fold belt can be obtained from balanced structure sections (Dahlstrom 1969*b*), in which the deep crustal structure is also considered, and is constrained by the relevant geophysical data. A balanced section is one in which there is geometric compatibility from one level to another. In a thrust faulted terrain the simplest application of this principle is the requirement that the rock units which occur on one side of a thrust fault must have matching counterparts on the other side. Provided that the deformation accompanying the thrust faulting has been accomplished mainly by flexure and interstratal slip, and that the penetrative strain within the beds is so small in comparison with the size of the folds and thrust faults that it can be ignored, a balanced section is one in which there is consistency of stratigraphic markers from one level to another, and in which the lengths of the stratigraphic markers in the deformed state are the same as they were in the undeformed state, only the shape having changed during the deformation. Under these circumstances one test of a balanced section is that it leads to a logical and reasonable palinspastic restoration (Dahlstrom 1969*b*).

Systematic regional geological mapping and many detailed studies of the stratigraphy and sedimentary petrography of the rocks of the exogeoclinal assemblage and the eastern carbonate facies of the platformal and miogeoclinal assemblages show that stratigraphic units do not change significantly in thickness between the limbs and hinge zones of folds, and also that even the finest details of original sedimentary and organic structures have survived the deformation without significant distortion. This leaves no doubt that intrastratal penetrative strain within these rocks is very low and can be neglected for purposes of constructing balanced sections. Accordingly, the structure sections through the carbonate facies of the platformal and miogeoclinal assemblages (Fig. 2) have been balanced, on the premise that the lengths of the stratigraphic contacts have not been altered significantly in the course of the folding and faulting; and the palinspastic sections have been constructed on the premise that the lengths of the beds in the deformed state are essentially the same as in the undeformed state. In the shale facies of the Lower Palaeozoic miogeoclinal assemblage, widespread development of penetrative cleavage and schistosity, and conspicuous distortion of primary structures (Balkwill 1972; Gardner 1977) preclude the application of this technique except at the highest stratigraphic levels (Middle Ordovician quartz sandstones and

Upper Ordovician and Lower Silurian dolomites), where interstratal penetrative strains are low, and shortening has been accomplished mainly by flexural-slip folding and associated thrust faulting (central parts of Sections NE–SW and E–W, Fig. 2).

In the Proterozoic rocks of the Windermere and Belt-Purcell assemblages, Penetrative intrastratal strain is variable, increasing from relatively low levels in the E to higher levels with progressively higher metamorphic grade toward the W. However, the general structure of the Purcell anticlinorium can be outlined satisfactorily by projecting what occurs at the surface along the northerly and northwesterly plunge into the plane of the sections. The Kootenay Arc, which forms the SW flank of the Purcell anticlinorium, is characterized by high, penetrative strains and complex tectonic overprinting associated with greenschist and amphibolite facies regional metamorphism and coaxial refolding about N and NW trending axes (Fyles 1964, 1967); and the general character of the structures within it can also be outlined on the basis of projections along the plunge (Fyles 1967; Höy 1977). There is a very large change in structural level across the Kootenay Arc, from the lowest part of the Belt-Purcell Supergroup in the core of the Purcell anticlinorium to an extensive tract of exposures of the Triassic–Jurassic eugeoclinal assemblage, which occurs adjacent to the Nelson batholith and the Shuswap Metamorphic Complex. The aggregate thickness of the succession of rock stratigraphic units which occur in superposition between these two horizons is more than 20 km. Lateral variations, involving a westerly decrease in the thickness of older units and an easterly decrease in the thickness of younger units, may account for some of this large change in stratigraphic level, but most of it is an expression of change in structural level. Accordingly, the Kootenay Arc, in spite of the complexity of the detailed structure within it, is basically a simple structure—a westerly facing monocline of crustal dimension. Geophysical investigations of deep crustal structure indicate that this monoclinal structural step in the supracrustal rocks marks the western edge of the thick (~40 km) slab of Hudsonian (>1750 Ma) continental crust that forms the basement of the North American craton in western Canada.

Deep drilling and seismic exploration for hydrocarbons show that the thrust faults flatten with depth within the supracrustal succession, above the Hudsonian basement, at least as far W as the central part of the Rocky Mountains, and probably beyond the Rocky Mountain Trench beneath the eastern flank of the Purcell anticlinorium (Bally *et al.* 1966). The distinctive pattern of broad, high-amplitude, NE trending magnetic anomalies that is characteristic of the Hudsonian basement, which extends under the cratonic platform from the Churchill Province of the Canadian Shield, can be followed across the whole of the Rocky Mountains into the eastern part of the Purcell anticlinorium, and perhaps as far W as the Kootenay arc (Fig. 3). These relationships corroborate the interpretation from the seismic reflection data. The Hudsonian basement does, indeed, extend under the thrust and fold belt and the Purcell anticlinorium without any apparent disruption of its characteristic NE trending tectonic fabric.

The results of seismic refraction experiments, using explosions in lakes in the interior of the Cordillera (Chandra & Cumming 1972; Berry & Forsyth 1975; Mereu *et al.* 1977), mine blasts in the coalfields of the southern Rocky Mountains and the porphyry copper mines of the interior of the Cordillera (Bennett *et al.* 1975; Spence *et al.* 1977; Cumming *et al.* 1979), show that the change in structural level that occurs in the supracrustal rocks across the Kootenay Arc coincides with a westerly decrease in crustal thicknesses, from a maximum thickness of 50–55 km in the vicinity of the western Rocky Mountains and the Purcell anticlinorium to about 30–40 km beneath the interior of the Cordillera (Monger & Price 1979). The monoclinal step at the Kootenay Arc also coincides with a westerly increase in crustal and upper mantle electrical conductivity as outlined by geomagnetic depth sounding experiments (Caner 1971; Dragert & Clarke 1976); and with a westerly change in the Bouguer gravity anomaly from about −200 mgal, where the crust is thickest, to about −140 mgal in the interior of the Cordillera (Fig. 4). It is worthy of note, that the lowest Bouguer gravity anomaly values occur within the widest part of the foreland fold and thrust belt, where the amount of tectonic thickening of the supracrustal wedge would be expected to be greatest. Morever, the amount of crustal thickening in this area, as indicated by the analysis of the Bouguer anomalies (Stacey 1973) and by the results of the seismic refraction experiments that were cited above, is about that which is required to accommodate the tectonic thickening of the supracrustal wedge that was calculated by Price (1973) from a comparison of structure sections and corresponding palinspastically restored sections by Bally *et al.* (1966) and Price &

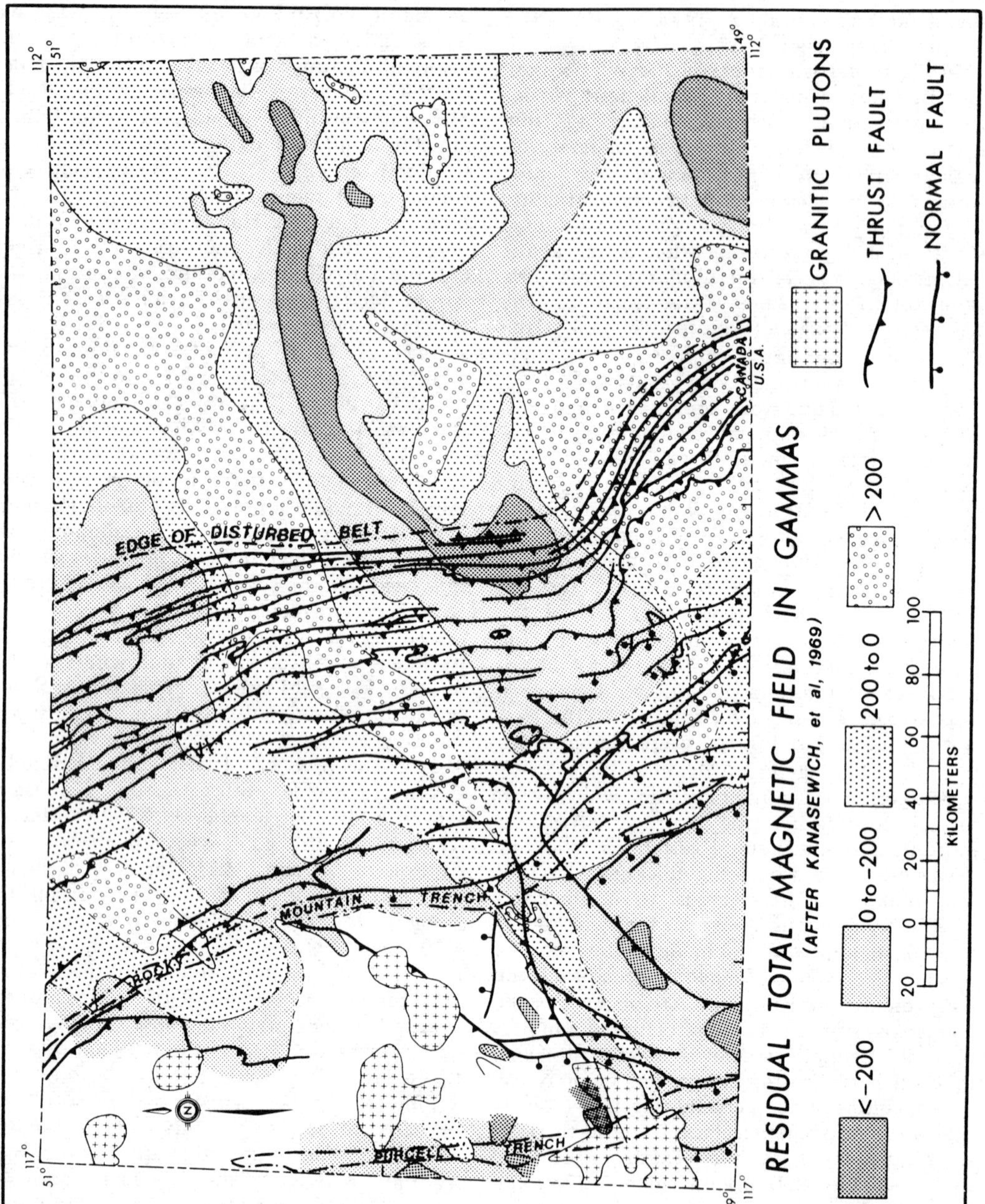
RESIDUAL TOTAL MAGNETIC FIELD IN GAMMAS
(AFTER KANASEWICH, et al, 1969)
<−200
0 to −200
200 to 0
>200
GRANITIC PLUTONS
THRUST FAULT
NORMAL FAULT
KILOMETERS
20 0 20 40 60 80 100
EDGE OF DISTURBED BELT
ROCKY MOUNTAIN TRENCH
PURCELL TRENCH
CANADA
U.S.A.
112°
51°
117°
49°

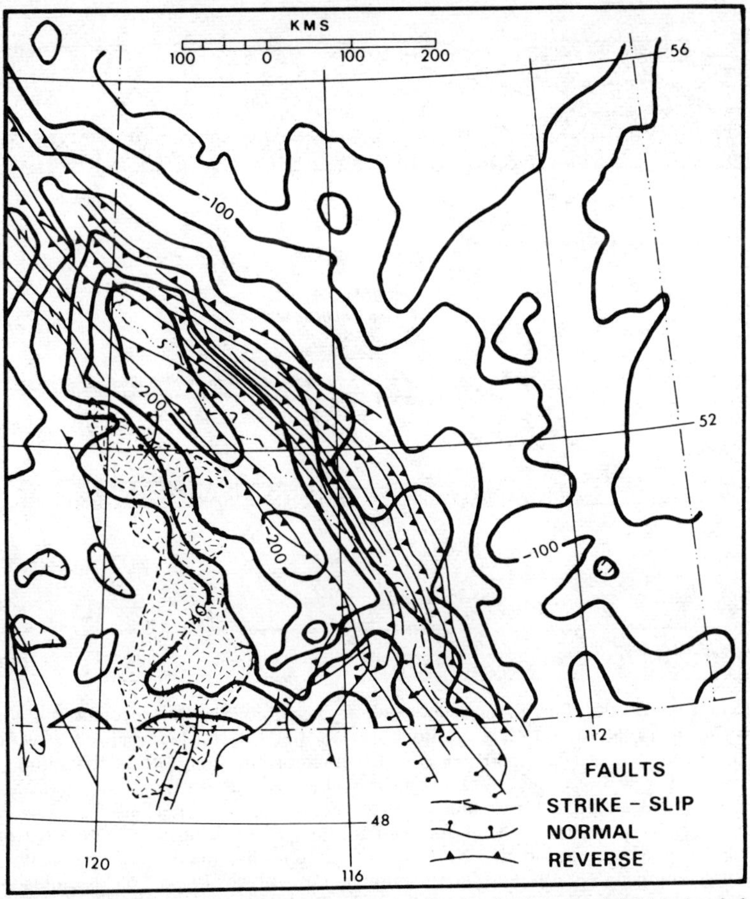

FIG. 4. Relationships between the structures in the foreland thrust and fold belt of the southern Canadian Rocky Mountains and the Bouguer gravity anomalies. The Bouguer gravity anomaly values are after the Earth Physics Branch of Canada (1974). The faults and the limits of the Shuswap Metamorphic Complex are from Fig. 1 and from King (1969).

Mountjoy (1970)—10 km of increase in crustal thickness, of which 2 km stand above sea level. However, it is not enough to balance, at a deep crustal level below the basement surface, the amount of tectonic shortening that occurs in even the eastern part of the thrust and fold belt, where deep drilling and seismic reflection data leave little scope for imagination in interpretations of tectonic shortening in the supracrustal wedge. Thus, a consideration of the deep crustal structure, based on the seismic refraction, magnetic and gravity data leads to the conclusion that the slab of continental crust, about 40 km thick, that underlies the cratonic platform in front of the thrust and fold belt, extends under the deformed wedge of supracrustal rocks, beneath the Rocky Mountains and the Purcell anticlinorium, to the Kootenay Arc, where the supracrustal rocks wrap around the edge of it (Fig. 5). This is contrary to the interpretation of Eisbacher *et al.* (1974) who suggested that the structure in the western Rocky Mountains and beyond is the result of uplift involving both the supracrustal wedge and its Precambrian crystalline basement, and that the flat thrust faults and undeformed basement are limited to the eastern Rocky Mountains. Their interpretation was based on structure sections which cannot be balanced between the supracrustal rocks and the basement because they involve relatively little shortening at the level of the basement,

FIG. 3. Relationships between the structures of the thrust and fold belt of the southern Canadian Rocky Mountains and the Hudsonian structural fabric of the Precambrian basement that extends under it. The magnetic anomaly map is after Kanasewich *et al.* (1969). The thrust faults and normal faults are a simplified portrayal of those in Fig. 1.

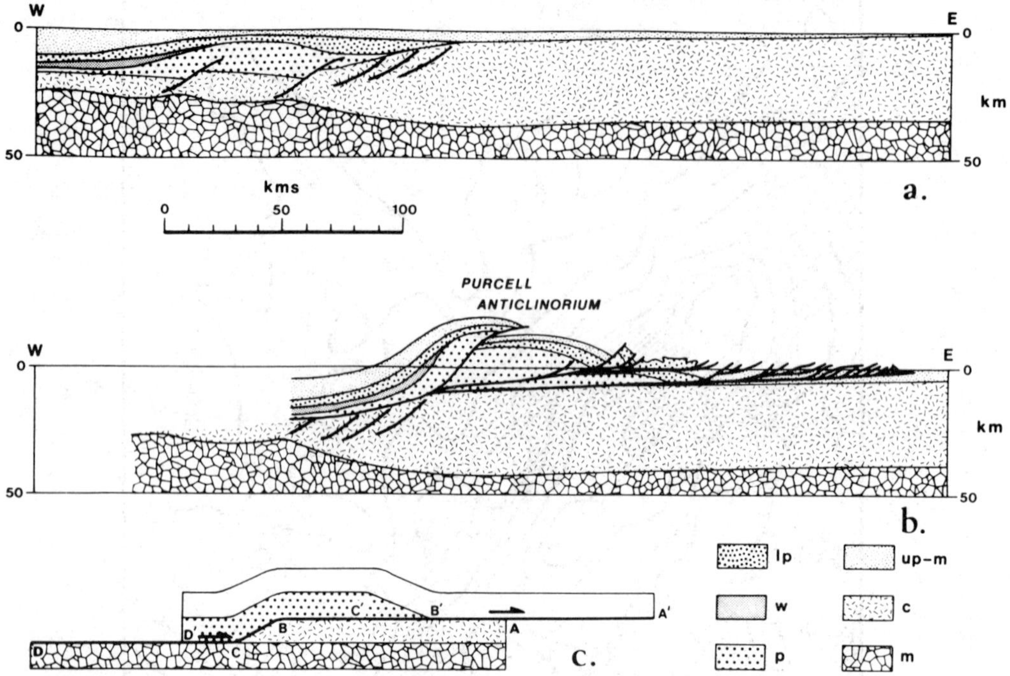

FIG. 5. Evolution of the Purcell anticlinorium a. Restored Section through the southeastern Canadian Cordillera along 49°45′N Latitude (as in Fig. 7). b. Generalized structure of the Cordilleran fold and thrust belt and the Purcell anticlinorium along 49°45′N Latitude (based on Section W–E of Fig. 2; but drawn to eliminate the effects of the erosion of supracrustal rocks that occurred during and after the thrust faulting). c. Schematic representation of the development of the Purcell anticlinorium (Points on the footwall and hanging wall of the zone of detachment, that were initially contiguous, are labelled with the same letters) m—mantle; c—continental crust; p—Belt-Purcell assemblage; w—Windermere assemblage; lp—Lower Palaeozoic assemblages; up-m—Upper Palaeozoic and Triassic–Jurassic assemblages.

whereas tectonic shortening in the supracrustal rocks to the E amounts to 65 km across the Foothills alone (Bally *et al.* 1966, plate 12), and all of this would have to be accommodated within the basement E of the boundary of their postulated basement uplift.

A conspicuous increase in the Bouguer gravity values occurs southward along the Purcell anticlinorium where the transverse northeasterly trending St Mary and Moyie Faults cut across it. Further S, in Montana, at about 48°15′N, Wynn *et al.* (1977), on the basis of their analysis of a gravity and audio-frequency magnetotelluric traverse across the Purcell anticlinorium, concluded that crystalline basement rocks occur in the core of the anticlinorium there. Transported basement rocks could be expected within the deformed wedge of supracrustal rocks at this structural position because suspected basement rocks occur in about the same position above the Purcell thrust fault, to the N, at about 52°30′N (Campbell 1968; Price & Mountjoy 1970).

Structure sections W–E and SW–NE (Fig. 2) have been drawn perpendicular to the local structural trend, in order to facilitate the projection of surface and well data, along the plunge of the structures, into the plane of the section. Errors in estimates of tectonic displacements, introduced by the fact that the structure sections may not be parallel with the direction of net tectonic displacement, vary with the cosine of the angle between the structure section and the direction of net tectonic displacement. The error introduced in this way is normally less than 15%, if the angle between the line of section and the average direction of net tectonic displacement is less than 30°. Both sections are based on the premise that the Hudsonian crystalline basement complex extends under the thrust and fold belt to the centre of the Purcell anticlinorium, and that all the thrust faults lie above it, in the supracrustal wedge, or at the basement surface (Bally *et al.* 1966). In both sections, the depth to the basement, as shown, is the minimum depth that is required to accommodate the total thickness of the supracrustal succession, as it is known from one place to another, and to meet the geometric requirements of a balanced section. The eastern part of Section W–E is relatively tightly constrained by information from deep drilling, and it illustrates several important principles, some of which are portrayed in more detail in Fig. 6. It is a well established empirical rule that the thrust faults cut up through the stratigraphic succession in the direction of relative displacement of the overthrust side, and that they carry older rocks over younger and produce repetitions of the normal stratigraphic succession. Thus, westerly verging thrust faults, such as the Waldron fault, cut up through the stratigraphic succession toward the W; and easterly verging thrust faults, even if folded and E-dipping, cut up through the stratigraphic succession toward the E and carry older rocks over younger. Certain stratigraphic zones are a preferred locus of décollement or bedding glide, and other thrust faults branch away from them as concave upward imbricate splays. The condition of geometric compatibility for a balanced section requires that an extensive bedding glide zone in the hanging-wall of a thrust fault must have its matching counterpart in the footwall. The bedding glide zones in the upper part of the Alberta Group (Kag) in the eastern part of Fig. 6, require matching footwall counterparts to permit the palinspastic reconstruction shown in the lower part of the figure. The thrust faults are commonly stepped, comprising extensive bedding glide zones that are linked by relatively narrow ramps. The external rotation and folding which is inherent in movement along such stepped fault surfaces affects all the overlying rocks, including any thrust faults which occur within them. Thrust faults may merge upwards as well as downwards, as shown by the tectonically thickened wedge of Alberta Group (Kag), Belly River Formation (Kbr) and Bearpaw Formation (Kbp) which is responsible for the easterly dip of the Alberta Syncline, and separates it from the relatively flat-lying strata below (Fig. 6). Elsewhere, as for example in the Coleman Thrust sheet, beneath where it is overlapped at the surface by the Lewis Thrust sheet (Fig. 2, Section W–E), sigmoidal imbricate thrust faults that merge upward with one fault and downward with another result in substantial tectonic thickening of an isolated lenticular mass of rock, as well as a significant transfer of displacement from the lower fault to the upper.

In the Foothills, along and beneath the Livingstone Thrust fault (Li in Section W–E, Fig. 2), extensive bedding glide zones occur in Jurassic marine shales (Fernie Group) and in or at the base of a thick unit of Upper Devonian micritic limestone (Palliser Formation). The condition of geometric compatibility for a balanced section requires that offset counterparts of these bedding glide zones be represented along the opposite side of the same thrust surfaces to the W; and this leads to the conclusion that autochthonous Upper Devonian and Cambrian strata lie on undeformed Hudsonian crystalline basement under the western Rocky Mountains, beneath the Bourgeau (Bo) and Bull River (BR) thrust sheets. It also implies that along this line of section the Lewis Thrust sheet (Le) cannot contain a thick succession of Belt-Purcell strata, as it does further S (Fig. 1), unless there is an abrupt change in the depth to the basement surface to provide the necessary space. Seismic reflection profiling shows no such change; instead, the basement in this area slopes gently W without disruption (P. L. Gordy pers. comm. 1979). Similar arguments, involving the Coleman thrust sheet (Co), the Lewis Thrust sheet (Le), and the Bourgeau Thrust sheet (Bo), indicate that the thick Lower Palaeozoic miogeoclinal shale facies and the underlying thick succession of Belt-Purcell strata that occurs between the Bull River Thrust fault and the Rocky Mountain Trench (RMT) first appears in the Bougeau Thrust sheet. The abrupt westerly increase in the thickness and facies of the supracrustal succession within the Bourgeau thrust sheet, which marks the change from platformal to miogeoclinal supracrustal rocks, must have had a matching counterpart in the 'autochthonous' cover succession on the basement below the Bourgeau thrust fault. Palinspastic reconstruction of only that part of the balanced section W–E (Fig. 2) which lies E of and includes, the Bourgeau Thrust sheet (eastern two-thirds of the Rocky Mountains) indicates that the locus of this matching 'autochthonous' counterpart is along the Kootenay Arc. This corroborates what was concluded from a consideration of the geophysical data—that the western edge of the thick slab of continental cratonic crust lies beneath the Kootenay Arc. It implies that all the thick miogeoclinal strata, including the Lower Palaeozoic shale facies and the underlying Windermere and Belt-Purcell assemblages, accumulated outboard from the edge of the craton on attenuated continental crust or oceanic crust.

The structure in section SW–NE differs in

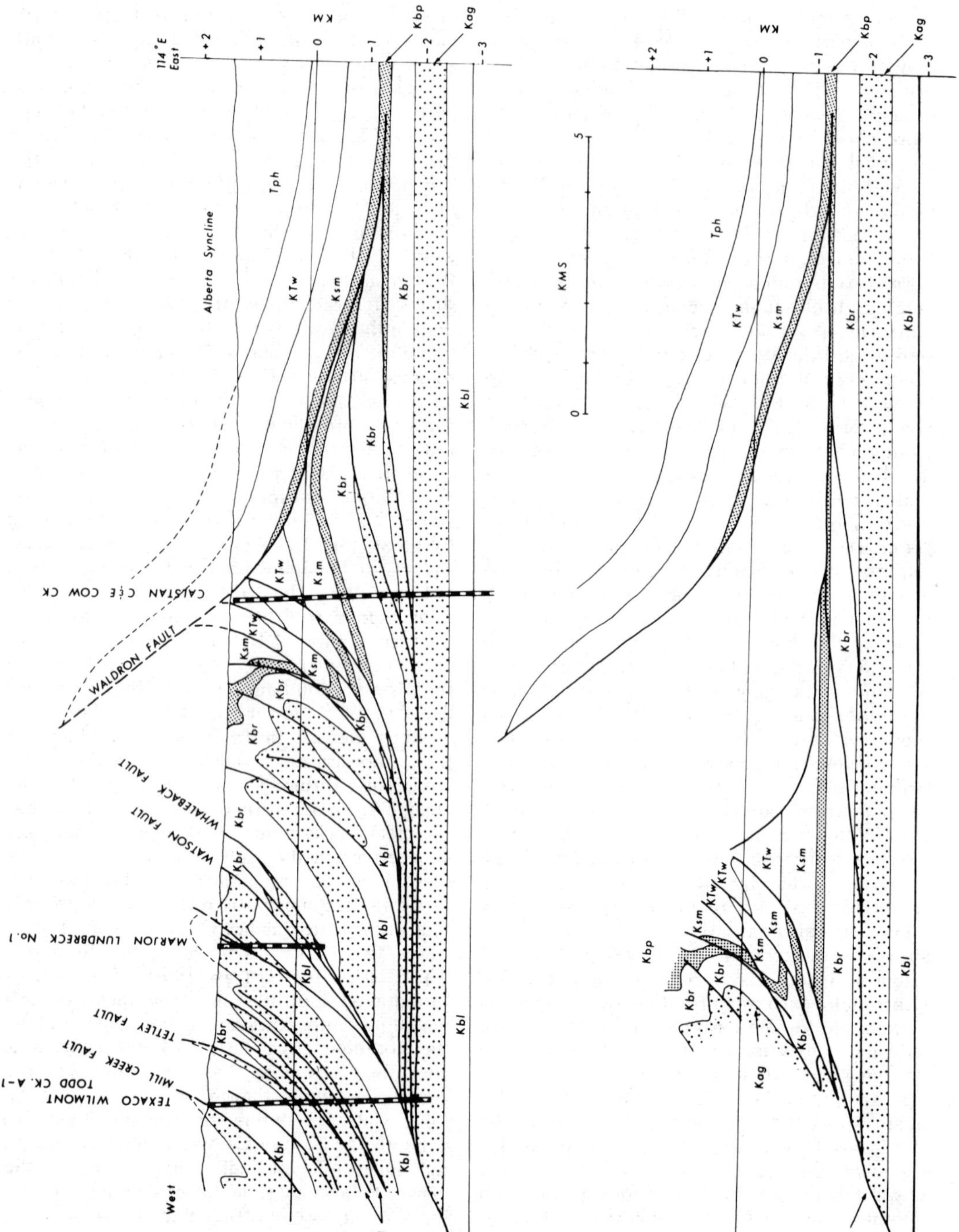

114°E East
KM
+2
+1
0
-1
-2
-3
Kbp
Kag
Tph
Alberta Syncline
KTw
Ksm
Kbr
Kbl
CALSTAN CEE COW CK
WALDRON FAULT
WHALEBACK FAULT
WATSON FAULT
MARJON LUNDBRECK No.1
TETLEY FAULT
MILL CREEK FAULT
TEXACO WILMONT TODD CK. A-1
West
KMS
0
5

detail from that in Section W–E because of changes in the character of the supracrustal succession, but also because of an increase in the amount of overall tectonic shortening or convergence across the fold and thrust belt. The Lower Palaeozoic platformal and miogeoclinal carbonate facies is substantially thicker and more complete beneath the sub-Devonian unconformity, and extends further W than in Section W–E. It forms a series of five moderately thick, relatively steeply dipping thrust sheets (Mc, La, In, Ru, and Sm in Section SW–NE of Fig. 2) beneath the Bourgeau Thrust fault (Bo); whereas in Section W–E there are only two in the same interval. As in Section W–E, the Bourgeau Thrust fault marks the approximate position of the relatively abrupt increase in thickness of the Lower and Upper Palaeozoic assemblages that defines the boundary between the platform and the miogeocline. It also marks the first appearance of Precambrian rocks, but these belong to the Windermere assemblage rather than the Belt-Purcell assemblage.

A broad, northerly plunging anticlinorium occurs W of the Bourgeau Thrust fault, in the centre of the Rocky Mountains near the Continental Divide. The Simpson Pass Thrust fault follows its crest, and several large thrust faults that develop as northeasterly diverging splays from the Simpson Pass Thrust fault are folded across the anticlinorium and adjacent syncline, and extend far to the NW (Price & Mountjoy 1970).

There are about 2 km of Windermere strata exposed above the Simpson Pass Thrust fault, and these, together with the thicknesses of the overlying formation, define a lower limit on the total thickness of supracrustal rocks that must be accommodated between the crystalline basement and the rocks that are exposed at the surface to the W of the Simpson Pass Thrust fault. This minimum estimate of the depth to the basement is what has been used to construct the section. Because the section is based on the premise that the Hudsonian basement is undeformed and slopes beneath the deformed supracrustal wedge to beyond the Rocky Mountain Trench, the space between the basement surface and the core of this northerly plunging anticlinorium is considered to be occupied by deformed supracrustal rocks, the specific structure of which is completely hypothetical. The particular interpretation shown in the section follows the style of the structures which occur in a similar setting beneath the Coleman fault and the Livingstone Thrust in Section W–E, where their existence is documented by deep drilling; and beneath the Lewis thrust sheet further S, where they are exposed (Fermor & Price 1976). The choice of the number of thrust faults that occur beneath the Bourgeau Thrust fault and merge upward with it may be arbitrary, but it does have implications regarding the rate at which the stratigraphic succession in the supracrustal wedge thickens westward between the Sulphur Mountain (Sm) and Bourgeau thrust faults. The interpretation shown implies that the thickening is gradual and relatively uniform. If the thickening is abrupt, fewer thrust faults and less shortening between the Sulphur Mountain and Bourgeau thrust sheets would be involved.

The carbonate bank margin, marking the boundary between the Lower Palaeozoic carbonate and shale facies, occurs W of the Simpson Pass fault and is, in part, offset by late listric normal faults (Cook 1975). W of this boundary, there is a profound change in structural style marked by the widespread development of slaty cleavage and high penetrative strain. A conspicuous fan structure is outlined by the orientation of the cleavage and the axial surfaces of the folds, which are northeasterly verging on the NE side and southwesterly verging on the SW side. Locally, it is situated over an anticlinal culmination in underlying Eo-Cambrian quartz sandstones (Balkwill 1972), but elsewhere (Section SW–NE) it coincides with a monoclinal flexure (see also Price & Mountjoy 1979*a* and *b*; Price *et al.* in press). The fan structure can be attributed to décollement above the Eo-Cambrian quartz sandstone succession, in Cambrian slates of the Lower Palaeozoic shale facies, in conjunction with displacement on the Purcell Thrust fault (as shown in Section SW–NE). Cross-cutting relationships amongst northeasterly and southwesterly verging thrust faults on the W flank of the fan structure, under the Purcell Thrust fault, outline two phases of tectonic overprinting. Older northeasterly verging thrust faults were rotated during continued folding, until overturned, together with the beds in which

FIG. 6. Structure of the eastern Rocky Mountain Foothills at 49°45'N Latitude. The line of section is the same as that of Section W–E in Fig. 2, and the sources of information on wells drilled for oil and gas are those given in Fig. 2. The interpretation in the upper part of the section is adapted, with some modifications, from Douglas (1950). The symbols identifying lithostratigraphic units are as follows: Kbl—Blairmore Group; Kag—Alberta Group; Kbr—Belly River Formation; Kbp—Bearpaw Formation; Ksm—St. Mary River Formation; KTw—Willow Creek Formation; Tph—Porcupine Hills Formation.

they occur (Balkwill 1972). They are cut by younger southwesterly verging thrust faults that involve displacements out of the core of the fan structure. These, in turn, are cut by still younger northeasterly verging thrust faults that are spatially related to the Purcell thrust fault which has overridden the W flank of the fan structure. The Lower Palaeozoic shale facies appears to have been 'scraped' off the underlying quartz sandstones, and piled up in front of the Purcell Thrust sheet. The same fan structure extends southward to Section W–E, beyond which it terminates against the northeasterly trending Moyie-Dibble Creek Fault (Moy) that follows the southern margin of the basin in which the Lower Palaeozoic shale facies accumulated. This asymmetric basin, which formed on top of the late Precambrian miogeoclinal prism, was probably controlled by deep-seated listric normal faulting. The steep, E flank forms a natural, local boundary between the miogeocline and platform.

In the Purcell anticlinorium, W of the Purcell fault, the Lower Palaeozoic miogeoclinal succession, which is about 8 km thick in the western Rocky Moutains, occurs as a condensed section that is overlain by Upper Devonian strata, and locally is less than 500 m thick (Reesor 1973). This contrast can be attributed to tectonic foreshortening, across the Purcell thrust fault, of the less steep W flank of the Lower Palaeozoic basin. The structure shown in the upper levels of the Purcell anticlinorium in Section SW–NE is based on projections along the plunge of structures exposed nearby; but the structure at depth is based on the hypothesis that the edge of the undeformed cratonic basement extends to the W flank of the Purcell anticlinorium, and that the Purcell Thrust fault and the décollement above the Eo-Cambrian quartz sandstone succession extends just about as far.

Palinspastic restorations

The tectonic displacements involved in the evolution of the thrust and fold belt can be analyzed most conveniently in terms of a horizontal datum defined by the transition from late Jurassic marine to non-marine deposits, which marks the onset of synorogenic sedimentation; and a reference frame fixed relative to the 'autochthonous' supracrustal rocks that are attached to the Hudsonian basement E of the thrust and fold belt (Price & Mountjoy 1970). The relative horizontal (and vertical) displacements of points in the thrust and fold belt, and the amount of tectonic shortening or convergence between them and the autochthon can be estimated by comparing their position in the restored section with their present position, provided that the angle between the section and the actual direction of net tectonic displacement is not more than about 30°. The minimum total convergence between the leading edge of the Bourgeau thrust sheet and the autochthon estimated in this way is about 170 km in Section SW–NE, and about 105 km in Section W–E. The estimated minimum total shortening up to and including the McConnell Thrust sheet, which stands above the Foothills in Section SW–NE, is about 75 km, whereas that up to and including the Lewis Thrust sheet, which stands above the Foothills in Section W–E, is about 90 km. However, the Lewis thrust fault dies out along the strike to the N within the Rundle Thrust sheet, and the estimated minimum net shortening up to and including the Rundle Thrust sheet in Section SW-NE is about 105 km. Although the total convergence across the fold and thrust belt probably does decrease southward, some of the difference in estimated tectonic convergence between the two sections can be attributed to errors arising from sections that are not parallel with the direction of net tectonic displacement.

The palinspastic restoration based on Section SW–NE, indicates that the Mesozoic exogeoclinal rocks occurring under and E of the McConnell Thrust fault have been horizontally compressed by at least 10 km more than the underlying Upper Palaeozoic platformal rocks from which they are separated by a regionally important bedding glide or décollement zone (Dahlstrom 1969*a*). A similar relationship in a nearby section has been described by Bally *et al.* (1966, p. 350), and attributed to cross-cutting relations involving a younger fault which developed in the Palaeozoic carbonate rocks, under the Mesozoic rocks, after the latter had been horizontally compressed above the décollement. In Section SW–NE, this younger fault is the McConnell Thrust fault, and the tectonic shortening that occurred above the décollement before it was offset by the McConnell Thrust must be related, at a deeper level, to other more westerly thrust faults.

Times of deformation

The record of deformation within the thrust and fold belt in this region spans an interval of almost 100 Ma from late Jurassic to Palaeocene time (Price & Mountjoy 1970).

The first stratigraphic record of deformation, uplift and erosion of the miogeoclinal assemblages occurs in the Upper Jurassic rocks (Kootenay Formation and upper part of Fernie Group). It is marked by an abrupt reversal in the provenance of clastic sediment, from the NE cratonic provenance, that characterized the miogeoclinal assemblages, to the SW Cordilleran provenance that characterized the exogeoclinal assemblage. Although the ages of individual thrust faults cannot be defined precisely, some broad limits can be established. Thrust faults in the western Rocky Mountains, particularly those along the W side of the fan structure that occurs in the Lower Palaeozoic shale facies, are linked spatially and kinematically to NE trending, right-hand reverse faults that cut across the Purcell anticlinorium (Fig. 1). Several of these NE trending, right-hand reverse faults are cut by early to mid-Cretaceous batholiths, and this establishes an upper limit for the times of last displacement on them and on the thrust faults that are linked to them. The Hall Lake fault, which is linked to faults in, and E of, the Rocky Mountain Trench that form part of the same thrust system as the Purcell fault, is cut by the White Creek batholith, for which Rb-Sr and K-Ar whole rock dates of 129 Ma and 111 Ma have been reported (Wanless *et al.* 1968). The St Mary fault, which is linked to faults in the western Rocky Mountains NE of Cranbrook, is cut by the Bayonne batholith, and K-Ar dates from the Bayonne batholith indicate that it was emplaced prior to 90 Ma (Archibald *et al.* 1977). Along the Kootenay Arc the late to mid-Cretaceous granitic plutons post-date most, or all, of the penetrative strain in the country rocks and have relatively low pressure thermal aureoles (Glover 1978; Sears 1979), in contrast with the late Jurassic plutons which are associated with intermediate pressure (Barrovian) regional metamorphism, and were deformed along with the enclosing strata (Gabrielse & Reesor 1974; Glover 1978). Thus, deformation in the eugeoclinal rocks W of the Kootenay Arc was underway in late Jurassic time, and thrust faults in the western Rocky Mountains had already developed by mid-early Cretaceous time.

The McConnell and Lewis thrust sheets both overlap Cenomanian to Santonian marine deposits and early Campanian non-marine deposits of the exogeoclinal assemblage (Alberta Group and Belly River and Brazeau Formations, respectively), and this establishes a lower limit for the time of displacement on these faults. An upper limit for the time of displacement on the Lewis Thrust is given by the late Eocene–early Oligocene age of the Kishenehn Formation, which unconformably overlaps the structures in the Lewis thrust sheet along the downthrown side of the Flathead fault (Price 1965), a SW dipping listric normal fault that merges at depth with the Lewis thrust fault (Bally *et al.* 1966). Thus, a minimum of about 100 km of the total horizontal convergence across the thrust and fold belt occurred between early Campanian and late Eocene time; and if the subsidence and sedimentation in the migrating foredeep was an isostatic response of the lithosphere to the loads imposed by the displacement of the thrust sheets (Price 1973), this amount of convergence must have occurred within less than 30 Ma, the time represented by the thick succession of early Campanian to Palaeocene fluvial sediments that makes up the youngest part of the exogeoclinal assemblage (McLean & Jerzykiewicz 1978). The implications of these conclusions warrant further consideration.

In early Campanian time, prior to displacement on the McConnell, Lewis and Rundle Thrusts, the Precambrian crystalline basement that now extends as an undeformed autochthonous slab under the Rocky Mountains and the eastern part of the Purcell anticlinorium was still covered by a thin sequence of autochthonous, Palaeozoic platformal carbonate rocks with an overlying easterly tapering blanket of autochthonous late Jurassic–Campanian exogeoclinal deposits (Fig. 2). At almost the same time, but perhaps somewhat earlier, prior to the displacement on the Bourgeau thrust fault, the hinge zone between the platform and the miogeocline lay above the edge of this slab of continental crust, and the tectonically compressed miogeocline was situated outboard from it, and was underlain by tectonically attenuated continental crust, and perhaps, in part, oceanic crust. At the end of Middle Jurassic time, prior to the displacements on the thrust faults in the western Rocky Mountains and the the eastern Purcell anticlinorium, and on the NE trending right–hand reverse faults that cut across the Purcell anticlinorium, the miogeocline was essentially undeformed and comprised a series of overlapping and interfingering, unconformity-bounded, tectono-stratigraphic assemblages, for which the aggregate thickness at any one place was about 15–20 km.

Tectonic models

The only obvious actualistic model for this situation seems to be a rifted continental mar-

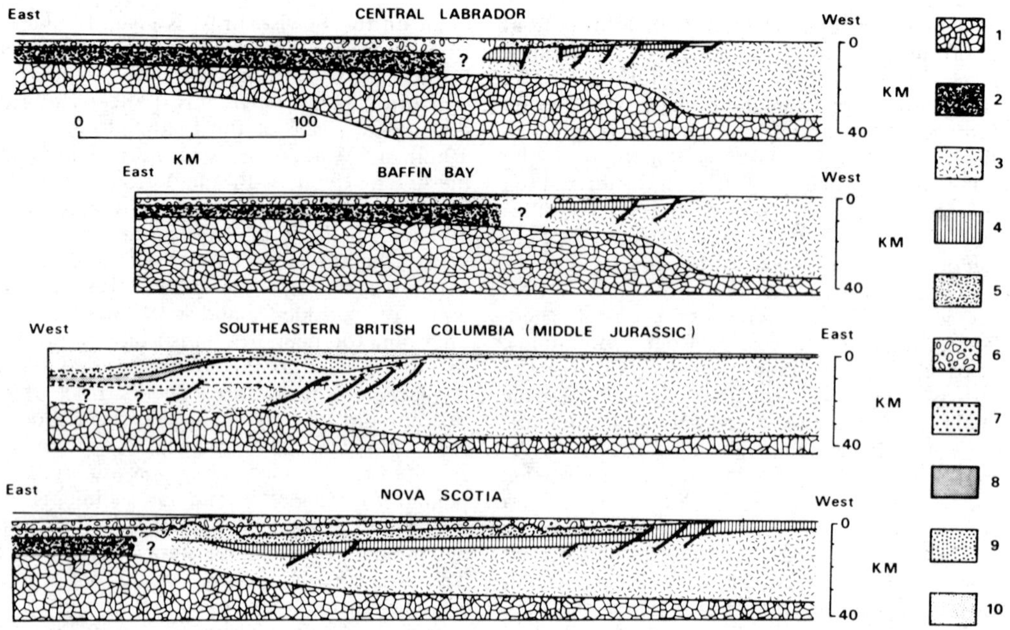

FIG. 7. Comparison of a palinspastically restored section through the foreland thrust and fold belt of southern Canada with crustal sections through the Atlantic margin of Canada. 1—Mantle; 2—oceanic crust; 3—continental crust; 4—continental basement rocks (pre-rifting); 5—early Jurassic and older sediments on the Nova Scotian shelf; 6—post-rifting sediments; 7—Belt-Purcell assemblage; 8—Windermere assemblage; 9—Lower Palaeozoic assemblages; 10—Upper Palaeozoic and Triassic–Jurassic assemblages.

The sections through the Atlantic continental margin are taken from Keen & Hyndman (1979). The palinspastically restored section for the Cordilleran margin in southern British Columbia and Alberta in Middle Jurassic time is based on the Section W–E of Fig. 2, and on the premise that the slab of continental crust approximately 40 km thick, which underlies the craton E of the foreland thrust and fold belt, extends under it to the vicinity of the Kootenay Arc, where it is tectonically attenuated as a result of rifting associated with the development of the western margin of North America in Proterozoic and early Palaeozoic time.

gin of the Atlantic-type. The results of recent investigations of the Atlantic continental margin of Canada (Keen & Hyndman 1979) provide some indication of the possible range in the configurations and relationships among continental and oceanic crust and an overlying, continental terrace wedge of supracrustal rocks (Fig. 7). The abrupt increase in thickness of the supracrustal wedge along section W–E can be compared with the abruptly attenuated continental margins that occur in Baffin Bay or Central Labrador, whereas the apparently more gradual increase in thickness along section SW–NE may be more comparable with gently tapering continental crust off Nova Scotia.

When the evolution of the thrust and fold belt is viewed from this perspective, the origin and significance of the Purcell anticlinorium and Kootenay Arc become obvious (Fig. 5). The Purcell anticlinorium formed in Campanian to early Palaeocene time, as a geometric consequence of the juxtaposition of the thick northeasterly tapering prism of sediment that had accumulated above the zone of abrupt crustal attenuation, outboard from the continental margin, with relatively flat, planar basement surface on the continental platform. The Kootenay Arc formed at the same time, as a W-facing monoclinal step beyond the crest of the Purcell anticlinorium where the sedimentary prism was draped over the edge of the continental platform. The zone of detachment, along which this juxtaposition occurred, reached the surface far to the E in the thrust and fold belt where it is expressed as a series of stacked imbricate thrust faults in the horizontally compressed supracrustal rocks. The overall structure is a crustal scale version of the stepped thrust faults which control the configuration of structures within the deformed supracrustal wedge (Fig. 5c).

The amount of displacement of the miogeoclinal prism relative to the cratonic basement can be visualized readily on even a regional geological map such as Fig. 1, provided that the map shows the relative distrubutions of the Palaeozoic shale and carbonate facies, and the location of the Kootenay Arc. The boundary between the Lower Palaeozoic shale and carbonate facies formed above, or close to, the edge of the continental craton, and the Kootenay Arc now follows the locus of the edge of this continental craton. Thus, the distance between the Kootenay Arc and the shale-carbonate facies boundary is a measure of how far this boundary has been displaced, and therefore, of the amount of convergence between the miogeocline and the autochthonous cover on the craton to the E. It is worth noting, that adjacent to the International Boundary, where there is a conspicuous right-hand deflection about 50 km long in the otherwise curvilinear N to NW trend of the Kootenay Arc, there is a matching right-hand deflection in the eastern and southern limit of the Lower Palaeozoic shale facies, which lies about 150 km NE of the Kootenay Arc. Moreover, S of 50°N Latitude, where the structural relief across the Kootenay Arc is greatest and the change in level across it is most abrupt, the change from a thin platformal facies of the Lower Palaeozoic assemblage to a very thick shale facies is also most abrupt. It seems clear that the miogeocline formed outboard from the continental craton, on tectonically attenuated continental crust, or on oceanic crust, and has subsequently been juxtaposed with the continental craton.

The idea that foreland thrust and fold belts may develop by lateral gravitational spreading is based on the premise that some other process operates to produce the topographic gradient, and thereby the gravitational potential, that drives the lateral spreading. Price & Mountjoy (1970) ascribed the gravitational potential in the southern Canadian Rockies to the buoyant upwelling of a hot, ductile infrastructure of high-grade metamorphic rocks, the exposed parts of which are represented by the intermediate-pressure (Barrovian) series metamorphic rocks of the Shuswap Metmorphic Complex. However, they left unspecified the processes that had carried the protolith of the metamorphic complex to the depths from which the subsequent upwelling occurred; and they also, mistakenly, placed too much emphasis on a conceptual model involving simple diapiric upwelling without any attendant convergence or crustal shortening. The dominantly flat foliation and mainly northeasterly oriented stretching lineation in the Shuswap Metamorphic Complex can be taken as evidence of large-scale vertical compression and horizontal northeasterly extension (Price & Mountjoy 1970). However, the fact that the less metamorphosed rocks in the suprastructure around the southern (Fig. 1) and northern (Campbell 1973) ends of the metamorphic culminations have a fabric indicative of horizontal compression, and moreover, show no indication of horizontal extension on the scale that is necessary to balance the more than 200 km of horizontal shortening that occurs between the metamorphic complex and the autochthonous cover on the craton, implies that there has been a large net horizontal shortening or convergence across the entire belt, and that the horizontal stretching in the deep infrastructure is a second-order effect associated with the special conditions which existed there. For example, tectonic slides are common around some of the margins of the metamorphic culminations. These are shallow-dipping extension faults (see Fig. 1) across which young rocks are juxtaposed on old rocks, and low-grade metamorphic rocks on high-grade metamorphic rocks. Faults of this type, marked by strong cataclasis, occur along the E side of the Shuswap Metamorphic Complex (Read 1977; Sears 1979), on the flanks of the Valhalla gneiss complex (Reesor 1965; Little 1960); on the W flank of a NNW-trending metamorphic culmination that extends along the Purcell Trench at about 116°30′ to 117°E Latitude (Fyles 1967); and around three-sides of a local structural and metamorphic depression within a U-shaped metamorphic culmination at Newport in northeastern Washington State (Miller 1974 *b–d*). There is a contrast in structural style across the faults from recumbent folds below, to upright folds above. Some of these faults are cut by early to mid-Cretaceous plutons (Read 1977); others cut early (?) Tertiary rocks (Miller 1974*b–d*). The basic structural relationships are similar to those around many metamorphic culminations further S in the Cordilleran belt of the United States. The faults around these metamorphic culminations have been interpreted (Davis & Coney 1979) as ductile, normal, growth faults related to Neogene thermal upwelling with horizontal stretching, pinch-and-swell, and mega-boudinage of the crust. However, they occur in the basin-and-range province where evidence of Neogene–Recent crustal extension abounds; and moreover, the displacements on them appear to be much younger than on the comparable faults in S Canada and NE Washington, where the shallow extension faults appear to be related to local diapiric movement involving the buoyant rise of culminations in the metamorphic infrastructure and relative subsidence of parts of the suprastructure. Lateral continuity in nearby rocks indicates that they do not represent a major regional horizontal extension of the crust.

Brown & Tippet (1978) have interpreted the evolution of the Selkirk fan structure, along the eastern margin of the metamorphic complex NW of Golden, as a product of two distinct phases of deformation; and Brown (1978) has concluded that the evolution of the SE Canadian Cordillera involved earlier 'eastward underthrusting from the W', and then 'westward underthrusting of North America from the E'. However, the Selkirk fan structure has also been interpreted as a fan fold (Price & Mountjoy 1970; Zwanzig 1973; Price 1979), like that in the Lower Palaeozoic shale facies of the western Rocky Mountains (Fig. 2), involving a single progressive deformation above a zone of regional décollement, and comprising a local reversal in the vergence of the structures, like that which occurs along the W side of the Alberta syncline (Fig. 6).

The conclusion that there has been a large net horizontal shortening or convergence across the entire tract from the E edge of the thrust and fold belt to the W edge of the Shuswap Metamorphic Complex constrains interpretations of the geotectonic significance of the foreland thrust and fold belt. Convergence between the W flank of the Shuswap Metamorphic Complex and the autochthonous cover on the continental craton appears to be approximately equal to the amount of crustal shortening across the thrust and fold belt—about 200 km (Bally *et al.* 1966; Price & Mountjoy 1970). The shortening must have been balanced by an equivalent convergence or crustal shortening at a deeper crustal level, and this constitutes a type of subduction. Bally (1975) has called this 'Alpinotype or A-subduction', and has contrasted it with 'Benioff or B-subduction', in which a slab of oceanic lithosphere dips deep into the mantle with décollement or scraping off of supracrustal rocks and imbrication of oceanic crust and upper mantle. He has suggested that it is the shallower conjugate homologue of B-subduction and that it involves the disposal of large amounts of continental crust. In the southern Canadian Rocky Mountains, this A-subduction may have been the conjugate homologue of B-subduction which occurred along the Pacific Margin of the Cordillera from mid-Jurassic to Palaeogene time (Monger & Price 1979). However, it apparently did not involve the disposal of any significant amount of continental crust, because it occurred outboard from the edge of the continental craton, in an area floored by attenuated continental crust or oceanic crust.

The following scenario is suggested. As the attenuated continental crust and its supracrustal cover of eugeoclinal and miogeoclinal rocks were carried into the subduction zone, they were metamorphosed to relatively high grades in an intermediate pressure (Barrovian) metamorphic facies series, became ductile, rose bouyantly and spread laterally into the overlying shallower suprastructure of less metamorphosed rocks, partly along ductile zones of extension faulting that are now represented by tectonic slides. Easterly verging imbricate thrust faults and related folds developed concurrently, at a higher level, farther to the E in the subduction zone, in a subduction complex that was tectonically prograded eastward up to and over the edge of the continental craton, in much the same way that subduction complexes above sediment-filled oceanic trenches have been tectonically prograded over the adjacent oceanic crust (Hamilton 1977). The topographic slope produced in this way was primarily due to the compression of the supracrustal cover that was scraped off the down-going slab, and therefore, was a second-order effect, rather than the primary cause of the thrust and fold belt (Chapple 1978; cf. Elliot 1976). Thus, the evolution of the foreland thrust and fold belt in the southern Rocky Mountains can be viewed as an example of intra-plate convergence, involving the collapse of a back-arc or marginal basin, behind the main easterly dipping Cordilleran subduction zone. As the supracrustal cover of miogeoclincal rocks was scraped off the down-going slab, it was tectonically prograded over an adjacent continental craton to form a very wide and shallow subduction complex.

ACKNOWLEDGMENTS. This work has been supported by the National Science and Engineering Research Council of Canada, and by the Canada Department of Energy Mines and Resources. It has been made possible by a Canada Council Killam Senior Research Fellowship. Mina Price assisted with the compilation of the maps and the preparation of the sections. Peter Fermor collaborated on the preparation of Section SW–NE in Fig. 2. Most of the figures were drafted by Christopher Peck. A. W. Bally, P. L. Gordy, E. W. Mountjoy, S. S. Oriel and R. I. Thompson reviewed an earlier draft of the manuscript and made helpful suggestions for improving it.

References

AALTO, K. R. 1971. Glacial marine sedimentation and stratigraphy of the Toby conglomerate (Upper Proterozoic) southeastern British Columbia, northwestern Idaho and northeastern Washington, *Can. J. Earth Sci.* **8,** 753–87.

AITKEN, J. D. 1971. Control of Lower Paleozoic sedimentary facies by the Kicking Horse rim, southern Rocky Mountains, Canada. *Bull. Can. Pet. Geol.* **19,** 557–69.

ARCHIBLAD, D. A., GLOVER, J. K. & FARRAR, E. 1977. K-Ar ages from the Bayonne batholith and some nearby plutons in SE British Columbia and their geological implications. *Geol. Assoc. Can. Progs with Abstr.* **2,** 5.

BALKWILL, H. R. 1972. Structural geology, lower Kicking Horse river region, Rocky Mountains, British Columbia. *Bull. Can. Pet. Geol.* **20,** 608–33.

——, PRICE, R. A. & MOUNTJOY, E. W. (in press *a*). Geology, Golden (east half), British Columbia. *Geol. Surv. Can. Map* **1496A.**

——, —— & —— (in press *b*). Geology, Golden (west-half), British Columbia. *Geol. Surv. Can. Map* **1497a.**

BALLY, A. W. 1975. A geodynamic scenario for hydrocarbon occurrences. *9th World Petrol. Cong. (Tokyo),* **2,** 23–44.

——, GORDY, P. L. & STEWART, G. A. 1966. Structure, seismic data and orogenic evolution of southern Canadian Rockies. *Bull. Can. Pet. Geol.* **14,** 337–81.

BENNETT, G. T., CLOWES, R. M. & ELLIS, R. M. 1975. A seismic refraction survey along the southern Rocky Mountain Trench, Canada. *Bull. seismol. Soc. Am.* **65,** 37–54.

BENVENUTO, G. L. 1978. *Structural Evolution of the Hosmer Thrust Sheet, South-Eastern British Columbia.* Queen's Univ. Ph.D. thesis (Unpubl.).

BERNOULLI, D., LAUBSCHER, H. P., TRÜMPY, R. & WENK, E. 1974. Central Alps and Jura Mountains. *In:* SPENCER, A. M. (ed), *Mesozoic-Cenozoic Orogenic Belts.* Spec. Publ. geol. Soc. London, **4,** 83–108.

BERRY, M. J. & FORSYTH, D. A. 1975. Structure of the Canadian Cordillera from seismic refraction and other data. *Can. J. Earth Sci.* **12,** 182–208.

BIELENSTEIN, H. U., PRICE, R. A. & JONES, P. B. 1971. Geology of the Seebe-Kananaskis area, Map. *In:* HALLADAY, I. A. R. & MATHEWSON, D. H. (ed), *A Guide to the Geology of the Eastern Cordillera along the Trans-Canada Highway between Calgary, Alberta and Revelstoke, British Columbia.* Alberta Soc. Petrol. Geol., Calgary.

BROWN, R. L. 1978. Structural evolution of the southeast Canadian Cordillera: A new hypothesis. *Tectonophysics,* **48,** 133–51.

—— & TIPPET, C. R. 1978. The Selkirk fan structure of the southeastern Canadian Cordillera. *Bull. geol. Soc. Am.* **89,** 548–58.

BURCHFIEL, B. C. & DAVIS, G. A. 1972. Structural framework and evolution of the southern part of the Cordilleran orogen, Western United States. *Am. J. Sci.* **272,** 97–118.

CAMPBELL, R. B. 1968. Canoe River, British Columbia. *Geol. Surv. Can. Map* **15–1967.**

—— 1973. Structural Cross-Section and Tectonic Model of the S.E. Canadian Cordillera. *Can. J. Earth Sci.* **10,** 1607–20.

CAMPBELL, I. & LOOFBOUROW, J. S., Jr. Preliminary geologic map and sections of the magnesite belt, Stevens county, Washington, *U.S. Geol. Surv. Map* **MF-117.**

CANER, B. 1971. Quantitive interpretation of geomagnetic depth-sounding data in western Canada. *J. geophys. Res.* **76,** 7202–16.

CHANDRA, N. N. & CUMMING, G. L. 1972. Seismic refraction studies in western Canada. *Can. J. Earth Sci.* **9,** 1099–109.

CHAPPLE, W. M. 1978. Mechanics of thin-skinned fold and thrust belts. *Bull. geol. Soc. Am.* **89,** 1189–98.

COOK, D. C. 1975. Structural style influenced by lithofacies, Rocky Mountain Main Ranges, Alberta-British Columbia. *Bull. geol. Surv. Can.* **233.**

CORBITT, L. L. & WOODWARD, L. A. 1973. Tectonic framework of cordilleran foldbelt in southwestern New Mexico. *Bull. Am. Assoc. Petrol. Geol.* **57,** 2207–16.

CUMMING, W. B., CLOWES, R. M. & ELLIS, R. M. 1979. Crustal structure from a seismic refraction profile across southern British Columbia. *Can. J. Earth Sci.* **16,** 1024–40.

DAHLSTROM, C. D. A. 1969*a*. The upper detachment in concentric folding. *Bull. Can. Pet. Geol.* **17,** 326–46.

—— 1969*b*. Balanced cross sections. *Can. J. Earth Sci.* **6,** 743–57.

—— 1970. Structural geology in the eastern margin of the Canadian Rocky mountains. *Bull. Can. Pet. Geol.* **18,** 332–406.

——, DANIEL, R. E. & HENDERSON, G. G. L. 1962. The Lewis thrust at Fording mountain, British Columbia. *J. Alberta Soc. petrol. Geol.* **10,** 373–95.

DAVIS, G. & CONEY, P. J. 1979. Geological development of the Cordilleran metamorphic core complexes. *Geology,* **7,** 120–4.

DE CSERNA, Z. 1971. Development and structure of the Sierra Madre Oriental of Mexico. *Geol. Soc. Am. (Abstracts with programs),* **3,** *Rocky mountain section,* 377–78.

DEWEY, J. A. & BIRD, J. M. 1970. Mountain belts and the new global tectonics. *J. geophys. Res.* **75,** 2625–47.

DIMROTH, E. 1970. Evolution of the Labrador geosyncline. *Bull. geol. Soc. Am.* **81,** 2717–42.

DOUGLAS, R. J. W. 1950. Callum creek, Langford creek and Gap map-areas, Alberta. *Mem. geol. Surv. Can.* **255.**

—— 1951. Pincher creek, alberta; preliminary map. *Pap. geol. Surv. Can.* **51–22.**

—— 1952. Waterton Alberta; preliminary map. *Pap. geol. Surv. Can.* **52–10.**

—— 1958. Mount Head map-area, Alberta. *Mem. geol. Surv. Can.* **291.**

—— 1970. in Douglas, R. J. W., Gabrielse, H., Wheeler, J. O., Stott, D. F. & Belyea, H. R. Geology of Western Canada, Chap. VIII of Geology and Economic Minerals of Canada. *In:* Douglas, R. J. W. (ed), *Geol. Surv. Can. Econ. Geology report,* **1,** 396.

Dragert, H. & Clarke, G. K. C. 1976. A detailed investigation of the Canadian cordillera geomagnetic transition anomaly. *J. geophys.* **42,** 373–90.

Earth Physics Branch. 1974. Bouguer anomaly map of Canada. *Dept. Energy Mines & Res., Can. Gravity map series* **74–1.**

Eisabacher, G. H., Carrigy, M. A. & Campbell, R. B. 1974. Paleodrainage pattern and late-orogenic basins of the Canadian Cordillera. *In:* Dickinson, W. R. (ed). *Tectonics and Sedimentation.* Spec. Publ. Econ. Paleon. & Min. **22,** 143–66.

Elliot, D. 1976. The motion of thrust sheets. *J. geophys. Res.* **81,** 949–63.

Fermor, P. R. & Price, R. A. 1976. Imbricate structures in the Lewis thrust sheet around the Cate Creek and Haig Brook windows. *Pap. geol. Surv. Can.* **76-1B,** 7–10.

Fox, F. G. 1959. Structure and accumulation of hydrocarbons in southern Foothills, Alberta, Canada. *Bull. Am. Assoc. Petrol. Geol.* **43,** 992–1025.

Fyles, J. T. 1964. Geology of the Duncan lake area, Lardeau district, British Columbia. *Bull. Br. Columbia Dep. Mines Petrol. Res.* **49.**

—— 1967. Geology of the Ainsworth-Kaslo area, British Columbia. *Bull. Br. Columbia Dep. Mines Petrol. Res.* **53.**

—— & Hewlett, C. G. 1959. Stratigraphy and structure of the Salmo lead-zinc area. *Bull. Br. Columbia Dep. Mines.* **41.**

Gabrielse, H. & Reesor, J. E. 1974. The nature and setting of granitic plutons in the central and eastern parts of the Canadian Cordillera. *Pacific Geology,* **8,** 109–38.

Gardner, D. A. C. 1977. *Structural Geology and Metamorphism of Calcareous Lower Paleozoic Slates, Blaeberry River-Redburn Creek Area, near Golden, British Columbia.* Queen's Univ., Ph.D. thesis (Unpubl.).

Gilman, R. A. 1972. Geology of the Clachnacudainn salient near Albert canyon, British Columbia. *Can. J. Earth Sci.* **9,** 1447–54.

Glover, J. K. 1978. *Geology of the Summit Creek Map Area, Southern Kootenay Arc, British Columbia.* Queen's Univ. Ph.D. thesis (Unpubl.).

Hamilton, W. 1977. Subduction in the Indonesian region. *In:* Talwani, M. & Pitman, W. C. (ed). *Island Arcs, Deep Sea Trenches and Back-Arc Basins.* Am. geophys. Union. Maurice Ewing series **I,** 15–31.

Harrison, J. E. 1972. Precambrian Belt basins of northwestern United States: its geometry, sedimentation, and copper occurrences. *Bull. geol. Soc. Am.* **83,** 1215–40.

—— & Jobin, D. A. 1963. Geology and ore deposits of the Clark Fork quadrangle, Idaho-Montana. *Bull. U.S. geol. Surv.* **1141-K.**

—— & —— 1965. Geologic map of the Packsaddle mountain quadrangle, Idaho. *U.S. geol. Surv. Map* **GQ-375.**

—— & Schmidt, P. W. 1971. Geologic map of the Elmira quadrangle, Bonner county, Idaho. *U.S. geol. Surv. Map* **I-603.**

Henderson, G. G. L. 1954. Geology of the Stanford range of the Rocky mountains. *Bull. Br. Columbia Dep. Mines* **35.**

Hoffman, P. 1973. Evolution of an early Proterozoic margin: the Coronation geosyncline and associated aulacogens of the northwestern Canadian Shield. *Philos. Trans. R. Soc. London* **A273,** 547–81.

Hoffman, P. F., St. Onge, M., Carmichael, D. M. & De Bie, I. 1978. Geology of the Coronation geosyncline (Aphebian) Hepburn lake sheet (86J), Bear province, district of Mackenzie; current research. *Pap. geol. Surv. Can.* **78-1A,** 147–51.

Höy, T. 1977. Stratigraphy and structure of the Kootenay arc in the Riondel area, southeastern British Columbia. *Can. J. Earth Sci.* **14,** 2301–15.

—— 1978. Geology of the Estella-Kootenay King area, southeastern British Columbia. *Br. Columbia Dep. Mines prelim. Map* **28.**

Huntting, M. T., Bennett, W. A. G., Livingston, V. E., Jr., & Moens, W. S. (Compilers). 1961. Geologic map of Washington. Washington Div. Mines and Geology.

Johns, W. M. 1970. Geology and mineral deposits of Lincoln and Flathead counties, Montana. *Bull. Montana Bur. Mines Geol.* **79.**

Kanasewich, E. R., Clowes, R. M. & McCloughan, C. H. 1969. A buried Precambrian rift in western Canada. *Tectonophysics,* **8,** 513–27.

Keating, L. F. 1966. Exploration in the Canadian Rockies and Foothills. *Can. J. Earth Sci.* **3,** 713–23.

Keen, C. E. & Hyndman, R. D. 1979. Geophysical review of the continental margins of eastern and western Canada. *Can. J. Earth Sci.* **16,** 712–47.

King, P. B. (Compiler) 1969. Tectonic map of North America. *U.S. geol. Surv.* Washington, D.C.

Leech, G. B. 1958. Fernie map-area, west half, British Columbia. *Pap. geol. Surv. Can.* **58–10.**

—— 1959. Canal Flats, British Columbia. *Geol. Surv. Can. Map* **24–1958.**

—— 1960. Fernie, west half, British Columbia. *Geol. Surv. Can. Map* **11–1960.**

—— 1962. Structure of the Bull river valley near latitude 49°35′. *J. Alberta Soc. petrol. Geol.* **10,** 396–407.

Lis, M. G. & Price, R. A. 1976. Large-scale block faulting during deposition of the Windermere supergroup (Hadrynian) in southeastern British Columbia. *Pap. geol. Surv. Can.,* **76-1A,** 135–6.

Little, H. W. 1960. Nelson map-area, west half, British Columbia. *Mem. geol. Surv. Can.* **308.**

McLean, J. R. & Jerzykiewicz, T. 1978. Cyclicity, tectonics and coal: Some aspects of fluvial sedimentology in the Brazeau-Paskapoo formations, Coal valley area, Alberta, Canada. *Mem. Can. Soc. petrol. geol.* **5,** *Fluvial Sedimentology*, 441–68.

MEREU, R. F., MAJUMDAR, S. C. & WHITE, R. E. 1977. The structure of the crust and upper mantle under the highest ranges of the Canadian Rockies from a seismic refraction survey. *Can. J. Earth Sci.* **14,** 196–208.

MILLER, F. K. 1974*a*. Preliminary geologic map of the Newport number 1 quadrangle, Pend Oreille county, Washington and Bonner county, Idaho. *Washington Dept. Natural Res. geol. Map* **GM-7.**

—— 1974*b*. Preliminary geologic map of the Newport number 2 quandrangle, Pend Oreille and Stevens counties, Washington. *Washington Dept. Natural Res. geol. Map* **GM-8**.

—— 1974*c*. Preliminary geologic map of the Newport number 3 quadrangle, Pend Oreille, Stevens and Spokane counties, Washington. Washington *Ddpt. Natural Res. geol. Map* **GM-9.**

—— 1974*d*. Preliminary geologic map of the Newport number 4 quadrangle, Spokane and Pend Oreille counties, Washington and Bonner county, Idaho. *Washington Dept. Natural Res. geol. Map* **GM-10.**

——, CLARK, L. D. & ENGELS, J. C. D. 1975. Geology of the Chewelah-Loon Lake area, Stevens and Spokane countries, Washington. *Prof. Pap. U.S. geol. Surv.* **806.**

MONGER, J. W. H. & PRICE, R. A. 1979. Geodynamic evolution of the Canadian Cordillera—progress and problems. *Can. J. Earth Sci.* **16,** 770–91.

——, SOUTHER, J. G. & GABRIELSE, H. 1972. Evolution of the Canadian Cordillera: a plate-tectonic model. *Am. J. Sci.* **272,** 577–602.

NGUYEN, E. K., SINCLAIR, A. J. & LIBBY, W. G. 1968. Age of the northern part of the Nelson batholith. *Can. J. Earth Sci.* **5,** 955–7.

NORTH, F. K. & HENDERSON, G. G. L. 1954. Summary of the geology of the southern Rocky Mountains of Canada. *Alberta Soc. petrol. Geol., Calgary, Guidebook, Fourth Annual Field Conference*, 15–81.

NORRIS, D. K. 1958*a*. Beehive mountain, Alberta and British Columbia. *Pap. geol. Surv. Can.* **58–5.**

—— 1958*b*. Livingstone river, Alberta. *Geol. Surv. Can. Map* **5–1958**.

—— & PRICE, R. A. 1966. Middle Cambrian lithostratigraphy of southeastern Canadian Cordillera. *Bull. Can. Pet. Geol.* **14,** 385–404.

NYE, J. F. 1952. The mechanics of glacier flow. *J. Glaciol. London*, **2,** 82–93.

OLLERENSHAW, N. C. 1972*a*. Geology, lake Minnewanka (east half), Alberta *Geol. Surv. Can. Map* **1347A**.

—— 1972*b*. Geology, Wildcat Hills (west half), Alberta. *Geol. Surv. Can. Map* **1351A.**

—— (Compiler), 1975. Rocky mountain Foothills and Front ranges; Calgary region. In guidebook, Savanna creek to Panther river, exploration update W5. *Can. Soc. Pet. Geol.*, Calgary.

—— (Compiler), 1978. Geology, Calgary, Alberta-British columbia. *Geol. Surv. Can. Map* **1457A.**

PARK, C. F., JR & CANNON, R. S., JR. 1943. Geology and ore deposits of the Metaline quadrangle, Washington, *Prof. Pap. U.S. geol. Surv.*, **202.**

PRICE, R. A. 1962. Fernie map-area, east half Alberta and British Columbia. *Pap. geol. Surv. Can.* **61–24.** (includes *Map* **35–1961).**

—— 1964*a*. The Precambrian Purcell System in the Rocky mountains of southern Alberta and British Columbia. *Bull. Can. Pet. Geol.* **12,** 399–426.

—— 1964*b*. Flexural-slip folds in the Rocky Mountains, southern Alberta and British Columbia. Seminars on Tectonics IV, Dept. Geol. Sci., Queen's Univ., Kingston, Ont. (Also as Geol. Surv. Can. reprint 78).

—— 1965. Flathead map-area, British Columbia and Alberta. *Mem. geol. Surv. Can.* **336.**

—— 1970. Geology, Canmore (east half) Alberta. *Geol Surv. Can. Map* **1265A.**

—— 1972. Rocky Mountain Trench—Golden to Cranbrook. *In:* PRICE, R. A., BALKWILL, H. R., CHARLESWORTH, H. A. K., COOK, D. G. & SIMONY, P. S. *The Canadian Rockies and Tectonic Evolution of the Southeastern Cordillera.* Guidebook, field excursion AC15, XXIV Inter. Geol. Congress. Montreal, 77–85.

—— 1973. Large-scale gravitational flow of supracrustal rocks, southern Canadian Rockies. *In:* DE JONG, K. A. & SCHOLTEN, R. (ed). *Gravity and tectonics.* Wiley, New York, 491–502.

—— 1979. The Selkirk fan structure of the southeastern Canadian Cordillera: Discussion. *Bull. geol. Soc.* Am. **90,** 695–8.

——, BALKWILL H. R. & MOUNTJOY, E. W. in press. Geology, McMurdo (east half), British Columbia. *Geol. Surv. Can. Map* **1501A.**

——, COOK, D. G., AITKEN, J. D. & MOUNTJOY, E. W. in press *a*. Geology, Lake Louise, (west half), Alberta-British Columbia. *Geol. Surv. Can. Map* **1483A.**

——, ——, —— & —— in press *b*. Geology, Lake Louise, (east half), Alberta-British Columbia. *Geol. surv. Can. Map* **1482A.**

—— & MOUNTJOY, E. W. 1970. Geologic structure of the Canadian Rocky Mountains between Bow and Athabasca rcvers—a progress report. *In:* WHEELER, J. O. (ed), *Structure of the Southern Canadain Cordillera.* Spec. Pap. geol. Assoc. Can. **6,** 7–25.

—— & —— 1972*a*. Geology, Banff, (east half), Alberta-British Columbia. *Geol. Surv. Can. Map* **1294A**.

—— & —— 1972*b*. Geology, Banff, (west half), Alberta-British Columbia. *Geol. Surv. Can. Map* **1295A.**

—— & —— 1978*a*. Geology, Hector Lake, (east half), Alberta-British Columbia. *Geol. Surv. Can. Map* **1463A.**

—— & —— 1978*b*. Geology, Hector Lake, (west half), Alberta-British Columbia. *Geol. Surv.*

Can. Map **1464A.**
—— & —— 1978*c*. Geology Siffleur River, (east half), Alberta. *Geol. Surv. Can. Map* **1465A.**
—— & —— 1978*d*. Geology, Siffleur River, (west half), Alberta. *Geol. Surv. Can. Map* **1466A.**
—— & —— 1979*a*. Geology, Mount Goodsir, (east half), Alberta-British Columbia. *Geol. Surv. Can. Map* **1476A.**
—— & —— 1979*b*. Geology, Mount Goodsir, (west half), Alberta-British Columbia. *Geol. Surv. Can. Map* **1477A.**
—— & in press. Geology, McMurdo, (west half), British Columbia. *Geol. Surv. Can. Map* **1502A.**
READ, P. B. 1977. Relationship of the Kootenay arc to the Shuswap Metamorphic Complex, Southern British Columbia. *Geol. Assoc. Can. Progs with Abstr.* **2,** 43.
—— & WHEELER, J. O. 1976. Geology of Lardeau west-half map area. *Geol. Surv. Can. open-file Report* **288.**
REESOR, J. E. 1965. Valhalla gneiss complex, British Columbia. *Bull. geol. Surv. Can.* **129.**
—— 1973. Geology of the Lardeau map-area, east half, British Columbia. *Mem. geol. Surv. Can.* **369.**
RICE, H. M. A. 1941. Nelson map-area, east half, British Columbia. *Mem. geol. Surv. Can.* **228.**
RODGERS, J. 1970. *The Tectonics of the Appalachians.* Wiley-Interscience, New York.
ROSS, C. P. 1959. Geology of Glacier National Park and the Flathead region, northwestern Montana. *Prof. Pap. U.S. geol. Surv.* **296.**
SEARS, J. W. 1979. *Tectonic Contrasts between the Infrastructure and Suprastructure of the Columbian Orogen, Albert Peak Area, Western Selkirk Mountains, British Columbia.* Queen's Univ., Ph.D. thesis (Unpubl.).
—— & PRICE, R. A. 1978. The Siberian connection: A case for Precambrian separation of the North American and Siberian cratons. *Geology,* **6,** 267–70.
SEELEY, D. R. 1977. The significance of landward vergence and oblique structural trends on trench inner slopes. *In:* TALWANI, M. & PITMAN, W. C. (eds), *Island Arcs, Deep Sea Trenches and Back Arc Basins.* Am. geophys. Union. Maurice Ewing Series **1,** 187–98.
SHAW, E. W. 1963. Canadian Rockies—orientation in time and space. *In:* CHILDS, O.E. (ed), *Backbone of the Americas.* Mem. Am. Assoc. Petrol. Geol. **2,** 231–42.
SIMONY, P. S. & WIND, G. 1970. Structure of the Dogtooth Range and adjacent portions of the Rocky Mountain Trench. *In:* WHELLER, J. O. (ed), *Structure of the Southern Canadian Cordillera.* Spec. Pap. geol. Assoc. Can. **6,** 41–53.
SPENCE, G. D., CLOWES, R. M. & ELLIS, R. M. 1977. Depth limits of the M discontinuity in the southern Rocky Mountain Trench, Canada. *Bull. seismol. Soc. Am.* **67,** 543–6.
STACEY, R. A. 1973. Gravity anomalies, crustal structure, and plate tectonics in the Canadian Cordillera. *Can. J. Earth Sci.* **10,** 615–28.
THOMPSON, R. I. 1972. *Geology of the Akolkolex River Area near Revelstoke, British Columbia,* Queen's Univ. Ph.D. thesis (Unpubl.).
WANLESS, R. K., LOVERIDGE, W. D. & MURSKY, G. 1968. A geochronological study of the White Creek batholith, southeastern British Columbia. *Can. J. Earth Sci.* **5,** 375–86.
WHEELER, J. O. 1963. Rogers pass map-area, British Columbia and Alberta. *Pap. geol. Surv. Can.,* **62–32.**
—— & GABRIELSE, H. 1972. The Cordilleran Structural Province. *In:* PRICE R. A. & DOUGLAS, R. J. W. (eds), *Variations in Tectonic Styles in Canada*; Spec. Pap. Geol. Assoc. Can., **11,** 1–81.
——, CHARLESWORTH, H. A. K., MONGER, J. W. H., MULLER, J. A., PRICE, R. A., REESOR, J. E., RODDICK, J. A. & SIMONY, P. S. 1974. Western Canada. *In:* SPENCER, A. M. (ed), *Mesozoic-Cenozoic orogenic belts.* Spec. Publ. geol. Soc. London, **4,** 592–623.
WILLIAMS, E. P. 1949. Preliminary map, Cardston, Alberta. *Pap. geol. Surv. Can.* **49–3.**
WYNN, J. C., KLEINKOPF, M. F. & HARRISON, J. E. 1977. Audio-frequency magnetotelluric and gravity traverse across the crest of the Purcell anticlinorium, northwestern Montana. *Geology,* **5,** 309–12.
YATES, R. G. 1964. Geologic map and sections of the Deep Creek area, Stevens and Pend Oreille counties, Washington. *U.S. geol. Surv. Map* **I-412.**
—— 1971. Geologic map of the Northport quadrangle, Washington. *U.S. geol. Surv. Map* **1-603.**
YOUNG, F. G., CAMPBELL, R. B. & POULTON, T. P. 1973. The Windermere Supergroup of the southeastern Canadian Rockies. *Belt Symposium* **1,** 181–203.
ZWANZIG, H. V. 1973. *Structural Transitions between the Foreland Zone and the Core Zone of the Columbian Orogen, Selkirk Mountains, British Columbia.* Queen's Univ. Ph.D. thesis (Unpubl.).

R. A. PRICE. Department of Geological Sciences, Queen's University, Kingston, Ontario, K7L 3N6, Canada.

Reprinted for private circulation from
THE AMERICAN ASSOCIATION OF PETROLEUM GEOLOGISTS BULLETIN
Vol. 60, No. 12, December, 1976

GEOLOGIC NOTES

Exposed Step in Laramide Thrust Fault, Southwest Montana[1]

D. S. BRUMBAUGH[2] and H. W. DRESSER[3]
Flagstaff, Arizona 86001, and Butte, Montana 59701

Abstract The Sandy Hollow thrust fault of southwestern Montana is a Laramide overthrust that possesses an unusually well-exposed step. In the step zone folding, fault-transported slices, splinter faults, and decollement have resulted from fault movement. The wide variety of both brittle and ductile structures commonly should be associated with steps because they are the inevitable result of step-fault mechanics in the restricted temperature/pressure environment in which thrusts form. Field evidence clearly indicates that preexisting folds controlled the development of this step zone in the Sandy Hollow thrust.

INTRODUCTION

Features associated with thrust faults include fensters, klippen, gouge, various kinds of folds, and steps. This report deals with the examination of structures related to an unusually well-exposed step, which is part of a Laramide overthrust fault of the northern Rockies.

From Dillon to Divide, Montana, Laramide fold axes and thrust faults bend eastward, defining a structural salient called the McCarthy Mountain salient (Brumbaugh, 1973). The salient has a chord length of 30 mi (48 km) and is 10 mi (16 km) wide from east to west (Fig. 1B).

This report concerns the exposed geometry of one fault of the McCarthy Mountain salient, the Sandy Hollow thrust (Figs. 1B, C, 2). Gale (1910) was apparently the first to map the Sandy Hollow thrust. Pardee (Richards and Pardee, 1925) used Gale's field notes and geologic map to describe the geology of the area. Peterson et al (1954) sampled and measured the phosphatic Permian rocks adjacent to the Sandy Hollow fault. Steuer (1956) mapped the area including the fault at a scale of 1:24,000.

SANDY HOLLOW THRUST AND RELATED STRUCTURES

Apparent stratigraphic displacement varies considerably along the trace of the Sandy Hollow thrust. Thicknesses of stratigraphic units in the area are given in Figure 3. The maximum stratigraphic displacement on the Sandy Hollow thrust is approximately 1,400 ft (427 m). This can be seen along the Burma Road (Fig. 2) where the Phosphoria Formation is thrust over the limestone in the middle part of the Kootenai Formation. The stratigraphic displacement decreases both to the north and south over an exposed trace of 6 mi (10 km), but the stratigraphic displacement probably bears little relation to the actual displacement of a thrust fault that cuts folded beds.

The most geometrically interesting exposures of the Sandy Hollow thrust are at the northern end of the fault trace. There, the fault trace swings abruptly from north-northeast to nearly due east (Figs. 1C, 2, 4). An obvious explanation for this change in trend would be erosional effects on an irregular low dip surface. However, field evidence does not support this. The fault as it is exposed at the Burma Road (Fig. 5) has a dip of 50°. Between the Burma Road and the locale where the change in strike occurs, the trace of the fault across the topography is straight, indicating that the high dip angles are maintained along this part of the fault. The change in fault trend is explained best if this part of the fault surface is envisioned as a plunging step (Fig. 6). Proof of the plunging character of the surface can be seen at an exposure of the fault plane in a dry wash east of the change in strike (at the circled *A* in Figs. 1C, 4). This outcrop clearly shows a northerly dip of the fault surface. Two mi (3 km) farther east the trend of the fault trace changes to north-northeast again at the base of Block Mountain (Figs. 1C, 2, 4). This is the base of a second step, the Block Mountain step.

[1]Manuscript received, September 29, 1975; accepted, April 19, 1976.

[2]Department of Geology, Northern Arizona University.

[3]Department of Geology, Montana College of Mineral Science and Technology.

This study was supported by grants from the Society of Sigma Xi and the Office of Research and Advanced Studies, Indiana University.

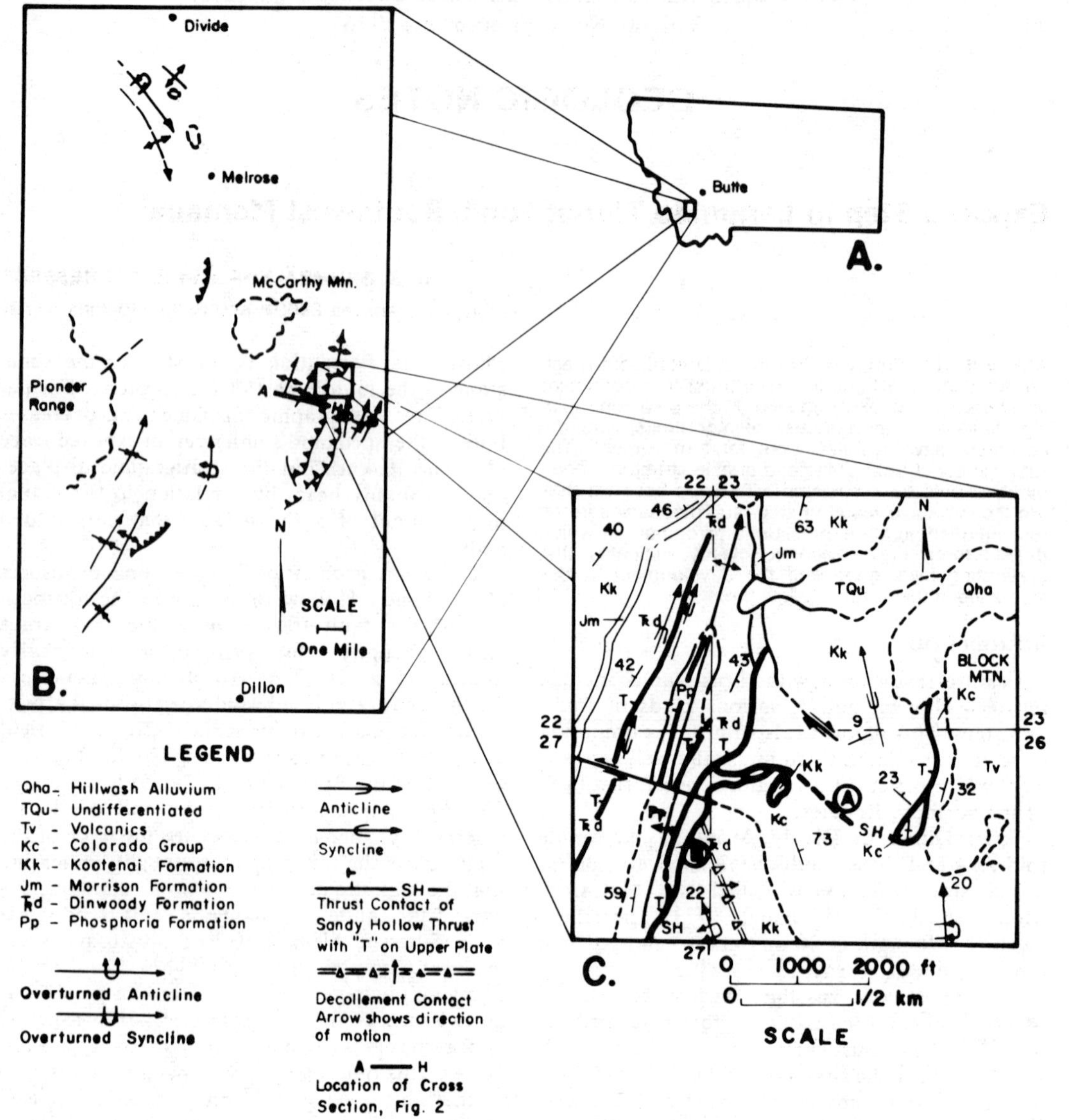

FIG. 1—**A,** Index map of Montana. **B**, Sketch map of McCarthy Mountain salient. **C**, Geologic map of northern part of Sandy Hollow thrust (T4S,R8W, Secs. 22, 23, 26, 27). Circled letters represent critical localities described in text.

The cross section of Figure 2 shows a subsurface step of the Sandy Hollow thrust along the line from *A* to *H* just north of the Big Hole River Valley. This hidden step is required to place the west flank of Ziegler anticline against the west flank of Sandy Hollow anticline.

It is instructive to examine this short stretch of the Sandy Hollow thrust to observe some of the structural phenomena associated with these steps and possibly characteristic of steps in general. Field observation of the Sandy Hollow thrust suggests that step zones are zones of difficulty of fault movement with the attendant production of a wide variety of structures. Both brittle and ductile structures abound at the site of the Sandy Hollow step (Figs. 4, 6).

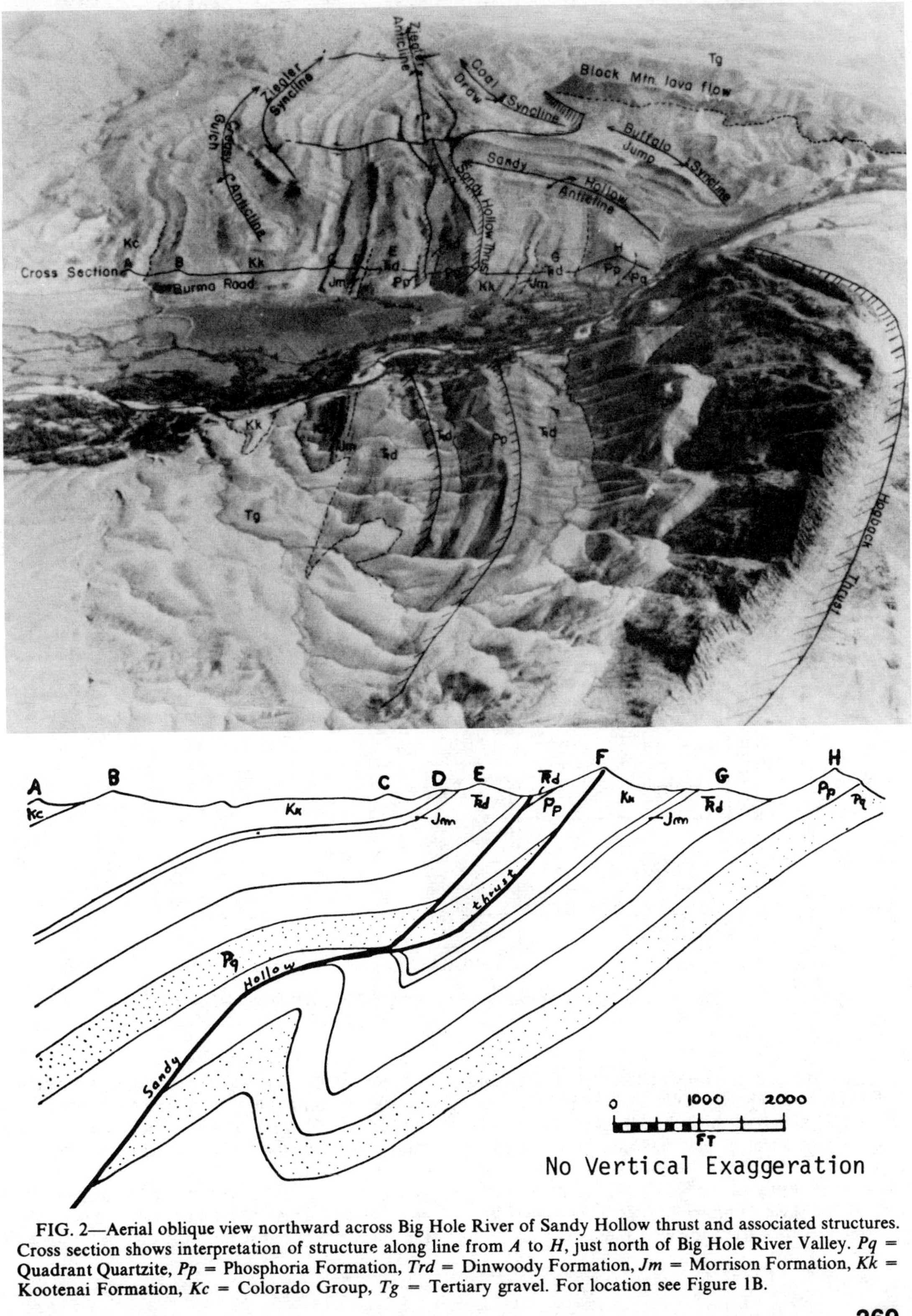

FIG. 2—Aerial oblique view northward across Big Hole River of Sandy Hollow thrust and associated structures. Cross section shows interpretation of structure along line from *A* to *H*, just north of Big Hole River Valley. *Pq* = Quadrant Quartzite, *Pp* = Phosphoria Formation, *Trd* = Dinwoody Formation, *Jm* = Morrison Formation, *Kk* = Kootenai Formation, *Kc* = Colorado Group, *Tg* = Tertiary gravel. For location see Figure 1B.

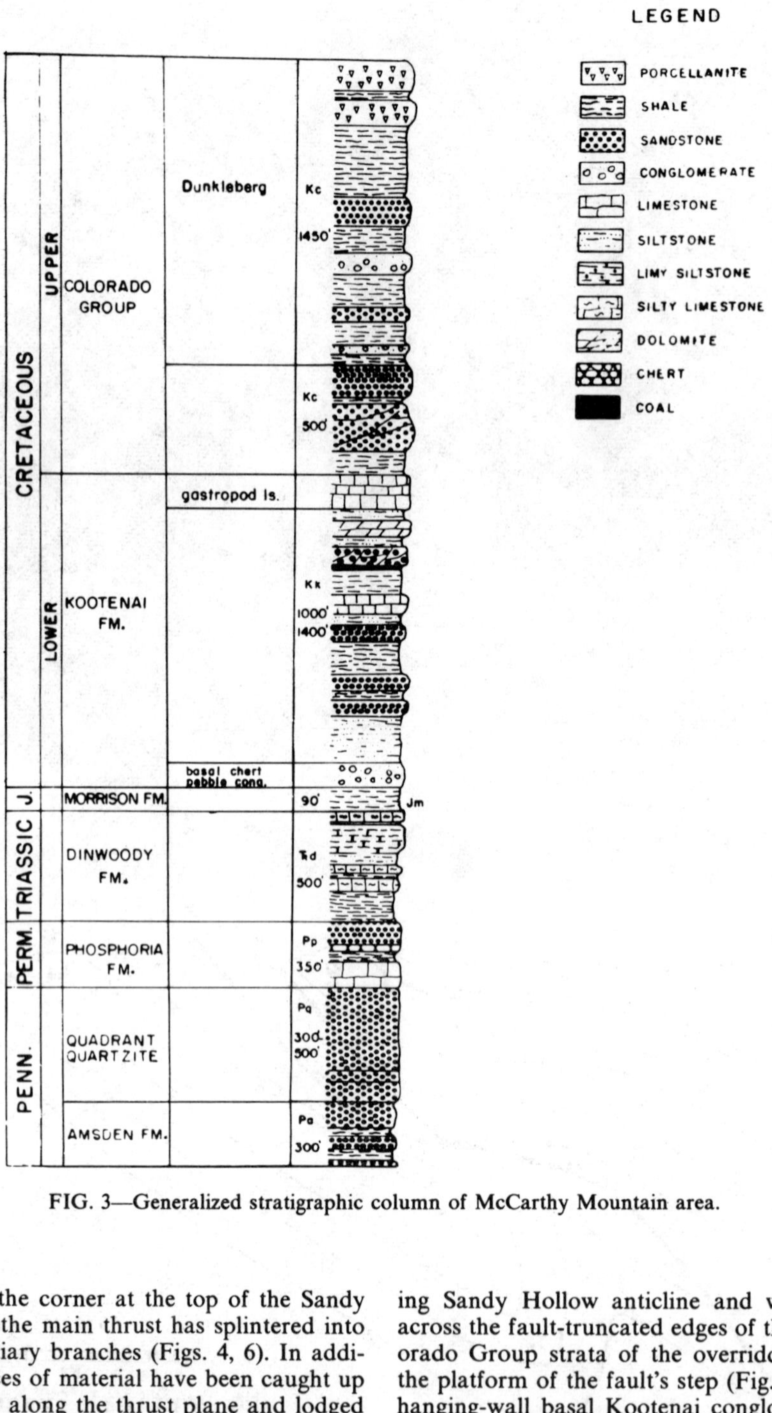

FIG. 3—Generalized stratigraphic column of McCarthy Mountain area.

In turning the corner at the top of the Sandy Hollow riser, the main thrust has splintered into several subsidiary branches (Figs. 4, 6). In addition, large slices of material have been caught up in and moved along the thrust plane and lodged on the platform of the Sandy Hollow step. A large irregular sheet of upper Kootenai gastropod limestone was torn from the west flank of the underlying Sandy Hollow anticline and was stretched across the fault-truncated edges of the lower Colorado Group strata of the overridden block on the platform of the fault's step (Fig. 4). Pieces of hanging-wall basal Kootenai conglomerate, broken from the fault-truncated edge of the steep east flank of Ziegler anticline as it moved across the platform, lie discordantly on the gastropod

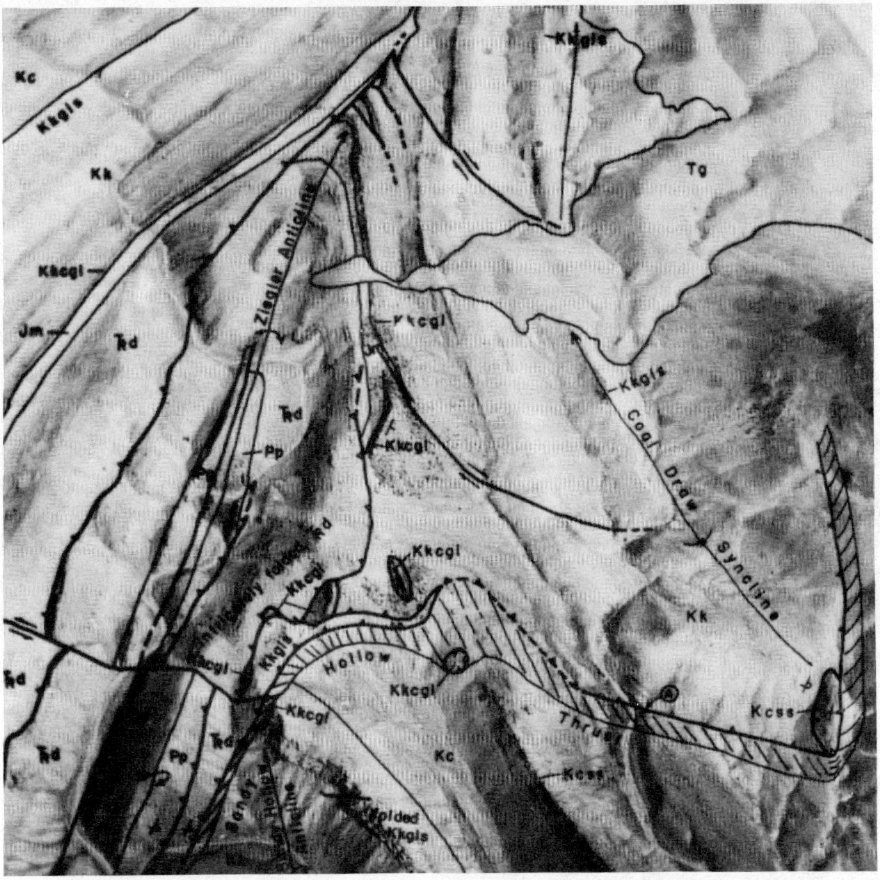

FIG. 4—Aerial oblique view northward of Sandy Hollow thrust in area between step in lower middle part of picture and Block Mountain step near right edge. *Pp* = Phosphoria Formation, *Trd* = Dinwoody Formation, *Jm* = Morrison Formation, *Kk* = Kootenai Formation, *Kkcgl* = basal chert pebble conglomerate member of Kootenai Formation, *Kkgls* = gastropod limestone member of Kootenai Formation at its top, *Kc* = Colorado Group, *Kcss* = prominent channel sandstone in lower part of Colorado Group, *Tg* = Tertiary gravel. Circled *A* location is same as that on Figure 1C. Length of side of photo is approximately 0.86 mi (1.4 km).

limestone and on the Colorado Group shale. A piece of Colorado Group sandstone torn from the upturned edge of a bed in the overridden block below the step platform was carried eastward and lodged against the riser of the Block Mountain step (Fig. 4). Tear faults (Figs. 2, 4) are confined to the overriding block. They represent velocity discontinuities that may result from local binding of the allochthon as it slides over the step.

In addition to these brittle structures, folds were produced in the overridden block at the top of the riser of the Sandy Hollow step (at circled *B* in Fig. 1C). A bedding surface within the gastropod limestone clearly served as a surface of decollement. The middle and upper parts of the gastropod limestone were folded strongly and pushed eastward over the lower part of the gastropod limestone (Figs. 4, 7). The dominantly upward movement of the overriding block at the step-fault riser changed to nearly horizontal movement over the step-fault platform. The overriding of the top edge of the step exerted a nearly horizontal stress against the upper part of the step, crumpling the gastropod limestone that here formed this part of the step (Fig. 7). Many fold structures also were produced in the thinly bed-

FIG. 5—View northward at outcrop of Sandy Hollow thrust and associated minor faults along Burma Road (for location see Fig. 2).

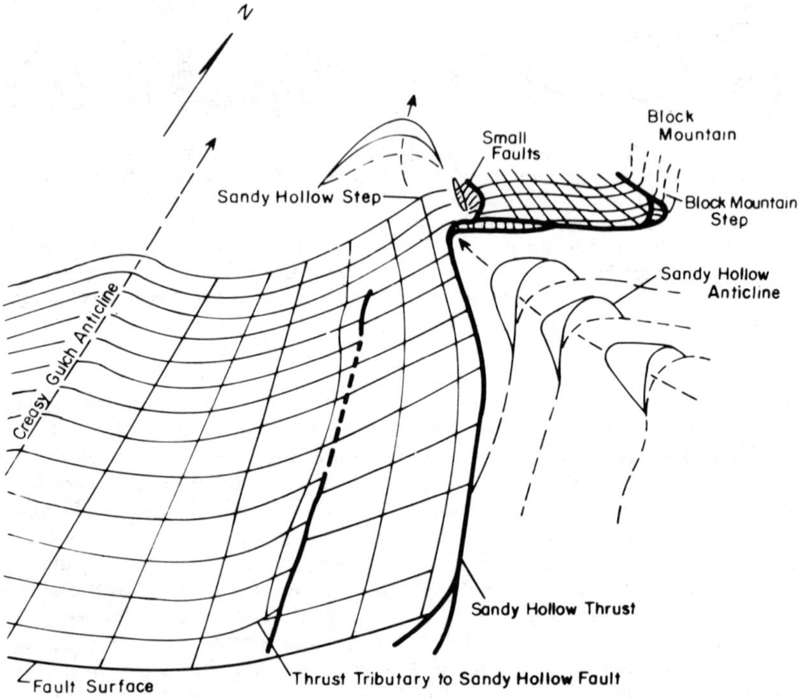

FIG. 6—Diagrammatic sketch of Sandy Hollow fault plane as it would appear if rocks of upper plate were removed. Sketch is based on tracing from oblique air photo.

FIG. 7—View eastward at *B* (Fig. 1c) of gastropod limestone at top of riser of Sandy Hollow step. Middle and upper parts of gastropod limestone crumpled and slid over its lower layers as overriding block of Sandy Hollow thrust surmounted top of riser.

ded Dinwoody Formation as that unit slid across the Sandy Hollow step above the splinter faults (Fig. 4).

This ductile response is in strong contrast to the more brittle splintering effect along the fault zone and to the tear faulting of the overriding block (Figs. 4, 6). Field evidence (Brumbaugh, 1973) from the area of the Sandy Hollow thrust clearly indicates shallow deformation—10,000 ft (3,048 m) or less. In such a tectonic environment, mechanical behavior becomes varied and dependent on lithology. Quartzite and dolomite respond in a more brittle fashion, whereas limestone behaves in a ductile manner. Thus, the wide variety of structures related to the Sandy Hollow step is more a function of the temperature and pressure than a necessary product of step mechanics. On the other hand, steps in general tend to be present in thin-skinned environments, so such an association of structures with steps probably is anything but unique.

Sequence of Faulting and Folding

The structural relations at the steps of the Sandy Hollow thrust indicate that the fault succeeded much of the folding of Sandy Hollow and Ziegler anticlines. The fault plane at best is folded only slightly. The fault was controlled by the folds, for the risers of its steps are located over underlying anticlinal structures, and the platforms of its steps across intervening synclines (Fig. 2). A further indication that the fault postdates much of the folding is shown by the relations of the fault plane to the beds it cuts. At the risers of the steps it dips more steeply than the beds it cuts; at the platforms of the steps it dips less steeply than the beds it cuts. In addition, its trace cuts at a slight angle across the axes of Ziegler and Sandy Hollow anticlines (Figs. 2, 6). These relations would be unlikely if the fault occurred first and then was folded with the beds.

Summary

This report has described a Laramide thrust or step fault and the structures produced by step-fault mechanics. Such structures include decollement, splinter faults, smaller scale folding, and fault transported slices. These step-fault structures should be associated quite commonly with steps everywhere because they are the predictable results of step-fault mechanics in the restricted temperature-pressure range in which thrust faults form, marginal to orogenic belts.

The sequence of faulting and folding for the Sandy Hollow fault is indicated quite clearly from field evidence. The classic model of thrust-fault steps is that of the Pine Mountain fault (Rich, 1934) where folding follows faulting. The case for the Sandy Hollow fault is one in which previously existing folds controlled the development of step zones in the subsequent thrust fault. Such a control by folds on the development and location of step zones provides an interesting alternative to the classic sequence of development of steps as envisioned by Rich for the Pine Mountain fault.

References Cited

Brumbaugh, D. S., 1973, Structural analysis of the complexly deformed Big Hole River area, Madison, Beaverhead, and Silver Bow Counties, Montana: PhD thesis, Indiana Univ., 96 p.

Gale, H. S., 1910, Rock phosphate near Melrose, Montana: U.S. Geol. Survey Bull. 470, p. 440-451.

Peterson, J. A., R. F. Gosman, and R. W. Swanson, 1954, Stratigraphic sections of the Phosphoria Formation in Montana, 1951: U.S. Geol. Survey Circ. 326, 27 p.

Rich, J. L., 1934, Mechanics of low-angle overthrust faulting as illustrated by Cumberland thrust block, Virginia, Kentucky, and Tennessee: AAPG Bull., v. 18, p. 1584-1596.

Richards, R. W., and J. T. Pardee, 1925, The Melrose phosphate field, Montana: U.S. Geol. Survey Bull. 780A, p. 1-32.

Steuer, F., 1956, Geology of the McCarthy Mountain area, Beaverhead and Madison Counties, Montana: Master's thesis, Univ. Utah, 71 p.

BULLETIN OF THE AMERICAN ASSOCIATION OF PETROLEUM GEOLOGISTS
VOL. 49, NO. 11 (NOVEMBER, 1965), PP. 1847-1866, 26 FIGS.

TECTONIC DEVELOPMENT OF IDAHO-WYOMING THRUST BELT[1]

FRANK C. ARMSTRONG[2] AND STEVEN S. ORIEL[2]
Denver, Colorado

ABSTRACT

Three stages are evident in the tectonic development of southeastern Idaho and western Wyoming. First are the changing patterns of tectonic elements during deposition; second, development of northward-trending folds and thrust faults; and third, development of block faults that produced horst ranges and graben valleys.

During Paleozoic time about 50,000 feet of marine sediments, mostly limestone and dolomite, were deposited in a miogeosyncline and about 6,000 feet of mixed marine sediments were deposited on the shelf to the east. Detritus came from both east and west from Cambrian time on. Starting in Mississippian time, the belt between shelf and miogeosyncline, where thicknesses increase markedly, shifted progressively eastward.

During Mesozoic time about 35,000 feet of marine and continental sediments were deposited in the western part of the region and about 15,000 feet in the eastern part, with terrestrial deposits becoming increasingly dominant. Western positive areas became the chief source of detritus. The belt of maximum thickening and the site of maximum deposition were relocated progressively eastward; maximum thicknesses of succeeding geologic systems are not superposed. In Late Triassic a belt on the west rose and the miogeosyncline started to break up. As Mesozoic time progressed the western high spread eastward, until by the end of the Jurassic the miogeosyncline gave way to intracratonic geosynclinal basins that received thick deposits, particularly in Cretaceous time. Cenozoic sedimentary rocks are products of orogeny in the region.

The second stage, which overlapped the first, produced folds overturned toward the east and thrust faults dipping gently west in a zone, convex to the east, 200 miles long and 60 miles wide. Stratigraphic throw on many larger faults is about 20,000 feet; horizontal displacement is at least 10 to 15 miles. Lack of metamorphism and mylonite along the faults is striking. From west to east, the thrust faults cut progressively younger beds, have progressively younger rocks in their upper plates, and are estimated to be successively younger. Thrusting started in the west in latest Jurassic and ended in the east perhaps as late as early Eocene time; detritus shed from emergent upper plates is preserved in coarse terrestrial strata of corresponding ages.

West of the thrust belt is a northwestward-trending area underlain mostly by lower Paleozoic rocks and flanked on east and west by upper Paleozoic and Mesozoic rocks. Scattered pieces of eastward-dipping thrust faults have been reported west of the older rocks. This central area of old rocks has been interpreted as: (1) part of a large continuous thrust sheet moved scores of miles from the west; or (2) an uplifted segment of the earth's crust from which thrust sheets on the east and west were derived. Both interpretations have defects; relative thrust ages are difficult to explain under the first; a large positive gravity anomaly, expectable under the second, is apparently absent.

Block faulting, the third stage of tectonic development, started in Eocene time. Faulting has continued to the Recent, as indicated by broken alluvial fans, displaced basalt flows less than 27,000 years old, and earthquakes. North-trending and east-trending fault sets are recognized. Old east-trending steep faults in the Bear River Range may be tear faults genetically related to thrusting. Movement along many faults has been recurrent. Patches of coarse Tertiary gravel on the flanks and crests of ranges, for which there is no provenance with present topography, may record reversed vertical movement along some north-trending faults. Present topographic relief of basins and ranges is tectonic.

INTRODUCTION

Many hundreds of man-years have been devoted to geologic studies of the thrust belt of southeastern Idaho and western Wyoming. The stimulus has been economic.

Classic reports resulted from the search for coal, oil, and gas (Veatch, 1907; Schultz, 1914), and phosphate (Mansfield, 1927). The region

[1] Publication authorized by the director of the United States Geological Survey. Read before the Rocky Mountain Section of the Association at Durango, Colorado, October 1, 1964. Manuscript received, April 26, 1965.

[2] United States Geological Survey.

Material for the paper has been assembled from many sources and we gratefully acknowledge this debt. We also thank many of our colleagues on the U.S. Geological Survey for their constructive discussion and criticism. Special thanks are due E. D. McKee, W. W. Mallory, A. E. Roberts, A. L. Benson, C. A. Sandberg, W. W. Rubey, and J. I. Tracey for use of unpublished material. Our greatest debt is owed William W. Rubey. Much of the information and many of the ideas on which this paper is based were originally set forth by him (Rubey, 1955; Rubey and Hubbert, 1959), and the paper was greatly improved as the result of his criticism. The paper also profited from the suggestions of G. D. Robinson and W. H. Hays.

contains the best deposits of the western phosphate field, one of the large phosphate fields of the world, and the phosphate is being mined at an increasing rate for fertilizer and chemicals. Along its eastern margin the region contains oil and gas fields that likely are but a small sample of potential reserves. Coal continues to contribute to the region's economy in the new steam-generating plants constructed by Utah Power and Light Company, near strip mines a few miles south of Kemmerer, Wyoming.

Geologic complexities of the region have made a sound scientific approach necessary for intelligent exploration. It is thus no accident that economically motivated studies have resulted in important contributions to geologic science. The region contains some of the earliest thrust faults described in North America; the Bannock and Absaroka thrust faults are widely known and discussed in many textbooks on structural geology.

Despite the amount of work already done, parts of the region have not been studied since Peale's (1879) reconnaissance for the Hayden Territorial Survey and answers to fundamental questions continue to be elusive. Accordingly, this paper is a progress report that summarizes some of the information and ideas that have emerged in recent years and some of the problems that remain.

SCOPE OF PAPER

The region discussed is southeast of the Snake River Plain, west of the Green River basin, and north of the Utah State line (Fig. 1). The discus-

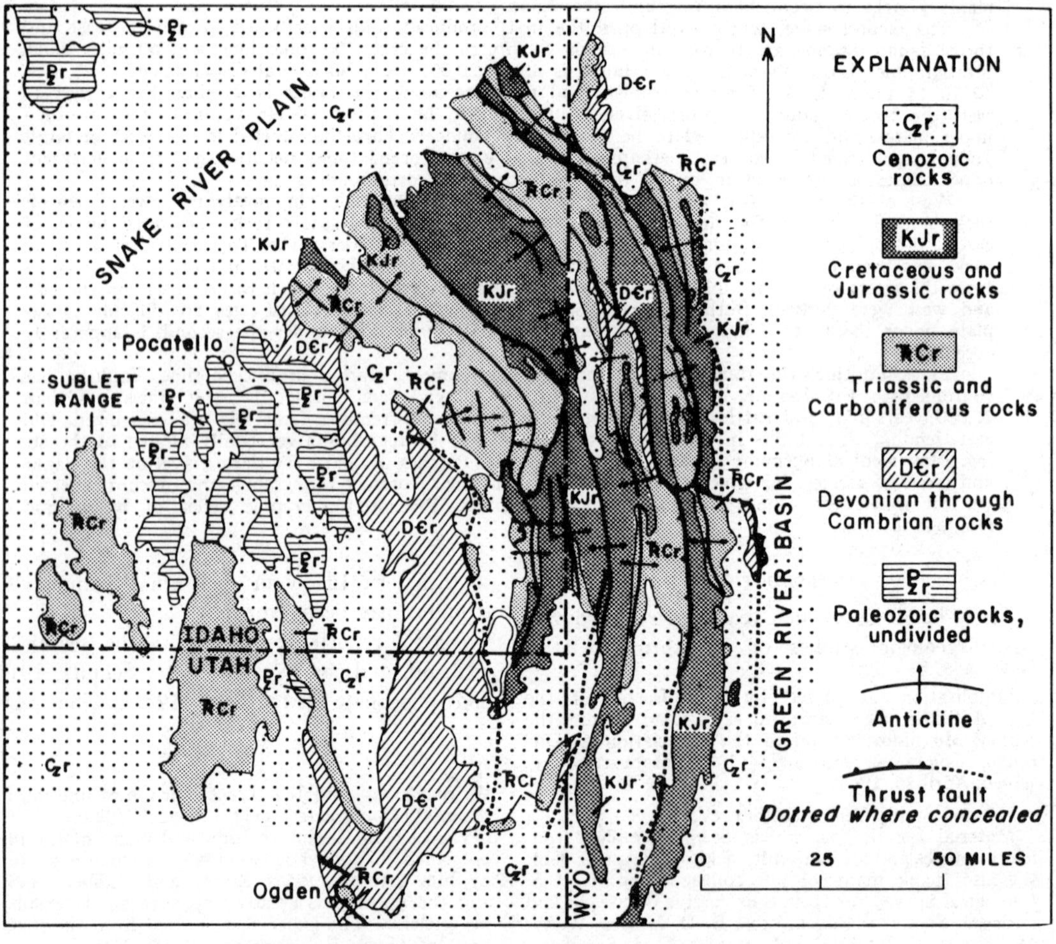

FIG. 1.—Generalized geologic map of Idaho-Wyoming thrust belt. Modified from "Geologic Map of the United States" (Stose, 1932).

sion is divided into three parts. The first deals with the changing patterns of deposition during the tectonic evolution in Paleozoic and early Mesozoic time of a miogeosyncline on the west and a shelf on the east; the second deals with regional deformation that formed northward-trending folds and large thrust faults in late Mesozoic and early Tertiary time; and the third deals with block faults, active since Eocene time, that have produced the present horst ranges and graben valleys.

Part 1. Depositional History

Aggregate thicknesses of sedimentary rocks deposited in southeastern Idaho and adjacent western Wyoming during Paleozoic and Mesozoic time are about 100,000 feet. The entire thickness was not deposited at any one place, however; rather, the site of maximum deposition changed with time. The changing patterns of deposition are shown in isopachous maps and diagrammatic cross sections for each of the geologic systems of the Paleozoic and Mesozoic Eras. Elements emphasized are: the platform on the east, where the sedimentary rocks are thin; the miogeosyncline on the west, where the rocks are thick; and a transitional area between. The position of this transitional area changed from system to system. Where data are available the sites of maximum deposition are shown; the positions of these also changed from system to system. The isopachous maps show the present distribution of thicknesses; the region is not yet sufficiently well understood to construct worthwhile palinspastic maps.

CAMBRIAN

The isopachous map of the Cambrian System excludes the basal quartzite (Fig. 2). A meaningful isopachous map of the quartzite is difficult to construct because different geologists have placed the Cambrian-Precambrian contact at different places in the thick sequence of conformable quartzite and argillite in the western part of the area. Accurate placement of this contact is one of the problems yet to be solved. The basal quartzite in the diagrammatic cross section is transgressive and is called Brigham on the west and Flathead on the east. Early Cambrian trilobites are present near the top of the Brigham Quartzite in the Portneuf Range (Oriel, 1964). The Flathead is Middle Cambrian.

Above the quartzite is a thick sequence of carbonate rocks, shale, and some arkosic quartzite; the sequence thins eastward. The isopachous map includes essentially the rocks deposited during Middle and Late Cambrian time. Westward thickening appears about 15 miles east of the Idaho-Wyoming State line and the site of maximum deposition was somewhere on the west. The Worm Creek Quartzite Member of the St. Charles Limestone contains abundant detrital cleavage fragments of potassium feldspar and some of the rock is arkose. The Worm Creek is not present in Wyoming; in Idaho it thickens westward and its feldspar content increases. This evidence indicates that during Late Cambrian time there was a source area at the west underlain by granite or granitic metamorphic rocks. During most of Paleozoic time major source areas both on the east and on the west (Rubey, 1955; Ross, 1962) shed detritus recurrently into the miogeosyncline.

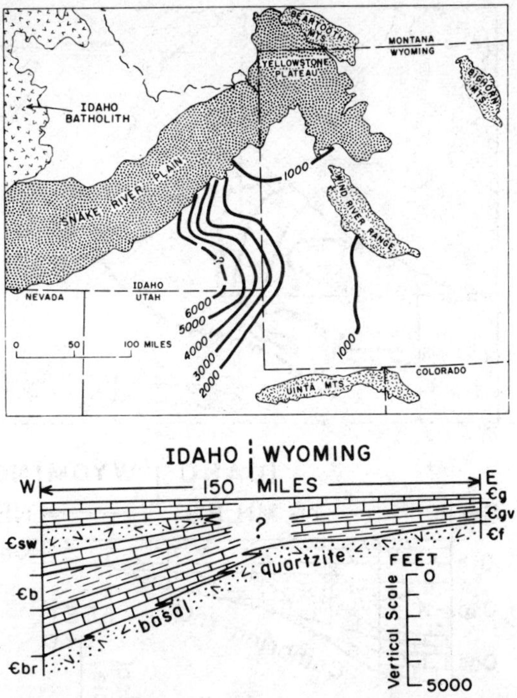

Fig. 2.—Isopachous map of Cambrian System and generalized section. Contoured thicknesses are of Cambrian rocks above basal quartzite. Thicknesses drawn in part from Lochman-Balk (1960) and Trimble and Carr (1962). Stratigraphic units shown in section: Ꞓsw, Worm Creek Quartzite Member of St. Charles Limestone; Ꞓb, Bloomington Formation; Ꞓbr, Brigham Quartzite; Ꞓg, Gallatin Limestone; Ꞓgv, Gros Ventre Formation; Ꞓf, Flathead Quartzite.

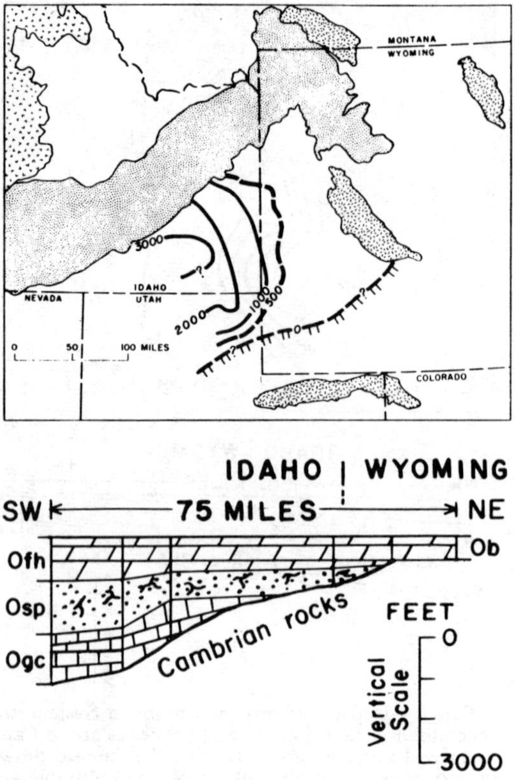

FIG. 3.—Isopachous map of Ordovician System and generalized section. Map drawn in part from Hintze (1960, p. 110). Stratigraphic units shown in section: Ofh, Fish Haven Dolomite; Osp, Swan Peak Quartzite; Ogc, Garden City Formation; Ob, Bighorn Dolomite.

ORDOVICIAN

The transitional area during Ordovician time was about 20 miles east of the Idaho-Wyoming State line; as during Cambrian time, the site of maximum deposition was on the west (Fig. 3). In Idaho, Lower and Middle Ordovician rocks are assigned to the Garden City Limestone and Swan Peak Quartzite, respectively. The entire thickness of Ordovician rocks in western Wyoming is included in the Upper Ordovician Bighorn Dolomite, which is correlative with the Fish Haven Dolomite in Idaho.

The Swan Peak is white, pure quartzite composed of uniformly medium-sized, frosted, well-rounded quartz grains of high sphericity. The grains are probably of second or higher cycle. In this part of Idaho, most of the sand may have been derived from the north, as suggested by Ross (1964, p. 1542, 1551), but it is difficult to visualize only a northern source for the large volume of quartzose detritus in the Ordovician quartzites that extend southwest through Utah and into Nevada.

SILURIAN

Silurian rocks, represented by the Laketown Dolomite in southeastern Idaho and adjacent Utah (Fig. 4), have not been found in western Wyoming (Berdan and Duncan, 1955). The absence of Lower and Middle Ordovician and Silurian rocks in Wyoming was formerly interpreted as suggesting non-deposition during those times. The recent discovery of fossiliferous exposures of Middle (?) and Upper Ordovician and Silurian rocks among the Precambrian crystalline rocks 22 miles southeast of Laramie, Wyoming (Chronic and Ferris, 1961), however, suggests a different interpretation. Consequently, the zero lines on the Ordovician and Silurian isopachous maps (Figs. 3 and 4) probably are erosional rather than depositional limits.

DEVONIAN

The isopachous map of the Devonian System shows local basins west of the Idaho-Wyoming State line in which thick deposits accumulated (Fig. 5).

Devonian rocks in western Wyoming have been mapped as the Darby Formation, which is divisible into a lower dolomite, comparable and correlative with the Jefferson Dolomite, and an upper siltstone and limestone unit that is correlated with the Three Forks Limestone. In southeasternmost Idaho the names Jefferson and Three

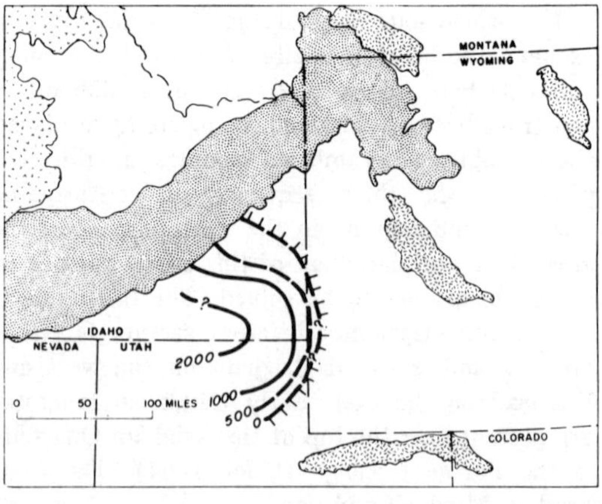

FIG. 4.—Isopachous map of Silurian System. Thicknesses in Utah from Hintze (*in* Rush, 1963, p. 17).

Forks have been used. In north-central Utah the Jefferson Formation is divided (Williams, 1948, p. 1140) into a lower member, the Hyrum Dolomite Member that correlates with the Jefferson Dolomite, and an upper member, the Beirdneau Sandstone Member that correlates with the Three Forks Limestone. Because fine-grained quartz sandstone is present in the top of the Jefferson in part of southeastern Idaho, and because farther northeastward the sandstone of the Beirdneau grades laterally into the siltstone and limestone of the upper division of the Darby, the sandstone in the Beirdneau likely was derived from the west or southwest.

Below the Hyrum in Utah (Williams, 1948, p. 1138; Williams and Taylor, 1964) and in extreme southeastern Idaho (Armstrong, 1953; Coulter, 1956, p. 30) is the Lower Devonian Water Canyon Formation, which consists of interbedded dolomite, argillaceous and silty dolomite, cross-bedded quartz sandstone, and calcareous quartz sandstone. Lower Devonian rocks of this type have not been recognized in western Wyoming. The presence of the Bighorn Dolomite on the east in Wyoming, blanketing underlying possible sources of quartz sand, requires that Osmond's (1962, p. 2052-54) interpretation of an eastern provenance for the quartz sand in the Water Canyon Formation and Sevy Dolomite in Utah and Nevada be modified for the sand in the Water Canyon in Idaho. The writers interpret the quartz sand in the Water Canyon in southeastern Idaho as having been derived from the southwest, transported by long-shore currents flowing northeastward along the eastern margin of the miogeosyncline.

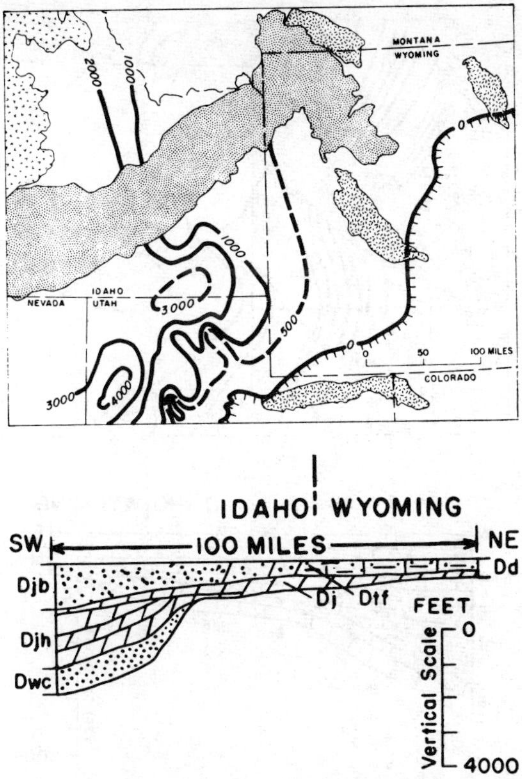

FIG. 5.—Isopachous map of Devonian System and generalized section. Thicknesses in Utah from Rigby (1963, p. 78); in Wyoming, partly from C. A. Sandberg (unpub. data). Stratigraphic units shown in section: Djb, Beirdneau Sandstone Member of Jefferson Dolomite; Djh, Hyrum Dolomite Member of Jefferson Dolomite; Dwc, Water Canyon Formation; Dtf, Three Forks Limestone; Dj, Jefferson Dolomite; Dd, Darby Formation.

MISSISSIPPIAN

Mississippian rocks thicken markedly westward from about 40 miles west of the Idaho-Wyoming State line (Fig. 6). During Cambrian through Devonian time this transitional area between miogeosyncline and shelf was east of its position in Missisippian time. From Mississippian time on, however, the position of the transitional area reversed with time and thenceforth it lay progressively farther to the east.

Because of facies changes and other stratigraphic complexities, Mississippian rocks in the region have been mapped in different ways and assigned different formational names. In the westernmost part of the region (not shown on Figure 6) Mississippian rocks have been called, in ascending order, Lodgepole Limestone, Deep Creek Formation, Great Blue Limestone, and Manning Canyon Shale (Carr and Trimble, 1961). In most of southeastern Idaho, however, Mississippian rocks have been mapped as two formations, the Madison Limestone and the overlying Brazer Limestone (Mansfield, 1927). In parts of adjacent western Wyoming the Mississippian consists of a basal carbonate unit and the lower part of the overlying Amsden Formation. In surface mapping the Madison and Brazer have not been differentiated; instead they have been mapped together as the basal carbonate unit, Madison and Brazer undivided (Rubey, 1958).

Mississippian nomenclature in the region has been modified in recent years. The name Brazer

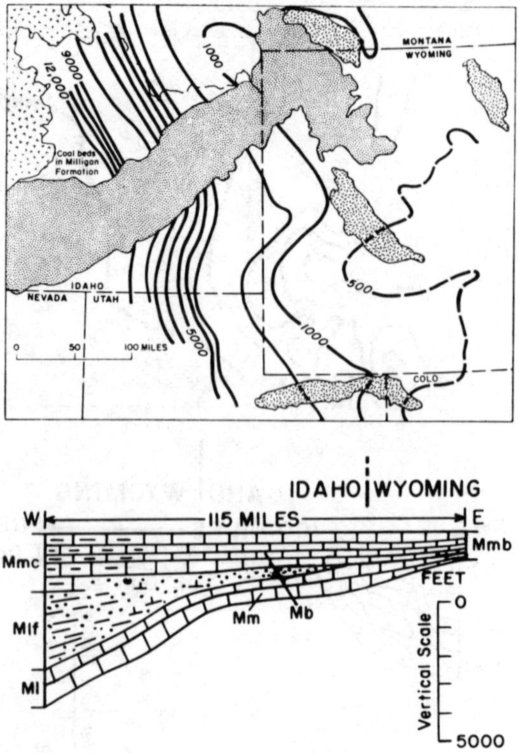

FIG. 6.—Isopachous map of Mississippian System and generalized section. Thicknesses partly from Strickland (1956, p. 52) and A. E. Roberts (unpub. data). Stratigraphic units shown in section: Mmc, Monroe Canyon Formation; Mlf, Little Flat Formation; Ml, Lodgepole Limestone; Mb, Brazer Limestone; Mm, Madison Limestone; and Mmb, Madison and Brazer Limestones, undivided. Lower part of Amsden Formation not shown in section.

was restricted in 1959 to its occurrence in the Crawford Mountains, Utah (Sando, Dutro, and Gere, 1959), and the Brazer Limestone of former usage in southeastern Idaho was subsequently redescribed as the Chesterfield Range Group, consisting of the Little Flat Formation and the overlying Monroe Canyon Limestone (Dutro and Sando, 1963). In the same article (Dutro and Sando, 1963) the name of the unit previously mapped as Madison Limestone (Mansfield, 1929, p. 23) was changed to Lodgepole Limestone. Fine-grained sandstone in the lower part of the Brazer, now named the sandstone member of the Little Flat Formation, is the stratigraphic equivalent of the Humbug Formation to the southwest, The Humbug is slightly thicker and coarser-grained than the sandstone member of the Little Flat, and contains detrital grains of microcline and plagioclase (Gilluly, 1932, p. 26-29); both minerals have been recognized recently in the Little Flat Formation in the Fish Creek Range. Accordingly, the sand in the Little Flat is interpreted as having been derived from a source area on the southwest.

The presence in Mississippian time of land on the northwest also is indicated by coal beds in the Milligen Formation (Umpleby, Westgate, and Ross, 1930, p. 26) in the Wood River region of Idaho north of the Snake River Plain (Fig. 6).

PENNSYLVANIAN

A large amount of westward thickening of Pennsylvanian rocks is apparent about 20 miles west of the Idaho-Wyoming State line (Fig. 7), and the site of maximum deposition is farther west, as shown. The upper part of the Amsden Formation and the lower and middle parts of the Wells Formation make up the Pennsylvanian in

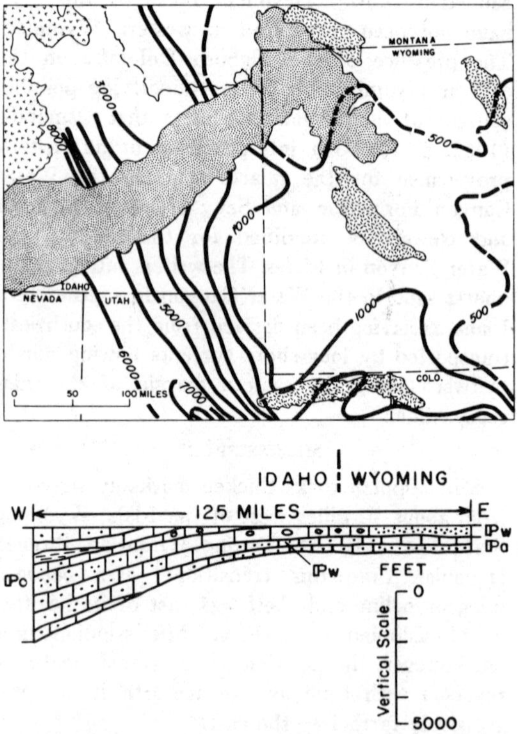

FIG. 7.—Isopachous map of Pennsylvanian System and generalized section. Thicknesses partly from Williams (1962, p. 181) and W. W. Mallory (unpub. data). Stratigraphic units shown in section: IPo, Oquirrh Formation; IPw, Wells Formation, lower and middle parts; and IPa, Amsden Formation, upper part.

western Wyoming. In southeasternmost Idaho the lower and middle parts of the Wells are the only Pennsylvanian rocks. Farther west in Idaho, Pennsylvanian rocks are assigned to the Oquirrh Formation (Carr and Trimble, 1961). The proportion of sandstone in Pennsylvanian rocks increases eastward, which suggests a source on the east. Westward from southeasternmost Idaho, however, Pennsylvanian rocks contain a large amount of sandstone and, although the writers do not have supporting quantitative data, their impression is that the total amount of the quartz sand fraction in the thick western sequence exceeds that in the thin eastern sequence. The large amount of sandstone in the Pennsylvanian rocks on the west, the presence of thick, coarse conglomerate beds in the Wood River Formation north of the Snake River Plain and east of the Idaho batholith (Umpleby, Westgate, and Ross, 1930, p. 32; Ross, 1937, p. 37), near the 8,000-foot isopachous line (Fig. 7), indicate a western source. Detritus in Pennsylvanian rocks of southeastern Idaho and adjacent Wyoming probably was derived both from the east and from the west.

PERMIAN

By Permian time, the eastern margin of the miogeosyncline was still farther east, close to the Idaho-Wyoming State line. The axis of maximum deposition lies between the 9,000-foot isopachous lines and trends northeastward (Fig. 8). A high area of similar trend was present farther to the west (McKee, Oriel, and others, in press).

The upper part of the Wells Formation, now assigned to the Grandeur Tongue of the Park City Formation (McKelvey and others, 1959, p. 12), and the Phosphoria Formation were deposited in Permian time. Pronounced thickening westward resulted almost wholly from thickening of the lower unit that is roughly equivalent to the top of the Wells. The Wells is interbedded sandstone and limestone that probably were deposited in shallow water. The sand may have been derived from both east and west. The Phosphoria is mudstone, chert, phosphorite, and limestone and was deposited in deep water.

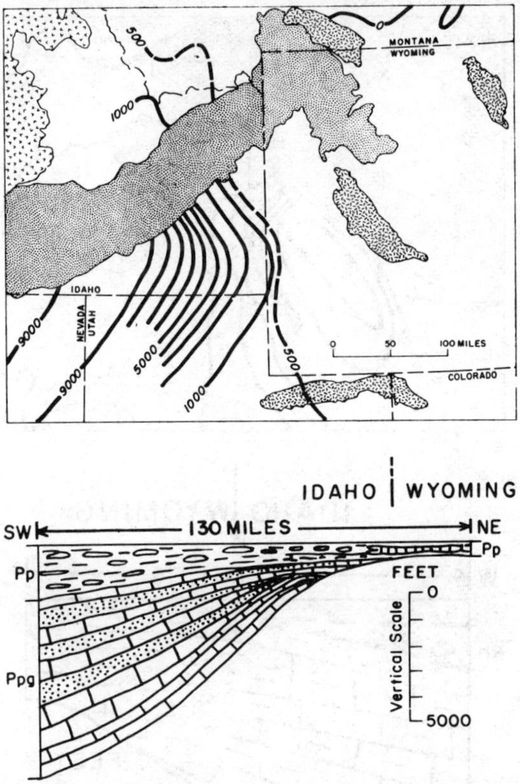

FIG. 8.—Isopachous map of Permian System and generalized section. Map generalized after McKee, Oriel, and others (in press). Stratigraphic units shown in section: Pp, Phosphoria Formation; and Ppg, Grandeur Tongue of Park City Formation which occurs at top of, and has been mapped with, upper part of Wells Formation.

TRIASSIC

All Paleozoic rocks in southeastern Idaho were deposited in a marine environment. Lower Triassic rocks are also marine, but the rest of the Triassic is non-marine.

Triassic sediments thicken markedly westward from about 30 miles east of the Idaho-Wyoming line, and the center of a deep basin is shown by the 8,000-foot isopach line (Fig. 9). In central Idaho west of the basin, a northeast-trending ridge rose early in Late Triassic time, and shed coarse detritus into the basin on the southeast to form the Higham Grit, which probably was deposited as coalescing fans along a mountain front (McKee and others, 1959, p. 17). The rise of this ridge, perhaps foreshadowed in the Permian, indicates the start of the breakup of the miogeosyncline.

JURASSIC

During Jurassic time the familiar pattern of shelf on the east and geosyncline on the west,

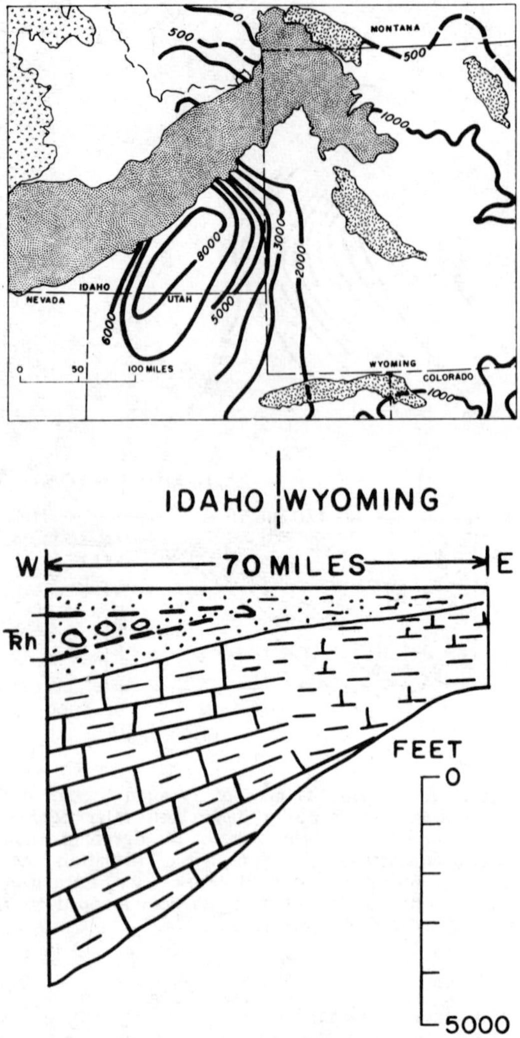

FIG. 9.—Isopachous map of Triassic System and generalized section. Map generalized from McKee and others (1959, pl. 5). Stratigraphic unit designated on section is Higham Grit, Trh.

though evident, is not as sharply defined as in older systems, and thick deposits of sedimentary rocks are present in western Wyoming (Fig. 10). The location of the western margin of the basin is not known, but it probably was not very far west. The non-marine conditions of Late Triassic persisted into Early Jurassic time. The Late Triassic northeast-trending high in central Idaho persisted into the Early Jurassic and supplied detritus for the Nugget Sandstone. During this time the high appears to have spread eastward and perhaps the relief was less. Subsequently the sea returned and the Twin Creek Limestone, Preuss Sandstone, and Stump Sandstone were deposited under marine conditions. In early Late Jurassic time the high on the west rose, again spread eastward, and shed detritus now found in the Preuss and Stump.

LOWER CRETACEOUS

During Early Cretaceous time the axis of a deep, north-trending basin was 5-10 miles west of the Idaho-Wyoming line (Fig. 11). In latest Jurassic or earliest Cretaceous time, parts of southeastern Idaho were raised and mountains were formed. Coarse detritus was deposited in

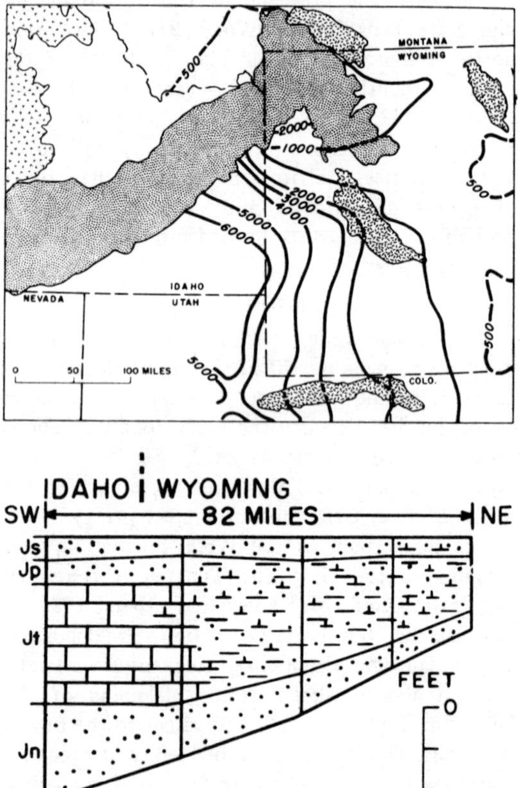

FIG. 10.—Isopachous map of the Jurassic System and generalized section. Map generalized from McKee and others (1956, pl. 3). Stratigraphic units shown on section are: Js, Stump Sandstone; Jp, Preuss Sandstone; Jt, Twin Creek Limestone; Jn, Nugget Sandstone.

coalescing alluvial fans to form the Gannett Group, and orogenic sediments, such as the Wayan Formation,[3] continued to be deposited throughout Early Cretaceous time. Because Lower Cretaceous sediments probably were not deposited west of the Bear River Range (Armstrong and Cressman, 1963, p. J11), the western parts of the isopachous lines may be shown incorrectly on the map. The Gannett and Wayan thin markedly eastward and become finer-grained.

UPPER CRETACEOUS

In Late Cretaceous time an axis of maximum deposition was 10-15 miles east of the Idaho-Wyoming line (Fig. 12). Farther east are other basins in which thick prisms of sediments were deposited. In the western basin probably more than two-thirds of the rocks were deposited in early Late Cretaceous (Colorado) time, whereas in the eastern basins most of the rocks were deposited in late Late Cretaceous (Montana) time. In a sense this reflects continued eastward movement with time of the site of maximum deposition. Detritus in sandy and partly conglomeratic tongues in the Frontier, Hilliard, and Adaville Formations was derived from the west and northwest.

PATTERNS OF DEPOSITION

The systemic isopachous maps show gross patterns, in addition to the features already mentioned. These gross patterns, however, because of insufficient information, may be more apparent than real. Isopachous lines on the Cambrian through Silurian maps are strongly concave westward, and the axis of concavity trends slightly north of east. The Devonian map is characterized by a subdued westward concavity and a northward trend for the eastern margin of the miogeosyncline. After Devonian time the eastern margin of the miogeosyncline had a well-defined northward to northwestward trend, and the axis of concavity had a northeastward trend. These changes in pattern in the transitional Devonian

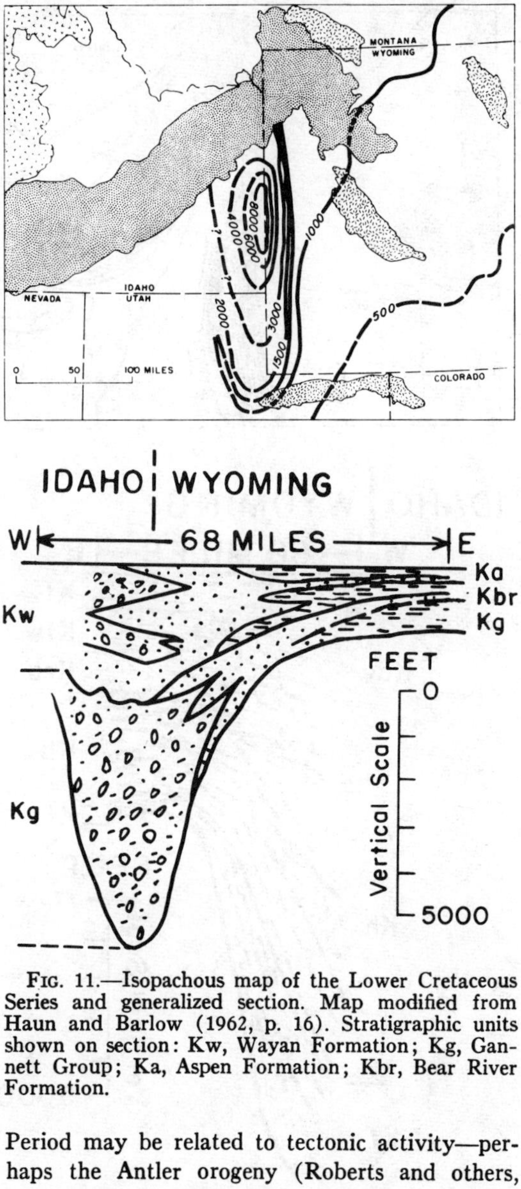

FIG. 11.—Isopachous map of the Lower Cretaceous Series and generalized section. Map modified from Haun and Barlow (1962, p. 16). Stratigraphic units shown on section: Kw, Wayan Formation; Kg, Gannett Group; Ka, Aspen Formation; Kbr, Bear River Formation.

Period may be related to tectonic activity—perhaps the Antler orogeny (Roberts and others, 1958)—on the west. These changes in pattern also may help explain the greater continuity northward, across the eastern part of the Snake River Plain, for upper than for lower Paleozoic stratigraphic units.

The northerly trend of the eastern margin of the miogeosyncline persisted into Cretaceous time and the isopachous lines lost their former westward concavity. During Jurassic time, however, the eastern margin of the miogeosyncline is not well defined. Can the Jurassic Period be compared

[3] Although the Wayan Formation has previously been assigned to the Lower(?) and Upper Cretaceous, it is here regarded as Lower Cretaceous. Recent studies in the type area, particularly by W. W. Rubey, have shown that many of the rocks mapped as Wayan belong to the Gannett Group. The younger rocks present are western facies of the Bear River and Aspen Formations of Early Cretaceous age and are arbitrarily depicted in the Figure 11 cross section to be about 3,000 feet thick.

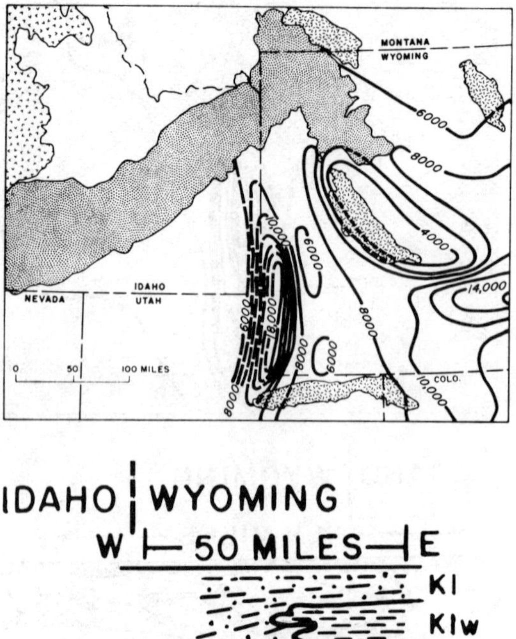

FIG. 12.—Isopachous map of the Upper Cretaceous Series and generalized section. Map generalized from Weimer (1961, p. 18). Stratigraphic units shown in section: Kav, Adaville Formation; Kh, Hilliard Shale; Kf, Frontier Formation; Kl, Lance Formation; Klw, Lewis Shale; Krs, Rock Springs Formation; Kb, Baxter Shale. Relation of western sequence to eastern sequence, shown as now commonly accepted, has not been established.

with the Devonian as an unsettled, transitional period after which a different pattern emerged in response to new forces? Were the deep, north-trending Cretaceous basins true descendants of the miogeosyncline? They may not have been, for these deep basins developed on the edge of a cratonic area that had been moderately stable for the previous 450 million years. Certainly during Late Cretaceous time a different pattern developed in Wyoming. During latest Cretaceous (Lance) time two deep eastward-trending basins began to form on the shelf in Wyoming, one across northern Wyoming and the other across southern Wyoming (Love and others, 1963, Fig. 4). The northern basin extended from the vicinity of Yellowstone Park east to the Black Hills and later developed into the Wind River and related basins; the southern one extended across south-central Wyoming and later developed into the Great Divide and related basins.

The sedimentary record thus indicates that during Paleozoic time there was a miogeosyncline on the west and a platform on the east. A thick accumulation of marine strata, mostly limestone and dolomite, was deposited across the area. Throughout Paleozoic time detritus came recurrently from both the east and the west. Beginning in Mississippian time the miogeosyncline was shifted eastward; at the same time the site of maximum deposition may also have shifted eastward, a trend which, with temporary reversal in the Permian, continued into the Mesozoic.

These shifts in the position of the miogeosyncline continued during the Mesozoic when marine and continental sedimentary rocks were deposited across the area. A "high" on the west became the principal source of detritus, and with the passage of time increasing proportions of continental strata were deposited, finally to the exclusion of marine sediments. As Mesozoic time progressed, the high on the west spread eastward until the miogeosyncline was destroyed in Late Jurassic time. In Late Cretaceous time basins developed on what had been the stable, cratonic platform on the east.

Destruction of the miogeosyncline in late Mesozoic time was accompanied by the development of large north-trending folds and by eastward movement on thrust faults along the eastern flank of the miogeosyncline. Thus the second chapter of tectonic history is entered.

PART 2. THRUST FAULTING

The thrust belt forms an eastwardly convex,

arcuate zone about 200 miles long and 60 miles wide, extending from the Snake River Plain southward into Utah (Fig. 13). It has been said (Rubey, 1955, p. 125) to consist of parts of Idaho and Utah that were pushed, or slipped, into Wyoming between the Teton Mountain and Uinta Mountain buttresses of Precambrian rock. The major thrusts to be discussed are, from west to east, the Paris, the Absaroka, the Darby, the Prospect, and the Hogsback.[4] The probable ages of the Meade, Crawford, and La Barge thrusts are shown in illustrations.

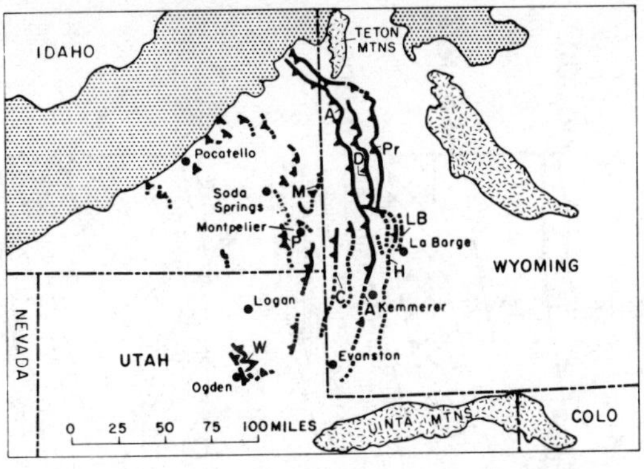

FIG. 13.—Mapped traces of major thrust faults. Solid line indicates exposed fault trace; broken line indicates concealed fault trace. Major faults shown: P, Paris; M, Meade; C, Crawford; A, Absaroka; D, Darby; Pr, Prospect; H, Hogsback; LB, La Barge; W, Willard. Modified from Rubey (1958); and Armstrong and Cressman (1963).

STRUCTURAL RELATIONS OF FAULTS

The thrust faults of the Idaho-Wyoming thrust belt are unlike many in the world in that strata involved are unmetamorphosed and no major fault breccia or mylonite is present. In structure sections (Fig. 14) prepared by Rubey (Rubey and Hubbert, 1959, p. 188), the Absaroka, Darby, and Prospect thrust faults are shown dipping slightly to moderately westward. Drilling has confirmed low westward dip for the Hogsback thrust (not shown in Figure 14) about 40 miles south of the cross sections; there the average dip of the Hogsback thrust, which has been penetrated by several wells across a downdip direction of 10 miles, is about 15° W. The Darby and Prospect faults mostly parallel bedding in the upper plates of the thrusts, although locally they cut across bedding. In the upper two cross sections the Absaroka fault cuts across bedding both above and below the fault surface, whereas in the lower two cross sections the Absaroka is like the Darby and Prospect faults and parallels bedding in the upper plate. Strata above and below all the faults are folded. Although folding formerly was thought to have preceded thrusting, it is now believed to have accompanied the faulting. Minimum horizontal displacements along the faults were 10-15 miles and may have been considerably more.

Stratigraphic throw on each of the major thrust faults is about 20,000 feet, as indicated by the vertical scale in Figure 15. Also illustrated is the fact that the upper plate of each successively more eastward thrust fault contains successively younger strata.

DATING MAJOR MOVEMENTS

Few of the thrust faults can be dated within narrow limits by usual geologic methods. One that can be, however, is the Prospect thrust (Fig. 16). The fault cuts the Hoback Formation of Paleocene and earliest Eocene age; the Eocene part of the Hoback is folded with the underlying strata, but only the Paleocene part is cut by the fault. South of where the fault cuts the Hoback, its trace is overlain by Wasatch strata of late early Eocene age. Major movement on the fault thus occurred in middle early Eocene time.

Somewhat similar evidence is used to date major movement on the Absaroka thrust north of Kemmerer. A short distance south of the area illustrated in Figure 17, the youngest unit cut by the Absaroka is the Adaville Formation, whose topmost part is of latest Cretaceous (Lance or late Montana) age. The Evanston Formation, which is of latest Cretaceous (Lance) and Paleocene age (Rubey, Oriel, and Tracey, 1961), crops out in a north-trending belt close to the trace of the Absaroka; and locally the basal beds of the Evanston, the Lance part, have been cut by late small movement on the Absaroka. Although the Evanston has not been found directly overlying the fault, it overlies rocks above and below the fault with angular unconformity. The Evanston thus was deposited in its present site after most

[4] Hogsback is a new name for the thrust fault south of Snider Basin called Darby by Schultz (1914, pl. 1) and subsequent workers, and Darby (?) by Oriel (1962, p. 2172). The fault is named for its excellent exposures along Hogsback Ridge west of La Barge, Wyoming.

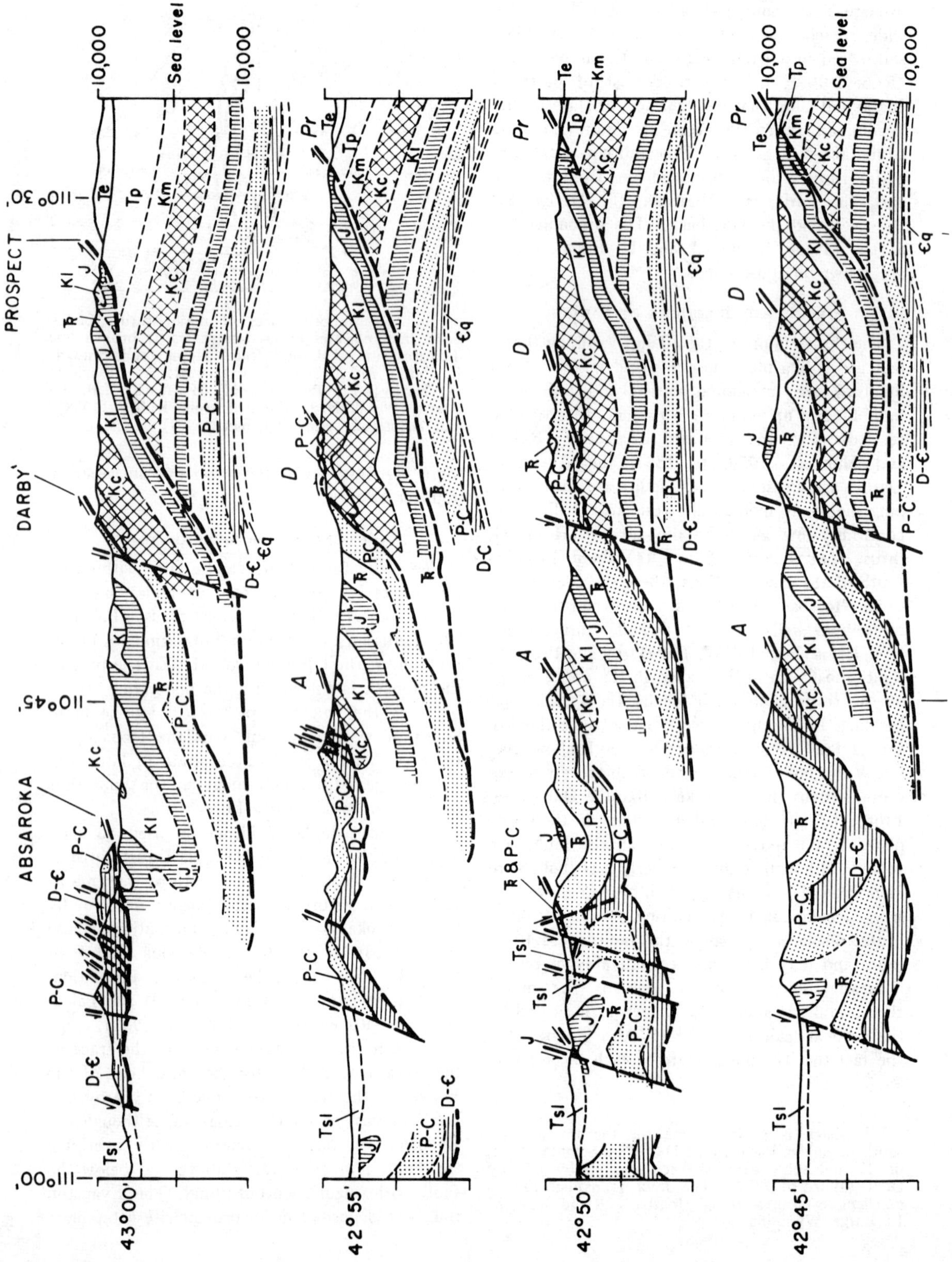
PROSPECT
DARBY
ABSAROKA
110°30'
110°45'
111°00'
43°00'
42°55'
42°50'
42°45'
10,000
Sea level
10,000

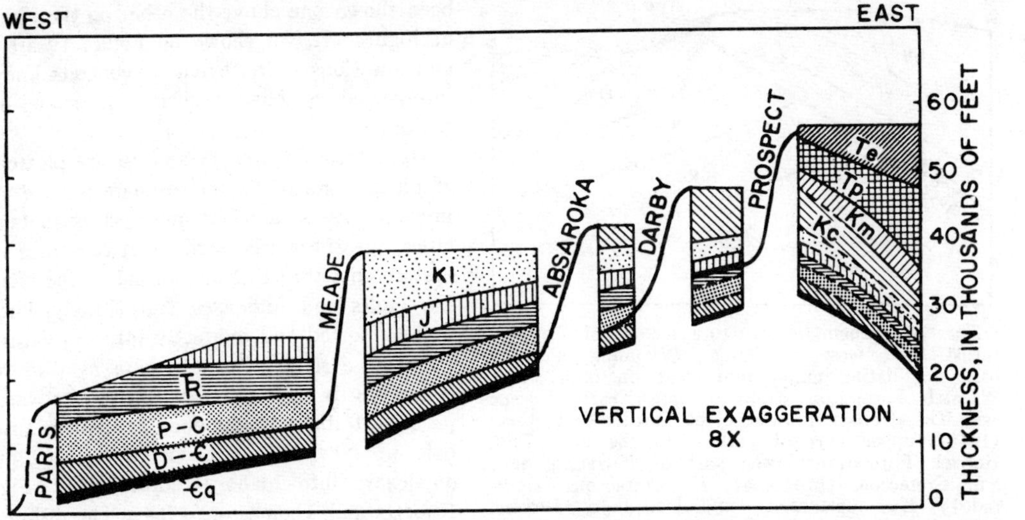

FIG. 15.—Stratigraphic sections in four large overthrust sheets and in Green River Basin. Diagram shows approximate constancy of stratigraphic throw along different faults and progressively younger age of involved rocks from west to east. Formational symbols as in Figure 14. Modified from Rubey and Hubbert (1959, p. 190).

of the movement; major thrust movement on the Absaroka is thereby dated as latest Cretaceous.

Another type of evidence was used to date first movement on the Paris thrust (Armstrong, 1962; Armstrong and Cressman, 1963, p. 48). At Red Mountain in the Gannett Hills (Fig. 18), coarsely conglomeratic redbeds, 5,000 feet thick, are a coarse facies of the lower part of the Gannett Group of earliest Cretaceous age whose basal beds probably are of latest Jurassic age. Examination of nearby Gannett stratigraphic sections shows that the conglomeratic unit coarsens westward and more than doubles in thickness in a distance of 8 miles, thereby indicating a western source for the detritus. A source area only 25-30 miles west is suggested by the rate of increase in pebble size (Rubey, *in* Moritz, 1953, p. 66) and by the survival of non-resistant rocks in the conglomerate (Plumley, 1948, p. 575). Pebbles of Paleozoic rocks in the conglomerate were derived from formations now exposed in the upper plate of the Paris thrust 20-30 miles west of Red Mountain. If the conglomeratic unit on Red Mountain is a synorogenic deposit (Eardley, 1960, p. 37), as the evidence suggests, then first movement on the Paris thrust is dated as latest Jurassic and earliest Cretaceous.

Evidence for accurately dating the other thrust faults shown on Figure 13 is meager. Nonetheless, in Figure 19 an attempt is made to date principal movement on each fault, with the realization that some dates may be slightly in error. Comparison

FIG. 14.—West-east structure sections, somewhat generalized, across Bedford, Blind Bull Creek, and part of Merna Butte Quadrangles, Wyoming. Horizontal and vertical scales same. €q, Cambrian quartzite; D-€, Devonian to Cambrian Gros Ventre Formation, inclusive; P-C, Permian to Carboniferous, inclusive; Tr, Triassic; J, Jurassic; Kl, Lower Cretaceous; Kc, Cretaceous (Colorado); Km, Cretaceous (Montana); Tp, Paleocene; Te, Eocene; Tsl, Pliocene Salt Lake Formation. A, Absaroka thrust fault; D, Darby thrust fault; Pr, Prospect thrust fault. From Rubey and Hubbert (1959, p. 188).

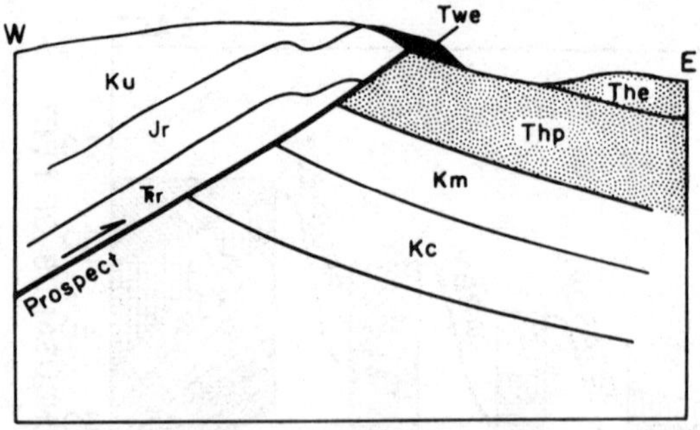

FIG. 16.—Schematic structure section of Prospect thrust fault west of Daniel, Wyoming, showing basis for dating major movement on fault. Twe, Wasatch Formation strata of latest early Eocene age; The, Hoback Formation of Eardley and others (1944), upper part of earliest Eocene age; Thp, Hoback Formation, lower part of Paleocene age; Km, Cretaceous (Montana); Kc, Cretaceous (Colorado); Ku, Cretaceous undivided; Jr, Jurassic rocks; Trr, Triassic rocks.

of Figures 13 and 19 shows that the oldest faults are on the west and the youngest on the east, a fact which has been recognized before. Because the Darby and Prospect thrusts probably are not the northward continuations of the Hogsback and LaBarge faults, these two pairs of faults have been shown one above the other on the right side of Figure 19. Not shown on Figure 19 are times and durations of recurrent movements known or interpreted to have occurred on some of the faults.

Data from Figures 13 and 19 are plotted as a graph in Figure 20[5]; the ordinate is time in millions of years, and the abscissa is distance in miles, measured westward from the trace of the easternmost thrust about normal to the trends of the faults and fold axes. Surprisingly, both the well-supported and indirectly inferred dates cluster along a straight line. Perhaps even more disconcerting is an implication of the westward projection of this straight line. The Antler orogenic belt of Nevada has been inferred to project northward into Idaho (Churkin, 1962, Fig. 11; Roberts and Thomasson, 1964). The distance between the northward projection of the belt and the easternmost fault is plotted in Figure 20; it intersects the line determined by the Idaho-Wyoming thrusts at about 380 million years, or at an age of about Middle Devonian (Holmes, 1959, p. 204). The result is not greatly discrepant from the Late Devonian or Early Mississippian to Early Pennsylvanian age deduced by Roberts

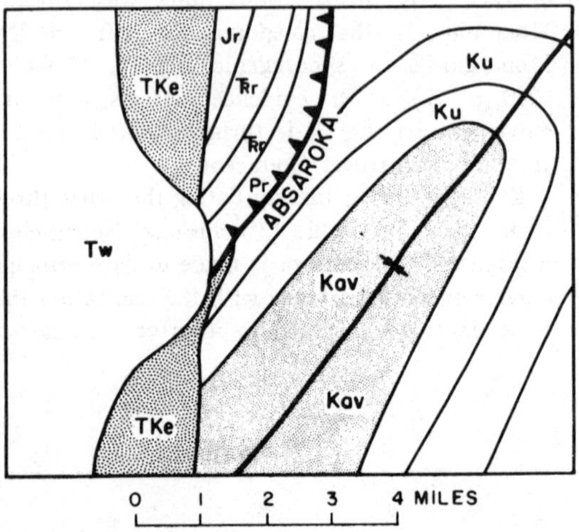

FIG. 17.—Schematic map of part of Absaroka thrust fault, showing the basis for dating major movement on fault in northern part of Kemmerer Quadrangle. Tw, Wasatch Formation (Eocene); TKe, Evanston Formation (Paleocene and uppermost Cretaceous); Kav, Adaville Formation (Upper Cretaceous); Ku, Cretaceous undivided; Jr, Jurassic rocks; Trr, Triassic rocks; Pr, Permian rocks. Modified from unpublished map by W. W. Rubey, J. I. Tracey, Jr., and S. S. Oriel.

[5] The graph was conceived initially to illustrate the futility of applying the term Laramide to deformations in the thrust belt. Use of the term masks, rather than reveals, temporal relations discovered in the region.

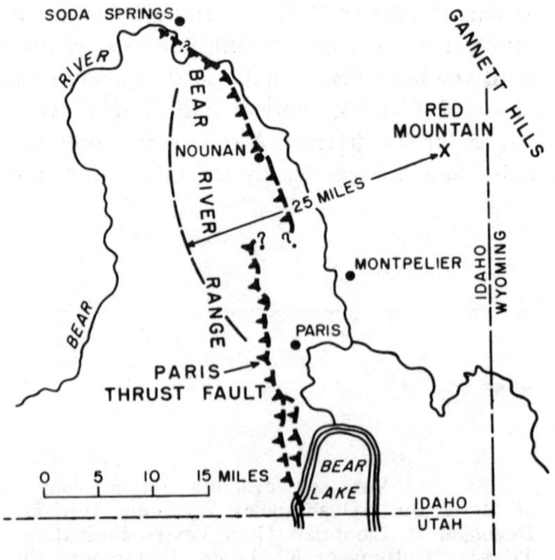

FIG. 18.—Map of localities cited in dating first movement on Paris thrust fault.

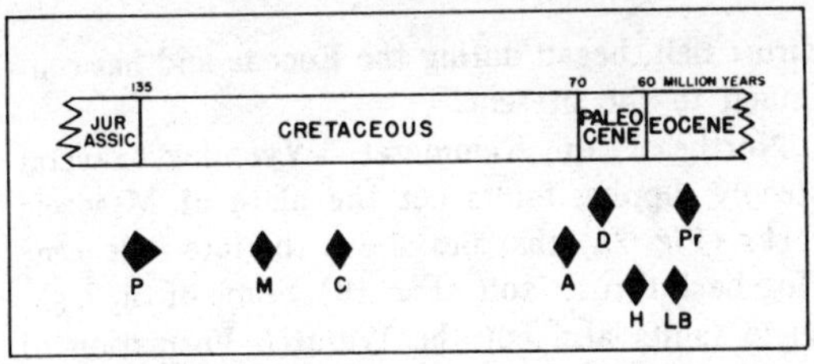

FIG. 19.—Estimated ages of principal movement on major thrust faults. Not shown are times of recurrent movements on some faults. Numerical ages from Holmes (1959, p. 204). Letter symbols are same as on Figure 13, with which this figure should be compared.

and others (1958, p. 2850) from other evidence. The graph may or may not have real meaning.[6]

GROSS STRUCTURE WEST OF THRUSTS

A question of some concern is, what is the gross crustal structure of the area of mainly lower Paleozoic rocks that extends northward from Ogden, Utah, and lies west of the Idaho-Wyoming thrust belt (Fig. 1)? The contrast in detail shown on the east and on the west in Figure 1 reflects contrasts in what is known about the area. Most of the area between Ogden and Pocatello is underlain by rocks of the Brigham through Ordovician units; upper Paleozoic and Mesozoic units are present on both sides of the area. Triassic rocks occur in the Sublett Range on the west and all Mesozoic systems are present in the thrust belt on the east.

East-dipping thrust faults (Fig. 13) near the western margin of the area of Paleozoic rocks have been reported by Blackwelder (1910, Figs. 4 and 6) and Eardley (1944, pl. 1) in the Ogden area, by Murdock (1961, p. 42) just north of the Utah-Idaho State line, by Anderson (1928, p. 8) and Ludlum (1943, Figs. 2 and 3) southeast of Pocatello, and by W. J. Carr and D. E. Trimble (oral communication, 1963) southwest of Pocatello.

Many geologists (Richards and Mansfield, 1912, p. 706, 707; Richardson, 1941, p. 39; Eardley, 1944, p. 869, 870; Crittenden, 1961) have inferred that the Willard thrust near Ogden connects with the Paris thrust to the east. On the basis of this interpretation (Fig. 21) the area

[6] An inference that could be drawn from the graph, for example, by projecting the line toward the right, would be the location of the next thrust faulting on the east.

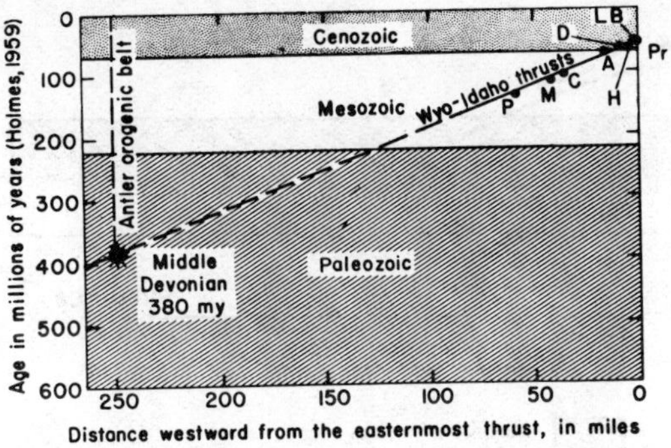

FIG. 20.—Relation of dates of principal movement to positions of present traces of major thrust faults. Curiously, extension of line in graph, based on Idaho-Wyoming thrust dates, yields possible age for northward projection of Antler orogenic belt on west surprisingly close to age inferred by Roberts and others (1958). Based on data in Figures 13 and 19. Letter symbols same as on Figure 13.

of Paleozoic rocks from Ogden northward is regarded as a remnant of the upper plate of a large thrust that moved scores of miles from the west and is therefore allochthonous. In addition to other objections (Armstrong and Cressman, 1963, p. J18), this interpretation is difficult to reconcile with the progressively younger ages of the thrust faults eastward from the Paris thrust, particularly if there is more than one major eastward-dipping thrust on the west. Moreover, if this block is allochthonous and did move scores of miles eastward, it is surprising that nowhere in southeastern Idaho or western Wyoming has a western facies of any stratigraphic unit yet been recognized east of an eastern facies.[7]

An alternate interpretation for the area of older Paleozoic rocks has been suggested (Eardley, 1944, p. 867-869). According to this interpretation (Fig. 22) the area was raised as a large wedge-shaped segment of the earth's crust from which the thrust plates were pushed or slid, and the area is thus autochthonous. If a segment of the earth's crust were raised as postulated, one might expect crystalline rocks of the basement to have been brought close enough to the surface so that their presence would be indicated, because of

[7] On the south in Utah, however, the Willard thrust fault separates a western thick sequence on the east from an eastern thin sequence on the west (Blackwelder, 1925; Crittenden 1961). Juxtaposition of western and eastern sequences is also evident still farther south, east of Provo, along the Charleston-Nebo thrust (Baker, 1959; Crittenden, 1959; 1961).

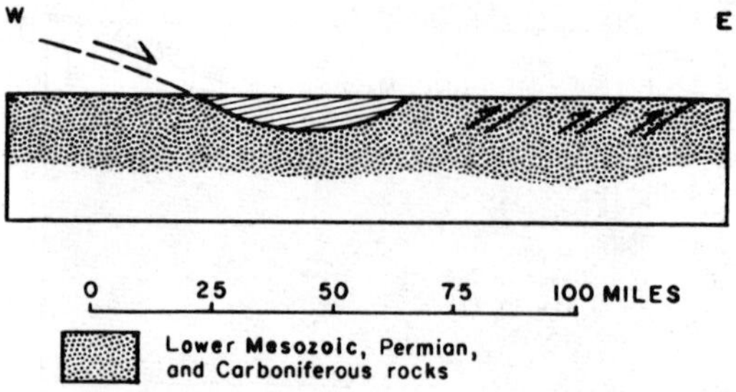

FIG. 21.—Schematic structure section illustrating overthrust origin inferred for area of Paleozoic rocks. Area extends north-northwestward from Ogden, Utah. See Figure 1.

their greater density, by a gravity high. The gravity map of Idaho (Bonini, 1963, pl. 1), however, does not show a gravity high southeast of Pocatello suitable to support this wedge hypothesis.

These two interpretations are based on the assumption that the westward-dipping eastern thrusts and the eastward-dipping western thrusts are connected—if not physically, then at least temporally and genetically. As yet, however, data are inadequate to date accurately the western thrusts and to demonstrate that they are the same ages as the eastern thrusts. Accordingly, a third possibility is that the faults are in fact unrelated, at least in the ways suggested thus far.

PART 3. BLOCK FAULTING

Block faulting, the third and latest chapter in the tectonic evolution of the Idaho-Wyoming thrust belt, began during the Eocene and has continued to the present.

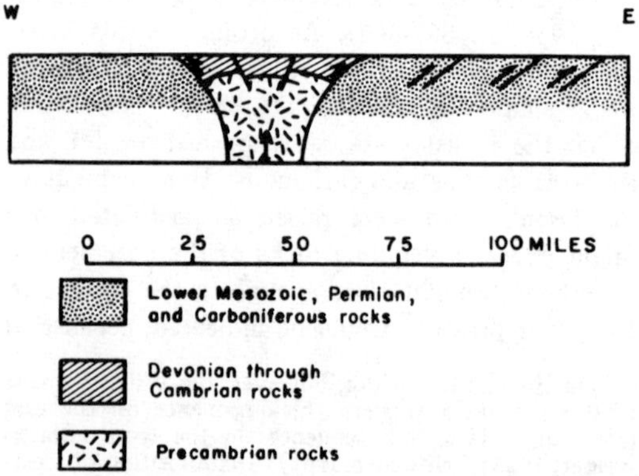

FIG. 22.—Schematic structure section illustrating uplifted wedge origin inferred for area of Paleozoic rocks.

Northeast of Kemmerer, Wyoming, several steeply dipping faults cut the plate of Mesozoic rocks (Fig. 23) that lies above the late Paleocene Hogsback thrust fault (Fig. 19). Some of the high-angle faults also cut the Wasatch Formation of late early Eocene age that unconformably overlaps the rocks above the thrust. Other steep faults, probably of early Eocene age, do not offset the Wasatch.

Eocene block faults have not been recognized in the western part of the region because Eocene rocks, used to date the faults, are absent where needed.

On the west, many block faults can not be dated within narrow limits. In a part of the Portneuf Range within the Bancroft Quadrangle, Idaho, three sets of block faults have been mapped (Fig. 24). The oldest trends northwest and does not cut the Salt Lake Formation; the next set trends northeast and cuts the Salt Lake; and the youngest set trends north to northwest, cuts the Salt Lake, and parallels present mountain fronts. Fossils have not been found in the Salt Lake in this area, and the formation is assumed to be Pliocene as it is at many places nearby.

In the next range to the east, the Bear River Range, similarly trending fault sets have been

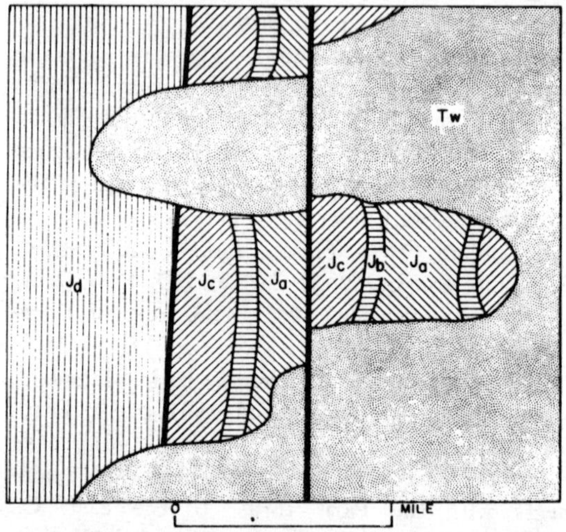

FIG. 23.—Schematic geologic map of part of Fort Hill Quadrangle, Wyoming, showing that Wasatch Formation, Tw, is cut by some steeply dipping faults but not by others. Ja, to Jd, Jurassic units, from oldest to youngest, in plate above Hogsback thrust fault.

mapped (Fig. 25). The oldest set, which trends east and northeast, may be tear faults related to thrusting (Armstrong, 1964).

The next set trends northwest, does not cut the Salt Lake Formation, and is cut by a younger northeast-trending set, which in turn is cut by a younger northwest-trending set that cuts the Salt Lake and parallels the mountain front.

In the western part of the region, the late northwest-trending faults served as conduits along which basalt flows rose during Pliocene to Recent time.

Some block faults have clearly moved more than once. Along one fault Tertiary strata have been offset about 1,000 feet vertically, whereas elsewhere along the same fault a modern alluvial fan has been offset only 50 feet (W. W. Rubey, J. I. Tracey, and S. S. Oriel, unpub. data). Vertical movement along some block faults may have been in opposite directions at different times. This conclusion is suggested by the presence, on the crests and flanks of present ranges, of coarse Tertiary gravel for which there is no possible source with the present topography. These gravels were derived from the west, an area that is now low, at a time when the source was higher than the site of deposition.

Movement on the northwestward-trending block faults was principally vertical. Gem Valley is a graben that separates the Fish Creek and Portneuf Ranges from the ranges on the east (Fig. 26). Gravimetric surveys by Mabey

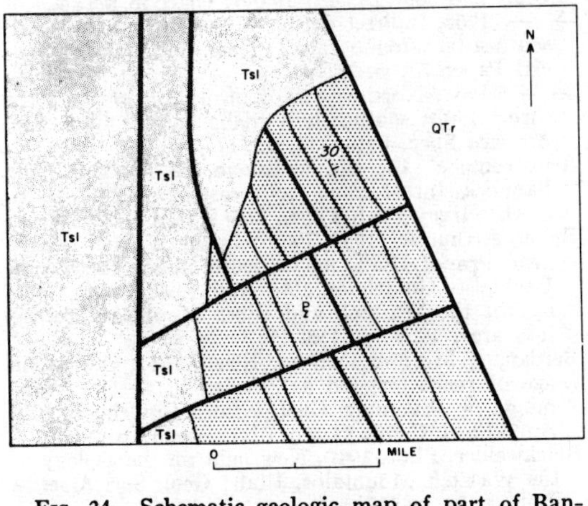

FIG. 24.—Schematic geologic map of part of Bancroft Quadrangle, Idaho, showing sets of faults mapped and their relations to rock units: QTr, Quaternary and uppermost Tertiary rocks; Tsl, Pliocene Salt Lake Formation; Pz, Paleozoic rocks.

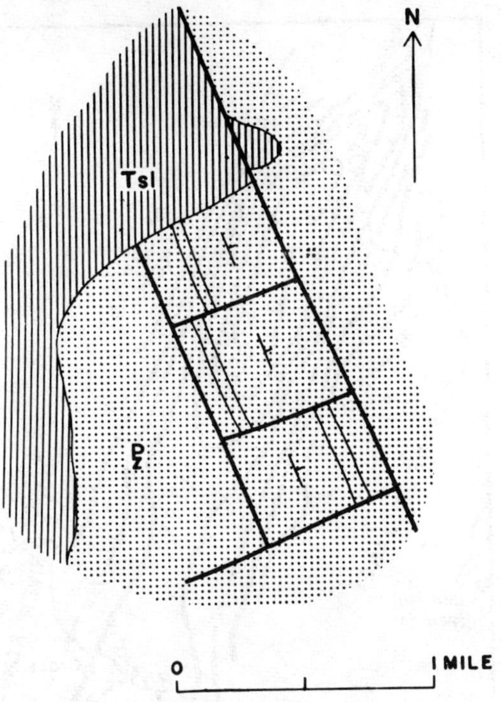

FIG. 25.—Schematic geologic map of part of Soda Springs Quadrangle, Idaho, showing sets of faults mapped and their relations to rock units: Tsl, Pliocene Salt Lake Formation; Pz, Paleozoic rocks.

(Mabey and Armstrong, 1962) reveal marked negative anomalies (Fig. 26) that are interpreted to indicate a relief of 7,000-10,000 feet between the crests of the flanking ranges and the bottom of the Tertiary fill in the valley. This relief is the result of movement on the block faults.

The young age of some block faulting hardly needs emphasis. Fault scarps are abundant, and modern alluvial fans have been cut by Recent faults. Basalt in Gem Valley dated as younger than 27,000 years (Bright, 1963, p. 30) has been cut by faults. Earthquakes continue in the area.

RELATION OF GEOLOGIC HISTORY TO PETROLEUM EXPLORATION

The writers have attempted to show the sequence of geologic events in part of an orogenic belt. Gross generalizations, as well as small geologic features, provide clues for petroleum exploration. Only a few can be mentioned here, for most lie beyond the scope of this paper.

The first broad episode in late Paleozoic and Mesozoic time was progressive eastward shift

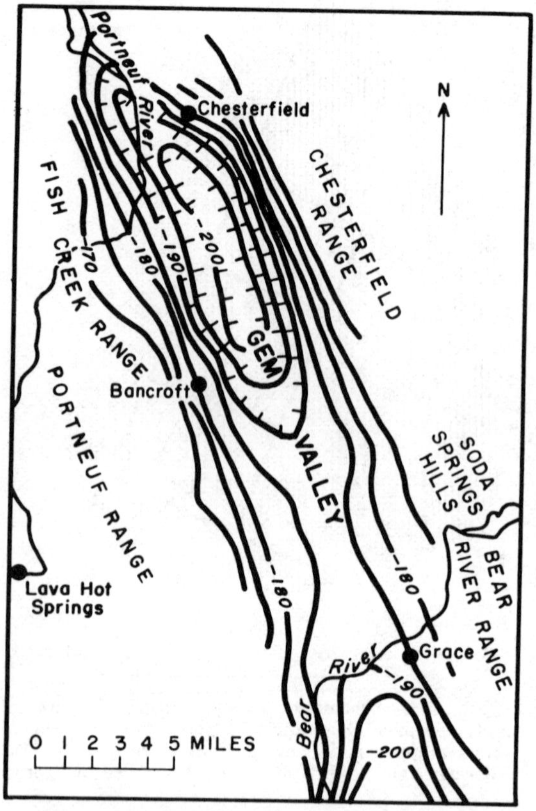

FIG. 26.—Bouguer-anomaly map of Gem Valley, Caribou County, Idaho. (Gravity contour interval 5 milligals.) After Mabey and Armstrong (1962, p. D73).

of a miogeosyncline encroaching on the craton that resulted in conditions favorable for the development of hydrocarbons and stratigraphic traps. Recurrently throughout deposition, both western and eastern sources shed detritus into the miogeosyncline to form eastward-thinning and westward-thinning detrital wedges, respectively. Precise delineation of regional facies changes and of pinch-outs of detrital wedges is critical to exploration.

The second broad episode, which overlapped late stages of the first, produced the folded and thrust belt along the eastern flank of the miogeosyncline; movement along the thrusts, measured in at least tens of miles, was progressively younger eastward. The times of hydrocarbon migration resulting from deformation may also have been progressively younger eastward. Thrust plates cover and conceal oil and gas fields, only a few of which have yet been found. Although some surface and near-surface allochthonous rocks may seem to be unfavorable prospects for drilling, they may conceal distinctly different rocks in underlying autochthonous sequences that are potential oil reservoirs. The structural history of stratigraphic traps is also important, for tilts and rotations accompanying folding during thrusting may have caused hydrocarbons to escape from some and to be concentrated in others.

The latest broad episode involves block faulting that apparently increased in intensity westward and is still active. Some minor dislocations and modifications of hydrocarbon fields are apparent in the east. Block faulting also resulted in the deposition of detrital wedges in Tertiary sequences that have yielded some gas and oil. Some areas in the western part of the region have been so intensely shattered by several sets of block faults that hydrocarbon concentrations accumulated prior to block faulting may have escaped late in the tectonic history. These same sets of faults, however, have provided channels for hydrothermal solutions that locally have deposited base-metal minerals.

REFERENCES CITED

Anderson, A. L., 1928, Portland cement materials near Pocatello, Idaho: Idaho Bur. Mines and Geology Pamph. 28, 14 p.

Armstrong, F. C., 1953, Generalized composite stratigraphic section for the Soda Springs Quadrangle and adjacent areas in southeastern Idaho, *in* Intermtn. Assoc. Petroleum Geologists Guidebook 4th Ann. Field Conf., Guide to the geology of northern Utah and southeastern Idaho: chart in pocket.

——— 1962, Indirect dating of the Paris thrust fault, southeastern Idaho (abs.): Geol. Soc. America Special Paper 73, p. 21.

——— 1965, Northwestward projection of the Paris thrust fault, southeastern Idaho (abs.): Geol. Soc. America Spec. Paper 82, p. 317.

Armstrong, F. C., and Cressman, E. R., 1963, The Bannock thrust zone, southeastern Idaho: U. S. Geol. Survey Prof. Paper 374–J, p. J1–J22.

Baker, Arthur A., 1959, Faults in the Wasatch Range near Provo, Utah, *in* Intermtn. Assoc. Petroleum Geologists Guidebook 10th Ann. Field Conf., Geology of the Wasatch and Uinta Mountains, transition area, 1959: p. 153–158.

Berdan, J. M., and Duncan, Helen, 1955, Ordovician age of rocks mapped as Silurian in western Wyoming, *in* Wyoming Geol. Assoc. Guidebook 10th Ann. Field Conf., Green River basin: p. 48.

Blackwelder, Eliot, 1910, New light on the geology of the Wasatch Mountains, Utah: Geol. Soc. America Bull., v. 21, p. 517–542.

——— 1925, Wasatch Mountains revisited (abs.): Geol. Soc. America Bull., v. 36, p. 132–133.

Bonini, W. E., 1963, Gravity anomalies in Idaho: Idaho Bur. Mines and Geology Pamph. 132, 10 p.

Bright, R. C., 1963, Pleistocene Lakes Thatcher and

Bonneville, southeastern Idaho: Minn. Univ., unpub. Ph.D. thesis, Minneapolis, Minn.

Carr, W. J., and Trimble, D. E., 1961, Upper Paleozoic rocks in the Deep Creek Mountains, Idaho, *in* Short papers in the geologic and hydrologic sciences: U. S. Geol. Survey Prof. Paper 424-C, p. C181–C184.

Chronic, John, and Ferris, C. S., Jr., 1961, Early Paleozoic outlier in southeastern Wyoming, *in* Rocky Mtn. Assoc. Geologists Guidebook 12th Ann. Field Conf., Lower and middle Paleozoic rocks of Colorado: p. 143–146.

Churkin, Michael, Jr., 1962, Facies across Paleozoic miogeosynclinal margin of central Idaho: Am. Assoc. Petroleum Geologists Bull., v. 46, no. 5, p. 569–591.

Coulter, H. W., 1956, Geology of the southeast portion of the Preston Quadrangle, Idaho: Idaho Bur. Mines and Geology Pamph. 107, 48 p.

Crittenden, M. D., Jr., 1959, Mississippian stratigraphy of the central Wasatch and western Uinta Mountains, Utah, *in* Intermtn. Assoc. Petroleum Geologists Guidebook 10th Ann. Field Conf., Geology of the Wasatch and Uinta Mountains, transition area, 1959: p. 63–74.

——— 1961, Magnitude of thrust faulting in northern Utah: U. S. Geol. Survey Prof. Paper 424-D, p. D128–D133.

Dutro, J. T., and Sando, W. J., 1963, New Mississippian formations and faunal zones in Chesterfield Range, Portneuf Quadrangle, southeast Idaho: Am. Assoc. Petroleum Geologists Bull., v. 47, no. 11, p. 1963–1986.

Eardley, A. J., 1944, Geology of the north-central Wasatch Mountains, Utah: Geol. Soc. America Bull., v. 55, p. 819–894.

——— 1960, Phases of orogeny in the fold belt of western Wyoming and southeastern Idaho, *in* Wyoming Geol. Assoc. Guidebook 15th Ann. Field Conf., Overthrust belt of southwestern Wyoming and adjacent areas: p. 37–40.

——— and others, 1944, Hoback-Gros Ventre-Teton [Range, Wyo.], field conference: Mich. Univ., geol. map, tectonic map, with sections, 2 sheets.

Gilluly, James, 1932, Geology and ore deposits of the Stockton and Fairfield Quadrangles, Utah: U. S. Geol. Survey Prof. Paper 173, 171 p.

Haun, J. D., and Barlow, J. A., Jr., 1962, Lower Cretaceous stratigraphy of Wyoming, *in* Wyoming Geol. Assoc. Guidebook 17th Ann. Field Conf., Early Cretaceous rocks of Wyoming and adjacent areas: p. 15–22.

Hintze, L. F., 1960, Ordovician tectonics of western Wyoming and vicinity, *in* Wyoming Geol. Assoc. Guidebook 15th Ann. Field Conf., Overthrust belt of southwestern Wyoming and adjacent areas: p. 110–115.

Holmes, Arthur, 1959, A revised geological time-scale: Edinburgh Geol. Soc. Trans., v. 17, pt. 3, p. 183–216.

Lochman-Balk, Christina, 1960, The Cambrian section of western Wyoming, *in* Wyoming Geol. Assoc. Guidebook 15th Ann. Field Conf., Overthrust belt of southwestern Wyoming and adjacent areas: p. 98–108.

Love, J. D., McGrew, P. O., and Thomas, H. D., 1963, Relationship of latest Cretaceous and Tertiary deposition and deformation to oil and gas in Wyoming, *in* The backbone of the Americas—Tectonic history from Pole to Pole, a symposium: Am. Assoc. Petroleum Geologists Mem. 2, p. 196–208.

Ludlum, J. C., 1943, Structure and stratigraphy of part of the Bannock Range, Idaho: Geol. Soc. America Bull., v. 54, p. 973–986.

Mabey, D. R., and Armstrong, F. C., 1962, Gravity and magnetic anomalies in Gem Valley, Caribou County, Idaho *in* Short papers in geology, hydrology, and topography: U. S. Geol. Survey Prof. Paper 450-D, p. D73–D75.

Mansfield, G. R., 1927, Geography, geology, and mineral resources of part of southeastern Idaho: U. S. Geol. Survey Prof. Paper 152, 453 p.

——— 1929, Geography, geology, and mineral resources of the Portneuf Quadrangle, Idaho: U. S. Geol. Survey Bull. 803, 110 p.

McKee, E. D., Oriel, S. S., Swanson, V. E., MacLachlan, M. E., MacLachlan, J. C., Ketner, K. B., Goldsmith, J. W., Bell, R. Y., Jameson, D. J., 1956, Paleotectonic maps of the Jurassic system, *with a separate section on* Paleogeography by Ralph W. Imlay: U. S. Geol. Survey Misc. Geol. Inv. Map I-175, 6 p.

McKee, E. D., Oriel, S. S., Ketner, K. B., MacLachlan, M. E., Goldsmith, J. W., MacLachlan, J. C., Mudge, M. R., 1959, Paleotectonic maps of the Triassic system: U. S. Geol. Survey Misc. Geol. Inv. Map I-300.

McKee, E. D., Oriel, S. S., and others, in press, Paleotectonic maps of the Permian system: U. S. Geol. Survey Misc. Geol. Inv. Map I-450.

McKelvey, V. E., and others, 1959, The Phosphoria, Park City, and Shedhorn Formations in the western phosphate field: U. S. Geol. Survey Prof. Paper 313-A, p. 1–47.

Moritz, C. A., 1953, Summary of the Cretaceous stratigraphy of southeastern Idaho and western Wyoming, *in* Intermtn. Assoc. Petroleum Geologists Guidebook 4th Ann. Field Conf.: p. 63–72.

Murdock, C. N., 1961, Geology of the Weston Canyon area, Bannock Range, Idaho: Utah State Univ., unpub. Master's thesis, Logan, Utah.

Oriel, S. S., 1962, Main body of Wasatch Formation near La Barge, Wyoming: Am. Assoc. Petroleum Geologists Bull., v. 46, No. 12, p. 2161–2173.

——— 1965, Brigham, Langston, and Ute Formations in Portneuf Range, southeastern Idaho (abs.): Geol. Soc. America Spec. Paper 82, p. 341.

Osmond, John C., 1962, Stratigraphy of Devonian Sevy Dolomite in Utah and Nevada: Am. Assoc. Petroleum Geologists Bull., v. 46, no. 11, p. 2033–2056.

Peale, A. C., 1879, Report on the geology of the Green River district: U. S. Geol. and Geog. Survey Terr. 11th Ann. Rept., p. 511–646.

Plumley, W. J., 1948, Black Hills terrace gravels; a study in sediment transport: Jour. Geology, v. 56, no. 6, p. 526–577.

Richards, R. W., and Mansfield, G. R., 1912, The Bannock overthrust; a major fault in southeastern Idaho and northeastern Utah: Jour. Geology, v. 20, p. 681–707.

Richardson, G. B., 1941, Geology and mineral resources of the Randolph Quadrangle, Utah-Wyoming: U. S. Geol. Survey Bull. 923, 54 p.

Rigby, J. K., 1963, Devonian System of Utah: Utah Geol. and Mineralog. Survey Bull. 54, p. 75–88.

Roberts, R. J., Hotz, P. E., Gilluly, James, and Ferguson, H. G., 1958, Paleozoic rocks of north-central Nevada: Am. Assoc. Petroleum Geologists Bull., v. 42, p. 2813–2857.

Roberts, R. J., and Thomasson, M. R., 1964, Comparison of late Paleozoic depositional history of northern Nevada and central Idaho, *in* Short papers in geology and hydrology: U. S. Geol. Survey Prof. Paper 475-D, p. D1-D6.

Ross, C. P., 1937, Geology and ore deposits of the Bayhorse region, Custer County, Idaho: U. S. Geol. Survey Bull. 877, 161 p.

——— 1962, Paleozoic seas, central Idaho: Geol. Soc. America Bull., v. 73, p. 769-794.

Ross, R. J., Jr., 1964, Relations of Middle Ordovician time and rock units in Basin Ranges, western United States: Am. Assoc. Petroleum Geologists Bull., v. 48, no. 9, p. 1526-1554.

Rubey, W. W., 1955, Early structural history of the overthrust belt of western Wyoming and adjacent states, *in* Wyoming Geol. Assoc. Guidebook 10th Ann. Field Conf., Green River basin, 1955: p. 125-126.

——— 1958, Geologic map of the Bedford Quadrangle, Wyoming: U. S. Geol. Survey Geol. Quad. Map GQ-109.

Rubey, W. W., and Hubbert, M. K., 1959, Overthrust belt in geosynclinal area of western Wyoming in light of fluid-pressure hypothesis, Pt. 2 *of* Role of fluid pressure in mechanics of overthrust faulting: Geol. Soc. America Bull., v. 70, p. 167-206.

Rubey, W. W., Oriel, S. S., and Tracey, J. I, 1961, Age of the Evanston Formation, western Wyoming *in* Short papers in the geologic and hydrologic sciences: U. S. Geol. Survey Prof. Paper 424-B, p. B153-B154.

Rush, R. W., 1963, Silurian strata in Utah: Utah Geol. and Mineralog. Survey Bull. 54, p. 63-73.

Sando, W. J., Dutro, J. T., Jr., and Gere, W. C., 1959, Brazer Dolomite (Mississippian), Randolph Quadrangle, northeast Utah: Am. Assoc. Petroleum Geologists Bull., v. 43, p. 2741-2769.

Schultz, A. R., 1914, Geology and geography of a portion of Lincoln County, Wyoming: U. S. Geol. Survey Bull. 543, 141 p.

Stose, G. W., 1932, Geologic map of the United States: U. S. Geol. Survey.

Strickland, J. W., 1956, Mississippian stratigraphy, western Wyoming *in* Wyoming Geol. Assoc. Guidebook 11th Ann. Field Conf., Jackson Hole: p. 51-57.

Trimble, D. E., and Carr, W. J., 1962, Preliminary report on the stratigraphy of the Paleozoic rocks southwest of Pocatello, Idaho: U. S. Geol. Survey open file rept., 18 p.

Umpleby, J. B., Westgate, L. G., and Ross, C. P., 1930, Geology and ore depositis of the Wood River region, Idaho: U. S. Geol. Survey Bull. 814, 250 p.

Veatch, A. C., 1907, Geography and geology of a portion of southwestern Wyoming: U. S. Geol. Survey Prof. Paper 56, 178 p.

Weimer, R. J., 1961, Uppermost Cretaceous rocks in central and southern Wyoming and northwest Colorado, *in* Wyoming Geol. Assoc. Guidebook 16th. Ann. Field Conf., Late Cretaceous rocks Wyoming and adjacent areas: p. 17-28.

Williams, J. Stewart, 1948, Geology of the Paleozoic rocks, Logan Quadrangle, Utah: Geol. Soc. America Bull., v. 59, p. 1121-1164.

——— 1962, Pennsylvanian System in central and northern Rocky Mountains, *in* Branson, C.C., ed., Pennsylvanian System in the United States: Tulsa, Okla., Am. Assoc. Petroleum Geologists, p. 159-187.

Williams, J. Stewart, and Taylor, M. E., 1964, The Lower Devonian Water Canyon Formation of northern Utah, *in* Contributions to geology: Wyo. Univ., v. 3, no. 2, p. 38-53.

BULLETIN OF THE AMERICAN ASSOCIATION OF PETROLEUM GEOLOGISTS
VOL. 50, NO. 12 (DECEMBER, 1966), PP. 2612-2621

DISCUSSION

TIME OF THRUSTING IN IDAHO-WYOMING THRUST BELT—DISCUSSION[1]

ERIC W. MOUNTJOY[2]
Montreal, Quebec

The American Association of Petroleum Geologists and the editors and authors of the recent issue concerning the Rocky Mountain Sedimentary Basins (A.A.P.G. *Bulletin,* v. 49, no. 11) are to be commended for publishing the papers of this symposium so soon after they were presented orally.

One of the papers of considerable interest is the excellent summary of the geology and tectonics of the Idaho-Wyoming thrust belt by Armstrong and Oriel (1965). Armstrong and Oriel have presented some important data concerning the dating and temporal relationships of the thrust-belt faults in this region in a clear and well-illustrated paper based not only on their own work but also on data summarized from several sources, especially the work of Rubey (1958; Rubey and Hubbert, 1959). They concluded mainly on stratigraphic evidence that thrusting began in latest Jurassic in the west and ended in the east about early Eocene time.

The purpose of this note is to consider carefully the evidence used to date thrusting in this region. Armstrong and Oriel (p. 1857) reported that "Few of the thrust faults can be dated within narrow limits by usual geologic methods." Nevertheless, they stated that three of the thrusts, Prospect, Absaroka, and Paris thrusts, can be dated. There is no doubt of the age of the Prospect thrust, the most easterly major thrust, because the trace of the thrust fault is overlain by late early Eocene Wasatch strata and cuts the Paleocene and earliest Eocene Hoback Formation. However, the evidence for the Absaroka and Paris thrusts is not nearly so convincing and this writer suggests that they can not be dated other than generally.

Armstrong and Oriel (1965, p. 1857–1859) concluded that the Evanston Formation of latest Cretaceous and Paleocene age was deposited after most of the movement along the Absaroka thrust, and the movement is thereby dated as latest Cretaceous. The significant relationships are illustrated in Figure 17 (p. 1860). In addition they state that ". . . the Evanston has not been found directly overlying the fault. . . ." (p. 1857). However, as shown in their Figure 17, the actual relationships are covered by the Eocene Wasatch Formation so that other interpretations are possible without more critical evidence to the contrary. Armstrong and Oriel suggested that only the basal beds of the Evanston Formation have been cut by the Absaroka thrust, but it could equally well cut through *all* of the Evanston Formation beneath the cover of Eocene Wasatch Formation. The unconformable base of the Evanston Formation need not be related to the thrusting. Consequently, the Absaroka thrust might be as young as early Eocene and may have formed at the *same* time or nearly the same time as the Prospect thrust.

The Paris thrust has been dated as latest Jurassic and earliest Cretaceous on the basis of the presence of coarsely conglomeratic redbeds of the Lower Cretaceous Gannett Group which contain Paleozoic pebbles of a non-resistant character. Armstrong and Oriel (p. 1859) infer that the source of the Gannett Group pebbles of the Gannett Hills was about 25–30 miles west of their present location and suggested that these pebbles were derived from erosion of Paleozoic rocks of the Paris thrust sheet. A more complete discussion is given in the report by Armstrong and Cressman (1963). However, again the evidence is not conclusive as these conglomerates could equally well have been derived before thrusting from some other uplifted source areas presently buried beneath the Meade or Paris thrust sheet. If the sequence of thrust development suggested by Armstrong and Oriel is correct, then surely

[1] Discussion received, December 6, 1965; accepted, May 13, 1966.

[2] Assistant Professor, Department of Geological Sciences, McGill University.

this conglomeratic unit should also be found in the Meade thrust sheet. Armstrong and Cressman (1963, p. J13) earlier concluded that the evidence regarding the Paris thrust sheet being the source of the conglomerate of the Gannett Group was not conclusive. Again the evidence is not compelling that the Paris thrust is Early Cretaceous and latest Jurassic; it could well be much younger, possibly as young as early Eocene. Consequently, much of the thrusting in the Idaho-Wyoming thrust belt may have taken place during the earliest Eocene or slightly earlier. Moreover, the sequence of thrusting from west to east can not be positively demonstrated, although it appears to be a reasonable inference from the present structural and stratigraphic information. Also applying the term Laramide to deformations in the thrust belt may not be so futile as suggested by Armstrong and Oriel (p. 1860).

Having worked in another well-known thrust belt for several years, the central part of the Alberta Rocky Mountains, the writer is interested in methods used for dating times of thrusting. In the Alberta Rockies near Jasper, the only evidence available is that the youngest beds (unnamed unit above the Brazeau Formation) cut by thrust faults or deformed in associated folds are of Late Cretaceous or Paleocene age. Geometric and fold relations suggest an order of development from west to east (Fox, 1959; Mountjoy, 1960, 1962) which obviously indicates that all of the thrust sheets did not form at the same time. Two scraps of information help date the thrusting of the southern Alberta Rockies. First, the upper limit of thrusting can be reasonably well dated near the International Boundary in the Flathead valley south of Crowsnest Pass on the basis of late Eocene and early Oligocene Kishenehn strata (Russell, 1954; Price, 1962). These strata were derived from progresive denudation of a local area of high relief and deposited in a basin bordered in part by normal faults which offset the Lewis thrust (Price, 1959, 1962). The second bit of information comes from the eastern Foothills where thrust faults appear to be covered by undeformed Paleocene Paskapoo and Porcupine Hills Formations (Bossart, 1957, p. 51; Fox, 1959, p. 996). However, this may be a situation in which thrust faults do not reach the present erosion surface as happens elsewhere in the Rockies (Fox, 1959) and also in the Appalachians (Gwinn, 1964). The stratigraphic evidence from western Canada in general indicates that the Canadian Rocky Mountain uplift and deformation culminated about the mid-point of the Eocene (Taylor *et al.*, 1964). By this statement the writer does not wish to imply that all the folding and thrusting of the Alberta Rockies took place simultaneously in mid-Eocene time, but that most of the thrust movements and uplifts appear to have occurred at this time. The available stratigraphic evidence does not rule out the possibility of somewhat earlier (especially Paleocene) uplift and thrust movements. It is interesting that the most accurate dating of thrusting in both the Alberta and Idaho-Wyoming thrust belts is approximately the same—Eocene.

The writer is not personally familiar with the structure of the Idaho-Wyoming thrust belt. This discussion has been written with the anticipation that Armstrong and Oriel will provide further evidence for time of thrusting and a more complete discussion of other ways in which the basic information may be interpreted, which they may not have had space to do in their excellent summary paper.

References

Armstrong, F. C., and E. R. Cressman, 1963, The Bannock thrust zone, southeastern Idaho: U.S. Geol. Survey Prof. Paper 374–J, 22 p.

——— and S. S. Oriel, 1965, Tectonic development of Idaho-Wyoming thrust belt: Am. Assoc. Petroleum Geologists Bull., v. 49, p. 1847–1866.

Bossart, D. O., 1957, Relationship of the Porcupine Hills to early Laramide movements: Alta. Soc. Petroleum Geologists, Guide Book, Seventh Ann. Field Conf., p. 46–51.

Fox, F. G., 1959, Structure and accumulation of hydrocarbons in southern Foothills, Alberta, Canada: Am. Assoc. Petroleum Geologists Bull., v. 48, p. 992–1025.

Gwinn, V. E., 1964, Thin-skinned tectonics in the Plateau and northwestern Valley and Ridge provinces of the central Appalachians: Geol. Soc. America Bull., v. 75, p. 863–900.

Mountjoy, E. W., 1960, Structure and stratigraphy of the Miette and adjacent areas, eastern Jasper National Park, Alberta: Univ. Toronto, unpub. Ph.D. thesis.

——— 1962, Mount Robson (southeast) map-area, Rocky Mountains of Alberta and British Columbia: Geol. Survey Canada Paper 61–31, 144 p.

——— in press, Significance of rotated folds and thrust faults, Alberta Rocky Mountains.

Price, R. A., 1959, Geology, Flathead, British Columbia and Alberta: Geol. Survey Canada Map 1-1959.

——— 1962, Fernie map-area, east half, Alberta and British Columbia: Geol. Survey Canada Paper 61–24, 65 p.

Rubey, W. W., 1958, Geology of Bedford quadrangle,

Wyoming: U.S. Geol. Survey Quad. Map GQ–109.
——— and M. K. Hubbert, 1959, Role of fluid pressure in mechanics of overthrust faulting, II. Overthrust belt in geosynclinal area of western Wyoming in light of fluid-pressure hypothesis: Geol. Soc. America Bull., v. 70, p. 167–206.
Russell, L. S., 1954, The Eocene-Oligocene transition as a time of major orogeny in western North America: Roy. Soc. Canada Trans., ser. 3, v. 48, sec. 4, p. 65–69.
Taylor, R. S., W. H. Mathews, and W. O. Kupsch, 1964, Chapter 13, Tertiary, geological history of western Canada: Alta. Soc. Petroleum Geologists, p. 190–194.

TIMES OF THRUSTING IN IDAHO-WYOMING THRUST BELT: REPLY[1]

STEVEN S. ORIEL[2] AND FRANK C. ARMSTRONG[2]
Denver, Colorado, and Spokane, Washington

Introduction

Professor Mountjoy's kind discussion, on the preceding pages, of our summary paper (Armstrong and Oriel, 1965) on the tectonic development of the Idaho-Wyoming thrust belt is most welcome for its own sake and because it has led the editors of the *Bulletin* to invite us to reply. We welcome the opportunity for more thorough discussion of the times and durations of thrust movements than was possible in the original paper.

Mountjoy points out, quite correctly, that in our 1965 paper unequivocal evidence is presented for dating the frontal Prospect thrust as Eocene but not for the greater age of thrusts farther west; he goes on to suggest that perhaps all the thrusts are Eocene and to note that thrusting in the Alberta Rockies also is Eocene. The plain, though no doubt unintended, implication of his suggestion is that folding and thrusting occurred during a very brief interval simultaneously throughout the Cordillera. Such a view is at variance with the experience of most modern structural geologists. It does not fit the evidence available from the Idaho-Wyoming thrust belt, which indicates that principal movements were not simultaneous on all the thrusts, and that some western thrusts have been inactive, or nearly so, since well before Eocene time. Before data supporting this firm statement are offered, it may be worthwhile to consider some general problems in dating thrust faults.

[1] Reply received, March 26, 1966; accepted, May 10, 1966. Publication authorized by the director of the United States Geological Survey.

[2] United States Geological Survey. We are indebted to our colleagues W. W. Rubey, J. I. Tracey, Jr., and L. B. Platt for critical review of this manuscript and for their permission to make use of unpublished information. The manuscript also profited greatly from suggestions by G. D. Robinson.

The origin of working ideas is commonly difficult to recall. The ensuing discussion no doubt makes use of many ideas first enunciated by our colleagues as well as a few of our own. Our debt to our colleagues, particularly Rubey and Tracey, is therefore considerable.

Problems in Dating Thrust Faults

paucity of precise dates

Thrust movements are difficult to date precisely in most of the world's thrust belts. A reason is that well-dated bracketing strata are preserved in very few places within orogenically active mountain systems. Along the mountain front and beyond, however, deposition predominates over erosion. Synorogenic and post-orogenic debris preserved along the mountain front may provide the critical stratigraphic record of major geologic events, and thus facilitate dating of frontal faults. Precise dates become far more scarce with increasing distance from the thrust-belt front. Of course, a precise date for a frontal fault does not establish the age for an entire mountain system.

Accordingly, coarse detrital units are being used increasingly to date tectonic events, but proving that a sediment is the product of a specific tectonic event is difficult. Does a particular conglomerate bed record movement along one thrust fault or along another or some other orogenic event?

Paucity of precise dates raises many questions, a few of which are discussed here.

duration of thrust movement

Discussions of times of thrusting commonly imply that the movement is brief. Yet movements on thrust faults that transport enormous plates of rock tens of miles laterally and that result in stratigraphic throws of several miles can not take place instantaneously. One estimate of the possible average rate of movement of a thrust sheet, based on ages of the youngest rocks exposed below successive thrust plates (Rubey and Hubbert, 1959, p. 190–191), is about a mile each million years (Rubey and Hubbert, 1959, p. 196) or 1.6 millimeters per year; the estimate may be low, for it envisions exceedingly slow, successive, eastward movements of the several thrust plates and is based on conservative estimates of minimum horizontal displacement. Far more rapid movement, about an inch a year or 16 miles per million years, has been measured briefly along an active thrust fault that has cut oil-well casings in the Buena Vista oil field of California (McMasters, 1943, p. 518; Gilluly, 1949, p. 564); this figure may be excessive over a long span of time. In the absence of better information, let us assume that the rate of thrust movement is between the two figures and is several miles per million years.

The time involved in the movement of a thrust sheet such as the Absaroka, then, is significant. If the sheet moved only 10–15 miles (Rubey and Hubbert, 1959, p. 187), approximately 3–5 million years may have been required; if the sheet moved as far as 30 miles, then as much as 10 million years may have elapsed during transport, a time comparable with the estimated 11- or 12-million-year duration of Paleocene time (Geol. Soc. London, 1964, p. 260).

If the estimated rates of thrusting previously mentioned are not greatly in error, then the span of time required for movement exceeds intervals of geologic time that we are now able to resolve on the basis of stratigraphic paleontology.

RECURRENT THRUST MOVEMENTS

Recurrent movements may seriously complicate the dating of thrust movements. Once a glide surface or thrust fault develops, continued application of forces which produce other thrust faults can, under some circumstances, result in renewed movement along the first surface of rupture. Not only do large thrust displacements require significant spans of time, but movement may have occurred during several such spans.

Thus, recognition that two distinct conglomerate units of different ages may have been derived from the upper plate of the Paris fault has led to the suspicion that this thrust plate has moved more than once. Some of the fragments in both conglomerate units are of rock types found only in the Paris thrust plate.

MOVEMENT OF PARTS OF SINGLE THRUST PLATE

Still another point not discussed in the original paper is the possibility that different parts of a single thrust sheet have moved at different times. Localities at which precise dates for fault movement are available are far too few to establish that a thrust sheet moved as a unit. A thrust plate as large as the Absaroka, which has been traced about 200 miles, could have moved by small increments at different times in different places without failing and breaking into separate plates.

The point is not entirely hypothetical, for it may explain apparent conflicts in time of thrusting of different parts of the Hogsback fault. Observations near La Barge, Wyoming, suggest one date for movement. Strata that overlie the fault here are believed to be involved in the faulting farther north (Michael, 1960, Fig. 2). One possible explanation is that different parts of the thrust plate moved at different times. Other possible explanations are that the critical strata at the two places may not be identical, or that structure at one of the localities may be more complex than now believed.

PRECISION IN DISCUSSING DATES OF THRUSTING

The long time probably required for movement of large thrust sheets imposes the need for precision in discussions of times of thrusting. By "time" or "date" of thrusting do we mean when thrusting began, when most of the movement took place, or when thrusting ceased?

Possible recurrent movement must also be considered in discussions of thrust dates. Times of recurrent movement were specifically excluded from diagrams in the original paper (Armstrong and Oriel, 1965, Figs. 19, 20) illustrating temporal relations among the thrust faults. Precise dates are insufficient to enable us to decipher the entire life history of any one thrust fault.

The absence of proof that all parts of a large thrust plate moved synchronously requires that a hard-earned precise date be applied only to a relatively small area. A precise date for Absaroka fault movement near Kemmerer, Wyoming, may not apply to the time of movement on the Absaroka 150 miles away along the edge of the Snake River Plain.

These considerations bear on the discussions to follow, and great care in analysis is necessary if we hope to gain an understanding of a complex assemblage of orogenic events. Now we can return to data that bear on faults in the Idaho-Wyoming thrust belt and answer points raised in Mountjoy's discussion.

Idaho-Wyoming Thrust Belt

Hogsback Thrust Fault

In illustrating the diversity of information used to date thrust faults, we ignored the Hogsback fault in the original paper. The basis for dating the Hogsback fault is the same as that for dating the Prospect fault.

On Hogsback Ridge near La Barge, Wyoming, movement along the Hogsback thrust ended before latest Paleocene time. The Hogsback fault lies west of and above the easternmost thrust known at this latitude, the La Barge or Hilliard, which was active in early Eocene time (Murray, 1960, p. 182–183).[3] The westward-dipping Hogsback fault, Cambrian rocks above the fault, and Upper Cretaceous rocks below it are overlain with angular unconformity by nearly horizontal beds along the crest of Hogsback Ridge (Oriel, 1963). The horizontal beds are distinctive and are assigned to the Chappo Member of the Wasatch Formation (Oriel, 1962), in which vertebrate remains of latest Paleocene (Clarkforkian) age have been found (Gazin, 1942, 1956a). The Hogsback fault ceased to move before the upper Paleocene strata were deposited. Major movement along the Hogsback fault at this locality, therefore, took place before latest Paleocene time. Even the very last stage of movement can not have been as recent as Eocene.

Absaroka Thrust Fault

The relations of the Absaroka fault, whose age Mountjoy questions, are similar near Kemmerer, Wyoming, to those of the Hogsback fault, although proof is more difficult. Critical evidence here is the relation of the Lazeart syncline, a part of which is shown on Figure 17 of Armstrong and Oriel (1965), to the Absaroka thrust and the age and areal distribution of the Evanston Formation relative to these two features.

The Lazeart syncline (Veatch, 1907, Pls. 3, 4) is a long, slightly sinuous, asymmetric chevron syncline with a steep west limb and a gentle east limb; at places along the west limb beds are overturned toward the east and consequently dip westward. The trace of the axial plane of the syncline roughly parallels the trace of the Absaroka fault for about 70 miles and lies 1–4 miles east of it (Veatch, 1907, Pl. 3). Fractures of the west limb of the syncline extend in places for some distance east of the Absaroka fault (Veatch, 1907, p. 110). Strata above the fault are folded similarly in both extent and degree of overturning and parallel with those below the fault. These relations suggest that the syncline and the thrust fault were formed by the same forces and at the same time.

The youngest unit east of the Absaroka that is involved in the gross folding of the syncline and that was presumably cut by the fault is the Adaville Formation, whose upper part is of Late Cretaceous (late Montana or early Lance) age.

The Evanston Formation is divisible into a lower and an upper part. The lower part consists of conglomerate and sandstone with some interbedded mudstone; it contains fossils of latest Cretaceous (late Lance, Hell Creek) age (Rubey *et al.*, 1961, p. B153). The upper part of the Evanston consists mainly of siltstone, mudstone, and sandstone; it contains fossils of early late Paleocene (Tiffany) age (Gazin, 1956b). Basal strata of the Evanston dip moderately westward but the dips decrease upward in the sequence; the upper beds of the formation are nearly flat.

The lower (Cretaceous) part of the Evanston lies with marked angular unconformity on upturned and eroded Adaville and older Cretaceous beds in the west limb of the Lazeart syncline about 9 miles north-northwest of Kemmerer and 1 mile northeast of Hams Fork (Veatch, 1907, Pl. 3). Farther south, however, basal beds of the Evanston lie parallel with strata in the east limb of the syncline along the east flank of the Fossil basin. Basal Evanston strata have been folded at one locality along the axis of the Lazeart syncline

[3] The La Barge or Hilliard fault cuts rocks that Murray assigned to the Evanston Formation. The upper part of Murray's Evanston is here assigned to the Chappo Member of the Wasatch Formation of latest Paleocene and earliest Eocene age.

on the southwest side of Hams Fork. At another place, 4½ miles north and along the trace of the Absaroka, basal Evanston beds are slightly overridden by the fault. Basal Evanston strata overlie units both above and below the Absaroka fault (Armstrong and Oriel, 1965, Fig. 17), although they have not been observed to overlie the fault trace itself; moreover, the linear belt of basal Evanston is not offset by the Absaroka fault.

The upper (Paleocene) part of the Evanston, in contrast, is mainly flat and only gently folded locally. In holes drilled through the Absaroka fault neither lower nor upper parts of the Evanston have been found beneath the fault. No faults have been observed during mapping by Rubey, Oriel, and Tracey in the upper part of the Evanston where, on the basis of subsurface information, it is believed to overlie the buried trace of the Absaroka fault about 6½ miles west of Kemmerer.

The foregoing relations are consistent with the inference that folding of the Lazeart syncline and movement along the Absaroka fault were essentially contemporaneous. The Absaroka plate near Kemmerer reached essentially its present site as the syncline formed and before the basal part of the Evanston was deposited. Major movement on the Absaroka thrust is thus post-Adaville and prebasal Evanston, or late Late Cretaceous.

Minor involvement of basal Evanston beds in the deformation indicates that a small orogenic pulse occurred later. This pulse involved lateral transport of only a few hundred or thousand feet, in contrast to the probable 15 miles or so of movement late in Cretaceous time. The absence of faulting in the upper part of the Evanston where it covers the Absaroka indicates that movement on the fault had ceased before deposition of these beds and that the late pulse took place between latest Cretaceous and early late Paleocene time.

If the inference that major movement on the Absaroka thrust and folding of the Lazeart syncline took place at the same time is not accepted, then perhaps most of the Absaroka movement took place after deposition of the basal Evanston. This alternative interpretation, however, calls for such special and fortuitous geologic events as to render it unlikely.

According to this alternative, the parallelism of folds above and below the Absaroka fault may reflect long persistence of a single stress field. The part of the basal Evanston now resting on the upper plate of the fault would have been transported piggyback to its present position. The amount of lateral transport must have been sufficient to move a western belt of basal Evanston (above the fault) just exactly the right distance to be aligned with an eastern belt of basal Evanston (below the fault). The credibility of this alternative is strained still further by absence of basal Evanston strata along most exposures of the Absaroka fault trace and in boreholes through the fault. This absence can be explained, under the assumption that basal Evanston deposits are older than the thrusting, only by spotty and patchy preservation of these beds. Thus, a nearly precise alignment of erratic patches of basal Evanston requires an even greater coincidence. The problems raised by the alternative make the first interpretation more palatable.

PARIS THRUST FAULT

The Paris fault lies considerably west of the thrust belt front and is more difficult to date precisely. The youngest rocks cut by the Paris fault in southeastern Idaho, where we have studied the relations in the field, are Lower Triassic. The oldest rocks directly above the fault are not Eocene as previously reported (Mansfield, 1927, Pl. 9), but more likely Pliocene where we have examined them; that is, we assign to the Salt Lake Formation patches of strata previously mapped as Wasatch Formation. The age span for Paris fault movement by the bracket method—Early Triassic to Pliocene—is so long as to be virtually meaningless.

However, if conglomerates in the Lower Cretaceous Gannett Group are the product of uplift and erosion related to movement along the Paris fault, they provide a means for dating some of this movement more closely. Mountjoy recognizes that evidence for dating movement on the Paris fault is not conclusive and suggests that Gannett conglomerates could have been derived from some other uplifted source areas now buried beneath the Meade or Paris thrust sheets. Pebbles in the conglomerates probably were not derived from uplifted source areas now concealed by the Meade and Paris thrust sheets because exposed rocks beneath both thrust sheets are thrust plates of uppermost Paleozoic and Mesozoic units. No middle to lower Paleozoic units are evident there now which would have provided a source for lithologic types recognized in the Gannett pebbles. These lithologic types are exposed only above and west of the Paris fault. The presence of an unknown source for these pebbles beneath the Paris or Meade faults is unsupported by present knowledge of the structure and is not required by any fact known to us.

That Gannett conglomerates record early movement on the Paris thrust may be unconvincing if considered alone, but other considerations lend support to this conclusion. First, no evidence has been found of pre-thrust block faulting. All of the uplift-bounding normal faults that we have been able to date formed after thrusting. Second, the Idaho-Wyoming thrust belt is characterized by an eastward progression of belts of coarse detritus. Conglomerates are thick, extensive, and rather coarse in units ranging from Upper Triassic to lower Eocene. The westernmost known occurrence and coarsest facies of each unit lie in a belt east of the coarsest facies of the next older unit.

A tenable inference that may be drawn is that the detrital units record tectonic events and that the progressively more eastward positions of successively younger coarse detrital units record progressively more eastward movements on the faults. This inference is supported by evidence of eastward progression in the development and eventual destruction of the Cordilleran miogeosyncline.

Additional support for the greater antiquity of the Paris fault is found along tectonic strike on the south in Utah, where Spieker (1946, p. 149–152) and Eardley (1944, p. 859–860, 865, Fig. 2) recognized evidence of a Late Jurassic or Early Cretaceous orogeny that involved folding and thrusting; neither they nor other geologists have recognized Eocene thrusting in that area.

Because the Paris thrust may have moved more than once (Armstrong and Cressman, 1963, p. J14), we have discussed only first movement on this fault in contrast to last or major movement on other faults. Lack of upper Mesozoic and lower Cenozoic strata in the area prevents accurate dating of major and last movements. Recurrent movement is suggested by the apparent absence of pebbles of distinctive, resistant Cambrian and upper Precambrian quartzites in the Lower Cretaceous Gannett Group and their abundance in the uppermost Cretaceous basal part of the Evanston Formation farther east. Although the middle part, stratigraphically, of the Paris thrust plate was exposed during Gannett deposition, the lower part of it apparently was not (Armstrong and Cressman, 1963, p. J11), for the resistant rock types are not preserved. No source other than the lower part of the Paris thrust plate in the Bear River Range is known for the Cambrian and upper Precambrian quartzite pebbles found in the Evanston Formation near Kemmerer. Thus, the sole of the Paris thrust plate had not broken out to the land surface during Gannett deposition, but the front of the plate composed of younger formations (Rubey and Hubbert, 1959, Fig. 9, p. 194) had; the basal part of the plate was exposed during Evanston deposition. The Evanston pebbles probably record recurrent movement along the Paris fault, or they may possibly suggest block-fault uplift of the ancestral Bear River Range; they probably were not recycled out of an older conglomerate because no such conglomerate has been found.

Several independent lines of evidence, none of which alone is conclusive, thus point to a Late Jurassic and Early Cretaceous age for the first, and perhaps the major, movement on the Paris thrust.

MEADE THRUST PLATE

Another point raised by Mountjoy is that, if the thrusting sequence is progressively younger eastward, then the Gannett Group should be found in the Meade thrust sheet. The Meade thrust zone extends eastward around the south end of Snowdrift Mountain into Crow Creek and thence north and northwestward to the southern edge of the Snake River Plain (Armstrong and Cressman, 1963, p. J14). In the northern part of this zone Gannett rocks do occur in the upper plates of thrust slices (Mansfield, 1927, Pl. 2, 4, 7, 12, sec. O–O″; 1952, Pl. 1). Recent mapping by Oriel and L. B. Platt in the Crow Creek part of the zone shows that Gannett rocks also occur in previously unrecognized thrust slices that likely are parts of the upper plate of the Meade thrust zone.

The absence of the Gannett Group in the western part of the Meade thrust plate is expectable. The longitude of the eastern part of the Meade thrust zone is the approximate western limit of known Lower Cretaceous rocks; because the source of the Gannett conglomerates was nearby, Lower Cretaceous strata probably were not deposited as far west as the area from which much of the Meade sheet was transported (Armstrong and Cressman, 1963, p. J11; Armstrong and Oriel, 1965, p. 1855). Moreover, the Meade thrust plate has been subjected to erosion during most of the time since the Late Cretaceous, and Lower Cretaceous rocks, if present earlier, have been stripped. In addition, Cenozoic, particularly late Cenozoic, block faults have raised the western part of Meade sheet exposures so that not only possible Cretaceous, but also thousands of feet of probable Jurassic, strata have been stripped away.

The Meade fault cuts the Gannett Group, indicating that movement on this fault was in latest Early Cretaceous time or later. If the date for first movement on the Paris fault is accepted, then here is another illustration of the eastward progression of tectonic events.

SOME GENERALIZATIONS

The foregoing evidence for dating the several thrust faults accomplishes two things. First, it establishes the pre-Eocene age of at least two major thrust faults. Second, it suggests but does not prove increasing ages for principal movements of the thrust faults from east to west.

Strata bracketing the Hogsback fault near La Barge and the Absaroka fault near Kemmerer establish, beyond doubt in our view, the pre-Eocene age of principal movements along these faults at these localities. Mountjoy's statement that the most accurately dated faults in the Idaho-Wyoming thrust belt are of Eocene age is incorrect. Only the faults along the front of the Idaho-Wyoming thrust belt, the youngest, are of this age.

The fairly well-dated thrust faults along the latitude of LaBarge and Kemmerer also indicate westward-increasing ages for principal movements. Major movement occurred in the early Eocene on the La Barge or Hilliard fault on the east, in the Paleocene on the Hogsback fault, and in latest Cretaceous on the Absaroka fault. The earliest movement on the Paris fault, on the west side of the Idaho-Wyoming thrust belt, was probably latest Jurassic and earliest Cretaceous.

Eastward progression of orogeny in the Cordilleran region has long been recognized. It is supported by the foregoing dates of thrust movements. It is supported by the progressively younger ages eastward of the youngest rocks present below the various thrust plates (Rubey and Hubbert, 1959, p. 190–191). It is also supported by information on the development and later the destruction of the miogeosyncline in which sediments of the Cordillera were deposited. Still other support comes from the distribution and ages of coarse detrital rock units along the eastern margin of the Cordilleran region. As orogeny in the thrust belt consisted of folding and thrust faulting, corroborative evidence of eastward progression of orogeny supports an eastward progression with time of thrust faulting.

The pattern of progressively younger faults eastward is apparently contradicted locally by geometric relations of some structures. In the Meade thrust plate, for example, along the Left Fork of Twin Creek northeast of Georgetown, Idaho, a younger western thrust slice overlies an older eastern thrust slice (Armstrong and Cressman, 1963, Pl. 3, p. J5-J6). There are other localities in the Idaho-Wyoming thrust belt (Staatz and Albee, 1963; Jobin and Soister, 1964; Albee, 1964) where folds and steeply dipping faults in eastern underlying thrust sheets do not affect western overlying thrust sheets; some eastern folds and faults apparently pre-date the western thrust fault. Fault traces at all such localities are closely spaced and no independent basis has been found for dating these faults except relative to one another. We interpret them as slices of the same major thrust fault. If they are, then the western slice might well have moved last, as inferred in the development of back-limb thrusts (Douglas, 1950, p. 93–95; Irish, 1965, p. 103). Late or recurrent movement of a western fault slice, late forceful squeezing and shattering of a small slice between major thrust faults, and different times of movement at different places along the same major thrust may explain local relations that apparently contradict the regional pattern. Whatever the explanation, the fact is that the few available precise dates indicate progressively younger movement eastward along the major thrust faults regionally.

CANADIAN DISTURBED BELT

Our knowledge of the literature on the Canadian Rocky Mountain thrust belt is meager and we are, regrettably, even less familiar with the rocks themselves. It would be risky to attempt to apply ideas gained in one region to another not known to us. However, we are puzzled by statements that the Canadian Rocky Mountain deformation culminated in mid-Eocene time. Information that contributes to our perplexity includes available evidence for Paleocene and older orogenic events (Bossart, 1957, p. 51; Taylor *et al.,* 1964, p. 192). Are these simply "precursors" of Laramide orogeny, or the main event? Also of interest are the well-reasoned inferences from structural data (Mountjoy, 1962, p. 62–64) that major thrust sheets were developed from southwest to northeast. If there was a progression in the development of the thrust faults, then how could thrusting have taken place almost everywhere in mid-Eocene time, particularly in view of the enormous amounts of time required for transport of each thrust plate?

Our intent in asking these questions is not to suggest that thrusting in the Canadian disturbed belt occurred at precisely the same times as thrusting in the Idaho-Wyoming thrust belt. Individual events no doubt overlapped but precise coincidence of major events in the two widely separated regions is unlikely. A general impression, that we are not prepared to defend, is that some of the latest events along the front of the Canadian belt may have occurred somewhat later than the last events along the front of the Idaho-Wyoming belt.

Dating of Orogenies and the Laramide

Despite the common interest which Mountjoy and we share in thrust faults, we may be very much apart on one issue. Neither we nor our colleagues working in the Idaho-Wyoming thrust belt (Rubey and Hubbert, 1959, p. 190–196; Eardley, 1960; Armstrong and Cressman, 1963, p. J14-J16) have believed it necessary to have orogeny or deformation culminate in a rousing crescendo for a region or mountain system. Precise dates for frontal faults of a mountain system do not date the system but merely the last of a whole sequence of tectonic events that include equally large or larger earlier episodes of tectonism that are more difficult to date. The local field evidence in the Idaho-Wyoming thrust belt, as in central Utah (Spieker, 1946), supports the concept of multiple, non-synchronous tectonic events in orogeny (Shepard, 1923; Spieker, 1946; Gilluly, 1949, 1963, 1965; Aubouin, 1965, p. 193, 211–213).

Acceptance of the concept of multiple tectonic events in orogeny may explain, in part, our not using the term Laramide, with which Mountjoy takes issue. The term Laramide was for decades a keystone in the theory of periodic world-wide diastrophism. The Laramide revolution formerly was used to mark the end of the Mesozoic and beginning of the Cenozoic Eras. The Laramide revolution has also been defined as: "A period of mountain building and erosion in Rocky Mountain region that began in late Cret. time and ended in early Tert. time" (Wilmarth, 1938, p. 1149). Mountjoy applies the term to mid-Eocene events. Our colleagues working in other parts of the continent apply the term to events ranging from mid-Mesozoic to mid-Cenozoic in age. It is difficult to tell what is meant by use of the term.

No doubt the term Laramide is just as useful and just as much needed in some discussions as the term Paleozoic. However, assignment of a stratigraphic unit to the lower Middle Devonian seems more meaningful to us than assignment to the Paleozoic. Moreover, however great their shortcomings, criteria are available for recognition of the beginning and the end of the Paleozoic. Tectonic events in many parts of the continent can not be dated more precisely than Laramide. If they can be, use of the term conceals the precision, for Laramide has been applied to episodes of orogeny extending through at least 60 million years (Gilluly, 1963, p. 150).

Semantic quibbling over the term Laramide, however, is not the real issue. The danger of use of the term is that it may imply synchroneity of tectonic events within a region or between regions. Mounting evidence in the western United States, in contrast, demonstrates that crustal deformation proceeds ". . . in a quasicontinuous, although doubtless episodic way" (Gilluly, 1965, p. 23). The concept has also been described as ". . . long lasting periods of tectonic deformation, at a most variable speed, with paroxysms, which may occur at different times in different places along the length of the range" (Goguel, 1965, p. 195).

The title of this reply, therefore, for all of the foregoing reasons, is *"Times"*—not "Time"—"of thrusting in the Idaho-Wyoming thrust belt."

REFERENCES CITED

Albee, H. F., 1964, Preliminary geologic map of the Garns Mountain NE quadrangle, Teton County, Idaho: U.S. Geol. Survey Mineral Inv. Field Studies Map MF-274.

Armstrong, F. C., and E. R. Cressman, 1963, The Bannock thrust zone, southeastern Idaho: U.S. Geol. Survey Prof. Paper 347-J, p. J1-J22.

——— and S. S. Oriel, 1965, Tectonic development of Idaho-Wyoming thrust belt: Am. Assoc. Petroleum Geologists Bull., v. 49, no. 11, p. 1847-1866.

Aubouin, Jean, 1965, Geosynclines, developments in geotectonics 1: Amsterdam, Elsevier, 335 p.

Bossart, D. O., 1957, Relationship of the Porcupine Hills to early Laramide movements, *in* Alberta Soc. Petroleum Geologists Guidebook, 7th Ann. Field Conf., p. 46-51.

Douglas, R. J. W., 1950, Callum Creek, Langford Creek, and Gap map-areas, Alberta: Geol. Survey Canada Mem. 255, 124 p.

Eardley, A. J., 1944, Geology of the north-central Wasatch Mountains, Utah: Geol. Soc. America Bull., v. 55, p. 819-894.

——— 1960, Phases of orogeny in the fold belt of western Wyoming and southeastern Idaho, *in* Overthrust belt of southwestern Wyoming and adjacent areas: Wyo. Geol. Assoc. Guidebook, 15th Ann. Field Conf., p. 37-40.

Gazin, C. L., 1942, Fossil Mammalia from the Almy Formation in western Wyoming: Wash. Acad. Sci. Jour., v. 32, no. 7, p. 217-220.

——— 1956a, The upper Paleocene Mammalia from the Almy Formation in western Wyoming: Smithsonian Misc. Colln., v. 131, no. 7, p. 1-18.

——— 1956b, The occurrence of Paleocene mammalian remains in the Fossil basin of southwestern Wyoming: Jour. Paleontology, v. 30, no. 3, p. 707-711.

Geological Society of London, 1964, The Phanerozoic time-scale, a symposium: Quart. Jour. Geol. Soc. London, v. 120, supplement, 458 p.

Gilluly, James, 1949, Distribution of mountain building in geologic time: Geol. Soc. America Bull., v. 60, p. 561-590.

——— 1963, The tectonic evolution of the western United States: Quart. Jour. Geol. Soc. London, v. 119, p. 133-174.

——— 1965, Volcanism, tectonism, and plutonism in the western United States: Geol. Soc. America Special Paper 80, 69 p.

Goguel, J., 1965, Tectonics and continental drift, *in* A symposium on continental drift: The Royal Society, London, Philos. Trans. no. 1088, p. 194-198.

Irish, E. J. W., 1965, Geology of the Rocky Mountain Foothills, Alberta: Geol. Survey Canada Mem. 334, 241 p.

Jobin, D. A., and P. E. Soister, 1964, Geologic map of the Thompson Peak quadrangle, Bonneville Co., Idaho: U.S. Geol. Survey Mineral Inv. Field Studies Map MF-284.

Mansfield, G. R., 1927, Geography, geology, and mineral resources of part of southeastern Idaho: U.S. Geol. Survey Prof. Paper 152, 453 p.

——— 1952, Geography, geology, and mineral resources of the Ammon and Paradise Valley quadrangles, Idaho: U.S. Geol. Survey Prof. Paper 238, 92 p.

McMasters, J. H., 1943, Buena Vista Hills area of the Midway-Sunset oil field: Calif. State Div. Mines Bull. 118, p. 517-518.

Michael, R. H., 1960, Hogsback and Tip Top units, Sublette and Lincoln Counties, Wyoming, *in* Overthrust belt of southwestern Wyoming and adjacent areas: Wyo. Geol. Assoc. Guidebook, 15th Ann. Field Conf., p. 210-216.

Mountjoy, E. W., 1962, Mount Robson (southeast) map area, Rocky Mountains of Alberta and British Columbia: Geol. Survey Canada Paper 61-31, 114 p.

——— 1966, Time of thrusting in the Idaho-Wyoming thrust belt: discussion: Am. Assoc. Petroleum Geologists Bull., v. 50, no. 12, p. 2612-2614.

Murray, F. E., 1960, An interpretation of the Hilliard thrust fault, Lincoln and Sublette Counties, Wyoming, *in* Overthrust belt of southwestern Wyoming and adjacent areas: Wyo. Geol. Assoc. Guidebook, 15th Ann. Field Conf., p. 181-186.

Oriel, S. S., 1962, Main body of the Wasatch Formation near La Barge, Wyoming: Am. Assoc. Petroleum Geologists Bull., v. 46, no. 12, p. 2161-2173.

——— 1963, Preliminary map of the Fort Hill quadrangle, Lincoln Co., Wyoming: U.S. Geol. Survey Oil and Gas Inv. Map OM-212.

Rubey, W. W., and M. K. Hubbert, 1959, Overthrust belt in geosynclinal area of western Wyoming in light of fluid-pressure hypothesis, Pt. 2 *of* Role of fluid pressure in mechanics of overthrust faulting: Geol. Soc. America Bull., v. 70, p. 167-206.

——— S. S. Oriel, and J. I. Tracey, Jr., 1961, Age of the Evanston Formation, western Wyoming, *in* Short papers in the geologic and hydrologic sciences: U.S. Geol. Survey Prof. Paper 424-B, p. B153-B154.

Shepard, F. P., 1923, To question the theory of periodic diastrophism: Jour. Geology, v. 31, p. 599-613.

Spieker, E. M., 1946, Late Mesozoic and early Tertiary history of central Utah: U.S. Geol. Survey Prof. Paper 205-D, p. 117-161.

Staatz, M. H., and H. F. Albee, 1963, Preliminary geologic map of the Garns Mountain SE quadrangle, Bonneville and Teton Counties, Idaho: U.S. Geol. Survey Mineral Inv. Field Studies Map MF-262.

Taylor, R. S., W. H. Mathews, and W. O. Kupsch, 1964, Chapter 13, Tertiary, *in* Geological history of western Canada: Alta. Soc. Petroleum Geologists, p. 190-194.

Veatch, A. C., 1907, Geography and geology of a portion of southwestern Wyoming: U.S. Geol. Survey Prof. Paper 56, 178 p.

Wilmarth, M. G., 1938, Lexicon of geologic names of the United States: U.S. Geol. Survey Bull. 896, 2396 p.

Geological Society of America Bulletin
1968, v. 79, p. 429-458

Sevier Orogenic Belt in Nevada and Utah

RICHARD LEE ARMSTRONG[1]

Abstract In Nevada and Utah, sedimentation in the Cordilleran miogeosyncline began before the appearance of Cambrian fossils and continued without orogenic interruption through the Triassic. During the Jurassic, deformation and regional metamorphism occurred in the western part of the miogeosyncline, and the area of sediment accumulation shifted onto the Colorado Plateau.

A major source of clastic material appeared along the eastern margin of the Cordilleran miogeosyncline in Early Cretaceous time; this source supplied the sediments that filled the Cretaceous to Paleocene Rocky Mountain geosyncline. Clasts in the Cretaceous conglomerates show an inverted stratigraphy, reflecting successive exposure of older and older rocks in an evolving orogenic belt along the eastern side of the Cordilleran miogeosyncline. This source area was the Sevier orogenic belt, which had a history of deformation through most of the Cretaceous (Sevier orogeny). Décollement thrusts with displacements of tens of miles are the characteristic structures of the belt, but several large folds are also known. The largest thrusts are overlain unconformably by uppermost Cretaceous conglomerates.

Thrusting in the Sevier orogenic belt had virtually ceased by the time the Laramide orogeny began east of the Sevier belt in latest Cretaceous time. Laramide mountains were the result of uplift of great blocks of crystalline basement along nearly vertical, reverse, and steep thrust faults. The Uinta arch, which intersects the Sevier orogenic belt almost at a right angle, is the only one of these basement uplifts closely involved with the deformation of the Cordilleran miogeosyncline.

North-south-trending regional normal faulting of post-Oligocene age has broken up the orogenic belt so that it is not immediately recognizable on geologic maps. Arch ranges, intrusive domes, and gravity slides are additional complications of the Tertiary geology, but widespread Tertiary deposits, particularly Oligocene ignimbrites, make a paleogeologic reconstruction possible; thus, the Sevier orogenic belt can be view as it existed before the normal faulting.

INTRODUCTION

The eastern Great Basin in eastern Nevada and western Utah is characterized by north-south-trending fault-block ranges composed of carbonate assemblage rocks of the Cordilleran geosyncline. The area under consideration is bounded on the west by the mid-Paleozoic Antler orogenic belt (Roberts and others, 1958), on the south by the Las Vegas shear zone (Longwell, 1960), and on the east by the Colorado Plateau. Although the Idaho boundary has been taken as an arbitrary northern limit, it should be emphasized that structures described in the Great Basin persist without significant modification northward into, and even past, central Idaho.

In order to understand the results of K-Ar dating studies of the region, a review of available knowledge of Great Basin geology was essential; the results of the K-Ar studies have been published elsewhere (Armstrong, 1963; 1966; Armstrong and Hansen, 1966). The only complete synthesis of Basin and Range geology (Nolan, 1943) has become a classic. Since the appearance of that report, a great amount of work has been done in the area, particularly as thesis projects. All Utah and approximately 80 percent of eastern Nevada have been mapped in enough detail to show the most significant structural features. Osmond (1960) discussed briefly the tectonic history of the Basin and Range province in Utah and Nevada; Misch (1960) discussed certain structural features of the eastern Great Basin; and Gilluly (1963) has reviewed the tectonic history of the western United States. No up-to-date detailed synthesis of the geology of the eastern Great Basin exists, however, and this led King (1959, p. 142) to say, after describing Longwell's discoveries in the Las Vegas region:

> To pursue details of the structures in other parts of the eastern Great Basin would probably only bewilder the reader without profit. Many folds and thrusts are known, but the larger pattern is for the most part undetermined. Not only have the fundamental structures been obscured over wide areas by Basin and Range structure, but many of the ranges have been little explored geologically.

It is the writer's opinion, however, that eventually we shall know more about the orogenic history of the Great Basin *because* of the faulting and volcanics, not in spite of them, for they provide exposures in the third dimension and key horizons for reconstructing the deformed orogen now exposed at varied structural levels.

The existence of thrust faults and folds of Mesozoic age along the eastern edge of the Cordilleran geosyncline is common knowledge (Eardley, 1962, 1963); the same area has been clearly recognized as a source of clastic material during the Cretaceous by

[1]Dept. Geology, Yale University, New Haven, Connecticut

Spieker (1946; 1949; 1956) and his students and by Harris (1959) who proposed the name, Sevier arch, for the clastic source. This paper is a review and analysis of the geology of this fold and thrust belt.

ACKNOWLEDGMENTS

I am indebted to the large number of geologists who contributed directly and indirectly to this project through their studies in the Great Basin. Without such previous work, this synthesis would be impossible. Kenneth F. Bick introduced me to the geology of the Great Basin in 1956. Pierre Biscaye and, later, Julia Armstrong assisted in the field studies during 1961. C. R. Longwell, Paul Williams, Robert Scott, T. B. Nolan, Keith Ketner, Harold Masursky, L. J. P. Muffler, J. C. Taylor, and Hoover Mackin provided hospitality and guidance in their respective field areas. During preparation of the original manuscript, John Rodgers, Edward Hansen, Keith Howard, Clark Burchfiel, D. H. Adair, Kenneth Pierce, and Pierre Biscaye provided helpful discussion. K. K. Turekian, John Rodgers, P. M. Orville, C. R. Longwell, R. J. Roberts, Peter Misch and D. H. Adair have read the present paper at various stages of preparation and provided helpful comments. Much of the drafting was done by Gary Audette. Field work was supported by National Science Foundation grant G14192. This research was done in major part while the writer was a National Science Foundation graduate fellow (1959-1962).

GEOLOGIC SETTING AND STRATIGRAPHIC HISTORY

Two principal parts of the Cordilleran geosyncline are recognized. The miogeosyncline in Nevada and Utah contains a thick section of Paleozoic rocks of the carbonate assemblage[2] (limestone, dolomite, clean sandstone, and little shale), and within, and west of the Antler orogenic belt Paleozoic rocks of the siliceous assemblage (shale, dirty sandstone, chert, and volcanic rocks) of the eugeosncline occur. East of the geosyncline in the central Wasatch Range and on the Colorado Plateau, rocks of the carbonate assemblage occur in a drastically thinned and incomplete Paleozoic section.

The relationships between the Paleozoic sections in the eugeosyncline, the miogeosyncline, and the adjacent shelf are obscured by major thrust faults with displacements of tens of miles. Eugeosynclinal rocks have been thrust over miogeosynclinal rocks in western and central Nevada, and miogeosynclinal rocks have been thrust over thin shelf facies in southeastern Nevada and western Utah. The present-day geographic distribution of the various rock assemblages, therefore, does not represent their distribution at the time of deposition.

Older Precambrian crystalline rocks that were metamorphosed approximately 1.5 b.y. ago underlie the shelf sections in Utah and southern Nevada and the miogeosynclinal rocks in the Death Valley, California, region. Within most of the geosyncline, however, no proven older Precambrian (> 1 b.y.) rocks are exposed. In the Uinta Mountains, Cottonwood Uplift, and Death Valley areas, thick sections of younger Precambrian sedimentary rocks unconformably overlie the older metamorphics and are in turn overlain unconformably by rocks of the Cordilleran geosyncline.

The Paleozoic history of areas west of a line extending south by southwest from northeastern Nevada was complex because two major Paleozoic orogenies occurred there (Roberts and others, 1958; Silberling and Roberts, 1962). In the miogeosyncline of eastern Nevada and western Utah, however, Eocambrian[3] through Triassic stratigraphic relations are relatively simple. Gentle truncation of units occurs, particularly along the eastern and western margins of the miogeosyncline, but only one distinctly angular unconformity is known (Stansbury anticline of Rigby, 1958), and this is only of local extent.

Deposition began in the Cordilleran miogeosyncline before the oldest Cambrian fossils appeared. Approximately one third of the stratigraphic section of the miogeosyncline is composed of a basal clastic sequence which includes Eocambrian, Lower, and Middle Cambrian quartzite and argillite and a widespread Eocambrian tillitic member. In southern Nevada and at scattered localities elsewhere, dolomite is present in this basal sequence. After Early Cambrian time, carbonate deposition became widespread. Middle and Upper Cambrian deposits are complexly intertonguing shale and carbonate rocks, more dolomitic toward the top. Lower Ordovician limestones with minor shale were succeeded in Middle Ordovician time by a distinctive, widespread, clean white quartz sand, which is absent only locally over the Tooele arch, a Cambrian and Ordovician positive element. Upper Ordovician, Silurian, and Lower Devonian deposits are almost exclusively dolomite, and Upper Devonian and later Paleozoic carbonates are predominantly limestone. Later

[2]"Carbonate assemblage" and "siliceous assemblage" are used for the contrasting geosynclinal facies as suggested by Silberling and Roberts (1962) and R. J. Roberts (1964, written commun.).

[3]The term "Eocambrian" is used to emphasize that no significant time gap separates the sediments referred to from overlying fossiliferous Cambrian strata. Usage of this term is the same as in the Caledonide area of Norway where Eocambrian was first proposed by W. C. Brogger in 1900 (Holtedahl, 1960, p. 111-112).

Devonian sedimentation was more varied because of tectonic activity in and near the miogeosyncline. A gentle arch formed in east-central Nevada between Middle and Late Devonian time; the Stansbury anticline rose in north-central Utah during Late Devonian time. From very late Devonian time until later Pennsylvanian time, the Antler orogeny affected sedimentation in Nevada and western Utah; a widespread uppermost Devonian-Lower Mississippian shale was succeeded by Lower Mississippian limestone, which was locally removed as a consequence of Early Mississippian warping and erosion. During the rest of Mississippian and Pennsylvanian time, a clastic wedge composed of material derived from the Antler orogenic belt extended into the miogeosyncline from the west. Subsidence of the Oquirrh basin in north-central Utah began in Mississippian time and continued into Permian time. From Late Mississippian through Permian time, clastics shed from the rising basement uplifts of the Ancestral Rocky Mountains accumulated in the Oquirrh basin which was bounded on the north by an east-west-trending monoclinal flexure. During Pennsylvanian time, most of eastern Nevada was the site of limestone deposition; in the Oquirrh basin, more than 20,000 feet of alternating limestone and quartzite accumulated. Thick lower and middle Permian deposits—limestone with much quartz sand, sandstone, siltstone, dolomite, and some evaporite—accumulated in the Arcturus basin in east-central Nevada, and thick limestone deposits accumulated in southern Nevada; in Utah, correlative strata consist of alternating quartzite and limestone with minor dolomite. Southeastward from the miogeosyncline, Permian marine strata intertongue with continental red beds.

Upper Permian deposits (Park City Group) are a widespread blanket of relatively uniform thickness and lithology (limestone and dolomite with minor chert and phosphate) over the entire region, including the Antler orogenic belt and much of the Colorado Plateau shelf. In the miogeosyncline, marine sedimentation continued without orogenic interruption into the Triassic over much of Nevada and Utah. Triassic (or earliest Jurassic) rocks were the last deposits of the Cordilleran miogeosyncline. In Middle Triassic time, marine waters withdrew from the eastern Great Basin in Nevada for the last time, marking the beginning of the orogenic chapter in the history of the region. The Triassic-Jurassic boundary probably lies within the widespread eolian sandstone (Navajo-Aztec-Nugget), which is the youngest pre-orogenic formation present in the eastern Great Basin west of the Mesozoic fold and thrust belt.

During the Jurassic, the region of thickest sediment accumulation shifted to central and eastern Utah, and the western part of the Cordilleran miogeosyncline became a source area (Stokes, 1960) in response to orogenic deformation taking place there. Continental clastic deposits of Upper Jurassic and lowermost Cretaceous age, derived from this western source, spread across the eastern edge of the Paleozoic miogeosyncline and the Colorado Plateau. In Early Cretaceous time, the eastern edge of the geosyncline became a source of clastic material, which accumulated during Cretaceous and Paleocene time in the Rocky Mountain geosyncline still farther east. At a few localities in eastern Nevada, continental Lower Cretaceous deposits are present, but over most of the region there is a great hiatus between deposits of the Cordilleran geosyncline and Tertiary deposits.

Tertiary strata of the eastern Great Basin and adjacent Colorado Plateau can be subdivided into three major groups. The oldest is composed of nonvolcanic continental sediments—scattered Eocene lacustrine deposits and undated conglomerates in Nevada and western Utah and Paleocene and Eocene fluviatile and lacustrine sediments that are well developed in central Utah and northward into Wyoming. The middle group consists of widespread intermediate to acidic volcanics, chiefly ignimbrites of latest Eocene, Oligocene, and early Miocene age. The youngest group, Miocene to Recent, is a heterogeneous collection of discontinuous clastic units, volcanic-rich sediments, volcanics (commonly basalts but also all other types), and lacustrine sediments, deposited during the development of the Basin and Range structure. Figure 1 is a correlation chart illustrating Tertiary stratigraphic relationships within, and adjacent to, the eastern Great Basin.

A more complete review of the Precambrian through Tertiary stratigraphic history, together with documentation not included in this paper, may be found in Armstrong (1968). Palinspastic isopach maps for all Paleozoic systems and three palinspatic paleostratigraphic profiles across the region are included in the review.

PRE-NORMAL FAULTING PALEOGEOLOGY—SEVIER OROGENIC BELT

Paleogeologic Map

The present-day structural pattern of the eastern Great Basin is dominated by the effects of Tertiary normal faulting. Geologic maps of the region cannot clearly portray the general features of the pre-Tertiary structures that are exposed in separated ranges. East individual exposure displays the older structures in a different attitude or aspect, and irregularly distributed Tertiary volcanics and sediments do not make things clearer. If we remove the effects of normal faulting and Tertiary sedimentation, we

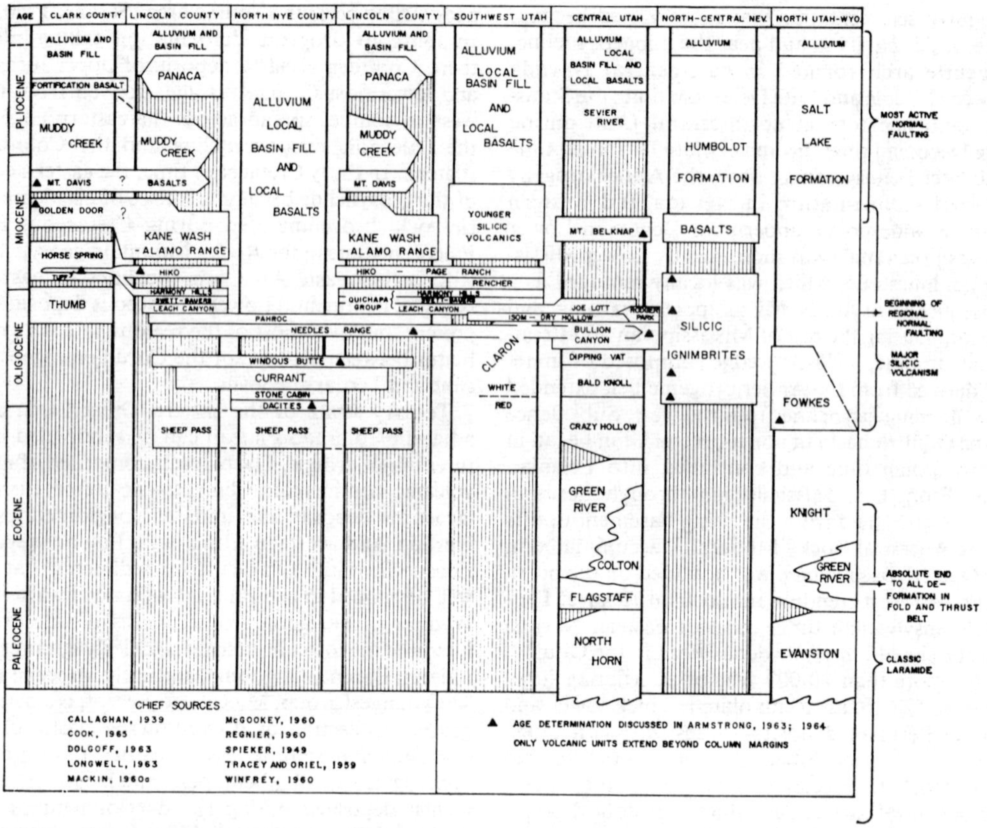

FIG. 1—Eastern Great Basin Tertiary correlation chart.

can view in a simple manner the broad features of the pre-Tertiary structures. This can be done by a paleogeologic reconstruction, as described by Levorsen (1960). All later effects of sedimentation and deformation are erased and the resultant map portrays the geology as it was at the time the unconformity was buried. This technique is applied here to the eastern Great Basin to display the pre-normal faulting paleogeology.

Plate 1, figure 1 shows the units which overlie the unconformity used for the reconstruction (*see* also Fig. 1). The range in age of the unconformity is latest Cretaceous to earliest Miocene. This undoubtedly has a somewhat distorting effect on the resultant paleogeology, but it does not alter the fundamental geologic pattern. In all areas, the unconformity postdates the main Mesozoic deformation, although locally in central Utah some minor folding and even thrusting may have occurred later than the unconformity. The unconformity, however, predates the major normal faulting. Early normal faulting may affect the pattern somewhat in eastern and southern Nevada but only to a relatively minor degree. The over-all structural pattern of the region was not significantly altered during the time spanned by the unconformity.

If the older structure is not enormously complex, the resultant paleogeologic map should display the broad features of the regional geology during early Tertiary time. Plate 1, figure 2 shows the distribution of points (~ 900) where information on rocks underlying the unconformity was recorded. The source maps consulted are given in Table 1. In addition to simple rock age information supplied by Tertiary–pre-Tertiary contacts, it is also often possible to construct paleo-strikes and dips by rotating the oldest Tertiary deposits (commonly ignimbrites) back to horizontal and determining the effect of the same rotation on the older rocks. In areas where a suitable unconformity is lacking, it is possible to put limits on the ages of rocks as those exposed in early Tertiary time must have been as young as, or younger than, those now present.

The normal faulting responsible for the present topography and much of the geologic complexity of the Great Basin occurred mostly during Miocene and Pliocene time. Pre-Miocene normal faults are known in many places in the region, but none

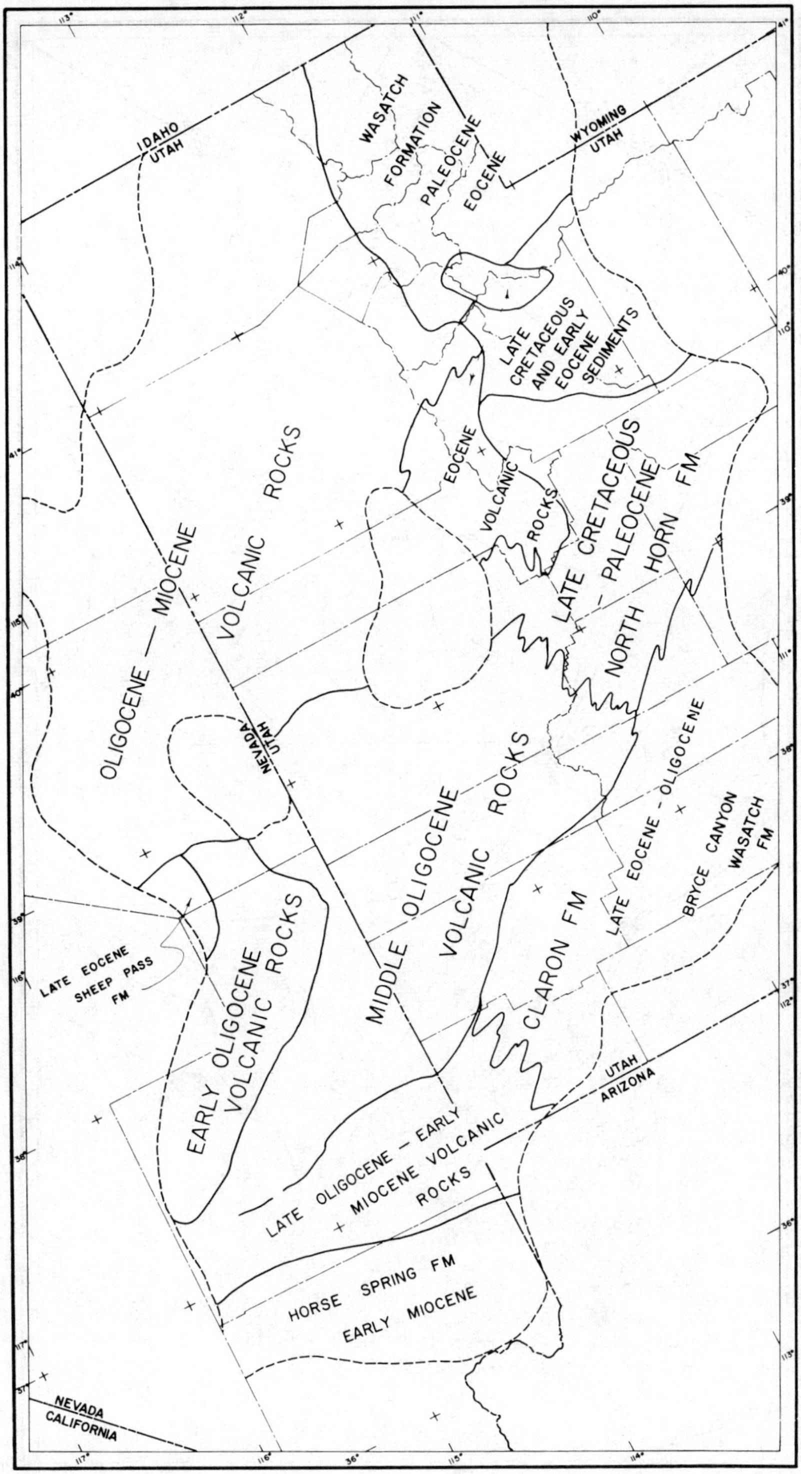

Plate 1 (fig. 1). Map showing units overlying unconformity studied, Eastern Great Basin.

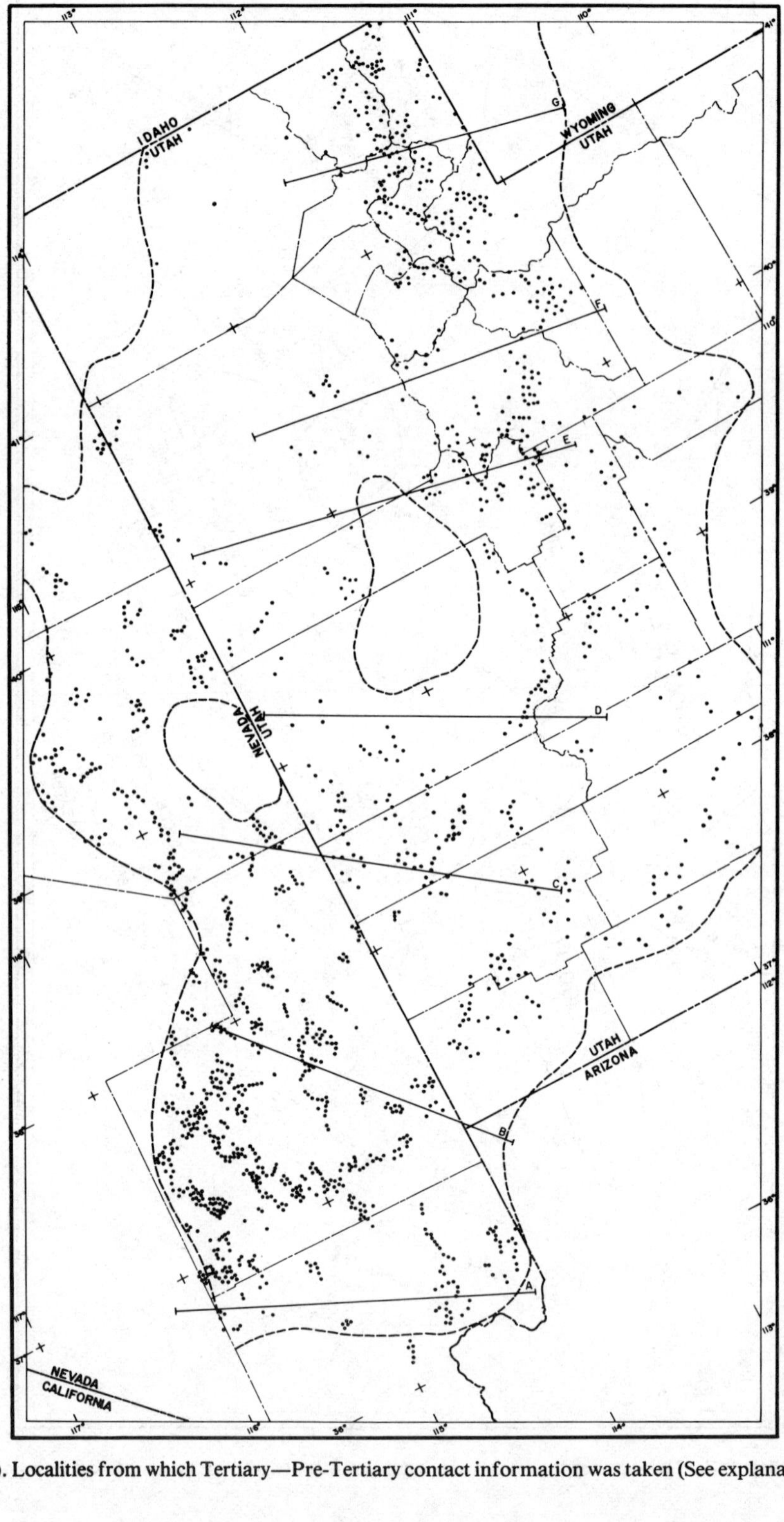

Plate 1 (fig. 2). Localities from which Tertiary—Pre-Tertiary contact information was taken (See explanation on **Plate 1**, fig. 4).

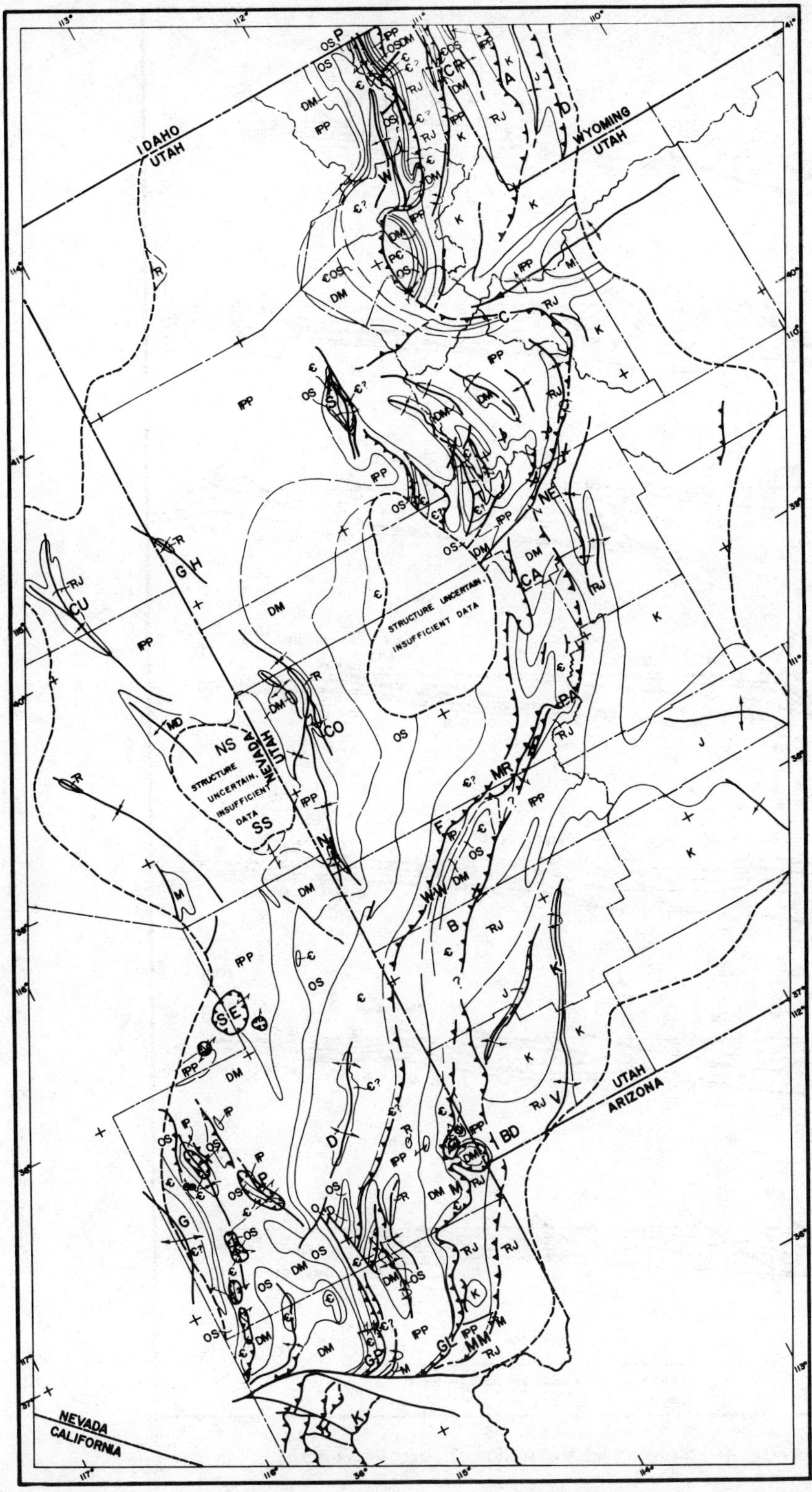

Plate 1 (fig. 3). Paleogeologic map, Eastern Great Basin (See explanation on **Plate 1,** fig. 4).

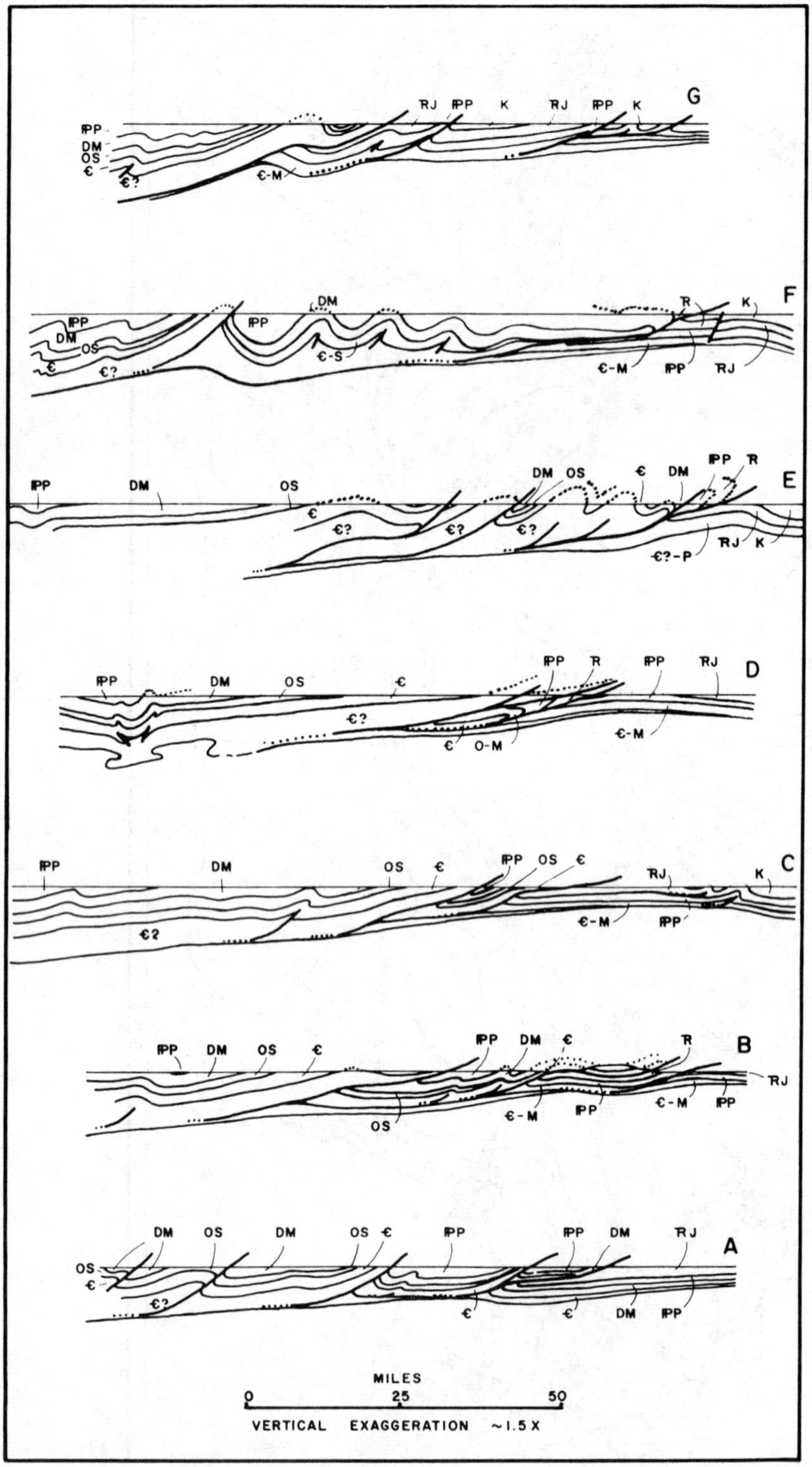

Plate 1 (fig. 4). Paleogeologic cross sections, Eastern Great Basin.

approach the magnitude of displacement of the faults formed later. No example can be cited where rocks differing in age by several geologic periods were juxtaposed along pre-Miocene normal faults. Thrusting accounted for the major discontinuities present.

A significant structural feature of the region is the widespread near-conformity of the Paleozoic sediments and Tertiary volcanics. Over large areas, the angularity of the unconformity is less than 5 degrees; only locally is it distinctly angular. Cook (1965, p. 54-55) states:

> pre-volcanism deformation was sharply localized along axes that trend east of north, leaving between the narrow belts of deformation broad areas of undeformed Paleozoic rocks, the ignimbrites in many sections are essentially parallel to the underlying sedimentary rocks...the attitude of the volcanics in many places reflects the attitude of the subadjacent sedimentary rocks; locally...the angularity of the unconformity is great.

Mackin (1961, oral commun.) made the same observation, and the writer also agrees. The relationship is of importance in reconstructing the pre-normal faulting paleogeology of the region. The generalization does not apply to more westerly areas near and in the Antler orogenic belt.

The observed near-parallelism of Tertiary and pre-Tertiary rocks in the region establishes that low dips were characteristic of pre-Tertiary rocks during early Tertiary time. Extrapolation between scattered data, therefore, should be safe at least for distances of a few miles. The near-parallelism of units, however, is only a generalization. In many areas, sharp angular unconformities occur, and the paleogeologic reconstruction must maintain consistency with these relationships. The paleogeology becomes increasingly complex westward into central Nevada. The reconstruction is limited to the area from which suitable data are available and where the paleogeology appears to have been fairly straightforward.

The pre-normal faulting paleogeology is shown on Plate 1, figures 3 and 4. The map agrees with all the data collected, but in many areas alternate interpretations of the data are possible; this does not imply, however, that the major structural features are in doubt. The map portrays the geology as it would have been mapped shortly after the end of the Cretaceous. The degree of definition (resolving power) is slightly better than that represented by more detailed parts of the 1932 edition of the U. S. Geological Survey geologic map of the United States.

The general features of the paleogeology are immediately evident. The structural trends are generally north to northeast. On the east side of the map is a broad, virtually undeformed, foreland basin area filled by Cretaceous deposits. Two broad arches which lie nearly perpendicular to the regional trend occur in this foreland. The middle of the map

EXPLANATION AND LIST OF LOCALITIES SHOWN ON MAP

- K — CRETACEOUS
- TRJ — TRIASSIC – JURASSIC
- PP — PENNSYLVANIAN – PERMIAN
- DM — DEVONIAN – MISSISSIPPIAN
- OS — ORDOVICIAN – SILURIAN
- € — CAMBRIAN SHALE - CARBONATE SEQUENCE
- €? — EOCAMBRIAN - CAMBRIAN BASAL CLASTIC SEQUENCE
- P€ — PRECAMBRIAN

- LIMIT OF COMPILATION
- LOCATION OF CONTACT DATA
- CONTACT
- THRUST FAULT
- FAULT
- KLIPPE
- SYNCLINE
- ANTICLINE
- OVERTURNED SYNCLINE
- OVERTURNED ANTICLINE
- STRIKE AND DIP
- STRIKE AND DIP, OVERTURNED BEDS
- LINE OF SECTION
- MONOCLINE

UTAH

A ABSAROKA THRUST
B BLUE MOUNTAIN THRUST
BD BEAVER DAM MOUNTAINS
C CHARLESTON THRUST
CA CANYON RANGE
CO CONFUSION RANGE
CR CRAWFORD THRUST
D DARBY THRUST
F FRISCO THRUST
K KANARRA FOLD
MR MINERAL RANGE THRUST
N NEEDLES THRUST
NE NEBO THRUST
P PARIS THRUST
PA PAVANT THRUST
S STANSBURY RANGE
T TINTIC DISTRICT
V VIRGIN ANTICLINE
W WILLARD THRUST
WW WAH WAH THRUST

NEVADA

CU CURRIE
D DELAMAR DISTRICT
G GROOM DISTRICT
GL GLENDALE THRUST
GP GASS PEAK THRUST
K KEYSTONE THRUST
M MORMON MOUNTAINS
MM MUDDY MOUNTAINS THRUST
NS NORTHERN SNAKE RANGE
P PAHRANAGAT RANGE
SE SOUTHERN EGAN RANGE
SS SOUTHERN SNAKE RANGE

0 50 100
SCALE OF MILES

Table 1. Geologic Maps of the Eastern Great Basin and Vicinity which Were Used for Construction of the Paleogeologic Map* (Pl. 1, fig. 3).

Nevada	
Clark County	Bowyer and others (1958)
Lincoln County	Kellogg (1963) Tschanz (1960) Tschanz and Pampeyan (1961)
White Pine County	Adair (1961) Bauer and others (1960) Douglass (1960) Drewes (1958; 1960) Fritz (1960) Langenheim and others (1960) Lloyd (1959) Nelson (1959) Playford (1962) Ward (1962) Whitebread and others (1962) Woodward (1964) Young, J. C. (1960)
Elko County	Harlow (1956) Nelson (1956) Schaeffer and Anderson (1960) Snelson (1955)
Utah	
State Geologic Map	
Northeast Quarter	Stokes and Madsen (1961)
Northwest Quarter	Stokes (1963)
Southwest Quarter	Hintze (1963)
Other Maps	Hintze (1962)
Wyoming	
State Geologic Map	Love and others (1955)
Other Data from	Cochran (1959) Schick (1959)

*A geologic map of Nevada (Webb and Wilson, 1962) is available but was not used for construction of Plate 1, figure 3. It includes data from the references in the table but in a more generalized form.

is occupied by a fold-and-thrust belt with eastward overturning and thrusting. The region farther west displays a deceptively simple structural pattern, in which, over broad areas, only gently folded upper Paleozoic rocks are shown. In southern Nevada, the pattern is complicated by an uplift exposing lower Paleozoic and Eocambrian rocks and surrounded by klippen of Paleozoic rocks overlying middle and upper Paleozoic strata. This pattern is the summation of all the effects of Mesozoic and early Tertiary deformation in the region.

Foreland

East of the belt of folds and thrusts is a broad area that was slightly deformed during Mesozoic time; later in the era, it was a basin in which orogenic debris from the area now occupied by the Great Basin accumulated. At the end of the Cretaceous and during early Tertiary time, broad arches formed in the foreland. These can be seen on the paleogeologic map as the Uinta arch in northern Utah and the Circle Cliffs upwarp in the southern part of the state.

Sevier Orogenic Belt

General statement. The Sevier arch was named by H. D. Harris (1959), who described it as a late Mesozoic positive area in western Utah and southeastern Nevada. The arch concept was based on paleogeology, and the arch was considered the source of the orogenic clastics shed to the east; thrusting was considered to be the climax of arching during late Cretaceous time. Harris (1959, p. 2646) says:

> There is no direct evidence of large-scale thrusting associated with the uplift of the Sevier arch. Deformation appears to have been generally limited to upwarping and development of major folds, some of which undoubtedly developed into belts of structural weakness that later became zones of thrusting.

The paleogeologic map of Plate 1, figure 3 is based on more than ten times the amount of data presented by Harris and does not support the concept of a simple late Mesozoic arch, for nowhere is an eastern limb evident. It shows that the exposure of old rocks in the area is indeed the result of thrusting, which was the deformation responsible for the clastics shed into the Rocky Mountain geosyncline during the latter half of the Mesozoic. It is suggested that "Sevier orogenic belt" is a better term for the belt of thrusts and folds originally described as the Sevier arch.

In several recent studies, paleogeographic arches or geanticlines have been discovered to be orogenic belts; for example, the Manhatten geanticline became the Antler orogenic belt (Roberts and others, 1958), and the Mesocordilleran geanticline was the site of Jurassic deformation (Misch, 1960; Armstrong and Hansen, 1966). It does not appear reasonable to expect enormous quantities of coarse clastics from simple arching. Orogenic deformation, including faulting, is necessary to account for the Mesozoic clastics.

In its earliest stages, the Sevier belt may well have been archlike. Permian isopachs offer the first faint suggestion of uplift along the locus of the belt. Further uplift in later Permian or early Triassic time may be indicated by the pre-Triassic unconformity in southern Nevada. The first conclusive evidence of orogenic deformation in the belt, however, is the appearance of lower Paleozoic clasts in early Colorado time in Utah and Nevada or as far back as the beginning of the Cretaceous in Idaho (F. C. Armstrong and Cressman, 1963, p. 10). The only possi-

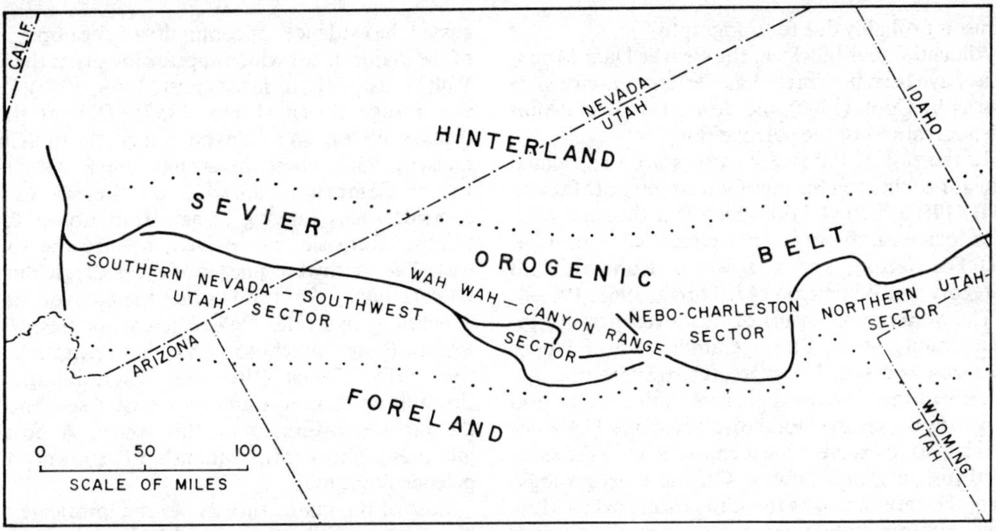

FIG. 2—Index map for Sevier orogenic belt, Nevada and Utah.

ble source of these clasts is the Sevier belt. Lower Cretaceous and Upper Jurassic sediments could have been derived from anywhere in the eastern Great Basin, but they may also have come, at least in part, from the Sevier belt during earlier stages of its development. The Eocambrian rocks of the belt occur almost exclusively in the sole of major thrusts. Their appearance, in abundance, as clasts approximately at the end of Colorado time is evidence that thrust displacements of tens of miles existed by then.

The relative ages of folds and thrusts in the belt normally cannot be determined, but in the Canyon Range folding definitely postdates displacement on the major thrust (Christiansen, 1952). In other mountain belts, such as the Appalachians, folding postdates movement along regional bedding-plane thrusts (Rich, 1935; Pierce and Armstrong, 1964), for after folding such thrusts are unable to develop. Probably, therefore, most of the Sevier thrusting represents an earlier stage of deformation than the folding. Some folding was probably also contemporaneous with, if not caused by, thrusting (Rich, 1934; Cressman, 1964). On the major thrusts of the belt, a minimum of 25 miles of total displacement has occurred; such a displacement is approximately equal to the thickness of the crust and cannot be merely the climax of deformation in a tight fold. To summarize, thrusting was prolonged and of great magnitude in the Sevier belt. It was not merely the climax to earlier folding and arching. Figure 2 provides an index map for the following detailed discussion of the Sevier orogenic belt.

Southern Nevada-Southwestern Utah Sector. The structural geology of Clark County, Nevada, has been discussed by Longwell (1949; 1952; 1952a; 1960; 1962), and a county map has been published (Bowyer and others, 1958). The paleogeology retains all features of Longwell's interpretation of the area. Modification of the structural pattern by later Tertiary normal faults is relatively minor.

The major thrust is terms of stratigraphic displacement is the Gass Peak thrust which brings the Lower Cambrian and Eocambrian quartzite over carbonates of the Pennsylvanian and Permian Bird Spring Formation. East of the major thrust is the Glendate thrust, which overrides the Muddy Mountain thrust; both thrusts apparently flatten out at depth in Cambrian shales and both override Jurassic sediments.

The absence of lower or middle Tertiary sediments prevents continuation of the paleogeology south of the Las Vegas shear zone, but similar thrusts appear there, although they are displaced approximately 25 miles to the west. The Wheeler Pass and Keystone thrusts would correspond to the Gass Peak and Glendale-Muddy Mountain thrusts, respectively.

The structural geology of southeastern Lincoln County has been discussed only in an abstract (Tschanz, 1960a), but a county map has been published (Tschanz and Pampeyan, 1961). In the Mormon Mountains in the southeast corner of the county, a thrust system, probably a continuation of the Glendale thrust, is present; the main thrust brings Cambrian over Mesozoic rocks which are locally imbricate. The structure as illustrated on the paleogeologic map undoubtedly is oversimplified; it represents one plausible interpretation consistent with the available data. The involute pattern of the thrust is probably due to topography.

Allochthonous blocks in the Beaver Dam Mountains have been interpreted as Tertiary gravity-slide blocks by Cook (1960) and Jones (1963) and thus are not features of the paleogeology.

To the east of the larger thrusts are two related structures: the Iron Springs Gap structure (Mackin, 1947; 1960a, p. 114-119) which is a thrusted anticline formed at the end of a décollement in the Carmel Formation, and the Virgin-Kanarra fold (Gregory and Williams, 1947; Threet, 1963, 1963a).

The westernmost thrust of the Sevier belt in Lincoln County places Lower Cambrian over Upper Paleozoic rocks. It has a greater stratigraphic displacement than the other thrusts in this sector and may have the greatest total displacement. There is a problem as to exactly how it connects with the similar thrust in Clark County. On the paleogeologic map, they are shown as the same thrust affected by later, but pre-middle Miocene, normal faulting. According to an alternative interpretation suggested by D. H. Adair (1962, oral commun.), the thrusts are *en echelon*, the Gass Peak dying out northward, the thrust in Lincoln County growing in the same direction, each compensating for the changing displacement on the the other. In the first interpretation, certain Upper Paleozoic outcrops in Lincoln County are autochthonous relative to the major thrust; in the second, they are allochthonous.

Approximately 15 miles west of the trace of the major thrust in Lincoln County is an elongate exposure of Lower Cambrian and Eocambrian (€?) quartzite. The quartzite-Tertiary volcanic contact is well exposed near Delamar, a ghost town. Except for this anticline, the structure above the thrust is a gentle homocline from Cambrian to Mississippian and younger rocks of the hinterland. The upper plate of the major thrust extends for miles westward virtually undeformed in contrast to the imbricate belt to the east. the stratigraphic section of the upper plate must have been relatively competent. The Delmar anticline might be related to a step in the underlying thrust surface in the same manner that surficial anticlines are related to steps in such thrusts as the Cumberland Plateau thrust (Wilson and Stearns, 1958).

Wah Wah-Canyon Range Sector. Between the Wah Wah Mountains and the north end of the Canyon Range, there are many exposures of the Sevier belt, and structural continuity of the various isolated pieces of the major thrust of the belt may be demonstrated with reasonable assurance.

The upper plate of the major thrust can be traced with structural continuity from Lincoln County to the Wah Wah Range, but the actual thrust trace and the imbricate faults to the east are obscured by volcanics. Miller (1958; 1963) described the geology of the southern end of the Wah Wah Range and discussed the evidence for continuity of the upper plate of the major thrust which appears locally as the Wah Wah thrust, Frisco thrust (East, 1956; 1957), Mineral Range thrust (Liese, 1957), Pavant thrust (Maxey, 1946), and Canyon Range thrust (Christiansen, 1952). These thrusts have the Eocambrian-Lower Cambrian quartzites on their soles and override rocks ranging in age from Lower Cambrian to Jurassic, successively, toward the southeast. The southwest quarter of the geologic map of Utah (Hintze, 1963) shows that the Canyon Range allochthon overrides Paleozoic carbonates of the Pavant Range allochthon, thus demonstrating that the major thrust (the one which brings the Eocambrian-Lower Cambrian clastic sequence to the surface) bifurcates in this sector. A possible interpretation of this relationship is shown on the paleogeologic map.

East of the major thrusts lies the imbricate belt, but it is not well exposed and lacks the structural continuity of the major thrusts. Southeast of the Wah Wah thrust is the Blue Mountain thrust, a structural twin of the Glendale or Muddy Mountain thrusts to the southwest, on which Cambrian carbonates override Jurassic(?) sandstone. The section between the Blue Mountain and Wah Wah thrusts is complexly faulted, but all the faults are relatively minor. The over-all pattern is highly generalized on the paleogeologic map. In the Frisco area, the imbricate belt is complex and so confused by later volcanics, intrusives, and faults, that little generalization is possible at present. Southwest of the Pavant thrust, Crosby (1959) worked out an example of an overturned sheared-off limb of an anticline. The entire section from Cambrian to Triassic dips gently northwestward but is upside down! The major thrust overrides the overturned block, which is itself in thrust contact with underlying strata. The relationship of the overturned block to structural units southwest of it is uncertain.

The Sanpete-Sevier anticline east of the Canyon Range has been described by Gilliland (1963). Jurassic shale and other clastics protrude through Cretaceous sediments. Diapiric phenomena are reported to be partially responsible for the structure.

The Sevier Desert west and northwest of the Canyon Range is a large area where geologic data are unavailable. It has furnished a name, but little supporting evidence, for the Sevier orogenic belt.

Nebo-Charleston Sector. A fault of uncertain character separates the Canyon Range structural block from the Gilson Mountains and other areas to the north at the only locality where the relationships are not covered. Costain (1960) considered this fault, the Leamington fault, to be a thrust with a dip of approximately 30° N. It is also possible that the movement on the fault may be normal or, in part, strike-slip. Morris and Shepard (1964) considered it to be a strike-slip fault and offered a slightly different interpretation than that shown on the paleogeologic map for the area between it and the Tintic district to the north.

Between Mount Nebo and the vicinity of Salt Lake City is a distinctly different sector of the Sevier orogenic belt. The rocks involved in deformation are those of the Oquirrh basin. The differences in structural style and appearance of the map in this sector are believed to be due to the difference in response to deformation of the thick sediments of the Oquirrh basin and not to differences in age of deformation or applied stress.

The major thrust in this sector continues out of the Great Basin into the Southern Wasatch Mountains; at the southern end is the Nebo thrust (Eardley, 1934). Overturned Triassic and Upper Paleozoic rocks are thrust over Middle Jurassic shales at Mount Nebo. The Nebo thrust is connected beneath a covered area with the Charleston and Strawberry thrusts at the northern end of the Southern Wasatch block (Bissell, 1959; Crittenden, 1959). The Charleston thrust was recognized by Baker and others (1949) and has been described by Baker (1959) and Baker and Crittenden (1961). Along this fault, the thick Oquirrh basin section has been juxtaposed against the much thinner shelf section. At the sole of the thrust, Cambrian and Eocambrian quartzites are thrust over rocks as young as Middle Jurassic.

West of the thrust, there is a wide fold belt. In this belt are the Tintic and Stansbury folds Eocambrian rocks exposed in their cores and several smaller folds such as the Bingham and Pole Canyon synclines. Several of the folds are overturned; a few have developed into thrusts. The Stansbury anticline (Rigby, 1958) is a reactivated older fold whose trend and location were predetermined by a buried Devonian anticline. The geology of the Tintic anticlines has been discussed by Morris (1957). At the northern end of the Oquirrh Range is the east-west-trending, north-dipping, North Oquirrh thrust (Roberts and Tooker, 1961).

The large folds are confined to the Oquirrh basin, but folding continues westward. Much of the hinterland in Utah is folded, but the folds are apparently of smaller magnitude as pre-Pennsylvanian rocks are not exposed in their cores.

The paleogeology becomes complex in the West Tintic (Groff, 1959) and Sheeprock (Cohenour, 1959) areas. On the Sheeprock thrust, which was first described by Loughlin (Butler and others, 1920, p. 436), Eocambrian quartzites are thrust over lower and middle Paleozoic rocks. The thrust rises in the section eastward and has always been considered a normal part of the belt of thrusts in Utah. Cohenour (1959) recognized an additional thrust dipping north and northeast, which he called the Pole Canyon thrust, the upper block of which was considered to have moved south and southwestward in contrast to eastward movement on the Sheeprock thrust. Figure 3 shows an alternative interpretation; the Pole Canyon and Sheeprock thrusts appear to be the same thrust with eastward movement. This interpretation would explain eastward overturning of the Eocambrian section below the Pole Canyon thrust. Total displacement on the Pole Canyon thrust is about 5 miles, which would easily account for the 2-3 miles of stratigraphic displacement on the Sheeprock thrust. The Government Creek fault of Cohenour (1959) may be a separate thrust west of the Sheeprock thrust. South and west of the Sheeprock area is an area left bare on the paleogeologic map (Pl. 1, fig. 3) because lack of exposure makes it impossible to determine what relationship the complex fold-and-thrust area has to the belt of relatively simple structure to the south.

Northern Utah Sector. No direct connection between the thrust belt in northern Utah and the Nebo-Charleston sector has been found. Autochthonous rocks of the Uinta arch extend to the edge of the Salt Lake basin; paleogeologic information in the basin and to the northwest is very limited. It is possible, however, to trace the Oquirrh Formation from central to northern Utah with only a few covered areas about 5 miles across; there is no evidence suggesting a major structural discontinuity between the two areas. Crittenden (1959; 1961) pointed out that the thrust sheets in the Nebo-Charleston and northern Utah sectors are parts of the same structural block, a conclusion supported by the similarity in age, direction, and amount of thrusting in both areas. Accordingly, the trace of the major thrust outlines a large re-entrant near Salt Lake city, connecting the Charleston with the Willard thrust. The fault surface is probably close to the base of the Eocambrian section. In all its present-day exposures, the Precambrian Farmington Canyon basement complex is autochthonous, or nearly so, when it is contrasted with the Willard thrust sheet.

In the Northern Wasatch Mountains, paleogeologic and structural information are abundant, and the interpretation shown on the map can be considered reasonably certain. Except for the Farmington Canyon complex and the immediately overlying Paleozoic strata between Salt Lake City and Ogden, all the Northern Wasatch Mountains are an immense allochthonous block.

The Willard and associated thrusts have been described by Eardley (1944). East of Ogden, a thick geosynclinal section, including 10,000 feet of Eocambrian, overlies (along the Willard thrust) a relatively thin shelf section with no Eocambrian. The thrust dips east and at one time was thought to have moved westward; this view is no longer gener-

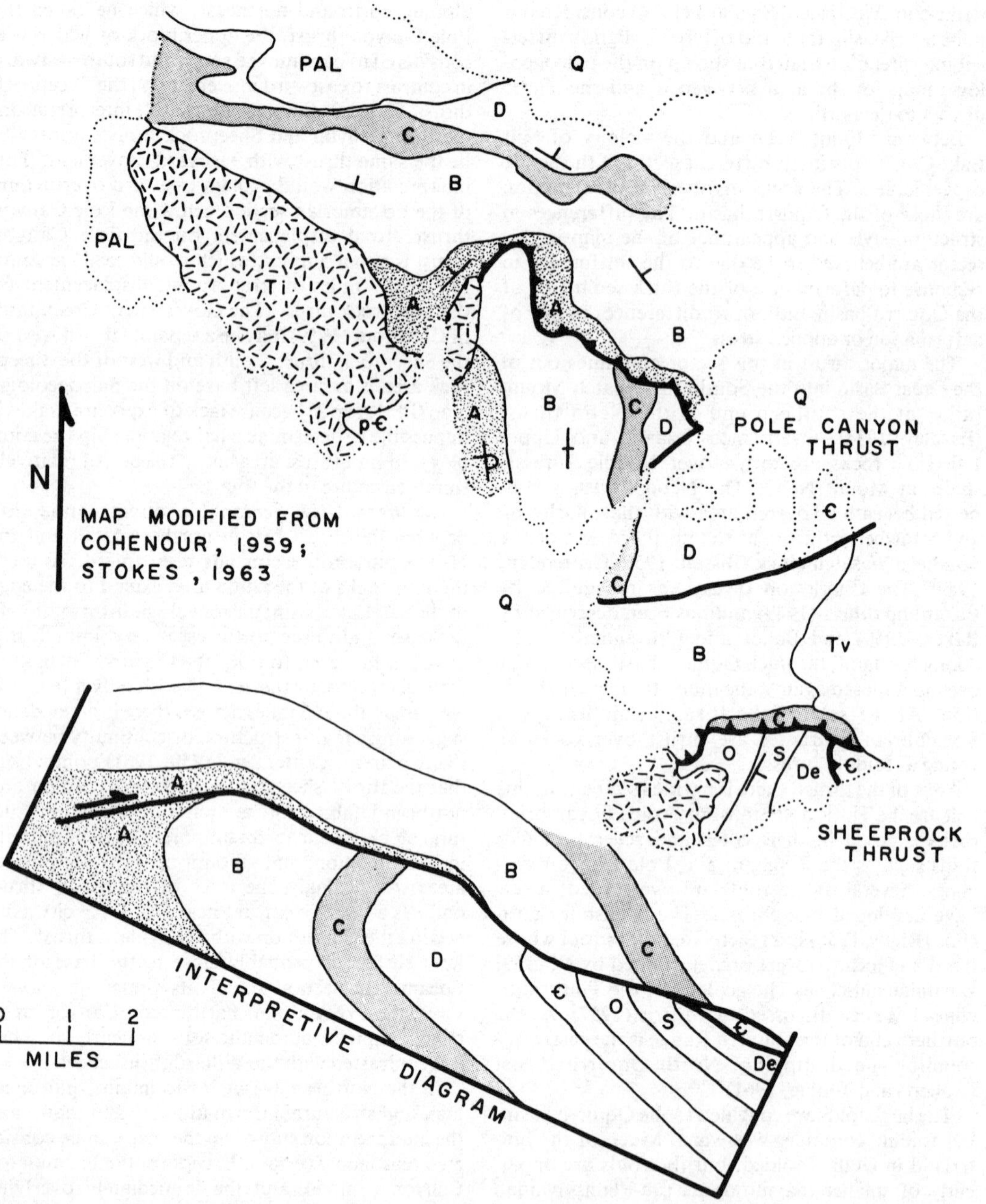

FIG. 3—Relationship of Pole Canyon thrust to Sheeprock thrust, Sheeprock-West Tintic area, Utah. Pre-Lower Cambrian stratigraphic units are indicated by *A, B, C,* and *D* from bottom to top, respectively; undifferentiated Paleozoic rocks, by PAL; Cambrian rocks, by Є; Ordovician rocks, by *O*; Silurian rocks, by *S*; Devonian rocks, by De; Tertiary plutons, by Ti; Tertiary volcanic rocks, by Tv; and Quaternary sediments, by Q.

ally accepted.

Drag features observable at Pineview Dam (including the gigantic Z fold) showing a down-dip (eastward) movement sense, and facies relationships requiring structural continuity between geosynclinal areas and the Willard block prove eastward movement. The thrust rises in the stratigraphic section to the east, which is consistent with eastward but not with westward movement. Indeed, no regional relationships support a western movement direction. Provable minimum displacement on the fault is 30 miles.

Structural continuity of the Willard block requires connection of the Willard thrust with the southern extension of the Paris thrust of Idaho. F. C. Armstrong and Cressman (1963, p. 18) did not accept this interpretation because of a presumed difference in age of the thrusts, but they were under the incorrect impression that the Willard is a Late Cretaceous or Paleocene thrust. They considered the Paris and Ogden-Taylor thrusts to be contemporaneous, and they also argued that the thrust near Woodruff Creek (west-dipping 30 miles east of Ogden) might be too insignificant to project 24 miles northward. The west-dipping fault at Woodruff Creek must have a *minimum* displacement of 30 miles. It *could not* die out before the Utah-Idaho boundary only 30 miles to the north. The Brigham Quartzite occurs on the sole of the thrust continuously from Paris, Idaho, to Woodruff Creek, Utah. In view of the lack of structural complexities between the two areas, it is reasonable to accept structural continuity between the Paris and Woodruff Creek and, hence, the Willard thrusts. The major thrust branches near the Idaho-Utah border, but this affects neither argument.

East of the major thrust is a series of thrusts that are part of the Idaho-Wyoming thrust belt. From west to east, these faults are the Crawford (Cambrian on Cretaceous), Absaroka (Upper Paleozoic on Cretaceous), and Darby (Jurassic on Cretaceous) thrusts. These thrusts appear to be successively younger eastward; each probably flattens out at depth into the same or successively higher stratigraphic horizons eastward. Rubey and Hubbert (1959) and F. C. Armstrong and Cressman (1963) discussed the evolution of the thrust belt north of Utah.

The Ogden and Taylor thrusts were described by Eardley (1944) as east-dipping with westward movement, in contrast to the eastward movement on the overlying Willard thrust (for a while all three thrusts were through to have moved west, but later the interpretation of the Willard was changed). Westward movement requires a separate episode of thrusting and a more complex structural history for northern Utah. This interpretation still is accepted by some geologists (Eardley, 1962; 1963; F. C. Armstrong and Cressman, 1963, p. 18-19). From the map pattern, a reinterpretation appears possible (Fig. 4). The Ogden and Taylor thrusts rise stratigraphically eastward relative to both upper and lower plates; this is awkward for a west-moving thrust and suggests that these thrusts moved eastward just as the overlying Willard thrust did. No large difference in age is necessary, and a separate orogeny is not required. The smaller thrusts could be simply peel thrusts in essentially autochthonous rocks overridden by the Willard block. It is noteworthy that basement crystallines are involved in these minor thrusts. The present dip of the faults must be the result of folding and of later eastward tilting of the Northern Wasatch Mountains; this tilting is already proven by the present dip of the Willard fault.

Amount of Shortening in Sevier Orogenic Belt

Several attempts have been made to estimate the amounts of shortening associated with parts of the Sevier orogenic belt. Rubey (Rubey and Hubbert, 1959, p. 190) gave an estimate of shortening by thrusting of about 75 miles for the Idaho-Wyoming belt north of Utah that was based on structural considerations. Crittenden (1961) estimated 40 miles for apparent displacement of isopachs by the Willard and Charleston thrusts. Along the Charleston thrust, a minimum of 20 miles can be concluded from eastward displacement of the Tintic Quartzite over Jurassic strata. Hintze (1960) gave an estimate of 12 miles for shortening by folding and thrusting at Mount Nebo. In the description of the Nebo-Charleston sector, 5 miles of displacement on the Sheeprock-Pole Canyon thrust was estimated. Hintze (1960) reported 8-10 miles of shortening in the Needles Range, which may be considered part of the Sevier belt. A minimum displacement of 20 miles is required for the Muddy Mountain and Glendale thrusts in southern Nevada in order to explain observable structural relationships (Longwell, 1961, oral commun.), not to mention the larger Gas Peak thrust in the same area.

In both northern and southern Utah, the sole of the major thrust is Eocambrian or lower Cambrian quartzite. From observable overriding relations, a minimum displacement of 25 miles in southern Utah and 30 miles in northern Utah is evident on this fault alone, and to this must be added the displacement on the faults farther east. The total displacement in the northern area must be greater than that estimated because the westernmost autochthonous and easternmost allochthonous sections are strikingly dissimilar. Additional intervening, now eroded, strata are required.

To summarize, it would seem that total shortening of 40 miles across the Sevier belt is a minimum value, and 60 miles is adequate to accommodate confortably any of the estimates. In the Nebo-Charleston sector much of the shortening is taken up by the folds and thrusts west of the major thrust so that the major thrust in that area may have a relatively smaller displacement (as little as 10-20 miles).

Structural Continuity of Thrust Belt

Between Gass Peak and Canyon Range, the evidence points to structural continuity of the upper

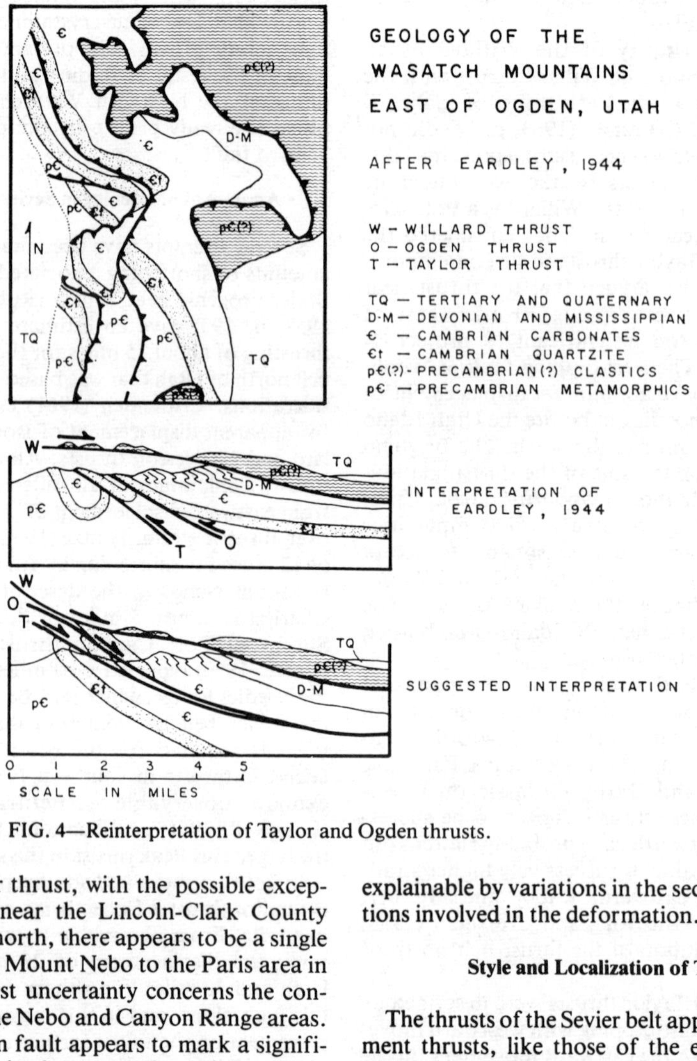

FIG. 4—Reinterpretation of Taylor and Ogden thrusts.

plate of the major thrust, with the possible exception of the area near the Lincoln-Clark County boundary. To the north, there appears to be a single major thrust from Mount Nebo to the Paris area in Idaho. The greatest uncertainty concerns the connection between the Nebo and Canyon Range areas.

The Leamington fault appears to mark a significant discontinuity in the upper plate of the major thrust of the Sevier belt; subsurface data may be required to understand the structural relations in this area.

The folds and thrusts east of the major thrust do not appear to be individually continuous. In places, structural complexities in this imbricate, highly deformed zone are unresolvable in the pre-normal faulting paleogeology, but the zone itself is continuous except where the thick sedimentary accumulation of the Oquirrh basin has been pushed eastward over the foreland.

In over-all view, the Sevier belt is a continuous entity from Nevada to Idaho; structural style, age of deformation, magnitude of shortening, and width of the highly deformed zone show no radical changes over a distance exceeding 500 miles. Admittedly, there are variations along strike, but these are explainable by variations in the sediment accumulations involved in the deformation.

Style and Localization of Thrusts

The thrusts of the Sevier belt appear to be décollement thrusts, like those of the eastern Canadian Rocky Mountains (O'Brien, 1960). The major thrust invariably appears to flatten out at depth in the Eocambrian quartzite-shale sequence (Pl. 1, fig. 4). There is no other way to explain the persistence along strike of the stratigraphic position of the thrust and the lack of apparent involvement of crystalline basement rocks in the major thrusts. If thrusts with displacements of 20 miles steepened westward, great uplifts of Precambrian would be inevitable. This sort of reasoning has led to similar conclusions for thrusts in Wyoming (Rubey and Hubbert, 1959, p. 187) and in southern Nevada (Longwell, 1950). The larger thrusts are restricted to the belt of rapid change of thickness of the entire geosynclinal prism. Under regional compression, the thick geosynclinal section of 30,000-40,000 feet of quartzite and carbonate with minor shale, has behaved with relative competence. The thinner section along the edge of the geosyncline has failed in one or more thrusts along which the geosynclinal strata are piled up on the transition zone. The shelf section was too thin to transmit stress and undergo significant deformation. The relatively broad transition zone in northern Utah and adjacent Wyoming resulted in a broader thrust-and-fold belt. A preexisting fold localized the Mesozoic Stansbury anticline. The Oquirrh basin was evidently less competent than other areas in the geosyncline, and, as a result, it experienced more intense folding and perhaps less shortening by thrusting than the rest of the belt during orogenic deformation.

The shaly zones in the Eocambrian and Cambrian clastic sequence provided the major zones of bedding-plane movement in the region; the major thrust rises from this level. Other important décollement zones are the Middle Cambrian shales, into which a number of the frontal thrusts apparently go downdip (Muddy Mountain, Glendale, Blue Mountain, Crawford), the various Mississippian shales, which are followed by important bedding-plane faults in the Southern Wasatch Mountains, and the Middle Jurassic shales and evaporites, into which several faults step up and flatten out (Blue Mountain, Pavant).

It is impossible to say definitely what happens to the décollement in the Eocambrian section as it continues westward from the Sevier thrust belt, for there is no direct evidence of its existence more than approximately 50 miles west of the trace of the major thrust. In northern Utah, the Taylor and Ogden thrusts have wedges of basement in their upper plates, but they are secondary thrusts about 30 miles west of the main fault trace. Along the Wasatch front near Santaquin, Utah, crystalline Precambrian rocks are shown on published maps in stratigraphic continuity with the upper plate of the Nebo thrust. If this is correct, then, the major thrust must be in crystalline rocks in that area. An alternative possibility is that the thrust is concealed as a bedding-plane fault along the mountain front.

In southern Nevada, the décollement can be observed 30 miles behind the frontal Wheeler Pass thrust as the Johnnie thrust described by Nolan (1929) and discussed by Burchfiel (1961, p. 129-131).

Eventually, the thrust must grade westward into mobile basement underlying the geosynclinal area. This concept returns to some degree to the idea Nolan (1929, p. 469-471) advanced to explain the Johnnie-Wheeler Pass thrust. Armstrong and Hansen (1966) proposed that a mobile basement underlay areas to the west of the thrust belt during the Mesozoic. During later stages of the deformation of the Sevier belt, particularly during some of the late postthrust folding, the basement within the fold and thrust belt may have become mobile enough to take part in the folding because of tectonic thickening of the overlying cover and the consequent rise in temperature of the basement. In earlier stages, apparently, the basement was relatively rigid, and deformation was confined to the sedimentary cover.

Hinterland

West of the Sevier orogenic belt on the paleogeologic map is a wide area of apparently simple structure that actually contains complex deformation of a different type than that in the Sevier belt. Most of the complexity lay deep below the surface in early Tertiary time, but fortunately, normal faulting has deeply exposed the region for geologic examination.

In the hinterland of southern Nevada, a major domal uplift is apparent in the Groom area. A number of exotic klippen occur near the dome; none has been mapped in detail, and little can be said about them. They contain lower and middle Paleozoic rocks and overlie middle and upper Paleozoic strata which are gently folded. D. H. Adair (1962, oral commun.) pointed out the close association of the apparent klippen with pre-normal faulting structural lows. They might be remnants of a large overthrust, but their chaotic character suggests that they are probably isolated blocks transported by gravity, most logically from the Groom dome. Similar blocks may occur west of the dome on the Nevada test site. Several thrust plates of upper Paleozoic rocks in the Southern Egan Range are shown by Tschanz (1960). These may be similar to the klippen in southern Nevada.

More work is needed to evaluate these klippen and their relationships. It should be stressed that later work may show the klippen, or some of them, to be nonexistent. Enough of the region is covered to allow a number of interpretations. The largest klippe in southern Lincoln County, the Pahranagat klippe, may be interpreted as a thrust-bound wedge without great transport, but the writer feels the klippe hypothesis is more likely on the present evidence.

East-central Nevada superficially appears to be a region of broad gentle folds. The angular unconformity between Paleozoic and Tertiary rocks in that region is small but increases westward. Cook (1965, p. 55) said, "the regional disconformity at the base of the volcanic sequence changes into an angular unconformity...in the central Egan Range, the Grant Range, the White Pine Mountains and the Pancake Range."

Many ranges in this area show a complex internal structure characterized by bedding-plane or near bedding-plane "shearing off" faults, i.e., faults where beds are consistently cut out instead of repeated (Misch, 1960). Misch (1960) discussed many examples of this type of structure; in a number of cases, he considered such faults to be parts of his regional décollement. In contrast to the complexity observed in many ranges, there are well-exposed ranges in the same area where the entire Paleozoic section can be observed intact without significant faulting. This contrast between simplicity and complexity of structure in the ranges was noted by Misc (1960) and is one of the remarkable features of the hinterland.

The near conformity of Tertiary volcanic strata with Upper Paleozoic strata in the same region containing this complex faulting has led some to conclude that most of the deformation is Tertiary (J. C. Young, 1960, p. 169-170). Much of the deformation may well be Tertiary and in some areas almost certainly is, but in the Snake Range "shearing off" faulting as well as folding and high-angle faulting predate Cretaceous plutons (Misch, 1960, Misch and Hazard, 1962). There appears, therefore, to have been fairly intense deformation, at least locally, before the time represented by the paleogeologic reconstruction. the paleogeologic method would be unable to detect bedding-plane faults that did not juxtapose rocks of contrasting age or that did not reach the early Tertiary surface. Deformation actually appears to have been greater at depth than near the surface. Most of the "shearing-off" faults occur in lower Paleozoic rocks, often in the Cambrian-Eocambrian part of the section (Misch, 1960).

From the available data, it seems necessary to conclude that in the hinterland of the Sevier belt, the surficial structure was fairly simple—mostly broad folds, with no complex tightly folded areas. Only upper Paleozoic and Mesozoic rocks were exposed at the surface, for nowhere do Oligocene volcanics lie on lower Paleozoic rocks. Even in Sacramento Pass between the central and southern Snake Ranges (in the blank area on Pl. 1, fig. 3), Oligocene volcanics lies on Pennsylvanian rocks disconformably. Much of the chaotic structure in that area is Tertiary (Misch, 1960). No major overthrusting involving telescoping of the Paleozoic section occurred in this region. Total shortening of the supercrustal rocks must have been relatively minor compared to that in the Sevier belt. In early Tertiary time, the complex structures of this part of the eastern Great Basin were deeply buried.

East of the Snake Range is the Confusion Range fold belt which is a narrow, deeply downfolded zone containing strata as young as Mesozoic. It is possible that this belt is the surface exposure of a zone sucked down between adjacent rising domal structures in a mobile basement. Alternatively, it could be a surficial structure formed by gravity-propelled gliding of upper Paleozoic rocks eastward off the rising Snake Range dome, possibly even in Tertiary time. In the Gold Hill area, the carefully worked out structural history of Nolan (1935) is in agreement with a simple paleogeology. In Eocene time, folded upper Paleozoic rocks were present at the surface. One fold was fairly tight and overturned to the east. The larger thrusts and most normal faults in the area are post-Eocene and, therefore, later than the time represented by the paleogeologic reconstruction.

A model for the deep-seated structure of the hinterland has been proposed by Armstrong and Hansen (1966). The present-day fault-block ranges expose different tectonic levels which may be interpreted as an *infrastructure*, a mobile deeply buried zone affected by regional metamorphism, and a *suprastructure*, an unmetamorphosed zone of broad open folds developed under a relatively thin cover. These two tectonic levels may grade into one another, but in most areas, they are separated by a zone of tectonic adjustment with disharmonic deformation and steep metamorphic gradients, an *Abscherungszone*.

The hinterland of the Sevier orogenic belt on the paleogeologic map exposes only the suprastructure—unmetamorphosed upper Paleozoic rocks, broadly folded but with little over-all shortening. Deeply buried in early Tertiary time but now exposed in areas such as the Snake Range, Ruby Mountains, and Raft River Range is the infrastructure—metamorphic rocks, in places, migmatites, with complex minor structures and large-scale recumbent folds. The transition between these two tectonic levels occurs in most areas in the upper part of the Eocambrian clastic section, but elsewhere, it rises into lower, and, in the Ruby Mountains area, even to middle Paleozoic rocks. This zone, in which adjustments were made for differential deformation of infrastructure and suprastructure, is the site of zones of intense shearing. In many areas, faults have been recognized in or closely related to this zone; they are the younger-on-older thrusts that characterize the hinterland. A note of caution is needed, however, because other structures also have been categorized as younger-on-older thrusts so that some confusion on this point is inevitable. Misch (1960) reviewed the occurrence of the structural complexities the *Abscherungs-zone*, although in a somewhat different context than the tectonic framework proposed by Armstrong and Hansen (1966).

Both Misch (1960) and Armstrong and Hansen (1966) concluded that the orogeny which affected the hinterland was pre-Lower Cretaceous, making the structures distinctly older than the Sevier orogenic belt. Lower Cretaceous plutons which cut across both orogenic structures and metamorphosed Eocambrian-Lower Cambrian rocks are unequivocal evidence of this. The regional décollement of Misch (1960) cannot be related directly to the Sevier orogenic belt as has been proposed by Miller (1963), for it is too high stratigraphically and distinctly older. Moreover, it does not appear possible to the writer or to Misch to consider that the Sevier belt is the result of gravity gliding eastward from central Nevada, the only possible area of "tectonic denudation." Rather, I feel compelled to seek the locus of shortening displayed by the orogenic belt in crustal shortening and deformation at depth within and west of the Sevier belt. The Paleozoic blanket of the hinterland was already deformed and "nailed down" by cross-cutting Lower Cretaceous plutons; only a very contrived explanation could specify a locus of gliding to bring the Sevier belt out of Nevada by gravity.

An entirely analogous situation is provided by the Rocky Mountains in Canada. There, regional metamorphism complex deformation, and emplacement of plutons of batholithic dimensions in the hinterland occurred before thrusting, with shortening exceeding 100 miles in the Rocky Mountains (White, 1959; Shaw, 1963; Gabrielse and Reesor, 1964). It appears necessary to invoke some sort of *Verschluckung* to explain the development of these later Mesozoic fold and thrust belts.

STRATIGRAPHIC EVIDENCE CONCERNING AGE OF DEFORMATION IN SEVIER OROGENIC BELT

Evidence for a Pre-Cretaceous Sevier Arch

It was in later Permian time, perhaps, that the orogeny began in the Sevier belt. Permian isopachs show a southwest-trending arch in southwestern Utah and adjacent Nevada that coincides with the belt of Mesozoic thrusting. Brill's maps (1963, p. 319) show that this arch affected the distribution of Leonardian strata but not the distribution of Wolfcampian strata. Data for later times are lacking, but possibly the arch is a precursor of later deformation in the area; it would be an ancestral Sevier arch.

Throughout the region, the basal contact of the Triassic is a disconformity above Permian strata. The time gap represented by the contact is somewhat greater in southern areas than in northern ones. In southern Nevada, erosional relief of more than 100 feet and coarse conglomerates are commonly found at the contact (Longwell, 1925; Secor, 1962). In the westernmost exposures, in the Spring Mountains, at least 1400 feet of Permian strata are missing below the disconformity, perhaps as a result of uplift and erosion on the ancestral Sevier arch.

Jurassic deposits, now occurring only in the easternmost part of the Sevier orogenic belt and on the Colorado Plateau shelf, are significant as sources of indirect information concerning events in the Great Basin. The shift of sediment sources during the Jurassic has been discussed by Wright and Dickey (1958, 1963, 1963a) and Stokes (1960, 1963a, 1963b). During Middle Jurassic time, the hinterland of the Sevier belt became a source of clastic sediment. In the southern Wasatch Mountains, Upper Jurassic and lowermost Cretaceous conglomerates occur conformably on both upper and lower plates of the major thrust of the Sevier belt (Bissell, 1959, p. 163; Stokes and Madsen, 1961), suggesting that uplift and deformation there began after the end of the Jurassic. In northern Utah, later Jurassic formations are truncated westward by erosion which preceded deposition of Lower Cretaceous Kelvin Conglomerate. In southern Idaho, movement on thrusts of the Sevier orogenic belt may have begun as early as the end of the Jurassic (F. C. Armstrong and Cressman, 1963, p. 8-16).

In summary, evidence for a pre-Cretaceous Sevier arch is suggestive but not conclusive. There is no evidence that extensive uplift or thrusting began in the belt before the end of the Jurassic; the story for the Cretaceous is an entirely different matter.

Cretaceous to Paleocene—Rocky Mountain Geosyncline

East of the Great Basin, an enormous accumulation of Cretaceous sediments provides a detailed record of the advances and retreats of the Cretaceous seas in response to eustatic changes, deformation, and repeated great floods of clastic materials from a westward source. Facies and thickness changes indicate a source west of the present outcrop areas.

The Sevier orogenic belt was the source of material that accumulated in the Cretaceous Rocky Mountain geosyncline (Fig. 5). This is proven by contemporaneity of deformation in the orogenic belt with sedimentation in the immediately adjacent area to the east, by the coarsening of clastics toward the orogenic belt, by the coarsening of clastics toward the westernmost deposits (some clasts being too large for transport of more than a few miles), and by conclusive provenance studies. The Sevier orogenic belt is the only possible source for the large quantity of carbonate and quartzite clasts derived from lower Paleozoic and Eocambrian rocks of the Cordilleran geosyncline. The prenormal faulting paleogeology shows that rocks of the same age and facies were not exposed elsewhere in the Great Basin during Cretaceous time.

In central Utah, a relatively complete sequence of Cretaceous strata have been studied by Spieker (1946, 1949, 1956) and his students (Schoff, 1951; Hardy and Zeller, 1953). The Cretaceous of Utah recently was reviewed by Burger (1963). Early Cretaceous rocks are absent in most westernmost exposures of Cretaceous formations because of Early Cretaceous uplift and erosion; however, Early Cretaceous rocks are present only a few miles to the

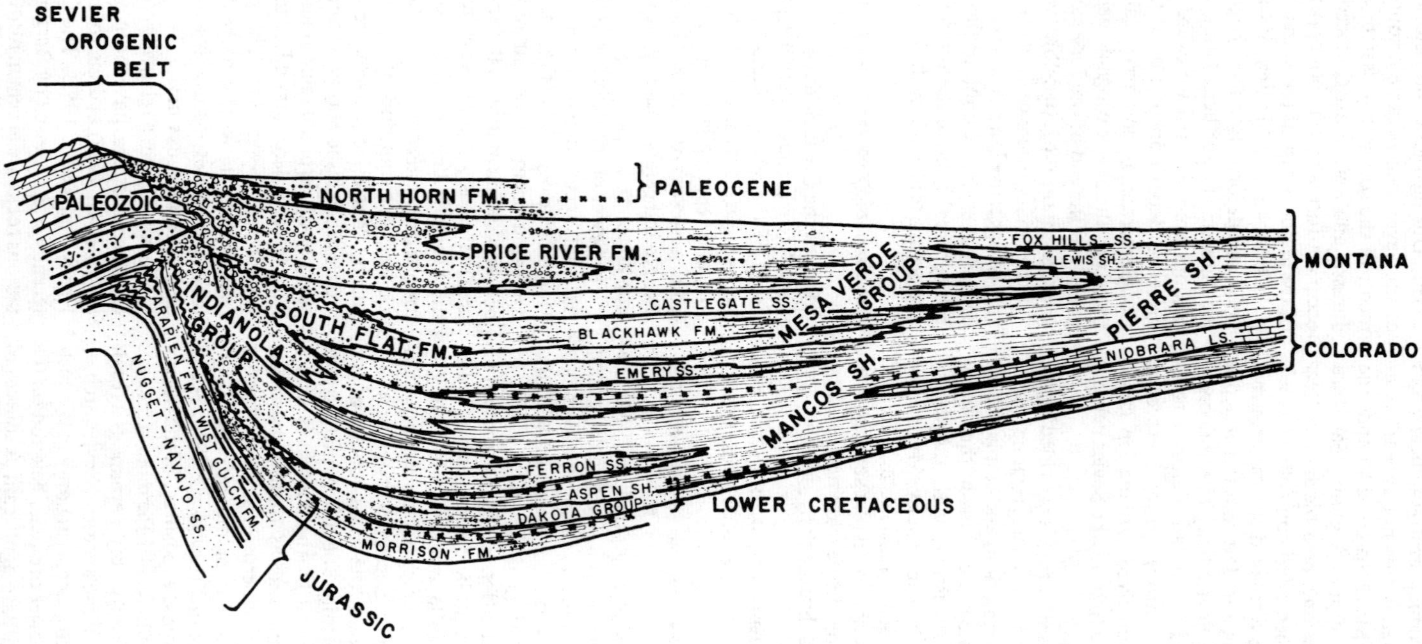

FIG. 5—Diagrammatic section across Rocky Mountain geosyncline in central Utah. (Principal sources: Cobban and Reeside, 1952; Risher and others, 1960; Spieker, 1949).

east. Numerous angular unconformities occur within the Cretaceous section along its western edge, indicating concurrent deformation and deposition. In central Utah, middle and upper Montana Price River Formation locally lies unconformably across major thrusts of the Sevier orogenic belt (Hintze, 1962). In southwestern Wyoming, a similar situation prevails in that upper Montana and Paleocene Evanston Formation lies unconformably across major thrust faults (Tracey and Oriel, 1959). In both areas, some deformation in the Sevier belt postdates the end of the Cretaceous.

The Cretaceous Rocky Mountain geosyncline deposits thicken gradually from east to west in Utah, reach a maximum near the westernmost exposures, and thin drastically toward the adjacent deformed belt. The greatest thickness of Lower Cretaceous deposits occurs in southern Idaho. The maximum thickness in any given east-west cross section decreases gradually from 15,000 feet in Idaho to a few hundred feet in southern Nevada. The Upper Cretaceous section is thick in central Utah, thins fairly rapidly southward toward southern Nevada, and thins slightly, and then thickens again northward into Idaho and Wyoming.

Review of information provided by clast provenance. The Sevier orogenic belt was the only possible source for the coarse lower Paleozoic clasts found in the Cretaceous deposits of the Rocky Mountain geosyncline. The sequential exposure of older and older units in the evolving orogenic belt resulted in an inverted stratigraphy of the lower Paleozoic clasts. The structures in the belt are large thrusts with a few folds; this type of deformation must have provided sources for the clastics.

The information available on clast composition is summarized in Figure 6. By the beginning of the Cretaceous, upper Paleozoic clastic sources were present within the orogenic belt so that at least locally the Mesozoic cover had been removed. Ordovician quartzites first appear in lower Colorado sediments in southern Nevada and in the Lower Cretaceous sediments in Idaho. The distinctive Eocambrian quartzites appear first in the upper part of the Indianola Group (Colorado) of central Utah, and they become widespread and abundant in Montana conglomerates from southern Nevada through Wyoming, at least as far north as the Teton area (Love, 1956). To a certain degree, the appearance of abundant Eocambrian clasts is almost coincident throughout the length of the Sevier orogenic belt. This would suggest that the evolutionary stages of the orogenic belt were similar over the entire region and that structures within the belt are of similar age and may be extrapolated and correlated along strike.

Problem of Canyon Range Fanglomerate. In the Canyon Range of central Utah (Christiansen, 1952), more than 10,000 feet of red-matrix boulder conglomerates, conglomerates, and sandstones, with local lacustrine limestones, occur. These clastic rocks, situated 20 miles west of the westernmost standard Cretaceous sections of the central Utah plateaus, will be referred to as the Canyon Range fanglomerate. The fanglomerate lies with great angular unconformity across a major thrust, which places Eocambrian quartzites of the Cordilleran geosynclinal section over Cambrian and younger carbonates. The fanglomerate strata have been tilted to near vertical attitudes, and locally minor movement along the major thrust has placed the Eocambrian quartzites in thrust contact with them.

Many writers have considered this area significant for the dating of orogeny in central Utah, but the quoted relationships here are discordant with those elsewhere in the fold-and-thrust belt. This is due to Christiansen's (1952) assignment of the Canyon Range fanglomerate to the Indianola (?) conglomerate. He made the Indianola correlation strictly on the basis of lithologic similarity. No alternatives were discussed, but at the time regional information was much more limited, and nothing about the age assignment seemed unreasonable. The time has come for reassessment of the age of the Canyon Range fanglomerate.

Variegated siltstone, conglomerate, sandstone, and light-colored limestone that overlie the fanglomerate with a slight angular ($\sim 10°$) unconformity were considered North Horn (?) Formation, thus establishing a Cretaceous age for the fanglomerate, which was assigned to the Indianola (?), although fossil evidence or tracing of lateral continuity were lacking. The Canyon Range fanglomerate was also correlated with red conglomerates in the Snake Range of eastern Nevada, which are now known to be Tertiary (Armstrong, 1964, p. 73). This is typical of how misleading lithologic correlations among clastics in this region may be. Spieker (1949, 1956) emphasized that similar facies are repeated in formations of different age.

The writer suggests that the Canyon Range fanglomerate may be a lateral equivalent of the Paleocene and early Eocene Flagstaff Limestone of the Utah Plateaus. This correlation can be supported by evidence as strong as that supporting an Indianola (?) age, although either correlation may be proven to be correct when fossil evidence is found. The important point is that the age of the fanglomerate is not well known.

Nowhere do rocks known to be older than upper Montana lie unconformably across major thrusts; hence, in terms of structural position, the fanglomerate should be Price River or younger. The basal conglomerates of the Canyon Range fanglomerate are composed mostly of Eocambrian quartzite, some boulders exceeding eight feet in diameter. An

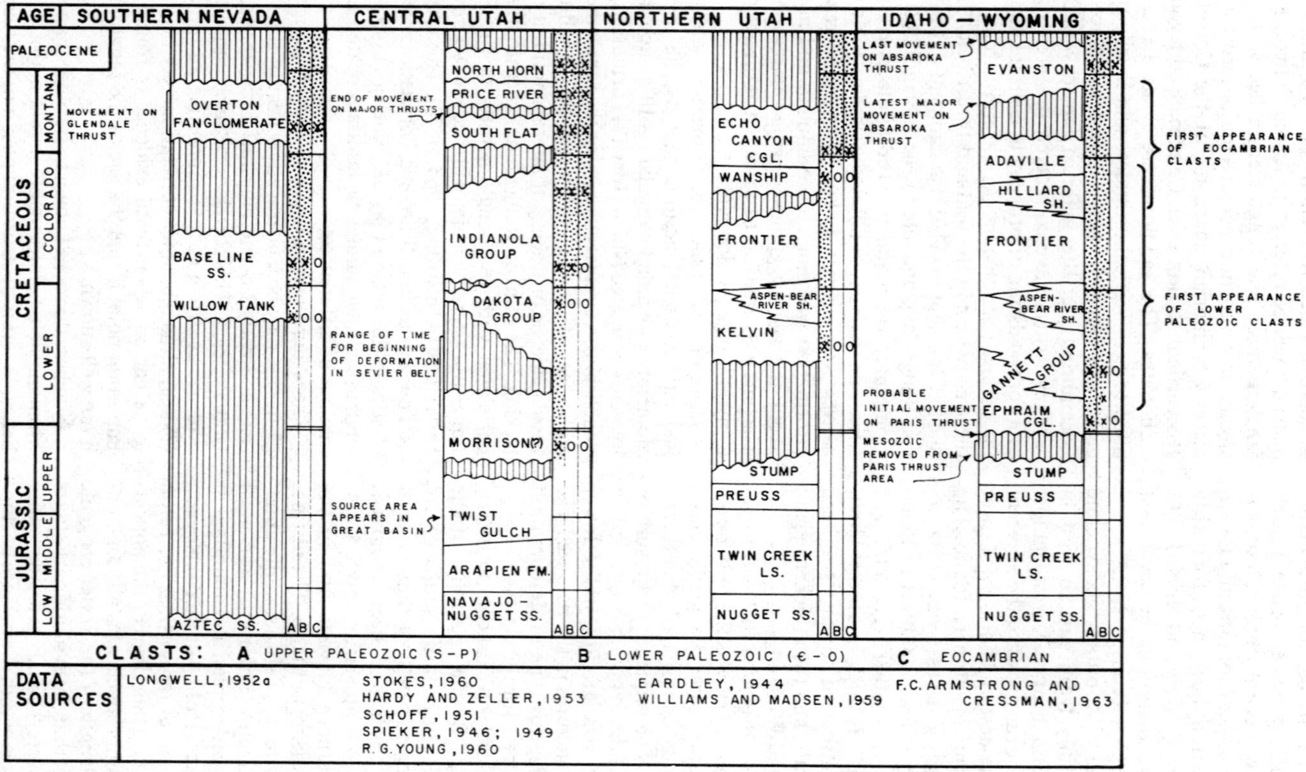

FIG. 6—Jurassic to Paleocene correlation chart showing inverted stratigraphy of clasts in Rocky Mountain geosyncline.

obvious local source is present in the upper plate of the Canyon Range thrust. In the Indianola strata of the Gunnison Plateau (due east of the Canyon Range), the lower beds contain none of the distinctive Eocambrian quartzite, but higher in the section it appears in modest quantities (5-10 percent locally). Only the Price River and later conglomerates contain the Eocambrian quartzites in abundance. Clast composition, therefore, would allow, at best, correlation of the Canyon Range fanglomerate with upper Indianola, and a Price River or later age would be more likely.

Where proven, Indianola beds are as closely associated with major thrusts as the Canyon Range fanglomerate, the deformation is much greater. Northeast of Nephi, Indianola conglomerates underlie an angular (almost 90°) unconformity below Price River conglomerates. The slight angular unconformity between Christiansen's Indianola (?) and North Horn (?) formations in an area much farther within the Sevier orogenic belt is further evidence against the age he assigned to the older conglomerates.

On Long Ridge, northeast of the Canyon Range, Muessig (1951) recognized a facies equivalent of the Flagstaff lacustrine deposits of the High Plateaus to the east that was composed of red matrix conglomerates and sands. This correlation is established by fossil evidence and intertonguing with other Tertiary formations. Flagstaff age conglomerates exposed near Mills, Utah, only 10 miles from the Canyon Range, are virtually identical to conglomerates of the Canyon Range fanglomerate. Rock types and clast compositions match so closely that on lithologic criteria correlation of Flagstaff Formation with Canyon Range fanglomerate would be stronger than the Indianola-Canyon Range correlation. This is by no means proof, but it emphasizes the fact that facies can be matched in rocks of widely different age. The Indianola of the Gunnison Plateau certainly is similar to the Canyon Range fanglomerate, but it is not as perfect a match as the Flagstaff of Long Ridge. As to later deformation and structural position, the Flagstaff on Long Ridge is reported by Muessig (1951) to be locally tilted to an almost vertical attitude, just like the Canyon Range fanglomerate.

Thus, much doubt can be cast on the correlation of the Canyon Range fanglomerate with the Indianola Group of central Utah, and it is at least equally possible to argue for a Flagstaff age although any age between Indianola and Flagstaff eventually may be established. The post-thrusting clastics of the Canyon Range should not be considered evidence of a distinct episode of pre-Upper Cretaceous thrusting in Utah, unless fossil proof of a Colorado age is found.

Summary

The evidence provided by the clastic rocks, unconformities, and structural relationships can be used to arrive at a summary of the timing of deformation in the eastern Great Basin. A faint suggestion of uplift along the Sevier belt is found in lessened Permian deposition and pre-Triassic erosion along a NE-SW-trending belt in southwestern Utah and southern Nevada. By Late Jurassic time, deformation was well under way within the Great Basin, and the area was sufficiently elevated to be a clastic source. By the end of the Jurassic, fairly coarse conglomerates were being shed.

In the southern Wasatch Mountains, faulting began just after the end of the Jurassic. In this sector of the Sevier orogenic belt, as elsewhere, deformation probably began earlier farther west (Sheeprock thrust, folds in the Oquirrh basin). Movement on the major thrust may have begun as early as the end of the Jurassic in northern parts of the Sevier orogenic belt (F. C. Armstrong and Cressman, 1963, p. 8-16). During most of Early Cretaceous time, thrusting must have been going on in the Sevier orogenic belt. By early Colorado time, lower Paleozoic clasts were being shed from the entire belt. Later in Colorado time, Eocambrian rocks became significant sediment sources in central Utah, and by Montana time clasts of these rocks were supplied in abundance from the entire belt. The Eocambrian rocks within the fold and thrust belt occur only in the soles of major thrusts. For the area to have been a source of these rocks, thrusting must have been of large magnitude by the beginning of Montana time. F. C. Armstrong and Cressman (1963, p. 8-16) consider that much of the movement on the Paris thrust in Idaho occurred during Lower Cretaceous time. Movement on the major thrusts of the Sevier belt was probably approximately coincident with movement on the Paris thrust.

Major thrusting in the Sevier orogenic belt ceased before the end of Montana time, for the upper beds of Price River of equivalent strata lie unconformably across the large thrusts. Later deformation consisted mostly of folding, at times with local thrusting. This period of waning deformation lasted into the Paleocene. During the Eocene, the orogenic belt was quiet, and erosion actively reduced the relief inherited from the deformation.

Within the broad continuous pattern of deformation of the eastern Great Basin, some fine structure can be recognized. In southern Nevada, earlier deformation in the Sevier orogenic belt began in the west (Longwell, 1952a; Secor, 1962). In the imbricate belt of Idaho and Wyoming, a series of thrusts become younger eastward (Rubey and Hubbert, 1959; F. C. Armstrong and Cressman, 1963, p. 8-

16). In Utah, it is possible, but not provable, that deformation likewise began in the west along the major thrusts. The clastic source evidence does indicate that the major thrusts were active during much of the deformation of the belt.

The clastics of the geosyncline in the Rocky Mountains can be subdivided into three great floods or complex tongues. The first was Upper Jurassic and Lower Cretaceous (Morrison, Dakota, Kelvin, Gannett). Diminished influx of clastics allowed the uppermost Lower Cretaceous shales (Aspen and equivalents) to spread over a wide area. The second clastic flood corresponds to the Indianola Group and Frontier Formation of Colorado age. This was followed by another time of widespread shale deposition in late Colorado time (Hilliard; part of Mancos). The flood of clastics in Montana time came with the final major deformation of the orogenic belt (Price River, South Flat, Echo Canyon, Mesaverde, Adaville). The retreat of the sea in latest Montana time after deposition of the "Lewis" Shale tongue above the Mesaverde Group in eastern Utah marked the end of the Rocky Mountain geosyncline.

The total volume of Cretaceous sediment eroded from the orogenic belt corresponds to a strip of material 25,000 feet thick and approximately 100 miles wide (estimated from isopachs of Weimer and Haun, 1960). More than half of this volume can be accounted for by thrusting of the geosynclinal section over the shelf with shortening corresponding to approximately 50 miles. The remainder would have to be derived from erosion of folds farther west and in the orogenic belt. Crudely, at least, the amount of sediment in the Cretaceous geosyncline equals that estimated on structural grounds to have been removed from the orogenic belt. Of the three major clastic floods, the second two are of greatest significance volumetrically. If amount of deformation and amount of debris could be correlated, then presumably most of the evolution of the Sevier belt took place in Late Cretaceous time. Such a correlation has not yet been established.

TERTIARY STRUCTURES

This discussion would be incomplete without mention of some of the Tertiary complications which have obscured the Sevier orogenic belt and its possible relation to them. The most evident structural development in the Tertiary was the breakup of the region by normal faults which are discussed in detail by Nolan (1943), Mackin (1960*a*), and Moore (1960). The major late Tertiary faults have a generally north-south trend and displacements commonly greater than 10,000 feet; some displacements exceed 30,000 feet. The origin of the normal faulting is a matter of debate. Thompson (1960) supported the concept of crustal stretching as the basic cause. Mackin (1960) has presented an eruptive-tectonic hypothesis that explains the faults as collapse features due to withdrawal of lateral support by eruption of 60,000 cubic miles of Tertiary volcanics. The time lag between volcanism and normal faulting presents this hypothesis with difficulties. It is too attractive an explanation to be rejected, however, and is probably an important part of the phenomenon, especially in areas like the Nevada Test Site where volcanism and faulting were synchronous. The regional pattern of the faulting and the timing anomaly support the crustal-stretching hypothesis. The regional faulting would have a deep tectonic control, and the volcanic eruptions would simply modify the pattern.

One strong suggestion in favor of over-all thinning of the crust in the Great Basin region is its present topographic altitude. It is, on the average, lower than surrounding areas. The Sierra Nevada on the west and the High Plateaus of Utah and the Wasatch Mountains on the east project far above the Great Basin region. Without crustal thinning, the region would probably now stand at altitudes of more than 10,000 feet as an Alpine chain undergoing deep dissection. Stretching has resulted in normal faulting, crustal thinning, and resulting lowering of average altitude. Consequently, structures that formed at shallow depths in the orogen have been preserved, and there is an unusual opportunity to observe here all depths of an orogenic belt simultaneously. The stretching of the crust in the Great Basin may be only part of a much larger tectonic system, the East Pacific Rise, that was described by Menard (1960). The strain appears to have taken the form of distributed faulting in contrast to rift valleys common elsewhere on oceanic rises or on the rise in Africa. Perhaps in the orogenic belt, even some tens of millions of years after deformation had ended, the crust would still be warmer and thereby weaker than elsewhere. That the lower crust and upper mantle under the Great Basin is relatively weak is shown by Crittenden's analysis (1963) of the isostatic rebound in the Bonneville basin which gave viscosities an order of magnitude smaller under the Great Basin than those found in Fennoscandia. Under "tension" (when the vertical principal stress exceeds the horizontal principal stresses in deeper parts of the crust), the softer zones of the crust would stretch like taffy, but shallow rocks would fail by faulting of the Basin and Range type. The Colorado Plateau was unaffected because it lacked a softened basement; thus, the weak basement hypothesis provides an explanation for the coincidence of the eastern margin of the Basin and Range province and the eastern edge of the Sevier orogenic belt.

Normal faulting is not the whole story, however.

Several ranges, particularly those most elevated during the Tertiary, appear to have been arched and may not have faulted margins (Misch, 1960, p. 19-20; Snelson, 1957; Woodward, 1964; Felix, 1956). These ranges occur in the hinterland of the Sevier belt and contain rocks metamorphosed during the Mesozoic; these rocks would have been warm and more plastic than the shallow rocks of the suprastructure. As a consequence, these ranges could have flexed during uplift, perhaps developing beneath horsts formed at shallower tectonic levels. Brittle deformation would characterize the suprastructural rocks; all suprastructure ranges appear to be simple fault blocks.

At a number of localities domal structures related to Tertiary plutons have formed (Mackin, 1960*a*; Blank, 1959; Cook, 1957; Wisser, 1960). Structural relief resulting from normal faulting, arching, and doming has resulted in gravity slides, some of large dimensions (Mackin, 1960*a*; Misch, 1960; Moores, 1963; Secor, 1962; Bissell, 1964; Cook, 1960; Jones, 1963; Armstrong, 1964). The structural style characteristic of these slides is pervasive brecciation. In addition to newly initiated faults, many older faults may have been reactivated during the Tertiary extension-gravity sliding regimen. The present complex geology represents the summation of Jurassic, Cretaceous, and Tertiary effects. With careful work, as many as five episodes of structural development have been worked out in some quadrangles (for example, Nolan, 1935).

NEVADAN, SEVIER, AND LARAMIDE OROGENIES

According to Wilmarth (1938), the Nevadian (Nevadan) revolution was a term applied to early Early Cretaceous and Late Jurassic diastrophic movements; the Laramide revolution was a mountain-building period in the Rocky Mountain region that began in Late Cretaceous time and ended in early Tertiary time. These terms, grounded in early conceptions of the nature of the orogenic record, are widely used today in the writing of Cordilleran geologists, and they have been so well established that new terms or concepts have not received acceptance. Although most recent reviews of regional geology in the west (King, 1959; Clark and Stearn, 1960) acknowledge an over-all continuity of orogeny in the Cordillera, the pictures presented indicate that confusion has not been eradicated. It is the universal practice to lump structures of the Sevier fold-and-thrust belt and the basement uplifts of the eastern Rocky Mountains under the term Laramide. Pre-Laramide orogenic activity in the Sevier belt is acknowledged, but the emphasis supplied by the two discrete names Nevadan and Laramide never allows a balanced picture to emerge. Attempts to extend, broaden, or blur the two concepts, or to refine subdivision within and between Nevadan and Laramide have only perpetuated confusion.

Basically a revision of the nomenclature is required. At best, the most that can probably be achieved is to add another term to the collections, something to allow at least a mental divorce of fundamentally different concepts now blurred into one.

In the Rocky Mountains, characterized by basement uplifts, movement began close to the end of the Cretaceous. Keefer and Love (1963) reviewed the evidence in Wyoming. Uplift began in Maestrichtian time, was intense during Paleocene and early Eocene time, and had ceased by middle Eocene time. The orogenic movements here are Laramide in the classic sense, as the Laramie Range in southeastern Wyoming was affected.

In the Sevier orogenic belt, orogenic deformation began approximately at the beginning of the Cretaceous, and major thrusting ended in Campanian time. Campanian and Maestrichtian Price River Formation unconformably overlies the thrusts in Utah, and Maestrichtian (?) Evanston Formation overlies those in Wyoming. Relatively minor movements continued as late as Paleocene in both areas. Thus the common implication that the Cordilleran thrusts are Laramide structures is erroneous. The orogenic structures of the Sevier belt are distinctly different in age from classic Laramide, and moreover, they represent a drastically different sort of deformation and tectonic regimen. General considerations on the nature of orogeny in the Cordillera are impossible unless this fundamental distinction between the Sevier belt and eastern Rocky Mountains is clearly recognized. If Nevadan and Laramide are terms that will continue to be used by geologists, a new term, of equal rank, is necessary. The term Sevier orogeny is suggested for the deformation which produced the structures of the Sevier orogenic belt, largely during Cretaceous time. This orogeny lies in the middle of a period of geologic time (Fig. 7). If it is used in the sense of the definition, it should help clarify discussion of the regional history. A proper summary then is that the central Cordillera of the western United States was affected during the Mesozoic by at least the Nevadan, Sevier, and Laramide orogenies. These orogenies, along with the Paleozoic Antler and Sonoma orogenies, are the principal Cordilleran orogenies which have been significant in the development of the central Cordillera of the western United States. Even this statement is an embarrassing oversimplification.

REFERENCES CITED

Adair, D. H., 1961, Geology of Cherry Creek District, Nevada: M. S. Thesis, Univ. of Utah, 112 p.

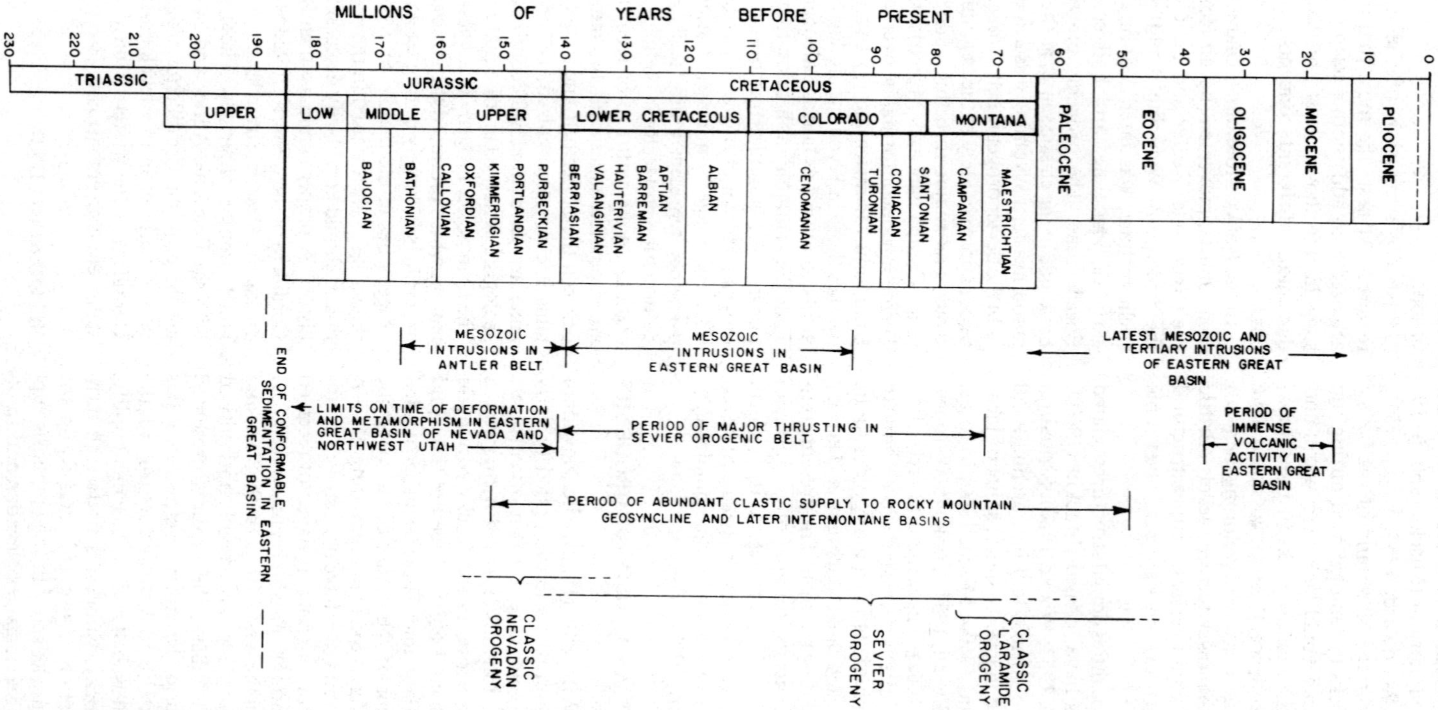

FIG. 7—Geologic time scale. Tertiary time scale *after* Evernden and others (1964); 120 m.y. to Tertiary *from* work of Folinsbee and others (1960); and 230-120 m.y. *modified from* data given by Kulp (1961), Baadsgard and others (1961); and Armstrong (1964).

Armstrong, F. C., and Cressman, E. R., 1963, The Bannock thrust zone, southeastern Idaho: U. S. Geol. Survey Prof. Paper 374J, 22 p.

Armstrong, R. L., 1964, Geochronology and geology of the eastern Great Basin in Nevada and Utah: Ph.D. Thesis, Yale University, 202 p.

——— 1963, K-Ar ages of volcanics in southwestern Utah and adjacent Nevada, *in* Guidebook to the geology of southwestern Utah: Intermountain Assoc. Petroleum Geologists Ann. Field Conf. Guidebook, no. 12, p. 79-80.

——— 1966, K-Ar dating using neutron activation for Ar analysis: granitic plutons of the eastern Great Basin, Nevada and Utah: Geochim. et Cosmochim, Acta, v. 30, p. 565-600.

——— 1968, The Cordilleran miogeosyncline in Utah and Nevada: Utah Geological and Mineralogical Survey Bulletin (in press).

Armstrong, R. L., and Hansen, Edward, 1966, Cordilleran infrastructure in the eastern Great Basin: Am. Jour. Sci., v. 254, p. 112-127.

Baadsgaard, Halfdan, Folinsbee, R. E., and Lipson, J. I., 1961, Potassium-argon dates of biotites from Cordilleran granites: Geol. Soc. America Bull., v. 72, p. 689-702.

Baker, A. A., 1959, Faults in the Wasatch Range near Provo, Utah: Intermountain Assoc. Petroleum Geologists 10th Ann. Field Conf. Guidebook, p. 153-158.

Baker, A. A., and Crittenden, M. D., Jr., 1961, Geology of the Timpanogos Cave Quadrangle, Utah: U. S. Geol. Survey Geol. Quadrangle Map, GQ 132.

Baker, A. A., Huddle, J. W., and Kinney, D. M., 1949, Paleozoic geology of north and west sides of Uinta Basin, Utah: Am. Assoc. Petroleum Geologists Bull., v. 33, no. 7, p. 1161-1197.

Bauer, H. L., Jr., Cooper, J. J., and Breitrick, R. A., 1960, Porphyry copper deposits in the Robinson mining district, White Pine County, Nevada: Intermountain Assoc. Petroleum Geologists 11th Ann. Field Conf. Guidebook, p. 220-228.

Bissell, H. J., 1959, North Strawberry Valley sedimentation and tectonics: Intermountain Assoc. Petroleum Geologists 10th Ann. Field Conf. Guidebook, p. 159-165.

——— 1964, Chaos near Ferguson Mountain, Elko County, Nevada (Abstract) p. 265-266: Geol. Soc. America Spec. Paper 76, 341 p.

Blank, H. R., 1959, Geology of the Bull Valley district Washington County, Utah: Ph.D. Thesis, Univ. of Washington, 173 p.

Bowyer, Ben, Pampeyan, E. H., and Longwell, C. R., 1958, Geologic map of Clark County, Nevada: U. S. Geol. Survey Mineral Invest. Field Studies Map, MF 138.

Brill, K. G., Jr., 1963, Permo-Pennsylvanian stratigraphy of western Colorado Plateau and eastern Great Basin regions: Geol. Soc. America Bull., v. 74, p. 307-330.

Burchfiel, B. C., 1961, Structure and stratigraphy of the Specter Range Quadrangle, Nye County, Nevada: Ph.D. Thesis, Yale University, 197 p.

Burger, J. A., 1963, The Cretaceous system of Utah: Utah Geol. and Mineralog. Survey Bull. 54, p. 123-139.

Butler, B. S., Loughlin, G. F., Heikes, V. C., and others, 1920, The ore deposits of Utah: U. S. Geol. Survey Prof. Paper 111, 672 p.

Callaghan, Eugene, 1939, Volcanic sequence in the Marysvale region in southwest-central Utah: Am. Geophys. Union Trans. 20th Ann. Meeting, pt. 3, p. 428-452.

Christiansen, F. W., 1952, Structure and stratigraphy of the Canyon Range, central Utah: Geol. Soc. America Bull., v. 63, p. 717-740.

Clark, T. H., and Stearn, C. W., 1960, The geological evolution of North America—a regional approach to historical geology: New York, Ronald Press Co., 434 p.

Cobban, W. A., and Reeside, J. B., Jr., 1952, Correlation of the Cretaceous formations of the Western Interior of the United States: Geol. Soc. America Bull., v. 63, p. 1011-1044.

Cochran, K. L., 1959, Results of Pre-Cretaceous exploration in the overthrust belt of southwestern Wyoming: Intermountain Assoc. Petroleum Geologists 10th Ann. Field Conf. Guidebook, p. 200-203.

Cohenour, R. E., 1959, Sheeprock Mountains, Tooele and Jaub Counties: Utah Geol. and Mineralog. Survey Bull. 63, 201 p.

Cook, E. F., 1957, Geology of the Pine Valley Mountains, Utah: Utah Geol. and Mineralog. Survey Bull. 58, 111 p.

——— 1960, Breccia blocks (Mississippian) of the Welcome Springs area, southwest Utah: Geol. Soc. America Bull., v. 71, p. 1709-1712.

——— 1965, Stratigraphy of Tertiary volcanic rocks in eastern Nevada: Nevada Bureau of Mines, Report 11, 61 p.

Costain, J. K., 1960, Geology of the Gilson Mountains and vicinity, Jaub County, Utah: Ph.D. Thesis, Univ. of Utah.

Cressman, E. R., 1964, Geology of the Georgetown Canyon-Snowdrift Mountain area, southeastern Idaho: U. S. Geol. Survey Bull. 1153, 105 p.

Crittenden, M. D., Jr., 1959, Mississippian stratigraphy of the central Wasatch and western Uinta Mountains, Utah: Intermountain Assoc. Petroleum Geologists 10th Ann. Field Conf. Guidebook, p. 63-74.

——— 1961, Magnitude of thrust faulting in northern Utah: U. S. Geol. Survey Prof. Paper 424-D, p. D128-D131.

——— 1963, Effective viscosity of the earth derived from isostatic loading of Pleistocene Lake Bonneville: Jour. Geophys. Research, v. 68, p. 5517-5530.

Crosby, G. W., 1959, Geology of the South Pavant Range, Millard and Sevier Counties, Utah: Brigham Young Univ. Research Studies Geology Series, v. 6, no. 3, 59 p.

Dolgoff, Anthony, 1963, Volcanic stratigraphy of the Pahranagat area, Lincoln County, southeastern Nevada: Geol. Soc. America Bull., v. 74, p. 875-900.

Douglass, W. B., Jr., 1960, Geology of the southern Butte Mountains, White Pine County, Nevada: Intermountain Assoc. Petroleum Geologists 11th Ann. Field Conf. Guidebook, p. 181-185.

Drewes, H. D., 1958, Structural geology of the southern Snake Range, Nevada: Geol. Soc. America Bull., v. 69, p. 221-240.

——— 1960, Bedding-plane thrust faults east of Connors Pass, Schell Creek range, eastern Nevada: U. S. Geol. Survey Prof. Paper 400-B, p. B270-B272.

Eardley, A. J., 1934, Structure physiography of the southern Wasatch Mountains, Utah: Mich. Acad. Sci.

Papers, v. 19, p. 377-400.
——— 1944, Geology of the north-central Wasatch Mountains, Utah: Geol. Soc. America Bull., v. 55, p. 819-894.
——— 1962, Structural geology of North America: New York, Harper and Row, 743 p.
——— 1963, Structural evolution of Utah: Utah Geol. and Mineralog. Survey Bull. 54, p. 19-29.
East, E. H., 1956, Geology of the San Francisco Mountains, western Utah: M. S. Thesis, Univ. of Washington.
——— 1957, Evidence of overthrusting in the San Francisco Mountains, Beaver County, western Utah (Abstract): Geol. Soc. America Bull., v. 68, p. 1825-1826.
Evernden, J. F., Savage, D. E., Curtis, G. H., and James, G. T., 1964, Potassium-argon dates and the Cenozoic mammalian chronology of North America: Am. Jour. Sci., v. 262, p. 145-198.
Felix, C. E., 1956, Geology of the eastern part of the Raft River Range, Box Elder County, Utah: Utah Geol. Soc. Guidebook to the Geology of Utah, no. 11, p. 76-97.
Fisher, D. J., Erdmann, C. E., and Reeside, J. B., Jr., 1960, Cretaceous and Tertiary formations of the Book Cliffs, Carbon, Emery, and Grand Counties Utah, and Garfield and Mesa Counties, Colorado: U. S. Geol. Survey Prof. Paper 332, 80 p.
Folinsbee, R. E., Baadsgaard, Halfdan, and Lipson, J. I., 1960, Potassium-Argon time scale: 21st Internat. Geol. Congress, Copenhagen, 1960, Rept., pt. 3, p. 7-17.
Fritz, W. H., 1960, Structure and stratigraphy of the Northern Egan Range, White Pine County, Nevada: Ph.D. Thesis, Univ. of Washington, 178 p.
Gabrielse, Hubert, and Reesor, J. E., 1964, Geochronology of plutonic rocks in two areas of the Canadian Cordillera: Royal Soc. Canada, Spec. Pub. 8, p. 96-138.
Gilliland, W. N., 1963, Sanpete-Sevier anticline of central Utah: Geol. Soc. America Bull., v. 74, p. 115-124.
Gilluly, James, 1963, The tectonic evolution of the western United States: Geol. Soc. London Quart. Jour., v. 119, p. 133-174.
Gregory, H. E., and Williams, N. C., 1947, Zion National Monument, Utah: Geol. Soc. America Bull., v. 58, p. 211-244.
Groff, S. L., 1959, Geology of the West Tintic Range and vicinity, Tooele and Juab Counties, Utah: Ph.D. Thesis, Univ. of Utah, 244 p.
Hardy, C. T., and Zeller, H. D., 1953, Geology of the west-central part of the Gunnison Plateau, Utah: Geol. Soc. America Bull., v. 64, p. 1261-1278.
Harlow, G. R., 1956, The stratigraphy and structure of the Spruce Mountain area, Elko County, Nevada: M. S. Thesis, Univ. of Washington.
Harris, H. D., 1959, A late Mesozoic positive area in western Utah: Am. Assoc. Petroleum Geologists Bull., v. 43, no. 11, p. 2636-2652.
Hintze, L. F., 1960, Thrust-faulting limits in western Utah (Abstract): Geol. Soc. America Bull., v. 71, p. 2062.
——— Editor, 1962 Geology of the Southern Wasatch Mountains: Brigham Young Univ. Geol. Studies, v. 9, 104 p.
——— Compiler, 1963, Geologic map of southwestern Utah: Utah Geol. and Mineralog. Survey.
Holtedahl, Olaf, Editor, 1960, Geology of Norway: Norges Geologiske Undersøkelse Nr. 208, 540 p.
Jones, R. W., 1963, Gravity structures in the Beaver Dam Mountains, southwestern Utah: Intermountain Assoc. Petroleum Geologists 12th Ann. Field Conf. Guidebook, p. 90-95.
Keefer, W. R., and Love, J. D., 1963, Laramide vertical movements in central Wyoming: Univ. Wyoming Contributions to Geology, v. 2, no. 1, p. 47-54.
Kellogg, H. E., 1963, Paleozoic stratigraphy of the southern Egan Range, Nevada: Geol. Soc. America Bull., v. 74, p. 685-708.
King, P. B., 1959, The evolution of North America: Princeton, N. J., Princeton Univ. Press, 190 p.
Kulp, J. L., 1961, Geologic time scale: Science, v. 133, p. 1105-1114.
Langenheim, R. L., Jr., Barr, F. T., Shank, S. E., Stensaas, L. J., and Wilson, E. C., 1960, Preliminary report on the geology of the Ely No. 3 Quadrangle, White Pine County, Nevada: Intermountain Assoc. Petroleum Geologists 11th Ann. Field Conf. Guidebook, p. 148-156.
Levorsen, A. I., 1960, Paleogeologic maps: San Francisco, W. H. Freeman and Co., 174 p.
Liese, H. C., 1957, Geology of the northern Mineral Range, Millard and Beaver Counties, Utah: M. S. Thesis, Univ. of Utah.
Lloyd, G. P., 1959, Geology of the Northern White River Valley, Nevada: M. S. Thesis, Univ. of California at Los Angeles.
Longwell, C. R., 1925, The pre-Triassic unconformity in southern Nevada: Am. Jour. Sci., 5th ser., v. 10, p. 93-106.
——— 1949, Structure of the northern Muddy Mountain area, Nevada: Geol. Soc. America Bull., v. 60, p. 923-968.
——— 1950, Tectonic theory viewed from the Basin Ranges: Geol. Soc. America Bull., v. 61, p. 413-434.
——— 1952, Basin and Range geology west of the St. George Basin, Utah: Utah Geol. and Mineralog. Survey Guidebook to the Geology of Utah, no. 7, p. 27-42.
——— 1952*a*, Structure of the Muddy Mountains, Nevada: Utah Geol. and Mineralog. Survey Guidebook to the Geology of Utah, no. 7, p. 109-114.
——— 1960, Possible explanation of diverse structural patterns in southern Nevada: Am. Jour. Sci., Bradley Volume, v. 258-A, p. 192-203.
——— 1962, Restudy of the Arrowhead fault, Muddy Mountains, Nevada: U. S. Geol. Survey Prof. Paper 450D, p. 82-85.
——— 1963, Reconnaissance geology between Lake Mead and Davis Dam, Arizona-Nevada: U. S. Geol. Survey Prof. Paper 374E, 51 p.
Love, J. D., 1956, Summary of geologic history of Teton County, Wyoming, during Late Cretaceous, Tertiary and Quaternary time: Wyo. Geol. Assoc. 11th Ann. Field Conf. Guidebook, p. 76-93.
Love, J. D., Weitz, J. L., and Hose, R. K., Compilers, 1955, Geologic Map of Wyoming: U. S. Geol. Survey.
Mackin, J. H., 1947, Some structural features of the intrusions in the Iron Springs district: Utah, Geol. Soc. Guidebook to the Geology of Utah, no. 2, 62 p.
——— 1960, Eruptive tectonic hypothesis for origin of Basin-Range structure (Abstract): Geol. Soc. American Bull., v. 71, p. 921.
——— 1960*a*, Structural significance of Tertiary volcanic

rocks in southwestern Utah: Am. Jour. Sci., v. 258, p. 81-131.
Maxey, G. B., 1946, Geology of part of the Pavant Range, Millard County, Utah: Am. Jour. Sci., 244, p. 324-356.
McGookey, D. P., 1960, Early Tertiary stratigraphy of part of Central Utah: Am. Assoc. Petroleum Geologists Bull., v. 44, p. 589-615.
Menard, H. W., 1960, The East Pacific Rise: Science, v. 132, p. 1737-1746.
Miller, G. M., 1958, Regional thrust relations of the Wah Wah Mountains, southwest Utah (Abstract) Geol. Soc. America Bull., v. 69, p. 1696.
——— 1963, Outline of structural-stratigraphic units of the Wah Wah Mountains, southwest Utah: Intermountain Assoc. Petroleum Geologists 12th Ann. Field Conf. Guidebook, p. 96-102.
Misch, Peter, 1960, Regional structural reconnaissance in central-northeast Nevada and some adjacent areas—Observations and interpretation, *in* Geology of east central Nevada: Intermountain Assoc. Petroleum Geologists 11th Ann. Field Conf., Guidebook, 1960, p. 17-42.
Misch, Peter, and Hazzard, J. C., 1962, Stratigraphy and metamorphism of late Precambrian rocks in central northeastern Nevada and adjacent Utah: Am. Assoc. Petroleum Geologists Bull., v. 46, p. 289-343.
Moore, J. G., 1960, Curvature of normal faults in the Basin and Range province of the western United States: U. S. Geol. Survey Prof. Paper 400B, p. 409-411.
Moores, E. M., III, 1963, Geology of the Current area, Nye County Nevada: Ph.D. Thesis, Princeton Univ., 208 p.
Morris, H. T., 1957, General geology of the East Tintic Mountains, Utah: Utah Geol. Soc. Guidebook to the geology of Utah, no. 12, p. 1-56.
Morris, H. T., and Shepard, W. M., 1964, Evidence for a concealed tear fault with large displacement in the central East Tintic Mountains, Utah: U. S. Geol. Survey, Prof. Paper 501-C, p. 19-21.
Muessig, S., 1951, Geology of a part of Long Ridge, Utah: Ph.D. Thesis, Ohio State Univ.
Nelson, R. B., 1956, The stratigraphy and structure of the region surrounding Currie, Elko County, Nevada: M. S. Thesis, Univ. of Washington.
Nelson, R. B., 1959, The stratigraphy and structure of the northernmost part of the northern Snake Range and the Kern Mountains in eastern Nevada and the southern Deep Creek Range in western Utah: Ph.D. Thesis, Univ. of Washington, 165 p.
Nolan, T. B., 1929, Notes on the stratigraphy and structure of the northwest portion of Spring Mountain, Nevada: Am. Jour. Sci., v. 17, p. 461-472.
——— 1935, The Gold Hill mining district, Utah: U. S. Geol. Survey Prof. Paper 177, 172 p.
——— 1943, The Basin and Range province in Utah, Nevada and California: U. S. Geol. Survey Prof. Paper 197-D, p. 141-196.
O'Brien, C. A. E., 1960, The structural geology of the Boule and Bosche Ranges in the Canadian Rocky Mountains: Geol. Soc. London Quart. Jour., v. 116, p. 409-436.
Osmond, J. C., 1960, Tectonic history of the Basin and Range province in Utah and Nevada: Min. Eng., v. 12, p. 251-265.
Pierce, K. L., and Armstrong, R. L., 1964, The Tuscarora fault, an Acadian(?) bedding-plane fault in the central Appalachian Valley and Ridge (Astract), p. 130 *in* Geological Society of America, Abstracts for 1963; Geol. Soc. America Special Paper 76, 341 p.
Playford, P. E., 1962, Geology of the Egan Range, near Lund, Nevada, Ph.D. Thesis, Stanford Univ., 294 p.
Regnier, J. P. M., 1960, Cenozoic geology in the vicinity of Carlin, Nevada: Geol. Soc. America Bull., v. 71, p. 1189-1210.
Rich, J. L., 1934, Mechanics of low-angle overthrust faulting as illustrated by Cumerland thrust block, Virginia, Kentucky, and Tennessee: Am. Assoc. Petroleum Geologists bull., v. 18, p. 1584-1596.
Rigby, J. K., 1958, Geology of the Stansbury Mountains, eastern Tooele County, Utah, Utah Geol. Soc. Guidebook to the geology of Utah, no. 13, p. 1-134.
Roberts, R. J., and Tooker, E. W., 1961, Structural geology of the north end of the Oquirrh Mountains, Utah: Utah Geol. Soc. Guidebook to the geology of Utah, no. 16, p. 36-48.
Roberts, R. J., Hotz, P. E., Gilluly, James, and Ferguson, H. G., 1958, Paleozoic rocks of north-central Nevada: Am. Assoc. Petroleum Geologists Bull., v. 42, p. 2813-2857.
Rubey, W. W., and Hubbert, M. K., 1959, Overthrust belt in geosynclinal area of western Wyoming in light of fluid-pressure hypothesis: Geol. Soc. America Bull., v. 70, p. 167-206.
Schaeffer, F. E., and Anderson, W. L., 1960, Geology of the Silver Island Mountains, Box Elder and Tooele Counties, Utah and Elko County, Nevada: Utah Geol. Soc. Guidebook to the geology of Utah, no. 15, 185 p.
Schick, R. B., 1959, Geologic sections from three deep wells, southwestern Wyoming: Intermountain Assoc. Petroleum Geologists 10th Ann. Field Conf. Guidebook, p. 193-199.
Schoff, S. L., 1951, Geology of the Cedar Hills, Utah: Geol. Soc. America Bull., v. 62, p. 619-646.
Secor, D. T., Jr., 1962, Geology of the central Spring Mountains, Nevada: Ph.D. Thesis, Stanford Univ., 152 p.
Shaw, E. W., 1963, Canadian Rockies—Orientation in time and space: Am. Assoc. Petroleum Geologists, Memoir 2, p. 231-242.
Silberling, N. J., and Roberts, R. J., 1962, Pre-Tertiary stratigraphy and structure of northwestern Nevada: Geol. Soc. America Special Paper 72, 58 p.
Snelson, Sigmund, 1955, The geology of the southern Peqop Mountains, Elko County, northeastern Nevada: M. S. Thesis, Univ. of Washington.
——— 1957, The geology of northern Ruby Mountains and the East Humboldt Range, Elko County, northeastern Nevada: Ph.D. Thesis, Univ. of Washington, 268 p.
Spieker, E. M., 1946, Late Mesozoic and early Cenozoic history of central Utah: U. S. Geol. Survey Prof. Paper 205-D, p. 117-161.
——— 1949, The transition between the Colorado Plateaus and the Great Basin in central Utah: Utah Geol. Soc. Guidebook to the geology of Utah, no. 4, 106 p.
——— 1956, Mountain-building chronology and nature of geologic time scale: Am. Assoc. Petroleum Geologists Bull., v. 40, p. 1769-1815.
Stokes, W. L., 1960, Inferred Mesozoic history of east-

central Nevada and vicinity, *in* Geology of east central Nevada: Intermountain Assoc. Petroleum Geologists 11th Ann. Field Conf. Guidebook, p. 117-121.

——— Compiler, 1963, Geologic map of northwestern Utah: Utah Geol. and Mineralog. Survey.

——— 1963*a*, Triassic and Jurassic periods in Utah: Utah Geol. and Mineralog. Survey Bull. 54, p. 109-121.

——— 1963*b*, Triassic and Jurassic formations of southwestern Utah: Intermountain Assoc. Petroleum Geologists 12th Ann. Field Conf., Guidebook, p. 60-64.

Stokes, W. L., and Madsen, J. H., Jr., Compilers, 1961, Geologic map of northeastern Utah: Utah Geol. and Mineralog. Survey.

Thompson, G. A., 1960, Problem of Late Cenozoic structure of the Basin Ranges: Copenhagen, 21st Internat. Geol. Cong., Proc., sec. 18, p. 63-68.

Threet, R. L., 1963, Geology of the Parowan Gap area, Iron County, Utah: Intermountain Assoc. Petroleum Geologists 12th Ann. Field Conf. Guidebook, p. 136-145.

——— 1963*a*, Structure of the Colorado Plateau margin near Cedar City, Utah: Intermountain Assoc. Petroleum Geologists 12th Ann. Field Conf. Guidebook, p. 104-117.

Tracy, J. I., Jr., and Oriel, S. S., 1959, Uppermost Cretaceous and Lower Tertiary rocks of the Fossil Basin: Intermountain Assoc. Petroleum Geologists 10th Ann. Field Conf. Guidebook, p. 126-130.

Tschanz, C. M., 1960, Geology of northern Lincoln County, Nevada: Intermountain Assoc. Petroleum Geologists, 11th Ann. Field Conf. Guidebook, p. 198-208.

——— 1960*a*, Thrust faults in southeastern Lincoln County, Nevada (Abstract): Geol. Soc. America Bull., v. 70, p. 1753-1754.

Tschanz, C. M., and Pampeyan, E. H., 1961, Preliminary geologic map of Lincoln County, Nevada: U.S. Geol. Survey, Mineral Inv. Field Studies Map, MF 206.

Ward, R., 1962, Geology of the northern half of the Ripetown Quadrangle, Nevada: M. S. Thesis, Univ. of Southern California.

Webb, Barbara, and Wilson, R. V., 1962, Progress geologic map of Nevada: Nevada Bureau Mines map 16.

Weimer, R. J., and Haun, J. D., 1960, Cretaceous stratigraphy, Rocky Mountain region, U. S. A.: 21st Internat. Geol. Cong., Copenhagen, Proc., Sect. 12, p. 178-184.

White, W. H., 1959, Cordilleran tectonics in British Columbia: Am. Assoc. Petroleum Geologists Bull., v. 43, p. 60-100.

Whitebread, D. H., Griggs, A. B., Rogers, W. B., and Mytton, J. W., 1962, Preliminary geologic map and sections of the Wheeler Peak Quadrangle, White Pine County, Nevada: U. S. Geol. Survey Mineral Inv. Field Studies Map, MF 244.

Williams, N. C., and Madsen, J. H., Jr., 1959, Late Cretaceous stratigraphy of the Coalville area, Utah: Intermountain Assoc. Petroleum Geologists, 10th Ann. Field Conf. Guidebook, p. 122-125.

Wilmarth, M. G., 1938, Lexicon of geologic names of the United States (including Alaska): U. S. Geol. Survey Bull. 896, 2396 p.

Wilson, C. W., Jr., and Stearns, R. G., 1958, Structure of the Cumberland Plateau, Tennessee: Geol. Soc. America Bull., v. 69, p. 1283-1296.

Winfrey, W. M., 1960, Stratigraphy, correlation, and oil potential of the Sheep Pass Formation, east-central Nevada: Intermountain Assoc. Petroleum Geologists 11th Ann. Field Conf. Guidebook, p. 126-133.

Wisser, E. H., 1960, Relation of ore deposition to doming in the North American Cordillera: Geol. Soc. America Memoir 77, 117 p.

Woodward, L. A., 1964, Structural geology of central northern Egan Range, Nevada: Am. Assoc. Petroleum Geologists Bull., v. 48, p. 22-39.

Wright, J. C., and Dickey D. D., 1958, Upper Jurassic strata of the Colorado Plateau as a record of tectonic history in the eastern Great Basin (Abstract): Geol. Soc. America Bull., v. 69, p. 1667.

——— 1963, Relations of the Navajo and Carmel formations in southwest Utah and adjoining Arizona: U. S. Geol. Survey Prof. Paper 450-E, p. 63-67.

——— 1963*a*, Block diagram of the San Rafael Group and underlying strata in Utah and part of Colorado: U. S. Geol. Survey Oil and Gas Inv. Map, OC-63.

Young, J. C., 1960, Structure and stratigraphy in north-central Schell Creek Range: Intermountain Assoc. Petroleum Geologists 11th Ann. Field Conf. Guidebook, p. 158-172.

Young, R. G., 1960, Dakota group of the Colorado Plateau: Am. Assoc. Petroleum Geologists Bull., v. 44, p. 156-194.

THE AMERICAN ASSOCIATION OF PETROLEUM GEOLOGISTS BULLETIN
Vol. 58, No. 12, December, 1974

Regional Extent and Age of Thrusts near Rockport Reservoir and Relation to Possible Exploration Targets in Northern Utah[1]

MAX D. CRITTENDEN, JR.[2]
Menlo Park, California 94025

Abstract Two east-northeast-striking steeply dipping faults involving rocks as young as the Frontier Formation (Upper Cretaceous) cross the Weber River at the Rockport Reservoir, 2.5-3.5 mi south of Wanship. Although they originally were interpreted as high-angle normal faults, several lines of evidence indicate that they are thrust faults, and that they owe their steep dips to folding: (1) overturning of beds in the Frontier Formation under the southernmost fault, (2) development of inverted Z-folds in the block between the faults, and (3) stratigraphic changes across the fault zone. Northeast of the reservoir, the faults are overlapped by Upper Cretaceous conglomerates of the Wanship Formation. These relations date the thrusting as early Late Cretaceous (early Carlisle to early Niobrara). Regional structure suggests that the faults are a northeastward extension of the Mount Raymond thrust mapped previously in the Wasatch Range on the west, and serves to date that episode of thrusting as well. Concealed extensions of these or related thrusts may help to account for structural complications at Coalville and may provide additional exploration targets along strike in southwestern Wyoming.

INTRODUCTION

Geologic mapping in the Cottonwood and Park City areas (Crittenden *et al.*, 1966; Bromfield and Crittenden, 1971) has been extended in reconnaissance to a small area adjoining the Rockport Reservoir on the Weber River, about 10 mi south of Coalville (Fig. 1). Attention was called to this area long ago by E. C. Morris (1953) who recognized that two steeply dipping faults in Dry Canyon and Crandall Canyon (Fig. 2) were truncated by an unconformity at the base of a unit later designated the Wanship Formation by Williams and Madsen (1959). Structural and stratigraphic evidence that these are thrust faults was recognized during study of the Kelvin Formation in the vicinity of Parleys Canyon, east of Salt Lake City (Crittenden, 1963).

STRATIGRAPHY

Stratigraphy of the area near Rockport has been described in earlier reports (Cobban and Reeside, 1952; Hale, 1960; Trexler, 1966) and is not repeated here. Pre-Tertiary units specifically involved in this study are listed in Table 1, and structurally significant features are noted briefly below. It is evident that (1) the Preuss Sandstone is less than half as thick in the block south of Crandall Canyon fault as it is on the north, and (2) the Kelvin Formation overlies the Morrison and Stump Formations in the block on the south but lies directly on the Preuss Sandstone in the block on the north. These stratigraphic differences cannot be explained by normal faulting but are consistent with the concept that the Crandall Canyon fault is a folded thrust whose upper plate—the block on the north—has moved eastward, carrying a thinner sedimentary section over a thicker section in the autochthon on the south.

The outcrops of Morrison Formation near Peoa, below Crandall Canyon fault, are the westernmost exposures known, as demonstrated by the presence of charophytes characteristic of Morrison in eastern Utah and Colorado (R. E. Peck, written commun., 1959). This identity is reinforced by the presence of glauconitic sands of the Stump Sandstone directly below. The unit formerly called Morrison(?) by Granger (1953) now is assigned to the Parleys Member of the Kelvin Formation. It includes brilliant red to pure white siltstone and marl with nodules and concretions of white limestone and local lenses of coarse conglomerate. In addition to the type section in Parleys Canyon (Granger, 1953), this unit forms spectacular exposures in roadcuts along the west edge of Rockport Reservoir and in the ridge just southwest of Wanship (Fig. 2). The extent of the genuine Morrison Formation in this area and the identity of these Morrisonlike rocks are discussed more fully elsewhere (Crittenden, 1963).

The structurally significant unconformity near the top of the Cretaceous section was recognized by Williams and Madsen (1959). Beginning at the edge of the alluvium north of the reservoir (Fig. 2), the base of the Wanship Formation of Williams and Madsen cuts gradually downward across the underlying sandstones of the Frontier Formation and truncates both the sharp folds within that unit and the thrusts that repeat part of the section (Fig. 2). The basal contact appears to be smooth and, except for one small fault, offset is unaffected by any of the underlying structures or beds. The unconformity strikes approximately west and dips about 60° N near the reservoir, but strikes N 60°

[1]Manuscript received, January 17, 1974; accepted, March 15, 1974. Publication authorized by the Director, U.S. Geological Survey.

[2]U.S. Geological Survey.

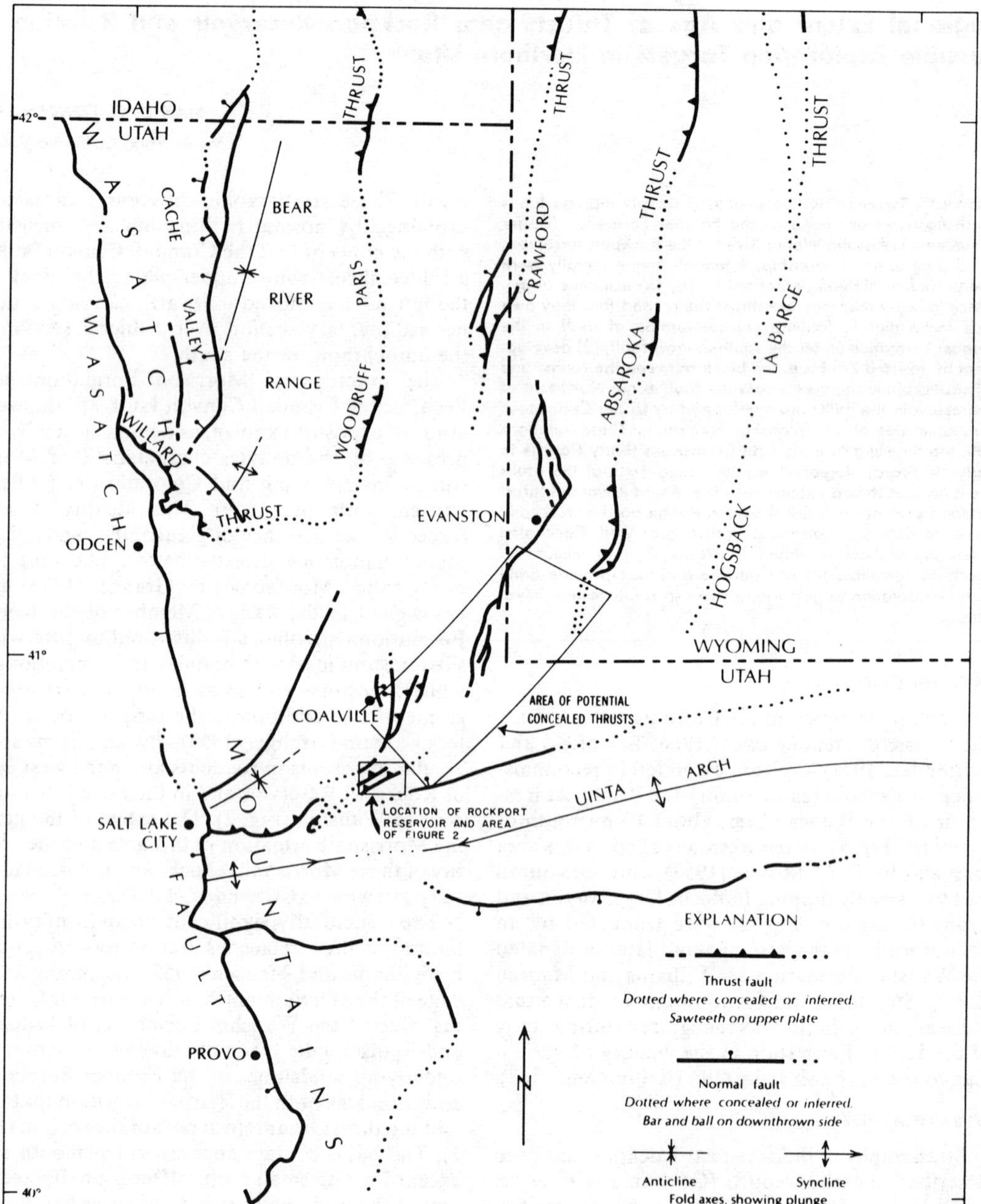

FIG. 1—Principal structural features, northeastern Utah and southwestern Wyoming, showing area of potential concealed thrust faults.

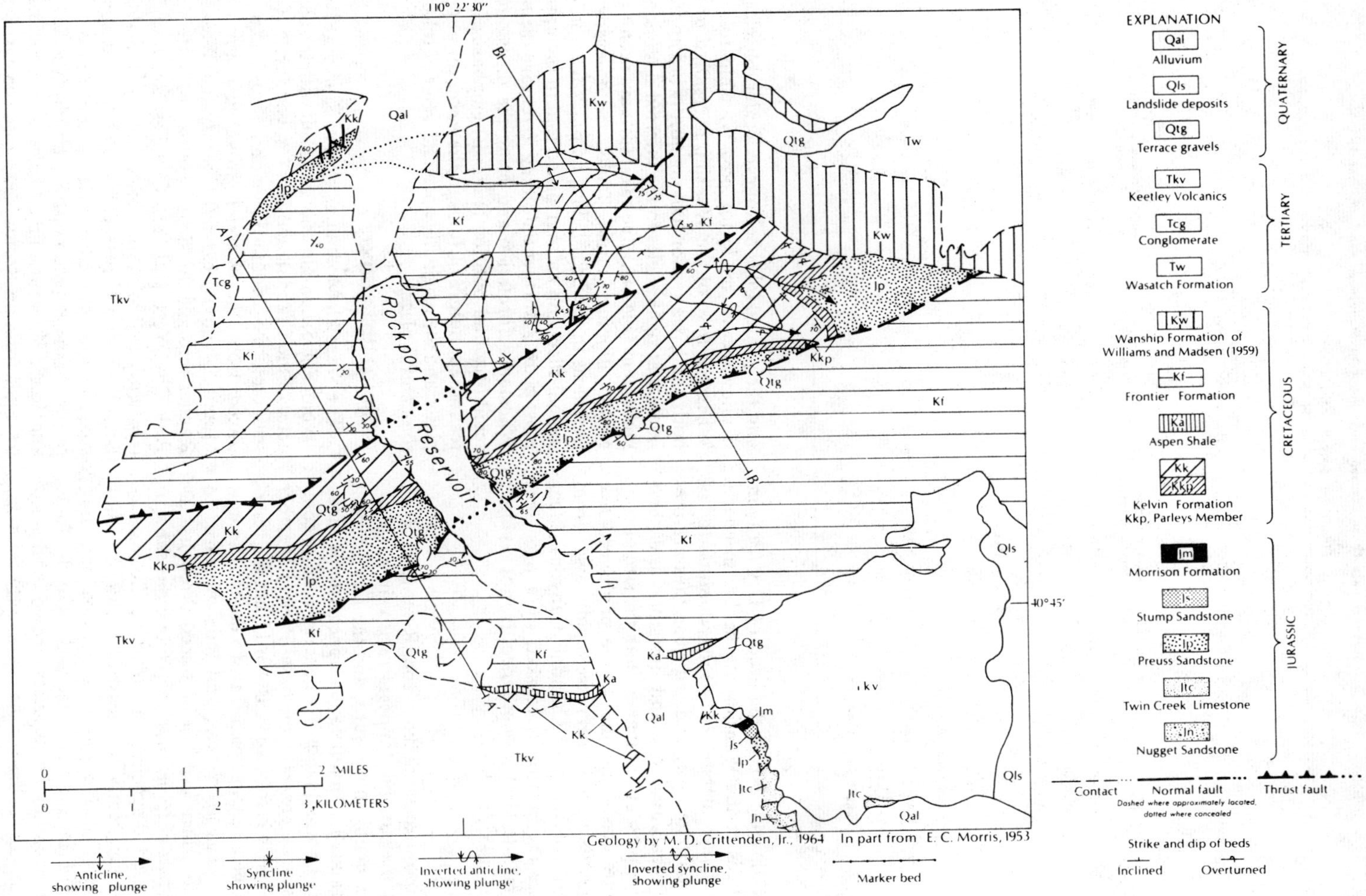

FIG. 2—Reconnaissance geologic map of area near Rockport Reservoir.

TABLE 1. Stratigraphic Units Exposed along Weber River between Peoa and Wanship

(Modified from Morris, 1953)

		Wanship Formation of Williams and Madsen (1959)	
		Unconformity	
	Below Crandall Canyon (Mount Raymond) thrust		Above Crandall Canyon (Mount Raymond) thrust
Frontier Formation	6,000 ft		5,000 + (no base)
Aspen Shale	525		Not exposed
Kelvin Formation	2,700		3,000 + (no top)
	UNCONFORMITY		UNCONFORMITY
Morrison Formation	258		Absent
Stump Sandstone	225		Absent
Preuss Sandstone	1,030		2,500 (no base)
Twin Creek Limestone	1,350		Not exposed
Nugget Sandstone	910		Not exposed

W and dips 45° NE at the head of Dry Canyon. Basal beds of the Wanship are pebble to coarse cobble conglomerate, consisting dominantly of white quartzite; gray and red quartzites are present in small amounts, and limestone and chert clasts are present but sparse. Both the size of conglomerate clasts and the degree of discordance with underlying units decrease rapidly away from the heads of Dry and Crandall Canyons. Near Coalville the horizon is marked by a persistent conglomerate, but discordance is not evident (Trexler, 1955, p. 30).

Structure

In exposures along the Weber River (Sec. *A-A'*, Fig. 3) the structure appears relatively simple—dips are mainly northward, and the Dry Canyon and Crandall Canyon faults appear to be steep reverse or normal faults. Evidence of a more complex history emerges as the basal Parleys Member of the Kelvin is followed eastward from the reservoir in the block between the two faults. The unit first steepens to vertical, and then overturns to about 80°; thence it can be traced through a Z-shaped pair of folds, which top directions show to be an inverted syncline and anticline. Both structures plunge gently southeastward and are truncated down section and up section by bounding faults. Still farther northeastward, both the folds and the bounding faults are truncated by the unconformity at the base of the Wanship Formation.

The nature of Dry and Crandall Canyon faults and the related folds becomes considerably clearer if allowance is made for post-Wanship folding. When the base of the Wanship is restored to the horizontal, in a section looking northward approximately at right angles to the inferred direction of movement (Fig. 4A), the two faults are seen to be gently dipping thrust faults, with eastward overturns in the beds below, complex folds in the medial slice, and an overturned fold in the uppermost plate. A further unfolding to remove the pre-Wanship deformation and bring the thrusts to a semihorizontal position (Fig. 4B) emphasizes the recumbent nature of folds in the lower and upper plates and the originally open character of folds in the central plate.

Regional Relations

The Wasatch Mountains west of Rockport Reservoir (Fig. 5) contain only one fault with the character of the Crandall and Dry Canyon faults. This structure, the Mount Raymond thrust, can be traced from the Wasatch front, just north of Mount Olympus, eastward across the Wasatch Range, through the Sugar House, Mount Aire, Park City West, and Park City East quadrangles (Crittenden, 1965a, b; Crittenden *et al.*, 1966; Bromfield and Crittenden, 1971). Its easterly strike in this interval is notably anomalous with respect to the general northerly trend of the Sevier and Idaho-Wyoming thrust belts, and is believed to result from postthrust folding on the east-west Uinta arch. The fault also rises stratigraphically in

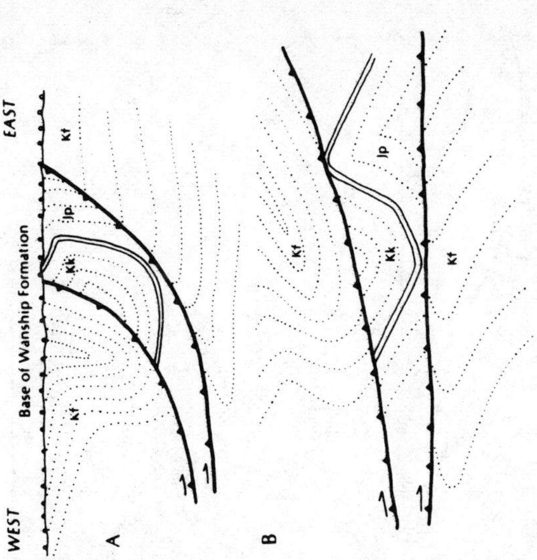

FIG. 4—Restored east-west longitudinal section through Dry Canyon and Crandall Canyon faults. A, During deposition of Wanship Formation of Williams and Madsen (1959). B, Shortly after thrusting. Lithologic units as on Figure 2.

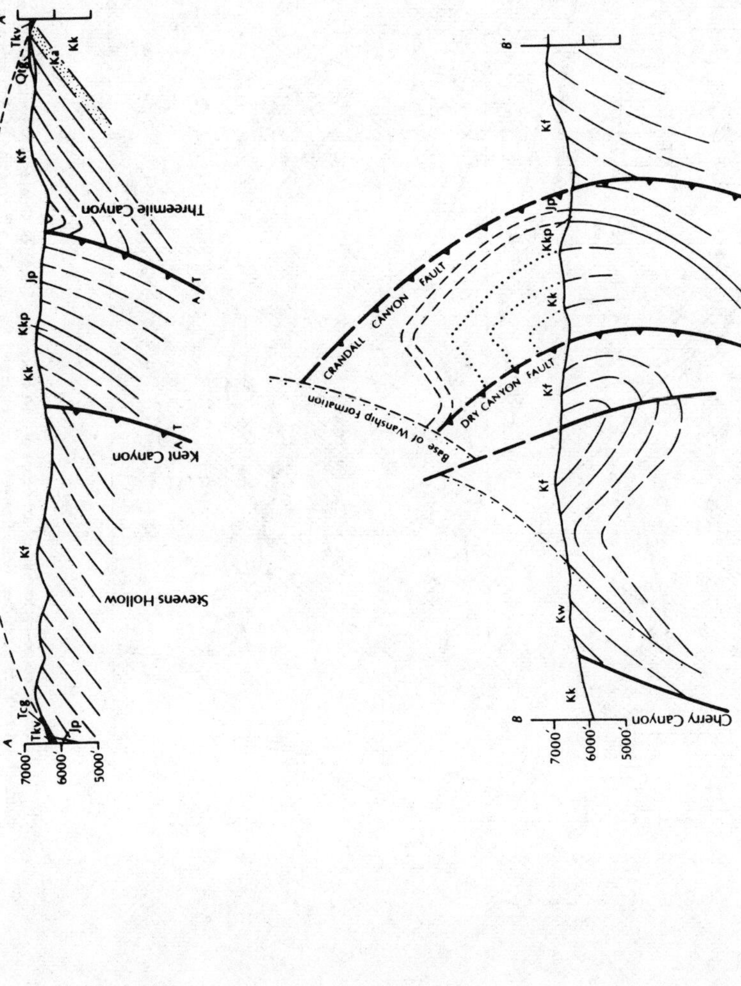

FIG. 3—Structure sections through area near Rockport Reservoir. Lithologic units and lines of sections shown on Figure 2.

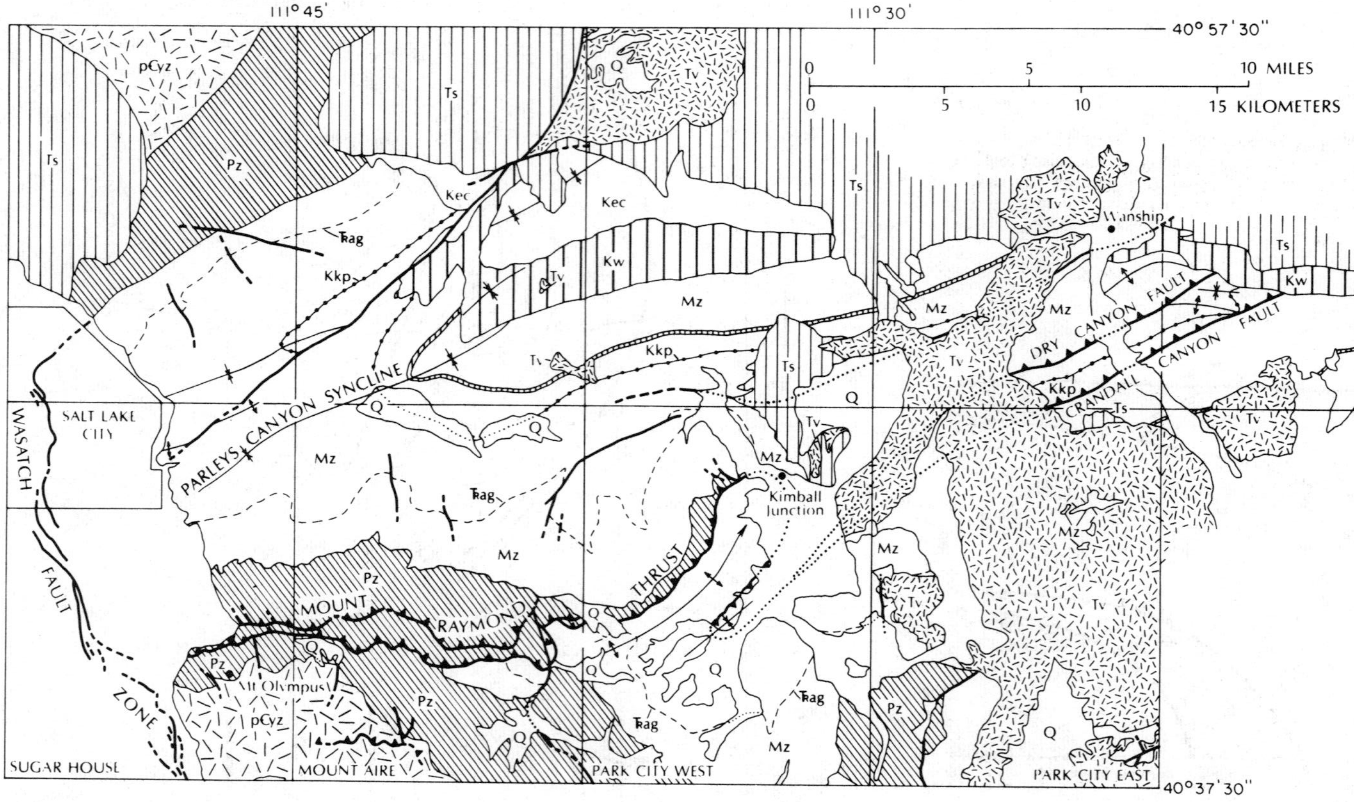

FIG. 5—Generalized geologic map of area east of Salt Lake City, Utah: *Q*, Quaternary rocks, mainly alluvium and gravel; *Ts*, Tertiary sedimentary rocks; *Tv*, Tertiary volcanic rocks, mainly Keetley Volcanics; *Kec*, Echo Canyon Conglomerate; *Kw*, Wanship Formation of Williams and Madsen (1959); *Mz*, Mesozoic rocks, includes two marker units: Parleys Member of Kelvin Formation (*Kkp*) and Gartra Grit Member of Ankareh Formation (*Ꞧag*); *Pz*, Paleozoic rocks; *pЄ*, undivided rocks of Precambrian age.

this interval from a position low in the Paleozoic section near the mountain front to a position low in the Mesozoic section at the place south of Kimball Junction (Fig. 5) where it disappears beneath a cover of Keetley Volcanics. In that area, the thrust is folded around adjoining east-northeast-trending anticlines and synclines and, as a consequence, has a remarkably sinuous trace. The Crandall and Dry Canyon faults appear on strike from beneath the volcanic rocks 5 mi northeast and are inferred to be branches of the Mount Raymond thrust. The alternative that they represent a single-folded thrust surface appears less probable because it requires much more complicated geometry.

Possible extent of the Mount Raymond (Crandall–Dry Canyon) thrust system beneath the cover of Cretaceous rocks on the northeast is also of interest. If the point of overlap northeast of Rockport Reservoir essentially marks the place where the thrusts broke through to the surface, they may not extend any farther in the direction of tectonic transport (inferred to be near east-southeast). On the other hand, they may extend for long distances parallel with the regional strike (north-northeast). This possibility is discussed later.

Age of Thrusting

The Mount Raymond thrust as extended cuts the Frontier Formation (restricted) and is overlapped by (hence is older than) the Wanship Formation. This episode of thrusting thus occurred during deposition of the classical Frontier Formation (Cobban and Reeside, 1952). The 90° unconformity east of Rockport Reservoir described in preceding paragraphs, used by Williams and Madsen as a basis for delimiting the Wanship Formation, is apparently very local. Nevertheless, both Trexler (1955, 1966) and Hale (1960) have recognized a persistent conglomerate at the base of the Dry Hollow Member of Hale, and believe the channel fills seen at the top of the Oyster Ridge Sandstone Member and minor erosion throughout the Coalville area represent this same unconformity. According to Hale (1960, p. 141)

> The age span of the unconformity cannot be precisely dated. The conglomerate immediately overlies rocks of early Carlisle age (*Collignoniceras woolgari* zone) and lies about 800 feet below rocks containing *Cardium curtum* and others of early Niobrara age. The intervening interval consists of fluvial sediments. . .not datable by faunal evidence, which represents either late Carlile or earliest Niobrara. The writer tentatively concurs with Williams and Madsen (1959) that the hiatus spans middle and late Carlile, and that rocks of early Niobrara rest on early Carlile.

This age assignment dates the Mount Raymond thrust as mid-Turonian to early Coniacian in terms of European stages. It thus appears significantly older than the Charleston-Nebo thrust, which cuts rocks of the Mesa Verde Group but is overlapped by rocks of the Price River or North Horn Formation. That episode of thrusting, though less precisely dated, is at least post-Santonian and may be as young as Maestrichtian. This sequence of events is consistent with structural relations, inasmuch as the Mount Raymond thrust is folded with the Uinta arch, and both the arch and thrust are truncated on the west by the Charleston-Nebo thrust (Crittenden, 1969).

Economic Implications

Both geometry and age relations (Fig. 1) suggest that the Mount Raymond thrust, as here extended, closely resembles some of the thrusts of southwestern Wyoming. The most likely appears to be the Absaroka (Fig. 1), which disappears southwestward beneath a cover of Tertiary and Cretaceous rocks southeast of Evanston. The two faults are approximately aligned and both are latest Cretaceous in age (Armstrong and Oriel, 1965, p. 1859). The subcrop of a connection, if present, should be at shallow depth (less than 1,000 ft) along the southeast edge of the area of concealed thrusts outlined on Figure 1. Farther downdip, the thrust would be expected at depths of several thousand feet. Similar northwest-dipping thrusts are described by Hale (1969, p. 106) along the eastern edge of the Coalville district. A connection at depth to the deeper Mount Raymond structure is possible and may help explain some structural complications revealed by deep drilling near Coalville (Shelley, 1959).

The possibility that this or related thrusts may produce structural traps in the covered interval between Wanship and Evanston should be considered in analyzing drilling and seismic records in this area. The shallowest potential reservoir rocks are sandstones of the Frontier and Kelvin Formations, whose equivalents have yielded both oil and gas in similar structural settings in Wyoming (Krueger, 1960). Deeper zones such as the Nugget and Weber Sandstones or sands of the Mississippian Humbug Formation are largely untested in this area (Skeeters and Hale, 1972). The Paleozoic section expected is essentially that exposed near Morgan, Utah (Mullens and Laraway, 1964; Mullens, 1967), in which a thin basal Cambrian unit (Tintic Quartzite) lies directly on gneissic or crystalline Precambrian basement.

References Cited

Armstrong, F. C., and S. S. Oriel, 1965, Tectonic development of Idaho-Wyoming thrust belt: Am. Assoc. Petroleum Geologists Bull., v. 49, no. 11, p. 1847-1866.

Bromfield, C. S., and M. D. Crittenden, Jr., 1971, Geologic map of the Park City East quadrangle, Summit and Wasatch Counties, Utah: U.S. Geol. Survey Geol. Quad. Map GQ-852, scale 1:24,000.

Cobban, W. A., and J. B. Reeside, Jr., 1952, Frontier Formation, Wyoming and adjacent areas: Am. Assoc. Petroleum Geologists Bull., v. 36, no. 10, p. 1913-1961.

Crittenden, M. D., Jr., 1963, Emendation of the Kelvin Formation and Morrison(?) Formation near Salt Lake City, Utah, *in* Geological Survey research, 1963: U.S. Geol. Survey Prof. Paper 475-B, p. B95-B98.

——— 1965a, Geology of the Mount Aire quadrangle, Salt Lake County, Utah: U.S. Geol. Survey Geol. Quad. Map GQ-379, scale 1:24,000.

——— 1965b, Geology of the Sugar House quadrangle, Salt Lake County, Utah: U.S. Geol. Survey Geol. Quad. Map GQ-380, scale 1:24,000.

——— 1969, Interaction between Sevier orogenic belt and Uinta structures near Salt Lake City, Utah (abs.): Geol. Soc. America Abs. with Programs, v. 1, pt. 5, p. 18.

——— F. C. Calkins, and B. J. Sharp, 1966, Geologic map of the Park City West quadrangle, Utah: U.S. Geol. Survey Geol. Quad. Map GQ-535, scale 1:24,000.

Granger, A. E., 1953, Stratigraphy of the Wasatch Range near Salt Lake City, Utah: U.S. Geol. Survey Circ. 296, 14 p.

Hale, L. A., 1960, Frontier Formation—Coalville, Utah, and nearby areas of Wyoming and Colorado, *in* Overthrust belt of southwestern Wyoming and adjacent areas: Wyoming Geol. Assoc. 15th Ann. Field Conf. Guidebook, p. 136-146.

——— 1969, Northern Utah, *in* Geologic guidebook of the Uinta Mountains, Utah's maverick range—Intermtn. Assoc. Geologists (and Utah Geol. Soc.), 16th Ann. Field Conf.: Utah Geol. Survey, p. 101-108.

Krueger, M. L., 1960, Occurrence of natural gas in the western part of Green River basin, *in* Overthrust belt of southwestern Wyoming and adjacent areas: Wyoming Geol. Assoc. 15th Ann. Field Conf. Guidebook, p. 194-209.

Morris, E. C., 1953, Geology of the Big Piney area, Summit County, Utah: M.S. thesis, Utah Univ.

Mullens, T. E., 1967, Preliminary geologic map of the Morgan 7½ quadrangle, Morgan County, Utah: U.S. Geol. Survey Open-File Rept.

——— and W. H. Laraway, 1964, Geology of the Devils Slide quadrangle, Morgan and Summit Counties, Utah: U.S. Geol. Survey Mineral Inv. Field Studies Map MF-290, scale 1:24, 000.

Shelley, C. T., 1959, Coalville anticline, Summit County, Utah: Intermtn. Assoc. Petroleum Geologists 10th Ann. Field Conf. 1959 Guidebook, p. 189-192.

Skeeters, W. W., and L. A. Hale, 1972, Southern Wyoming, *in* Geologic atlas of the Rocky Mountain region: Rocky Mountain Assoc. Geologists, p. 274-276.

Stokes, W. L., and J. H. Madsen, Jr., compilers, 1961, Geologic map of Utah—northeast quarter: Utah Geol. and Mineralog. Survey, scale 1:250,000.

Trexler, D. W., 1955, Stratigraphy and structure of the Coalville area, Utah: Ph.D. thesis, Johns Hopkins Univ.

——— 1966, The stratigraphy and structure of the Coalville area, northeastern Utah: Colorado School Mines Prof. Paper, Contr. 2, 69 p.

Williams, N. C., and J. H. Madsen, Jr., 1959, Late Cretaceous stratigraphy of the Coalville area, Utah: Intermtn. Assoc. Petroleum Geologists 10th Ann. Field Conf. Guidebook, p. 122-125.

The American Association of Petroleum Geologists Bulletin
V. 60, No. 3 (March 1976), P. 379-388, 6 Figs.

Complex Fault Structures in Veracruz Province of Mexico[1]

R. W. MOSSMAN[2] and FRANCISCO VINIEGRA[3]
Tulsa, Oklahoma 74102, and Mexico City, Mexico

Abstract Tertiary rocks of the Veracruz basin overlap westward onto overthrust Cretaceous limestones and infolded lower Tertiary clastic rocks, which comprise the eastern flank of the Sierra Madre Oriental. Evaluation of the structural relations of the buried foothills was accomplished through close integration of the extensive seismic and borehole information available. This involved deciphering details of the complex fault system by study of the vast number of diffractions associated with the seismic data, as well as the resolution of the occasionally ambiguous relations of the well data. A middle Eocene age is established for the orogeny. The concepts derived from this investigation have significant tectonic implications as well as potential in guiding economic development.

INTRODUCTION

The Veracruz province of Mexico encompasses approximately the south-central part of the state of Veracruz (Fig. 1). It is separated from the old productive area of Poza Rica on the north by a zone of Tertiary volcanism consisting of the Santa Ana–Teziutlan massif. On the southeast there is a less well-defined separation between the Veracruz province and the Reforma area, where recent petroleum development has occurred in the states of Chiapas and Tabasco. Such demarcation, however, appears to be approximately coincident with the San Andres massif, also a zone of Tertiary volcanic rocks.

The province is often referred to as the Veracruz basin although the basinal part actually consists only of the area on the east containing thick Tertiary sedimentary rocks. The Tertiary basin is characterized by north-south-trending anticlines, bounded on the seaward side by extensive, down-to-the-east growth faults. On the west (Fig. 2), the Tertiary beds overlap the buried foothills of the Sierra Madre Oriental which are productive locally of oil and gas. These complex, buried structures thus are of particular economic interest. They comprise extensively overthrust, imbricate plates of Cretaceous limestones with associated, locally infolded lower Eocene and Paleocene shales.

About 2,000 km of seismic profiles, on approximately a 2-km grid, provides detailed seismic coverage of an area of more than 4,000 sq km within the limestone zone. The seismic data were recorded using either digital or analog magnetic-recording methods, and most such recording involved common-depth-point techniques. Figure 3 shows the extent of the seismic grid in the foothills zone.

In addition to the seismic control, nearly 100 wells have been drilled in the area, about one-half of which represent exploitation drilling, principally in Angostura field. Most of the wells were drilled after 1950, and the available data include standard electric logs, gamma ray-neutron logs, acoustic logs, detailed sample logs, lithologic and mineral analyses, paleontologic analyses of the microfauna, and core analyses. The locations of most of these tests also are shown on Figure 3, although some wells are omitted where several are in close juxtaposition.

It is difficult to determine structural and chronological relations in an area of complex faulting, and hence to assay any economic potential that may exist. Characteristically, in the Veracruz province the seismic data are replete with diffractions and usually are obscure; also the relations of the wells not always are defined clearly. However, the data have been integrated into positive geologic concepts which are pertinent to the regional tectonics of the area and provide guidance for commercial exploitation.

STRATIGRAPHY

The stratigraphic relations within the province are shown by the columnar section in Figure 4. The Miocene to recent strata consist primarily of shales with locally interbedded fine sandstones, whereas coarse clastic rocks are present locally in the basal Miocene, the Oligocene, and the upper Eocene. The middle and lower Eocene and Paleocene consist principally of shales and sandy shales, with coarse clastic rocks rarely present.

In the western part of the Veracruz province, that part which composes the foothills of the Sierra Madre Oriental, an extensive sequence of Upper and middle Cretaceous limestones and shaly limestones is present beneath the Tertiary. The true sedimentary thickness of these limestones is

[1]Manuscript received, May 7, 1975; accepted, September 5, 1975. Presented at the 60th Annual Meeting of the Association, Dallas, Texas, April 7, 1975.

[2]Seismograph Service Corporation.

[3]Petroleos Mexicanos.

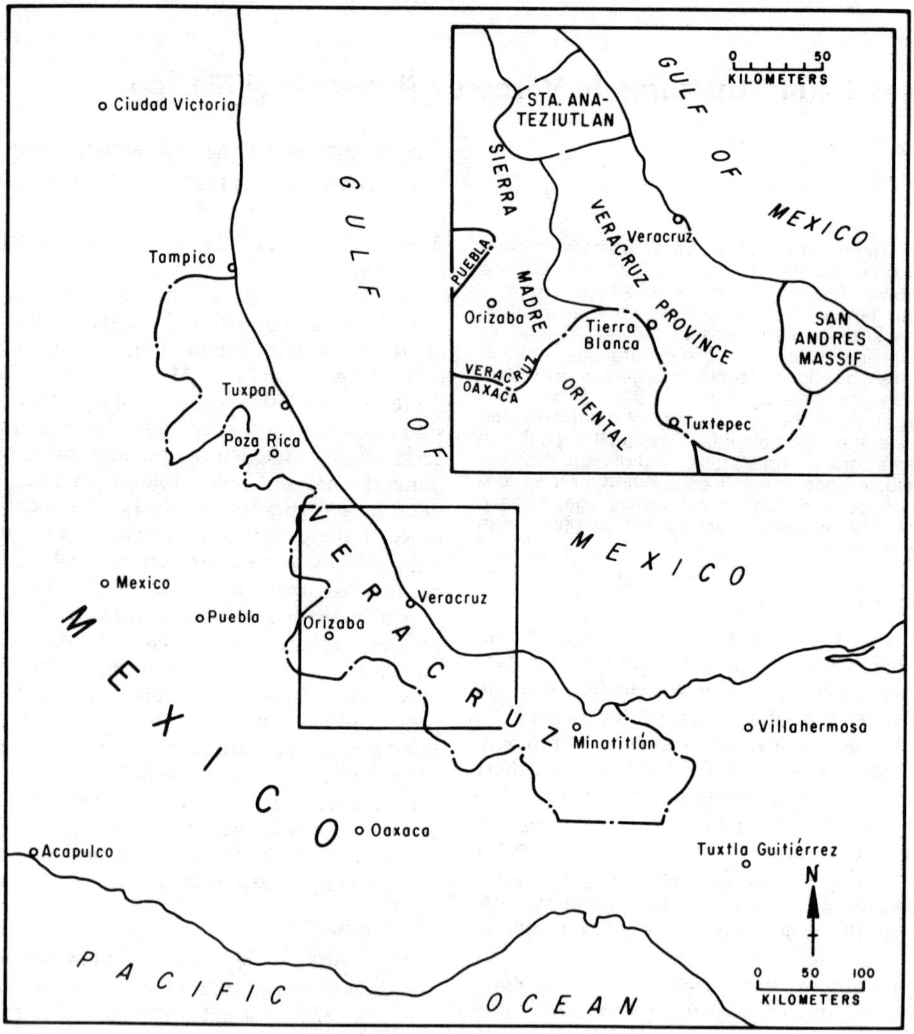

FIG. 1—Map of eastern Mexico with inset of Veracruz province.

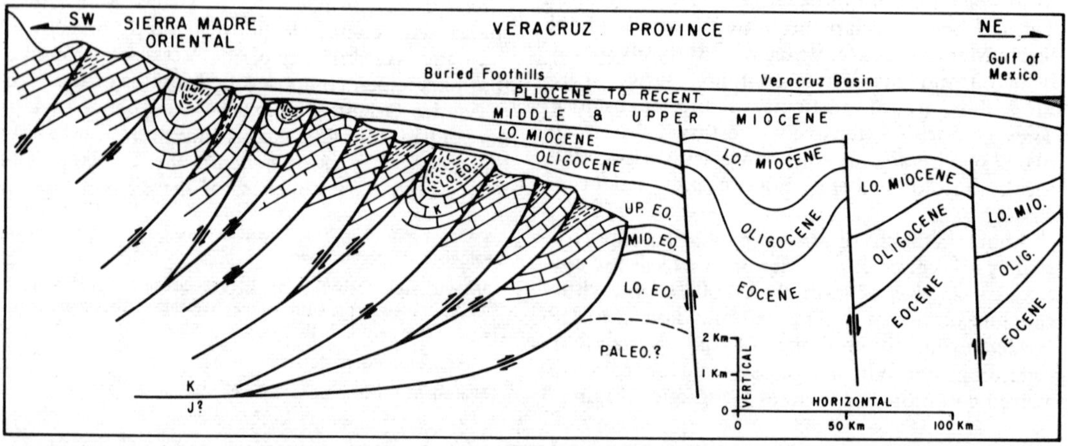

FIG. 2—Schematic cross section across Veracruz province.

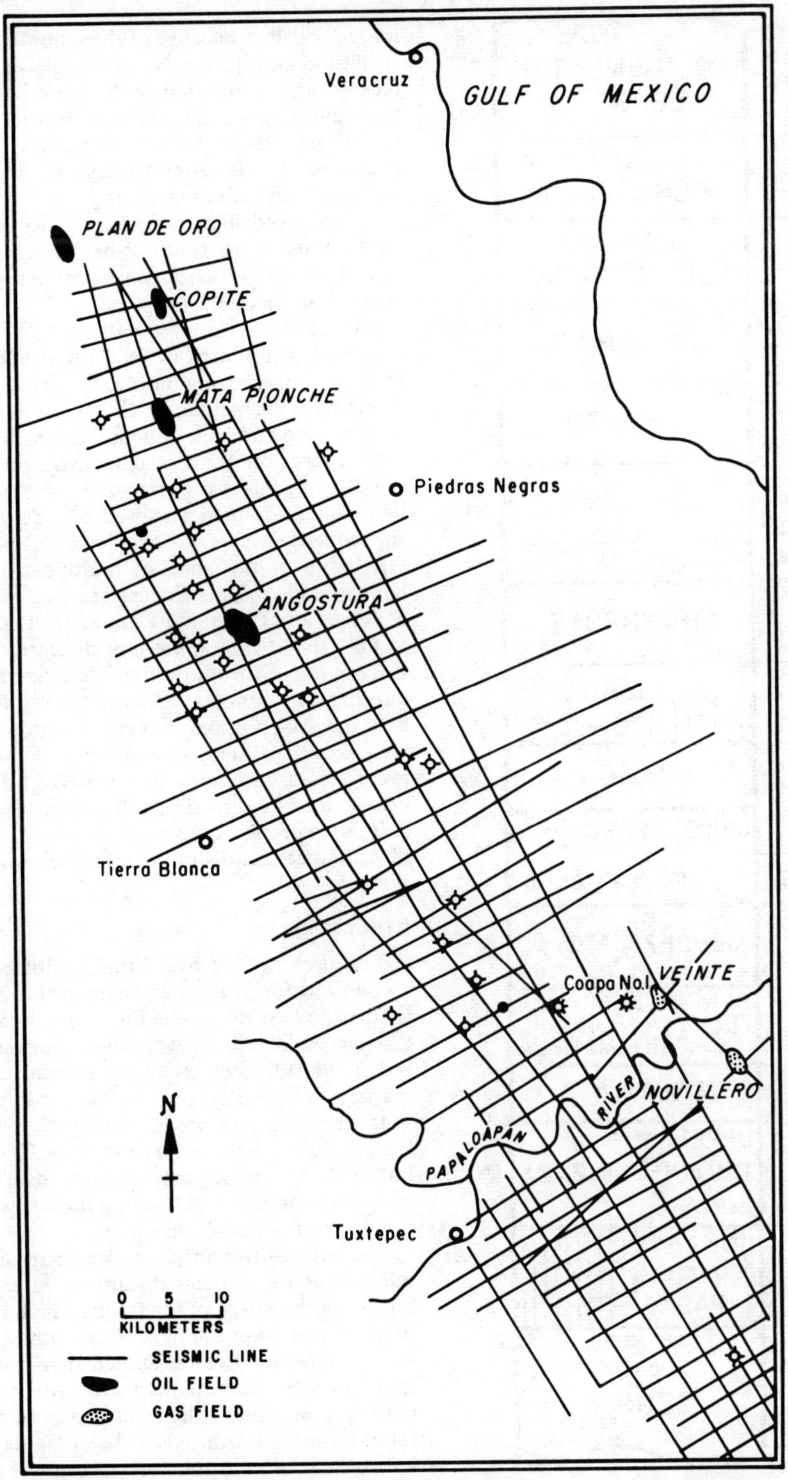

FIG. 3—Location of seismic lines, oil and gas production, and principal dry holes in zone of foothills structure.

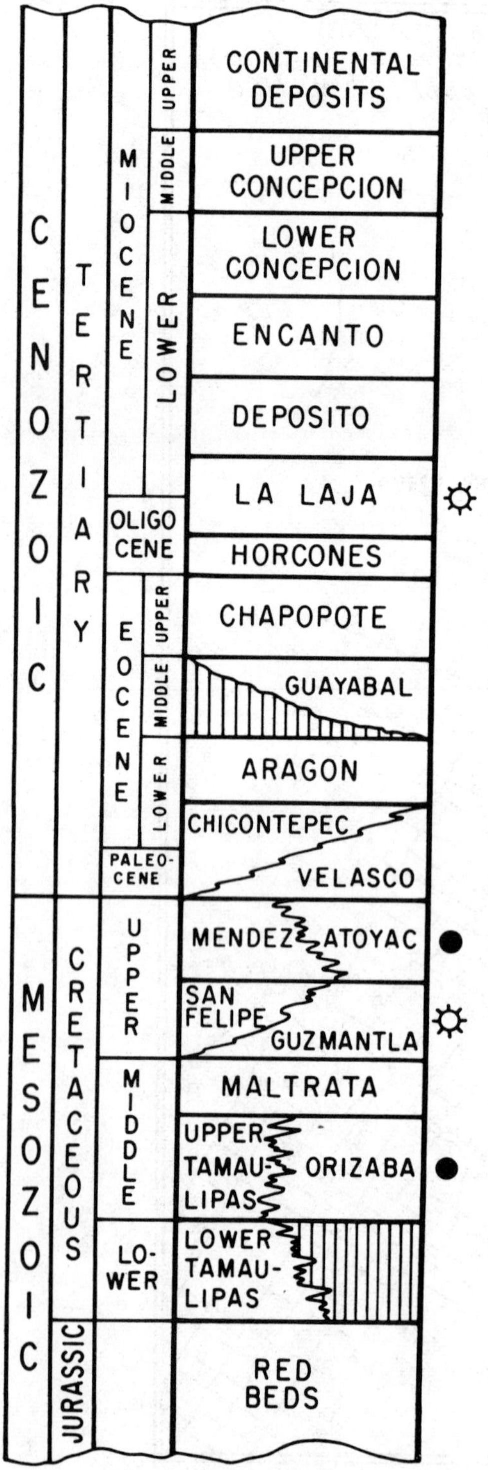

FIG. 4—Columnar section, Veracruz province.

not established because of the complex, imbricate structural conditions. Nor are rocks of Early Cretaceous age recognized in the buried foothills, except about 20 km northwest of Fortin in the Sierra Madre where Lower Cretaceous limestones equivalent to the lower Tamaulipas are present in a limited outcrop area.

Jurassic red beds are believed to underlie the Cretaceous in the foothills belt, but this postulation is based primarily on extrapolation of relations from areas on the west.

Of particular interest within the stratigraphic sequence in the area of foothills structures is the middle Eocene unconformity. This hiatus is thought to be primarily the result of structural disturbance and lack of deposition rather than extensive erosion. This is demonstrated readily in the deeper, eastern parts of the overthrust-fault belt, but is defined less well on the west where subsequent erosion has occurred. East of the eastern limits of the zone of faulting a completely conformable middle Eocene section, consisting of shales of the Guayabal Formation, is present.

The first stratigraphic unit beneath the middle Eocene unconformity differs from one location to another as a function of structural conditions. In extreme cases upper Eocene Chapopote shales, which overlie the unconformity in most of the area, are in contact with inverted middle Cretaceous limestones. The stratigraphic sequence (Fig. 4) represents the normal vertical continuity of sediments, undistorted by structural deformation.

Structure

The presence of overthrust faulting has been recognized for at least 25 years in this area, since before the first oil production in the province was discovered in 1953 at Angostura, but only recently has the full intensity of the overthrusting been realized. Originally one or two, no more than four, thrust plates were postulated, but present studies have identified as many as 18 sequential imbricate plates in parts of the subsurface area, with additional thrust faulting farther west where outcrops of the rocks are present.

The zone of overthrusting was mapped by seismic means for a linear distance greater than 130 km along the strike of the faults, which is generally north-northwest. A part of this zone of faulting near Angostura field is shown in Figure 5. The lines on the subcrop map are fault traces, not contours, and mark the intersection of the thrust faults with the northeastward-sloping unconformity surface. The distance between the faults as they intersect the unconformity ranges from ½ to 3 or 4 km, and they appear to coalesce at depth.

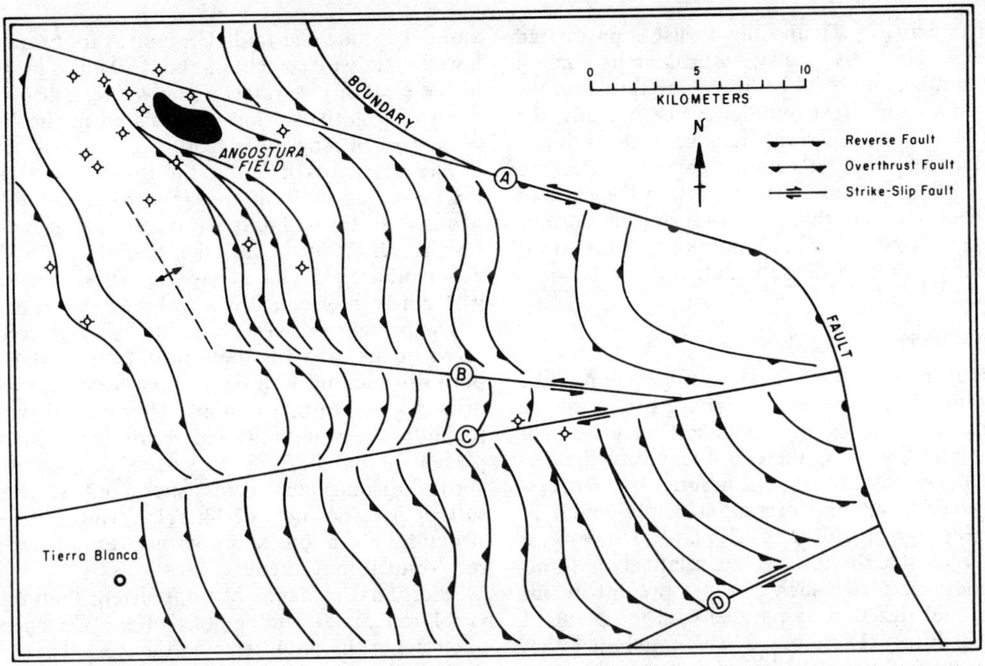

FIG. 5—Map of part of area near Angostura oil field showing traces of faults on middle Eocene unconformity.

The deepest and most eastern of the fault blocks is more than 3,500 m in depth at its uppermost limit against the unconformity.

AGE OF FAULTING

The existence of the faulting and of the overlying unconformity, and the characteristics of these features, are defined amply by the many boreholes and the extensive seismic data in the area (Fig. 3). The unconformity is established as being of middle Eocene age, with several criteria supporting this conclusion. Sedimentary rocks of this age are not present anywhere in the Cretaceous overthrust belt, yet directly east of the zone of thrusting about 500 m of middle Eocene Guayabal shale is present. The Coapa 1 well, which is just a few hundred meters, or less, east of the fault marking the eastern limit of the Cretaceous overthrust belt, penetrated 495 m of Guayabal shale with no coarse clastic rocks present in any of the Eocene section, nor with any evidence of a depositional or erosional hiatus within the Eocene. Because the overlying upper Eocene Chapopote shale extends continuously across the bounding fault, and the underlying lower Eocene Aragon beds are essentially identical both beneath the Guayabal and beneath the unconformity, the hiatus is established as being the age equivalent of the Guayabal Formation.

Upper Eocene to recent sedimentary deposits lie directly on the unconformity and progressively overlap it toward the west; the unconformity surface truncates various Cretaceous limestones as well as lower Eocene and Paleocene shales. The pre-unconformity rocks are folded and overturned locally, yet no rocks younger than early Eocene have been discovered beneath the unconformity.

In Alberta, Canada, the "foothills structures" described by various authors present geologic conditions somewhat similar to those in Veracruz. Although the thrusting in Alberta involves Paleozoic as well as Cretaceous rocks, and there are other dissimilarities, the Mesozoic part of the Alberta foothills as described by Dahlstrom (1970) and others appears to be closely related both in form and age to the Veracruz thrust belt.

BOUNDARY FAULT

Figure 5 shows the fault lines to be subparallel, with their eastern limit marked by a high-angle reverse fault. The latter delineates the eastern boundary of Cretaceous rocks in the area. It is interpreted to be the upturned edge of the lowermost thrust (Fig. 2), and this thrust is postulated to be a sole fault, nearly vertical at its eastern limit and becoming a bedding-plane thrust on the west at depth. Horizontal movement along the fault plane may represent slippage of the indurated rocks of the Cretaceous along the softer red beds of the Jurassic. Farther west, in the state of Oaxaca, a similar thrust is suggested by surface geology. However, in the Veracruz area the stratigraphic position of the sole fault at depth is largely speculative.

Transcurrent Faults

From purely theoretical considerations, the horizontal forces necessary to have produced the extensive overthrusting would not have been of equal intensity along the entire length of the disturbed area. Such variations in intensity of necessity would result in differential displacement of different segments of the fault plates. However, it is not plausible that broad, flat, relatively indurated limestone plates such as those present in this area could flex sharply parallel with the plane of their bedding. Therefore it was concluded that strike-slip faulting must be present to relieve the stresses produced. The seismic data confirm that such tear faults did develop. Designated as transcurrent faults, these are vertical faults which strike normal to the overthrust plates and are contemporaneous with the overthrusting. Similar faults are encountered in the Alberta foothills and elsewhere.

Fault A, along the north side of Angostura field (Fig. 5), is an example of this type of faulting, representing a large offset in the traces of the thrust faults. Fault A is a left-lateral strike-slip fault which does not appear to involve basement rocks and probably extends no deeper than the sole fault with which it is associated closely. It is one of 12 such tear faults which were recognized within the area.

The compressional forces producing the extensive overthrusting appear to have been at a maximum in the area south of Angostura field, between transcurrent faults A and D (Fig. 5). In this locality 18 imbricate plates are present, which is the greatest number present anywhere in the area. In addition, northwest of transcurrent fault B all the strike-slip faults mapped are left-lateral, and from fault C southeast all such faults are right-lateral, providing additional evidence that the maximum development of eastward-thrusting energy was in this central zone.

These transcurrent faults have approximately zero hade, yet exhibit little evidence of vertical movement. Horizontal slip commonly is extensive and quite varied between different fault plates along the same tear fault. On fault A for example, horizontal displacement of about 20 km is present in the eastern lowermost plate, whereas displacement of less than 5 km is indicated in the fault plate west of Angostura field.

The transcurrent faults, the thrust faults, and the bounding fault all appear to be contemporaneous. The thrust faults always are offset by the transcurrent faults, and the boundary fault is either offset, exhibits a change in strike, or both, where it is intersected by a transcurrent fault.

The presence of vertical faults has been recognized in the area for some time from studies of both seismic and well data. They were described originally as normal faulting. However, they apparently are tear faults representing lateral displacement but little or no vertical motion. Any vertical displacement across such faults is the result of juxtaposition of blocks having different structural elevations and does not result from direct vertical movement.

Additional evidence of transcurrent faulting is visible on aerial photographs of the Cretaceous outcrops in the mountains on the west. These reveal similar, transversely oriented faults, at least some of which appear to be projections or continuations of transcurrent faults identified in the subsurface.

In an extensive paper on the general paleogeography and tectonics of the Mesozoic of this area, Viniegra (1966) noted the existence of several very straight U-shaped valleys, particularly Acultzingo Valley west of Orizaba, Veracruz. These valleys are unique in that they have cross sections similar to those of glacial valleys, and are oriented normal to the structural and topographic grain of the area. Some investigators in the past considered them to be evidence of glaciation. However Viniegra suggested that they were the result of chemical erosion, that is, of solution along fractures. This hypothesis is substantiated by the present study. The orientation of the valleys, their appearance, even the inability to trace faults across the valleys, from one valley wall to the opposite, all are explained adequately if they are considered to be representative of erosional development along transcurrent faulting.

Frontal Trough

East of the fault which marks the eastern limit of the overthrust belt, and extending southeast along that fault for at least 60 km, is a graben averaging about 10 km in width. The east side of the graben is formed by a high-angle reverse fault which, like the normal faults which characterize the Tertiary basin on the east, appears to be a growth fault extending upward into the late Miocene or younger beds and with throw increasing with depth. Contrary to the other Tertiary faults in the basin, which are normal faults downthrown to the east, this is a reverse fault downthrown to the west. The Loma Bonita anticline on the eastern upthrown side of the fault may represent a shale diapir.

The strike of the reverse fault is more north-south than the average strike of the bounding fault marking the west edge of the trough, resulting in a narrowing of the graben from north to south. Whether this reverse fault dies out southward or ultimately intersects the fault separating the Cretaceous overthrusts and the graben is not known. However, activity involving the fault on the east of the graben continued over a long period of time, with the probability that movement along this alignment was present at least as early as the beginning of the Tertiary, as suggested by seismic evidence. Apparently the graben and the associated reverse fault on the east were present during the middle Eocene when the major thrust faulting was going on in the west, and may have acted as a buttress to limit the eastward thrust of the Cretaceous limestone plates.

Within the graben at Novillero and Veinte fields (Fig. 3), gas production is obtained from Oligocene conglomerates which are folded gently into anticlinal structures. Seismic evidence suggests that more than 6,000 m of Tertiary sedimentary rocks are present in this trough. However, the deepest hole drilled in the graben is the Coapa 1; this well penetrated only 4,500 m of Tertiary shale containing but a few fine-grained sands from which gas is produced. It is bottomed in uppermost lower Eocene. It encountered no coarse clastic rocks but, as previously noted, did encounter 495 m of middle Eocene Guayabal shale. The shale, or any equivalent thereof, is not present in the west as shown by the many wells including several located only 5 to 6 km west of the Coapa 1.

Seismic evidence suggests that at least 2,000 m of lower Eocene and Paleocene sedimentary rocks are present below the depth reached by the Coapa well, but gives no indication of either the depth or type of the pre-Tertiary rocks which underlie the lower Tertiary strata.

Genetic Environment

The well data and to some extent the seismic data provide information bearing on the environmental conditions in the middle Eocene during the time of orogenic movement. No single criterion of itself presents irrefutable evidence of the ambient conditions at such time, but the consensus of data strongly indicates that at least the eastern part of the imbricate zone and the accompanying unconformity developed beneath sea level, probably in deep water.

The faunal assemblages which are associated with both the overlying upper Eocene and underlying lower Eocene rocks are typical of a deep-water environment, including no shallow-water fauna except pelagic forms. The Foraminifera identified include predominantly such bathyal species as *Cibicides robertsoniana, Planulina wuellerstorfi, Anomalina dorri, Cyclomina cancellata,* as well as a few other similar genera. It is evident that movement of the sea floor from deep ocean to dry land thence back to deep water would have been required if the unconformity surface representing middle Eocene time had developed above sea level, and all this would have had to take place very rapidly, for the time span of the middle Eocene is geologically relatively short. It appears more probable that the thrusting occurred on the sea floor with the unconformity developing in deep water.

Additional evidence for such submarine genesis of the unconformity is present where the Cretaceous limestones are directly beneath the hiatus. Drilling has disclosed that the Cretaceous rocks consist, in places, of breccias containing massive blocks, often over 10 m thick and of unknown lateral extent, but much too large to have been transported any distance. They apparently have not been subjected to subaerial erosion, only to the less drastic action of marine currents. Such ocean currents also would have served to remove the crumpled debris of lower Eocene shales involved in the thrusting, and to redeposit this fine-grained material in the trench in front of the rising thrust plates.

The absence of subaerial erosion is demonstrated further by the angularity of the edges of some of the smaller limestone fragments present in cores taken in the breccia, and by the total absence of coarse clastic rocks in the middle Eocene deposits east of the zone of faulting. Although not necessarily positive indications of submarine conditions, these criteria do strongly suggest such an ambiance.

Coarse sedimentary rocks, nevertheless, are present locally in the upper Eocene and Oligocene both east of the overthrust zone and above it. They reach their greatest extent in the Oligocene where they act as reservoir rocks to produce gas at Novillero and Veinte fields. The conglomerates generate strong seismic reflections because of porosity and density contrast with the enveloping shales and on this basis can be traced westward by seismic means. Such reflections do not coalesce with the reflection representing the unconformity surface, indicating that the conglomerates do not impinge on that surface. This suggests that these upper Eocene and Oligocene clastic rocks are derived from source rocks farther west which were eroded as a result of vertical movement of the Sierra Madre Oriental subsequent to the overthrusting. They thus are not directly representative of the specific orogenic movement which produced the middle Eocene unconformity.

Today the middle Eocene unconformity in Veracruz exhibits an eastward dip ranging between 10 and 25°, but the configuration of the unconformity surface suggests that its original attitude may well have been more horizontal. A low-slope angle would be compatible with the theory that at least the eastern part of the unconformity developed below sea level, in deep water.

From these criteria it is possible to postulate a sequence of events during Eocene time in which expanding forces in the Sierra Madre Oriental, or west of it, caused extensive thrusting and imbrication of the Cretaceous and lower Eocene rocks during middle Eocene time while the rocks were under a considerable depth of water. Such orogeny, as it progressed, caused tilting of the area toward the east-northeast and elevation of the western part of the overthrust foothills belt. During very late Eocene and in Oligocene time this resulted in rapid erosion and deposition of conglomerates in the bordering seas east of or overlying the eastern part of the overthrust belt.

Subsequent subsidence is indicated by the gradual overlap of upper Tertiary beds toward the west onto the unconformity, but the uniformly fine-grained character of the clastic rocks which form the upper Tertiary throughout the Veracruz basin suggest that this tilting, uplift, and subsidence took place very gradually, continuing throughout most of Tertiary time. More recently, the area has been uplifted, accompanied by late Tertiary and Holocene volcanism in peripheral areas. Present-day earthquake activity suggests that related orogenic processes still continue.

Hydrocarbon Entrapment

Gas production is obtained in the Tertiary basin from basal Miocene and Oligocene sandstones and conglomerates, and both oil and gas are produced from the intensely overthrust Upper and middle Cretaceous limestones which compose the foothills of the Sierra Madre Oriental.

Until recently the only commercial production in the area of overthrusting was at Angostura, discovered in 1952, which has yielded prolific amounts of oil from breccias in the Upper Cretaceous Mendez limestone. At Angostura the fault plate which provides the oil reservoir exhibits reversal to the north, undoubtedly associated with major horizontal movement along the transverse fault bounding the north edge of the field. Additionally, rollover into the edge of the overthrust fault is present, and low-porosity, low-permeability upper Eocene shales provide an unconformable, impervious cover for the reservoir rocks. Although these shales could be source beds, the oil is believed to be of Cretaceous origin, not Tertiary, as in the Veracruz basin the Tertiary appears to generate gas exclusively.

For most of the fault-plate structures, the seismic data or drill data disclose an absence of one or more prerequisite trapping mechanisms. Rollover is lacking, inversion of plunge along strike is not present, or the sedimentary deposits overlying the unconformity are not sufficiently impervious to form a seal, thus explaining the dearth of hydrocarbon accumulation. A few small, isolated zones of oil or gas accumulation, such as at Nopaltepec, are present but these have such limited areal extent that they are not commercially attractive. They even may represent lateral migration of Tertiary gas into Cretaceous limestone reservoirs, in some cases.

One particular fault plate just west of Angostura (Fig. 5) exhibits wide and extensive anticlinal structure suggesting good productive possibilities. Unfortunately, all of the wells drilled on this feature have encountered fresh water (5,000 ppm or less of salt) rather than saline water, with only vestiges of oil, leading to the conclusion that this feature has been flushed out by invasion of meteoric water.

Most recently, oil and gas have been discovered north of Angostura at Mata Pionche, Copite, and Plan de Oro. Except for the last which produces from Cretaceous rocks directly underlying the unconformity at depths of about 500 m, production appears to be coming from various zones of secondary porosity well below the top of the Cretaceous, at depths of 2,500 to 2,800 m primarily from the middle Cretaceous Orizaba limestone. Fracture porosity may be a factor, but anticlinal structure, affecting the thrust plates and possibly predating the ovethrusting, is of significant importance.

Seismic Data

Much evidence bearing on the complex structural conditions in the Veracruz province was provided by the many exploratory holes in the area, as previously noted. However, detailed evaluation of the structural conditions was accom-

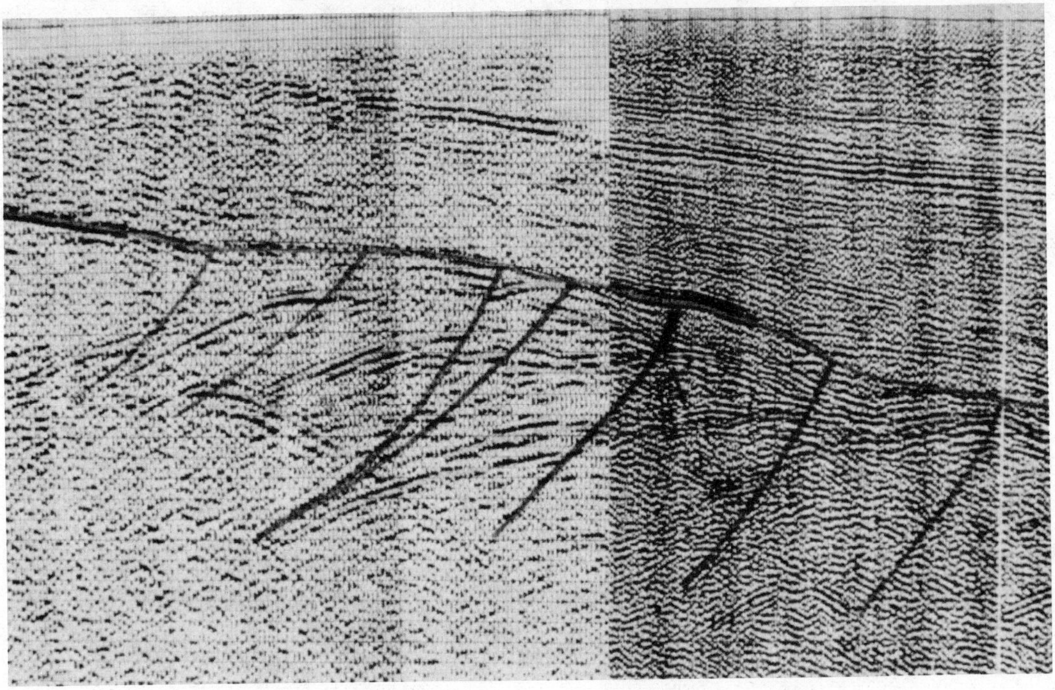

FIG. 6—Representative seismic section near Angostura field.

plished primarily through study of the seismic data. These data vary from single-coverage analog sections to the most sophisticated, digitally recorded and processed presentations, and provide very detailed coverage (Fig. 3).

Although not particularly evident from the sample section shown in Figure 6, Tertiary information is usually good on the seismic sections, but reflections representing Cretaceous bedding planes are limited in extent or are absent. However all sections, like the illustration, show a great many events which are identified as diffractions.

Diffracted energy is defined as energy which is reflected from a subsurface source acting as a single point relative to the plane of the section. In this area conditions are ideal to produce diffractions, for the thrusting has resulted in high-velocity, high-density limestones subsequently being enveloped in low-velocity, low-density shales, yielding reflecting points of little lateral extent as seen by seismic lines normal to the strike, but having great acoustic contrast.

As a consequence of the preponderance of diffracted events and the scarcity of conventionally reflected data, the locations and characteristics of these many diffractions were used in preparing much of the interpretation. Not only did most of the diffraction sources line up in an ordered and meaningful manner but, with the help of the well-data available, the thrust plates which they bounded could be correlated from line to line. Use of diffracted energy as a means of developing a structural interpretation is unconventional.

CONCLUSIONS

The evaluation and analysis of the Cretaceous structures have demonstrated that careful integration of geologic and seismic data can resolve highly complex structural conditions even when the seismic data are obscure and the relations of the wells far from obvious. The precise dating of the middle Eocene orogeny in the Veracruz area may be of significance for other regions where comparable orogenic activity has occurred but has left a less well defined historical record.

Certainly the work involved has proved economically rewarding to Petroleos Mexicanos. The recent discoveries of new oil and gas production in the overthrust zone suggest that the economic potential of the area may be considerable and demonstrate that the effort required to decipher such complex structural relations can be worthwhile.

Selected References

Dahlstrom, C. D. A., 1960, Structural geology in the eastern margin of the Canadian Rocky Mountains: Bull. Canadian Petroleum Geology, v. 18, p. 332-406.

Douglas, R. J. W., H. Gabrielse, J. O. Wheeler, O. F. Stott, and H. R. Belyea, 1970, Geology of western Canada, Chap. 8, *in* Geology and economic minerals of Canada, 5th ed.: Canada Geol. Survey Econ. Geology Rept. 1, p. 365-488.

Link, T. A., 1949, Interpretations of Foothills structures, Alberta, Canada: AAPG Bull., v. 33, p. 1475-1501.

Price, R. A., and E. W. Mountjoy, 1970, Geologic structure of the Canadian Rocky Mountains between Bow and Athabasca Rivers, a progress report, *in* Structure of the Southern Canadian Cordillera: Geol. Assoc., Canada Spec. Paper 6, p. 7-25.

Viniegra, F., 1963, Geologia del Macizo de Teziutlan y de la Cuenca Cenezoica de Veracruz: Asoc. Mexicana Geologos Petroleros Bol., v. 15, p. 101-163.

——— 1966, Paleogeografia y tectonica del Mesozoico en la Provincia de la Sierra Madre y Macizo de Teziutlan: Asoc. Mexicana Geologos Petroleros Bol., v. 18, p. 145-171.

BULLETIN OF THE AMERICAN ASSOCIATION OF PETROLEUM GEOLOGISTS
VOL. 46, NO. 11 (NOVEMBER, 1962), PP. 2019-2032, 9 FIGS.

MOUNTAIN FLANK THRUSTING IN ROCKY MOUNTAIN FORELAND, WYOMING AND COLORADO[1]

ROBERT R. BERG[2]
Denver, Colorado

ABSTRACT

The principal mountain ranges of the Wyoming and Colorado foreland were raised asymmetrically by Laramide uplift which began in latest Cretaceous time as dominantly vertical movement along arcuate trends. Uplift continued into the Paleocene, and the steeper flanks of some ranges developed into large overturned folds by local compressive forces marginal to the main uplifts. Along segments of maximum uplift, overturned folds were broken and thrust far over the basin synclines in latest Paleocene or earliest Eocene time. Throughout the long period of uplift the adjacent basins were downwarped continuously and received sediment from the rising mountains.

The process of uplift by folding and thrusting better explains observed structure of mountain flanks than the older ideas of block uplift along high-angle faults or thrust uplift by regional compression. Fold-thrust structures are best known along the major Wyoming thrust zones bordering, on the south, the Wind River and Granite Mountains and the Washakie and Owl Creek Mountains. Other examples of faulted overturned folds occur in Colorado along the Golden thrust of the Front Range and Willow Creek thrust in western Colorado. In all these areas thrusts are well documented by subsurface control which includes both deep wells and seismic data.

INTRODUCTION

For many years it was believed that the mountain ranges of the Rocky Mountain foreland originated primarily as vertical block uplifts bounded by high-angle faults. This idea was gradually replaced by the concept of thrust uplift by regional compression after many low-angle reverse faults were recognized in surface mapping. However, in recent years additional data from the subsurface lead to the conclusion that the thrust margins of uplifts originated by a process of deformation which began as vertical uplift but resulted in some places in the growth of overturned folds and culminated in thrusting. This fold-thrust idea gives a better understanding of the genetic relationships between mountain flank structures and adjacent basins and makes easier the interpretation of local structure as well as regional tectonic patterns. However, for some geologists the old idea of vertical block uplift persists today, as an exclusive process of foreland mountain building, to the detriment of tectonic understanding. Some recently published papers have concluded that thrusting is not common or significant in the Wyoming foreland. Therefore, this paper is a defense of thrusting and presents details of known thrust zones which are significant mountain flank structures. It is concluded that folding and thrusting on a grand scale were major factors in mountain flank deformation.

Exploration for oil and gas in the Rocky Mountains, especially in Wyoming, often has extended into the folded and thrust-faulted mountain flanks. In these areas structural interpretation is not simple, and in most places is exceedingly difficult because of the common occurrence of post-Laramide sediments which mask structure in the older rocks. A wider knowledge of mountain flank structure will aid exploration in these complex belts. Furthermore, the study of flank structure undoubtedly will lead toward a better understanding of Laramide history and the influence of uplifts on related sediments of adjacent basins, in which increasing amounts of oil and gas are now being found.

As used in the Rocky Mountain area, "thrust fault" is a low-angle reverse fault, especially one which has formed by compression, and "thrust zone" is a fault zone believed to consist of thrust faults as its major structure. It is shown that the large thrust zones of the Wyoming foreland have fault plane dips which range from an average low dip of about 20° to nearly vertical. Also, parts of these thrust zones appear to have originated by folding; that is, great vertical and horizontal displacements were attained primarily by overturn-

[1] Manuscript received, November 22, 1961.

[2] Embar Oil Company. The writer is grateful to Cosden Petroleum Corporation for the use of geophysical data essential to the interpretation of the Wind River thrust, and to F. A. Romberg of Texas Instruments, Inc., and to E. B. Wasson of Embar Oil Company for assistance in the structural interpretation of these data. Many others have given helpful advice. Among them are G. G. Anderman, C. J. Gudim, J. M. Parker, and T. C. Wilson. The manuscript has benefited from the criticism of D. L. Blackstone, Jr., and G. R. Downs.

ing of mountain flanks with minor amounts of displacement achieved by faulting alone. Therefore, in the following discussion, the use of "thrust" and "thrust zone" admittedly is not invariably precise, but no other names are suitable. However, there is the deliberate attempt to describe relative uplift in terms of total displacement of the Precambrian (sub-Cambrian) surface, and the use of the terms throw and heave is avoided except where specifically applied to faults.

Wyoming Flank Structure

MAJOR THRUST ZONES

Zones of thrusting are significant mountain flank structures in the Wyoming foreland. All are the result primarily of deformation of Precambrian crystalline rocks which makes them distinctly different from the thrust slices of sediments in the Overthrust Belt of western Wyoming.

The two major thrust zones of central Wyoming are dominantly west- to northwest-trending complexes of multiple thrust faults which form the south and southwest flanks of principal asymmetric mountain ranges (Fig. 1). The individual thrusts within these zones are arcuate in plan, convex toward the south or southwest, and form an *en échelon* pattern. The longest of these is the Wind River-Seminoe thrust system which bounds the central mountain masses from the Gros Ventre Range in the northwest, past the Wind River Mountains, and along the Granite Mountains in the southeast. Maximum vertical displacement is on the order of 40,000 feet and maximum horizontal displacement is estimated to be about 50,000 feet. A second zone is the Washakie-Owl Creek thrust system which extends along the south flanks of the Washakie and Owl Creek Mountains to the southern end of the Bighorn Mountains where the zone turns southeast and may continue along the west margin of the Casper arch. Maximum displacements along this zone are probably equal to those of the Wind River-Seminoe thrust zone, but here displacements are estimated because more precise data are not available.

These two major thrust zones are better known than others in Wyoming because of available detailed surface maps and a greater amount of subsurface information from seismic surveys, mountain flank test holes, and deep wells in the adjacent basins.

Two similar thrust zones are known in adjacent areas. One is the flanking thrust system of the Uinta Mountains in northeast Utah and northwest Colorado, which consists of the Henrys Fork, Uinta, and Sparks thrusts. The other is the Beartooth thrust zone extending from southwest Montana into northwest Wyoming, which is perhaps continuous with the steep west flank of the Bighorn Basin as far south as Oregon Basin. Both zones include thrust faults, but the nature of these faults at depth is still a matter of argument. Recent diagrams of both the Uinta thrust (Ritzma, 1959) and the Beartooth thrust (Foose et al., 1961) show them as markedly steepening with depth, essentially high-angle faults.

Other zones of less significant thrusting in Wyoming are located along the eastern flanks of the Bighorn and Medicine Bow Mountains, whereas the steep monoclines on the west side of the Rock Springs and the Rawlins-Sierra Madre uplifts are believed to be faulted at depth. However, the monoclinal zones are covered by a great thickness of younger Cretaceous and Tertiary sediments, and the exact nature of deep structure is unknown.

WIND RIVER THRUST

The Wind River Mountains are a northwest-trending asymmetric anticline that has an extensive exposed core of Precambrian crystalline rocks (Fig. 1). At the northwest end of the range significant thrust faults are exposed (Richmond, 1945), but for a distance of about 100 miles the southwest flank is overlapped by Eocene sediments which extend mountainward from the Green River Basin and obscure structure in the older rocks.

For many years it was apparent that great structural relief exists between the southwest flank and the adjacent syncline of the Green River Basin, but the nature of flank structure was unknown. It was suggested that the mountain flank is bounded by a reverse fault (Coffin, 1946; Eardley, 1951), and Love (1950) inferred that the flank is bounded by a thrust fault having 20,000 feet or more of throw. Then more definite evidence became available from seismograph surveys in the Big Sandy area (Berg, 1961b). From this data the Wind River thrust is interpreted to have an average dip of 20° NE., a vertical displacement of 30,000 feet or more, and a horizontal displacement approaching 50,000 feet (Fig. 2). The seismic survey does not extend far enough mountainward to show the relationships in the root zone of the

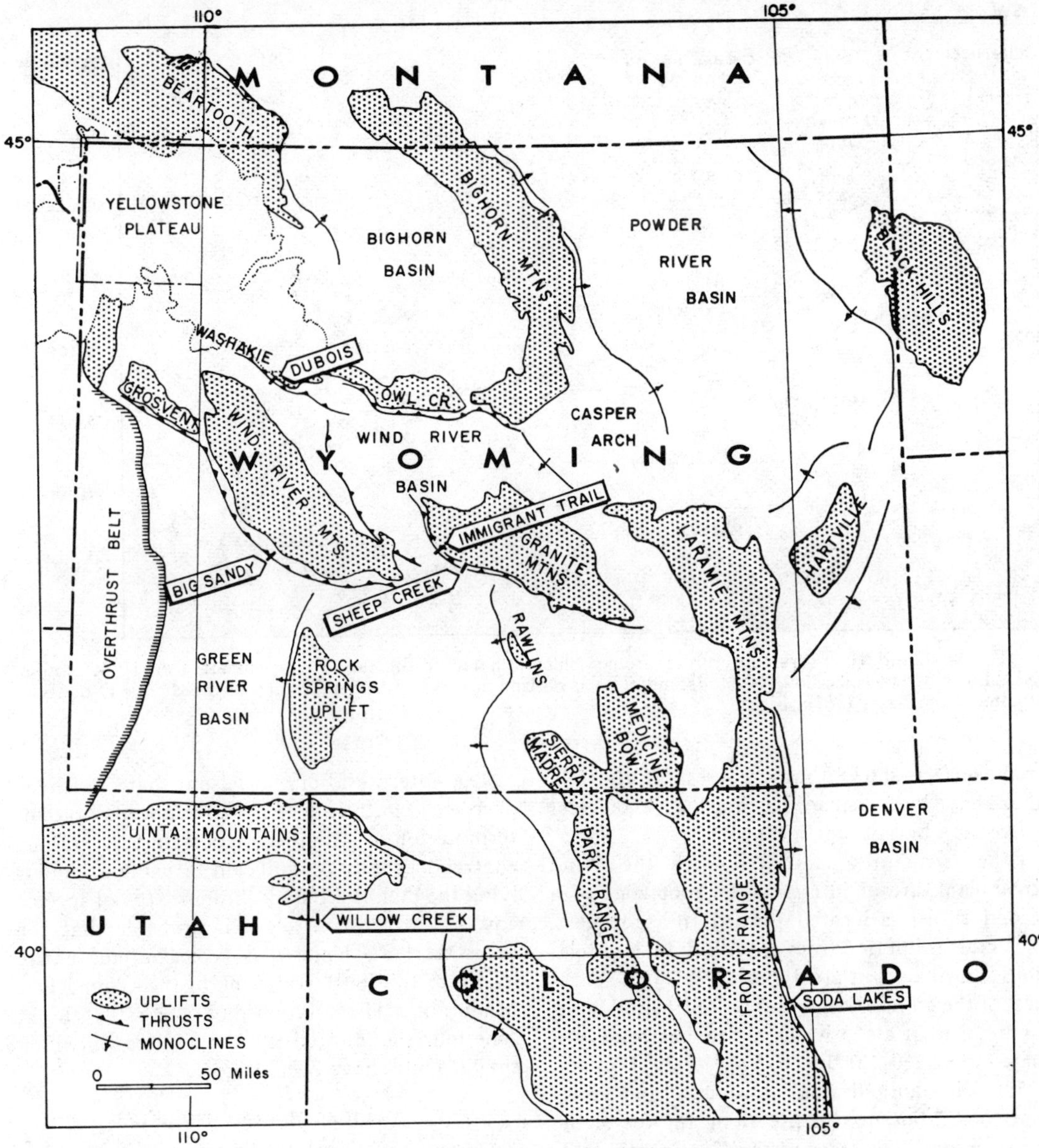

FIG. 1.—Index map of foreland, Wyoming and adjacent areas, showing principal Laramide uplifts, major thrust zones, and location of cross sections.

thrust, but steepening of the fault plane here is expected, based on analogy with other thrust faults at the surface (Berg, 1961b, p. 74).

A prominent feature of the seismograph profile is a strong reflection band that dips northeast and is believed to represent the base of the Precambrian thrust wedge (Berg, 1961b). The thickness, uniformity, and persistence of this band suggest that the thrust is not a simple plane but is a zone 1,000–2,500 feet thick which consists of layered rocks. By comparison with similar thrusts that have been drilled, it is likely that the thrust zone consists of a sheet of sediments, largely Paleozoic in age, deformed by folding and minor faults, but essentially in regular but inverted stratigraphic sequence. Therefore, the Wind River thrust appears to be the faulted limb of a huge, overturned fold. Projection of the possible original Precam-

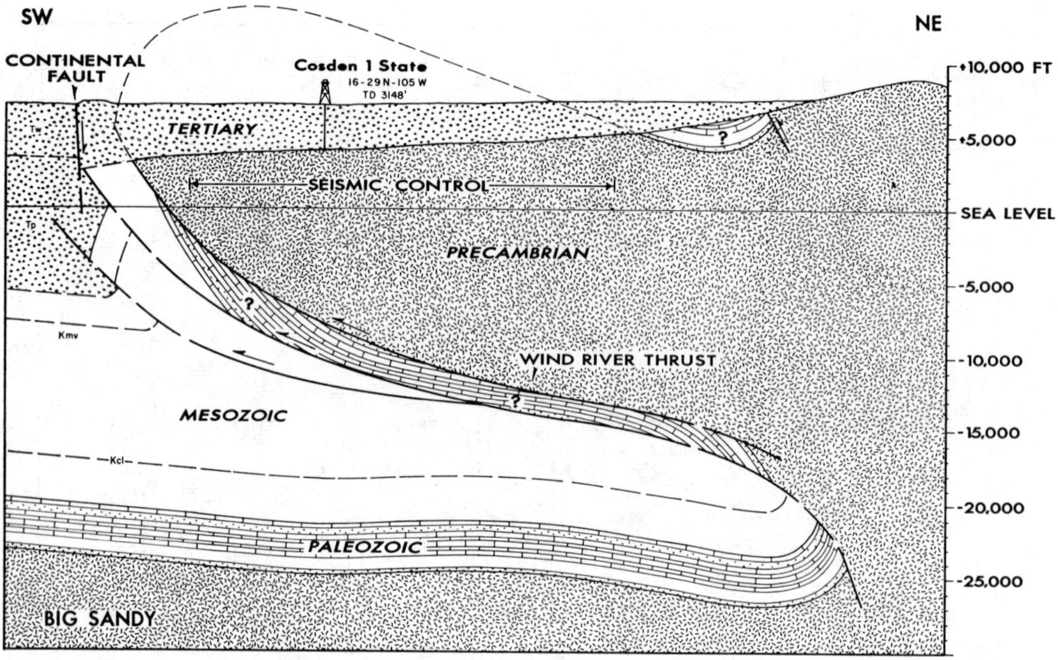

FIG. 2.—Wind River thrust interpreted from seismograph data in Big Sandy area, Sublette County, Wyoming. Kcl is Lower Cretaceous Cloverly Sandstone, Kmv is horizon near top of Upper Cretaceous Mesaverde Formation. Modified from Berg, 1961b, Fig. 3.

brian surface shows that the crest of the fold might have been displaced as much as 45,000 feet above the adjacent syncline.

Other structures associated with the Wind River flank are of interest. The root zone of a second thrust is located toward the northeast. This fault is inferred from gravity data, and relations shown along it are hypothetical. Just southwest of the granite wedge is the Continental fault, a normal fault along which throw has been estimated to be 250–1,000 feet (Nace, 1939; Berman, 1955). The fault dies out downward and represents post-Miocene collapse along the toe of the thrust wedge. The trace of the Continental fault zone along the south flank of the Wind River Range approximately parallels the pre-Wasatch trace of the thrust.

The subthrust sediments of the Green River Basin syncline are remarkably uniform in dip and show no evidence of strong deformation. A gentle, low relief fold is shown in the subthrust section, but this fold may be merely a velocity anomaly resulting from laterally increasing velocity in the subthrust section.

The interpretation of seismic data, without deep well control, does not give an infallible interpretation of structural details, but the seismic control does show a significant thrust fault underlying the mountain flank. From scattered surface and subsurface control, it is probable that the thrust as here depicted is typical of the entire length of the southwest flank of the Wind River Mountains. The relationships along the thrust zone must be inferred from other areas in which similar faults have been drilled.

IMMIGRANT TRAIL THRUST

In other areas of central Wyoming, drilling has revealed overturned sections beneath thick Precambrian thrust wedges. One such area is Immigrant Trail where a Carter Oil Company test well drilled through the Immigrant Trail thrust which forms the southwestern bounding fault of the Granite Mountain uplift. The Carter No. 1 Unit (sec. 32, T. 30 N., R. 93 W.) penetrated 1,200 feet of flat Tertiary beds then 1 900 feet of Precambrian granite, and 880 feet of overturned rocks ranging in age from Cambrian to Jurassic before encountering a normal sequence of Upper Cretaceous sandstone and shale which dip 13° NE. (Fig. 3). The overturned section itself is partly repeated by minor faults, and the Paleozoic formations are represented by limited thicknesses of only the most competent beds. In the cross section, no attempt is made to show the structural complexity of this section. The Immigrant Trail thrust dips 20° NE., an estimate based on all available data; these include a published cross section drawn from seismic data (Wyoming Geological Association, 1951), shallow core holes in the vicinity of the pre-Eocene trace of the thrust, and a second well recently drilled through the same thrust zone, The Atlantic Refining Company No. 1 Unit (sec. 9, T. 29 N., R. 93 W.). Total vertical displacement of the Precambrian surface has been about 13,000 feet, and horizontal displacement is estimated to be 16,000 feet. The known extent of the inverted section below the granite is nearly half of the horizontal displacement; therefore, it appears that a significant part of this displacement has been accomplished by overturning. The Precambrian surface has been restored in the structure section to show the outline of an overturned fold.

Osterwald (1961, p. 229) interprets the Immigrant Trail fault as dominantly high-angle despite the relatively abundant subsurface control to the contrary. His interpretation rests largely on the fact that one well, the Stanolind's Scarlett Ranch Unit No. 1 (sec. 22, T. 31 N., R. 94 W.), drilled 2,600 feet of Tertiary and Paleozoic sediments and 5,400 feet of Precambrian granite without penetrating the fault. This well is located 9 miles northwest of the Carter No. 1 Unit, generally along strike of the thrust, and only 1½ miles from the pre-Eocene fault trace. The position of the Scarlett Ranch well relative to the Immigrant Trail thrust is indicated by its projected location in Figure 3.

Rather than disproving a thrust, the Scarlett Ranch well shows that the fault is deep at this location and that it is overlain by a greater thickness of Precambrian in the superthrust fold. Therefore, both the overlying fold and the subthrust syncline plunge northwestward between the two wells. These conclusions are supported by other subsurface data from nearby wells and seismic surveys.

SHEEP CREEK THRUST

The Sheep Creek thrust, another major fault of central Wyoming, bounds the Granite Mountains on the south from the vicinity of Crooks Gap to Muddy Gap. The Immigrant Trail thrust may be the northwestward extension of the Sheep Creek thrust, but the relation between the two is not clearly established. Osterwald (1961, p. 230) presents a cross section in the vicinity of the Sheep Creek oil field which shows the Sheep Creek thrust as a high-angle fault at depth, an interpretation based only on surface exposures and incompatible with available subsurface data. The exposed thrust dips 5°–18° NE. (Reva, 1959), and seismic control indicates that this dip persists at least 1 mile north (Fig. 4). Underlying the granite thrust wedge are steeply overturned Paleozoic and Mesozoic sediments, bounded below by a second thrust.

Immediately in front of the thrust zone is the strongly asymmetric and faulted Sheep Creek

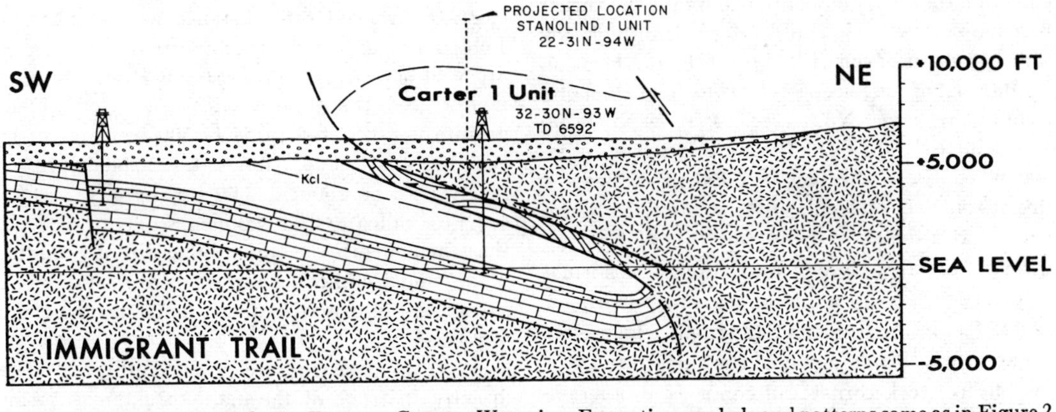

FIG. 3.—Immigrant Trail thrust, Fremont County, Wyoming. Formation symbols and patterns same as in Figure 2.

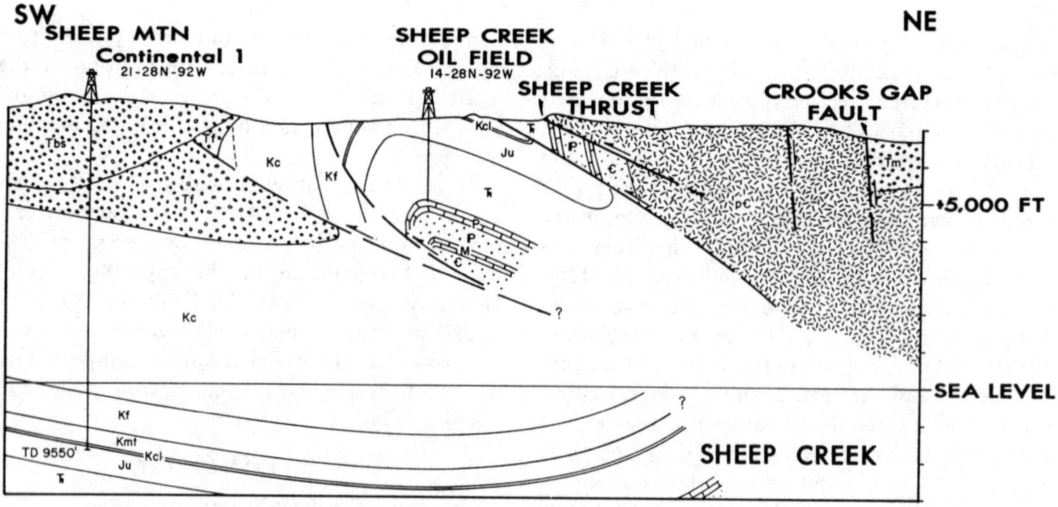

FIG. 4.—Sheep Creek thrust, Fremont County, Wyoming. Surface geology from Reva (1959); structure of Sheep Creek anticline simplified.

anticline. This fold may lack a basement core. Its limited areal extent, sharp asymmetry, and absence of deep folding all indicate that it is truncated below by another relatively shallow thrust. Therefore, the fold lies within a strongly deformed slice in front of and below the main thrust. Not shown on the section (Fig. 4) is another major thrust, located below and south of the Continental's Sheep Mountain well No. 1 (sec. 21, T. 28 N., R. 92 W.). This unnamed fault separates the Crooks Gap-Happy Springs folds from the deep syncline on the south in which the top of the Precambrian surface lies at an elevation of about −15,000 feet.

The root zone of the Sheep Creek thrust is unknown because of limited subsurface control, but enough data are available to show that the thrust is a major low-angle fault. More definitely, the same thrust has been drilled recently at a location 8 miles east of the Sheep Creek field. The Sinclair's Cooper Creek well No. 1 (sec. 19, T. 28 N., R. 90 W.) drilled 2,100 feet of Precambrian granite below which were near-vertical Paleozoic beds, a thrust zone section similar to that shown at Sheep Creek. At a depth of 3,444 feet the well passed through the thrust zone and entered a normal Upper Cretaceous section to total depth of 12,225 feet in the Frontier Sandstone. This well is situated only 1 mile north of the surface trace of the Sheep Creek thrust and confirms an average low dip of about 25° NE. for the fault plane.

EA THRUST

Another drilled thrust is located near Dubois in the northwestern Wind River Basin (Fig. 1). Here a Shell Oil Company test (sec. 9, T. 42 N., R. 105 W.) penetrated the EA thrust, one of the bounding faults of the Washakie Mountain uplift (Fig. 5). This well drilled 7,400 feet of Precambrian schist and found an inverted section 800 feet thick which consisted of 500 feet of deformed and sheared rocks of Mississippian, Pennsylvanian, and Permian age underlain by 300 feet of Triassic redbeds. Below this was about 700 feet of steeply dipping Cretaceous Mowry shale, possibly dragged along the fault, and then a normal section of Lower Cretaceous and Jurassic rocks which dip 15° NE. Here again is an inverted section which exists for a considerable distance beneath a thrust. Folded Precambrian rocks above the EA thrust are well shown by the overlying Paleozoic sediments, and the total vertical elevation of the granite is estimated to be 17,500 feet above the sub-thrust syncline. Horizontal displacement is believed to be about 22,000 feet because seismic surveys indicate that sediments extend at least that far beneath the thrust. The dip of the fault plane is not definitely known, but projection of all subsurface information indicates an average dip of 20° NE.

A subsidiary structure is the additional slice directly in front of the main thrust mass. This frontal zone was drilled nearby and consists of

Cretaceous shales and sandstones, in part folded and overturned. It is similar to the Sheep Creek frontal slice (Fig. 4) but more strongly deformed.

OTHER THRUSTS

Few areas in Wyoming show the nature of mountain flank thrusts as well as those described where wells have actually penetrated a Precambrian thrust wedge. However, similar thrusts of smaller size are known, and one of these was drilled at Sage Creek anticline in the southwest Wind River Basin. The Amerada's Tribal No. 1 (sec. 21, T. 1 N., R. 1 W.) penetrated a Paleozoic section on the steep west flank of the fold. Near the base of the Orodovician the well passsed through a thrust fault and encountered an overturned section of Pennsylvanian and Permian rocks nearly 1,200 feet thick. Then it drilled a second thrust and entered a normal Paleozoic section to bottom in the Tensleep Sandstone at a total depth of 7,390 feet. This is another low-angle fault with dip estimated to be 30° E. and with a typical overturned section in the thrust zone. Here total vertical displacement is only 6,500 feet. The thrust penetrated is part of a fault zone which extends northward along the entire Sage Creek-Steamboat Butte line of folding on the west flank of the Wind River Basin (Fig. 1).

Colorado Flank Structure

GOLDEN THRUST

The Front Range of the Southern Rocky Mountains west of Denver is bounded by a complex reverse fault zone, the Golden thrust. Along much of its extent fault relationships must be inferred, for nowhere is it well exposed so that dip may be accurately determined. The Golden thrust previously was thought to be a high-angle reverse fault (Zeigler, 1917), but later it was believed to be a low-angle thrust based largely on sparse seismic data (Stommel, 1951). Osterwald (1961, Fig. 3) used a single well for subsurface control to show that the fault dips westward at a rather high angle, about 60°. Recently Harms (1961) has further concluded that "the major structures outlining the flank of the range south of Denver are high-angle reverse faults whose dips steepen with depth." However, when all well control is used, the Golden thrust appears to be less a simple thrust and more a faulted and overturned fold (Fig. 6). The overturned limb is complexly thrust faulted, but the individual thrusts have relatively small throws of approximately 1,000 feet or less, whereas the total vertical displacement of the Precambrian surface is more than 13,000 feet. The thrusts dip 35°–50° W. between the wells. They disappear upward in the thick Cretaceous Pierre Shale (Kp) and are not seen in the near-vertical beds at the surface.

Unfortunately, the Golden fault must be interpreted only from the two deep wells and their relation to outcrops, because other data, such as adequate seismic control and dipmeter surveys, are lacking. It appears, however, that the overturned fold is the primary mountain flank structure and that the faults are a secondary feature of the overturned limb. At the town of Golden, 7 miles north of Soda Lakes, the thrust has greater throw, for the Fountain Formation (IPf) is faulted against Pierre Shale (Kp) (Van Horn, 1957). Therefore, the fold at Soda Lakes passes northward into a more prominent thrust fault.

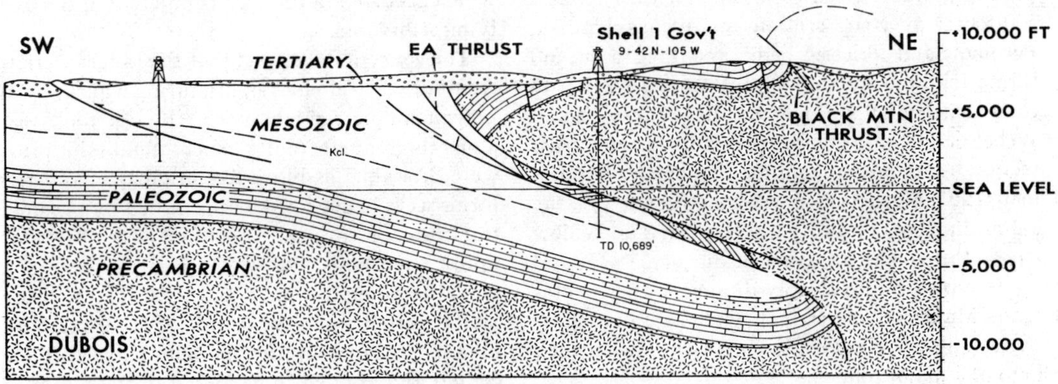

Fig. 5.—EA thrust near Dubois, Fremont County, Wyoming. Kcl is Lower Cretaceous Cloverly Sandstone; surface geology from Love (1939).

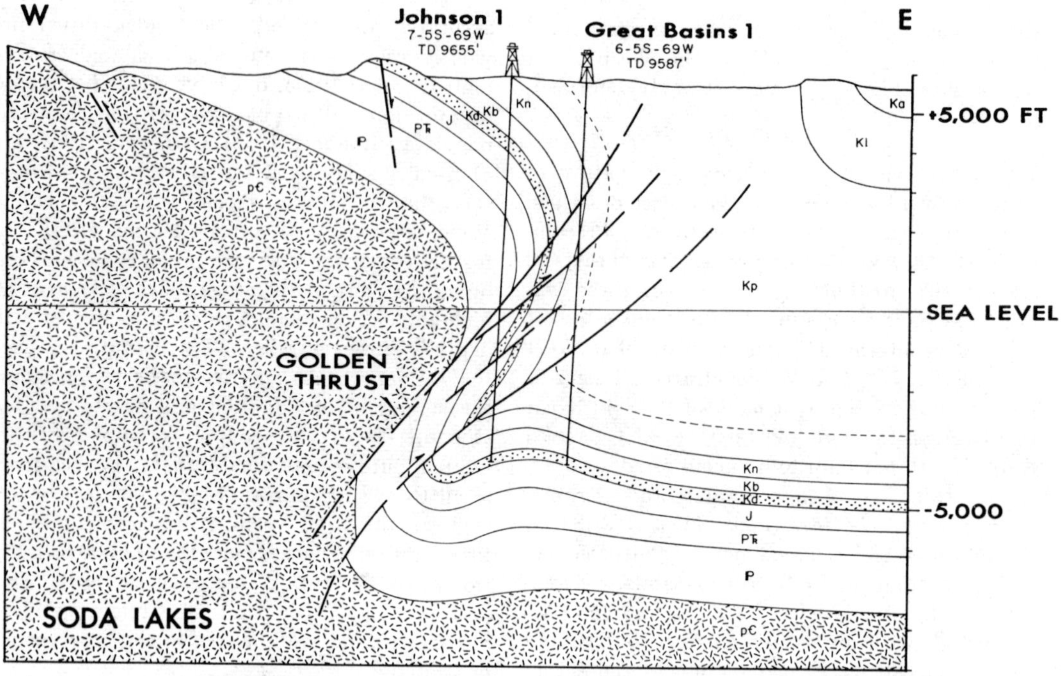

FIG. 6.—Golden thrust in the Soda Lakes area, Jefferson County, Colorado. From Berg (1962).

WILLOW CREEK THRUST

A smaller but better known thrust was drilled in western Colorado on the south flank of the Blue Mountain anticline, a subsidiary fold nearly parallel with the east plunge of the major Uinta Mountain uplift. Here, at Willow Creek, subsurface control shows a major overturned fold with thrust-faulted limb (Fig. 7). The Tennessee's Gov't. well No. 1-A (sec. 3, T. 3 N., R. 103 W.) drilled the overlying sediments and then penetrated 2,000 feet of Precambrian rock, the Uinta Mountain quartzite. Below the first thrust was a 500-foot inverted section of Pennsylvanian, Permian, and Triassic beds, and then a second thrust followed by an 800-foot normal section to total depth of 9,371 feet in the Pennsylvanian Weber Sandstone. Figure 7 is adapted from a cross section by Anderman and DeChadenedes (Anderman, 1961, Fig. 2) which was constructed with the aid of dipmeter and core dip control. The Willow Creek thrust has an average dip of 25° N. and grades upward into steeply dipping Upper Cretaceous Mancos Shale (Kmc). Here again is a low-angle fault zone associated with the overturned limb of a major fold.

ORIGIN OF THRUSTS

BLOCK UPLIFT

The mountain ranges of Wyoming and Montana have long been the subjects of tectonic study, but strong differences of opinion still exist on the interpretation of flank structure. Geologists responsible for early geologic mapping did not recognize thrust faulting. Later when extensive low-angle faults were mapped many geologists persisted in the belief that these were not significant flank structures but only secondary phenomena which occurred in the least competent of the overlying sediments.

The concept developed that the ranges were the result of vertical block uplift during which upward movement of rigid basement blocks took place along high-angle faults. Such mountain ranges were described as block-like in outline, and their forms were believed to be controlled by basement fractures, or conjugate sets of joints. The block uplift hypothesis was widely used to explain regional tectonic patterns, and its foremost proponents were Thom (1923) and R. T. Chamberlin (1945). As block uplift gradually came to be accepted and applied to nearly all of the major uplifts of the foreland, marginal thrusting was relegated to a minor role in deformation. Thus, the thrust structures of the Gros Ventre Range were described as "trap door" uplifts, an asymmetric variation of block uplift, by Horberg, Nelson, and Church (1949). Later, Bengtson (1956) applied the same interpretation and offered a mathematical explanation for thrust faults related to vertical movement. Recently, Osterwald (1961) has presented the extreme view that thrust faults are rare, and he has applied vertical block uplift principles to structural interpretation throughout the entire Cordilleran foreland.

The principles of block uplift are illustrated in Figure 8A. Great vertical displacement is attained by movement along an essentially high-angle reverse fault. The major conclusions of this idea are that the mountain ranges moved upward as rigid basement blocks; that uplift was along vertical fractures developed from pre-existing conjugate joint patterns in the basement; and that thrusts are not major flank structures.

According to the "block-uplifters," all features not compatible with vertical uplift are considered to be of minor importance. Thus Richmond (1945), in discussing thrust faults at the northwest end of the Wind River Mountains, states that "they do not extend beneath the range for any great distance . . . and that more likely they represent local planes of lateral relief" formed during vertical uplift. Osterwald (1961, p. 234) says that "structures that superficially resemble overthrusts are probably subsidiary breaks resulting from rigid basement blocks crowding aside thick sequences of relatively non-resistant sedimentary rocks." These views seem to be extensions of the older idea expressed by Chamberlin (1945, p. 110) that "in a later episode of the orogeny, further easement by horizontal thrust movements became increasingly prominent." In other words, Chamberlin and other proponents of vertical block uplift did not deny the occurrence of thrusts as many have done, but merely assigned their formation to a late period of mountain uplift.

THRUST UPLIFT

Beginning in the 1930s, large-scale thrust faults were recognized more widely in central Wyoming, and it became generally accepted that low-angle

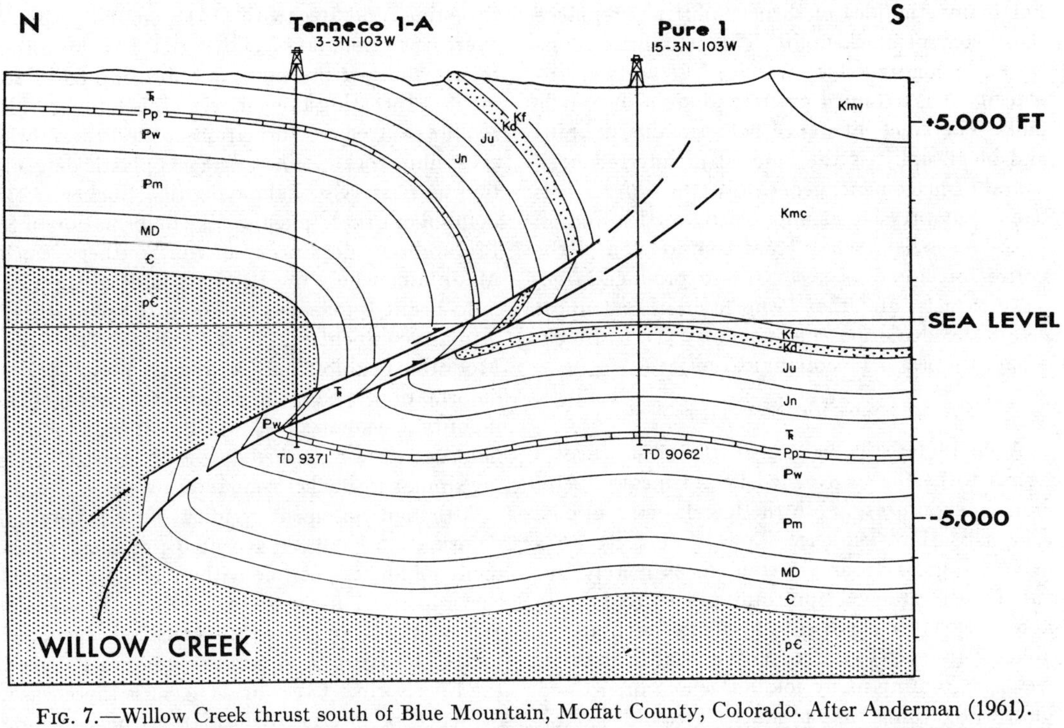

FIG. 7.—Willow Creek thrust south of Blue Mountain, Moffat County, Colorado. After Anderman (1961).

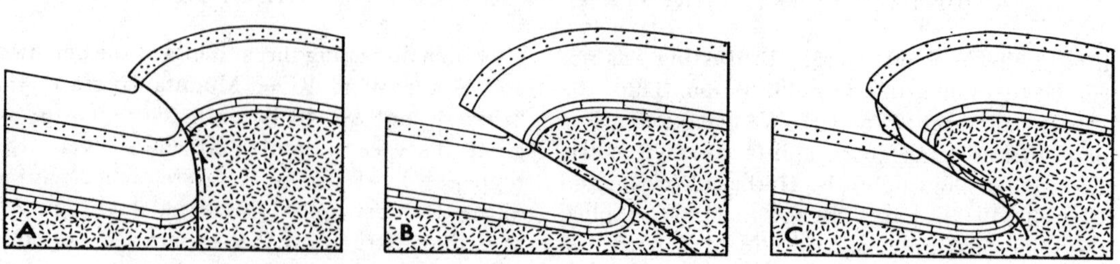

FIG. 8.—Diagrams illustrating hypotheses of mountain flank deformation: A—Block uplift, B—thrust uplift, C—fold-thrust uplift.

faulting observed or inferred at the surface was actually an indication of significant thrust structure at depth. Thrust faults of great magnitude were believed to occur on the flanks of the arcuate ranges which are asymmetric toward the south and southwest, such as the Owl Creek Mountains (Fanshawe, 1939), the Washakie Range (Love, 1939), and the ranges bordering the Granite Mountain uplift (Lovering, 1929, Carpenter and Cooper, 1951, Heisey, 1951). Extensive thrusting was also mapped in other areas, such as along the east flank of the Medicine Bow Range (Beckwith, 1938), and in the Beartooth Mountains. The general character of the thrust-bounded uplift now is known rather well, although the details of the thrust zones at depth are obscure in most places. For many, the idea of thrust uplift has replaced the theory of block uplift. Thrust uplift is illustrated in Figure 8B, in which great vertical displacement is attained entirely along a low-angle fault. The chief difference between thrust uplift and block uplift is the amount of inferred horizontal displacement; genetically, the difference is one of compressive as opposed to vertical forces. Many geologists believe that compression of the entire foreland was necessary to produce large-scale thrusts, and those who favor thrust uplift retain the idea of basement rigidity (D. L. Blackstone, Jr., personal communication).

FOLD-THRUST UPLIFT

A characteristic feature of all major thrusts seems to be the overturned beds beneath them, both in exposures and in the few drilled sections. The subsurface evidence shows that these beds have too great an areal extent to be merely isolated blocks dragged along fault planes, and therefore they must represent the distorted limbs of folds. This leads to the conclusion that thrusts developed from uplift by folding, which progressed into overturning that produced great horizontal displacement, and was followed by thrusting to give an additional but minor amount of horizontal displacement. This mode of mountain flank deformation has been termed fold-thrust uplift (Berg, 1961a, b) and is illustrated in Figure 8C, where great vertical and horizontal displacements are attained primarily by overturning.

Fold-thrust uplift explains an extensive overturned sequence beneath the fault plane whereas uplift by thrusting alone does not. It suggests moderate to extreme deformation of the Precambrian crystalline rocks as opposed to the rigid basement of both block and thrust uplift.

The idea of related overturning and thrusting is not new in Rocky Mountain structure, but its significance has not been recognized. In fact, folding is commonly found in association with major thrusting, and Beckwith (1941) actually showed overturned fold axes above the Elk Mountain thrust while attributing major horizontal displacement to thrusting alone. Large-scale folding which requires extreme deformation of the Precambrian crystalline rocks seems to suggest plastic deformation at relatively shallow depths. Bucher (1933) pointed out that "plastic"-like deformation of the Precambrian does occur in the Southern Rocky Mountains where the Ute Pass fault of the Colorado Front Range was observed to pass laterally into a recumbent fold in the granite. The Rocky Mountains exhibit only a mild form of "plastic" deformation when compared with other great mountain systems of the world, but the mountain flanks do show a marked degree of "plastic" yielding similar to Bucher's mountain welts.

Although basement rigidity during Laramide deformation has been accepted previously, basement folding is possible without plastic flow. On the northwest flank of the Bighorn Mountains, Wilson (1934) observed slickensided, secondary joints in Precambrian granite along a flexure in the Five Springs Creek area. Because there was no evidence of internal deformation, he concluded that granite folding was accomplished by small movements along a large number of closely spaced joints. This may be the manner of deformation of crystalline rocks along thrust zones.

A possible objection to large-scale overturned folding is the missing volume of sediment on the overturned limb. Some of this sediment was eroded during an early stage of uplift when movement was dominantly vertical. Some of it was pushed ahead of the fold in thick frontal slices and partly eroded. Also, there was probably compaction of younger sediments beneath the heavy Precambrian thrust wedge. However, there seems to be a remaining bulk of sediments not accounted for beneath the thrust zones such as in the Wind River thrust (Fig. 2). This may be a failure in structural interpretation rather than in mechanics of folding.

FOLD-THRUST CHARACTERISTICS

The features of the major foreland thrust zones are particularly useful in local structural studies as well as in regional tectonic interpretation. Besides the thrust faults themselves and great displacements, these characteristics may be important, individually or severally, in diagnosing the significance of local structure in exploration along the flanks of major uplifts.

1. Multiple fault planes are common, but usually two faults are prominent, one between Precambrian and Paleozoic rocks, and another between Paleozoic rocks and Late Cretaceous or early Tertiary beds.

2. The fault zone consists of overturned Paleozoic and early Mesozoic sediments, generally a reduced thickness of only the most competent rocks. The inverted section may be relatively undisturbed, somewhat folded and faulted, or highly deformed and jumbled. Its thickness ranges from about 500 feet to several thousand feet.

3. Dip of the main fault plane is variable, and its amount at any place depends on the position of the exposed or drilled part of the thrust. Dips range from very high to vertical at the point of greatest horizontal displacement and in the root zone, to only a few degrees at a location along the middle of the overturned limb. The observed angle of dip may be a real help in determining local position along a fault plane.

4. Stratigraphic throw is variable when measured in different places along the thrust, and apparent throw is least in the root zone.

5. One or more frontal thrusts are known in some places. These frontal slices involve chiefly Mesozoic sediments, are up to 5,000 feet or more in thickness, and do not include Precambrian rocks. They may be strongly folded and faulted, defying a simple structural interpretation even after drilling.

6. Transverse shear faults of major horizontal displacement up to several miles may offset the thrust zones. These result from breaking and lateral movement in a late stage of fold-thrust development. Examples of lateral shearing occur along the Beartooth thrust near Red Lodge, Montana (Foose et al., 1961).

7. A large anticlinal fold involving basement rocks may suggest the presence of deep-seated thrusts. Conversely, the presence of significant thrust faulting may indicate that major Laramide folding is present in the exposed Precambrian core, a feature usually not mapped but often suggested by structures along mountain fronts. Such folding can be observed in Precambrian sediments along a portion of the Uinta thrust (Hansen, 1957).

8. The overlying major fold may have backthrusts on its flank, but these are relatively minor when compared with the total displacement of the mountain flank.

9. Significant normal faults are associated with those fold-thrust zones that have greatest horizontal displacement, such as the Continental fault at the toe of the Wind River thrust (Fig. 2). These faults represent post-Miocene collapse of the thrust wedges. Other normal faults occur behind the thrusts and are downthrown toward the mountains, such as the Crooks Gap fault north of the Sheep Creek thrust (Fig. 4) and the Pathfinder fault north of the Seminoe thrust (Carpenter and Cooper, 1951). Significant normal faults are not recognized where horizontal displacements are small (Figs. 6, 7).

10. Subthrust sediments of the deep basin synclines are not strongly deformed. In most cases the subsurface evidence indicates only more or less uniform dip into the root zones, thereby arguing against the possibilities of subthrust anticlinal prospects for drilling.

Time of Deformation

The time of deformation in the Wyoming Mountain ranges seems to be largely Laramide, that is, uplift began in late Cretaceous time, Lancian or even earlier, and ended in latest Paleocene or earliest Eocene time. Evidence from the basins suggests continuous downwarping and more significant, a continuously growing uplift. This is indicated by the influx of coarse clastics of increasing age which were derived from the rising mountain flank. The sedimentary evidence leads to the conclusion that uplift of the mountains progressed uniformly with downwarp of the basin (Berg, 1961, a, b). This conclusion stands in opposition to some former interpretations of Laramide history derived from areas of uplift. These have stressed numerous repeated uplifts followed by long periods of widespread erosion. Locally, tectonic events may have been episodic and deformation may have been limited and intense, but these interpretations give an erroneous regional history. In considering the basin-mountain relationship on a larger scale, a continuous sequence of uplift and downwarp is suggested.

As defined here, "Laramide" deformation is restricted to the period in which greatest uplift was achieved; it does not include subsequent minor movements of the middle and late Tertiary. Some smaller thrusts are post-early Eocene, even post-Oligocene in age. For example, the Piney Creek thrust on the east flank of the Bighorn Mountains pushed Paleozoic rocks over the early Eocene Wasatch conglomerates. Such faults are not major flank structures but are the result of final lateral push toward the basin following the main period of uplift. Perhaps the motivation for some of these thrusts was gravity. In the case of the Piney Creek thrust, it does not effect great vertical displacement, but rather it forms a cap on the monoclinal fold which is the main flank structure.

Although fold-thrust structure may be present along the margins of some ranges, it is an extreme condition for the foreland as a whole. Great horizontal displacement is not a universal condition, and thrusting is limited or absent in parts of many uplifts such as the Bighorn Mountains, Laramie Range, Hartville Uplift, Black Hills, and the Colorado Front Range. Therefore, fold-thrust may represent the final phase of mountain flank deformation opposite only the areas of maximum downwarp of the intermontane basins. Maximum uplift probably was attained only along relatively limited parts of the major thrust zones. It is assumed that less intensely deformed structures are present laterally from areas of maximum deformation.

Uplift began during the late Cretaceous with dominant vertical movement, not of blocks but along linear and arcuate trends. This resulted in asymmetry of basin and range and possibly some high-angle reverse faults at points of great stress (Fig. 9A). Vertical displacement of the Precambrian surface probably attained 10,000–15,000 feet relative to the adjacent basins. During this early period compressional features were weakly developed.

Upwarp continued during the Paleocene. Moderate to high-angle reverse faulting and perhaps gentle folding in the basin sediments accompanied strong arching in the Precambrian rocks of the uplift (Fig. 9B). Vertical relief of 15,000–25,000 feet was attained, and moderate compression of the mountain flank resulted in overturning of major folds.

Uplift culminated in latest Paleocene or earliest Eocene time with dominant compression of the flank which resulted in strong overturning and finally thrusting of the overturned limb (Fig. 9C). Maximum foreland relief ranging from 25,000 to 40,000 feet was achieved. Later Tertiary uplift was largely epeirogenic.

Some uplifts did not progress beyond the first stage while others reached the second stage, and along the major Wyoming thrust zones the mountain flanks reached the final stage of maximum overturning and structural elevation. As shown in Figure 9, erosion of the rising uplifts was continuous, and deposits in the adjacent basins consisted of three facies. Nearest the uplift were conglomeratic sandstones of the flank facies containing pebbles derived first from the sediments, and then from the Precambrian core of the rising mountains. In the basin syncline was a fine-grained shale and siltstone facies representing a non-marine, quiet-water environment of basinal deposition. This facies constitutes the major part of the Hoback Formation in the Green River Basin and Waltman shale member of the Fort Union Formation in the Wind River Basin (Keefer, 1961). Toward the gentle flank of the basin were the interbedded sandstones and coal-bearing shales of the marginal facies that was deposited in a fluvial and fresh-water swamp environment which completely surrounded the non-marine basin. Overturning and thrusting of the mountain flank in the final stage of uplift brought older rocks over the Paleocene sediments that were deposited earlier.

Summary

The structures observed on the mountain flanks of the foreland show that there are thrust zones along the major asymmetric uplifts. Great struc-

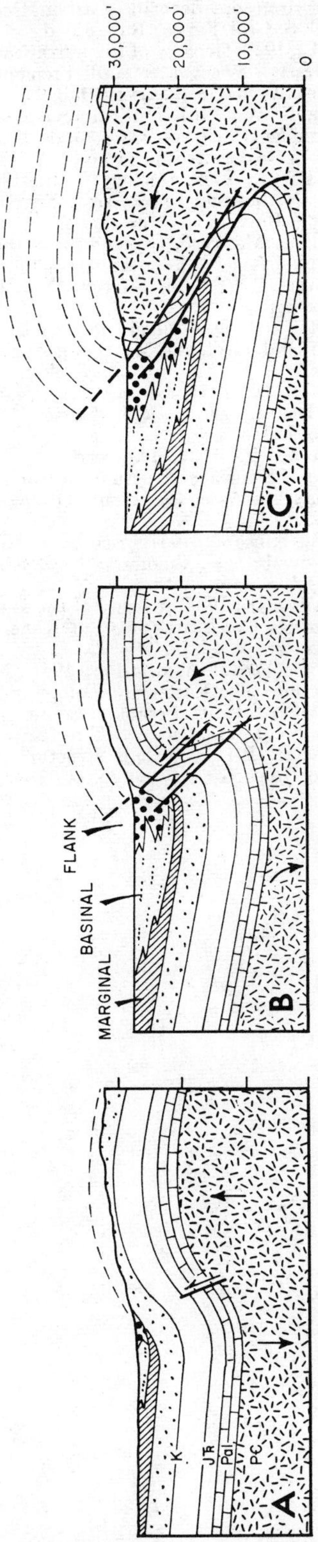

FIG. 9.—Diagrams illustrating stages in mountain flank deformation: A—latest Cretaceous or earliest Paleocene, B—middle Paleocene, C—earliest Eocene.

tural elevation of the uplifts was attained by folding and along significant low-angle faults. There is no subsurface evidence for an exclusive block uplift theory with its rigid basement rocks and dominantly high-angle faults. Neither is compressive thrusting of Precambrian rock wedges an adequate answer, for it does not explain the extensive overturned sequence beneath the thrust planes. Rather, it appears that Laramide basin and mountain flank deformation began with vertical movement along arcuate trends and developed linear upwarps and downwarps with local compressive features. In some places uplift culminated in overturned folding and thrusting. All of this was accompanied by an increasing degree of deformation of the crystalline crust. The entire process was carried on continuously throughout the long duration of Laramide time, from late Cretaceous into the early Tertiary with no significant interruptions either of uplift or of deposition in the adjacent basins.

REFERENCES

Anderman, G. G., 1961, Structure contour map of Colorado, *in* Symposium on Lower and Middle Paleozic rocks of Colorado: Rocky Mtn. Assoc. Geologists, 12th Ann. Field Conf. Guidebook, p. 101–105.

Beckwith, R. H., 1938, Structure of the southwest margin of the Laramie Basin, Wyoming: Geol. Soc. America Bull., v. 49, p. 1515–1544.

——— 1941, Structure of the Elk Mountain District, Carbon County, Wyoming: Geol. Soc. America Bull., v. 52, p. 1445–1486.

Bengtson, C. A., 1956, Structural geology of the Buffalo Fork area, northwestern Wyoming, and its relation to the regional tectonic setting: Wyo. Geol. Assoc., 11th Ann. Field Conf. Guidebook, p. 158–168.

Berg, R. R., 1961a, Laramide sediments along the Wind River thrust, Wyoming (abs.): Am. Assoc. Petroleum Geologists Bull., v. 45, p. 416.

——— 1961b, Laramide tectonics of the Wind River Mountains, *in* Symposium on Late Cretaceous rocks: Wyo. Geol. Assoc., 16th Ann. Field Conf. Guidebook, p. 70–80.

——— 1962, Subsurface interpretation of Golden fault at Soda Lakes, Jefferson County, Colorado: Am. Assoc. Petroleum Geologists Bull., v. 46, p. 704–707.

——— and Wasson, E. B., 1960, Structure of the Wind River Mountains, Wyoming (abs.): Geol. Soc. America Bull., v. 71, p. 1824.

Berman, J. E., 1955, Geology of the Tabernacle Butte area, Sublette County, Wyoming: Wyo. Geol. Assoc. 10th Ann. Field Conf. Guidebook, p. 108–111.

Bucher, W. H., 1933, The deformation of the earth's crust: p. 161–171, Princteon, N.J., Princeton Univ. Press.

Carpenter, L. C., and Copper, H. T., 1951, Geology of the Ferris-Seminoe Mountain area: Wyo. Geol. Assoc. 6th Ann. Field Conf. Guidebook, p. 77–79.

Chamberlin, R. T., 1940, Diastrophic behavior around the Bighorn Basin: Jour. Geology, v. 48, p. 673–716.

——— 1945, Basement control in Rocky Mountain deformation: Am. Jour. Sci., v. 243-A, p. 98–116.

Coffin, R. C., 1946, Recent trends in geological-geophysical exploration: Am. Assoc. Petroleum Geologists Bull., v. 30, p. 2031–32.

Eardley, A. J., 1951, Structural geology of North America: fig. 203, p. 348, New York, Harper and Brothers.

Fanshawe, J. R., 1939, Structural geology of Wind River Canyon area, Wyoming: Am. Assoc. Petroleum Geologists Bull., v. 23, p. 1439–92.

Foose, R. M., Wise, D. U., and Garbarini, G. S., 1961, Structural geology of the Beartooth Mountains, Montana and Wyoming: Geol. Soc. America Bull., v. 72, p. 1143–1172.

Hansen, W. R., 1957 Structural features of the Uinta Arch: Intermtn. Assoc. Petroleum Geologists 8th Ann. Field Conf. Guidebook, p. 35–39.

Harms, J. C., 1961, Laramide faults and stress distribution in Front Range, Colorado (abs.): Am. Assoc. Petroleum Geologists Bull., v. 45, p. 413.

Heisey, E. L., 1951, Geology of the Ferris Mountains-Muddy Gap area: Wyo. Geol. Assoc. 6th Ann. Field Conf. Guidebook, p. 71–76.

Horberg, C. L., Nelson, V. E., and Church, H. V., Jr., 1949, Structural trends of central western Wyoming: Geol. Soc. America Bull. v. 60, p. 183–216.

Keefer, W. R., 1961, Waltman shale and Shotgun members of Fort Union formation (Paleocene), Wyoming: Am. Assoc. Petroleum Geologists Bull., v. 45, p. 1310–1323.

Love, J. D., 1939, Geology along the southern margin of the Absaroka Range, Wyoming: Geol. Soc. America Special Paper 20.

——— 1950, Paleozoic rocks on the southwest flank of the Wind River Mountains near Pinedale, Wyoming: Wyo. Geol. Assoc. 5th Ann. Field Conf. Guidebook, p. 25–27.

Lovering, T. S., 1929, The Rawlins, Shirley, and Seminoe iron ore deposits, Carbon County, Wyoming: U. S. Geol. Survey Bull. 811-d.

Nace, R. L., 1939, Geology of the northwest part of the Red Desert, Sweetwater and Fremont Counties, Wyoming: Wyo. Geol. Survey Bull. 27.

Osterwald, F. W., 1961, Critical review of some tectonic problems in Cordilleran foreland: Am. Assoc. Petroleum Geologists Bull., v. 45, p. 219–37.

Reva, J. P., Jr., 1959, Geology of the Sheep Creek-Middle Cottonwood Creek Area, Fremont County, Wyoming: Univ. Wyo. M.S. thesis.

Richmond, G. M., 1945, Geology of northwest end of the Wind River Mountains, Sublette County, Wyoming: U. S. Geol. Survey Map 31, Oil and Gas Inv. Ser.

Ritzma, H. R., 1959, Geologic atlas of Utah, Daggett County: Utah Geol. and Mineralog. Survey Bull. 66, p. 71–75.

Stommel, H. E., 1951, Seismic investigation in the Golden-Denver area: Colo. School Mines D.Sc. thesis no. 729.

Thom, W. T., Jr., 1923, The relation of deep-seated faults to the surface structural features of central Montana: Am. Assoc. Petroleum Geologists Bull., v. 7, p. 1–13.

Van Horn, Richard, 1957, Bedrock geology of the Golden quadrangle, Colorado: U. S. Geol. Survey Map GQ 103.

Wilson, C. W., Jr., 1934, A study of the jointing in the Five Springs Creek area, east of Kane, Wyoming: Jour. Geology, v. 42, p. 498–522.

Wyoming Geological Association, 1951, North-south cross section, The Carter Oil Company No. 1, Immigrant Trail Unit, Fremont County, Wyoming: 6th Ann. Field Conf. Guidebook, p. 122.

Zeigler, Victor, 1917, Foothills structure in northern Colorado: Jour. Geology, v. 25, p. 728–733.